Aquatic and Wetland Plants of India

Aquatic and Wetland Plants of India

A reference book and identification manual for the vascular plants found in permanent or seasonal fresh water in the subcontinent of India south of the Himalayas

CHRISTOPHER D. K. COOK
Institut für Systematische Botanik
der Universität Zürich

Oxford New York Delhi
OXFORD UNIVERSITY PRESS

This book has been printed digitally and produced in a standard specification in order to ensure its continuing availability

OXFORD
UNIVERSITY PRESS

Great Clarendon Street, Oxford OX2 6DP

Oxford University Press is a department of the University of Oxford.
It furthers the University's objective of excellence in research, scholarship, and education by publishing worldwide in

Oxford New York

Auckland Cape Town Dar es Salaam Hong Kong Karachi
Kuala Lumpur Madrid Melbourne Mexico City Nairobi
New Delhi Shanghai Taipei Toronto

With offices in

Argentina Austria Brazil Chile Czech Republic France Greece
Guatemala Hungary Italy Japan South Korea Poland Portugal
Singapore Switzerland Thailand Turkey Ukraine Vietnam

Oxford is a registered trade mark of Oxford University Press
in the UK and in certain other countries

Published in the United States
by Oxford University Press Inc., New York

Reprinted 2011

ISBN 978-0-19-854821-8

Printed and bound in Great Britain by CPI Antony Rowe,
Chippenham and Eastbourne

Preface

The importance of wetlands to life on this planet is nowadays generally accepted. In spite of the internationally accepted resolutions to conserve wetland biodiversity agreed at Ramsar in 1971 they are not always treated with the respect they deserve. A difficult problem presented to biologists is the setting of priorities for the conservation for a given selection of wetlands or indeed any other habitats. Even worse, biologists may be expected to rank several wetlands and suggest a triage. Or they may be asked to present management programmes to conserve, 'improve' or rescue a wetland. What criteria should be used? One is biological diversity. It is a frequently used and important gauge to decide the value of a particular habitat. Changes in diversity indices are used as an indication of how a habitat is responding to a chosen kind of management or mismanagement. Biological diversity is a scientifically acceptable and objectively measurable parameter. In many areas of this world the taxonomy of the organisms present is so poorly known that even species diversity indices are of little scientific value, although inadequate ones may be of some political use; see 'Directory of Indian wetlands 1993'. Even if appropriate literature exists, it is often not available to the workers in the field. A more serious problem is that today very few people are trained to use the existing literature. Even fewer are trained to write the needed works.

During my career as a teacher of botany I have tended to preach the message that writing a Flora, although enjoyable, is not science and should be left to amateur botanists or elderly professionals. My aim was, as a last fling before going into retirement, to write a Flora. I have now realized how very wrong I was on nearly all counts. Originally, I started to write a local Flora for aquatic plants. It was Dr Brij Gopal of the Jawaharal Nehru University, New Delhi who made it clear to me that India has many local Floras but very few for larger areas. The major problem he and his students encounter is the identification of wetland rather than aquatic plants. With time I have come to realise just how right he was with the latter statement. It has also become clear that many Indian local Floras are highly inadequate and exclude many and often critical, species. I have tried to write this Flora in a style that requires only little botanical training and all except a very few recently described local endemic species are illustrated. I hope it will be possible with this book to, at least, document the diversity of higher plants in any freshwater aquatic or wetland habitat in the subcontinent of India.

Zürich
March 1995

C. D. K. C.

Contents

Introduction	1
Conventions and abbreviations	6
References	8
Excluded species	11
Identification keys	13
Descriptions	22
Index to plant names	371

Introduction

THE PLANTS

The definition of 'aquatic' used in this book remains the same as presented by Cook (1990). Vascular aquatic plants are interpreted as all Pteridophytina (ferns and fern allies) and Spermatophytina (seed-bearing plants) whose photosynthetically active parts are permanently or, at least, for several months each year partly or wholly submerged in water or which float on the surface of water.

The definition of a wetland plant is rather more difficult and perhaps potentially dangerous to the future of wetlands; recent legislation in the USA defines a wetland as a place where 80 per cent or more of the plants growing there are wetland plants. The subject of what a wetland is and what a wetland plant is has been reviewed by Gopal (1990) for India and Tiner (1991) for the USA. Without going into too much detail I consider a wetland to be a place where inundation must occur for at least 14 days and saturation for at least 60 consecutive days. This is rather less stringent than the rules in the USA but India is a tropical land with many climates and with very heavy monsoon rains. In some places in India almost everything is under water following a monsoon storm. The water usually recedes relatively quickly. Plants found in places which may be flooded each year but drain quickly so the soil does not remain waterlogged for very long after the floods have receded are not considered to be wetland species. Such plants are listed on page 11.

I have excluded plants that are confined to marine or highly brackish water. These species are listed on page 11.

THE REGION

The history of botanical discovery in the subcontinent of India is well described by Burkill (1965) and Desmond (1992).

The geographic delimitation of the subcontinent of India is still very much influenced by historical and political invention. Today one usually recognizes it as a region incorporating Afghanistan, Pakistan, India, Sri Lanka, Nepal, Bhutan, Bangladesh and Myanmar, an area once known, incorrectly, as British India. The arc of the Himalayas and Burmese Highlands is considered to represent the northern boundary. Botanically this mountainous northern arc is populated largely by plants that are widespread in central Asia and Eurasia or which, at least, have their nearer relatives there. Of these, the greater majority have never established themselves in or south of the plains of the Rivers Indus, Sutlej, Ganges and Brahmaputra. The Himalayan flora has ecologically and historically little to do with the flora of the rest of India (Mani 1974). Aquatic and wetland plants which occur in the Himalayas but do not extend south to the Indus-Gangetic Plain or into Peninsular India are excluded from this Flora but are mentioned in the text after the family or genus description. The families that otherwise have no wetland species in India are listed on page 11.

For this book I have chosen to set my northern boundary of India at the foothills of the Himalayas and not at their summits. My altitudinal limit along this northern border is set at 1000 m. Starting at the west of the region this Flora incorporates the plain of the Indus River, mostly in Pakistan, the western limit being a line drawn north of Karachi. The boundary follows the northern limit of the Ganges and Brahmaputra plains, following the 1000 m contour along the Himalayas; this excludes most of Jammu and Kashmir, Himachal Pradesh, Nepal, Sikkim, Bhutan and Arunchal Pradesh. The eastern boundary is the border of Myanmar with China, Laos and Thailand. A map of the region is shown in Figure 1.

In early botanical works Assam, Bengal, Punjab and Sikkim occupied larger areas than they do today. I have tried to confine the cited distributional records to the modern States as shown in Figure 1. I am aware that sometimes records given as Assam may refer to collections made in Meghalaya or even northern parts of Bangladesh, also that Sikkim may include parts of northern West Bengal and that Punjab may include parts of Haryana or eastern Pakistan. This is unfortunate but I feel a small geographic error is better than writing nothing. I have used a question mark when I am aware of doubt.

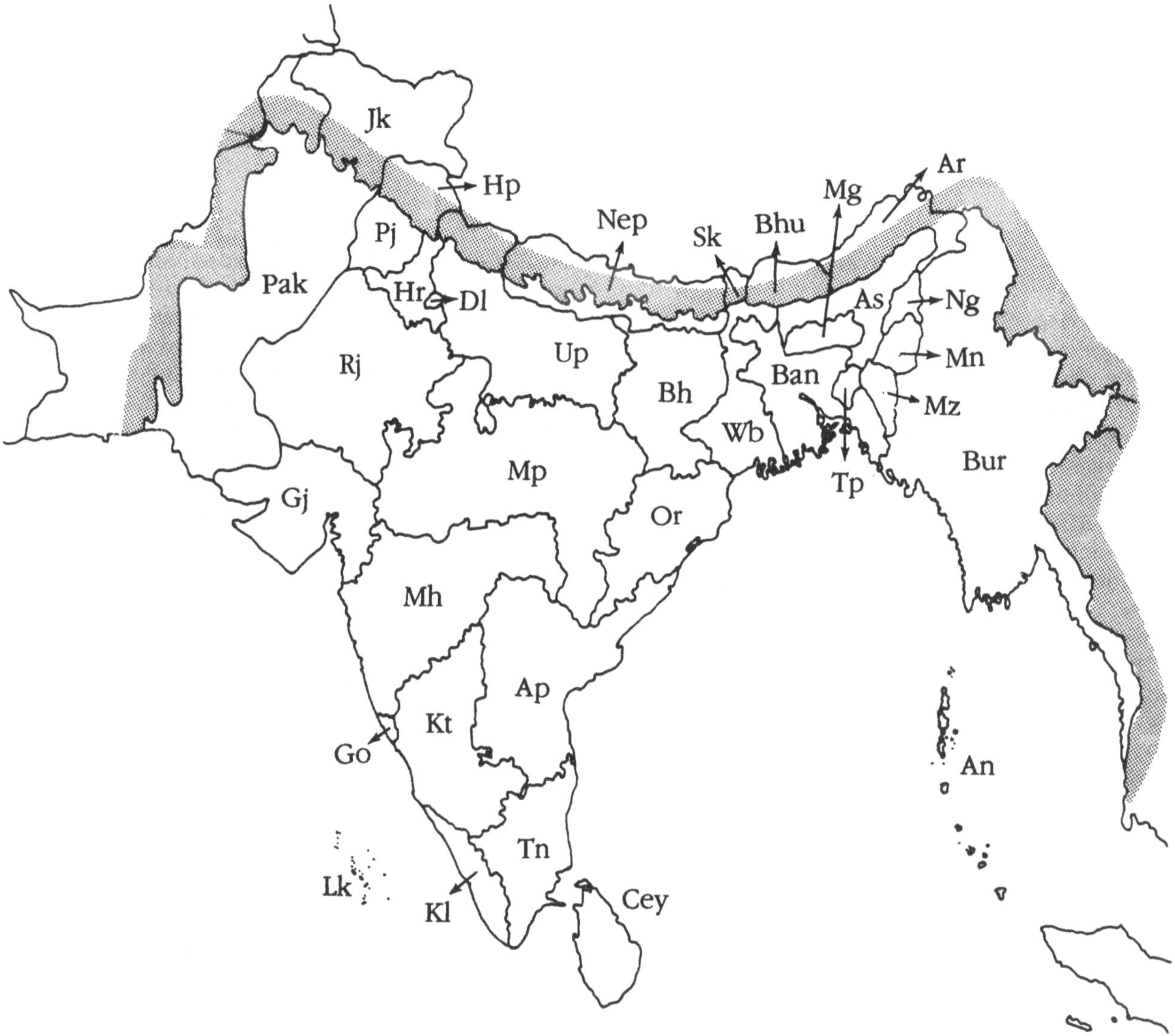

Fig. 1. Map of the Indian subcontinent showing the area covered in this work. Abbreviations: Ban = Bangladesh; Bhu = Bhutan; Bur = Myanmar (Burma); Cey = Sri Lanka (Ceylon); Nep = Nepal; Pak = Pakistan; An = Andaman and Nicobar; Ap = Andra Pradesh (including Yanam); Ar = Arunachal Pradesh; As = Assam; Bh = Bihar; Dl = Delhi; Gj = Gujarat (including Daman and Diu); Go = Goa; Hp = Himachal Pradesh; Hr = Haryana; Jk = Jammu and Kashmir; Kl = Kerala (including Mahe); Kt = Karnataka; Lk = Lakshadweep; Mg = Meghalaya; Mh = Maharashtra (including Dadra and Nagar Haveli); Mn = Manipur; Mp = Madhya Pradesh; Mz = Mizoram; Ng = Nagaland; Or = Orissa; Pj = Punjab; Rj = Rajasthan; Sk = Sikkim; Tn = Tamil Nadu (including Karaikal and Pondicherry); Tp = Tripura; Up = Uttar Pradesh; Wb = West Bengal.

ILLUSTRATIONS

The illustrations have been drawn with the aim of helping the identification of the plants. They are not meant to be plant portraits and are often restricted to the diagnostic features. Nearly all the illustrations are based on herbarium specimens. The outline of the plant was traced directly from the specimen or drawn through a camera lucida. Some were drawn by Marta Seitz and published in Cook (1990). The remainder are drawn by myself. I worked on a table that was reputed to have belonged to the painter Dante Gabriel Rosetti who founded the Pre-Raphaelite Brotherhood in 1848. I make no comparison but the knowledge that he might have faced similar problems of transferring living objects onto a flat surface on the same table has, at least, helped me. Those illustrations prepared from other sources are mentioned in the text.

DISTRIBUTIONAL RECORDS

It has not been possible to confirm all the distributional records by examining herbarium specimens in Indian institutions. Many lists and accounts of aquatic plants in India or in regions within India have been published. Several of them are based almost entirely on previously published records and have, therefore, not been used as primary sources in this work. Others clearly contain records based on incorrectly determined material; such works have been used with caution. There are, however, notable exceptions where herbarium specimens are cited. A list of references is presented in the bibliography on p. 8. It is my hope that the distributions in this book are, at least, accurate but it is clear they are a long way from being complete. The most common and widespread species are often the poorest known; I have not accepted statements such as 'common in all regions'.

Introduced plants that have become naturalized have '<' as a prefix; if the date of introduction is known it is cited. N, E, S and W refer to compass points and not political divisions; 'warm' is taken to mean from subtropical to warm temperate.

GROWTH FORM

There have been many attempts to classify aquatic plants according to their growth form. Cook (1990) summarized them and suggested a classification based on the response of the plant to milieu for growth and development rather than directly on its morphology or the way it survives adverse conditions. For this work I have slightly elaborated this earlier classification of mine but have tried to keep it simple.

There are many plants, particularly in the tropics, which grow on rocks or trees by streams, rivers and waterfalls. After rain they may be submerged in swiftly flowing water, sometimes for only a few hours or minutes at a time. These plants are called rheophytes and, although specialized for this kind of habitat, are excluded from this book as their growth and establishment takes place in the terrestrial milieu. Steenis (1981, 1987) has written extensively on rheophytes.

Key to growth forms

1. Plants not physiologically bound to water but tolerating longer periods of submergence **Helophytes**

1. Plants physiologically bound to water with at least part of the generative cycle taking place submerged in or floating on water

  2. Plants attached to but not penetrating a solid substrate, as in the Podostemaceae **Haptophytes**

  2. Plants with parts penetrating a substrate or lying totally free, submerged in or floating on water

    3. Plants with the juvenile phase submerged in or floating on water and the adult (flowering) phase terrestrial **Tenagophytes**

    3. Plants developing flowers while at least partly submerged in or floating on water

      4. Plants with all photosynthetic parts submerged

        5. Plants occupying the zone between the bottom and the lower surface of the water **Plankton**

        5. Plants rooted in the substrate

          6. Leaves borne in a rosette **Rosulate**

          6. Leaves arranged along elongated stems **Vittate**

      4. Plants with some photosynthetic parts in contact with air

        7. Plants free-floating on the surface, not attached to or penetrating the substrate **Pleustophytes**

        7. Plants with roots penetrating the substrate

          8. Leaves and/or stems floating on but not arising above water surface **Epihydates**

          8. Leaves and/or stems emerging above the above water surface **Hyperhydates**

Many plant species change their dependence on water in regions with different climates or with different hydrological regimes. This is quite normal and such plants are give more than one life form in the text.

GROWTH AND FLOWERING

In India aquatic plants with floating or submerged photosynthetic parts tend to grow and flower throughout the year as long as there is sufficient water and warmth; the same applies to plants which grow in permanent swamps. Most of India has a very marked seasonal difference in rainfall and permanent swamps are relatively rare. Most of the wetland plants of India are adapted to habitats that dry out after a period of inundation and waterlogging. Germination and growth usually take place in the wet phase; flowering and fruiting starts as the water recedes. For the whole of the Indian subcontinent over 70 per cent of the annual rainfall falls during the southwest monsoon; in the central plain of India this is more extreme and over 90 per cent falls during the southwest monsoon. The development of aquatic vegetation is dependent on this monsoon. In Sri Lanka, the south of India and Myanmar the southwest monsoon starts towards the end of May and withdraws as late as December or even in January, while in the northeast it starts as late as mid July and withdraws by the beginning of September. There are many local variations depending on the topography and there are frequent variations from year to year. At any particular locality most plant species have well-defined growth and flowering rhythms, see Ghosh *et al.* (1993). Because of the large scale climatic gradient over the whole subcontinent and the local variations in rainfall it is futile or even misleading to cite flowering and fruiting times for widespread species. For local endemic species this is different and when reliable data exist they are cited. To all biological generalizations there are exceptions. There are aquatic and wetland species which develop their flowers at the height of the dry season while others have their flowering time correlated with low temperatures. These exceptions are mentioned.

ECOLOGY

Following each species description I have attempted to write a short ecological diagnosis. This has proved rather difficult. Not all the species are known to me in the field. The essential information is often lacking in Indian botanical literature or is reduced to statements like 'wet places' or 'weed'!

TERMINOLOGY

All the plants are given Latin names. No attempt has been made to list ethnic or local names because I have no way of knowing if they are correct. Most plants today, unfortunately, have more than one Latin name. One name alone is correct; the others are synonyms. Only the synonyms which appear in widely used Indian floristic works are cited. Above the level of species there are sometimes differences in opinion as to where to draw the limits of the various taxonomic categories. I have taken the, I hope, practical solution and tried to adopt the system used in the most widely used local Floras unless there are convincing reasons for doing otherwise. Sometimes the choice is difficult and occasionally I have gone against the widespread Indian usage. For example, I have chosen to accept the genus *Scirpus* (Cyperaceae) in the very restricted sense because it seems to me more important to be botanically correct than to lump a lot of not closely related bisexually flowered sedges into one genus.

I have attempted to keep the botanical terminology as simple as possible. Wherever it is reasonable I have chosen Anglo-Saxon words: hair not trichome, woolly instead of lanate, sepals not calyx, petals not corolla and many others. When describing leaf surfaces and bilaterally symmetrical flowers the terms 'upper/lower', 'ventral/dorsal' and 'posterior/anterior' are widely used and usually considered interchangeable. They become more or less meaningless when the leaves are upright or the flowers are hanging. I have used the words 'adaxial' for surfaces that arise towards the axis and 'abaxial' for surfaces arising away from the axis.

ACKNOWLEDGEMENTS

This book has been written without any 'research grants' or any paid technical assistance. With the exception of my own salary and the University infrastructure, this book has consumed no taxpayers' money. The 'Georges und Antoine Claraz-Stiftung' has supported field work in India. I have, however, received great amounts of unpaid assistance. I would like to thank the following people for their help: Dr R. Ansari, G. A. College, Kozhikode, Kerala; Dr N. P. Balakrishnan, Botanical Survey of India, Coimbatore, Tamil Nadu; Prof. T. N. Bhardwaja, Kota Open University, Rajasthan; Dr K. Gopalakrishna Bhat, Poornaprajna College, Udupi, Karnataka; Mr Josef Bogner, Botanischer Garten, München, Germany; Prof. Subin K. Ghosh, University of Kalyani, West Bengal; Dr Brij Gopal and his research students, Jawaharal Nehru University, New Delhi; Prof. E. Govindarajalu, University of Madras, Tamil Nadu; Dr K. T. Joseph, University of Calicut, Kozikode, Kerala; Ms Ursula Kradolfer, University of Zürich, Switzerland; Prof. K. S. Manilal, University of Calicut, Kozikode, Kerala; Mr P. Thomas Mathew, NBPGR, Thrissur (Trichur), Kerala; Father K. M. Matthew, St Joseph's College, Tiruchchirappalli, Tamil Nadu; Dr K. T. Mathew, Environmental Resources Research Centre, Thiruvananthapuram (Trivandrum), Kerala; Mrs Indu Mehta, Chambal Project, Kota, Rajasthan; Prof. H. Y. Mohan Ram, University of Delhi, Delhi; Dr Marian Ørgaard, Royal Vetinary and Agricultural University, Fredricksberg, Denmark; Prof. Mohammad Qaiser, University of Karachi, Pakistan; Dr Manjula K. Saxena, University of Rajasthan, Jaipur, Rajasthan; Dr Sylvia Phillips, Royal Botanic Gardens, Kew, UK; Dr R. Rutishauser, University of Zürich, Switzerland; Dr M. Sivadasan, University of Calicut, Kozhikode, Kerala;

Prof. V. V. Sivarajan, University of Calicut, Kozhikode, Kerala; Prof. Thomas Stützel, Ruhr-Universität, Bochum, Germany; Dr R. Vasudevan Nair, Palakkad (Palghat), Kerala. Special thanks go to the librarians and herbarium curators of: Conservatoire et Jardin botaniques de la Ville de Genève, Switzerland; The Royal Botanic Gardens, Kew, UK; Institut für Systematische Botanic der Universität Zürich, Switzerland.

Conventions and abbreviations

Names in square brackets are, in the widest sense, synonyms.
auct. Ind. = of authors concerned with Indian plants
nom. illeg. = a name not in accordance with the international code of botanical nomenclature
orth. mut. = an altered spelling not in accordance with the international code of botanical nomenclature
subsp. = subspecies
var. = variety (varietas)

LIFE FORMS

⊙ = annual, (⊙) = essentially annual but sometimes perennating, otherwise all plants assumed to be perennial.

GROWTH FORMS

Eph = Ephydate — Bottom-rooted with floating leaves.
Hap = Haptophyte — Attached to but not penetrating the substrate.
Hel = Helophyte — Terrestrial plants which tolerate submergence.
Hyp = Hyperhydate — Emergent aquatic (lower parts almost always in water).
Pla = Plankton — Free-swimming below the water surface.
Ple = Pleustophyte — Free-floating (at the water surface).
Ros = Rosulate — Submerged, bottom rooted, leaves in a rosette.
Ten = Tenagophyte — Juvenile submerged, adult usually terrestrial.
Vit = Vittate — Submerged, bottom rooted, leaves cauline.

When the abbreviation is in brackets it indicates that the plant only occasionally has this kind of growth form.

DISTRIBUTION

♦ denotes endemic (used for countries).
< denotes introduced in historical times, the date and place of introduction is given when known.

GEOGRAPHICAL ABBREVIATIONS

The borders of the countries and states are taken from Muthiah, S. (ed.) An Atlas of India. Oxford University Press, Delhi (1990).
Ban = Bangladesh
Bhu = Bhutan
Bur = Myanmar (Burma)
Cey = Sri Lanka
Ind = India
Nep = Nepal
Pak = Pakistan.

An = Andaman and Nicobar
Ap = Andra Pradesh (including Yanam)
Ar = Arunachal Pradesh
As = Assam
Bh = Bihar
Dl = Delhi
Gj = Gujarat (including Daman and Diu)
Go = Goa
Hp = Himachal Pradesh
Hr = Haryana
Jk = Jammu and Kashmir
Kl = Kerala (including Mahe)
Kt = Karnataka
Lk = Lakshadweep
Mg = Meghalaya
Mh = Maharashtra (including Dadra and Nagar Haveli)
Mn = Manipur
Mp = Madhya Pradesh
Mz = Mizoram
Ng = Nagaland
Or = Orissa

Pj = Punjab
Rj = Rajasthan
Sk = Sikkim
Tn = Tamil Nadu (including Karaikal and Pondicherry)
Tp = Tripura
Up = Uttar Pradesh
Wb = West Bengal.

References

Numerous records have been published in the following Indian periodicals (not all are cited in the text): Bulletin of the Botanical Society of Bengal; Bulletin of the Botanical Society of India; Journal of the Bombay Natural History Society; Journal of the Indian Botanical Society; Journal of Economic and Taxonomic Botany (Jodhpur).

The Botanical Survey of India has published numerous local Floras they are not all cited individually.

Taxononomic monographs and revisions are cited in the text after the family or generic description. The following are more general works ('*n.v.*' indicates those that I have not been able to see).

Ali, S. I. and Nasir, Y. J. or Qaiser, M., 1991–1993. Flora of Pakistan. Issued in numerous parts.

Bhadri, B., Singh, B. and Desai, B. L., 1962. Water plants. Indian Council of Agricultural Research, New Delhi. pp. ? *n.v.*

Bhaskar, V. and Razi, B. A., 1973. Hydrophytes and marsh plants of Mysore City. Prasaranga, University of Mysore. i–viii, 1–102.

Bhat, K. G., (my copy is undated). Grasses and sedges of Coorg and South Kanara Districts of Karnataka — A taxonomic study. PhD Thesis, University of Mysore. 1–622.

Biswas, K. and Calder, C. C., dated 1954, published 1955. Handbook of common water and marsh plants of India and Burma, 1936. Health Bulletin No. 24, Malaria Bureau No. 11 (Government Press, Calcutta); ed. 2, revised and enlarged by K. Biswas. i–xvi, 1–216.

Burkill, I. H., 1965. Chapters in the history of botany in India. Botanical Survey of India, Calcutta. i–xi, 1–245.

Chavan, A. R. and Sabnis, S. D., 1961. A study on the hydrophytes of Baroda and environs. Journal of the Indian Botanical Society 40: 121–130.

Cook, C. D. K., 1990. Aquatic plant book. SPB Academic Publishing, The Hague. 1–228.

Cooke, T., 1901–1908. The flora of the Presidency of Bombay. Taylor and Francis, London. 1: i–ix, 1–645; 2: i–iv, 1–1083.

Dalgado, D. G., 1898. Flora de Goa e Savantvadi. Imprensa Nacional, Lisboa. i–xvii, 1–290.

Dan, P. K. and Mandal, S., 1990. Aquatic Plants of Durgapur Industrial Belt, West Bengal with reference to water environment. Recent Researches in Ecology, Environment and Pollution 4: 271–284. *n.v.*

Dassanayake, M. D., 1980–1994. A revised handbook to the flora of Ceylon. Amerind Publishing Co. Pvt., New Delhi. 1: i–vii (1980), 1–508; 2: i–vii, 1–511 (1981); 3: i–vii, 1–499 (1981); 4: i–vii, 1–532 (1983); 5: i–vii, 1–476; 7: i–vii, 1–439 (1991); 8: i–v, 1–458 (1994).

Dassanayake, M. D., 1988. A revised handbook to the flora of Ceylon. A. A. Balkema, Rotterdam. 6: i–vii, 1–424.

Deb, D. B., 1976. A study on the aquatic vascular plants of India. Bulletin of the Botanical Society of Bengal 29: 155–170.

Deb, D. B. and Dutta, R. M., dated 1987, published 1988. Contributions to the flora of Mizoram. Journal of Economic and Taxonomic Botany 10(1): 21–61.

De Thabrew, W. H. and De Thabrew, W. V., 1983. Water plants of Sri Lanka. Suhada Press, England [sic] *n.v.* cited by — Khan and Halim (1987).

Dey, A. K., Mandal, S., Rahaman, C. H. and Dan P. K., 1988. A contribution to the aquatic angiosperm [sic] of Santiniketan W. B. with reference to water environment. Proceedings Environmental Management and Planning 164–174. *n.v.*

Desmond, R., 1992. The European discovery of the Indian flora. Oxford University Press, Oxford. i–xii, 1–355.

De Wit, H. C. D., 1982. Aquariumplanten. Uitgeverij Hollandia, Baarne. 1–463.

Directory of Indian wetlands 1993, 1993. World wide fund for nature, India, New Delhi and Asian wetland bureau, Kuala Lumpur. i–xvi, 1–263, 32 maps.

Duthie, J. F., 1903–1929. Flora of the upper Gangetic plain and of the adjacent Siwalik and sub-himalayan tracts. Calcutta (Superintendent of Government Printing). 3 volumes issued in 6 parts. 1: 1–500 (1903–1905); 2: 1–266 (1911); 3: 1–371 (1915–1929).

Gamble, J. S. with Fischer, C. E. C. and Dunn, S. T., 1915–1936. Flora of the Presidency of Madras. Adlard and Son, London. i–lxiv, 1–2017.

Ghoshi, S. K., Santra, S. C. and Mukherjee, P. K., 1993. Phenological studies in aquatic macrophyte plants of lower Gangetic Delta, West Bengal, India. Feddes Repertorium 104 (1–2): 93–111.

Gopal, B. (ed.), 1990. Ecology and management of aquatic vegetation in the Indian subcontinent. Kluwer Academic Publishers, Dordrecht. i–xiii, 1–257.

Grierson, A. J. C. and Long, D. G., 1983–1991. Flora of Bhutan including a record of plants from Sikkim. Royal Botanic Garden, Edinburgh. 1(1): 1–186 (1983), 1(2): 197–462 (1984), 1(3): 463–834 (1987); Flora of Bhutan. Royal Botanic Garden, Edinburgh. 2(1): 1–426 (1991); 3(1): 1–456 (1994).

Gupta, R. S., 1966. A study of hydrophytes and marsh plants of Kota and environs. Tropical Ecology 7: 153–162.

Haines, H. H., 1921–1025. The botany of Bihar and Orissa. Adlard and Son, London. Issued in 6 parts. i–ix, 1–1350.

Hara, H., Stearn, W. T. and Williams, L. H. J., 1978. An enumeration of the flowering plants of Nepal. British Museum, London. 1: 1–154.

Hara, H. and Williams, L. H. J., 1979. An enumeration of the flowering plants of Nepal. British Museum, London. 2: 1–220.

Hara, H., Chater, A. O. and Williams, L. H. J., 1982. An enumeration of the flowering plants of Nepal. British Museum, London. 3: 1–226.

Hooker, J. D., 1872–1897. The Flora of British India. Reeve and Co., London. 1: i–viii, 1–740 (1872–1875); 2: 1–792 (1876–1878); 3: 1–712 (1880–1882); 4: 1–780 (1883–1885); 5: 1–910 (1886–1890); 6: 1–792 (1890–1894); 7: 1–842 (1896–1897).

Islam, M., 1989. Aquatic weeds of north-east India. International Book Distributors, Dehra Dun. 1–155.

Islam, M., dated 1998 published ?1990. Aquatic and marshland flora of Assam. Journal of Economic and Taxonomic Botany 13(2): 461–491.

Jha, U. N., 1965. Hydrophytes of Ranchi. Tropical Ecology 6: 98–105. *n.v.*

Kachroo, P., 1959. Aquatic vegetation of the Damodar valley I. Phanerogamic flora of fresh water ponds and marshy lands with particular reference to Eden canal areas of West Bengal. Journal of the Asiatic Society of Bengal 1: 271–298. *n.v.*

Kak, Majeed, M., 1990. Aquatic and wetland vegetation of the Kashmir Himalaya. Journal of Economic and Taxonomic Botany 14(1): 1–14.

Karthikeyen, S., Kumar Anad and Sharma, B. D., 1982. Aquatic angiosperms of Maharashtra. Journal of Economic and Taxonomic Botany 3(2): 423–445.

Karthikeyan, S., Jain, S. K., Nayar, M. P. and Sanjappa, M., 1989. Florae Indicae Enumeratio: Monocotylonae. Flora of India Series 4, 1–435.

Khan M. S. and Rahman, M. M., 1972–1990. Flora of Bangladesh. Bangladesh Agricultural Research Council. Issued in numerous parts.

Khan, M. S. and Halim Mahbuba, M., 1987. Aquatic angiosperms of Bangladesh. National Herbarium of Bangladesh. i–vi, 1–120.

Leach, G. J. and Osborne, P. L., 1985. Freshwater plants of Papua New Guinea. University of Papua New Guinea Press. i–xv, 1–254.

Maheshwari, J. K., 1960. The vegetation of marshes, swamps and riverside [sic] in Khandwa district (M. P.). Journal of the Bombay Natural History Society 57: 371–388.

Maheshwari, J. K., 1963. The Flora of Delhi. CSIR, New Delhi. 1–447.

Majumdar, N. C., 1965. Aquatic and semiaquatic flora of Calcutta and adjacent localities. Bulletin of the Botanical Society of Bengal 19: 10–17. *n.v.*

Majumdar, N. C., Ghosh Barin and Singh, G. P., dated 1988, published 1990. Observations on the wetland ecosystem of Kabar Lake in Begusarai, Bihar, with special reference to vegetation. Bulletin of the Botanical Society of India 30(1–4): 134–139.

Malik, K. C. and Safui, B., dated 1987, published 1988. Additions to the flora of Manipur State. Journal of Economic and Taxonomic Botany 10(1): 1–19.

Mani, M. S. (editor), 1974. Ecology and biogeography in India. Dr W. Junk b. v. Publishers, The Hague i–xv, 1–773.

Manilal, K. S. and Sivarajan, V. V., 1975. A contribution to the study of hydrophytic flora of Kerala. Proceedings of the National Academy of Sciences, India 45B, IV: 225–231.

Matthew, K. M., 1981–1988. The Flora of the Tamilnadu Carnatic. The Rapinat Herbarium, St Joseph's College, Tiruchirapalli. Vol. 1 (Materials for the Flora of the Tamilnadu Carnatic): i–x, 1–472 (1981); Vol. 2 (Illustrations on the flora of the Tamilnadu Carnatic): i–xlvi, 1–1027 (1982); Vol. 3, part 1: i–lxxxiv, 1–688 (1983); Vol. 3, part 2 (Gamopetalae and Monochlamydeae): lxxxvi–lxxxvii, 689–1539 (1983); Vol. 3, part 3 (Monocotyledones): 1541–2154 (1983); Vol. 4 (Further illustrations on the flora of the Tamilnadu Carnatic): i–xxxvi, 1–915 (1988).

Mirashi, M. V., 1954. Studies in the hydrophytes of Nagpur. Journal of the Indian Botanical Society 33: 299–308.

Mirashi, M. V., 1957 Studies in the hydrophytes of Umred. Journal of the Indian Botanical Society 36: 396–407.

Mukhopadhyay, C. R., 1987. Aquatic and semi-aquatic plants of Birbhum District, West Bengal. Journal of Economic and Taxonomic Botany 9(1): 230–238.

Nair, N. C., 1978. Flora of the Punjab Plains. Records of the Botanical Survey of India 21(1): i–xx, 1–326.

Naithani, H. B., 1990. Flowering plants of India, Nepal and Bhutan not recorded in Sir J. D. Hooker's Flora of British India. i–viii, 1–711.

Nasir, E. and Ali, S. I., 1970–1979. Flora of West Pakistan. Issued in numerous parts.

Nasir, E. and Ali, S. I., 1980–1989. Flora of Pakistan. Issued in numerous parts.

Naskar, K. R., 1990. Aquatic and semi-aquatic plants of the lower Ganga delta. Daya Publishing House, Delhi. i–xiv, 1–408.

Nayar, M. P., 1984–1986. Key works to the taxonomy of flowering plants of India. Botanical Society of India. 1: 1–462 (1984); 2: 1–279 (1984); 3: 1–169 (1985); 4: 1–268 (1984); 5: 1–268 (1986).

Noltie, H. J., 1994. Flora of Bhutan including a record of plants from Sikkim and Darjeeling. 3(1): i–vi, 1–456.

Paul, S. R., 1973. On the aquatic and marsh plants of Monghyr, Bihar. Botanique 4(2): 140–151. *n.v.*

Raizada, M. B., 1976. Supplement to Duthie's Flora of the Upper Gangetic Plain and of the adjacent Siwalik and sub-Himalayan tracts. Bishen Singh Mahendra Pal Singh, Dehra Dun. i–vii, 1–358.

Rao. R. S., 1985, 1986. Flora of Goa, Diu, Daman, Dadra and Nagarhaveli. Botanical Survey of India, Flora of India series 2, 1: i–xxxxi, 1–198 (1985); 2: 199–546 (1986).

Saldhanha, C. J. and Nicolson, D. H., 1976. Flora of the Hassan district, Karnataka, India. Amerind Publishing, New Delhi. i–viii, 1–915.

Seerwani, A. B., 1962. A study of the hydrophytes and plants of low-lying habitats in Jabalpur. Bulletin of the Botanical Society of India 14: 227–231.

Sen, D. N. and Chatterjee, U. N., 1959. Ecological studies on aquatic and swampy vegetations [sic] of Gorakhpur. A survey. Agra University Journal of Research (Science) 8: 17–29.

Sharma, B. D., Balkrishnan, N. P., Rao, R. R. and Hajra, P. K., 1993. Flora of India, Botanical Survey of India, Calcutta 1 (Ranunculaceae — Barclayaceae): i–xiv, 1–467.

Sharma, B. D. and Balkrishnan, N. P., 1993. Flora of India, Botanical Survey of India, Calcutta 2 (Papaveracaeae — Caryophyllaceae): i–xi, 1–625.

Sharma, B. D. and Sanjappa, M., 1993. Flora of India, Botanical Survey of India, Calcutta 3 (Portulaceae — Ixonanthaceae): i–xi, 1–639.

Shetty, B. V. and Pandey, R. P., 1983. Flora of Tonk District. Botanical Survey of India, Delhi. 1–253.

Singh Hirjam Bikramit, 1990. Aquatic angiosperms of Manipur (monocotyledonous plants). Journal of Economic and Taxonomic Botany 14(3): 659–661.

Singh, K. K. and Tomar, R. P. S., 1982. The aquatic and marshland flora of the Kheri district, Uttar Pradesh. Journal of the Indian Botanical Society 79(2): 261–274.

Singh, M. P., 1990. Hydrophytes of Ranchi, Bihar. Journal of Economic and Taxonomic Botany 14(3): 373–577.

Singh, N. P., 1988. Flora of eastern Karnataka. Mittal Publications, Delhi 1: i-xiv, 1–422; 2: 423–794.

Soerjani Mohamad, Kostermans, A. J. G. H. and Tjitrosoepomo Gembong (eds.), 1987. Weeds of rice in Indonesia. Balai Pustaka, Jakarta. i–xvi, 1–716.

Srivastava, R. C. and Kumar Anad, 1987. Aquatic and marshy angiosperms of Madhya Pradesh. Journal of Economic and Taxonomic Botany 9(2): 433–458.

Steenis, C. G. G. J. van, 1981, 1987. Rheophytes of the world: an account of the flood-resistant flowering plants and ferns and the theory of autonomous evolution. Sijthoff and Noordhoff, Alphen aan den Rijn. i–xv, 1–408 (1981); supplement: Allertonia 4: 267–330 (1987).

Stewart, R. R., 1972. An annotated catalogue of the vascular plants of West Pakistan and Kashmir. i–xvii, 1–1028.

Subramanyam, K., 1962. Aquatic angiosperms: a systematic account of common Indian aquatic angiosperms. CSIR, New Delhi, Botanical Monographs 3: i–vi, 1–190.

Tiner, R. W., 1992. The concept of a hydrophyte for wetland identification. Bioscience 41(4): 236–247.

Tiwari, D. K., 1966. Hydrophytes of Raipur. Proceedings of the 53rd Indian Science Congress 3: 407–408. *n.v.*

Trivedi, B. S. and Sharma, P. C., 1965. Studies on the hydrophytes of Lucknow and environs. 1. Distribution and habit, with reference to ecological classification. Proceedings of the National Academy of Sciences, India 35: 1–14. *n.v.*

Unni, K. S., 1967. Studies on the vegetation of ponds, swamps and river banks of Raipur (M.P.). Journal of the Bombay Natural History Society 64: 95–102.

Vasudeva Rao, M. K., 1986. A preliminary report on the angiosperms of Andaman–Nicobar islands. Journal of Economic and Taxonomic Botany 8(1): 107–184.

Vyas, L. N., 1964. A study of the hydrophytes and marsh plants of Alwar and environs. Journal of the Indian Botanical Society 43: 17–30.

Yan, Su-Zhu, 1983. Higher water plants of China (in Chinese). Science Press, i–v, 1–335.

Excluded species

The following aquatic plants are excluded because they are confined to the marine environment:
CYMODOCEACEAE: ***Cymodocea serrulata*** (R. Brown) Ascherson et Magnus; ***C. rotundata*** Ehrenberg et Hempricht ex Ascherson; ***Halodule uninervis*** (Forsskål) Ascherson; ***Syringodium isoetifolium*** (Ascherson) Dandy.
HYDROCHARITACEAE: ***Enhalus acoroides*** (Linnaeus fil.) Royle; ***Halophila beccarii*** Ascherson; ***H. decipiens*** Ostenfeld; ***H. ovalis*** (R. Brown) J. D. Hooker; ***H. ovata*** Gaudich; ***H. stipulacea*** (Forsskål) Ascherson; ***Thalassia hemprichtii*** (Ehrenberg) Ascherson.

Mangrove vegetation is regularly flooded in sea water by the tide. These plants are consequentially very specialized and restricted to this habitat. The mangrove vegetation of India is relatively well known and well documented and is, therefore, excluded from this work. For information see: Banerjee, L. K., Sastry, A. R. K. and Nayar, M. P. Mangroves in India: identification manual. Botanical Survey of India. 1–113 (1989); Banerjee, L. K. and Rao, T. A. Mangroves of the Orissa coast and their ecology. Bishen Singh Mahendra Pal Singh, Dehra Dun. 1–119 (1990).

Aquatic and wetland plants which occur in the Himalayas but do not extend south to the Indus-Gangetic plain or into Peninsular India are excluded from this Flora but are mentioned in the text after the family or genus description. The following are in families that otherwise have no wetland species in India:
BUTOMACEAE: ***Butomus umbellatus*** Linnaeus
CRASSULACEAE: ***Crassula schimperi*** Fischer et C. A. A. Meyer var. ***schimperi*** [*Tillaea pentandra* Royle ex Edgeworth]
HIPPURIDACEAE: ***Hippuris vulgaris*** Linnaeus
JUNCAGINACEAE: ***Triglochin maritima*** Linnaeus; ***T. palustre*** Linnaeus
PARNASSIACEAE: ***Parnassia laxmannii*** Pallas ex Schultes; ***P. nubicola*** Wallich ex Royle; ***P. palustris*** Linnaeus
PRIMULACEAE: ***Glaux maritima*** Linnaeus; ***Lysimachia candida*** Lindley; ***L. lobeliodes*** Wallich; ***Samolus valerandi*** Linnaeus (the records from Bijoli Mandalagarth (Rj) and Ratnagiri (Mh) are surely based on missidentifactions – the illustration in the Journal of the Bombay Natural History Society 77: 360 (1980) does not much resemble *Samolus*.
SAURURACEAE: ***Houttuynia cordata*** Thunberg; ***Saururus chinensis*** (Loureiro) Baillon
SPARGANIACEAE: ***Sparganium emersum*** A. Rehmann; ***S. erectum*** Linnaeus subsp. ***stoloniferum*** (F. Hamilton ex Graebner) C. D. K. Cook et M. S. Nicholls

Many weedy plants grow in places where water stood during and shortly after the monsoon. These plants are not wetland species but sometimes they are found growing with wetland species as the water recedes after the monsoon. In the dry season they then replace the wetland species in this kind of habitat. These species are not described in this Flora but are mentioned in the text after the family or genus description. The following are in families that otherwise have no wetland species in India:
AIZOACEAE [MOLLUGINACEAE]: ***Glinus lotoides*** Linnaeus [*Mollugo hirta* Thunberg]; ***G. oppositifolius*** (Linnaeus) A. P. de Candolle; ***Mollugo disticha*** (Linnaeus) Seringe; ***M. pentaphylla*** Linnaeus
CAPPARIDACEAE: ***Cleome chelidonii*** Linnaeus fil.
CARYOPHYLLACEAE: ***Polycarpon prostratum*** (Forsskål) Ascherson et Schweinfurth [*Pharnaceum depressum* Linnaeus; *Hapalosa loeflingiae* Wallich ex Wight et Arnott]; ***P. tetraphyllum*** (Linnaeus) Linnaeus
CHENOPODIACEAE: ***Chenopodium ambrosioides*** Linnaeus
EUPHORBIACEAE: ***Euphorbia prostrata*** Aiton; ***E. thymifolia*** Linnaeus
HYPERICACEAE: ***Hypericum laxum*** (Blume) Koidzumi
IRIDACEAE: ***Cipura paludosa*** Aublet
LOGANIACEAE [SPIGELIACEAE]: ***Mitrasacme indica*** Wight [*M. alsinoides* R. Brown]
NYCTAGINACEAE: ***Boerhavia erecta*** Linneaus
POLYGALACEAE: ***Polygala elongata*** Klein ex Willdenow; ***Salomonia ciliata*** (Linnaeus) A. P. de Candolle
PRIMULACEAE: ***Anagallis pumila*** Swartz; ***Lysimachia obovata*** J. D. Hooker
RESEDACEAE: ***Oligmeris linifolia*** (Vahl) Macbride
ROSACEAE: ***Potentilla supina*** Linnaeus
SAXIFRAGACEAE [VAHLIACEAE]: ***Vahlia*** nom. cons. [*Bistella*]: ***Vahlia dichotoma*** (Murray) O. Kuntze [*Oldenlandia pentandra* Retzius; *V. oldenlandioides* Roxburgh]; ***V. digyna*** (Retzius) O. Kuntze [*V. viscosa* Roxburgh]

STERCULIACEAE: ***Melochia cochorifolia*** Linnaeus; ***M. concatenata*** Linnaeus; ***Pentapetes phoenicea*** Linnaeus
TILIACEAE: ***Corchorus aestuans*** Linnaeus; ***C. capsularis*** Linnaeus
VIOLACEAE: ***Hybanthus suffruticosa*** Baillon ex Lanessy

Large perennial plants found in wet or moist places in forests but not in wetlands:
MARANTACEAE: ***Donax arundastrum*** Loureiro; ***Phrynium*** sp.; ***Schumanianthus*** [***Clinogyne***] ***dichotoma*** (Roxburgh) Gagnepain

Trees and shrubs tolerating seasonal flooding:
ARECACEAE [PALMAE]: ***Calamus erectus*** Roxburgh: ***C. tenuis*** Roxburgh
BARRINGTONIACEAE [LECYTHIDACEAE]: ***Barringtonia acutangula*** (Linnaeus) Gaertner
COMBRETACEAE: ***Terminalia arjuna*** (Roxburgh) Wight et Arnott
DILLENIACEAE: ***Dillenia indica*** Linnaeus
EUPHORBIACEAE [BISCHOFIACEAE]: ***Bischofia javanica*** Blume; ***Hormonia retusa*** Müller Argoviensis; ***H. riparia*** Loureiro
FABACEAE [LEGUMINOSAE]: ***Cassia alata*** Linnaeus
MELASTOMACEAE: ***Blastus cochinchensis*** Loureiro; ***Melastoma malabarthricum*** Linnaeus; ***Osbeckia rostrata*** Don
MYRTACEAE: ***Syzygium cumini*** (Linnaeus) Skeets
RUBIACEAE: ***Cephalanthus occidentalis*** Linnaeus
SALICACEAE: ***Salix tetrasperma*** Roxburgh
TAMARICACEAE: ***Tamarix dioica*** Roxburgh; ***T. ericoides*** Rottler; ***T. troupii*** Hole
VERBENACEAE: ***Clerodendron indicum*** (Linnaeus) Kuntze; ***Vitex negundo*** Linnaeus

Identification keys

The identification keys are biased in favour of easily seen vegetative characters. This has the consequence that many taxa appear several times, the keys are thus lengthened but it is hoped that this will enable plants to be identified in the field without the use of a microscope. Also the user need not worry too much when confronted with difficult to determine alternatives because the plant may appear in both.

Key to major groups

1. Vegetative parts not differentiated into stems with leaves, (leaves, scales or bracts may be found just below flowers) **Key 1 p. 13**
1. Vegetative parts clearly differentiated into stems with leaves or scales
  2. Leaves not photosynthetic, reduced to scales (stems green and photosynthetic, bladed or laminate leaves absent) **Key 2 p. 13**
  2. Leaves photosynthetic, not scale-like (non-photosynthetic scales may be present on underground parts or just below the flowers)
    3. Leaves jointed between sheath and blade (the base of the leaf is a cylindrical sheath, free or united at the edges, enveloping the stem; above the sheath the blade is free; between the blade and sheath is a joint usually bearing a membranous ligule but sometimes a swelling or row of hairs) **Key 3 p. 14**
    3. Leaves without a joint between sheath and blade
      4. Leaf blades peltate (attached to their petioles inside the margin), non-peltate leaves may be present on other parts of the same plant **Key 4 p. 14**
      4. Leaf blades never peltate
        5. Leaves or whorls or clusters of leaves cauline (dispersed along elongated stems at more or less regular intervals)
          6. Leaves 3 or more at most nodes **Key 5 p. 14**
          6. Leaves 1 or 2 at each node
            7. Stems creeping, rooting at most nodes (free-floating or bottom-rooted) **Key 6 p. 15**
            7. Stems erect (emergent, floating or submerged) usually rooting only at lower nodes
              8. Leaves in opposite pairs **Key 7 p. 16**
              8. Leaves 1 at each node **Key 8 p. 17**
        5. Leaves or whorls or clusters of leaves not regularly dispersed along elongated stems, most in basal or terminal rosettes or clusters (stems often truncated, corm-like or rhizomatous)
          9. At least some foliage leaves clearly differentiated into petiole and blade; blades flattened and laminate **Key 9 p. 19**
          9. Leaves without a distinct petiole; blades flattened or not **Key 10 p. 19**

Key 1: Vegetative parts not differentiated into stems with leaves (leaves, scales or bracts may be found just below flowers)

1. At least the basal portion of the plant firmly attached to and flattened against a hard substrate (usually on rocks in flowing water) **Podostemaceae p. 316**
1. Plants not attached to and flattened against a hard substrate, free-swimming, free-floating or attached and penetrating a soft substrate
  2. Vegetative parts filamentous, repeatedly and regularly branched; roots absent; with animal catching bladders (mostly submerged) **Utricularia p. 231**
  2. Vegetative parts not filamentous, frond-like, either globose or flattened and discoid to elongate; roots absent or present (mostly free-floating, some submerged but then found near the surface) **Lemnaceae p. 226**

Key 2: Plants with stems and leaves but the leaves reduced to non-photosynthetic scales, the stems green and photosynthetic (bladed or laminate leaves absent).

1. Scales united below, arranged in regular whorls
  2. Lateral branches in regular whorls or absent **Equisetum p. 23**
  2. Lateral branches in clusters at nodes **Cyperaceae p. 94**
1. Scales not united below, arranged spirally or in pairs
  3. Scales in opposite pairs
    4. Petals united to a tube below; stamens not exerted; fruit a many-seeded capsule **Dopatrium p. 342**

4. Petals free to base, scale-like; stamens exerted; fruit of 4 mericarps **Myriophyllum p. 213**
3. Scales spirally arranged, usually at base of stem or solitary
5. Fruit a 1-seeded nut; perianth of hairs, bristles or absent; stem solid **Cyperaceae p. 94**
5. Fruit a many-seeded capsule; perianth of 6, scarious scales; stem hollow **Juncus p. 222**

Key 3: Leaves jointed between the sheath and blade (the base of the leaf is a cylindrical sheath enveloping the stem, free or united at the edges; above the sheath the blade is free; between the blade and sheath is a joint usually bearing a membranous ligule but sometimes a swelling or row of hairs)

1. Leaves pinnately nerved, with straight and parallel nerves diverging from the midrib (large emergent herbs) **Canna p. 83**
1. Leaves not pinnately nerved or if pinnate then nerves not straight and parallel
2. Leaf inserted at the base of the sheath; nerves irregularly branched **Polygonum p. 325**
2. Leaf inserted at or towards the apex of the sheath; nerves indistinct or parallel
3. Leaves in 3 rows (stems sometimes twisted) **Cyperaceae p. 94**
3. Leaves in 2 rows or spirally arranged
4. Perianth present (all segments petaloid, petaloid and sepaloid or appearing as scales attached to stamens)
5. Perianth entirely petal-like **Pontederiaceae p. 328**
5. Perianth with both petals and sepals or entirely sepal-like
6. Petals showy; anthers stalked, all fertile or some staminodial (without fertile anthers) **Commelinaceae p. 85**
6. Petals absent or scale-like; anthers sessile or nearly so, all fertile **Potamogeton p. 331**
4. Perianth absent or reduced to hairs, bristles or scales
7. Fruits sessile in leaf axils or in long-stalked terminal racemes; leaves without silica bodies; stamens 1 or 2 (totally submerged)
8. Fruits in terminal umbels, borne on long flexible stalks; stamens 2 **Ruppia p. 338**
8. Fruits sessile or nearly so in leaf axils; stamen 1 **Zannichellia p. 370**
7. Fruits in spikes, racemes or heads; leaves with silica bodies; stamens 3–6; (flowers in the axils of dry, scarious bracts, grass- or sedge-like plants)
9. Flowers usually each enclosed by 1 bract (glume), variously arranged in spikelets; stems usually solid and triangular in transverse section, without swollen nodes; leaves mostly basal and in 3 rows **Cyperaceae p. 94**
9. Flowers each enclosed by 2 bracts (lemma outside, palea inside), arranged in spikelets each usually subtended by 1 or 2 empty bracts (glumes); stems usually hollow and round in transverse section, with solid and swollen nodes; leaves mostly cauline and in 2 rows **Poaceae p. 280**

Key 4: Leaf blades peltate (attached to their petioles inside the margin), plants may also bear non-peltate leaves on other parts of the stem

1. Leaf blades 3-lobed (usually sagittate or hastate) or elongate-elliptical
2. Leaves mostly radical (borne near to the root); peltate leaf blades borne on rigid petioles and held above the water surface; finely divided submerged leaves absent **Araceae p. 50**
2. Leaves cauline (borne along the stem); peltate leaf blades borne on flexible petioles, floating; finely divided submerged leaves present **Cabomba p. 81**
1. Leaf blades entire, with or without a sinus (the sinus if present not reaching the petiole attachment point) orbicular or nearly so
3. Petioles bearing spines; leaf blades usually more than 10 cm in diameter
4. Leaf blades without spines; plants with milky latex **Nelumbo p. 272**
4. Leaf blades with spines on the under surface; plants without milky latex **Nymphaeaceae p. 273**
3. Petioles without spines; leaf blades less than 10 cm in diameter
5. Plants bearing bladder-like, animal catching traps; flowers bilaterally symmetrical; petals united below **Utricularia p. 231**
5. Plants not bearing bladders; flowers radially symmetrical; petals absent or free
6. Flowers single, conspicuous, borne above the leaf blades; most underwater parts covered in mucilaginous jelly **Brasenia p. 80**
6. Flowers in umbels, heads or whorls, inconspicuous, borne below the leaf blades; underwater parts without mucilaginous jelly **Hydrocotyle p. 45**

Key 5: Leaves whorled (3 or more at most nodes) and cauline (dispersed along elongated stems at regular intervals, not in terminal or basal rosettes or clusters)

1. Whorls with 2 different kinds of leaf (2 floating, entire, green and laminate–the third submerged, divided, brown and root-like) **Salvinia p. 29**

1. Whorls with all leaves alike or nearly so
2. Plants bearing bladder-like, animal catching traps **Utricularia p. 231**
2. Plants not bearing bladders
3. Leaves terminating in an orbicular lobe hinged along the midrib (catching animals) and 4–6 bristles **Aldrovanda p. 182**
3. Leaves not terminating in an orbicular and hinged lobe, terminal bristles absent or less than 4
4. Leaves compound or simple divided into secondary linear or capillary segments
5. Leaves pinnately divided
6. Leaves 2-pinnate or 1-pinnate with at least the lower segments forked; petals tubular towards the base **Limnophila p. 343**
6. Leaves 1-pinnate, all segments simple; petals free, caducous **Myriophyllum p. 213**
5. Leaves repeatedly forked or their segments repeatedly forked
7. Ultimate segments with small, marginal thorn-like projections and terminal bristles, all divisions 2-fid (simply forked) **Ceratophyllum p. 84**
7. All segments smooth, lower divisions 5- or 3-fid **Cabomba p. 81**
4. Leaves simple and entire
8. Leaves interspersed with hard and sharp spines; seeds inserted on hardened, hook-like outgrowths from the central placenta (retinacula) **Hygrophila p. 30**
8. Leaves not interspersed with spines; seeds inserted directly on the placenta or seeds solitary
9. Leaves lanceolate to ovate (submerged, dark green or reddish, somewhat translucent with marginal spines) **Hydrocharitaceae p. 215**
9. Leaves linear to capillary
10. Leaves not equally spaced around the nodes (leaves in clusters at each node)
11. Leaves flat, translucent, with toothed margins **Najas p. 267**
11. Leaves capillary, not translucent, with entire margins
12. Flowers numerous, subtended by spirally arranged glumes, in ovoid heads **Cyperaceae p. 94**
12. Flowers sessile, 1 male and 1 female in leaf axils **Zannichellia p. 370**
10. Leaves equally spaced around the nodes (there is a remarkable amount of convergent evolution among aquatics with this particular growth form, identification without flowers is not practical)
13. Flower buds enclosed within a spathe **Hydrilla p. 218**
13. Flower buds not enclosed within a spathe
14. Petals united below into a tube
15. Inflorescence very dense, spike-like, mostly terminal; fruit of 4 nutlets **Pogostemon p. 224**
15. Inflorescence lax and racemose, if spike-like then not dense or flowers solitary; fruit a capsule **Scrophulariaceae p. 338**
14. Petals free or absent
16. Sepals free, not appendaged between lobes; petals and stamens free from the sepals; fruit of 4 mericarps **Myriophyllum p. 213**
16. Sepals tubular, often appendaged between lobes; petals and stamens inserted on the inner surface of the sepal tube; fruit a capsule **Rotala p. 252**

Key 6: 'Creepers'; Leaves 1 or 2 at each node (alternate or opposite), stems creeping and rooting at most nodes (plants free-floating or bottom-rooted).

1. Plants free-floating (pleustophytes); floating leaves sessile or nearly so, in 2 rows
2. Floating leaves opposite, 1-lobed, aerial surface with multicellular hairs **Salvinia p. 29**
2. Floating leaves alternate, 2-lobed, the lower lobe in contact with the water, aerial surface without multicellular hairs **Azolla p. 22**
1. Plants bottom-rooted or if floating then floating leaves with petioles and not in 2 rows
3. Leaves elongate, ± terete, without a flattened blade, not differentiated into petiole with blade
4. Leaves 2 at each node, in opposite pairs; plants without bladders **Scrophulariaceae p. 338**
4. Leaves 1 at each node; plants with bladder-like, animal catching traps **Utricularia p. 231**
3. Leaves with flattened blades, usually clearly differentiated into petiole with blade
5. Leaves compound (4 or more leaflets on each petiole)
6. Leaves 1- or 2-pinnate; each leaflet with midrib and lateral veins; stems usually somewhat woody but when in water spongy **Fabaceae p. 206**
6. Leaves with 4 terminal leaflets; each leaflet without midrib but with repeatedly forked veins; stems not woody and not spongy **Marsileaceae p. 25**
5. Leaves simple (sometimes lobed to the midrib)
7. Leaves 2 at each node, in opposite pairs; leaf blades usually sessile or nearly so
8. Flowers in dense terminal or axillary heads
9. Heads not surrounded by enlarged bracts, usually white and paper-like; ovaries superior **Amaranthaceae p. 40**

9. Heads surrounded by enlarged bracts, not white and paper-like; ovaries inferior **Asteraceae p. 66**

8. Flowers solitary in leaf axils or in loose cymes or racemes

10. Petals united to a tube below

11. Flowers radially symmetrical; ovaries inferior **Rubiaceae p. 335**

11. Flowers bilaterally symmetrical; ovaries superior

12. Seeds inserted on hardened, hook-like outgrowths from the central placenta (retinacula); capsules usually elongate **Acanthaceae p. 30**

12. Seeds inserted directly on the central placenta, hard, hook-like outgrowths absent; capsules usually globose **Scrophulariaceae p. 338**

10. Petals free to the base

13. Ovaries inferior; petals usually showy, yellow or white **Ludwigia p. 276**

13. Ovaries superior; petals either inconspicuous or if showy then pink, red or purple

14. Sepals free to the base; petals free from the sepals **Elatinaceae p. 184**

14. Sepals united into a tube surrounding the ovaries; petals borne at the top of the sepal tube between the sepal lobes **Lythraceae p. 248**

7. Leaves 1 at each node, alternate; leaf blades sessile or with petioles

15. Stems rhizomatous, covered with brownish scales; leaves rolled spirally in bud, at maturity bearing sporangia; ferns **Microsorum p. 27**

15. Stems or rhizomes without brownish scales; leaves not rolled spirally in bud, not bearing sporangia; flowering plants

16. Stems woody throughout; fruits soft orange to red drupes (rocky river beds) **Rotula p. 78**

16. Stems herbaceous or sometimes somewhat woody at the base; fruits hard (nuts, mericarps or capsules), not orange or red

17. Leaves inserted at the base of a tubular sheath which envelops the stem (ochrea) **Polygonum p. 325**

17. Leaves not inserted at the base of a sheath

18. Leaves crisped, asymmetrical, with sharp teeth; petals and stamens usually 4 **Coldenia p. 75**

18. Leaves flat, not crisped, symmetrical or nearly so, entire or toothed; petals and stamens either 3 or more

19. Flowers inconspicuous; ovaries inferior; fruit of 2 mericarps **Apiaceae p. 44**

19. Flowers conspicuous; ovaries superior; fruits capsules

20. Perianth entirely petaloid **Pontederiaceae p. 328**

20. Perianth differentiated into sepaloid sepals and petaloid petals

21. Petals 4, free to base; stamens 6 **Nasturtium p. 78**

21. Petals 4–9 united into a tube below: stamens as many as the petal lobes

22. Petals 5, the lobes much shorter than the tube; leaf blades with midrib and pinnate nerves, usually aerial, the margins usually lobed **Convolvulaceae p. 91**

22. Petals 4–9, the lobes much longer than the tube; leaf blades without a dominant midrib, palmately nerved, usually floating, the margins entire **Nymphoides p. 263**

Key 7: Leaves 2 at each node, in opposite pairs regularly dispersed along erect stems (stems emergent, floating or submerged, usually rooting only at the lower nodes)

1. Plants firmly attached to rocks in swiftly flowing water **Podostemaceae p. 316**

1. Plants rooting in a soft substrate or floating

2. Leaves divided into linear segments, the sinuses reaching the midrib or leaf base

3. Leaves palmately divided; sepals 3, free to base; petals 3, free to base; stamens 6 **Cabomba p. 81**

3. Leaves pinnately divided; sepals 4 or 5, free or united at the base; petals 5, free or united at the base; stamens 2 or 4

4. Fruit breaking into 4, 1-seeded mericarps; petals inconspicuous, free to the base, often caducous **Myriophyllum p. 213**

4. Fruit breaking into 4, 1-seeded mericarps; petals showy, united into tube below

5. Leaf segments distinctly flattened, the margins notched or toothed **Acanthaceae p. 30**

5. Leaf segments filiform, terete or slightly flattened, the margins entire **Scrophulariaceae p. 338**

2. Leaves entire, sometimes lobed but the sinuses not reaching the midrib

6. Leaves capillary, with sheathing stipular bases (totally submerged herbs)

7. Flowers stalked, borne in spike-like inflorescences, bisexual **Potamogeton p. 331**

7. Flowers sessile or subsessile, borne in leaf axils, unisexual **Zannichellia p. 370**

6. Leaves not capillary, without sheathing stipular bases (submerged or emergent herbs)

8. Flowers in heads subtended by an involucre of bracts **Asteraceae p. 66**
8. Flowers not in heads or if in heads then not subtended by an involucre of bracts
9. Flowers naked (without sepals and petals or perianth) **Callitriche p. 81**
9. Flowers with sepals and petals or a perianth
10. Ovaries completely inferior
11. Petals free to the base; petals showy, yellow or white **Ludwigia p. 276**
11. Petals united to a tube below; relatively small, whitish **Rubiaceae p. 335**
10. Ovaries superior or perigynous
12. Plants with milky latex; capsules swelling and inflated at maturity; seeds with hairs **Oxystelma p. 66**
12. Plants without latex; fruit not a capsule or if a capsule then not swelling at maturity; seeds without hairs
13. Perianth of 4, dry, membranous segments; flowers usually in silver-white heads **Amaranthaceae p. 40**
13. Perianth of sepals and petals, never dry and membranous; flowers not in silver-white heads
14. Petals free to the base
15. Sepals free or united only at base **Elatinaceae p. 184**
15. Sepals united into a perigynous tube **Lythraceae p. 248**
14. Petals united and tubular towards the base
16. Fruits breaking into 4, 1-seeded nutlets or mericarps **Pogostemon p. 224**
16. Fruits 2-valved capsules
17. Petals radially symmetrical, twisted in bud; petal lobes 4 **Hoppea p. 212**
17. Petals usually bilaterally symmetrical, 2-lipped, not twisted in bud; petal lobes 5
18. Seeds inserted on hardened, hook-like outgrowths from the central placenta (retinacula) **Acanthaceae p. 30**
18. Seeds inserted directly on the central placenta, hard, hook-like outgrowths absent **Scrophulariaceae p. 338**

Key 8: Leaves solitary at nodes (alternate) regularly dispersed along erect stems (stems emergent, floating or submerged, usually rooting only at the lower nodes)

1. Leaves thistle-like (with pinnately arranged marginal spines); flowers in ± spherical heads **Sphaeranthus p. 71**
1. Leaves not thistle-like; flowers in heads or not
2. Leaf blades of 4, ± equally sized leaflets
3. Each leaflet with midrib and lateral veins; stems usually somewhat woody but when in water spongy; flowers borne in the axils of bracts **Geissaspis p. 208**
3. Each leaflet without midrib but with repeatedly forked veins; stems not woody and not spongy; flowers absent, spores borne in sporocarps towards the base of the petioles **Marsileaceae p. 25**
2. Leaf blades never of 4, ± equally sized leaflets
4. Leaves compound or leaf blades lobed
5. Leaves palmately divided or palmately lobed
6. Flowers inconspicuous, borne below the leaf blades; ovaries inferior; fruits mericarps **Apiaceae p. 44**
6. Flowers conspicuous, borne above the leaf blades; ovaries superior; fruits heads of nutlets **Ranunculus p. 334**
5. Leaves pinnately divided or pinnately lobed
7. Leaves 2- or more-pinnate with regular, ovate leaflets **Neptunia p. 209**
7. Leaves 1-pinnate or pinnately lobed
8. Flowers solitary in leaf axils, rather inconspicuous; fruits of 4 mericarps **Myriophyllum p. 213**
8. Flowers in inflorescences, naked or in the axils of bracts; fruits not of 4 mericarps
9. Flowers in dense heads, each head with an involucre of bracts **Asteraceae p. 66**
9. Flowers in lax racemes or umbels, without involucres of bracts
10. The leaflets or pinnae entire and very constant in size and shape in any single leaf; stems usually somewhat woody but when in water spongy **Fabaceae p. 206**
10. The leaflets or pinnae lobed and varying in size and shape in any single leaf; stems herbaceous, not spongy
11. Inflorescences racemose; fruits dehiscent capsules **Nasturtium p. 78**
11. Inflorescences umbellate; fruits of 2 mericarps **Oenanthe p. 45**
4. Leaves simple and entire
12. Leaves inserted at the base of a tubular sheath which envelops the stem (ochrea) **Polygonum p. 325**
12. Leaves not inserted at the base of a sheath
13. Flowers crowded in dense, ± globose heads; heads subtended by an involucre of sterile bracts
14. Peduncles (culms) enclosed in a sheath at the base; flowers unisexual; roots transversely septate **Eriocaulaceae p. 187**
14. Peduncles not enclosed in a sheath at the base; flowers female or bisexual; roots not septate

15. Leaf blades laterally flattened, with a distinct midrib, without transverse septae; perianth segments not dry and not scale-like **Asteraceae p. 66**

15. Leaf blades vertically flattened (unifacial), without a midrib, with transverse septae; perianth segments dry and scale-like **Juncus p. 222**

13. Flowers not in dense, globose heads; involucre absent

16. Leaves linear, not differentiated into petiole and blade

17. Totally submerged herbs; leaves flaccid, almost transparent

18. Fruit a many seeded capsule **Hydrocharitaceae p. 215**

18. Fruit a nut or head of nutlets

19. Perianth of 4, clawed scales, each attached to a stamen **Potamogeton p. 331**

19. Perianth absent

20. Fruits distinctly stalked, in umbels **Ruppia p. 338**

20. Fruits sessile or subsessile in leaf axils **Zannichellia p. 370**

17. Emergent herbs; leaves self-supporting, not transparent

21. Leaf blades bearing sensitive, sticky, gland-tipped tentacles, spirally rolled in bud; inflorescences spirally rolled (helicoid) **Drosera p. 183**

21. Leaf blades not bearing gland-tipped tentacles, not spirally rolled in bud; inflorescences erect, not rolled

22. Flowers borne in loose cymes; bract margins entire; flowers radially symmetrical **Juncus p. 222**

22. Flowers borne in elongate spikes; bract margins ciliate; flowers bilaterally symmetrical **Zeuxine p. 279**

16. Leaves not linear, either differentiated into petiole and blade or blade flattened and laminate

23. Leaf blades bearing sensitive, sticky, gland-tipped tentacles, spirally rolled in bud; inflorescences spirally rolled (helicoid) **Drosera p. 183**

23. Leaf blades not bearing gland-tipped tentacles, not spirally rolled in bud; inflorescences erect, not rolled

24. Inflorescence a simple, fleshy spike (spadix) of small and inconspicuous flowers subtended by a thick bract (spathe) **Araceae p. 50**

24. Inflorescence without a spadix, spathe absent or if present then of 2, thin segments

25. Perianth not clearly differentiated into sepals and petals, of 4, clawed scales, each attached to a stamen **Potamogeton p. 331**

25. Perianth segments clearly differentiated into sepals and petals, not scale-like, not attached to stamens

26. Floral parts in whorls of 3 or 6 (monocots)

27. Sepals and petals sepaloid; stamens petaloid; leaf blades pinnately nerved with straight, parallel nerves arising from the midrib (large emergent herbs) **Cannaceae p. 83**

27. Either sepals and petals petaloid or sepals sepaloid and petals petaloid; stamens not petal-like; leaf blades with nerves longitudinally parallel from base to tip

28. Perianth of 6 petaloid segments **Pontederiaceae p. 328**

28. Perianth of 3 sepals and 3 petals

29. Ovaries superior; leaves sheathing at the base; usually emergent herbs; some stamens usually staminodial **Commelinaceae p. 85**

29. Ovaries inferior; leaves not sheathing; submerged herbs; all stamens fertile or flowers unisexual **Hydrocharitaceae p. 215**

26. Floral parts in whorls of 4 or 5 (leaf blades with pinnate or reticulate nerves or apparently lacking nerves except for a midrib; dicots)

30. Ovaries inferior

31. Flowers bilaterally symmetrical, adaxially split; anthers united in a ring around the style; petals blue to pink or purple **Lobelia p. 82**

31. Flowers radially symmetrical, not split; anthers free from each other; petals yellow or white **Ludwigia p. 276**

30. Ovaries superior or plants dioecious (female flowers might be lacking)

32. Fruits indehiscent, splitting into 1-seeded nutlets or mericarps

33. Flowers borne in elongate, coiled cymes; petals united into a tube below; flowers bisexual **Boraginaceae p. 75**

33. Flowers solitary in leaf axils or borne in upright racemes; petals free, caducous; flowers unisexual or bisexual (if male then flowers 4-merous) **Myriophyllum p. 213**

32. Fruits dehiscent capsules or berries, few- to many-seeded

34. Flowers with spurs; leaf bases bearing paired glands; fruits purplish red **Hydrocera p. 75**

34. Flowers without spurs; leaf bases without glands; fruits not purplish-red

35. Petal tube much longer than the petal lobes; flowers trumpet-like **Convolvulaceae p. 91**

35. Petals free or petal tube shorter than the petal lobes; flowers never trumpet-like

36. Inflorescence a densely packed spike; capsule opening by an apical lid **Sphenoclea p. 363**
36. Inflorescence a panicle of loose axillary cymes or flowers solitary in leaf axils; capsule opening by longitudinal slits or valves **Hydrolea p. 221**

Key 9: Leaves petiolate and not regularly dispersed along the total length of the stem, most leaves in basal or terminal rosettes or clusters; stems often truncate, corm-like or rhizomatous

1. Leaves compound or at least some leaves lobed to the base or to the midrib
2. Leaves irregularly pinnatisect near the sinus; petioles spiny; flowers borne on a spadix surrounded by a spathe **Araceae p. 50**
2. Leaves not irregularly pinnatisect; flowers not borne on a spadix, not surrounded by a spathe
3. Leaves palmately lobed; fruit of several, free nutlets **Ranunculus p. 334**
3. Leaves pinnately lobed; fruit not of free nutlets or fern without fruit
4. Young leaves not spirally rolled; mature leaves not bearing sporangia; flowers in umbels (leaves ± triangular or irregular in outline; most pinnae pinnately divided or repeatedly forked) **Oenanthe p. 45**
4. Young leaves spirally rolled; mature leaves bearing sporangia on the abaxial surface
5. Leaflets very constant in shape from top to bottom of each leaf, margins entire; leaves erect, up to 3 m or more long **Acrostichum p. 27**
5. Leaflets variable in shape from top to bottom of each leaf, margins usually lobed; leaves submerged, floating or erect, rarely more than 50 cm long **Ceratopteris p. 27**
1. Leaves simple, entire or lobed only at the margin
6. Leaf blades forming a circular mosaic of leaves floating on the surface
7. Petioles not inflated; fruits flattened and winged, without lateral horns; flowers unisexual, without perianth; male flowers with 1 stamen **Callitriche p. 81**
7. Petioles inflated; fruits with 2 or 4, lateral, hard horns; flowers bisexual, with sepals and petals **Trapa p. 363**
6. Leaf blades not forming a circular mosaic of whorled leaves floating on the surface
8. Inflorescence a simple, fleshy spike (spadix) of small and inconspicuous flowers subtended by a thick bract (spathe) **Araceae p. 50**
8. Inflorescence without a spadix, spathe absent or if present then of 2, thin segments
9. Carpels 2 or more, free
10. Leaf nervature almost regularly rectangular; flowers in simple or branched spikes **Aponogetonaceae p. 46**
10. Leaf nervature not regularly rectangular; flowers not in spikes
11. Leaf margins lobed or notched but not sagittate; flowers not 3-merous (petals and sepals 5, stamens numerous) **Ranunculus p. 334**
11. Leaf margins entire or sagittate; flowers 3-merous
12. Fruits 1-seeded nutlets; flowers borne in whorls **Alismataceae p. 37**
12. Ovaries containing many seeds, the ovules scattered over the inner surface of the carpels; flowers borne in simple umbels **Limnocharitaceae p. 247**
9. Carpels 1 or more, united
13. Styles united into a disk with radiating stigmas; stamens more than 6 **Nymphaeaceae p. 273**
13. Styles not united into a disk, without radiating stigmas; stamens 6 or less
14. Ovaries inferior; male flowers if present 3-merous **Hydrocharitaceae p. 215**
14. Ovaries superior; male flowers if present 5-merous
15. Perianth differentiated into sepals and petals; bisexual flowers 4–8-merous **Nymphoides p. 263**
15. Perianth either sepaloid or petaloid; flowers 3-merous
16. Perianth sepaloid or bract-like, not showy; leaves stiff and distinctly ribbed; petioles not inflated (Sri Lanka) **Hanguana p. 214**
16. Perianth petaloid, showy; leaves not stiff and not distinctly ribbed; petioles often inflated (widespread) **Pontederiaceae p. 328**

Key 10: Leaves without distinct petioles and not regularly dispersed along the total length of the stem, most leaves in basal or terminal rosettes or clusters; stems usually truncate, corm-like or rhizomatous

1. Leaves flattened throughout most of their length (parallel-sided in transverse section, sometimes 'V'-shaped or curved, not terete, not semicircular, not triangular, not rectangular)
2. Leaves widest towards the apex (spathulate, obovate or obtriangular)
3. Leaf blades bearing sensitive, sticky, gland-tipped tentacles, spirally rolled in bud; inflorescences spirally rolled (helicoid) **Drosera p. 183**
3. Leaf blades not bearing gland-tipped tentacles, not spirally rolled in bud; inflorescences erect, not rolled

4. Leaves densely pubescent; plants free-floating rosettes **Pistia p. 65**
4. Leaves glabrous; plants attached to the bottom
5. Fruits splitting into 2 or 4 mericarps
6. Mericarps 4, flattened and winged; flowers unisexual; perianth absent **Callitriche p. 81**
6. Mericarps 2, ± hemispherical; flowers bisexual; perianth of 5 sepals and 5 petals **Phyla p. 366**
5. Fruits capsules; petals united into a tube below
7. Plants with bladder-like, animal catching traps; petals with abaxial spurs; leaves alternate **Utricularia p. 231**
7. Plants without animal catching traps; petals without spurs; leaves opposite **Scrophulariaceae p. 338**
2. Leaves widest at or towards the base or leaves parallel-sided
8. Midrib running asymmetrically towards one side of the leaf; inflorescence cylindrical apparently arising near the base of the leaf **Acorus p. 50**
8. Midrib running symmetrically up the middle of the leaf or midrib absent
9. Leaves with distinct spiny margins
10. Stems emergent, elongate, hard, woody; ovaries superior **Pandanus p. 279**
10. Stems submerged, corm-like or if elongate then soft, flexible, never woody; ovaries inferior **Hydrocharitaceae p. 215**
9. Leaves with entire or minutely toothed margins
11. Flowers in compact, spherical to discoid heads
12. Heads sessile surrounded by and embedded in foliage leaves (minute herbs) **Trithuria p. 215**
12. Heads borne on stalks or culms, held above the foliage leaves
13. Flowers unisexual; petals small, not showy, white or black; seeds solitary **Eriocaulon p. 187**
13. Flowers bisexual; petals showy, yellow; seeds 2 or more in each loculus **Xyris p. 367**
11. Flowers not in compact, spherical to discoid heads
14. Flowers densely packed in simple cylindrical, superposed spikes on a single axis; flowers unisexual **Typha p. 364**
14. Flowers not densely packed, not in cylindrical inflorescences
15. Perianth united into tube below
16. Perianth blue to purple; stamens 3; small annuals without a rootstock **Burmannia p. 79**
16. Perianth white tinged with red; stamens 6; rootstock a bulb **Crinum p. 43**
15. Perianth or free sepals and free petals
17. Perianth of scarious, bract-like segments; leaves usually with transverse septae **Juncus p. 222**
17. Perianth of sepaloid sepals and petaloid petals (petals sometimes reduced or absent); leaves without septae
18. Carpels free; fruit a head of 1-seeded nutlets **Alismataceae p. 37**
18. Carpels united; fruit a many-seeded capsule
19. Ovary superior; stems usually emergent with self-supporting leaves **Commelinaceae p. 85**
19. Ovary inferior; stems submerged with submerged or floating leaves **Hydrocharitaceae p. 215**
1. Leaves not flattened throughout most of their length (circular, semicircular, ellipsoidal, triangular or rectangular in transverse section)
20. Leaves capillary (hair-like and flaccid)
21. Flowers tightly packed in globose heads; heads subtended by an involucre of bracts
22. Perianth dry, papery, inconspicuous, greyish or blackish; flowers unisexual; roots septate **Eriocaulon p. 187**
22. Perianth petaloid, showy, yellow, blue or rarely white; flowers bisexual; roots not septate **Xyris p. 367**
21. Flowers not tightly packed in globose heads or if so then the heads not subtended by an involucre of bracts
23. Flowers radially symmetrical, bisexual; perianth of 6 ± equally sized dry and scale-like segments **Juncus p. 222**
23. Flowers not radially symmetrical, unisexual or bisexual, each subtended by 1 glume-like bract; perianth absent or of scales or bristles **Cyperaceae p. 94**
20. Leaves not capillary (awl-like or elongate, self-supporting)
24. Leaves with 4, longitudinal, septate gas canals (visible in transverse section); leaf bases swollen and usually bearing spores **Isoetes p. 23**
24. Leaves without 4, longitudinal, gas canals; leaf bases not swollen, never bearing spores
25. Perianth absent or of scales or bristles
26. Leaves in 2 rows, equitant; inflorescence of 2, superposed spikes (male above and female below) **Typha p. 364**
26. Leaves in 3 rows or arranged spirally, not equitant; inflorescence not of superposed spikes
27. Perianth absent or of scales or bristles **Cyperaceae p. 94**

27. Perianth of 6 scarious segments **Juncus p. 222**

25. Perianth sepaloid and/or petaloid, not of scales or bristles

28. Flowers radially symmetrical; stamens 6; rootstock a bulb **Crinum p. 43**

28. Flowers bilaterally symmetrical; stamen 1; rootstock a corm **Philydrum p. 280**

Descriptions

PTERIDOPHYTINA (ferns and fern allies)

Plants with an alteration of free-living generations. Sporophytes herbaceous, with vascular tissue, reproducing by spores which give rise to small, short-lived gametophytes (prothalli) bearing archegonia and antheridia.

AZOLLACEAE: 1 genus

Azolla: 6 species: through introductions almost cosmopolitan, 1 species in India

Azolla pinnata R. Brown subsp. ***asiatica*** R. M. K. Saunders et K. Fowler [*A. imbricata* (Roxburgh ex Griffith) Nakai, *A. pinnata* var. *imbricata* (Roxburgh ex Griffith) Bonaparte]: Fig. 2

Roots simple, bearing numerous root hairs; root hairs in obliquely arranged groups of (3–) 5–6. Stems alternately branched; branches arranged pinnately or bipinnately (the whole plant appearing deltoid or trapezoid in outline), up to 4 cm long but usually less. Leaves alternate, lax or overlapping towards the tips of the branches, ovate, obtuse, each divided into 2 lobes; the adaxial (upper) lobe thick, green and usually somewhat reddish, borne above the water, with 2-celled papillae arranged in regular rows, the margin hyaline and very narrow; the abaxial (lower) lobe 1 cell thick, colourless, short and broad (length to width ratio ± 1.2:1), in contact with the water on the abaxial surface only, with cavities usually inhabited by a cyanobacterium (*Anabaena azollae* Strasburger). Sporocarps borne in pairs in the axils of the first leaves of branches on older parts of the plant; megasporangial sporocarps small, each producing 1 megasporangium with 1 megaspore; megaspores each with 9 floats. Microsporangial sporocarps larger, producing up to 100 microsporangia, each containing 32 or 64 microspores; microsporangia free floating, held together in groups (massulae) by hardened frothy mucilage, each with hairs or hooks (glochidia). Glochidia on the microsporangial massulae without barbs, thread-like, borne only on the inner surface.

Annual or sometimes perennial in permanent water. Free-floating on permanent and temporary water, in inundated rice fields, canals and ponds. It rarely reaches pest proportions in India. The disseminules are sporocarps or sporangia in groups (massulae), dispersed by water and perhaps also by animals. The megaspores germinate at the water surface and produce minute prothalli. *Azolla* is used as a fertilizer as it contains nitrogen-fixing cyanobacteria. It can replace 25–30 kg N ha^{-1} as applied through ammonium sulphate. It can also become a weed and may be controlled by diesel distillates 550–1100 l ha^{-1} or by Diquat 0.25–1 ppmw or 3 kg ha^{-1} applied as foliar spray or injected into confined waters. Saunders, R. M. K. and Fowler, K. A morphological taxonomic revision of *Azolla* Lam. section *Rhizosperma* (Mey.) Mett. (Azollaceae). Botanical Journal of the Linnean Society 109(3): 329–357 (1992).

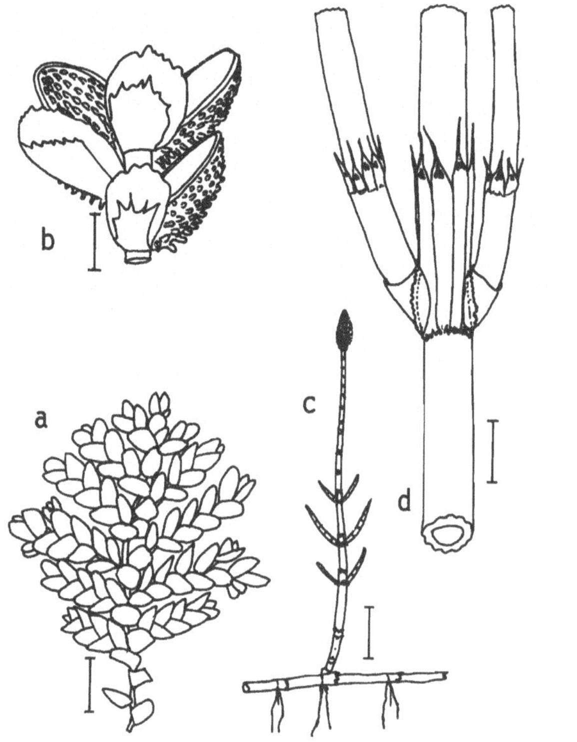

Fig. 2. *Azolla pinnata* subsp. *asiatica*: a, plant viewed from above (2 mm); b, shoot with three leaves viewed from below (0.5 mm). *Equisetum ramosissimum* subsp. *debile*: c, habit (2 cm); d, portion of a shoot (2 mm).

{(⊙) Ple}. Tropical and subtropical Asia: Ban, Bur, Cey, Ind, Nep, Pak: Ap, As, Bh, Kl, Kt, Mh, Mn, Or, Rj, Sk, Tn, Tp, Up, Wb.

EQUISETACEAE: 1 genus

Equisetum: 29 species: almost cosmopolitan, mainly north temperate zone

Hauke, R. A taxonomic monograph of the genus *Equisetum*. Part 1, subgenus *Hippochaete*. Beihefte zur Nova Hedwigia 8: 1–123 (1966); part 2, subgenus *Equisetum*. Nova Hedwigia 30: 385–455 (1978).

Excluded: the following are found in the Himalayas but do not extend to the Indus–Gangetic plain or into Peninsular India: ***Equisetum diffusum*** D. Don; ***E. fluviatile*** Linnaeus [*E. limosum* Willdenow]; ***E. palustre*** Linnaeus; ***E. ramosissimum*** Desfontaines subsp. ***ramosissimum***.

Equisetum ramosissimum Desfontaines subsp. ***debile*** (Roxburgh) Hauke [*E. debile* Roxburgh]: Fig. 2

Rhizomes creeping, giving rise to erect stems. Erect stems jointed, smooth, evergreen, 4–40 cm tall, 2.5–8 mm in diameter, usually freely branched; grooves 10–32; central hollow half to two-thirds the diameter of the stem; stomata in 1 line on each side of the groove. Branches hollow, 6- to 10-angled; the lowest internode much shorter than the sheath on the main stem. Leaves whorled, scale-like, as many as grooves, united into a sheath below, green; teeth of sheaths breaking off, with narrow scarious border elongated into an aristate or cuspidate tip. Cones borne at the stem tip, consisting of numerous, whorled, peltate sporangiophores. Sporangia borne several together around the adaxial surface of the sporangiophores. Spores all alike, overlaid by 2 or 4 spiral bands (elaters). Gametophytes unisexual, with cushion-like base and lobed tip.

Perennial, usually found in moist soil. It is essentially a helophyte and sometimes grows in regions that are regularly flooded and may even be found in shallow water. It is often locally dominant and can form extensive stands. The disseminules are spores dispersed in clumps by wind; local spread is by rhizomes.

{Hel}. S and E Asia, extending to New Caledonia and Fiji: Bur, Cey, Ind, Nep: Dl, Hr, Jk? Rj, Tn, Up.

ISOETACEAE: 1 genus

Isoetes: ± 130 species (there are some higher estimates): cosmopolitan

Stems short, tuber-like, 2- or 3-lobed; roots forked. Leaves in a rosette, linear to subulate, each with 4 transversely septate, longitudinal gas channels and a central vascular strand; leaf base spoon-shaped, with a membranous margin and a delicate ligule on the adaxial surface where the leaf narrows. Sporangia develop near the leaf base, embedded, naked or covered with thin tissue (velum). Spores polymorphic or dimorphic, mostly in separate sporangia (mega- and microsporangia). Gametophytes remain within the spore wall.

Submerged, emergent or seasonally submerged perennials. Heterosporous: disseminules mega- and microspores or fertilized megagametophytes, dispersed by water and in mud by animals (the feet of birds, in excreta of earthworms, etc.). Some species are sensitive indicators of water quality. The stems are readily eaten by waterfowl and rodents. Srivastava, G. K., Pant, D. D. and Shukla, P. K. The Genus *Isoetes* L. in India. American Fern Journal 83(4): 105–119 (1993).

Thirteen endemic species have been described in India. Most seem to intergrade with each other and Indian authors do not agree about their specific limits. I feel that none have been adequately studied in the field and have, therefore, listed them circumscribed under the widespread species *I. coromandelina* giving no more than their published 'diagnostic features'. I suspect there are no more than two or three species with perhaps some hybrids or agamospermous races and most of these names will later be shown to be no more than 'paper' species.

Isoetes coromandelina Linnaeus fil. [*I. indica* J. G. König ex J. E. Smith, *I. brachyglossa* A. Braun]: Fig. 3

Leaves (5–) 30–60 (–80) cm long, with 4 strong and several small peripheral fibrous strands; gas channels with internal hairs. Sporangia orbicular on the outer side becoming obovoid towards the centre, 9–14 mm in diameter; velum absent. Megaspores white when dry, grey when wet, di- or polymorphic, (300–) 330–560 (–660) μm in diameter, distally tuberculate; tubercles blunt and rounded; the triradiate ridges straight and simple. Microspores sometimes rare, reddish or dull, smooth or rough with papillae or spines.

Perennial. Usually found at the edges of pools and lakes. The leaves sprout as the plants become flooded and wither when the water recedes and substrate dries out, often remaining green for a few weeks only; dormant corms survive the dry season. They grow solitary or gregarious, in open or closed vegetation. Panigrahi recognized *I. brachyglossa* as a distinct subspecies.

{Ros/Ten}. Distribution uncertain depending on the taxonomic delimitation either widespread in tropical Asia, N Australia and N Italy or endemic to the Indian subcontinent: Cey, Ind: Ap, Bh, Gj, Kl, Kt, Mh, Mp, Or, Rj, Tn, Up, Wb.

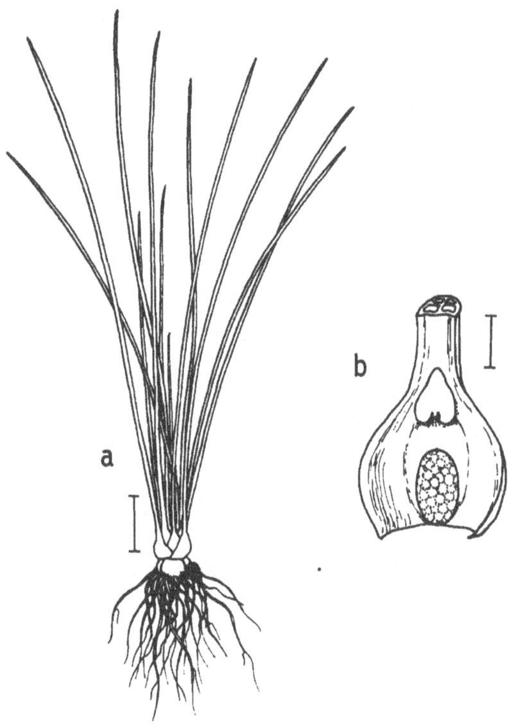

Fig. 3. *Isoetes coromandelina*: a, whole plant (2 cm); b, adaxial surface of the leaf base showing sporangium and ligule (3 mm).

Isoetes bilaspurensis G. Panigrahi

Like *I. coromandelina* but leaves 5–10 cm long; velum covering two-thirds to three-quarters of the sporangium; megaspores 250–440 μm in diameter, with pustules of almost equal size.

{Ros/Ten}. Endemic to India: ♦Ind: Mp.

Isoetes dixitei D. V. Shende

Like *I. coromandelina* but leaves usually less than 25 cm long; tubercles of the megaspores unequal in size; megaspores 320–580 μm in diameter; megasporangia ovate, 2–6 mm long, 2–2.5 mm wide; microspores echinate.

{Ros/Ten}. Endemic to India: ♦Ind: Mh.

Isoetes mahadevensis G. K. Srivastava, D. D. Pant et P. K. Shukla

Like *I. coromandelina* but megaspores reticulate; anticlinal height of muri not exceeding the tangential thickness of the muri; velum rudimentary.

{Ros/Ten}. Endemic to India: ♦Ind: Mp.

Isoetes panchananii D. D. Pant et G. K. Srivastava

Like *I. coromandelina* but leaves usually less than 25 cm long; megaspores reticulate on all surfaces; velum covering ± half the sporangial surface; megaspores 240–410 μm in diameter, the muri generally forming areoles; megasporangia oval, 3–5 mm long, 2–3 mm wide..

{Ros/Ten}. Endemic to India: ♦Ind: Mh, Mp.

Isoetes panchganiensis G. K. Srivastava, D. D. Pant et P. K. Shukla

Like *I. coromandelina* but megaspores with reticulations on the central part of the distal face only; velum covering more than three-quarters of the sporangium

{Ros/Ten}. Endemic to India: ♦Ind: Mh.

Isoetes panchganiensis subsp. ***kemangundiensis*** G. K. Srivastava, D. D. Pant et P. K. Shukla

Like *I. panchganiensis* but has larger megaspores with incomplete areoles.

{Ros/Ten}. Endemic to India: ♦Ind: Kt.

Isoetes pantii H. K. Goswami et B. S. Arya

Like *I. coromandelina* but with a tilted rhizomorph (3-lobed but 1 lobe smaller than the other 2); leaves 15–33 cm long; megasporangia circular to oblong-elongate, 3–7 mm long, 2–4 mm broad; megaspores (?70–) 280–600 μm in diameter, tuberculate; megaspore tubercles large and small intermixed; microspores 15–54 μm in diameter.

{Ros/Ten}. Endemic to India: ♦Ind: Mp.

Isoetes rajasthanensis C. B. Gena et T. N. Bhardwaja

Like *I. coromandelina* but leaves 5–16 cm long; megaspores reticulate; velum covering three-quarters or more of the sporangia surface; megaspores 300–350 μm in diameter, trimorphic, exine with branched ridges; megasporangia ovate to elongate, 3–6.5 mm long, 2–4.5 mm wide.

{Ros/Ten}. Endemic to India: ♦Ind: Rj.

Isoetes reticulata C. B. Gena et T. N. Bhardwaja

Like *I. coromandelina* but leaves 4–10 cm long; megaspores tuberculate on the proximal surface and reticulate on the distal surface; velum covering two-thirds or more of the sporangia surface; megaspores 160–245 μm in diameter, dimorphic; megasporangia oval to circular 2.5–3.5 mm in diameter.

{Ros/Ten}. Endemic to India; ♦Ind: Rj.

Isoetes sahyadrii T. S. Mahabalé [*I. sahyadriensis* G. Panigrahi orth. mut.]

Like *I. coromandelina* but leaves usually less than 25 cm long; megasporangia almost round, ± 2 mm in diameter; velum almost complete, with slit-like aperture at the base; tubercles of the megaspores discrete but unequal in size; megaspores 325–430 μm in diameter; microspores echinate.

{Ros/Ten}. Endemic to India: ♦Ind: Kt, Mh.

Isoetes sampathkumaranii L. N. Rao

Like *I. coromandelina* but leaves usually less than 25 cm long; velum incomplete (covering half to two-thirds); tubercles of the megaspores confluent at their bases, giving rise to irregularly branched ridges; megaspores 256–458 μm in diameter; megasporangia oval, 2–6 mm long, 2–2.5 mm wide.

{Ros/Ten}. Endemic to India: ♦Ind: Ap, Kt.

Isoetes tuberculata C. B. Gena et T. N. Bhardwaja

Like *I. coromandelina* but megaspores trimorphic.

{Ros/Ten}. Endemic to India; ♦Ind: Rj.

Isoetes unilocularis J. E. Smith [*I. indica* D. D. Pant et G. K. Srivastava non J. G. König ex J. E Smith, *I. mirzapurensis* G. Panigrahi et Dixit]

Like *I. coromandelina* but the megaspores with the triradiate ridges mostly curved and often branched, the tubercles tapering; megaspores 90–640 μm in diameter; megasporangia elongate, 6–19 mm long, 4–9 mm wide.

{Ros/Ten}. Endemic to India: ♦Ind: Mp, Tn, Up.

MARSILEACEAE: 3 genera

Marsilea: ± 45 species: tropical to warm regions of the world

Stems creeping or floating, rooting at nodes. Leaves spirally rolled when young, compound; petioles long, terminating in 4 leaflets; leaflets arranged symme trically cross-wise, each obdeltate or obovate, the margin entire when growing in water usually becoming toothed or crenate when dry, the veins repeatedly forked. Sporangia in closed sporocarps, borne on short stalks arising from the petioles or in their axils. Spores of 2 kinds, usually found together in the same sporocarp; megaspores solitary; microspores numerous.

Juveniles submerged, adults floating, emergent or terrestrial. Plants usually require an emergent period to develop sporocarps. The sporocarps may remain viable for 100 years; they also may germinate after passing through a duck. The disseminules are sporocarps, dispersal by birds (endozoochor), other animals, by water and probably also in mud. *Marsilea* is sometimes considered to be troublesome in rice fields and in irrigation ditches. It is used locally in medicine. The sporocarps are eaten by the aborigines in the Andaman Islands. Gupta, K. M. *Marsilea.* Botanical Monographs (CSIR, New Delhi) 2: i–vi, 1–113 (1962), this work has not found acceptance; Launert, E. A monographic survey of the genus *Marsilea* Linnaeus. I. The species of Africa and Madagascar. Senckenbergiana Biol. 49(3/4): 273–315 (1968); Johnson, D. M. Systematics of the New World species of *Marsilea.* Systematic Botany Monographs 11: 1–87 (1986).

Excluded: ***Marsilea quadrifolia*** Linnaeus is frequently confused with other species in the Indian literature. It occurs in temperate Eurasia and is naturalized in N America; in India it is confined to the NW Himalayas and does not extend further south in India.

1. Sporocarp stalks arising from the petiole well above its axis, solitary or united above the base (appearing paired or dichotomously branched) (NW Himalayas) ***M. quadrifolia***

1. Sporocarp stalks arising at or near the base of the petiole, free or united below (often in clusters but appearing unbranched) (Peninsular India)

  2. Leaflets with pellucid sclerenchymatic interstitial streaks (visible when held against the light); sporocarps with 4 or more, lateral ribs ***M. coromandelina***

  2. Leaflets without pellucid streaks; sporocarps smooth or with less than 4, rather faint, lateral ribs

    3. Sporocarps almost square in lateral view (almost a cube), distinctly truncate, with a distinct abaxial (dorsal) and frontal furrow ***M. aegyptiaca***

    3. Sporocarps longer than high (oval or bean-shaped), without a distinct abaxial (dorsal) and frontal furrow ***M. minuta***

Marsilea aegyptiaca Willdenow [*M. ballardii* K. M. Gupta, *M. condensata* Baker, *M. rajasthanensis* K. M. Gupta nom. illeg.]: Fig. 4

Plants mat-forming; rhizomes slender, wiry, repeatedly branched. Internodes slightly pilose or glabrous, 0.5–4 cm long on land, up to 30 cm in water; nodes densely pilose. Petioles (1–) 3 (–4) cm long on land or (4–) 5–6 (–10) cm in water. Leaflets very variable 2–25 (–30) mm long, 2–20 (–25) broad; terrestrial leaflets with outer margins entire, bi-lobed, many-lobed, to deeply crenulate; floating leaflets with margins entire. Sporocarp stalks free, simple, (2–) 3–8 (–12), arising at the base of the petiole. Sporocarps usually crowded or rarely solitary, square to rectangular in lateral view, sometimes adaxially curved, abaxially straight or saddle-shaped, lateral sides with a shallow or deep curved vertical groove with the frontal furrow usually present and often expanded over the abaxial (dorsal) side, densely pilose at first, when mature, 2–3 (–5) mm long, 2–3 (–4) mm thick; raphe present but often indistinct; teeth 2, the superior one prominent.

Perennial. Found in shallow water in seasonally flooded areas but it can withstand considerable desiccation and is sometimes found in dry habitats.

{Ten/(Ple)}. NW India, N and S Africa and Madagascar: Ind: Gj, Or, Rj.

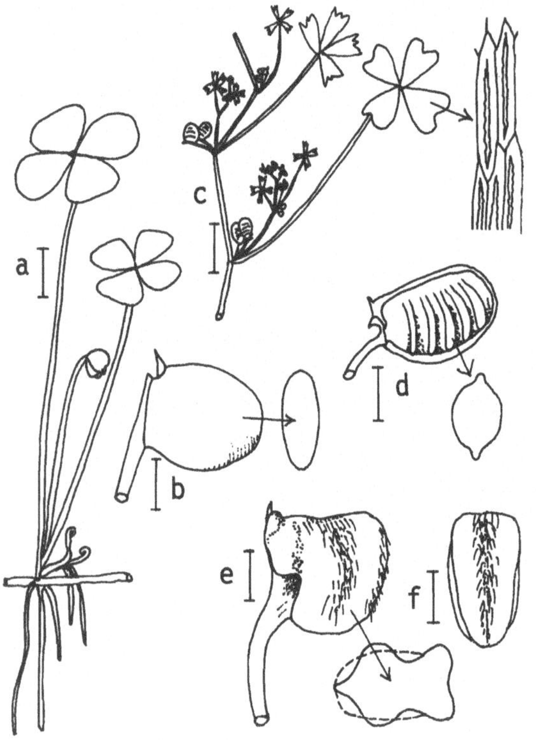

FIG. 4. *Marsilea minuta*: a, aquatic plant (1 cm); b, lateral view of sporocarp (1.5 mm) with a dorsi-ventral transverse section (inset). *M. coromandelina*: c, terrestrial plant (1 cm) with (inset) a piece of leaf with pellucid streaks; d, lateral view of sporocarp (1.5 mm) with a dorsi-ventral transverse section (inset). *M. aegyptica*: e, lateral view of sporocarp (1.5 mm) with a dorsi-ventral transverse section (inset); f, sporocarp viewed from the front (1.5 mm).

Marsilea coromandelina Willdenow ['*coromandeliana*' orth. mut., the name *M. coromandelica* N. L. Burman is invalid]: Fig. 4

Delicate plant; rhizome very fine, thread-like, repeatedly branched, at maturity growing upright in a zigzag, glabrous or villous at the nodes. Petioles very slender, filiform, flexible, 1–7 (–12) cm long. Leaflets small, (1.5–) 2.8 (10) mm long, 1.2–6 (–8) mm broad, with pellucid sclerenchymatic streaks of various lengths between the nerves. Sporocarp stalks simple, free, 5–15 (25) mm long, slender, wiry, usually glabrous. Sporocarps solitary, very rarely crowded, subcircular to broadly elliptic in outline, lemon-shaped in dorsoventral section, 2–3.75 (–4) mm long, 1.5–2.5 (–2.7) mm high, up to 1.75 mm thick, always distinctly bordered all round, with lateral walls conspicuously bulging and when fully mature, pilose when young, soon becoming glabrous; lateral ribs 4 or more, almost always visible; raphe often short covering half to two-thirds the length of the sporocarp; teeth 2, prominent, the inferior usually downward pointed.

Perennial or perhaps sometimes annual. Found at the edges of small pools and buffalo wallows and in seasonally swampy areas. When the habitat dries out the sporocarp-bearing stems grow upright. Bhardwaja, T. N. and Sharma, P. Distribution, intraspecific variation and epidermal studies of *Marsilea coromandeliana* in India. Nova Hedwigia 28(2): 451–462 (1974).

{(⊙)? Ten}. India, Madagascar, Socotra, Mauritania, Senegal, Sudan, Tanzania, Burundi, Angola, Namibia and South Africa: Cey, Ind: Ap, Kl, Kt, Mh, Rj, Tn, Up.

Marsilea minuta Linnaeus [*M. poonensis* Kolhat]: Fig. 4

An extremely variable plant: cushion-like on dry land, spreading and creeping in water. Internodes 0.5–20 cm long. Petioles 2–8 cm long on land or up to 25 cm or more in water. Leaflets 0.4–2.5 (–3) cm long, 0.3–2.25 (–2.75) mm broad, in air entire or with crenulate outer margins, in water entire. Sporocarp stalks simple, numerous, arising one above the other at or near the base of the leaf stalk, sometimes 2–3 (–4) united at their bases; sporocarps usually crowded or rarely in groups of 2 or 3, very rarely solitary, broadly oblong to broadly elliptic in lateral view, distally rounded, elliptical in dorsoventral transverse section, (2–) 3–7 mm long, (0.5–) 0.8–1.4 (–2) mm thick, when mature usually glabrous and dark brown to black; raphe semi-terete, at least two-thirds the length of the sporocarp; teeth prominent, the inferior usually shorter and blunter than the superior.

Perennial or sometimes annual. Found in shallow pools, at the edges of rivers, canals and ditches and in rice fields. It is most abundant in temporarily flooded places where it may form large and dense colonies which can become locally dominant. It tolerates considerable organic pollution. Sporocarps develop as the water recedes.

{(⊙) Ten}. Tropical Africa and Asia, naturalized in the New World: Ind: An, Ap, As, Bh, Dl, Gj, Kl, Kt, Mh, Mp, Or, Pj, Rj, Tn, Up, Wb.

It is a highly plastic species. In China it intergrades with *M. quadrifolia*. The following Indian species are within the acceptable range of variation of *M. minuta* in Africa and are perhaps no more than environmentally induced phenotypes.

Marsilea brachycarpa A. Braun

Like *M. minuta* but sporocarps not more than 2 mm long.

Endemic to Myanmar and India: ♦Bur, ♦Ind: Tn, Kt.

Marsilea brachypus A. Braun

Like *M. minuta* but sporocarps ribbed.

Endemic to India. ♦Ind: Pj (Amritsar), Tn (Nilgiris).

Marsilea gracilenta A. Braun

Like *M. minuta* but sporocarps almost round.

Endemic to India: ♦Ind: Mh.

Marsilea maheshwari B. Gopal

Like *M. minuta* but leaves remaining entire when dry; sporocarps borne in groups on the petiole base.

Endemic to India: ♦Ind: Tn (Pondicherry).

POLYPODIACEAE: 29 genera: 1 with wetland species

Microsorum ['*Microsorium*' orth. mut.]: ± 60 species, mostly epiphytic: tropical Asia

Bir, S. S. and Trikha Chander K. Taxonomic revision of the polypodiaceous genera of India — I: *Microsorium* Link. Bulletin of the Botanical Survey of India 10(2): 133–148 (dated 1968, published 1969).

Microsorum pteropus (Blume) Copeland [*Polypodium tridactylon* Wallich]: Fig. 5

Rhizomes creeping, up to 7 mm thick, infrequently and irregularly branched, with numerous roots, densely clothed in triangular to linear, brown scales. Leaves erect, developing at almost regular intervals along the rhizomes, usually simple up to 15 cm long or more and 2 cm wide, sometimes 3- or rarely 5-lobed up to 30 cm long and 3 cm wide; petioles usually short, winged and with some brown scales; leaf blades or lobes linear-lanceolate to lanceolate, glabrous but with some scattered, almost transparent, early deciduous scales on both surfaces; margins entire; apex rounded. Sporangia in circular or elongate sori scattered irregularly on the adaxial surface of the leaves; sori without indusia.

It is usually a rheophyte growing at the edges of mountain streams but some plants may grow almost permanently submerged. The disseminules are spores. The spores develop only on terrestrial fronds and are presumably dispersed by wind but perhaps also by water; rhizome fragments may be dispersed by water. *Microsorum pteropus* and perhaps other species are cultivated for decoration in aquaria; a dwarf race, usually called *M. pteropus* var. *minor* or 'Java fern', is often offered for sale.

{Hel/Rheophyte}. Tropical Asia: Ban, Bhu, Bur, Cey, Ind: As, Kl, Sk.

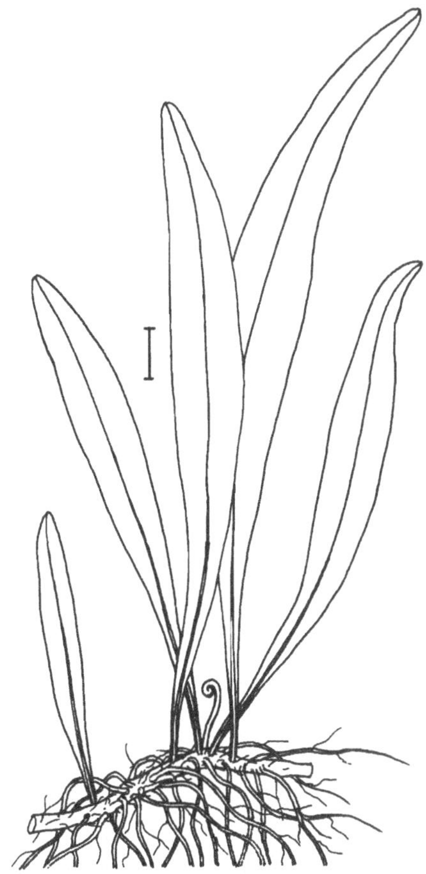

FIG. 5. *Microsorum pteropus*: whole plant (1 cm).

PTERIDACEAE [PARKERIACEAE]: 34 genera

Acrostichum: 2 or 3 species: warm and tropical regions

Acrostichum aureum Linnaeus: Fig. 6

Rhizomes short, woody, creeping to erect, clothed in large, entire-edged scales. Leaves erect, developing in terminal rosettes; petioles well developed but shorter than the blades; blades 1-pinnate (juveniles sometimes simple); up to 3 m or more long; leaflets tongue-shaped, up 40 cm long, 3–7 cm wide, leathery, base asymmetric, rounded to cuneate, margin entire, apex abruptly rounded to truncate or retuse, shortly mucronate; veins reticulate. Sporangia not arranged in sori but borne on veins as well as on leaf tissue, with club-shaped hairs scattered regularly on the adaxial surface of the leaves; sori without indusia.

Perennial. Usually in brackish and saline marshes, in mangrove vegetation, but occasionally also found inland in freshwater swamps, marshes and around lakes. It is eaten in the Nicobar Islands.

{Hel}. Warm and tropical regions of the world: Cey, Ind: An, As, Go, Kl, Mh, Or, Pj.

Ceratopteris: 3 or 4 species: tropical to warm regions of the world

Stems very reduced, bearing numerous roots and a few scales. Leaves in a rosette, entire to 1- or more-

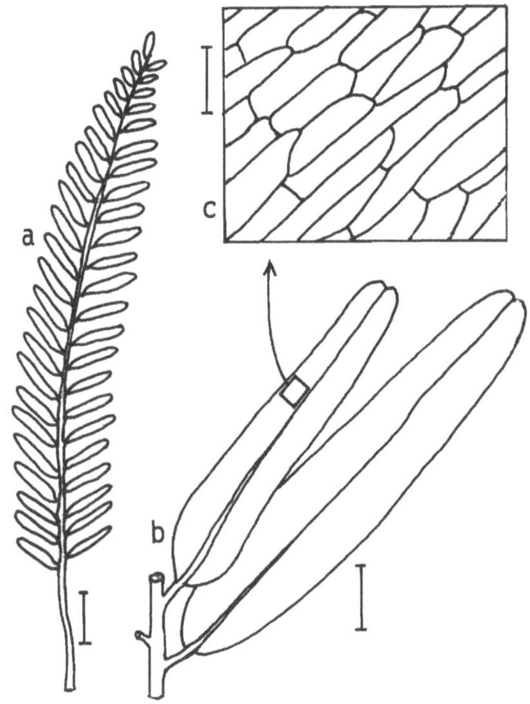

FIG. 6. *Acrostichum aureum*: a, complete leaf (25 cm); b, two leaflets (4 cm); c, venation (2 mm).

pinnate; the lower ones sterile and usually less divided than the upper ones, often bearing buds which may develop into new plants; fertile leaves finely divided, erect; sporangia borne on veins on the adaxial surface; margins of the fertile leaves revolute, partly covering the sporangia. Leaves transitional in form between sterile and fertile ones often develop.

Juveniles free-floating or submerged, adults submerged, free-floating, partly emergent or bottom-rooted and emergent, mostly annual. The disseminules are spores or rooting vegetative shoots, dispersal presumably by water. *Ceratopteris* is cultivated in E Asia as a vegetable. It is also reported to be a weed in rice fields and man-made lakes. It is also used for ornament in aquaria. Lloyd, R. M. Systematics of the genus *Ceratopteris*. Brittonia 26: 139–160 (1974).

1. Lower sterile leaves simple with or without broad basal lobes, deltate in outline; petioles short, thick, usually inflated, broader above than below; ultimate segments of mature fertile leaves with margin revolute only part of the way to the vein, covering only the edges of the lower surface ***C. pteridoides***
1. Lower sterile leaves pinnately divided, ± triangular in outline; petioles long, slender, not inflated, usually broader below than above; ultimate segments of mature fertile leaves with margin revolute the whole way to the vein, covering the whole lower surface. ***C. thalictroides***

Ceratopteris pteridoides (Hooker) Hieronymus [*Ceratopteris cornuta* (Palisot de Beauvois) Leprieur]: Fig. 7

Petioles short and thick, often inflated, broader above than below. Sterile leaves often with adventitious buds, usually floating, rarely submerged or emergent, the lower ones entire or divided into 2 or 3 lobes, deltate in outline; the upper sterile leaves pinnatifid, 1 or rarely 2 times pinnate, deltate to triangular in outline, with broad ultimate segments, when divided usually becoming emergent. Fertile leaves emergent, 1–3 (or 4) times pinnate; the margins of the mature ultimate segments revolute only part of the way to the vein, leaving a central portion uncovered.

Annual. Usually free floating but sometimes rooted in the substrate. Found in fish ponds, water reservoirs and rice fields. In India I suspect it is more widely distributed than previously thought and has perhaps been misidentified as *C. thalictroides*. There is some confusion among botanists as to whether *C. cornuta* and *C. pteridoides* are distinct species or not.

FIG. 7. *Ceratopteris pteridoides*: a, complete young plant (2 cm); b, apical part of a fertile leaf (5 mm). *C. thalictroides*: c, complete young plant (2 cm); d, portion of a fertile leaf (2 cm).

{⊙ Ple/(Eph/Hyp)}. Almost throughout the tropics and subtropics; usually considered to be native in the Old Word: <Ban, <Bur, Cey, <Ind: <As.

Ceratopteris thalictroides (Linnaeus) Brongniart: Fig. 7

Petioles long and slender, not inflated, usually broadest at the base. Sterile leaves submerged, floating or emergent, (1) 2 or 3 times pinnate; the ultimate leaf segments linear to oblong, trapezoid or triangular. Fertile leaves 2–4 (–5) times pinnate; the ultimate segments elongate, up to ± 4 cm long and 0.2 cm wide; the margins of the mature ultimate segments revolute the whole way to the vein, covering the whole lower surface.

Annual or perhaps sometimes perennial. Usually found in permanent or semi-permanent water in large reservoirs and tanks, but also in rice fields and in mats of floating vegetation. Often growing partially submerged but sometimes becoming detached and floating. The floating plants are sometimes found in canals and ditches. It is sometimes cooked as a vegetable.

{(⊙)? Ple/Hyp/(Ten)}. Pantropics: Bur, Cey, Ind, Nep: An, As, Bh, Gj, Go, Kl, Kt, Or, Rj, Up.

SALVINIACEAE: 1 genus

Salvinia: ± 10 species, plus hybrids: through introductions almost cosmopolitan, absent from colder regions

Stems floating, irregularly forked, without roots. Leaves in whorls of 3, 2 of them floating and 1 submerged; floating leaves photosynthetic, entire with complex unwettable hairs on the adaxial (upper) surface and wettable hairs on the abaxial surface; submerged leaves not photosynthetic, finely divided into filamentous segments which bear a striking resemblance to roots. Sporocarps borne on modified segments of submerged leaves; megasporangial sporocarps developed first, producing up to 25 megasporangia, each containing 1 megaspore; microsporangial sporocarps developed later, producing numerous microsporangia, each containing usually 64 microspores. Prothalli develop inside floating spores.

Perennial or annual. Free floating mat-building plants. Heterosporous: disseminules sporocarps or groups of sporangia and perhaps also megaprothalli. It can be a very serious pest, spreading by vegetative fragments.

Excluded: ***Salvinia natans*** (Linnaeus) Allioni is widespread in the warm temperate Old World. It is frequently confused with other species in Indian literature. It occurs in the Himalayas but does not extend into Peninsular India.

1. Hairs on the adaxial surface of the floating leaves multiaxillary at the base dividing into 4 uniaxillary filaments above, each then bending outwards below and joining above (the whole looking like an egg beater) ***S. molesta***

1. Hairs on the adaxial surface of the floating leaves uniaxillary (erect, unbranched and multicellular)

  2. Hairs on the adaxial surface of the floating leaves solitary, scattered without any obvious pattern on the surface ***S. cucullata***

  2. Hairs on the adaxial surface of the floating leaves in tufts of 3 or 4, the tufts then clearly arranged in rows along the surface ***C. natans***

Salvinia cucullata Roxburgh: Fig. 8

Like *S. molesta* but very much smaller, floating leaves 1–1.8 cm wide; the adaxial surface of the floating leaves bearing simple hairs; hairs unbranched, multi-

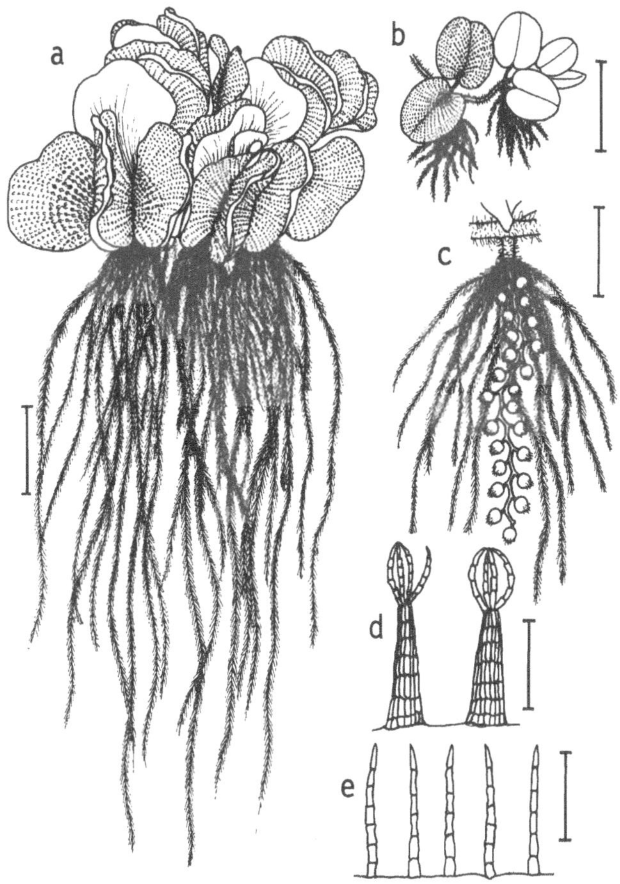

Fig. 8. *Salvinia molesta*: a, whole plant in 'closed' or 'mature' phase (2 cm); b, whole plant in 'open' or 'pioneer' phase (2 cm); c, submerged leaf with sporocarps (2 cm); d, two trichomes on the adaxial surface of a floating leaf (0.5 mm). *S. cucullata*: e, five trichomes on the adaxial surface of a floating leaf (0.5 mm).

cellular, transparent, scattered without any obvious pattern on the surface. Sporocarps unknown.

Perennial or perhaps annual. This species is very poorly known and would appear to be rather rare.

{(⊙?) Ple}. Perhaps endemic to Myanmar and NE India: Ban? Bur, Ind: As, Bh, Mg, Mn, Or, Wb.

Salvinia molesta Mitchell: Fig. 8

The main stem fragments easily, but is often up to 30 cm long. Floating leaves 1–4 (–6) cm long, 1–5 cm wide, flat in open water becoming folded inwards when in a closed mat; the adaxial surface with complex hairs; hairs multiaxillary at the base dividing into 4 uniaxillary filaments above, each then bending outwards below and joining above (the whole looking like an egg beater); the base green, the apical filaments transparent. Sporocarps born on submerged leaves, ± globose, usually less than 1 mm long, up to 55 produced in 2 rows along a segment, the first 2 or 3 with macrosporangia, the remainder with microsporangia.

Perennial. Free floating in a wide variety of aquatic habitats. This species is usually considered to be a hybrid between two S American species. It grows best in high nutrient levels and may form very large stands. In many parts of India, particularly in Kerala, it is a very serious pest. It is today controlled by a S American beetle *Cyrtobagous singularis* Hustache. Abbasi, S. A. and Nipaney, P. C. Worlds' [sic] worst weed (Salvinia) — its impact and utilization. International Book Distributors, Dehradun i–x, 1–128 (1993) — this book does not live up to its title.

{Ple}. Sterile hybrid, the parents native in S America, naturalized in Old World tropics: <Ban, <Cey, <Ind: <Ap, <Dl, <Kl ± 1960, <Kt, <Tn 1980, <Wb.

SPERMATOPHYTINA (seed bearing plants)

Plants reproducing by seeds (megagametophytes retained within ovules).

MAGNOLIOPSIDA [ANGIOSPERMAE]

Ovules enclosed in an ovary. Ovary usually crowned by a style and stigma. Pollen grains (microspores) adhering to the stigma. Fertilization affected by means of a pollen tube.

ACANTHACEAE: ± 350 genera

Herbs or shrubs. Leaves opposite. Inflorescences terminal or axillary or flowers solitary. Flowers bisexual, bilaterally symmetrical. Sepals united below, 4- or 5-lobed. Petals united below, 4- or 5-lobed or 2-lipped. Stamens 2 or 4. Ovary superior, 2-locular; ovules 2 or more in each loculus, produced on hook-like projections (retinacula); fruit a 2-valved capsule.

There is some confusion concerning generic delimitation in this family which is reflected in the synonymy. I have taken the genera in the older classical sense.

Excluded: ***Acanthus ilicifolius*** Linnaeus is a woody perennial with spiny margined leaves, which grows in brackish water in mangrove swamps and in the tidal zone of coastal marshes.

The following may occasionally be found in moist or damp places but are not considered to be wetland plants: ***Asystasia gangetica*** (Linnaeus) T. Anderson [*A. coromandelina* Wight ex Nees]; ***Hemigraphis dura*** T. Anderson; ***H. hirta*** (Vahl) T. Anderson; ***Rungia repens*** Nees.

1. Stamens 4 (sometimes 2 with fertile anthers and 2 staminodial); anther sacs at the same level, parallel, without tail-like appendages; capsule with 6 (rarely less) or more seeds; nectariferous scale inconspicuous ***Hygrophila***

1. Stamens 2 (both with fertile anthers, staminodes absent); anther sacs superposed, with tail-like appendages; capsule with 4 or less seeds; nectariferous scales conspicuous, cup-shaped with 2, 3 or 5 lobes; bracts, bracteoles and sepal segments glabrous or nearly so and with scarious margins ***Justicia quinqueangularis***

Hygrophila [*Adenosma* pro parte, *Asteracantha, Cardanthera, Dyschoriste, Hemidelphis* pro parte, *Nomaphila, Plaesianthera* ('*Plaesiantha*' orth. mut.), *Physichilus, Ruellia* pro parte, *Synnema*]: ± 85 species (probably less): tropical to warm regions of the world.

Leaves decussate, either homomorphic simple and entire or heteromorphic with pinnately divided basal or submerged leaves with filiform or linear, simple or forked segments and entire cauline or emergent leaves. Inflorescence axillary of subsessile clusters of 1–many flowers or rarely in terminal heads, with or without spines; bracts and bracteoles large, often covering the sepals usually hairy. Sepals with 4 or 5, hairy segments. Petals white to purple, either almost equally 5-lobed or 2-lipped; adaxial lip, when present, erect, concave, 2-lobed or 2-toothed; abaxial lip 3-lobed. Stamens 4, occasionally 2 staminodial, not exerted, adnate to petal tube above the middle; filaments united in pairs by a membrane at the base; anthers 2-locular; anther locules at the same level and parallel, without spurs or tails; pollen 3- or rarely 4-colporate. Nectariferous disk inconspicuous. Seeds 6 (rarely less) to many.

Mostly perennial. They range from submerged aquatics with emergent flowers to helophytes or terrestrial plants. Flowers entomophilous, usually self-incompatible. The disseminules are relatively heavy seeds actively ejected from the capsules. Many are troublesome weeds in rice fields and in irrigation and drainage ditches; others are used in medicine in India and several species are cultivated as decorative aquarium plants. Most species are very plastic and there is considerable taxonomic confusion. This account is superficial and not based on critical work; the tribe Hygrophileae is very much in need of revision. Sreemadhaven, C. P. The taxonomic status of *Cardanthera*. Bulletin of the Botanical Society of India 10(2): 222–223 (dated 1968, published 1969).

Excluded: the following species may sometimes be found in damp or moist places but they are not considered to be wetland plants: ***Hygrophila erecta*** (N. L. Burman) Hochreutner [*H. pusilla* Blume, *Ruellia erecta* N. L. Burman, *Dyschoriste erecta* (N. L. Burman) O. Kuntze]; ***H. incana*** Nees: ♦Ind; ***H. phlomoides*** Nees: ♦Bur, ♦Ind; ***H. thwaitesii*** (Bentham) Heine [*Cardanthera thwaitesii* Bentham, *Plaesianthera thwaitesii* (Bentham) Livera]: ♦Cey.

The following is confined to the Himalayas: ***H. griffithii*** (T. Anderson) Sreemadhavan [*Cardanthera griffithii* Bentham ex C. B. Clarke].

The following are problematic: ***H. biplicata*** (Nees) Sreemadhaven [*Cardanthera avana* Wallich ex Bentham ex C. B. Clarke]: ♦Bur; ***Nomaphila parishii*** T. Anderson: ♦Bur; ***N. stricta*** Nees: probably not found in India.

1. Plants with hard and sharp axillary spines ***H. schulli***
1. Plants without spines
  2. Flowers in discrete, bracteate, terminal spikes; bracts not like foliage leaves, ovate to orbicular, in 4 ranks and overlapping ***H. heinei***
  2. Flowers in axillary clusters of 2–5 or solitary in leaf axils; bracts leaf-like, not overlapping
    3. Blades of lower or submerged leaves pinnately divided into linear segments or pinnatifid; blades of emergent leaves pinnatisect to toothed; retinacula ± conical and straight, rather small
      4. Sepals linear with pinnate lobes or if entire then wider above than below ***H. pinnatifida***
      4. Sepals linear and entire, narrower above than below
        5. Leaf blades of emergent stems (subtending flowers or flower clusters) lanceolate, mostly 4–5 times as long as wide; emergent stems and leaves densely clothed in short, equally sized, glandular hairs; flowers usually in clusters of 2–5 ***H. balsamica***
        5. Leaf blades of emergent stems (subtending flowers or flower clusters) elliptic to ovate rarely more than one and a half times as long as wide; emergent stems and leaves densely clothed in long and short glandular hairs; flowers often solitary or 2 or 3 in a cluster ***H. difformis***
    3. Blades of lower or submerged leaves simple and entire; blades of emergent leaves simple and entire; retinacula long and curved, mostly hooked
      6. Stamens 2 with fertile anthers, the adaxial (upper) pair reduced to minute teeth; flowers usually in dense terminal spikes with imbricate bracts ***H. polysperma***
      6. Stamens 4 with fertile anthers; flowers usually in axillary clusters or solitary in leaf axils, if in terminal spikes then spikes interrupted below and the bracts not imbricate
        7. Sepal tube split to not more than two-thirds; leaves subtending the flowers or axillary flower clusters all leaf-like and longer than the flowers; stems usually erect ***H. quadrivalvis***
        7. Sepal tube split almost to the base, much more than two-thirds; leaves subtending the flowers or axillary flower clusters leaf-like and longer than the flowers only at the base of the inflorescence, becoming shorter and bract-like above; stems usually creeping ***H. serpyllum***

Hygrophila balsamica (Linnaeus fil.) Rafinesque [*Cardanthera balsamica* (Linnaeus fil.) Bentham ex C. B. Clarke, *Adenosma thymus* Nees, *Cardanthera verticillata* (Nees) Bentham ex C. B. Clarke]: Fig. 9

Stems up to 60 (–80) cm tall, branched, when submerged swollen and glabrous with scabrid nodes, when emergent densely covered with short, ± equally long, glandular hairs. Leaves mostly decussate, sessile or shortly petiolate; petiole up to 1.5 cm long; blades at base of stem or submerged, pinnately divided into simple or branched linear segments or pinnatifid, usually glabrous, up to 10 cm long; blades on emergent stems simple, lanceolate, 3–8 cm long, 0.5–2.5 cm wide, densely clothed in short, ± equally long glandular hairs, base cuneate, margins serrulate, tips acute. Flowers axillary, in clusters of 2–5, rarely solitary; bracts elongate-elliptical to obovate ± 4 mm long; bracteoles like bracts but smaller, ± 3.5 mm long. Sepal tube unequally 5-lobed, the longest lobe linear-lanceolate, ± 0.7 cm long. Petals purple to violet; petal tube ± 4 mm long; petal lobes up to 5 mm long; adaxial (upper) lip glabrescent, ciliate, subacute. Stamen filaments in 2 pairs, 1 pair ± 2 mm long, the other pair ± 4 mm long; anthers unequal, 1 pair ± 1 mm long, the other pair ± 1.5 mm long. Capsules linear-oblong, ± 6 mm long, hairy. Seeds numerous, ± 0.4 mm long, ± 3 mm in diameter; retinacula minute, straight.

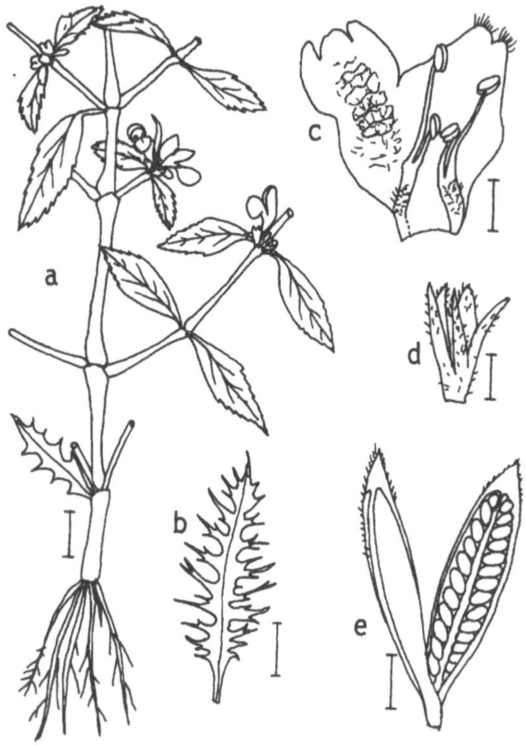

FIG. 9. *Hygrophila balsamica*: a, base of plant (1 cm); b, submerged leaf (1 cm); c, opened petal tube (2 mm); d, sepal tube (5 mm); e, capsule with seeds (1 mm).

Annual or perennial. In shallow water, in moist clayey ground by watercourses, irrigation canals and in rice fields. A very aromatic plant apparently poisonous to fish when grown in aquaria. Kasselmann, Christel, *Hygrophila balsamica* (Linné fil.) Rafinesque: giftig für Fische! Aqua-Planta 83(3): 6–9 (1983).

{(⊙) Hyp/Ten/(Hel)}. Endemic to Sri Lanka and S India: ♦Cey, ♦Ind: Kl, Kt, Mh, Tn.

Hygrophila difformis (Linnaeus fil.) Blume [*Cardanthera triflora* (Roxburgh ex Nees) F. Hamilton ex Voigt or Nees, *Synnema triflora* (Roxburgh ex Nees) O. Kuntze]: Fig. 10

Like *H. balsamica* but leaf blades of emergent stems (subtending flowers or flower clusters) elliptical to ovate, rarely more than one and a half times as long as wide; emergent stems and leaves densely clothed in long and short glandular hairs; flowers often solitary or 2 or 3 in axillary clusters.

Perennial or perhaps sometimes annual. Usually found in shallow water where it may be bottom-rooted or straggling over the surface. A very popular aquarium plant.

{(⊙)? Hyp/Ten}. Probably endemic to the northern part of Indian subcontinent, naturalized in the south: ♦Ban, ♦Bur, ♦Ind, ♦Nep: As, Bh, <Kl, Up, Wb. 0–200 m.

Hygrophila heinei Sreemadhavan [*Adenosma uliginosa* (Linnaeus fil.) R. Brown ex Nees, *Cardanthera uliginosa* (Linnaeus fil.) F. Hamilton ex C. B. Clarke, *Hygrophila heloides* Heine]: Fig. 11

Stems erect or decumbent, 3–40 cm tall, branched, emergent parts with some hairs. Leaves mostly decussate, sessile or nearly so; leaf blades of emergent stems oblong to elongate-elliptical, simple, 1–5 cm long, 3–8 mm wide, glandular and sometimes also hairy, margins crenate, serrate or entire. Flowers in terminal spikes, (1.5–) 3–8 (–12) cm long, scarcely interrupted at the base even in fruit; bracts in 4 ranks, broadly ovate to orbicular, 4–6 mm long, glabrous, glandular and/or hairy; bracteoles ovate broadly elliptical, 3–4 mm long. Sepals 5, ± equal, 4–5 mm long, ± 0.6 mm wide at the base, attenuate above, hairy. Petals purple to blue, 2-lipped; petal tube funnel-shaped, ± 4 mm long, yellow below, hairy within; abaxial (lower) petal lip ± 3 mm

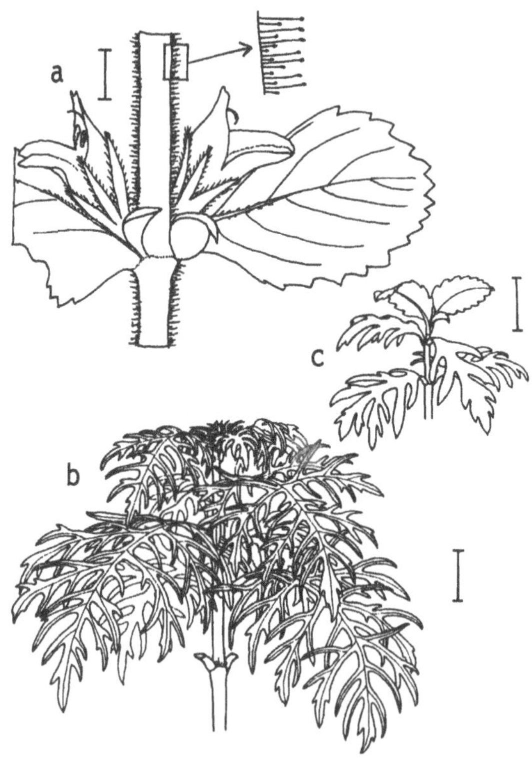

FIG. 10. *Hygrophila difformis*: a, inflorescence (4 mm) with marginal glandular hairs (inset); b, apical part of a submerged shoot (2 cm); c, emergent shoot (2 cm).

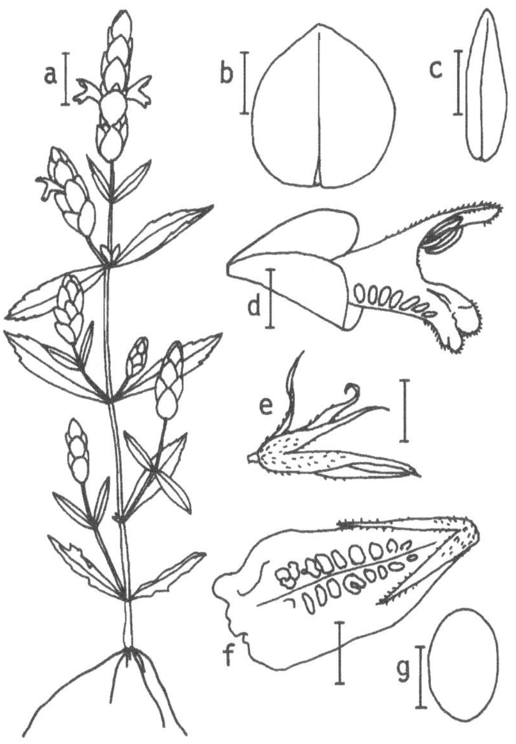

Fig. 11. *Hygrophila heinei*: a, whole plant (1 cm); b, bract (2 mm); c, capsule valve (2 mm); d, flower viewed from the side with bracteoles (2 mm); e, sepals (2 mm); f, flower viewed from below with sepals (2 mm); g, seed (0.5 mm).

long, 3-lobed, hairy; adaxial (upper) petal lip 2-lobed, ± 3.5 mm long, hairy. Stamens 4, 2 unequal pairs; adaxial (upper) anthers ± 1.2 mm long, the abaxial (lower) pair shorter. Capsules linear-oblong, 5–6 mm long. Seeds 25–30, oblong-ovoid, ± 0.5 mm long, smooth; retinacula minute, straight.

Annual. In shallow water, in swamps and rice fields, also reported from sugar cane fields.

{⊙ Ten/Hel}. Probably endemic to the Indian subcontient: ♦?Cey, ♦?Ind: An, Ap, Mp, Or, Sk, Tn.

Hygrophila pinnatifida (Dalzell) Sreemadhavan [*Nomaphila pinnatifida* Dalzell, *Cardanthera pinnatifida* Bentham ex C. B. Clarke]: Fig. 12

Like *H. balasmica* but leaves and bracts deeply pinnatifid, sometimes becoming crenate towards the apex of old shoots; sepal tube split almost to the base; sepal lobes linear with pinnately lobed segments or if segments entire then wider above than below; flowers often borne on lax axillary spikes

Poorly known, probably annual or perennial. Probably found in shallow water or in wet places.

{(⊙) Hyp/Ten/(Hel)}. Endemic to Peninsular India: ♦Ind: Kt, Mh, Tn, Up.

Hygrophila polysperma (Roxburgh) T. Anderson [*Hemidelphis polysperma* (Roxburgh) Nees, *Hemidelphis polysperma* var. *joshianus* Rao et Biswas]: Fig. 13

Stems creeping and rooting at nodes, ascending or erect, branched, 20–50 (–100) cm tall, when submerged not noticeably swollen. Submerged leaves usually distinctly stalked, usually glabrous, elliptical to oblong, with entire margins, usually larger than the emersed ones; emergent leaves sessile or shortly stalked, glabrous or pubescent, elliptical to obovate-elliptical, 1.2–4 cm long, with entire margins. Flowers mostly in distinct terminal spikes, occasionally in axillary clusters or solitary in the axils of leaf-like bracts; spikes dense, oblong or cylindric, up to 15 cm long. Bracts broadly elliptical, 0.6–1 cm long, pubescent, usually imbricate. Bracteoles lanceolate to linear, nearly as long as the sepals. Sepals 5, the lobes linear with ciliate margins. Petals blue or white, pubescent, up to ± 12 mm long; palate of the abaxial (lower) lip glabrous. Stamens 4, 2 with fertile anthers; adaxial (upper) filaments with an appendage towards the base; the adaxial (upper) stamens reduced to minute teeth.

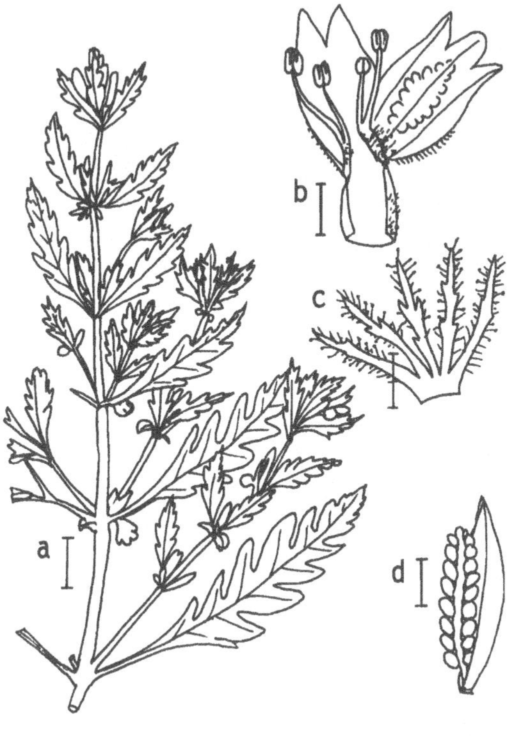

Fig. 12. *Hygrophila pinnatifida*: a, part of inflorescence (1 cm); b, opened petal tube (4 mm); c, sepals (4 mm); d, capsule with 1 valve removed (5 mm).

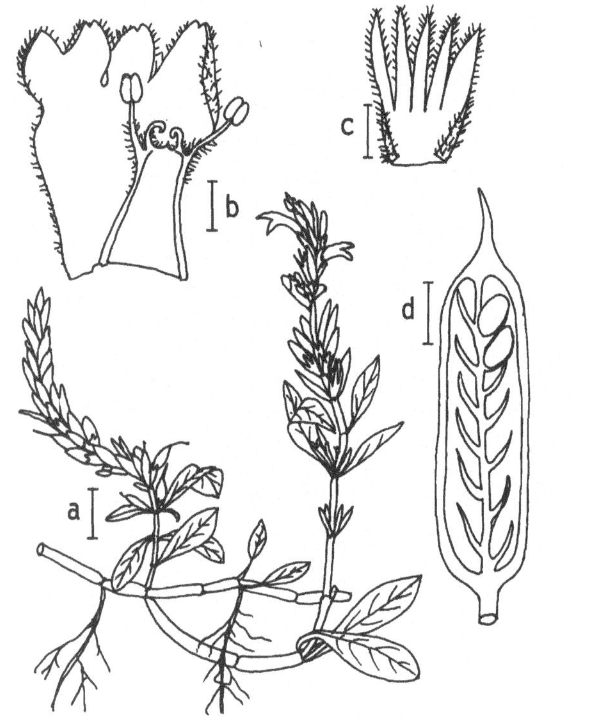

FIG. 13. *Hygrophila polysperma*: a, habit (1 cm); b, opened petal tube (2 mm); c, sepals (2 mm); d, capsule with 1 valve removed (2 mm).

Fruit narrow, up to 15 mm long. Seeds 15–25; retinacula hard, slender and curved.

Perennial or perhaps rarely annual. Found in marshy places, drains, temporary pools and in slowly moving water in streams and canals.

{(⊙)? Hyp/Vit}. India to Indochina, introduced and established in some warmer parts of N America: Bhu, Cey, Ind, Nep, Pak: Ap, As, Bh, Dl, Gj, Kt, Mh, Mp, Pj, Rj, Up, Wb. 0–1200 m.

Hygrophila quadrivalvis (F. Hamilton) Nees [*H. obovata* Wight, *H. salicifolia* (Vahl) Nees, *H. angustifolia* auct. Ind., non R. Brown]: Fig. 14

Stems often creeping, often somewhat woody below, up to 1 m or more tall, aerial parts usually hairy. Leaves petiolate; petiole up to 2.5 cm long; blades simple, lanceolate to elliptical-ovate or obovate, 2.5–8 (–10) cm long, (1–) 1.5–3.5 cm wide, base rounded or attenuate, margins entire to somewhat crenate, tips obtuse or acute. Flowers in axillary whorls or occasionally solitary; bracts scale-like, lanceolate to obovate, 5–9 mm long; bracteoles like bracts but smaller, 4–6 mm long. Sepals ± 9 mm long, the lower quarter to third united below, equally 5-lobed. Petals white with pink lines or pink to violet; petal tube 6–8 mm long; adaxial (upper) petal lip ± 10 mm long, glabrescent, ciliate, sub-acute; abaxial (lower) lip ± 12 mm long, glandular hairy outside. Stamen filaments in 2 pairs, 1 pair 2–3.5 mm long, the other pair 5.5–6 mm long; anthers unequal, 1 pair 1.5–2 mm long, the other pair 2–2.5 mm long. Capsules linear-oblong, longer than the sepals. Seeds 8–12, globose, 0.7–1.5 mm in diameter; retinacula long and curved.

Perennial or perhaps sometimes annual. In shallow water or in marshy habitats and along the banks of streams. It is often gregarious and forms dense single-species clumps. It is a very plastic species with long, thin and almost glabrous leaves in wet conditions and short, rounded, very hairy ones in dry conditions. In many Indian floristic works plants with long elliptical leaves and bracts and deeply divided sepal tubes are called *H. salicifolia*; while plants with obovate leaves and bracts and shortly divided sepal tubes are called *H. obovata*. As there are so many intermediate states, sometimes even on the same plant, it is felt wiser to incorporate them all into *H. quadrivalvis*. Two additional species, *Ruellia ringens* Linnaeus and *R. erecta* N. L. Burman, may prove to be conspecific with *H. quadrivalvis*; the earliest acceptable name in the genus *Hygrophila* is *H. ringens* (Linnaeus) R. Brown ex Steudel.

{⊙? Hyp/Ten}. S and SE Asia: Ban, Cey, Ind, Nep, Pak: An, Ap, Kl, Kt, Mh, Mp, Tn. 0–1300 m.

FIG. 14. *Hygrophila quadrivalvis*: a, flowering shoot (1 cm); b, fruit (2 mm); c, opened sepal tube (4 mm); d, opened petal tube (4 mm).

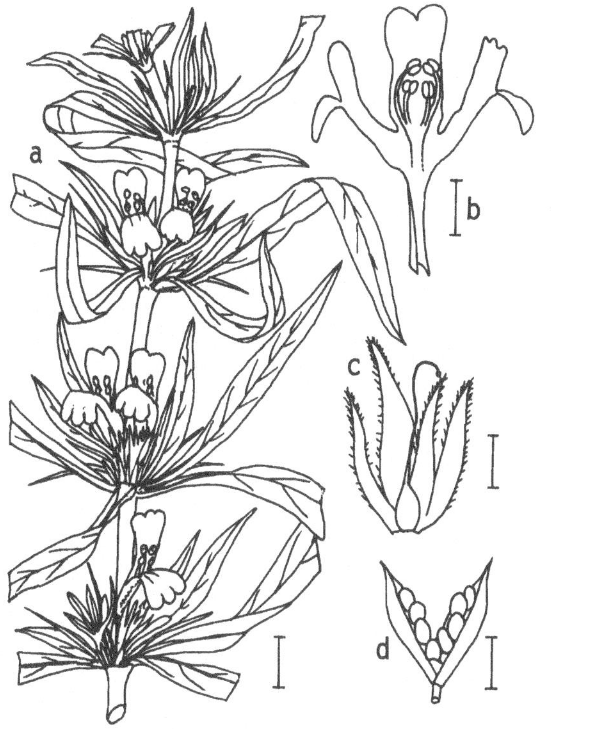

FIG. 15. *Hygrophila schulli*: a, part of inflorescence (2 cm); b, opened petal tube (5 mm); c, opened sepal tube with ovary (5 mm); d, capsule with seeds (5 mm).

Hygrophila schulli (F. Hamilton) M. R. et S. M. Almeida [*H. spinosa* T. Anderson, *Asteracantha longifolia* (Linnaeus) Nees, *H. auriculata* (K. Schumacher) Heine]: Fig. 15

Subshrubby herbs with axillary spines up to 4 cm long. Stems up to 1 m or more tall, mostly unbranched, ± hispid, thickened at the nodes. Leaves sessile or nearly so, simple, in spurious whorls of 6 thorns and 8 leaves, the outermost leaves longest, lanceolate to linear-lanceolate, 5–10 (–15) cm long, 1.5–2.5 cm wide, hairy when aerial, base cuneate, margins minutely dentate, tips acute; thorns up to 3 cm long. Flowers in axillary whorls, usually surrounded by spines; bracts and bracteoles leafy, ± 1.5 cm long. Sepal tube unequally 4-lobed, the longest lobe linear-lanceolate, ± 1 cm long. Petals blue, purple or sometimes pink, glabrous, up to 2.5 cm long; petal lobes rounded to acute, the middle lobe of the abaxial (lower) lip with a yellow spot. Stamen filaments in 2 pairs, 1 pair ± 5 mm long, the other pair ± 10 mm long; anthers unequal, up to 3 mm long. Capsules linear-oblong, ± 1 cm long. Seeds 4–8 orbicular, ± 3 mm in diameter; retinacula long and curved.

Perennial. Found in swamps, in temporary pools, at the edges of tanks, canals and ditches and in rice fields. Often abundant and rather difficult to remove as it has very sharp spines. It has diuretic properties and is used locally for diseases of the urino-genital tract and for dropsy. Seed sold commercially in markets.

{Ten/Hyp/Ten}. S and SE Asia: Ban, Cey, Ind, Nep, Pak: As, Bh, Gj, Go, Kl, Kt, Pj, Rj, Tn, Up, Wb. 0–700 m.

Hygrophila serpylla (Nees) T. Anderson [‘*serphyllum*’ orth. mut., *Physichilus serpyllum* Nees, *Hygrophila stocksii* T. Anderson]: Fig. 16

Stems creeping and rooting at nodes, with numerous ascending or erect flower-bearing shoots; creeping stems branched, often woody or wire-like, not noticeably swollen when submerged, mostly glabrous or nearly so; erect shoots usually simple up to ± 25 cm tall but usually less, usually densely clothed in whitish hairs above. Submerged leaves usually distinctly stalked; petiole 0.3–1.5 (–2.5) cm long; blades of submerged leaves elliptical to oblong or orbicular, up to 2.5 cm long and ± 2 cm wide, usually glabrous, with entire margins; blades of emergent leaves sessile or shortly stalked, lanceolate to elliptical or ovate, 1–2 (–4) cm long, rarely glabrous, mostly pubescent, with long

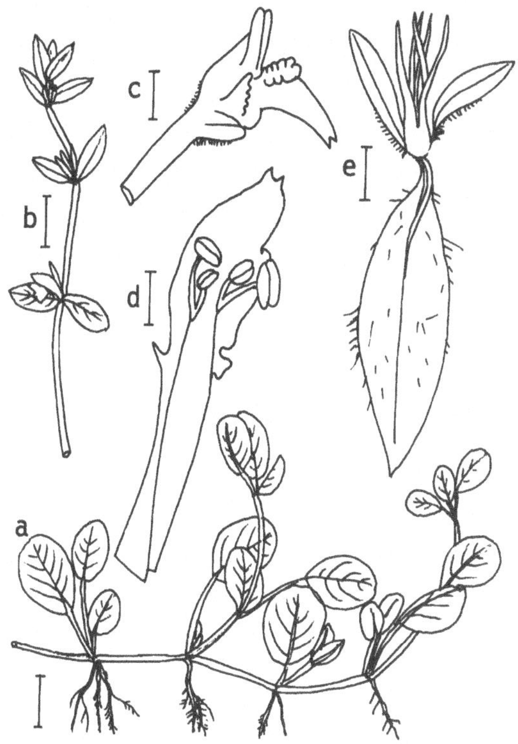

FIG. 16. *Hygrophila serpylla*: a, creeping shoot (1 cm); b, flowering shoot (1 cm); c, petal tube viewed from the side (2 mm); d, opened petal tube (1 mm); e, sepals with bracteoles and bract (2 mm).

white hairs and short glandular hairs, margins entire. Flowers mostly in distinct terminal spikes, at the beginning of the flowering period developing solitarily or in axillary clusters in the axils of leaf-like bracts; spikes distinctly interrupted below, becoming dense above. Bracts leaf-like to elliptical, 0.5–1 cm long, pubescent with long white hairs and short glandular hairs, imbricate only at the tip of the spike. Bracteoles oblanceolate to linear, ± as long as the sepals. Sepals 5, divided almost to the base; the lobes all alike, linear with ciliate margins, 4.5 mm long. Petals bright blue; tube ± 5 mm long, pubescent; abaxial (lower) lip ± 4 mm long, the palate bright blue with whitish swellings, glabrous. Stamens 4, all with fertile anthers; anthers deep purple to black, the adaxial (upper) pair ± 7 mm long, the abaxial (lower) pair ± 4 mm long. Fruit narrow, ± 6 mm long. Seeds 10–16; retinacula hard, slender and curved.

Perennial. In marshy places, drains, temporary pools and in slowly moving water in streams and rivers. It is a very plastic species. During and shortly after the monsoon it is prostrate with largish almost orbicular, ± glabrous, petiolate leaves; this phase is dark green and develops some flowers in the axils of leaf-like bracts; it has been named *H. serpylla* var. *hookeriana* C. B. Clarke. Towards winter as the water recedes the shoots become upright with narrower, shortly petiolate or sessile leaves, covered with whitish hairs and developing flowers in terminal spikes; the whole plant becomes whitish-green and looks very unlike the wet season plants. In Rajasthan it is often locally dominant along the banks of streams and rivers.

{Ten/Hel}. Endemic to Peninsular India: ♦Ind: Gj, Kt, Mp, Rj.

Justicia [*Adhatoda* pro parte, *Rostellularia*, '*Rostellaria*' orth. mut., *Ecbolium, Nicoteba, Rhaphidospora*]: ± 600 species: temperate to tropical regions of the world

Graham, Victoria, A. W. Delimitation and infrageneric classification of *Justicia*. Kew Bulletin 43(4): 551–624 (1988); Ellis, J. L. *Rostellularia diffusa* (Willd.) Nees (*Justicia diffusa* Willd.) complex. Bulletin of the Botanical Survey of India 30(1–4): 126–133 (dated 1988, publ. 1990).

Excluded: the following may sometimes be found in damp or moist places but they are not considered to be wetland plants: ***Justicia betonica*** Linnaeus; ***J. glabra*** J. G. König ex Roxburgh [*Rhaphidospora abyssinica* Nees]; ***J. simplex*** D. Don [*Rostellularia procumbens* (Linnaeus) Nees var. *simplex* D. Don, *Ecbolium procumbens* (Linnaeus) O. Kuntze].

Justicia quinqueangularis J. G. König ex Roxburgh [*J. peploides* (Nees) T. Anderson, *Rostellaria peploides* Nees, *J. quinqueangularis* var. *peploides* (Nees) C. B. Clarke, *R. quinqueangularis* (J. G. König ex Roxburgh) Nees var. *peploides* (Nees) Raizada]: Fig. 17

Stems decumbent or erect, swollen if in water, often rooting at the lower nodes, freely branched above, up to ± 45 cm tall, grooved, glabrous or with a few downwardly directed hairs. Leaves shortly petiolate; blades simple, elliptical to oblong or obovate up to 4 (–8) cm long, up to ± 2 cm wide, margin entire, attenuate at base, rounded at the apex; cystoliths arising from the midrib, ± 20 μm long. Inflorescence a simple spike, 2–6 (–14) cm long. Flowers paired at nodes, each subtended by 1 bract and 2 bracteoles; bracts and bracteoles alike, linear to linear-oblong or spathulate, up to ± 7 mm long, ± 1 mm wide, glabrous or nearly so, the margin scarious. Sepal lobes 4 large and 1 reduced, linear, ± 6 mm long, ± 1 mm wide, glabrous or nearly so, the margin scarious. Petals pink or purplish-pink, 6–8 mm long, tubular below, 2-lipped, the adaxial (upper) lip shallowly 2-lobed, the abaxial (lower) lip 3-lobed. Stamens 2, laterally attached at the sinuses, joined to the petal tube above the middle; staminodes absent; filaments united in pairs by a membrane at the base; anthers 2-locular; anther sacs 2-locular, the lower sacs

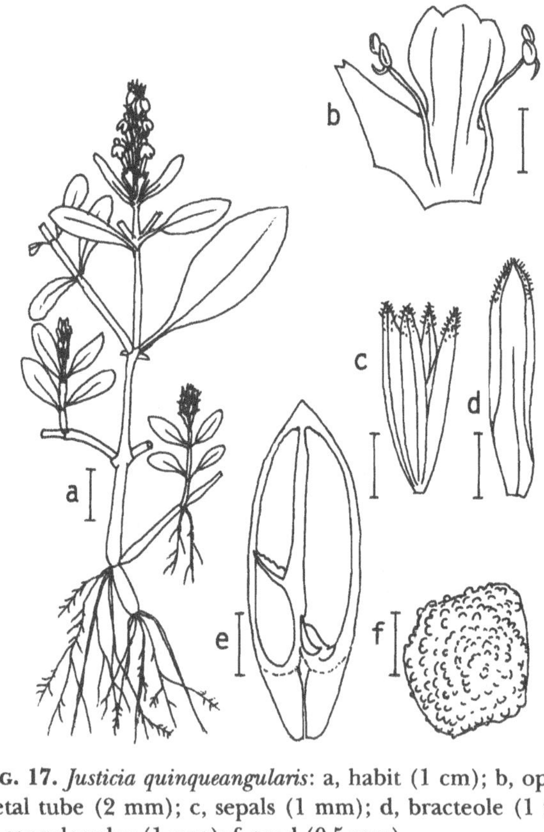

FIG. 17. *Justicia quinqueangularis*: a, habit (1 cm); b, opened petal tube (2 mm); c, sepals (1 mm); d, bracteole (1 mm); e, capsule valve (1 mm); f, seed (0.5 mm).

each with a tail-like appendage. Nectariferous scales conspicuous, cup-shaped with 2, 3 or 5 lobes. Ovary glabrous. Capsule 4-seeded, ± 7 mm long, ± 2 mm wide, glabrous, the lower part sterile, solid and stalk-like for ± a quarter of its length. Seeds supported by retinacula, flattened, 1–1.5 mm across, almost black tuberculate.

Perennial or annual. In wet and marshy places, at the edges of ponds and streams and in rice fields; it is not confined to wetlands and is also found in fields and waste places after the monsoon. Flowering after the rains. Plants with small roundish leaves are often recognized as *J. peploides* or as var. *peploides*; there are rather a lot of intermediate plants so I have recognized just one taxon.

{(⊙) Hel}. India to Indochina: Ban, Cey, Ind, Nep, Pak: Ap, Bh, Dl, Kt, Mh, Rj, Up. 0–200 m.

ACORACEAE see ARACEAE

ALISMATACEAE: 11 genera

Annual or perennial. Leaves basal; juvenile leaves not differentiated into petiole and blade, usually linear and submerged; adult leaves occasionally linear but usually with petiole and blade. Inflorescences usually compound or rarely simple, of whorls of branches, rarely umbel-like or with solitary flowers. Flowers radially symmetrical, bisexual or unisexual. Sepals 3, free. Petals 3, free, caducous or rarely absent. Stamens 3, 6, 9 or more. Carpels superior, free or united at the base, 3–numerous, in 1 whorl or spiral. Fruit a head of nutlets or in *Damasonium* dehiscent follicles.

Excluded: ***Alisma*** is found in the Himalayas but does not extend south in India (***A. gramineum*** C. G. Gmelin; ***A. lanceolata*** Withering; ***A. orientale*** (G. Samuelsson) Juzepczuk; ***A. plantago-aquatica*** Linnaeus).

1. Inflorescence spike-like; flowers sessile or nearly so, whorled; stamens 3 ***Wiesneria***

1. Inflorescence not spike-like; flowers distinctly stalked, whorled or not; stamens 6 or more

 2. Fruiting heads globose and smooth (head made up of very many flattened nutlets fitting together giving a smooth outline) ***Sagittaria***

 2. Fruiting heads either globose and spiny or flattened or ± disk-like

  3. Nutlets flattened, when ripe fitting together to form a circular or triangular whorl ***Alisma***

  3. Nutlets subglobose, when ripe forming an irregular, loose and flattened head

   4. Leaf blades glabrous, elliptical to broadly ovate or suborbicular, with base cuneate to deeply cordate; all flowers bisexual; nutlets laterally somewhat flattened, without lateral gas-filled chambers ***Caldesia***

   4. Leaf blades bearing stellate hairs, sagittate, with base deeply lobed; upper flowers male; nutlets swollen, with 2 lateral gas-filled chambers ***Limnophyton***

Caldesia: 2–4 species: Old World

Leaves submerged or floating, rarely emergent; petioles transversely septate; leaf blades elliptical to broadly ovate or suborbicular; base truncate to deeply cordate with broad lobes; tips acute or rounded to obtuse. Inflorescence a pyramidal panicle, with whorls of (2 or) 3 (–5) branches or flowers. Flowers bisexual. Sepals persisting in fruit. Petals white or pinkish, with claw at base. Stamens 6 (–11). Carpels 2–9 (–20), crowded on a small flat receptacle; style lateral. Nutlets laterally somewhat flattened, smooth, ridged, warty or spiny.

Usually perennial. Mostly found growing in shallow water. The flowers are pollinated by insects. The disseminules are ovoid nutlets (or in cooler regions inflorescence bulbils) dispersed in water or perhaps also by animals. *Caldesia* shows great variability in form and size of the leaf blade, sculpturing of the nutlets and in number of stamens and ovaries. The delimitation of species is uncertain; it is in need of a worldwide taxonomic revision.

Excluded: ***Caldesia grandis*** G. Samuelsson, it is found in the eastern Himalayas but not in Peninsular India (like *C. parnassiifolia* but generally larger, leaf blades leathery, mostly broader than long, up to 10 cm wide; stamens 8; carpels (8–) 10–12).

1. Bracts 1–8 cm long; transverse secondary veins 1–2 mm apart, connecting the main veins at ± 90°; nutlets with warty or spiny ribs ***C. oligococca***

1. Bracts not more than 0.7 cm long; transverse secondary veins 0.3–0.5 mm apart, connecting the main veins at 60°; nutlets with smooth ribs ***C. parnassiifolia***

Caldesia oligococca (F. von Mueller) Buchenau [*Alisma glandulosum* Thwaites]: Fig. 18

Petioles up to 70 cm long; leaf blades submerged or floating, membranous, ovate-cordate, 5–15 cm long, 4–8 cm wide, with (5–) 11–17 veins, the transverse secondary veins 1–2 mm apart, connecting the main veins at ± 90° the base deeply cordate with a narrow sinus, the lobes obtuse or acute. Bracts elongate-lanceolate with acute tips, very large at base of inflorescence up to 8 cm long, becoming smaller above. Scapes 40–115 cm long; panicle compound, pyramidal, 15–22 cm long and broad; pedicels 1–3.5 cm long. Sepals ovate, 2–4 mm long, persistent and reflexed in fruit. Petals ovate-lanceolate, 2–6 mm long. Nutlets (1–) 2–6 (–10), kidney-shaped, 3–6 mm long, 2–4 mm wide, dorsally 4-ribbed, the ribs usually warty, endocarp thick and hard; style inserted ventrally almost halfway down, very short, deciduous.

Perennial or perhaps occasionally annual. In standing permanent or temporary water in lakes, pools and very slowly flowing rivers. A very variable species, particularly in the sculpturing of the nutlets.

{⊙? Hyp}. W Africa, SE Asia and Australia: Ban, Cey, Ind: Kl, Pj, Up, Wb.

Caldesia parnassiifolia (Bassi ex Linnaeus) Parlatore var. ***major*** (Micheli) Buchenau [‘*parnassifolia*’ orth. mut., *C. reniformis* (D. Don) Makino]: Fig. 18

Petioles up to 25 (–50) cm long; leaf blades submerged or floating, leathery, elliptical to suborbicular or reniform, (2–) 5–8 (–12) cm long, (2–) 4–8 (–18) cm wide, with 11–15 (–17) veins, the transverse secondary veins 0.3–0.5 mm apart, connecting the main veins at ± 60°, base cuneate to cordate with rounded lobes. Bracts oblong-lanceolate up to 1 cm long. Scapes 30–100 cm long; panicle compound or rarely simple; pedicels 1.5–4 cm long. Flowers in whorls of 3 sometimes proliferating. Sepals ovate, 3.5–4.5 mm long, erect or spreading in fruit. Petals ovate-lanceolate, 4–6 mm long. Nutlets 5–8, ellipsoid, weakly flattened, 2–4 mm long, 1.5–2.5 mm wide, dorsally 3- to 5-ribbed, the ribs smooth; style subterminal, persistent forming an awn in fruit.

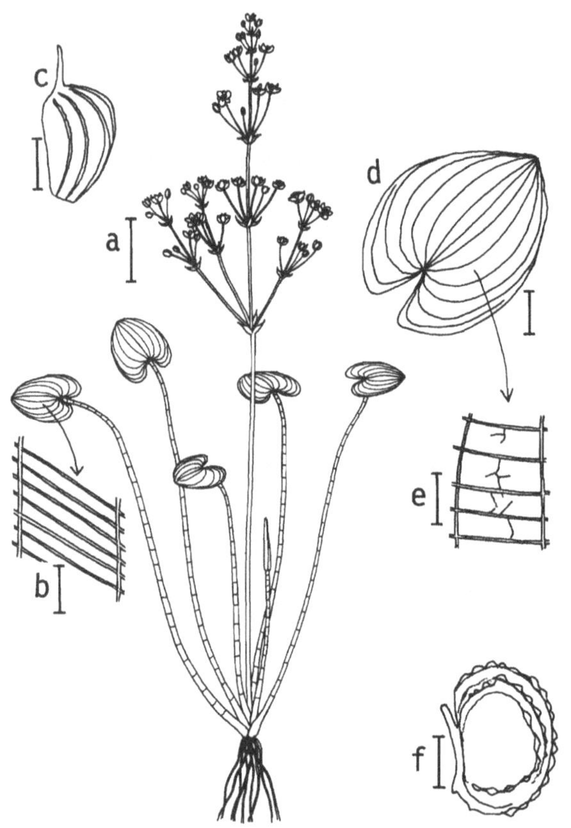

Fig. 18. *Caldesia oligococca*: d, mature leaf blade (1 cm); e, venation (1 mm); f, nutlet (1 mm). *C. parnassiifolia* var. *major*: a, plant (1 cm); b, venation (2 mm); c, nutlet (1 mm).

Perennial or sometimes annual. Usually in permanent shallow water, particularly along the margins of lakes, on river banks and in swamps. Most Indian authors have recognized *C. reniformis* as a separate species but on a world-wide scale it seems to merge with *C. parnassiifolia.*

{(⊙) Eph/(Hyp)}. Warm and tropical Old World: Ban, Ind, Nep: Bh, Kt, Mn, Tn, Up, Wb. 0–1500 m.

Limnophyton: 4 species: tropical Africa, SE Asia and Australia

Limnophyton obtusifolium (Linnaeus) Miquel: Fig. 19

Plants with milky latex. Leaves erect; petioles up to 1 m long; leaf blades ovate to sagittate or triangular, 4–20 (–30) cm long, 3–15 (–20) cm wide, glandular, with stellate hairs; lobes spreading, tapering, many-nerved. Scapes 1–3 m long, stout, angled. Inflorescences with 2–7 whorls of flowers; the lowest whorls with up to 15 flowers, sometimes branched and bearing female or bisexual flowers; upper whorls mostly with male flowers. Petals white, obovate, larger than the sepals. Stamens 6. Carpels 10–30, crowded on a relatively small receptacle; style ventral, forming a short beak. Nutlets reticulately ribbed and swollen, ± 4 mm long, with 2 lateral gas-filled chambers.

Annual or perennial. Juveniles submerged, adults submerged, emergent or seasonally terrestrial. In shallow water in tanks, ponds, ditches and swamps, sometimes near the coast. Pollinated by insects. The disseminules are swollen nutlets with lateral air chambers; probably dispersed in water. Economic importance unknown.

{(⊙) Hyp/(Eph)}. Old World tropics: Ban, Cey, Ind: Kt, Mp, Tn, Wb.

Sagittaria [*Lophotocarpus*]: ± 20 species: cosmopolitan, most species in the New World, some New World species are spreading in the Old World

Leaves submerged, floating or emergent; blades of the submerged leaves usually linear, sometimes lanceolate, mostly flat but sometimes inflated and spongy; blades of the floating leaves linear to ovate, cuneate or cordate at the base; emergent blades linear to ovate or sagittate. Inflorescences erect or sometimes floating, with 1–12 whorls, each whorl usually 3-flowered, the lower whorls bisexual or female, the upper whorls bisexual or male. Petals white, pink, yellowish or white with a purple or brown spot near the base. Stamens 7 to numerous. Carpels very numerous, spirally

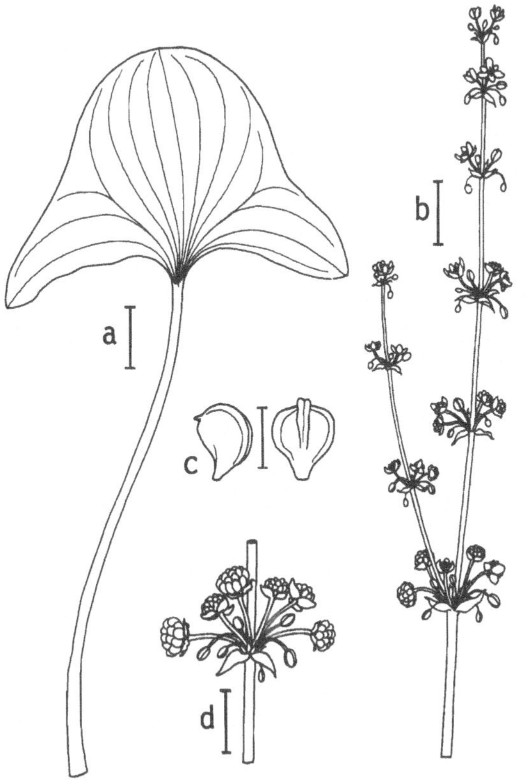

FIG. 19. *Limnophyton obtusifolium*: a, mature leaf (6 cm); b, inflorescence (6 cm); c, nutlet viewed from side and back (2 mm); d, whorl of fruits (1.5 cm).

arranged on a convex receptacle; style central. Nutlets laterally compressed, winged and beaked.

Juveniles submerged, adults submerged, floating, emergent or occasionally terrestrial, mostly perennial but some are annual. The flowers of most species unisexual; they are pollinated by insects. The disseminules are laterally flattened, winged and beaked, often very numerous nutlets; they are dispersed by water and by surface adhesion to animals. Some species, particularly *S. sagittifolia*, are cultivated for human food (tubers and leaves) and for ornament; some species are used for animal fodder and green manure; others are valuable for wildlife; some are potential weeds. Bogin, C. Revision of the genus *Sagittaria*. Memoirs of the New York Botanical Garden 9: 179–233 (1955); Rataj, K. Revision of the genus *Sagittaria*. Annotationes Zoolologicae et Botanicae (Bratislava) 76: 1–30 (1972); 78: 1–61 (1972). Neither of these revisions have found general acceptance.

Excluded: the following have been reported from the Himalayas but do not extend south into Peninsular India: ***Sagittaria greggii*** J. G. Smith [?*S. longiloba* Engelmann ex Torrey]; ***S. latifolia*** Willdenow: <Jk; ***S. pygmaea*** Miquel; ***S. sagittifolia*** Linnaeus sensu stricto; ***S. tengtsungensis*** H. Li.

1. Sepals in fruit reflexed; pedicels in fruit erect, ascending or spreading, not thickened; petals entirely white, without a purple blotch at the base ***S. sagittifolia*** subsp. ***leucopetala***
1. Sepals in fruit appressed or spreading, never reflexed; pedicels in fruit recurved and thickened; petals white, with a purple blotch at the base
   2. Mature leaves heart-shaped and floating; nutlets with a jagged wing ***S. guayanensis*** subsp. ***lappula***
   2. Mature leaves sagittate and held above the water; nutlets smooth, unwinged ***S. montevidensis***

Sagittaria guayanensis Humbolt, Bonpland et Kunth subsp. ***lappula*** (D. Don) Bogin ['*guyanensis*' orth. mut., *Lophotocarpus guayanensis* (Humbolt, Bonpland et Kunth) J. G. Smith]: Fig. 20

Stolons absent. Mature leaves differentiated into petiole and blade; petioles flexible up to 50 cm long; the blades floating, heart-shaped, 1.5–8 cm long, nearly as long as wide, obtuse or acute at the tips, the lobes 1–3 cm long with obtuse or acute tips; emergent leaves never develop. Scapes erect or floating, up to 45 cm long, bearing 3–5 whorls of flowers. Bracts somewhat united at the base, 6–20 mm long, 3–8 mm wide. Flowers usually bisexual or male or rarely female. Pedicels of the bisexual or female flowers, 1.5–4 cm long, thickened and recurved in fruit, of male flowers 2–6.5 cm long, filiform and deciduous. Sepals triangular, 1–1.5 cm long, appressed in bisexual flowers. Petals white usually with a purple blotch at the base, as long as or longer than the sepals. Stamens 6–9 in bisexual flowers, 9–12 in male flowers. Aggregate fruit globose, 1.2–1.5 cm in diameter. Nutlets flattened and winged, 2.7–4.5 mm long, 1.8–3.5 mm wide, the wings jagged or toothed to sometimes almost spiny.

Annual or perhaps sometimes perennial. Found in shallow water in tanks and at the edges of ditches and canals in slowly flowing water. It is sometimes found in temporary water but cannot withstand long periods of emergence and never develops erect emergent leaves. Frequently found in rice fields. The flowers wither by afternoon.

{⊙ (⊙)? Eph}. Old World tropics but absent from Australasia and Oceania (subsp. *guayanensis*: New World tropics): Ban, Ind, Nep: As, Bh, Dl, Kl, Mn, Mp, Or, Pj, Rj, Tn, Tp, Up, Wb. 0–1300 m.

Sagittaria montevidensis Chamisso et Schlechtendal subsp. ***montevidensis***: Fig. 20

Like *S. guayanensis* but mature leaves erect; petiole 30–80 cm long; blades sagittate, 10–20 cm long,

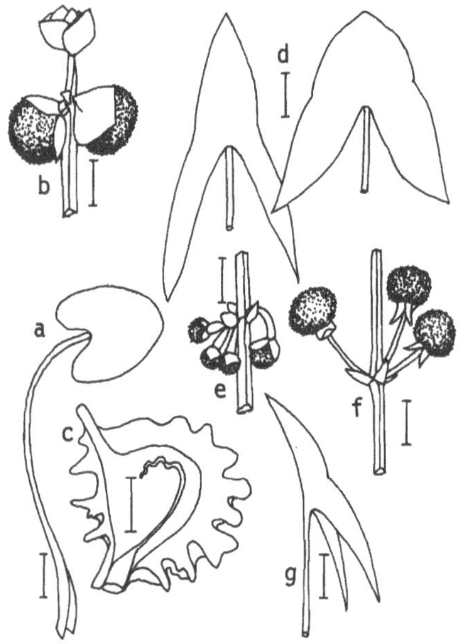

Fig. 20. *Sagittaria guayanensis*: a, mature leaf (2 cm); b, fruiting heads (1 cm); c, nutlet (1 mm). *S. montevidensis*: d, mature leaf blades (4 cm); e, fruiting heads (3 cm). *S. sagittifolia* subsp. *leucopetala*: f, fruiting heads (1 cm); g, mature leaf blade (4 cm).

5–25 cm wide; lobes as long as blade; fruiting pedicels, 3–7 cm long, recurved, thickened; aggregate fruit 1.5–2.5 cm in diameter; nutlets 2–3 mm long, 1–1.5 mm wide, smooth and unwinged.

Perennial. Wet places or in shallow water.

{Hyp}. Native to warm temperate S America, reported from Bangladesh: <Ban 1844.

Sagittaria sagittifolia Linnaeus subsp. ***leucopetala*** (Miquel) Hartog [*S. triflora* auct. Ind., non Linnaeus, *S. hastata* D. Don, *S. sinensis* Sims, '*chinensis*' orth. mut.]: Fig. 20

Stolons developing corms or tubers. Mature leaves differentiated into petiole and blade; petioles erect (20–) 50–80 (–120) cm long; the blades emergent, sagittate, 3–25 cm long, 0.4–6 cm wide, acute at the tips, the lobes always divergent with acute apices. Scapes erect, 30–100 cm tall, bearing 2–10 whorls of flowers, usually simple. Bracts somewhat united at the base, 3–15 mm long. Flowers usually unisexual. Pedicels of the female flowers 0.5–1.5 cm long, erect or spreading and not thickened in fruit; pedicels of male flowers longer, filiform and deciduous. Sepals triangular to ovate, 4–9 mm long, reflexed in fruit. Petals white, without a purple blotch at the base, 10–15 mm long, 14–20 mm wide, ± twice as long as the sepals. Stamens numerous; filaments glabrous, 1–2 mm long, yellow; anthers yellow. Aggregate fruit globose, 1–2 cm in diameter. Nutlets flattened, obovate, 4–5.5 mm long, 2.5–3.5 (–5) mm wide, the faces unadorned; the apical beak erect, 0.2–0.8 mm long.

Perennial. In rice fields and ditches. A cultivated race (sometimes called *S. edulis* Schlechtendal) develops edible corms up to 7 cm long (the wild form develops much smaller corms), it is widely cultivated in China; in India the corms are eaten by poor people; they are also used in medicine as an antiseptic.

{Hyp}. Tropical and warm Asia and Pacific, in India it is not clear where it is native and where it is an escape from cultivation; in S India it is found only on higher ground: Ban, Bhu, Bur, Ind, Nep, Pak: An, As, Bh, Jk, Kl, Kt, Mn, Mp, Pj, Tp, Up, Wb.

Wiesneria: 3 species: tropical Africa and India

Wiesneria triandra (Dalzell) Micheli: Fig. 21

Leaves basal, grass-like when juvenile, differentiated into petiole and blade when mature; blade and petiole usually separated by a constricted and curved hinge; petiole erect, 15–60 (–100) cm long, often inflated; blades linear to linear-lanceolate, 9–16 cm long, 4–10 mm wide, usually 3-angular at the base. Inflorescence a spike; scape shorter than the leaves, 10–50 cm

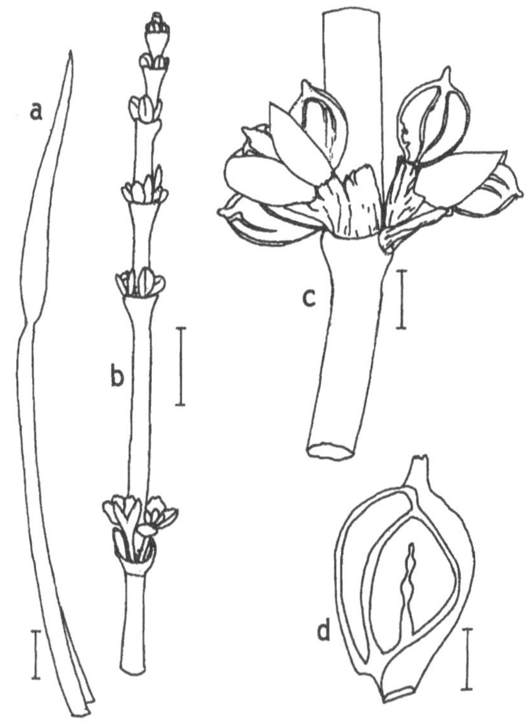

Fig. 21. *Wiesneria triandra*: a, mature leaf (3 cm); b, young inflorescence (1 cm); c, fruiting heads (2 mm); d, nutlet (1 mm).

long at anthesis, the lower part elongating in fruit. Flowers sessile or subsessile, in whorls of 5–8, the lower 2 or 3 whorls female (3rd whorl sometimes with bisexual flowers), upper whorls male (sometimes sterile at the tip). Sepals ± 3 mm long, ± 2 mm wide, cream-coloured. Petals shorter and narrower than the sepals, cream-coloured. Stamens 3. Carpels 3–4 (–6); styles terminal. Nutlets ovoid to globose, swollen, ± 5 mm long, ± 3 mm in diameter, with 2 lateral gas-filled chambers inside.

Annual. Often gregarious on flooded land, in temporary pools and marshes. The juveniles are submerged, the adults are partly submerged with emergent leaves. The flowers are mostly unisexual and probably pollinated by insects, although pollination by wind is possible. The disseminules are very hard, ovoid nutlets with 2 gas-filled chambers, dispersal probably by water or perhaps endozoochor. Flowering and fruiting after the SW monsoon. Sivadasan, M. *Wiesneria triandra*, eine sehr seltene und wenig bekannte Alismatacee aus Indien. Aqua-Planta 86: 159–161 (1986).

{⊙ Hyp}. Endemic to SW India: ♦Ind: Kl, Mh; records from NE India need confirmation.

AMARANTHACEAE: 70 genera

Herbs. Leaves opposite, entire, without stipules. Flowers inconspicuous, in axils of thin bracts, each subtended by 2 membranous bracteoles. Perianth of 3–5, free or nearly so, dry and membranous segments. Stamens (2–) 5, usually alternating with short pseudostaminodes; filaments united at the base into a short tube. Ovary superior, 1-locular; ovule solitary, pendulous from a long funicle; fruit 1-seeded capsule or nut.

Excluded: the following are not wetland plants but may be found in places after the flood water has receded or in dry rice fields: ***Allmania nodiflora*** (Linnaeus) R. Brown ex Wight [*A. longipedunculata* (Trimen) Gamble, *Chamissoa albida* Moquin-Tandon, *C. aspera* Moquin-Tandon, *C. dichotoma* Moquin-Tandon] not to be confused with *Alternanthera nodiflora* R. Brown; ***Celosia argentea*** Linnaeus; ***Digera muricata*** (Linnaeus) Martius [*D. alternifolia* (Linnaeus) Ascherson et Schweinfurt, *D. arvensis* Forsskål]; ***Gomphrena celosioides*** Martius: S America, <Ind; ***Nothosaerva brachiata*** (Linnaeus) Wight; ***Psilotrichum ferrugineum*** (Roxburgh) Moquin.

1. Flowers in dense, sessile, axillary heads; stems not ribbed ***Alternanthera***
1. Flowers in elongate, pedunculate spikes, with flowers and fruits increasingly distant towards the base of the inflorescence; stems ribbed ***Centrostachys***

Alternanthera [*Achyranthes* pro parte, *Gomphrena* pro parte, *Telanthera*]: ± 80 species: tropical to warm regions of the world, most in tropical S America

Stems not grooved. Leaves lanceolate, ovate or obovate. Inflorescences axillary or rarely terminal, sessile or stalked heads or short spikes. Perianth segments 5, glabrous or hairy, equal or unequal, the outer one not terminating in a subulate tip. Stamens 2–5; anthers 2-locular.

Herbs. Probably entomophilous. Disseminules are seeds or stem fragments

Excluded: not considered to be wetland plants: ***Alternanthera ficoidea*** (Linnaeus) R. Brown ex Grisebach: America, <Ind 1939; ***A. pungens*** Kunth; ***A. tenella*** Colla: <?Ind; ***A. nodiflora*** R. Brown: Australia <?Ind.

1. Flower heads distinctly pedunculate, either terminal or single at nodes; stems hollow with age ***A. philoxeroides***
1. Flower heads sessile or nearly so, either in clusters or pairs; stems not hollow
   2. Leaves and flower heads clustered; perianth segments unequal, each 3-nerved; fertile stamens 5 ***A. paronychioides***
   2. Leaves and flower heads paired; perianth segments equal, each 1-nerved; fertile stamens 3 ***A. sessilis***

Alternanthera paronychioides A. St. Hillaire: Fig. 22

Stems reddish, prostrate, creeping, branched, mat-forming, ± white-villous when young. Leaves clustered, subsessile or petiolate; blades oblanceolate-elliptical to obovate-rhomboid, (0.5–) 1–1.5 (–4) cm long, 0.2–1.2 cm wide, white hairy when young, base attenuate, tip obtuse to subacute. Flower heads sessile, solitary or 2–3 in clusters, subglobose to oblong, sometimes to 1.5 cm long, 0.5 cm in diameter. Bracts ovate-acuminate, mucronate with the excurrent midrib, ± 2.5 mm long; bracteoles like bracts but narrower, deciduous. Perianth oblong-lanceolate, 3-nerved, pilose on back, unequal, the outer 2 larger, 4–4.5 mm long, the inner 3.5–4 mm long. Stamens 5, all fertile, alternating with short, subulate pseudostaminodes. Capsule compressed orbicular-obcordate, ± 2 mm in diameter; seeds discoid, ± 1.2 mm in diameter.

Perennial with a stout vertical rootstock. It is not confined to wetlands but is found in sandy places subjected to periodical inundation, such as rice fields, river banks, irrigation ditches. It also found in brackish water. Often mistaken for *A. sessilis*.

{Hel}. Tropical America, naturalized in N America and the Old World: <Cey, <Ind, <Pak: <An, <Ap, <Dl, <Gj, <Kt, <Mg, <Mh, <Mp, <Pj, <Rj, <Tn, <Up, <Wb 1961.

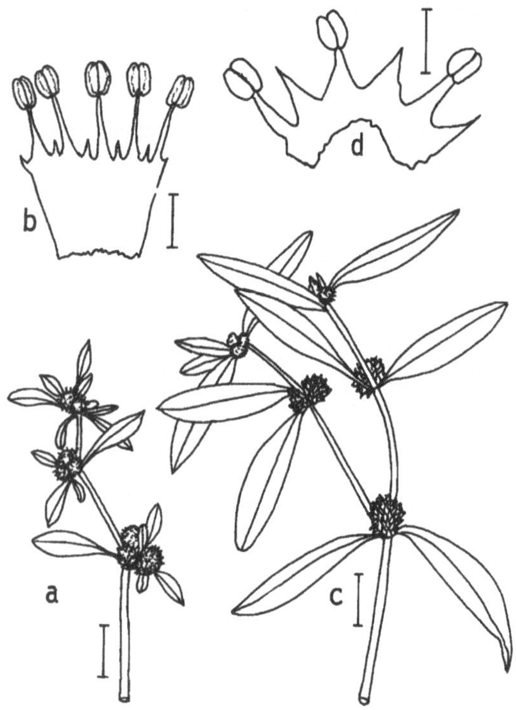

Fig. 22. *Alternanthera paronychioides*: a, part of flowering shoot (1 cm); b, staminal tube (0.3 mm). *A. sessilis*: c, part of flowering shoot (1 cm); d, staminal tube (0.3 mm).

Alternanthera philoxeroides (Martius) Grisebach: Fig. 23

Stems decumbent, up to 10 m or more long, up to ± 8 mm thick, becoming hollow and flattened with age, usually much branched below, forming mats, glabrous except for a band or strip of multicellular hairs within the leaf bases. Leaves paired; blades linear-elliptical, 2–7 (–9) cm long and 1–2 cm wide, acute at the tips, tipped with a tiny spine, tapering at the base to form a short petiole-like base which clasps the stem. Flower heads single in leaf axils or terminal, white, globose, ± 1 cm in diameter, pedunculate; peduncles up to 5.5 cm or more long, up to ± 1.5 mm thick. Bracts and bracteoles 1–2 mm long. Perianth segments ± equal, ± 4 mm long, glabrous, acute. Stamens 5, alternating with fringed petaloid staminodes; filaments united below into a short tube or replaced by pseudo-ovaries. Mature fruit and seeds not seen in India.

Perennial. The stems often form tangled masses in water (often partly floating) or along the shore. Seeds have not been reported outside its native range. It is often considered to be a very noxious weed. Two insects (*Agasicles hydrophila, Vogtia malloi*) feed exclusively on this species and are used for its control. In China, however, it is regarded as a valuable food for livestock, particularly pigs. It is widely used as a green manure; it is also used in medicine; in West Bengal it is sold in the markets as a vegetable. Majumdar, N. C. and Banerjee, R. N. The distribution and economic uses of *Alternanthera philoxeroides*. Bulletin of the Botanical Society of Bengal 30(1–2): 147–148 (1976).

{Hyp/Eph}. Native in S America, naturalized in N America, Asia and Australia: <Ban, <Bur 1932, <Ind 1940: <As, <Bh, <Dl, <Jk, <Kl, <Kt 1971, <Mg, <Mn 1961, <Mp 1976, <Or, <Up, <Wb 1940.

Alternanthera sessilis (Linnaeus) R. Brown ex A. P. de Candolle [*Alternanthera triandra* Lamarck, *Achyranthes prostrata* D. Don]: Fig. 22

Stems prostrate, usually much branched, forming mats, fistular when floating, with clusters of whitish roots at nodes, with a narrow line of whitish hairs along each side of the stem, the nodes pilose. Leaves paired, sessile or nearly so; blades lanceolate to spathulate or oval, 1–9 (–15) cm long, 0.2–2 (–3) cm wide, rounded to acute at the tips, cuneate at the base. Flower heads paired or rarely in clusters of up to 5, white, globose and sessile at anthesis, ± 7 mm in diameter, becoming cylindrical in fruit. Bracts deltoid-ovate,

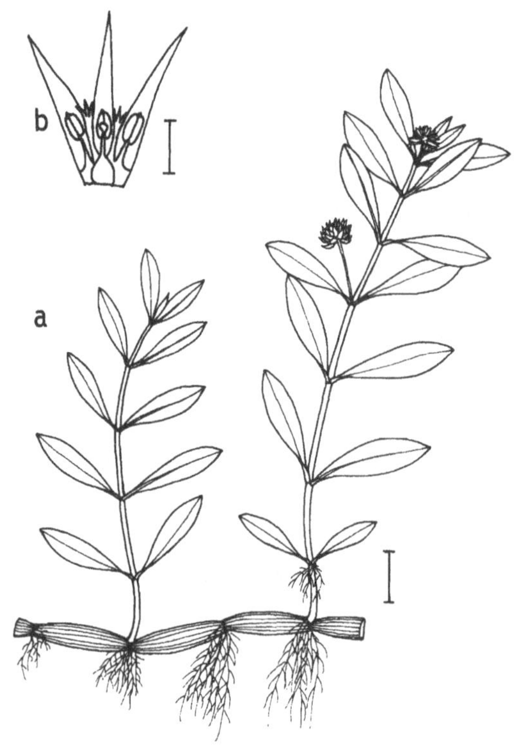

Fig. 23. *Alternanthera philoxeroides*: a, part of a floating shoot with inflorescences (1 cm); b, longitudinal section of a flower (2 mm).

± 1 mm long; bracteoles like bracts but shorter and folded, persistent. Perianth segments ± equal, oval-elliptical, 1.5–2.5 mm long, glabrous, 1-nerved, acute to blunt. Stamens 5, 3 fertile alternating with 2 filamentous staminodes; filaments ± 0.5 mm long; anthers ± 0.3 mm long. Capsule compressed heart-shaped, 1.2–2.5 mm long, ± 2 mm wide. Seeds discoid, 0.75–1 mm in diameter.

Annual or sometimes perennial. Essentially a terrestrial plant of seasonally waterlogged soil, but particularly common at the edges of tanks, rivers, canals and ditches. It is sometimes found growing in water (up to 1 m or more deep, where it is usually partly floating and partly emergent). It is also found in rice fields and occasionally in mats of floating vegetation. It may grow in brackish water near the coast. Flowers and fruits develop throughout the year if there is sufficient water. It is eaten as a vegetable and cultivated as such, also used in medicine.

{(⊙) Hel/(Ple)}. Pantropics: Bhu, Cey, Ind, Nep, Pak: An, Ap, As, Bh, Dl, Gj, Go, <Hp, Kl, Kt, Mh, Mp, Mz, Ng, Or, Pj, Rj, Sk, Tn, Tp, Up, Wb. 0–2000 m.

Centrostachys: 1 species

Centrostachys aquatica (R. Brown) Wallich ex Moquin-Tandon [*Achyranthes aquatica* R. Brown]: Fig. 24

Stems prostrate, straggling or erect, up to 1.5 m long, usually much branched, spongy and inflated below, up to 2 cm in diameter, conspicuously grooved above. Leaves opposite; petioles 0.4–4 cm long; blades lanceolate to ovate, (2.5–) 7–18 cm long; (0.8–) 2–6 cm wide, pilose beneath with appressed white hairs when young. Inflorescences usually terminal, erect, spikes, 4–12 cm long in flower, elongating to 25 (–45) cm in fruit. Flowers at first crowded, later remote. Bracts persistent, deltoid-lanceolate, 3–4 mm long, hyaline; bracteoles orbicular, up to 2 mm long. Perianth segments 5, glabrous, unequal, 5–8 mm long, the outer one terminating in a firm and slightly recurved subulate tip. Stamens 5; anthers 4-locular; staminodes with a fimbriate dorsal appendage. Nut thin-walled, falling with bract and hardened perianth.

Perennial or perhaps sometimes annual. Emergent in shallow water or straggling over the water surface. Probably entomophilous. The disseminules are seed-like nuts, dispersal unknown. The leaves are cooked as a vegetable. Prasad, V. P. and Vijayan, L. *Centrostachys aquatica* (R. Br.) Wallich ex Moq. — a rare species in India and its occurrence in Rajasthan. Journal of Economic and Taxonomic Botany 14(3): 612–614 (1990).

{(⊙) Hel/Hyp}. Tropical Africa and Asia eastwards to Norfolk Islands: Ban, Bur, Cey, Ind, Nep: As, Bh, Gj, Or, Rj, Up, Wb. 0–200 m.

FIG. 24. *Centrostachys aquatica*: a, upper part of shoot with inflorescences (3 cm); b, flower (5 mm).

AMARYLLIDACEAE [excluded from LILIACEAE]: 50 genera

Crinum: 110 species: pantropical

Crinum viviparum (Lamarck) R. Ansari et V. J. Nair [*Amaryllis vivipara* Lamarck, *C. asiaticum* Roxburgh non Linnaeus, *C. ensifolium* Roxburgh, *C. defixum* Ker-Gawler]: Fig. 25

Herbs with tunicated, bulbous rootstocks with fusiform, stoloniferous base; bulbs up to 10 cm in diameter. Leaves basal, linear, thick and fleshy, inrolled, 30–60 cm or more long, 3–6 cm wide, with an obscurely scabrid margin. Inflorescence an umbel subtended by a thick involucre which splits into 2 spathes. Flowers 6–12, erect at anthesis. Bracts up to 8 cm long. Perianth inserted above the ovary, tubular at base, divided into 6 segments above, without a corona, showy, white tinged with red, the tube 6–13 cm long, the lobes ± 8 cm long. Stamens 6, inserted at the mouth of the tube opposite the segments; filaments free to the base, bright red or purple, two-thirds to three-quarters as long as the perianth lobes; anthers

Fig. 25. *Crinum viviparum*: a, inflorescence (3 cm); b, habit.

versatile, ± 1.3 cm long. Ovary inferior, sessile or subsessile, 3-locular; fruit a capsule with irregular dehiscence; ovules few. Styles purple. Capsules up to ± 5 cm in diameter.

Bulbous perennial. Leaves submerged or emergent or partly submerged. It may flower in water up to 40 cm deep. Mostly found in clear streams and rivers but also recorded from swamps and mangroves. The bulbs are usually firmly imbedded in pebbles and may remain permanently submerged or may dry out each year. The flowers are pollinated by insects, they open at night and are fragrant; in the morning they wither. The disseminules are seeds and also bulbils, dispersal unknown but probably by water.

{Vit/Hyp/Ten}. Perhaps endemic to the Indian subcontinent: ♦?Bur, ♦?Cey, ♦?Ind: Go, Kl, Kt, Mh, Mp, Or, Rj.

APIACEAE [UMBELLIFERAE]: 275 genera

Herbs. Leaves alternate, often heterophyllous, compound or simple. Inflorescences of compound or simple umbels, sometimes in heads or occasionally proliferous; umbels usually subtended by bracts. Flowers small, bisexual, unisexual or sterile, mostly radially symmetrical. Sepals 5, conspicuous, inconspicuous or absent. Petals 5, usually with inflexed tip. Stamens 5, inserted on the disk above the ovary; anthers versatile. Ovary 2-locular; 1 ovule in each loculus; styles 2, at base united and often swollen; fruit dry, of 2, 1-seeded mericarps, contiguous along the inner surface (commissure), usually with 5 longitudinal ribs on the exposed outer (dorsal) surface; dark, oil-containing tubes frequently found between the ribs on both the dorsal and the commissural sides.

Mukherjee, P. K. and Constance, L. Umbelliferae (Apiaceae) of India. Oxford and IBH Publishing. New Delhi: i–viii, 1–279 (1993).

Excluded: aquatics growing in the Himalayas but not extending southwards: ***Berula erecta*** (Hudson) Coville; ***Cicuta virosa*** Linnaeus; ***Sium latijugum*** C. B. Clarke.

Found in damp ground but not a wetland species: ***Kedarnatha sanctuarii*** P. K. Mukherjee et Constance.

1. Stems stout, erect or ascending; leaf blades pinnately divided; flowers in compound umbels, borne above the leaf blades; fruits terete or nearly so ***Oenanthe***

1. Stems filiform, stolon-like, rooting at nodes; leaf blades entire or palmately lobed; flowers in simple umbels or clusters, inconspicuous, borne below the leaf blades; fruits distinctly flattened

  2. Leaves in clusters of 1–6 on short shoots at nodes; leaf blades entire to dentate; petioles sheathing at the base; stipules absent; involucral bracts conspicuous ***Centella***

  2. Leaves solitary or sometimes paired at nodes; leaf blades palmately lobed; petioles not sheathing at the base; stipules present; involucral bracts inconspicuous ***Hydrocotyle***

Centella: ± 20 species: pantropical

Centella asiatica (Linnaeus) Urban: Fig. 26

Stems creeping, stolon-like, usually reddish, hairy when young. Leaves in rosettes, 1–6 clustered on short shoots at each node, distinctly petiolate; petioles 1–10 (–40) cm long, sheathed below; blades simple, reniform to orbicular, 0.5–2 (–7) cm long, 1–3 (–8) cm broad; margins entire or crenate to crenate-dentate. Umbels borne opposite leaves, simple, solitary or 2–5 together; involucral bracts 2, ovate, 2–3 mm long; peduncles 0–2 (–5) cm long (much shorter than the leaves). Flowers usually 3, the middle one usually sessile; pedicels 0.5–1 mm long. Petals white to red. Fruit ovoid to ellipsoid, flattened, 2–3 mm long, 3–4 mm broad; mericarps 7- to 13-ribbed, ribs connected by transverse veins; oil tubes absent.

Perennial. Floating, emergent or terrestrial. Pollinated by insects or perhaps autogamous. Disseminules flattened mericarps, dispersal unknown probably by water or animals. It is often reported as a weed but is probably not very troublesome. It is sometimes grown as a cover crop because it has a great capacity for holding earth against erosion. Leaves used as a pot herb and as medicine. Flowers and fruits develop throughout the year.

{Hel}. Pantropics: Ban, Bhu, Cey, Ind, Nep, Pak: < An, Ap, Ar, As, Bh, Dl, Gj, Go, Hp, Kl, Kt, Mg, Mh, Mp, Or, Pj, Rj, Tn, Up, Wb. 0–2500 m.

Hydrocotyle: 75 species: almost cosmopolitan

Hydrocotyle sibthorpioides Lamarck [*H. rotundifolia* Roxburgh, *H. tenella* F. Hamilton ex D. Don, *H. nitidula* A. Richard, *H. ranunculoides* var. *incisa* Blume, *H. zollingeri* Molkenboer, *H. perexqua* Hance, *H. puncticulata* Miquel]: Fig. 27

Stems filiform, creeping, branching, rooting at most nodes, sparsely hairy or glabrous. Leaves alternate, borne singly or in pairs, distinctly petiolate, with stipules; stipules ovate or obovate, 0.5–1.5 mm long, acute, entire or fringed at the tip; petioles filiform, (0.5–) 0.8–4 (–6) cm long; blades almost orbicular in outline, 0.2–1.3 (–2.5) cm long, 0.5–1.5 (–3) cm broad, palmately cleft halfway down into 5 or 7 lobes; lobes broad, in contact but not overlapping, glabrous and shining above, with scattered coarse hairs beneath, margins crenate, tips obtuse. Umbels inconspicuous, solitary at nodes, simple, 4- to 8- (to 15)-flowered, compact, globular; involucral bracts minute; peduncles 0–1 (–3) cm long (usually shorter than the leaves). Flowers sessile or nearly so. Petals greenish-white. Fruit suborbicular, strongly flattened, 0.7–1.3 mm long, 1.2–1.5 mm broad; mericarps glabrous, ribs 3, subequal, blunt, oil tubes absent.

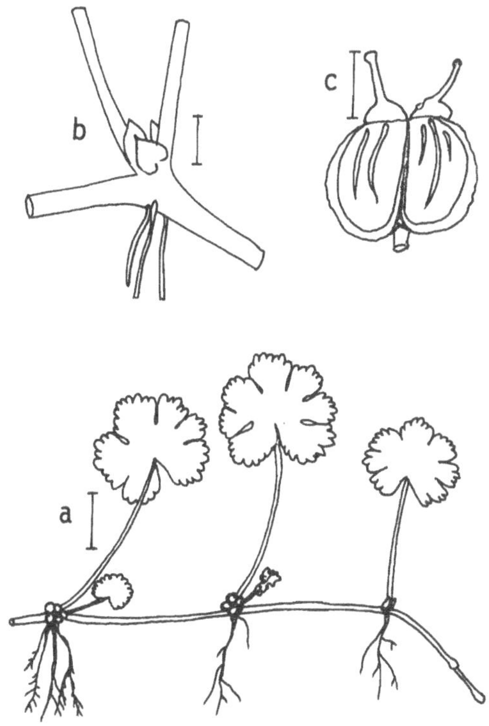

Fig. 27. *Hydrocotyle sibthorpioides*: a, flowering shoot (1 cm); b, node showing stipules (2 mm); c, fruit (0.5 mm).

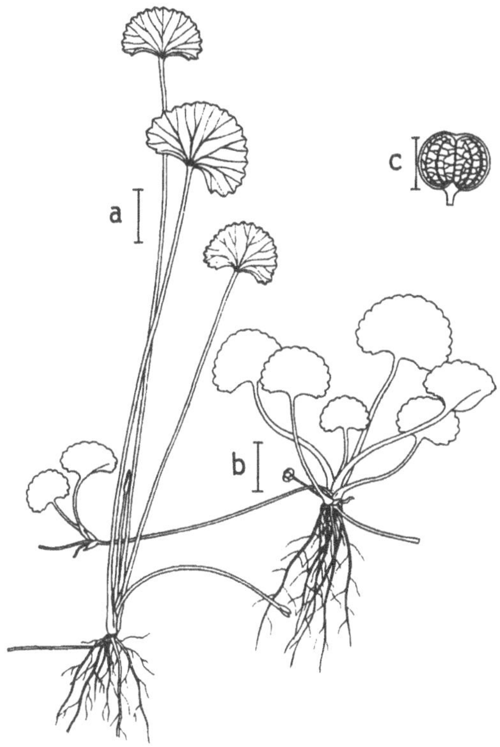

Fig. 26. *Centella asiatica*: a, aquatic shoot (2 cm); b, terrestrial shoot (2 cm); c, fruit (2 mm).

Perennial or perhaps sometimes annual. In damp, sunny or shady localities, in river beds, sometimes floating when flooded. Flowering throughout the year. It is occasionally eaten as a vegetable and used in medicine. Pollination by insects or perhaps autogamous. The disseminules are flattened mericarps, probably dispersed in mud, by water or perhaps by animals. It is very variable and plastic in form; numerous poorly defined species have been described. I have followed a 'lumping' policy.

{⊙? Hel}. Warm and tropical Asia, introduced in the New World: Ban, Bhu, Cey, Ind, Nep, Pak: Ar, As, Jk, Kl, Kt, Mg, Mp, Sk, Tp, Up, Wb. 0–4000 m.

Oenanthe: ± 40 species: mostly temperate and tropical N Hemisphere: 1 wetland species in India

Excluded: found in the Himalayas but not extending southwards: ***Oenanthe hookeri*** C. B. Clarke; ***O. thomsonii*** C. B. Clarke.

Oenanthe javanica (Blume) A. P. de Candolle [*O. stolonifera* (Roxburgh) Wallich ex A. P. de Candolle, *O. linearis* Wallich ex A. P. de Candolle, *O. bengalensis* (Roxburgh) S. Kurz, *O. schlechteri* Wolff, *O. rivularis* Dunn, *Sium laciniatum* Blume]: Fig. 28

Stems floating, ascending or erect, 10–100 cm long, internodes often swollen, rooting at lower nodes. Leaves borne singly at nodes, petiolate; petioles up to 10 cm long, sheathing at the base; stipules absent; blades 1-, 2- or rarely more-pinnate; segments ovate to narrowly oblong, serrate or pinnatifid. Umbels once compound, terminal and lateral; peduncles 1–20 cm long; rays 5–15, 0.5–3 cm long; pedicels 10–25, 2–5 mm long; involucral bracts inconspicuous, narrow, entire or absent. Sepals up to 0.5 mm long, acute, usually persistent in fruit. Fruit ellipsoid to almost cylindrical, terete or nearly so, 2–3 mm long, 0.5–1 mm wide, not constricted at the commissure, style base conical, ribs corky, inconspicuous; oil tubes solitary between ribs.

Perennial. In swamps, wet muddy places, wet grasslands and along streams and water channels. Pollinated by insects. The disseminules are corky mericarps, dispersed by water or perhaps also by animals. Sometimes cultivated as a vegetable (most other species are poisonous). It is very variable and plastic in form; numerous poorly-defined species have been described. I have followed a 'lumping' policy.

{Hyp/Hel}. Warm to tropical SE and E Asia and Australia: Ban, Bhu, Bur, Ind, Nep, Pak: As, Dl, Mg, Ng, Pj, Rj, Sk, Up, Wb. 200–2500 m.

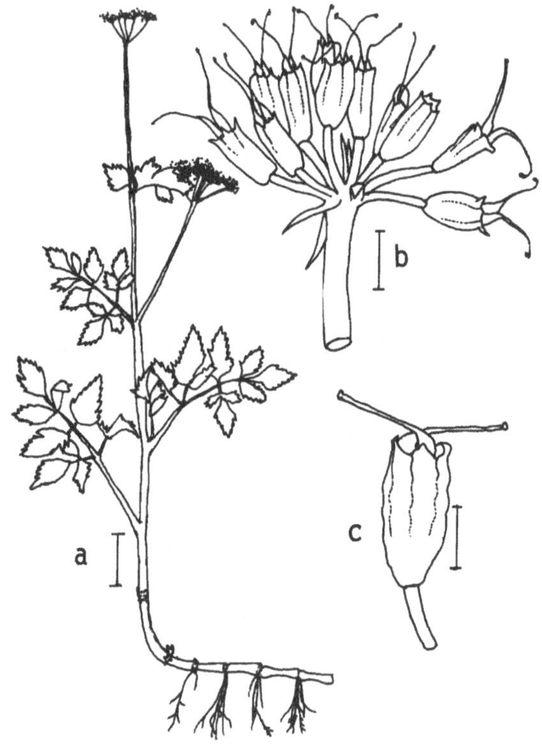

Fig. 28. *Oenanthe javanica*: a, flowering shoot (1 cm); b, umbel with young fruits (2 mm); c, fruit (1 mm).

APONOGETONACEAE: 1 genus

Aponogeton: 43 species: tropical to warm regions of the Old World

Stems tuberous or rhizomatous. Leaves radical, in basal rosettes, linear or petiolate; blades linear to ovate; margins entire, sometimes crisped or undulate. Flowers unisexual or bisexual, sessile, borne in spikes. Spikes stalked, usually elongate, solitary or two or more, emerging above the water surface, when immature enveloped by a caducous or persistent spathe. Perianth segments (0–) 2 (–6), petal-like or bract-like, mostly persistent. Stamens 6 or more. Ovary superior; carpels 3, 4 or rarely 5, free, mostly sessile; each ovary narrowed into a style with an adaxial stigmatic ridge; ovules 2–12 in each carpel. Fruits follicles.

Perennial. Juvenile plants submerged, the adults submerged, floating or sometimes temporarily terrestrial. Mostly in freshwater. Pollination by insects, autogamous or perhaps some agamospermous, one species is dioecious. The disseminules are seeds, bulbils or small plantlets, dispersed in water or in mud. The seeds do not survive desiccation and often germinate while still on the mother plant. The plants mostly survive the dry period as a tuber or rhizome. Several species are cultivated for decorative purposes; a few sterile artificial hybrids have been made. The tubers of some species eaten by humans and their livestock. Bruggen, H. W. E. van. Monograph of the genus *Aponogeton*. Bibliotheca Botanica, 137: 1–76 (1985).

1. Plants developing bulbils or plantlets instead of flowers ***A. undulatus***

1. Plants not developing bulbils or plantlets instead of flowers

2. Leaves semi-erect, emergent; leaf blades thick and brittle (ovate to cordate, 2–7.5 cm long, 1.8–3.3 cm wide. Maharashtra, Sindhurgurg District) ***A. bruggenii***

2. Leaves submerged or floating; leaf blades thin and flexible

3. Inflorescence with 2 spikes; flowers unisexual; plants dioecious, not developing submerged leaves (floating leaves lanceolate, 3.5–8.2 cm long, 0.5–1.4 cm wide. W India, Maharashtra) ***A. satarensis***

3. Inflorescence a single spike; flowers bisexual; plants monoecious, developing submerged leaves
4. Spathe persisting during anthesis; perianth segments falling off after flowering *A. undulatus*
4. Spathe falling off before anthesis; perianth segments persisting after flowering
5. Submerged leaves triangular to ovate; plumule attached at the middle of the embryo (Sri Lanka) *A. jacobsenii*
5. Submerged leaves narrowly ovate to linear; plumule basal or terminal
6. Plumule terminal; germinating embryo warty and crowned with a whorl of bent, linear or conical, ± 1 cm long appendages (brackish lagoons, coastal S and E India) *A. appendiculatus*
6. Plumule basal; germinating embryo smooth or if rough then not crowned with terminal appendages
7. Rhizomes cylindrical, creeping, up to 15 cm long, up to 1 cm in diameter; submerged leaves rigid, band-shaped, with flat margins; floating leaves absent (Sri Lanka) *A. rigidifolius*
7. Rhizomes tuber-like, scarcely longer than wide; submerged leaves not rigid, narrowly ovate to linear, with undulate or crisped margins; floating leaves often present
8. Perianth segments yellow; fruit less than 4 mm long; anthers yellow (NE India) *A. lakhonensis*
8. Perianth segments white, pink, violet or purple: fruit more than 4 mm long; anthers violet
9. Follicles with 2 seeds, up to 12 (–18) mm long, up to 5 (–7) mm wide, smooth or with irregular excrescences, with a short terminal beak; seeds with a simple coat (Sri Lanka) *A. crispus*
9. Follicles with ± 8 seeds, up to 4 mm long, up to 2.25 mm wide, smooth, with a very long (up to 1 mm) terminal beak; seeds with a double coat (India and Sri Lanka) *A. natans*

Aponogeton appendiculatus H. Bruggen: Fig. 29

Tuber imperfectly known, up to 2 cm in diameter. Submerged leaves petiolate; petioles 13–25 cm or more long; blades narrowly elliptical, 25–40 cm or more long, up to ± 3 cm wide, base and tip cuneate, margins flat or undulate, midrib wide with 2–4 parallel veins on each side. Floating leaves absent. Inflorescence a single spike, up to 15 cm long; spathe caducous, 1.5–6 cm long. Flowers bisexual, turned in all directions. Perianth segments 2, white, petal-like, obovate, 0.75–1.25 (–2) mm long, 0.75–1.5 (–2) mm wide, often wider than long. Filaments white, strongly widened towards the base; anthers pale yellow. Ovaries 2–4. Follicles 1- or 2-seeded, ± 12.5 mm long, ± 5 mm wide, with a short, terminal beak. Seeds with a simple coat; germinating embryo ± 10 mm long, ± 5 mm wide, warty and crowned with a whorl of bent, linear or conical appendages; plumule terminal.

Perennial. Near the coast in fresh or brackish lakes and backwaters, in water up to 4 m deep during the rains when the brackish water is flushed out by the monsoon rains. Flowering from August to December. The rhizomes are harvested in January and February for food in Lake Vembanad, Kl.

{Ros}. Endemic to S India: ♦Ind: Kl, Mh, Or? Tn, Wb?

Aponogeton bruggenii S. R. Yadav et R. S. Govekar

Like *A. crispus* but leaves nearly erect, emergent; petioles acutely 3-angled; leaf blades oblong-ovate to cordate, thick, dark green, aerial, brittle, at most two and a half times as long as wide; ovaries blue; germi-

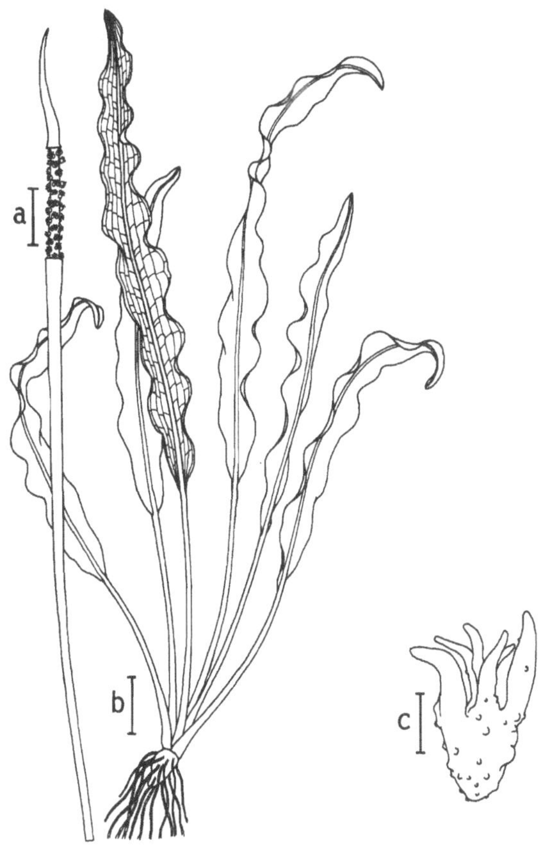

Fig. 29. *Aponogeton appendiculatus*: a, inflorescence (6 cm); b, habit (6 cm); c, germinating embryo (6 mm).

nating embryo cylindrical, three times as long as thick, smooth, without a longitudinal groove.

Perennial, surviving the dry period as tubers. Known only in rice fields. It is described and illustrated by Yadav and Govekar in Rheedea 4(1): 34–36 (1994) and Aqua-Planta 95(1): 3–11 (1995).

{Hyp/Ten?}. Endemic to W India: ♦Ind: Mh.

Aponogeton crispus Thunberg [*A. echinatum* Roxburgh]: Fig. 30

Tuber obovoid, up to 5 cm in diameter. Submerged leaves petiolate; petioles 10–35 cm long; blades linear to narrowly ovate, 10–25 cm long, 0.8–4.2 cm wide, usually more than two and a half times as long as wide, flexible, base and tip cuneate or rounded, margins undulate, midrib wide with 2–4 parallel veins on each side. Floating leaves occasionally absent; blades lanceolate to ovate, up to 20 cm long, up to 5 cm wide, base rounded or cordate, tip rounded or cuneate, midrib wide, with 7–11 secondary veins. Inflorescence a single spike, up to 13 cm long; spathe caducous, up to 25 mm long. Flowers bisexual, turned in all directions. Perianth segments 2, petal-like, white, pink or pale violet, up to 2.25 mm long, up to 1.75 mm wide. Filaments white; anthers violet. Ovaries 3. Follicles 2-seeded, up to 12 (–18) mm long, up to 5 (–7) mm wide (± twice as long as wide), smooth or with irregular excrescences, with a short terminal beak. Seeds with a simple coat; germinating embryo cylindrical, at most two times as long as thick, with a wide, longitudinal groove.

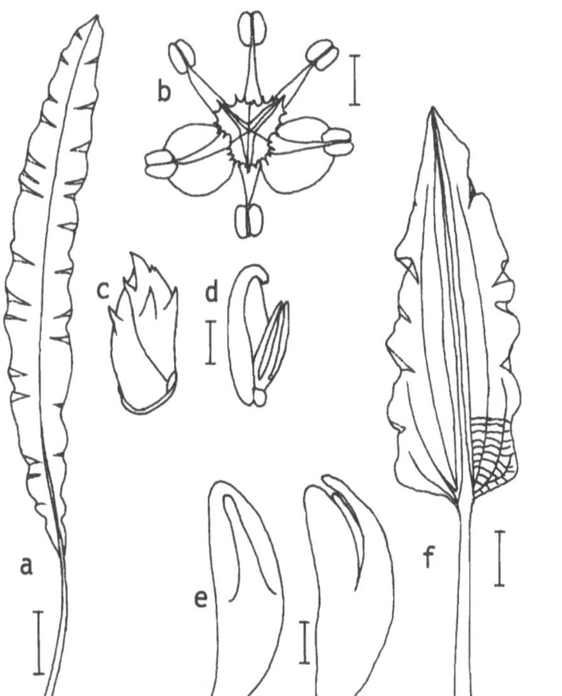

Fig. 30. *Aponogeton crispus*: a, submerged leaf (2 cm); b, flower (2 mm); c, fruit (2 mm); d, germinating embryo (2 mm). *A. jacobensii*: e, submerged leaf (1 cm); f, germinating embryo viewed from back and side (1 mm).

Perennial. In permanent and temporary water in ponds, tanks and lakes. Often forming large populations. The tubers are harvested for food or fodder, pigs dig them out; they may reach large dimensions, Bruggen (1985) mentions a tuber that measured 8 cm by 5 cm and weighed 105 g. It differs from *A. undulatus* in having floating leaves without cordate bases, submerged leaves with crisped margins, caducous spathes and does not proliferate at the inflorescence.

{Ros/Eph}. Endemic to S India and Sri Lanka: ♦Cey, ♦Ind: Ap, Kl, Kt, Mh, Pj? Tn, Wb?

Aponogeton jacobsenii H. Bruggen: Fig. 30

Like *A. crispus*, but the blades of submerged leaves are triangular to ovate, up to 20 cm long, up to 6 cm wide; floating leaves very rarely develop but when they do so then the base of the blade rounded or truncate; the plumule develops laterally at the side of the embryo above the middle.

Perennial. In shallow or deep water in stagnant ponds or in flowing water. Sometimes growing emergent on banks or shores. Locally frequent. Flowering almost throughout the year.

{Ros/(Eph)}. Endemic to Sri Lanka (Uva and Central Provinces): ♦Cey. 1650–2300 m.

Aponogeton lakhonensis A. Camus: Fig. 31

Like *A. natans*, but perianth segments yellow and obovate; follicles with a short beak (less than 1 mm long); the leaves usually larger, submerged blades up to 25 cm long and 6 cm wide, floating blades up to 17 cm long and 4 cm wide

Perennial. In slowly flowing streams, ponds and flooded rice fields. Flowering, at least, from February to May and from August to October.

{Ros/Eph}. Indochina extending eastwards to Assam: Ind: As. 230–800 m.

Aponogeton natans (Linnaeus) Engler et K. Krause [*A. monostachyon* Linnaeus fil., *A. indicus* sensu Roth ex Roemer et Schultes, non Roxburgh]: Fig. 31

Tuber up to 2 cm in diameter. Submerged leaves petiolate; petioles up to ± 5 cm long; blades lanceolate, up to 6.5 cm long, up to 1.5 cm wide, at base and tip cuneate, midrib wide with 2–3 parallel veins on each side. Floating leaves up to 11.5 cm long and 3 cm wide, at base cordate, at tip cuneate, with 5 or 7 secondary

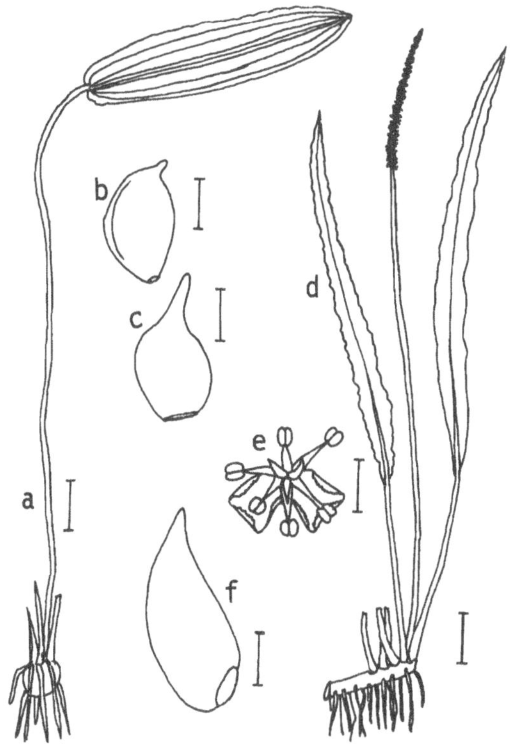

FIG. 31. *Aponogeton lakhonensis*: a, plant with corm and floating leaf (2 cm); b, fruit (7 mm). *A. natans*: c, fruit (7 mm). *A. rigidifolius*: d, plant (5 cm); e, flower (7 mm); f, fruit (3 mm).

veins. Inflorescence a single spike, up to 7 cm long, elongating in fruit; spathe caducous, up to 15 mm long. Flowers bisexual, turned in all directions. Perianth segments 2, petal-like, white, pink or purple, spathulate, 1.5–2 mm long, 0.5–1 mm wide. Filaments white; anthers yellow. Ovaries 3. Follicles with ± 8 seeds, up to 4 mm long, up to 2.25 mm wide, with a very long terminal beak (up to 1 mm). Seeds with a double coat.

Perennial. Mostly in temporary ponds, rice fields and marshes. Often abundant. Probably the commonest species in India. Flowering throughout the year. The tubers are gathered for food. Seed set is always very good in this species, Bruggen (1985) suspects agamospermy.

{Ros/Eph}. Endemic to India and Sri Lanka: ♦Cey, ♦Ind: Ap, Bh, Hr, Kl, Kt, Mp, Or, Pj, Rj, Tn, Tp, Up, Wb. 0–800 m.

Aponogeton rigidifolius H. Bruggen: Fig. 31

Like *A. crispus*, but it has a cylindrical, creeping rhizome, up to 15 cm long, up to 1 cm in diameter; floating leaves absent; blades of submerged leaves rigid, band-shaped, up to 60 cm long, up to 3 cm wide, with usually flat margin (may become undulate when cultivated in still water); anthers brownish.

Perennial. In slow or fast running streams and rivers in forest. It may grow temporarily emergent on banks as the water recedes. Flowers almost throughout the year.

{Ros}. Endemic to southwestern Sri Lanka: ♦Cey. 0–200 m.

Aponogeton satarensis Raghavan, Kulkarni et Yadav: Fig. 32

Plants dioecious. Tuber globose or ellipsoidal, 1.4–1.6 cm in diameter. Submerged leaves absent. Floating leaves petiolate; petioles 4–12 cm long; blades lanceolate, 3.5–8.2 cm long and 0.5–1.4 cm wide, base cuneate or rounded, tip acute, midrib wide, with 2 or 3 secondary veins. Inflorescences of 2 spikes, male inflorescence up to 7 cm, female 2.5–3.5 cm long, elongating in fruit, up to 12 cm; spathe caducous, up to 8–12 mm long. Flowers turned in all directions. Perianth segments in male flowers 2, petal-like, purple, obovate to almost orbicular, ± 1.6 mm long, ± 1.3 mm wide; in female flowers absent. Filaments in male flowers purple; anthers brown. Ovaries in male flowers rudimentary, sterile, in female flowers 3. Follicles 2-seeded, 6–7 mm long, ± 4 mm wide, with a terminal curved beak. Seeds with a simple coat.

Perennial. In temporary pools at 1200 m. Male plants are more abundant than females. Flowering from June to August. Yadav, S. R. and Mujawar, I. *Aponogeton satarensis* Raghavan, Kulkari et Yadev. Aqua-Planta 17(2): 50–58 (1992).

{Eph}. Endemic to W India: ♦Ind: Mh (Mavashi Plateau, Satara District).

Aponogeton undulatus Roxburgh [*A. microphyllum* Roxburgh]: Fig. 32

Tuber globose or elongate, up to ± 2.5 cm in diameter, smooth and whitish. Submerged leaves invariably present; petioles 10–35 cm long; blades elongate, 10–25 cm long, 0.8–4.2 cm wide, alternately transparent or opaque between the veins in an irregular pattern, at base and tip cuneate or rounded, margins undulate, midrib wide with 2–4 parallel veins on each side. Floating leaves occasionally absent; petioles up to 70 cm long; blades lanceolate, up to 20 cm long and 3.5 cm wide, at base rounded or cordate, at tip cuneate, midrib wide, with 5 or 7 secondary veins. Young plants and bulbils develop at the end of peduncle-like stalks, the young plants may also proliferate and develop more plantlets. Inflorescence a simple spike up to 10 cm long, elongating to 16 cm in fruit; spathe persisting during anthesis, up to 17 mm

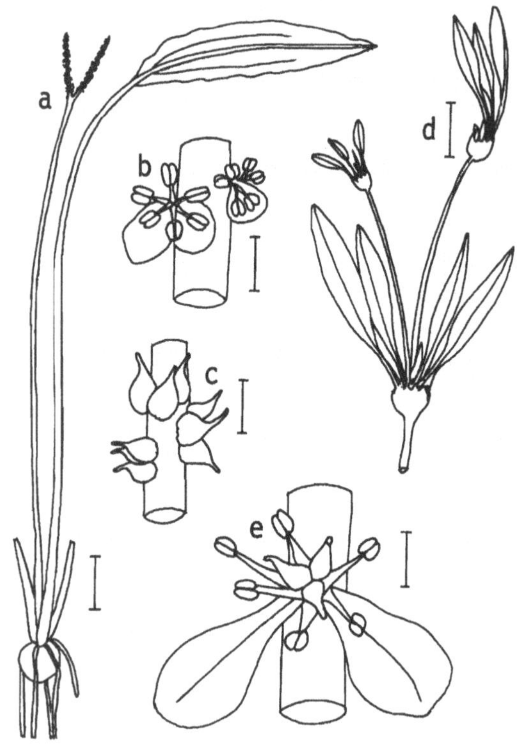

FIG. 32. *Aponogeton satarensis*: a, plant with corm and one floating leaf (2 cm); b, two male flowers (7 mm); c, three female flowers (7 mm). *A. undulatus*: d, proliferating bulbils (1 cm); e, flower (7 mm).

long. Flowers bisexual, turned in all directions. Perianth segments (1–) 2 (–3), petal-like, white or pinkish, 5–6 (–12) mm long, 1–2.5 (–4) mm wide, falling after flowering. Filaments white; anthers yellow (or sometimes blue?). Ovaries 3 or 4, pale pink. Follicles 2-seeded, smooth, 5–7 mm long, ± 4 mm wide, with a short, terminal curved beak. Seeds with a simple coat.

Perennial. Found in temporary ponds, ditches and tanks. It flowers from July to November but mostly the flowers are replaced by bulbils which may, in turn, proliferate into more bulbils. Bruggen, H. W. E. van, *Aponogeton undulatus* Roxburgh. Aqua-Planta 85(3): 20–24 (1985).

{Ros/(Eph)}. S and SE Asia: Ban, Bur, Ind: As, Bh, Mh, Pj, Rj, Up, Wb. 0–300 m.

ARACEAE [including ACORACEAE]: 107 genera

Mostly perennial herbs. Leaves mostly in basal or apical rosettes, with a membranous sheath at the base of the petiole. Inflorescence a fleshy spike (spadix), mostly enclosed in a large bract (spathe). Flowers small to minute, usually densely crowded on the spadix, bisexual, unisexual or sterile; female flowers confined to the base of the spadix; male flowers above the females; sterile flowers when present between the females and the males or on the apical part of the spadix. Perianth absent or of 4–6 (–8), free or united segments. Stamens 1–8, hypogynous, free or united into a synandrium, opposite the perianth segments, staminodes often present in sterile or female flowers. Ovary superior or immersed in the spadix; fruit a spongy, fleshy or leathery berry.

1. Plants free-floating; leaves subsessile and cuneate, in rosettes, hairy; inflorescence with a single female and a few male flowers ***Pistia***
1. Plants bottom-rooted or floating and creeping; leaves linear or differentiated into petiole and blade, never cuneate, usually glabrous; inflorescence with several female and male flowers or flowers bisexual
   2. Leaves unifacial, reed-like, not clearly differentiated into petiole and blade; spadix apparently borne directly on a leaf (stalk fused to the leaf) ***Acorus***
   2. Leaves bifacial, not reed-like, differentiated into petiole and blade; spadix borne on its own stalk, not fused to the leaf
      3. Petioles, lower leaf veins and peduncles bearing spines; flowers bisexual; spathe open only at the base ***Lasia***
      3. Petioles, lower leaf veins and peduncles without spines; flowers unisexual or sterile; spathe open above the base
         4. Leaf blades peltate at the base; female and male flowers separated, many sterile flowers crowded on a narrower part of the spadix, not surrounded by a fold of the spathe; the spathe with overlapping but free margins below ***Colocasia***
         4. Leaf blades ovate to linear-lanceolate; female and male flowers separated by a slender, naked axis of the spadix; the spathe united below (forming the so-called kettle)
            5. One half of the adaxial surface of the leaf blade rolled into the other half in bud (convolute); fruits united, forming a syncarpium opening from the apex ***Cryptocoryne***
            5. Abaxial surfaces of the leaf blades rolled inwards towards the midrib in bud (involute); fruits free, opening from the base ***Lagenandra***

Acorus: 2 species: N America and Asia, introduced in Europe and S America

Excluded: ***Acorus gramineus*** Solander is found in the Himalayas but does not extend into tropical India; it is like *A. calamus* but smaller; rhizomes relatively tough and wiry; leaves mostly arched, without a midrib, but with ribs along the margins; transition zone between sheath and blade suddenly narrowed and then widened above.

Acorus calamus Linnaeus: Fig. 33

Rhizomes creeping, 1–2 cm thick, relatively soft, aromatic. Leaves equitant (in 2 rows, each 'riding' the other), unifacial, erect, (50–) 60–120 (–185) cm long, (0.5–) 2–2 (–3) cm wide, with a ± prominent midrib, but without ribs along the margins, aromatic; sheaths usually shorter than blades; transition zone between sheath and blade not suddenly narrowed or widened above. Spadix appears to be borne laterally on a foliage leaf (the stalk is fused to the leaf), (4–) 6–8 (–10) cm long, 0.6–1.5 cm in diameter, slightly curved, with 700–800 flowers. Spathe absent (the prolongation of the fertile axis is sometimes interpreted as the spathe). Flowers bisexual, 3-merous. Perianth of 6 segments (2 whorls of 3), each ± 1 mm long, yellowish-green. Stamens 6; filaments linear. Ovary 2- or 3-locular; stigmas sessile or nearly so, round; ovules numerous. Fruit a few-seeded, gelatinous berry.

Perennial. Reed-like, emergent or temporarily terrestrial. In marshes, swamps and other wet and open places. It is pollinated by insects. The disseminules are gelatinous, few-seeded berries, presumably dispersed by animals and perhaps also by water; it is also dispersed by rhizome fragments. In the warm to hot regions of India it does not flower and some cultivars even if they do flower are sterile. It is probably not native or even naturalized in Peninsular India but is cultivated for medicine, as an insecticide and for perfume.

There are good grounds for recognizing the Acoraceae as a distinct family but I prefer to follow the usage found in most Floras. Röst, L. C. M. Biosystematic investigations with *Acorus*, 4. A synthetic approach to the classification of the genus. Planta Medica 37: 289–307 (1979). For evidence and arguments supporting the removal of *Acorus* from the Araceae see Grayum, M. Taxon 36: 723–729 (1987); Bogner, J. and Nicolson, D. H. Willdenowia 21: 35–50 (1991).

{Hyp/(Hel)}. N America and Asia, naturalized in Europe and S America: Bhu, <Cey, Ind, Nep, Pak: <Ap, As, <Go, Hp, Jk, <Kl, <Kt, Mn, Sk, Up, Wb. ?–2300 m.

Colocasia: ± 6 species plus numerous cultivated races: pantropics

Colocasia esculenta (Linnaeus) Schott [*C. antiquorum* Schott, *Arum colocasia* Linnaeus, *Caladium nymphaeaefolia* Ventenat]: Fig. 34

Rhizomes tuberous, starch-filled, sometimes developing stolons. Stems thick, up to 1 m high. Petioles up

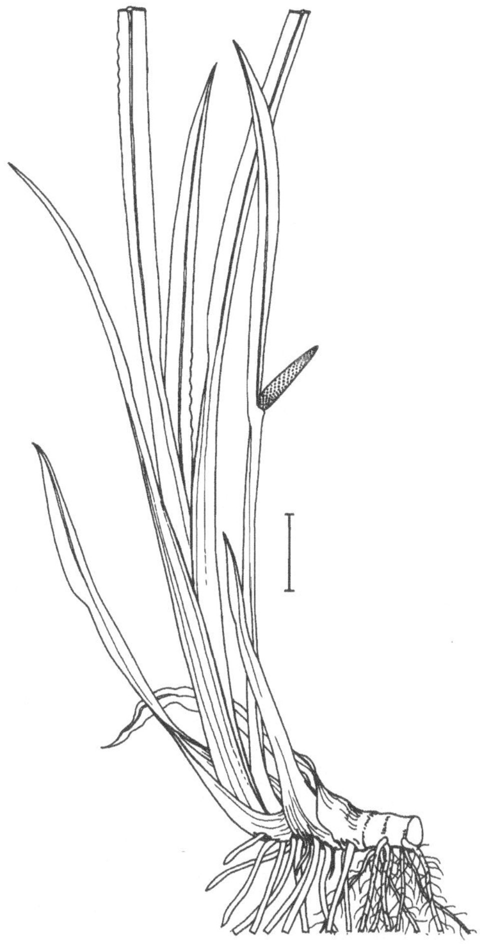

FIG. 33. *Acorus calamus*: whole plant (10 cm).

to 1 (–2) m long; blades peltate for 3 cm or more, ovate with a cordate base to sagittate in outline, up to 40 cm long and 30 cm wide. Inflorescences shorter than the petioles. Spathes ovate-lanceolate, constricted below the middle, up to 35 cm long, green below, bright yellow to orange above. Spadix up to 20 cm long; female flowers on the lower 2.5–4 cm, sterile for 1–1.5 cm and male for 3–4 cm, with an apical appendix 2–4 cm long (absent in some cultivated races). Perianth absent. Stamens united into a peltate mass (synandrium). Ovary 1-locular; ovules numerous, with parietal placentation; style short. Berries oblong to obconic, enclosed in the persistent tube of the spathe.

Perennial. Marshy places, but often standing in water in streams, rivers and ponds. In deep water it may develop floating leaves. It is pollinated by small flies but in India it apparently does not set fruit. It is widely cultivated for its starchy rhizomes (taro) and edible leaves or for ornament. It is propagated by tubers; there are numerous cultivars which vary in size,

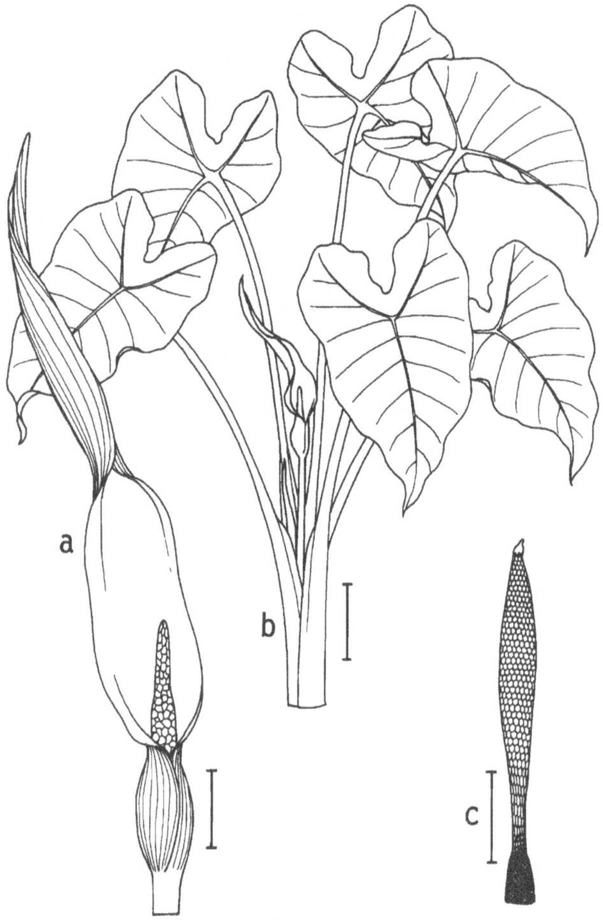

FIG. 34. *Colocasia esculenta*: a, spathe and spadix (4 cm); b, habit (20 cm); c, inflorescence (4 cm).

shape and colour. It is sometimes recorded to be a weed in irrigation ditches. Hill, A. F. The nomenclature of Taro and its varieties. Botanical Museum Leaflets (Harvard University) 7: 113–118 (1939).

{Hel/Eph}. Native in tropical Asia, widely naturalized in the tropics: <Bhu, Cey, Ind, Nep, <Pak: <An, As, <Dl, Go, Kl, Kt, Mh, Mn, Mz, <Rj, <Sk, Wb. 0–1200 m.

Cryptocoryne: ± 50 species: tropical Asia

Rhizomes creeping, thick or stoloniferous, often developing runners. Leaves often heterophyllous, linear or distinctly petiolate; blades convolute (one half of adaxial surface rolled into the other half) in bud, submerged or emergent, linear-lanceolate to ovate, cuneate to cordate at the base, sometimes bullate; cataphylls (like bladeless sheathes but homologous to prophylls) develop between some of the upper leaves. Spathes tubular, limb ovate or elliptical to elongate-lanceolate, sometimes spirally twisted, usually shorter than the tube. Spadix very short, enclosed in the dilated base (kettle) of the spathe; the female part below, separated by a slender sterile axis of the spadix; male part conic, wrapped in an involute fold or flap of the spathe. Male flowers very short, reduced to single stamens. Female flowers few, united in a single whorl. Ovary flask-shaped, 1-locular; ovules numerous. Fruit capsule-like, of several, united berries (syncarpium) which open from the tips. Seeds usually numerous.

Submerged, emergent or temporarily terrestrial. Pollinated by small flies. The fruits are syncarpia of united berries which open from the tips; the disseminules are seeds or seedlings, dispersed by water and perhaps also by animals. Many species are cultivated as decorative aquarium plants. Jacobsen, N. Cryptocorynen. Kernen Verlag 1–112 (1982); de Wit, H. C. D. Aquariumplanten (Hollandia) ed. 4, 1–463 (dated 1982 published 1983). Jacobsen, N. Die schmalblättigen Cyptocorynen des asiatischen Festlandes. Aqua-Planta 16(1): 2–33 (1991). Most figures are based on drawings from de Wit (1983).

1. Limb of spathe ciliate along the margin; the cilia mostly more than 5 mm long, simple or branched (brackish water) ***C. ciliata***

1. Limb of spathe smooth or warty or if ciliate then cilia less than 5 mm long and unbranched

2. Leaf blade gradually running into the petiole or leaves linear without a clear distinction between blade and petiole

3. Spathe without a tube, opening by a longitudinal slit extending down to the kettle, the limb elongate and remaining partly closed and curling below but not spirally twisted above (S India) ***C. spiralis***

3. Spathe with a tube between the kettle and limb, the limb opening and spreading or if remaining partly closed then spirally twisted above

4. Spathe tube gradually changing into a limb, without a clearly demarcated throat between them; the spathe limb spirally twisted at least at the tip

5. Spathe limb opening by a narrow slit, ± evenly spiral from base to tip, the markings reddish and ± round; stigmas borne almost horizontal (India) ***C. retrospiralis***

5. Spathe limb open and somewhat spreading at the base, the base less tightly spiral than the tip, the markings brown to purple, round and in longitudinal stripes; stigmas borne vertically to almost horizontal

6. Spathe limb rarely longer than 2.5 cm, the base open, widened and somewhat spreading, the tip bent backwards (often spiral and with a short tail), the markings brown to somewhat reddish; leaf blades margins flat, untoothed (Myanmar) ***C. albida***

6. Spathe limb usually more than 3 cm long, the base opening by a slit, not widened, but sometimes spreading, the tip spiralling upwards like a corkscrew or almost untwisted, the markings red to purple; leaf blades margins flat, undulate or bullate, entire or finely toothed (NE India) ***C. crispatula***

4. Spathe tube abruptly changing into a limb, with a clearly demarcated throat between them; the lower part of the spathe limb spreading not spirally twisted; the upper part of spathe flat or sometimes spiral at the tip

7. Rhizomes knotty, partly elongate and partly corm-like; male flowers 100 or more (Myanmar and India, not in Sri Lanka)

8. Spathe limb with a tail 7 cm or more long; kettle not wider than and somewhat longer than the tube (SW India) ***C. cognata***

8. Spathe limb with a tail less than 6 cm long; kettle not wider than and shorter than the tube

9. Spathe tube ± 10 cm long; kettle 2.5–3.5 cm long; submerged band-shaped leaves with undulate margins (S India) ***C. consobrina***

9. Spathe tube 2.5–5 cm long; kettle ± 1 cm long; submerged band-shaped leaves with flat margins (Myanmar) ***C. cruddasiana***

7. Rhizomes ± even, elongate and creeping; male flowers 40–60 (small plants with grass green leaves. Sri Lanka)

10. Spathe 10–20 cm long; spathe limb 2–3 cm long; collar between spathe tube and limb protruding 0.5–1 mm above the surface of the limb; leaf blades (0.5–) 1–2 cm wide ***C. nevillii***

10. Spathe usually less the 4 cm rarely up to 8 cm long; spathe limb usually less than 1 cm long; collar between spathe tube and limb never protruding more than 0.3 mm above the surface of the limb; leaf blades usually considerably less than 1 cm wide ***C. parva***

2. Leaf blade abruptly running into the petiole (the demarcation between blade and petiole short and clear)

11. Spathe limb without an abruptly differentiated collar at the base but sometimes with a diffuse collar zone

12. Spathe limb with an apical tail, white to pink or red, with or without red spots

13. Spathe limb white or rarely pink or red, without spots; spathe tube usually less than 1 cm long; leaf blades ovate to narrowly ovate ***C. alba***

13. Spathe limb white with red spots; spathe tube usually more than 2 cm long; leaf blades subelliptical ***C. thwaitesii***

12. Spathe limb without an apical tail, yellow to greenish, without any spots

14. Inner surface of the spathe limb rough ***C. bogneri***

14. Inner surface of the spathe limb smooth or very nearly so ***C. walkeri***

11. Spathe limb with an abruptly differentiated and prominent collar at the base

15. Collar between spathe tube and limb protruding 0.5–1 mm above the surface of the limb; upper part of the spathe throat with large and discrete purple spots ***C. nevillii***

15. Collar between spathe tube and limb not protruding more than 0.3 mm above the surface of the limb; upper part of the spathe throat without spots or if spotted then spots small and irregular

16. Collar between spathe tube and limb brown to blackish-purple; spathe limb usually paler than the collar

17. Spathe limb suberect, twisted to recurved, apical point short, never tail-like ***C. beckettii***

17. Spathe limb ± obliquely twisted, apical point sometimes tail-like ***C. wendtii***

16. Collar between spathe tube and limb green to yellow or cream sometimes becoming brownish; spathe limb the same colour as the collar

18. Spathe limb yellow to green, ± flat or recurved ***C. walkeri***

18. Spathe limb yellowish to cream or sometimes brownish, ± obliquely twisted ***C. undulata***

Cryptocoryne alba de Wit: Fig. 35

Like *C. thwaitesii* but spathe limb white or occasionally pink or red; spathe tube shorter (0.5–1.5 cm long); leaf blades subovate, almost smooth, with a flat or rarely wavy margin.

Perennial. Along or in small streams in the forest. It is perhaps only a variant of *C. thwaitesii.* Bastmeijer, J. D., Christensen, C. and Jacobsen, N. *Cryptocoryne alba* und ihre Variationsbreite. Aqua-Planta 84(1): 18–22 (1984).

{Ten/Hel}. Endemic to southwestern Sri Lanka: ♦Cey.

Cryptocoryne albida Parker [*C. retrospiralis* var. *albida* (Parker) Rataj, *C. costata* Gagnepain, *C. korthausae* Rataj]: Fig. 35

Like *C. retrospiralis* but spathe limb rarely more than 2.5 cm long; base of the spathe limb open and loosely spiral, towards the tip it is more tightly spiral and usually bends back on itself, the margins usually have brown or reddish stripes; stigmas are vertical; leaf blades green, green mottled with brown or brown; male flowers 80–120.

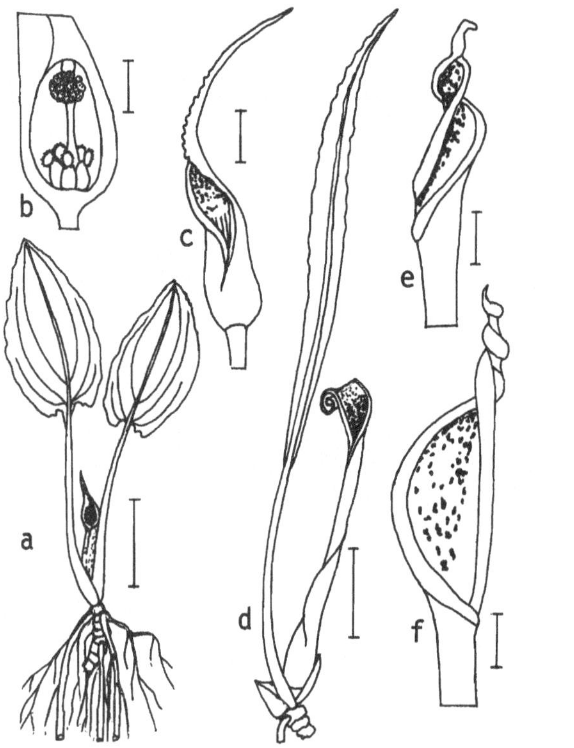

Fig. 35. *Cryptocoryne alba*: a, whole plant (4 cm); b, kettle (5 mm); c, spathe (1 cm). *C. albida*: d, leaf and spathe (2 cm); e, spathe limb (5 mm); f, spathe limb from another plant (5 mm).

Perennial. Grows on banks and gravel islands in slowly flowing rivers.

{Ros/Ten}. Myanmar and Thailand: Ban? Bur.

Cryptocoryne beckettii Trimen [*C. petchii* Alston]: Fig. 36

Like *C. walkeri* but spathe limb twisted and upright to somewhat recurved and twisted, the inner surface yellowish to brownish with a chocolate brown to almost black collar; spathe throat whitish or irregularly purple-spotted; leaf blade green and marbled with reddish-brown on the upper surface, the lower surface usually violet or at least tinged red; spathe outer surface purplish-brown with minute spots.

Perennial. Along and in swiftly flowing streams and rivers, frequently submerged. Frequently cultivated in aquaria. It is variable and sometimes rather similar to *C. wendtii* and *C. undulata*.

{Ros/Ten}. Endemic to central and southwest Sri Lanka: ♦Cey.

Cryptocoryne bogneri Rataj: Fig. 36

Rhizomes up to 7 mm thick, rough, with runners 5–10 cm long. Leaves ± upright; petioles 3–12 cm long; leaf blades ovate, 3–8 cm long, 2–4 cm wide, ± smooth, upper surface green to purple-green or reddish, lower surface paler often reddish, veins conspicuous, margin entire and sometimes wavy, base subcordate, tip acute. Spathes (2–) 4–8 cm long, outside yellowish below becoming reddish-purple above; kettle up to 1 cm long, slightly warty on the upper wall; tube 1–3 cm long; collar absent but the collar zone smooth; the throat yellowish to almost white; limb narrowly ovate, (1–) 2–5 cm long, 0.5–1 cm wide at the base, gaping and widened below, pointed but not tailed above, the inner surface rough, warty, yellow. Male flowers 20–30. Ovaries 4–6; stigmas circular, borne vertical.

Perennial. In small and large streams in forest, mostly submerged.

{Hyp/Ten}. Endemic to southwestern Sri Lanka: ♦Cey.

Cryptocoryne ciliata (Roxburgh) Schott: Fig. 36

Rhizomes runner-like, 40 cm or more long. Leaves erect; petioles up to 20 (–40) cm long, green, with wide sheath; leaf blades elongate-lanceolate to elliptical, 12–40 (–80) cm long, 6–10 (–15) cm wide, both surfaces uniformly pale green, base cuneate or

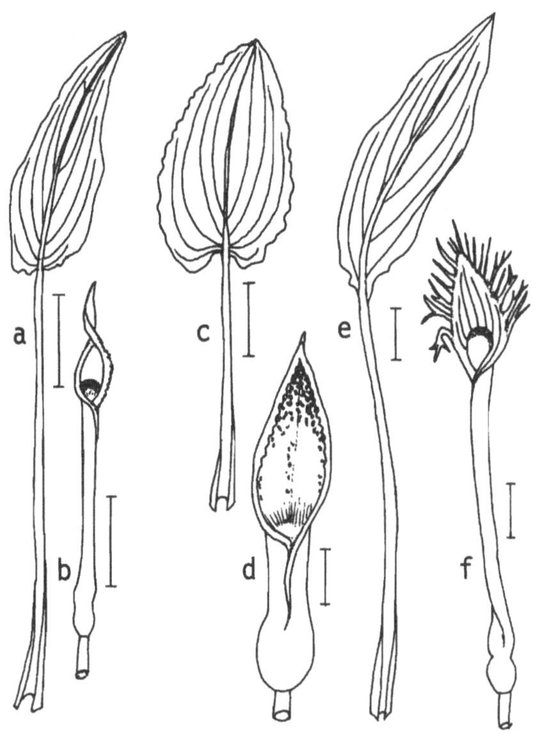

Fig. 36. *Cryptocoryne becketii*: a, leaf (2 cm); b, spathe (2 cm). *C. bogneri*: c, leaf (2 cm); d, spathe (1 cm); *C. ciliata*: e, leaf (3 cm); f, spathe (2 cm).

rounded, tip acuminate, becoming leathery when emergent. Spathes 20–45 cm long; kettle ± 2 cm long, with inner wall rough; tube 10–30 cm long; throat light yellow, dotted purple, with a narrow ligule; limb narrowly ovate to ovate, 5–7 cm long, the centre yellow, the remainder purplish-red or sometimes pink or yellowish-green, bearing 3–5 somewhat darker longitudinal ridges, the margin fringed with simple or branched, often curling cilia 5–15 mm long. Male flowers 60–80, arranged irregularly. Ovaries 5–7; stigmas elliptical. Seeds 11–13, developing numerous claw-like protuberances as they germinate.

Perennial. Usually found in tidal estuaries near the coast or in mangrove swamps. It prefers direct sunlight and usually grows along muddy banks where it is flooded twice a day with brackish water.

{Ros/Ten}. S and SE Asia to New Guinea: Ban, Ind: An, Or, Wb.

Cryptocoryne cognata Schott: Fig. 37

Rhizomes up to 1 cm thick and knotty, with stolons, bearing long thick contractile roots and thin absorbing roots. Leaves petiolate; petioles 3.5–27 cm long, the sheaths 1.5–10 cm long; leaf blades elongate-lanceolate or elongate-elliptical, 7–32 cm long, 2.6–7.6 cm wide, both surfaces shiny and uniformly green or pale green, with a thick midnerve and fine lateral nerves, margin entire and somewhat undulate, base cuneate running into the petiole, tip acuminate. Spathes 9–33 cm long, with 3–4 mm long stalks; kettle 0.7–6.5 cm long, cylindrical, 0.4–1.3 cm in diameter; tube 1.5–4 cm long, ± same in diameter as the kettle; throat ring purple above and yellow with purple flecks in the tube; limb 7–23 cm long, 2–3 times spiralled, drawn above into a long tail, the inner surface purple, warty. Male flowers 60–110. Ovaries 3–5 (–6); stigmas circular to broad-elliptical, borne vertical. Seeds ellipsoid, sometimes bent, 3–5 mm long, smooth, brown.

Perennial. Found rooted in sand or gravel in shady streams. Known from only 2 localities in Maharashtra (Ratnagiri and Sinhudurd Districts). Patil, K. S., Yadav, S. R., Dixit, G. B. and Bogner, J. Eine lang verschollene *Cryptocoryne* aus Indien: *Cryptocoryne cognata* Schott (Araceae). Aqua-Planta 17(2): 59–65 (1992).

{Ros/Ten/Hyp}. Endemic to S India: ♦Ind: Mh.

Cryptocoryne consobrina Schott [*C. wightii* Schott]: Fig. 37

Rhizomes up to 1 cm thick and knotty, with stolons, bearing long thick contractile roots which sometimes become detached and develop new plants and thin absorbing roots. Leaves heterophyllous; submerged (monsoon) leaves not clearly differentiated into petiole and blade, terete or ribbon-like, up to 60 cm or more long, less than 1 cm wide, when bladed then with undulate margin; emergent leaves either linear or with petiole and blade; petioles 10–18 cm long, the sheaths 7–8 cm long, gradually widening above and running into the blade; leaf blades elongate-lanceolate or elongate-elliptical, 16–20 cm long, 1.5–2.5 cm wide, flat or wavy, both surfaces uniformly green or sometimes purplish, with a thick and pale midnerve and fine lateral nerves, margin entire, base cuneate running into the petiole, tip acuminate (plants near Calicut flower without developing petiolate leaves). Spathes 15–25 cm long, shorter than the leaves, with 1–2 cm long stalks at anthesis (elongating up to ± 10 cm in fruit); kettle 2.5–3.5 cm long, cylindrical, 6–10 mm in diameter; tube 10–15 cm long, 2–4 mm in diameter; throat ring densely purple flecked; limb 2.5–7.5 cm long, gaping and widened below, with an erect or backwardly bent, long, twisted tail, the inner surface warty, yellowish-green or sometimes purple, the margin rough below. Male flowers ± 60. Ovaries 3–5 (–7); stigmas circular, borne vertical.

Perennial. Today it is known only from three man-maintained streams in a 10 km radius of the campus of the University of Calicut (Kohzikode, N Kerala). It is

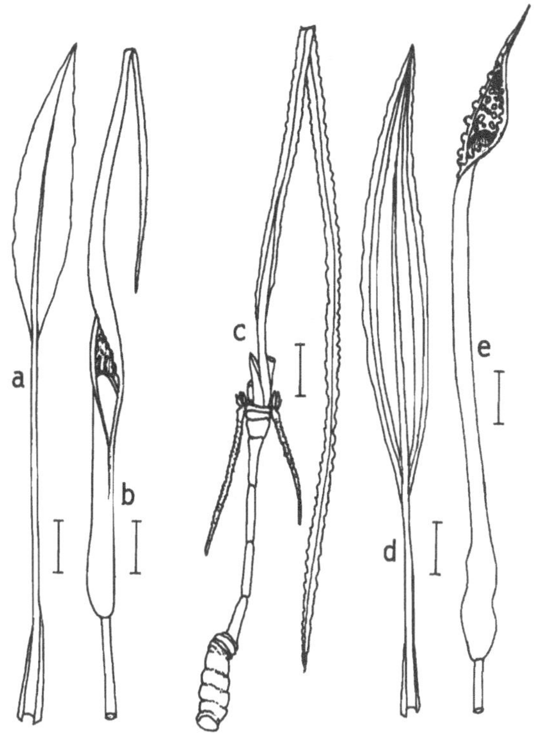

FIG. 37. *Cryptocoryne cognata*: a, leaf (2.5 cm); b, spathe (1 cm). *C. consobrina*: c, rhizome with two budding contractile roots and one submerged leaf (2 cm); d, emergent leaf (1 cm); e, spathe (1 cm).

probably extinct in earlier recorded habitats. It is in cultivation. During the monsoon it is submerged and develops long, grass-like leaves. It flowers in the winter from December to March after the water has receded. The plants near Calicut never seem to develop laminate leaf blades (even during the dry season they are grass-like); these plants may thus, perhaps, be better recognized as *C. wightii* but more field observations are needed. Sivadasan, M. *Cryptocoryne consobrina*, eine seltene Art aus Südindien. Aqua-Planta 85(2): 3–7 (1985); Jacobsen, N., Sivadasan, M. and Bogner, J. Ungewöhnliche vegetative Vermehrung bei der Gattung *Cryptocoryne*. Aqua-Planta 89(3): 83–88, (4): 127–132 (1989).

{Ros/Ten/Hyp}. Endemic to S India: ♦Ind: Kl, perhaps extinct in Kt and Tn.

Cryptocoryne crispatula Engler var. ***flaccidifolia*** N. Jacobsen [*C. retrospiralis* var. *crispatula* (Engler) de Wit, *C. sinensis* Merrill, *C. balansae* Gagnepain, *C. tonkinensis* Gagnepain]: Fig. 38

Like *C. retrospiralis* but spathe limb grey to yellowish with purple to purple-black spots and short and rather irregular stripes; stigmas vertical to horizontal; leaf blades sometimes brownish, usually undulate in India but sometimes flat or bullate, sometimes larger (reaching 70 cm long and 4 cm wide, but mostly less than 1.2 cm wide in India), with more distinct lateral nerves; male flowers 90–130.

Perennial. Grows in swiftly flowing rivers. It is very variable in form of the leaves; several varieties are recognized, in India only var. *flaccidifolia* N. Jacobsen is found.

{Ros/Ten}. NE India, Myanmar, Indochina and S China: Ban? Bur, Ind: As, Wb. The record from Bh is probably erroneous.

Cryptocoryne cruddasiana Prain [*C. burmensis* Rataj]: Fig. 38

Like *C. consobrina* but kettle ± 1 cm long, very slightly constricted at the middle; spathe tube 2.5–5 cm long; spathe limb ± 4 cm long, 1–2 times twisted, with irregular transverse ribs and a short apical tail, reddish inside with a dark throat; male flowers 100–120; submerged band-shaped leaves 25–35 cm long, 0.5–1 cm wide, with flat margins.

Perennial. Ecology unknown, presumably in rivers.

{Ros/Ten/Hyp}. Endemic to Upper Myanmar: ♦Bur.

Cryptocoryne nevillii Trimen ex J. D. Hooker: Fig. 38

Rhizomes up to 8 mm thick and knotty, stolons and runners unknown. Leaves ± upright; petioles 5–10 (–15) cm long; leaf blades elliptical to narrowly ovate, 2–8 cm long, (0.5–) 1–2 cm wide, smooth or somewhat rough, both surfaces uniformly green, not shiny, veins conspicuous, margin entire, base cuneate running into the petiole or truncate, tip subobtuse. Spathes 10–20 cm long, outside whitish, becoming brownish-green towards the tip; kettle 0.8–1.5 cm long, inside somewhat constricted below the male flowers, upper part ± red-spotted; tube 8–15 cm long; collar between spathe tube and limb protruding 0.5–1 mm above the surface of the limb; the upper part of the throat with large and discrete purple spots; limb narrowly ovate, 2–3 cm long, 0.5–0.8 mm wide at the base, gaping and widened below, with an erect or backwardly bent and spirally twisted tail, smooth, the inner surface maroon. Male flowers 60–80. Ovaries 4–6; stigmas circular, borne vertical.

Perennial. In seasonally inundated areas, marshes and swamps and sometimes along the coast. It grows in the dry region of Sri Lanka. The leaves die completely during the dry season. As the rains come it sometimes flowers before the leaves are fully developed.

{Ten}. Endemic to central eastern Sri Lanka: ♦Cey.

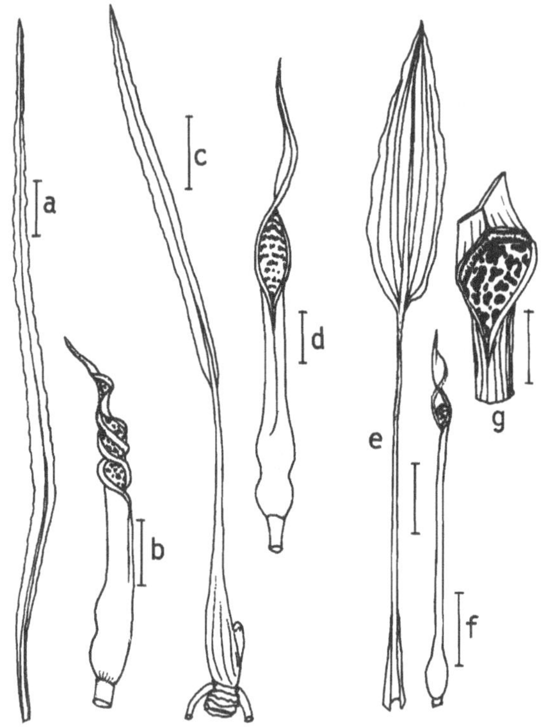

FIG. 38. *Cryptocoryne crispatula* var. *flaccidifolia*: a, leaf (2 cm); b, spathe (2 cm). *C. cruddasiana*: c, leaf (2 cm); d, spathe (1 cm). *C. nevilii*: e, leaf (2 cm); f, spathe (2 cm); g, throat of spathe (1 cm).

Cryptocoryne parva de Wit: Fig. 39

Rhizomes up to 5 mm thick, often with rather long internodes, with runners 1–5 (–10) cm long, the shorter ones somewhat rough. Leaves ± upright; petioles up to 10 cm long; leaf blades narrowly ovate-elliptical, 0.8–1.5 cm long, 0.2–0.8 cm wide, smooth, both surfaces uniformly green, veins inconspicuous, margin entire, base attenuate to obtuse running into the petiole, tip acute. Spathes 1.5–2.5 (–3.5) cm long, outside redish-purple, lower part whitish; kettle 0.5–0.8 cm long; tube 0.7–1.5 cm long; collar between spathe tube and limb purple to dark purple, not protruding more than 0.3 mm above the surface of the limb; the throat purple to unevenly purple-spotted; limb subovate, 0.4–0.9 (–1.2) cm long, 0.3–0.5 cm wide at the base, gaping and widened below, subobliquely twisted, the inner surface rough to warty, purple. Male flowers 30–50. Ovaries 4–6; stigmas circular, borne vertical.

Perennial. Along streams and rivers, often forming dense mats, particularly in swift water. It sometimes grows with and hybridizes with *C. beckettii* and *C. walkeri.* These hybrids are fertile and show considerable variation but are virtually indistinguishable. This hybrid complex is usually named *Cryptocoryne* × *willisii* Reitz [*C. lucens* de Wit] by the cultivators of aquarium plants; *C.* × *willisii* differs from *C. parva* in having rather larger spathes, the limb is suberect and twisted.

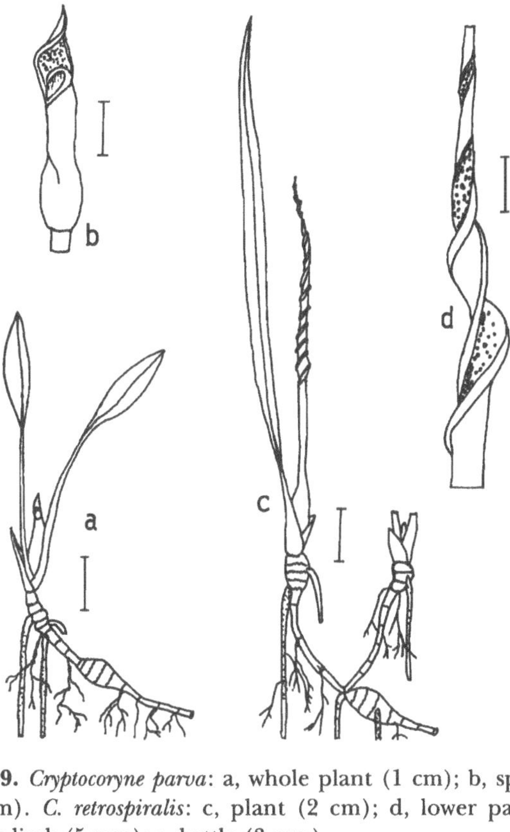

Fig. 39. *Cryptocoryne parva*: a, whole plant (1 cm); b, spathe (5 mm). *C. retrospiralis*: c, plant (2 cm); d, lower part of spathe limb (5 mm); e, kettle (3 mm).

Cryptocoryne parva probably also forms hybrids with *C. undulata.*

{Ros/Ten}. Endemic to central Sri Lanka: ♦Cey.

Cryptocoryne retrospiralis (Roxburgh) Kunth [*C. uniloculare* (Roxburgh) Kunth, *C. roxburghiana* Voigt, *C. roxburghii* Schott, *C. dalzellii* Schott]: Fig. 39

Rhizomes 0.5–1.5 cm thick and knotty, with stolons, bearing long thick contractile roots which sometimes form corm-like structures above the ground and thin absorbing roots. Leaves erect or spreading; petioles shorter than the blades, the lower half to two-thirds with a rolled sheath, gradually widening above and running into the blade; leaf blades linear to elongate-lanceolate or elongate-elliptical, 10–40 cm long, 0.3–1 (–1.5) cm wide, flat or wavy, both surfaces uniformly green or sometimes brownish, with a thick and pale midnerve and very thin, indistinct lateral nerves, margin sometimes finely toothed, base cuneate running into the petiole, tip acuminate, becoming leathery when emergent; monsoon leaves awl-like, ± terete, 10–15 cm long. Spathes (7.5–) 10–30 cm long, shorter than the leaves, with up to 2.5 cm long stalks; kettle (1–) 1.5–2.5 cm long, cylindrical, with inner wall contracted about halfway or more up; tube 5–20 cm long, gradually changing into a limb; throat closed or a narrow curved split; limb (1–) 3–8 cm long, usually tightly spiralled with up to 7 or more turns (resembling a corkscrew), rarely short and open, olive green to yellowish with red spots not streaks, the margin smooth. Male flowers 100–140. Ovaries 4–7; stigmas circular, somewhat funnel-shaped, borne horizontal or nearly so. Seeds 12–20.

Perennial. Mostly in and along rivers, where it often forms turf-like communities. It is usually in sand or gravel; the petioles, sheaths and lower part of the spathes are often buried. Used locally in medicine. It has often been confused with other species; in this work it is considered to be distinct from *C. albida* and *C. crispatula.* In SW India it is the only species with a long tube.

{Ros/Ten}. Endemic to India and perhaps Bangladesh, in the south on higher land: ♦Ban? ♦Ind: Ap, As, Bh, Go, Kl, Kt, Mh, Mp? Tn, Wb? 0–1000 m.

Cryptocoryne spiralis (Retzius) Fischer ex Wydler [*C. huegelii* Schott, *C. cognatoides* Blatter et McCann, *C. tortuosa* Blatter et McCann]: Fig. 40

Rhizomes up to 1 cm thick, up to 30 cm long, bearing long thick contractile roots and thin absorbing roots. Leaves erect or spreading; petioles 5–8 (–15) cm long, green, with wide sheath; leaf blades elongate-elliptical to linear, (7–) 10–20 (–30) cm long, (0.5–) 1–2 (–3) cm wide, both surfaces uniformly pale green

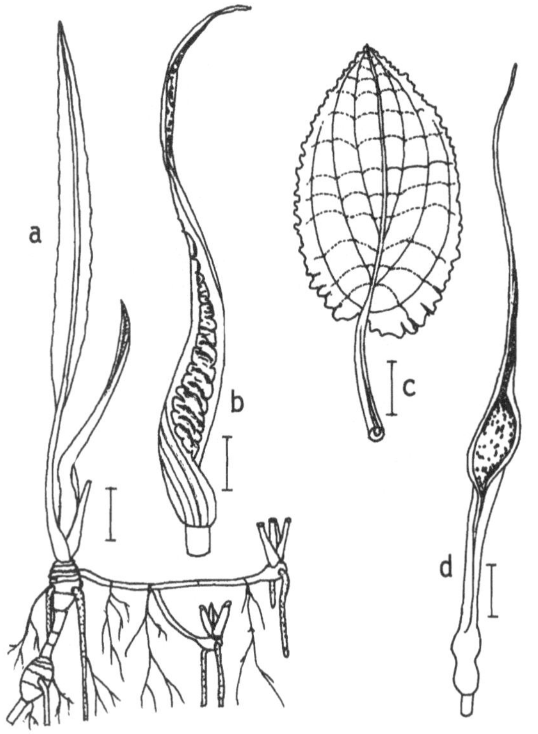

FIG. 40. *Cryptocoryne spiralis*: a, plant (2 cm); b, spathe (1 cm). *C. thwaitesii*: c, leaf (1 cm); d, spathe (1 cm).

sometimes with a brownish shimmer, with a thick and pale midnerve and thin indistinct lateral nerves, base cuneate running into the petiole, tip acuminate, becoming leathery when emergent. Spathes 8–17 cm long, shorter than the leaves, with up to 1 cm long stalks; kettle 1.5–2 (–3) cm long, ellipsoid, with inner wall contracted about halfway or more up; tube virtually absent; limb usually twisted, narrowing from a 1–1.5 cm wide base to an acuminate tip, 7–15 cm long, the outer surface greenish-brown; the inner surface purple to red or sometimes yellowish with wart-like transverse ridges, the margin with warty tooth-like projections. Male flowers 60–70. Ovaries 5; stigmas irregularly circular, somewhat funnel-shaped.

Perennial. Found in seasonally inundated places along rivers, roadside ditches and wet sandy places along the coast. It is often abundant and by no means rare in Kerala.

A highland variety with very long spathes, as long or longer than the leaves (15–35 cm long), has been called *Cryptocoryne spiralis* var. *cognatoides* (Blatter et McCann) Yadiv, Patil et Bogner. Yadiv, S. R., Patil, K. S. and Bogner, J. Kritische Bemerkungen über die Identität von *Cryptocoryne cognatoides* Blatter et McCann (Araceae). Aqua-Planta 18(2): 62–67 (1993).

{Ten/Hel}. Endemic to SW India, records from Ban and Wb probably referrable to *C. retrospiralis*: ♦Ind: Kl, Kt, Mh.

Cryptocoryne thwaitesii Schott: Fig. 40

Rhizomes up to 1 cm thick, rough, with runners 1–5 cm long, mostly short, giving the plant a tufted appearance. Leaves ± upright; petioles 3–10 (–15) cm long; leaf blades elliptical to ovate, 3–8 cm long, 1.5–4 cm wide, ± rough, upper surface green to marbled green or marbled purple, lower surface usually pale green, veins conspicuous, margin entire but finely waved appearing toothed, base cordate, tip obtuse. Spathes 6–14 cm long, outside greyish to brownish; kettle up to 1 cm long, red-spotted on the inside; tube 2–4 cm long, red-spotted on the inside; collar absent; the throat red-spotted; limb 4–7 cm long, 0.6–1 cm wide at the base, gaping and widened below, developed into a long tail above, the inner surface smooth, white and ± densely red-spotted. Male flowers 20–30. Ovaries 5 or 6; stigmas circular, borne vertical.

Perennial. Along or in small streams in the forest, mostly emergent or sometimes even terrestrial. Very similar to *C. alba* but has a longer spathe tube with a red-spotted limb and rather rough leaf blades with a distinctly wavy, almost tooth-like margin.

{Ten/Hel}. Endemic to southwestern Sri Lanka: ♦Cey.

Cryptocoryne undulata Wendt [*C. axelrodii* Rataj]: Fig. 41

Like *C. walkeri* but spathe usually smaller (4–7 (–9) cm long); spathe limb narrowly ovate, subobliquely twisted to spiral, inner surface smooth to slightly rough; cream to yellow or brownish; collar not clearly distinguishable from the limb; throat whitish; leaf blades green to green-marbled, brown-marbled or brown on the upper surface, often wavy.

Perennial. In and along swiftly flowing streams and rivers in forests, often growing submerged. It is very similar to *C. beckettii* and *C. wendtii*.

{Ros/Ten}. Endemic to central Sri Lanka: ♦Cey.

Cryptocoryne walkeri Schott [*C. lutea* Alston, *C. legroi* de Wit]: Fig. 41

Rhizomes up to 8 mm, when submerged thinner and with longer internodes, with slender runners (5–) 10–15 cm long. Leaves ± upright; petioles 5–10 (–15) cm long; leaf blades narrowly ovate to ovate, 3–9 (–15) cm long, (1–) 1.5–3.5 (–6) cm wide, ± smooth, upper surface dark green, lower surface paler green or reddish, veins conspicuous and usually tinged with red, margin entire and sometimes finely

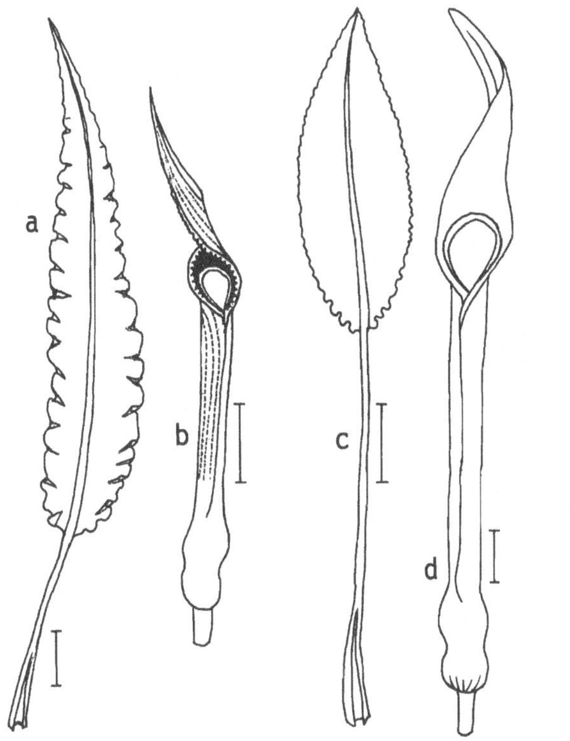

Fig. 41. *Cryptocoryne undulata*: a, leaf (1 cm); b, spathe (1 cm). *C. walkerii*: c, leaf (2 cm); d, spathe (1 cm).

waved, base obtuse to cordate, tip acute-acuminate; submerged leaves usually smaller and with a violet tinge. Spathes 4–15 (–35) cm long, outside purplish to greenish with small spots; kettle up to 2 cm long, warty inside; tube 3–10 (–20) cm long; collar either absent or present and green to yellow or at late anthesis brownish; the throat whitish-yellow to purplish; limb 1.5–4 (–6) cm long, 0.5–1 cm wide at the base, gaping and widened below, subrecurved to backwardly twisted, the inner surface smooth or slightly rough, green to yellow or at late anthesis brownish. Male flowers 40–90. Ovaries 4–7; stigmas circular, vertical.

Perennial. Along or in swiftly flowing streams and rivers, mostly emergent. When the collar is well-developed it is sometimes difficult to distinguish from *C. beckettii*, *C. wendtii* and *C. undulata*.

{Ten/Hel}. Endemic to central Sri Lanka: ♦Cey.

Cryptocoryne wendtii de Wit: Fig. 42

Like *C. walkeri* but spathe limb subobliquely twisted, elongated into a long almost tail-like point, outer surface brownish, inner surface yellowish to purplish-brown or purple; collar between spathe tube and limb chocolate brown to blackish-purple; spathe throat whitish; leaf blades ± smooth to somewhat bullate, the upper surface green to brownish-red and slightly marbled; vegetative parts usually larger (rhizome up to 1 cm thick, petioles up to 25 cm long, leaf blade up to 20 cm long).

Perennial. In and along swiftly flowing rivers and in ditches, often growing submerged. It is very variable in leaf colour and form. It is rather similar to *C. beckettii* and *C. undulata*.

{Ros/Ten}. Endemic to central, west and northwest Sri Lanka: ♦Cey.

Lagenandra: 14 species: all endemic to the Indian subcontinent

Rhizomes erect or creeping. Leaves evergreen, up to 120 cm long; sheathing at the base; petioles usually longer than the blades; blades linear-lanceolate to ovate, acute at the apex, involute (margins rolled inwards towards the abaxial midrib) in bud. Cataphylls (resembling bladeless sheaths, homologous to prophylls) develop between some of the upper leaves, usually with 2 keels or ribs. Inflorescences usually shorter than the leaves. Spathes tubular with edges united at the base, the limb usually about equal to the

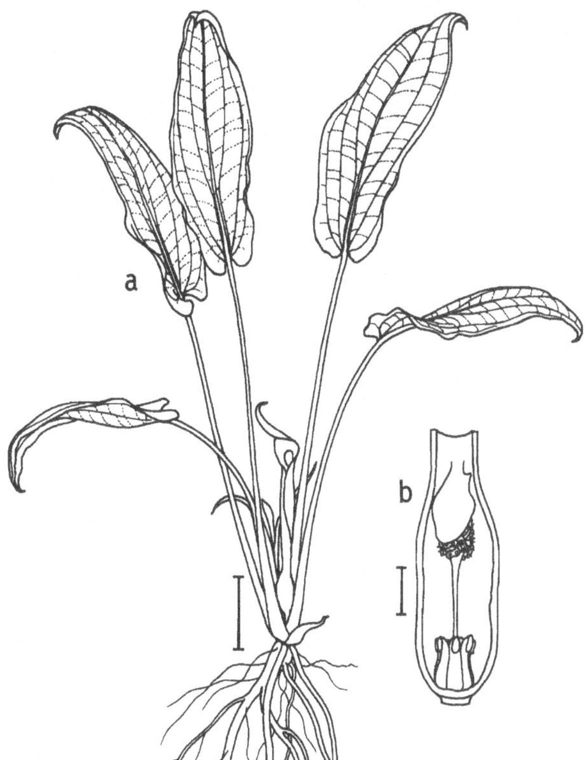

Fig. 42. *Cryptocoryne wendtii*: a, whole plant (2 cm); b, kettle (2 mm).

tube, usually twisted, sometimes inflated. Spadix very short, enclosed within the inflated base of the spathe (kettle); the male part conic, wrapped in an involuted flap or fold of the spathe. Stamens 1 or 2. Female flowers usually many, free, at the base of the spadix, separated from the males by a slender sterile axis. Ovary 1-locular, widest at the top. Berries free, opening from the base, 1- to several-seeded.

Perennial. All species are ecologically very much alike; they are usually found in and along steams and rivers in forests and plantations but also in irrigation ditches around rice fields. In the wet season they may be submerged. Flowering takes place as the water recedes during the winter. Pollination is entomophilous. The disseminules are berries, dispersed by animals and also by water. Many are cultivated for decoration. Natural and artificial hybrids are known. Most figures are based on drawings from de Wit (1983).

1. Female flowers at one level, arranged side by side in a row around the spadix
  2. Petioles and leaf blades glabrous; spathes as long or longer than the petioles; ovaries smooth — ***L. gomezii***
  2. Petioles and under surface of the leaf blades hairy; spathes much shorter than the petioles; ovaries covered in warts — ***L. nairii***
1. Female flowers arranged one above the other spirally along the spadix
  3. Spathes with a pale green outer surface
    4. Leaf blades ovate-lanceolate, up to 5 times as long as broad; leaf margins entire; spathes long-stalked (8–18.5 mm long); inner surface of spathe pale greenish-white, without flecks — ***L. bogneri***
    4. Leaf blades narrowly lanceolate to almost linear, ± 10 times longer than broad; leaf margins crispate-erose; spathes short stalked (1–2 cm) to almost sessile; inner surface of spathe pale green with purple flecks — ***L. erosa***
  3. Spathes with a dark greenish-purple to purple, red, brown or pinkish outer surface
    5. Spathes widely gaping, the limb very much wider than the kettle
      6. Limb of spathe strongly twisted to one side and opening subhorizontally — ***L. jacobsenii***
      6. Limb of spathe erect or nearly so — ***L. meeboldii***
    5. Spathes opening by a long slit, never gaping, the limb not or swollen above the kettle
      7. Ovaries smooth, without warts; spathes up to ± 2 cm long — ***L. undulata***
      7. Ovary with warty outgrowths; spathes more than ± 2 cm long
        8. Spathes elongate, with a smooth outer surface; stigmas sessile or nearly so
          9. Leaf blades narrowly elliptical to lanceolate, rarely more than 2 cm wide; spathe with irregular transverse ribs inside — ***L. koenigii***
          9. Leaf blades ovate, 7–10 cm wide; spathe almost smooth inside — ***L. toxicaria***
        8. Spathes swollen, with a warty, longitudinally ribbed or rarely smooth outer surface; stigmas borne on styles
          10. Spathes up to 3.5 cm long, often with an up to 2 cm long tail, the outer surface smooth or warty — ***L. lancifolia***
          10. Spathes usually more than 5 cm long, with a 5–25 cm long tail, the outer surface bearing warts or papillae
            11. Leaves spreading, the blades ± horizontal; plants usually less than 50 cm tall
              12. Spathes swollen above the kettle, bearing warts on the outer surface — ***L. dewitii***
              12. Spathes not swollen above the kettle, bearing papillae on the outer surface — ***L. thwaitesii***
            11. Leaves erect; plants 1 m or more tall
              13. Spathes abruptly globose above the kettle, the warts 1.5–3 mm long — ***L. ovata***
              13. Spathes subcylindrical, not or only slightly swollen above the kettle, the warts up to 1 mm long — ***L. praetermissa***

Lagenandra bogneri de Wit: Fig. 43

Rhizomes up to 1.5 cm thick. Leaves erect and spreading; petioles 4–10 (–20) cm long; leaf blades ovate to narrowly ovate, 6.5–13 cm long, 1–6.5 cm wide. Cataphylls up to 10 cm long. Peduncle 8–18.5 cm long. Spathes 8–13 cm long; limb up to 4 cm long, slightly twisted and swollen above the kettle, opening by a narrow slit; outer surface light green or yellowish, minutely papillate on the vertical striae; inner wrinkled; tail 3–7 cm long. Ovaries with protuberances below the stigma; stigma dome-shaped.

{Hel/Ten}. Endemic to SW Sri Lanka: ♦Cey.

Lagenandra dewitii Crusio et de Graaf: Fig. 43

Like *L. ovata* and *L. praetermissa* but leaves spreading with subhorizontal blades, the blades narrowly ovate, up to 25 cm long and 6 cm wide, the margins silvery or reddish-brown and wavy; spathes very shortly stalked, up to 12 cm long, twisted and somewhat swollen above the kettle, warty on the outside. Seeds elongate ovoid, ± 8 mm long, ± 2 mm wide, with many longitudinal ribs.

Crusio, W. and de Graaf, A. *Lagenandra dewitii* Crusio et de Graaf, eine neue Art aus Sri Lanka. Aqua-Planta 11(2): 56–59 (1986).

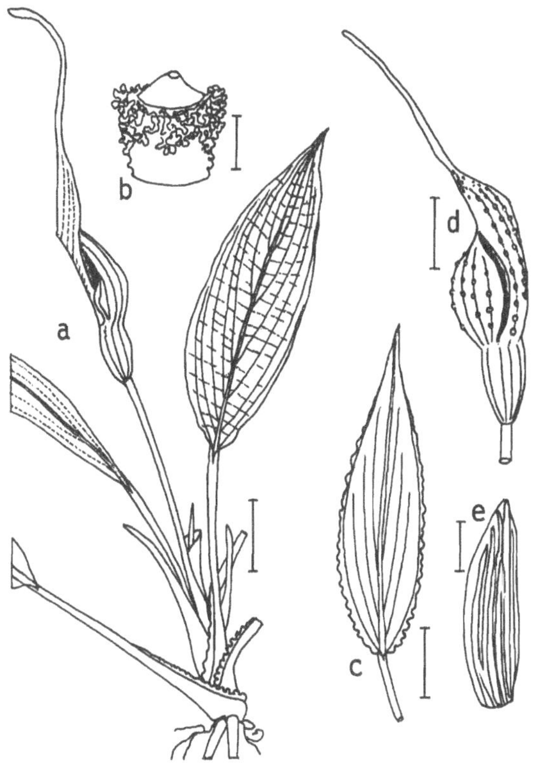

Fig. 43. *Lagenandra bogneri*: a, habit (2 cm); b, ovary (0.5 mm). *L. dewitii*: c, leaf (4 cm); d, spathe (2 cm); e, seed (2 mm).

{Hyp/Ten}. Endemic to SW Sri Lanka (Kanneliya Ela): ♦Cey.

Lagenandra erosa de Wit: Fig. 44

Rhizomes up to 1 cm thick. Leaves spreading; petioles 4–5 (–16) cm long; leaf blades linear-lanceolate to almost linear, 8–10 (–20) cm long, 0.5–1.5 cm wide, veins on lower blade surface bearing very small branched outgrowths, margins crispate-toothed (erose) towards the base, long-tapered at the tip. Cataphylls 4–10 cm long. Peduncle very short or 1–2 cm or more long. Spathes 10–23 cm long, slender, not swollen above the kettle; limb up to 20 cm long, somewhat twisted, opening by a relatively narrow slit; outer surface smooth, green to purple; inner surface purple, spongy, irregularly ribbed with protuberances; tail 5–10 cm long. Ovaries with warty outgrowths below the stigma; stigma sessile, dome-shaped. Seeds unknown.

{Hyp/Ten}. Endemic to Sri Lanka, exact locality not known: ♦Cey.

Lagenandra gomezii (Schott) Bogner et N. Jacobsen [*Cryptocoryne gomezii* Schott]: Fig. 44

Rhizomes up to 5 mm thick. Leaves erect; petioles 6–12 cm long; leaf blades ovate, cordate at the base, 5–7 cm long, 2.5–3.5 cm wide. Cataphylls up to ± 4 cm long. Peduncle up to ± 2.5 cm long. Spathes up to ± 6 cm long, slightly contracted and then expanded above the kettle; limb up to ± 3 cm long, not twisted, probably opening with a wide gape; outer surface smooth; inner surface with bristle-like hairs (crystals); tail up to ± 2 cm long. Ovaries 7–9 at 1 level, with smooth styles; stigma domed. Seeds unknown.

Bogner J. and Jacobsen N. Die systematische Stellung von *Lagenandra gomezii* (Schott) Bogner et Jacobsen, comb. nov. Aqua-Planta 12(2): 43–50 (1987).

{Hyp/Ten}. Endemic to Bangladesh and/or NE India (collected in 'Sylhet' in 1828 and never re-collected): ♦Ban/Ind: Mg?

Lagenandra jacobsenii de Wit: Fig. 45

Rhizomes up to 4 cm thick. Leaves erect; petioles 15–30 cm long; leaf blades ovate to narrowly ovate, 15–30 cm long, 7–10 cm wide. Cataphylls 12–17 cm long. Peduncle 4–5 cm long. Spathes up to 20 cm long, much expanded above the kettle; limb 4–6 cm long, with a pronounced half twist and held horizontally, opening with a wide gape, surrounded by an almost

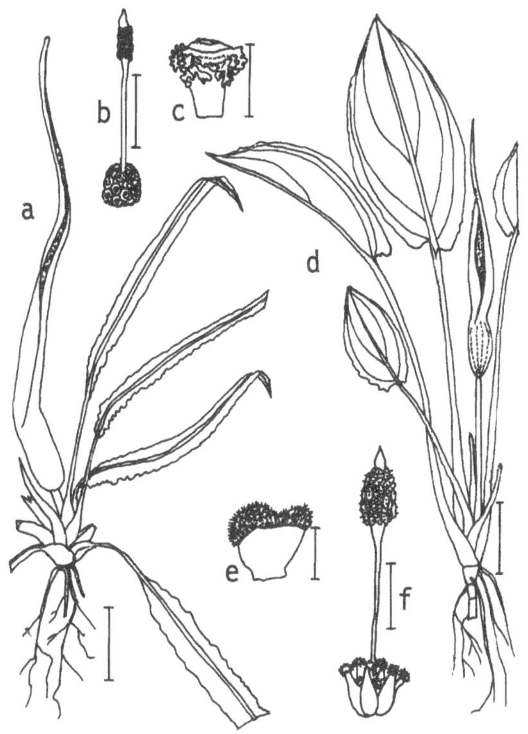

Fig. 44. *Lagenandra erosa*: a, habit (2 cm); b, spadix (1 cm); c, ovary (1 cm). *L. gomezii*: d, habit (2 cm); e, ovary (1 mm); f, spadix (1 cm).

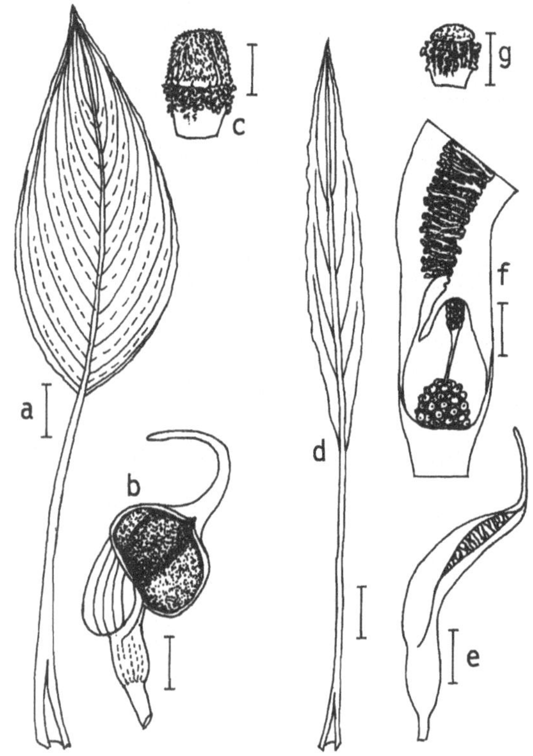

Fig. 45. *Lagenandra jacobsenii*: a, leaf (3 cm); b, spathe (2 cm); c, ovary (1 mm). *L. koenigii*: d, leaf (2 cm); e, spathe (1 cm); f, longitudinal section through the kettle (5 mm); g, ovary (0.7 mm).

orbicular extension of the limb; outer surface granular, greenish-brown; inner surface deep purple, irregularly horizontally ribbed with scattered white-tipped and branched protuberances; tail 5–9 cm long. Ovaries with elongate, warty outgrowths below the stigma; stigma sessile, dome-shaped. Seeds unknown.

{Hyp/Ten}. Endemic to SW Sri Lanka: ♦Cey.

Lagenandra koenigii (Schott) Thwaites: Fig. 45

Rhizomes up to 2 cm thick. Leaves erect; petioles 12–35 cm long; leaf blades narrowly lanceolate, 17–37 cm long, 1–2 cm wide, marginal veins bearing very small branched outgrowths, long-tapered at the tip. Cataphylls up to 17 cm long. Peduncle 6–18 cm long. Spathes 6–12 cm long, fleshy, suberect to curved, not swollen above the kettle; limb 4–5 cm long, not twisted, opening by a relatively narrow slit; outer surface smooth, blackish-purple to pale red; inner surface dark purple, horizontally ribbed; tail up to 6 cm long. Ovaries with warty outgrowths; stigma sessile, dome-shaped.

{Hyp/Ten}. Endemic to SW Sri Lanka: ♦Cey.

Lagenandra lancifolia (Schott) Thwaites [*L. blassii* de Wit]: Fig. 46

Rhizomes ± 1 cm thick. Leaves erect; petioles 5–35 cm long; leaf blades ovate to narrowly ovate, 6–14 cm long, 2.5–5 cm wide. Cataphylls up to 6 cm long. Peduncle 1–2 cm long. Spathes 2–3.5 cm long, twisted and swollen above the kettle, opening by a narrow slit; outer surface greenish-purple, longitudinally ribbed, the ribs usually warty; inner surface finely striate, dark purple, outer surface with white-tipped branching protuberances near the collar; tail 0.5–2 cm long. Ovaries with long protuberances below the stigma; stigma sessile, dome-shaped. Seeds 2 or 3.

{Hyp/Ten}. Endemic to SW Sri Lanka: ♦Cey.

Lagenandra meeboldii (Engler) C. E. C. Fischer: Fig. 46

Rhizomes up to ± 1 cm thick. Leaves erect; petioles 8–15 cm long; leaf blades ovate, cordate at the base, 6.5–10 cm long, 3.5–6 cm wide. Cataphylls 4–6 cm long. Peduncle up to 12 cm long. Spathes 8–11 cm long, much expanded above the kettle; limb 3–6 cm long, not twisted, upright, opening with a wide gape, the gape not more than 2.5 cm long and surrounded by an

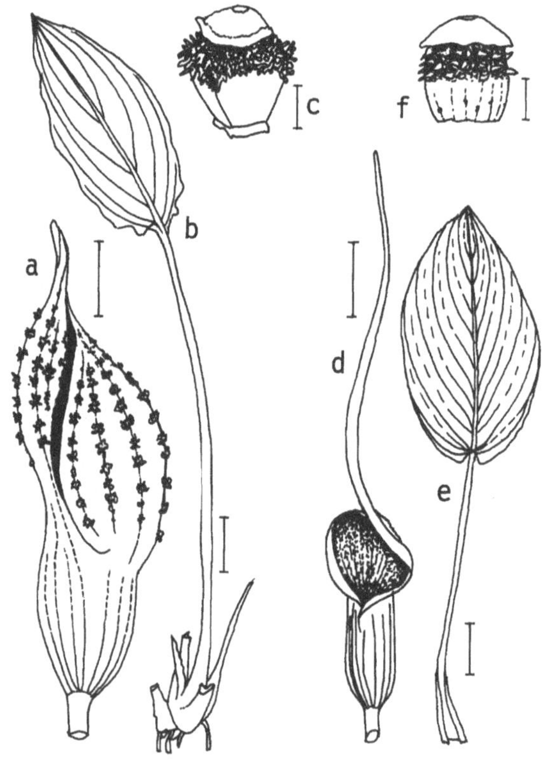

Fig. 46. *Lagenandra lancifolia*: a, spathe (5 mm); b, leaf (3 cm); c, ovary (0.5 mm). *L. meeboldii*: d, spathe (2 cm); e, leaf (2 cm); f, ovary (0.5 mm).

almost orbicular extension of the limb; outer surface smooth, purple; inner surface deep purple, irregularly horizontally ribbed with scattered white-tipped and branched protuberances; tail 8–12 cm long. Ovaries with elongate, spiny outgrowths below the stigma; stigma sessile, dome-shaped. Seeds unknown.

Usually found in shady situations in marshes, around tanks, along streams and in ditches at the edges of rice fields.

{Hel/Ten/Hyp}. Endemic to SW India: ♦Ind: Kt, Kl.

Lagenandra nairii K. Ramamurthy et R. Rajan: Fig. 47

Rhizomes 1.5–2 cm thick, occasionally developing stolons. Leaves erect; petioles up to 28 cm long, densely clothed in short hairs; leaf blades elliptical-lanceolate to ovate, cordate or rounded at the base, 10–17 cm long, up to 9 cm wide, lower surface clothed in short hairs. Cataphylls 4–9 cm long. Peduncle 2–5 cm long. Spathes 4–5 cm long, hardly expanded above the kettle; limb 2.5–3.5 cm long, ear-like, ± 1.4 cm wide, opening with a wide gape; outer surface smooth but with short hairs, purple with greenish-yellow collar; inner surface deep purple, irregularly horizontally ribbed; tail absent or rudimentary. Ovaries 10–15 in more than 1 whorl but in 1 level, with small wart-like outgrowths on the style; stigma head-like. Seeds 2–5, elongate, slightly curved, 5–7 mm long, ± 2 mm in diameter, brown, with longitudinal ribs.

Found partly submerged in the Chalakudy river near the Adirappaly waterfalls. Sivadasan, M. *Lagenandra nairii*, eine ungewöhnliche Art aus Indien. Aqua-Planta 11(2): 60–64 (1986).

{Hyp/Ten}. Endemic to SW India: ♦Ind: Kl (Trichur/Thrissur District).

Lagenandra ovata (Linnaeus) Thwaites [*L. insignis* Trimen, *Lagenandra toxicaria* auct. Ind., non Dalzell]: Fig. 47

Rhizomes up to 8 cm thick. Leaves erect; petioles up to 50 cm long; leaf blades elliptical to ovate, (10–) 25–60 cm long, (5–) 7–15 (–20) cm wide. Cataphylls 8–15 (–30) cm long. Peduncle up to 25 cm long. Spathes 15–25 cm long, twisted and swollen above the kettle, opening by a narrow slit, the slit usually more than 5 cm long, running into the tail; outer surface dark purple, warty, the warts 1.5–3 (–10) mm long; inner surface finely striate, dark purple, with white-tipped branching protuberances near the collar; tail up to ± 8 cm long. Ovaries bearing short spiny protuberances below the stigma; stigma subsessile, dome-shaped. Seeds 3–5.

Along streams and ditches in open and shady situations, also at the edges of rice fields; usually submerged during the rains. It is mostly found in coastal regions. Flowers and fruits the whole year but with a peak from August to February.

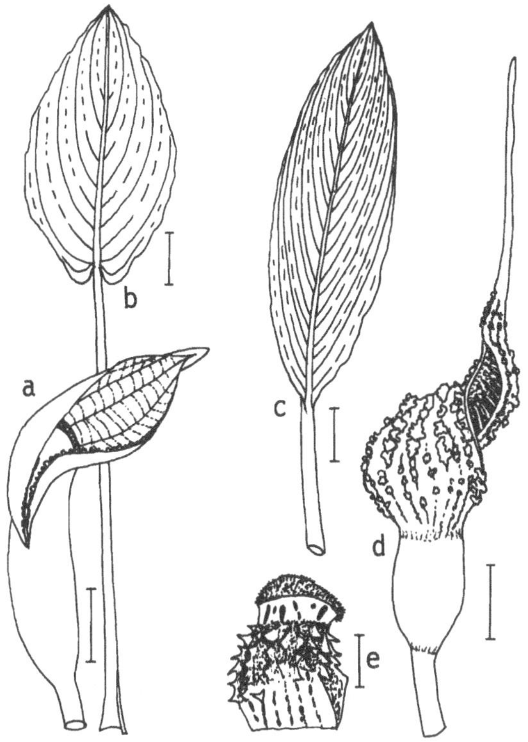

FIG. 47. *Lagenandra nairii*: a, spathe (1 cm); b, leaf (2 cm). *L. ovata*: c, leaf (5 cm); d, spathe (2 cm); e, ovary (1 mm).

{Hel?/Ten/Hyp}. Endemic to SW India and SW Sri Lanka: ♦Cey, ♦Ind: Go, Kl, Kt, Tn.

Lagenandra praetermissa de Wit: refer to Fig. 47

Like *L. ovata* but spathes 15–20 cm long; limb 6–9 cm long, subcylindrical, not or only slightly swollen above the kettle, dark red, the warts on the outer surface up to 1 mm long; tail 5–8 cm long; leaf blades longer than their petioles.

It is a rather unsatisfactory species, perhaps of hybrid origin. Further investigations are needed.

{Hyp/Ten}. Endemic to SW Sri Lanka: ♦Cey.

Lagenandra thwaitesii Engler [*L. schultzei* de Wit]: Fig. 48

Rhizomes up to 1 cm thick. Leaves spreading; petioles up to 20 cm long; leaf blades ovate to lanceolate, usually undulate, up to 20 cm long, up to 4 cm wide, the margins usually greyish-white to silver, long tapered and acute at the tip. Cataphylls up to 9 cm long. Peduncle 1–7 cm long. Spathes up to 10 cm long, twisted and swollen above the kettle, opening by a narrow slit; outer surface greenish-brown

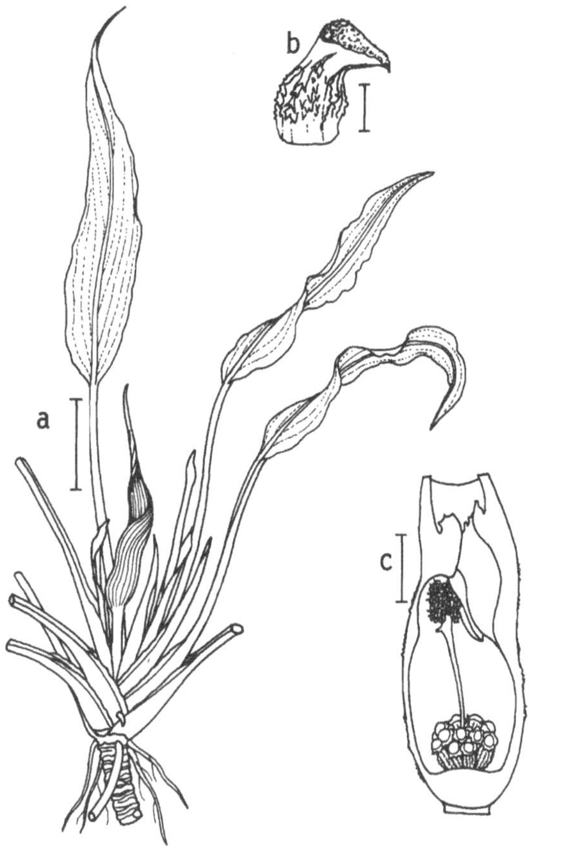

Fig. 48. *Lagenandra thwaitesii*: a, habit (4 cm); b, ovary (0.5 mm); c, longitudinal section of the spathe showing the 'kettle' (2 cm).

to purple, without warts, at most finely roughened by vertical striae; inner surface finely striate, dark purple, with white-tipped branching protuberances near the collar; tail up to 4 cm long or occasionally absent. Ovaries warty below the style; stigma subsymmetrical. Seeds 3–4, brownish, ± 6 mm long, ± 2 mm wide.

{Hel/Ten}. Endemic to SW Sri Lanka: ♦Cey.

Lagenandra toxicaria Dalzell: Fig. 49

Rhizomes up to 3 cm thick. Leaves erect; petioles up to ± 40 cm long; leaf blades elliptical-ovate, 12–20 cm long, 5–12 cm wide. Cataphylls up to 10 cm long. Peduncle up to ± 3 cm long. Spathes up to 18 cm long, ± cylindrical, not swollen above the kettle, opening by a gaping slit; outer surface light brown, smooth; inner surface deep purple; tail up to 9 cm long. Ovaries with warty outgrowths; stigma subsessile, dome-shaped. Seeds 3–5.

Often abundant in marshes and streams; usually submerged during the monsoon rains.

{Hyp/Ten}. Endemic to SW India: ♦Ind: Ap, Go, Kl, Kt, Mh, Tn.

Lagenandra undulata A. R. K. Sastry: Fig. 49

Rhizomes up to ± 5 cm thick. Leaves erect; petioles up to ± 6 cm long; leaf blades elliptical-lanceolate, undulate, 1.5–9 cm long, 0.5–1.5 cm wide. Cataphylls ± 2 cm long. Peduncle 2–3 cm long. Spathes ± 2 cm long, not expanded above the kettle; limb ± 3 cm long, ovate, ± 1 cm wide, opening by a narrow slit; outer surface smooth but with longitudinal ribs; inner surface deep purple flecks; tail absent or rudimentary. Ovaries ± 30 in 3–4 rows, smooth but flecked below the style; stigma peltate.

{Hyp/Ten}. Endemic to NW India: ♦Ind: As? Ap.

Lasia: 2 species

Lasia spinosa (Linnaeus) Thwaites [*L. heterophylla* Schott, *L. aculeata* Loureiro]: Fig. 50

Rhizomes creeping and robust, up to 4 cm thick and 1.5 m long. Petioles spiny, up to 1.6 m long; juvenile blades sagittate, mature blades usually pinnatifid, spiny on the main nerves of the abaxial surface. Inflorescences about equalling the leaves; peduncles spiny.

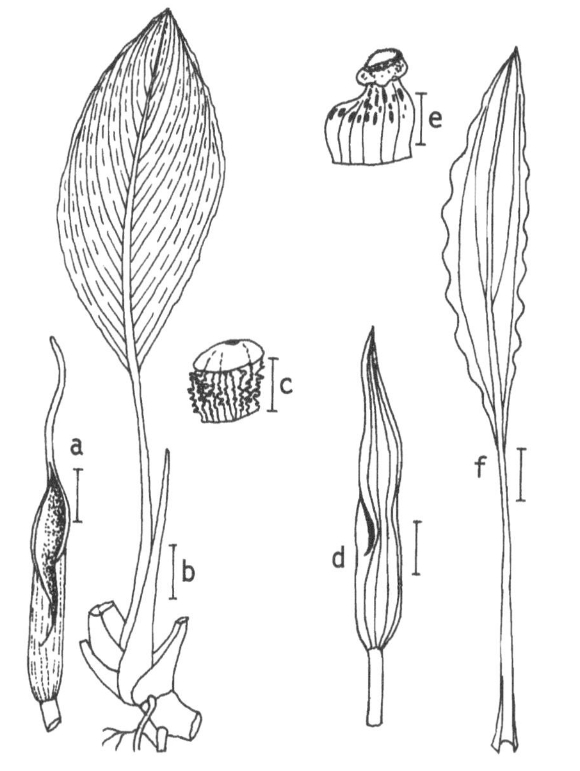

Fig. 49. *Lagenandra toxicaria*: a, spathe (5 mm); b, leaf (3 cm); c, ovary (0.5 mm). *L. undulata*: d, spathe (2 cm); e, ovary (0.5 mm); f, leaf (1 cm).

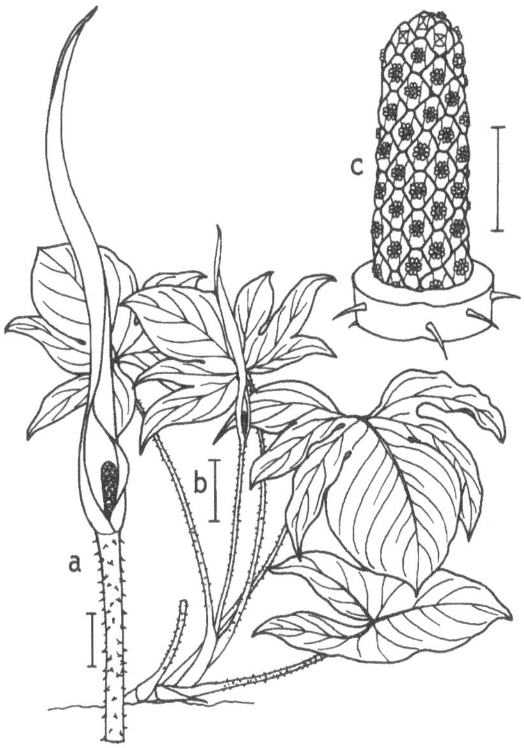

Fig. 50. *Lasia spinosa*: a, inflorescence (3 cm); b, leafy shoot (2 cm); c, spadix (1 cm).

Spathes slightly twisted, open only at the base, up to 55 cm long, greenish-brown to purple. Spadix completely covered in fertile flowers, at anthesis up to 7.5 cm long, in fruit up to 18 cm long, green and fleshy. Flowers bisexual. Perianth segments 4–6, ovate-cuneate. Stamens 4 or rarely 6, filaments broad. Ovary 1-locular; stigma small, sessile. Berries 1-seeded, warty on the exposed side.

Perennial. Marshy areas, sometimes in brackish water. Pollinated by insects. The disseminules are warty (or rarely smooth) berries, dispersal unknown. It is very variable and is sometimes cultivated; the insides of the stems and tender leaves are used as pot herbs. Hay, A. *Cyrtosperma* and its Old World allies. Blumea 33: 427–469 (1988).

{Hel/Hyp}. S and SE Asia: Ban, Bhu, Bur, Cey, Ind, Nep: As, Mn, Mp, Or, Sk, Tn, Wb.

Pistia: 1 species

Pistia stratiotes Linnaeus: Fig. 51

Stoloniferous, free-floating rosettes with numerous roots. Leaves subsessile, obovate to ovate-cuneate, up to 15 cm long and 5 cm wide, with (3–) 5–9 (–12) parallel veins, densely pubescent on both surfaces. Inflorescences axillary, shortly stalked, much shorter than the leaves. Spathe (0.7–) 2–4 cm long, greenish-yellow, hairy, convolute at the base, spreading above, with a white ciliolate margin, constricted between the female and male parts. Spadix shorter than and partly adnate to the spathe at the base, with 1 female flower below and a 2–8 male flowers above. Flowers unisexual, perianth absent, the males with 2 united stamens. Ovary flask-shaped, 1-locular. Berries many-seeded, green, crowned by the persistent stigma.

Perennial or annual. Free-floating rosettes with emergent leaves, sometimes annual. In tanks, lagoons and rice fields. Usually in fresh water but sometimes in brackish or even salt water. Flowers mostly during the hot weather but occasionally throughout the year. Pollinated by insects, self-compatible and probably frequently self-pollinated. Disseminules berries, seeds or detached rosettes, mostly dispersed by water but fruits may be eaten by animals. A serious pest in many regions but used in medicine in India as an antiseptic, antidysenteric, insecticide and cure for asthma. It is fed to ducks and pigs in Bengal.

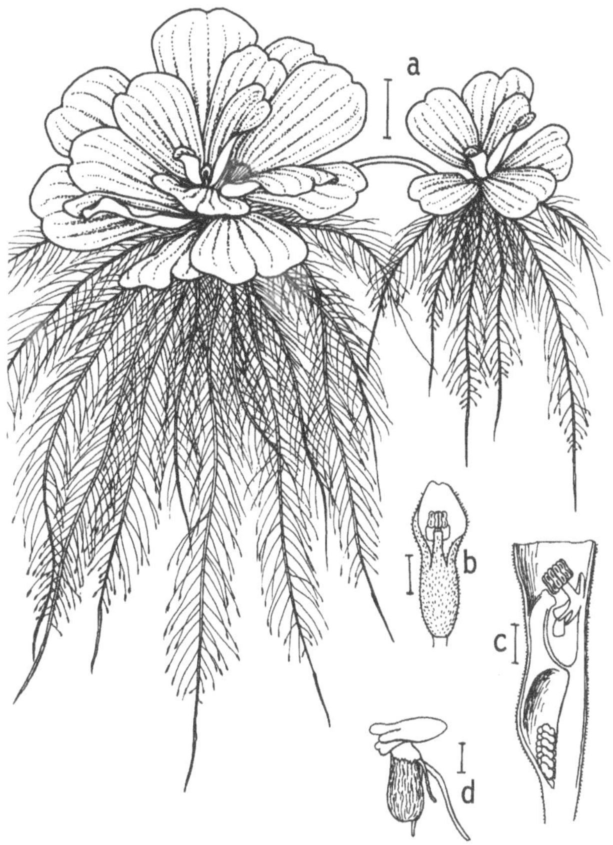

Fig. 51. *Pistia stratiotes*: a, two floating rosettes (4 cm); b, spathe (4 mm); c, longitudinal section of an inflorescence (8 mm); d, seedling (1 mm).

{(⊙) Ple}. Pantropics: Ban, Bur, Cey, Ind, Pak: As, Bh, Gj, Go, Kl, Kt, Mn, Mp, Rj, Wb. 0–1000 m.

ASCLEPIADACEAE: 348 genera

Twining or straggling herbs and shrubs, with milky latex. Leaves simple, often reduced, opposite, without stipules. Flowers radially symmetrical, bisexual, 5-merous. Petals campanulate. Stamens and style united to a column (gynostegium). Pollen united into pollinia (each derived from the anther cells of 2 adjacent anthers); pollinia paired, borne on spoon-shaped translators. Fruit a pair of follicles. Seeds tufted by a crown of hairs.

Oxystelma: 4 species: Old World tropics

Oxystelma esculentum (Linnaeus fil.) R. Brown ex J. A. Schultes [*O. secamone* auct. Ind., non Linnaeus, *Sarcostemma esculentum* (Linnaeus fil.) Holm]: Fig. 52

Branched straggler. Leaves oblong to linear, 4–8 cm long, 0.3–2 cm wide. Flowers usually paired, in lax racemes. Sepals cup-like, the lobes 2–3.5 mm long. Petals purplish-pink to pink or white, 1.8–2.5 cm across, lower half united; lobes triangular. Corona double; the inner staminal. Gynostegium ±7 mm long. Follicles ellipsoidal, inflated, 3.8–7 cm long, ±2 cm in diameter. Seeds ± 3.5 mm long, bearing hairs up to 1.8 cm long.

Perennial. In swamps, lakes and ponds, often rooting in water then straggling over emergent aquatic plants such as *Typha* and various sedges. The flowers are very beautiful but short-lived; pollinated by insects. The seeds have tufts of hairs and are dispersed by wind. Used as a gargle and a cure for jaundice.

{Hel}. Extending from Egypt and Iraq to SE Asia: Cey, Ind, Nep, Pak: Dl, Gj, Kt, Mh, Pj, Rj, Tn, Up, Wb

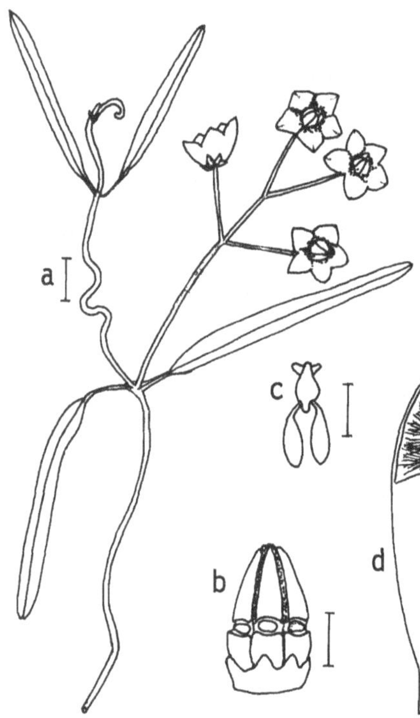

Fig. 52. *Oxystelma esculentum*: a, habit (1 cm); b, staminal column (1 mm); c, pollinium (1 mm); d, fruit with top part opened to show seeds and hairs (1 cm).

ASTERACEAE [COMPOSITAE]: ± 1300 genera

Annual or perennial. Flowers arranged on a common receptacle in heads; heads surrounded by an involucre of bracts. Sepals absent, replaced by a pappus of hairs, bristles or scales or absent. Petals 5, united, either radially symmetrical (tubular florets) or bilaterally symmetrical and tongue-like (ligulate florets). Stamens 5; anthers loosely joined around the style. Ovary inferior, 1-locular; ovule 1, erect on the base of the loculus. Fruit a cypsela (a specialized nut, sometimes called achene).

Shukla, B. K. The family Asteraceae in Madhya Pradesh, India. Journal of Economic and Taxonomic Botany 8(2): 373–418 (1986).

Excluded: many composites are weedy annuals that grow in places where water stood during the monsoon but has dried out. These plants are not wetland species but are sometimes found growing together with wetland species shortly after the monsoon; in the dry season they then replace the wetland species in this habitat. The following species come into this category: ***Acanthospermum hispidum*** A. P. de Candolle; ***Adenostemma angustifolium*** Arnott; ***Ageratum conyzoides*** Linnaeus; ***Bidens bipinnata*** Linnaeus; ***B. biternata*** (Loureiro) Merrill et Sherff; ***B. pilosa*** Linnaeus; ***B. radiata*** Thuillier [*B. foliosa* Willdenow]; ***Blainvillea acmella*** (Linnaeus) Philipson [*B. rhomboidea* Cassini, *Verbesina acmella* Linnaeus, *Spilanthes acmella* (Linnaeus) Murray, *Eclipta latifolia* Linnaeus fil.]; ***Blumea bifoliata*** (Linnaeus) A. P. de Candolle [*B. oligocephala* A. P. de Candolle]; ***B. mollis*** (D. Don) Merrill [*Erigeron molle* D. Don]; ***Dichrocephala integifolia*** (Linnaeus fil.) O. Kuntze [*Cotula bicolor* Roth, *Grangea latifolia* Lamarck]; ***Cotula anthemoides*** Linnaeus [*Pleiogyne cardiosperma* Edgeworth]; ***C. hemisphaerica*** (Roxburgh) Wallich ex C. B. Clarke; ***Crassocephalum crepidioides*** (Bentham) A. H. Moore; ***Gnaphalium hypoleucum*** A. P. de Candolle; ***G. indicum*** Linnaeus; ***G. polycaulon*** Persoon; ***G. pulvinatum*** Delile; ***Pseudognaphalium luteo-album*** (Linnaeus) Hillard et Burtt [*Gnaphalium luteo-album* Linnaeus, *G. affine* D. Don]; ***Pulicaria foliosa*** A. P. de Candolle; ***Sonchus brachyotus*** A. P. de Candolle; ***Struchium***

sparganophorum (Linnaeus) O. Kuntze; ***Synedrella nodiflora*** (Linnaeus) Gaertner; ***Wedelia chinensis*** (Osbeck) Merrill [*W. calendulacea* (Linnaeus) Lessing]; ***Xanthium indicum*** J. G. König [*X. strumarium* Linnaeus].

The following wetland species are found in the Himalayas but do not extend south: ***Bidens cernua*** Linnaeus; ***B. tripartita*** Linnaeus; ***Cotula coronopifolia*** Linnaeus: <Up.

1. Leaves alternate, mostly cauline but sometimes forming basal rosettes
2. Stems, branches and peduncles winged; involucral bracts scarious or papery
3. Flowering heads stalked, single or in corymbs but never crowded on a common receptacle, less than 7 mm in diameter ***Epaltes***
3. Flowering heads sessile,crowded on a common receptacle into ± globose terminal compound heads, more than 1 cm in diameter ***Sphaeranthus***
2. Stems, branches and peduncles without wings; involucral bracts not scarious or papery
4. Involucre of 2 (3) bracts, each orbicular or broadly ovate with a sharply pointed tip; stems rooting and creeping at the base, not obviously arising from a single taproot ***Caesulia***
4. Involucre of many more than 3 bracts, the bracts never orbicular or broadly ovate; stems spreading from a single tap root
5. Leaves spathulate and tapering at the base, never pinnately lobed, glabrous or nearly so; flowering heads sessile ***Centipeda***
5. Leaves not spathulate, pinnately divided or pinnately lobed, densely clothed in soft, multicellular hairs; flowering heads mostly pedunculate
6. Flowering heads leaf-opposed and solitary or 2 (3) together, most overtopped by leaves; cypselas without a pappus; receptacle convex or conical ***Cyathocline***
6. Flowering heads in terminal clusters, borne above the leaves; cypselas with tubular pappus with a fimbriate mouth; receptacle concave, sometimes with a conical projection at the centre ***Grangea***
1. Leaves in opposite pairs, never forming rosettes
7. Involucral bracts 4, leaf-like, very wide and enclosing the heads ***Enhydra***
7. Involucral bracts more than 4, scale-like, not enclosing the heads
8. Stems, leaves and bracts clothed in white, appressed, almost bristle-like hairs; receptacles ± flat, not elongate in fruit; cypselas dimorphic but alike in each head, either straw-coloured, flattened and corky or blackish, cylindrical and smooth; petals cream or white ***Eclipta***
8. Stems, leaves and bracts glabrous or slightly pubescent; receptacles elongate, accrescent in fruit; cypselas all alike or slightly dimorphic within each head, never corky; petals yellow ***Spilanthes***

Caesulia: 1 species
Caesulia axillaris Roxburgh [*Meyera orientalis* D. Don]: Fig. 53

Glabrous, sometimes somewhat succulent. Stems creeping, ascending or erect, 15–30 (–45) cm tall, with violet to brown longitudinal stripes, rather swollen at the nodes. Leaves alternate, sessile, lanceolate to almost linear, up to 15 cm long, acute, margin distantly serrulate, somewhat narrowed towards the base but expanded and sheathing at the point of attachment to the stem. Flower heads sessile, axillary, up to 1.8 cm in diameter; involucre of 2 (3) bracts, each orbicular to broadly ovate, crenate or denticulate, with a sharply pointed elongate tip. Ligulate florets absent; tubular florets ± 1.5 mm long, 5-toothed, pale mauve to blue or white, the outer florets subtended by scales, the inner without scales. Anthers dark violet to black. Cypselas obovoid, 2–3 mm long, flattened and laterally

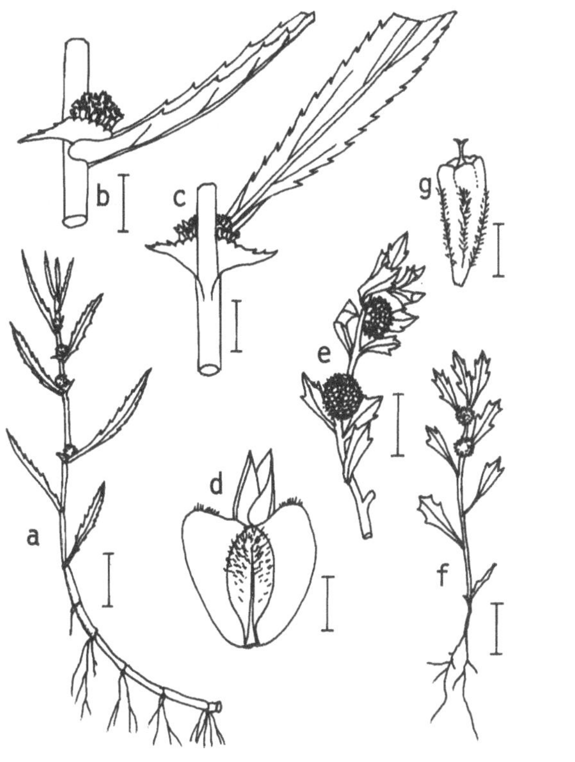

FIG. 53. *Caesulia axillaris*: a, whole plant (5 cm); b, c, flowering head from side and back (1 cm); d, cypsela (1 mm). *Centipeda minima*: e, shoot with two flowering heads (4 mm); f, whole plant (1 cm); g, cypsela (0.3 mm).

winged, dark brown, sparsely hairy; pappus reduced to 2 scales, united below into a ± 2 mm long tube.

Usually annual but in very wet conditions it may perennate by rooting at the nodes. It is sometimes found floating but usually grows rooted in the substrate in shallow water at the edges of tanks, but is also found in marshes, along the banks of irrigation canals and ditches and also sometimes in rice fields where it is not considered to be a pest. Pollinated by insects or perhaps sometimes autogamous. The disseminules are flattened and winged cypselas with a pappus of 2 scales forming a tube, the dispersal mechanism is unknown but some are probably carried by water while others may be transported by surface adhesion to animals, particularly water fowl. It is used in medicine.

{(⊙) Hel/Hyp}. Endemic to the Indian subcontinent but lacking in Sri Lanka: ♦Ban, ♦Bur, ♦Ind, ♦Nep, ♦Pak: Ap, Bh, Dl, Gj, Go, Kt, Mh, Mp, Or, Pj, Rj, Up, Wb. 0–1500 m.

Centipeda: 4 species

Centipeda minima (Linnaeus) A. Braun et Ascherson [*C. orbicularis* Loureiro, *Dichrocephala schmidtii* Wight]: Fig. 53

Stems erect or spreading from a single tap root, very variable, 1–20 (–30) cm long, slender, leafy, glabrous or thinly pilose. Leaves alternate, numerous, subsessile, 4–12 mm long, 2–4 mm wide, oblanceolate or spathulate, with usually 2 teeth on each side, base tapering gradually to point of attachment. Flowering heads sessile, solitary, axillary but sometimes turning to the side away from the axil, ± hemispherical, 2.5–4 (–6) mm in diameter; involucral bracts in 2 rows, subequal, small, oblong, obtuse, with membranous margins; receptacle glabrous. Ligulate florets yellow, female, many-seriate, the limbs 2- or 3-lobed; tubular florets relatively few, yellow, bisexual, 4-toothed. Cypselas minute, 0.75–1 mm long, 4-angled, bristly on the angles; pappus absent or of 4, thin scales forming a crown.

Short-lived annual. Wet places, marshy banks, sometimes found submerged in shallow water, flowering as the water recedes, also sometimes in rice fields but not considered a nuisance. It promotes sneezing and thus relieves nasal congestion, also used to treat swellings and inflammation.

{⊙ Hel}. Tropical Africa, Asia, Australia and many Pacific Islands: Cey, Ind, Nep, Pak: <An, As, Kl, Kt, Mp, Mz, Pj, Rj, Tn, Tp, Up. 0–1600 m.

Cyathocline: 3 species: tropical Asia

Stems, leaves and bracts clothed in white, soft, multicellular hairs, also bearing minute glands. Stems erect, simple or branched. Leaves mostly basal, alternate, sessile, 1- or 2-pinnate, leaf bases auriculate. Flowering heads in, shortly stalked, terminal clusters. Receptacle concave, sometimes with a conical projection at the centre. Ligulate florets absent but the outer tubular florets, female, in several rows, the limbs 2- or 3-lobed; tubular florets 12–20, 5-toothed. Cypselas not more than 0.7 mm long, ovoid, glabrous, shiny; pappus absent.

Fayed, A. Revision der Grangeinae (Asteraceae — Astereae). Mitteilungen der Botanischen Staatssammlung München 15: 425–576 (1979).

1. Flowering heads in dense, rounded clusters; style cylindrical — ***C. jacquemontii***
1. Flowering heads in loose groups or flattened corymbs; style club-shaped
 2. Stems smooth; petals yellow — ***C. lutea***
 2. Stems with longitudinal ribs; petals purple-red — ***C. purpurea***

Cyathocline jacquemontii Gagnepain: refer to Fig. 54

Like *C. purpurea* but flowering heads in dense clusters; style cylindrical; leaves more densely hairy; involucre felty and white.

Ecology unknown.

{⊙? Hel?}. Endemic to central India: ♦Ind: Mp (Prov. Malwa, Mhandou).

Cyathocline lutea Law ex Wight [*C. flava* C. B. Clarke]: refer to Fig. 54

Like *C. purpurea* but stems smooth; petals yellow.

Ecology unknown.

{⊙? Hel?}. Endemic to the central part of the west coast of India: ♦Ind: Kt, Mh, Tn?

Cyathocline purpurea (F. Hamilton ex D. Don) O. Kuntze. [*C. lyrata* Cassini, *Tanacetum viscosum* Wallich ex A. P. de Candolle, *Dichrocephala minutifolia* Vaniot]: Fig. 54

Stems erect, up to 60 cm tall, with longitudinal ribs, simple or sparsely branched. Rosette leaves 1- or 2-pinnate, with or without small segments between, 3–8 (–14) cm long, up to 4 cm wide; the lobes almost alternate, oblong-linear, bluntly toothed. Flowering heads pedunculate, in terminal corymbs, hemispherical, ± 5 mm in diameter; involucral bracts in 2 or 3 rows, equal, up to 3 mm long, obtuse, with reddish to white hairs and glands outside, 1-nerved; receptacle concave, with a small conical projection at the centre. Outer florets female, ± 2.5 mm long, numerous; petals dark purple-red (florets towards centre somewhat paler), 3-lobes, glandular; central florets bisexual or male, 5-lobed, purple-red, glandular. Anthers whitish-yellow, with obovate appendages. Styles red to white, club-shaped. Cypselas ovate, ± 0.6 mm long, glabrous and shiny; pappus absent.

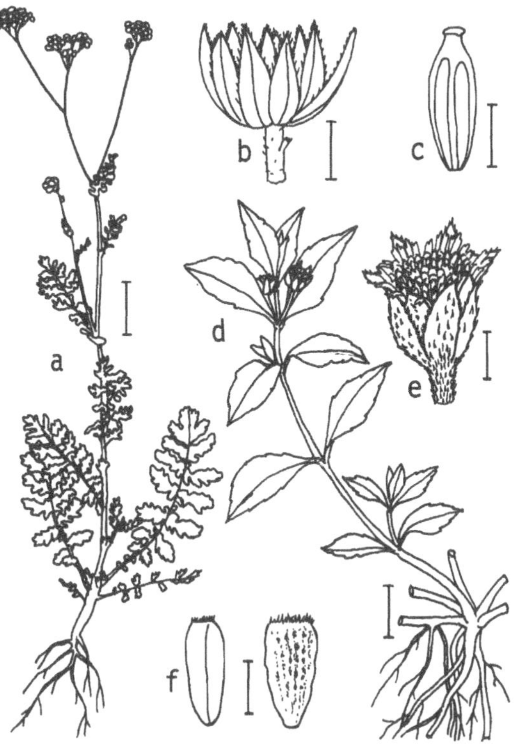

FIG. 54. *Cyathocline purpurea*: a, whole plant (2.5 cm); b, involucre (1 mm); c, cypsela (0.2 mm). *Eclipta alba*: d, plant (1 cm); e, flowering head (4 mm); f, two cypselas, left smooth, right corky (1 mm).

Annual or occasionally perennial. Found in waterlogged soil, places subject to seasonal inundation and along water courses. Pollinated by insects or autogamous. Aromatic, used in medicine against headaches.

{(⊙) Hel}. India to Indochina and S China: Ban, Bhu, Bur, Ind, Nep, Pak: As, Bh, Gj, Go, Hp, Jk, Kt, Mg, Mh, Mn, Mp, Or, Pj, Up, Wb. ?–1800 m.

Eclipta: ± 4 species: cosmopolitan in tropical to warm regions

Eclipta alba (Linnaeus) Hasskarl [*E. prostrata* (Linnaeus) Linnaeus, *Verbesina alba* Linnaeus]: Fig. 54

Stems, leaves and bracts clothed in white, appressed, almost bristle-like hairs. Stems creeping, ascending or erect, 20–30 (–75) cm long, often reddish, rather hard, usually branched, swollen when submerged, often rooting at internodes. Leaves opposite, sessile or nearly so, lanceolate or narrowly ovate-lanceolate, 1–6 (–8) cm long, 0.3–2 cm wide, margins entire to serrate. Flower heads pedunculate, most axillary, up to 1 cm in diameter; peduncles 4–7 cm long; involucral bracts in 2 rows, few, ovate-lanceolate, overlapping. Ligulate florets white to cream; tubular florets white to cream, 4-toothed. Cypselas 3–3.5 mm long, dimorphic but alike in each head; most heads with straw-coloured, tuberculate, slightly flattened cypselas; other heads with dark grey to black, smooth cypselas; pappus reduced to 2, minute and united scales, with or without a few hairs.

Usually annual but in very wet conditions it may perennate by rooting at the nodes. It is common in and along the edges of pools, tanks, irrigation canals and ditches, sometimes found floating in mats, also often found in rice fields but rarely considered to be a serious weed. Pollination entomophilous or perhaps sometimes autogamous. The disseminules are cypselas with a minute pappus, dimorphic: either corky, flattened and floating or hard, cylindrical and sinking, at least some dispersed by water. It is used for many medical purposes in India and it is a source of a blackish dye. It is also a source of thiophene derivatives which are active against nematodes. Plants with smooth cypselas are sometimes distinguished as *E. alba* while those with marginal tubercles are called *E. prostrata*; there seem to be numerous intergrading plants.

{(⊙) Hel(Ple/Hyp)}. Cosmopolitan in warm and tropical conditions: Bur, Cey, Ind, Nep: <An, Ap, As, Bh, Dl, Gj, Go, Hp, Kl, Kt, Lk, Mp, Mz, Or, Pj, Rj, Tn, Tp, Up, Wb. 0–1200 m.

Enydra [*Enhydra* orth. mut.]: ± 10 species: cosmopolitan in warm regions, most in America

Enydra fluctuans Loureiro: Fig. 55

Stems creeping and rooting at internodes or ascending, up to 30 cm tall. Leaves in opposite pairs, subsessile, linear to lanceolate, 1.2–4 (–7) cm long, 0.5–1 cm wide, glabrous, sometimes auriculate at the base, margins slightly toothed. Flower heads sessile or subsessile in one of a pair of leaves, ± 8 mm in diameter; involucral bracts 4, very wide and leaf-like, enclosing the heads, the outer pair obtuse, ± 1.5 cm long, the inner pair rounded at the apex, ± 6 mm long. Florets all ligulate, with very short tongues, mostly appearing tubular. Cypselas ± 1 mm long, brown, enclosed in scales; scales persistent, rigid and hardened, with glandular hairs at the tip; pappus absent.

Annual or sometimes perennial. Erect helophytes found in marshy places, tanks, ditches or sometimes also floating. Pollinated by insects. The disseminules are cypselas enclosed in hardened scales, the dispersal is unknown. Almeida, S. M. and Daruwalla, A. R. Massive occurrence of *Enydra fluctuans* in a city pond. Journal of the Bombay Natural History Society 84: 463–465 (1987).

{(⊙) Hel/Hyp?}. S and SE Asia, perhaps not native in W India: Ban, Bur, Ind, Pak: As, Bh, Dl, Jk, <?Mh, Mp, Wb.

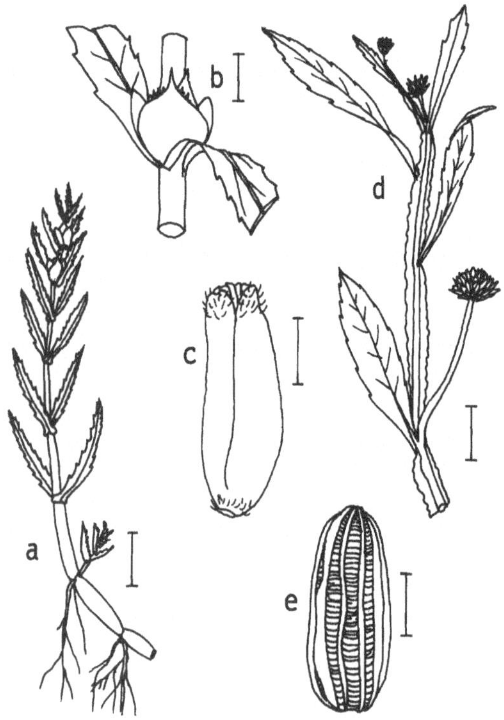

Fig. 55. *Enydra fluctuans*: a, whole plant (2 cm); b, flowering head (1 cm); c, receptacular scale (1 mm). *Epaltes divaricata*: d, upper part of flowering shoot (1 cm); e, cypsela (0.25 mm).

Epaltes: 9–15 species: pantropics

Epaltes divaricata (Linnaeus) Cassini [*Ethulia divaricata* Linnaeus]: Fig. 55

Stems spreading, much branched, 3–30 (–50) cm tall, glabrous, 3- or 4-winged; wings decurrent from leaf bases, entire or dentate. Leaves alternate, narrowly oblanceolate or spathulate, 0.7–10 cm long, 1–13 mm wide, somewhat fleshy, glabrous, base somewhat attenuate, margins minutely toothed or entire, tip acute or obtuse but usually mucronate. Heads globose, ± 1 cm in diameter, 1 or more at branch ends; peduncles up to ± 5 cm long. Involucral bracts in 5 or 6 series, ovate-lanceolate, 3–5 mm long, 1–1.5 mm wide, cartilaginous, acuminate, recurved at the tip, longer than the flowers. Flowers all tubular, pink to purple; female flowers numerous, tubular, the petals 3- or 4-lobed, 1.2–1.5 mm long; bisexual flowers fewer (± 20–30), the petals 5-lobed, 2–3 mm long, the ovary usually abortive. Cypselas of female flowers obovoid, ± 1 mm long brown, with 10 paler wing-like ribs; pappus absent.

Annual. Often very abundant, it is not confined to wetlands but is often found in wet places, even in standing water in ditches, marshes and rice fields. The roots are bitter and astringent and used as a tonic. A second species, *Epaltes pygmea* A. P. de Candolle, is sometimes found in dry rice fields; it differs from *E. divaricata* in having smaller heads (± 4 mm in diameter); involucral bracts appressed and shorter than the flowers; cypselas ellipsoid, ± 1 mm long, smooth and glossy black.

{⊙ Hel}. S and SE Asia: Bur, Cey, Ind: Go, Kl, Kt, Or, Tn.

Grangea: 10 species: tropical and warm Africa and Madagascar, 1 species extending into Asia

Fayed, A. Revision der Grangeinae (Asteraceae—Astereae). Mitteilungen der Botanischen Staatssammlung München 15: 425–576 (1979).

Grangea maderaspatana (Linnaeus) Poiret: Fig. 56

Stems, leaves and bracts clothed in white, soft, multicellular hairs, also bearing minute glands. Stems arising from a single tap root, ribbed, creeping, ascending or erect, sometimes reduced to a single rosette, usually branched, the branches 5–30 cm long. Leaves alternate, numerous, sessile, up to 10 cm long and 4 cm wide, oblong-ovate to obovate, usually

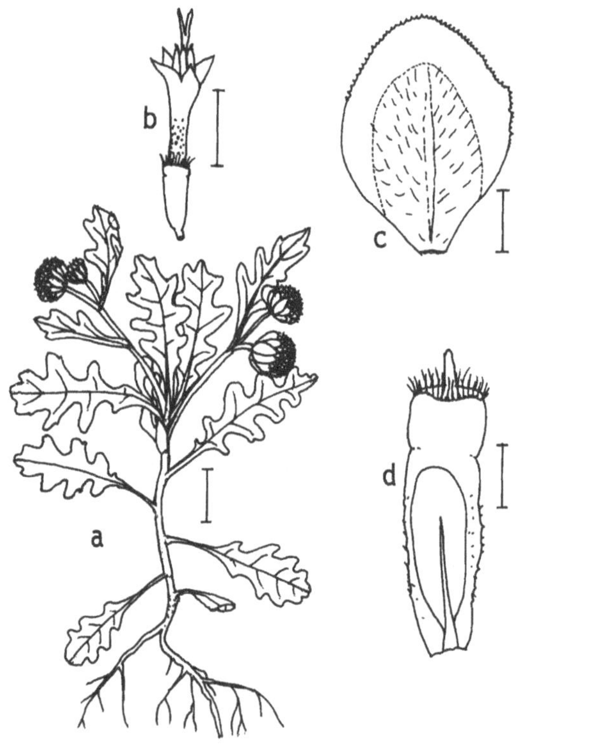

Fig. 56. *Grangea maderaspatana*: a, whole plant (1 cm); b, bisexual flower (1 mm); c, involucral bract (1 mm); d, cypsela (0.5 mm).

pinnatifid or at least pinnately lobed, the lobes opposite or nearly so, in 2–6 pairs, smaller towards the base, each entire, toothed or the larger ones sometimes pinnately lobed, leaf base subauriculate. Flowering heads pedunculate or rarely sessile, solitary and leaf-opposed or sometimes 2 or 3 together, globose, 0.75–1.3 cm in diameter; peduncles up to 2.5 cm long; involucral bracts ± in 3 rows, subequal, the outer ones up to 7 mm long and 2.5 mm wide, linear-oblong to elliptical, obtuse, with membranous and ciliate margins; receptacle convex, glabrous. Ligulate florets yellow, in ± 3 rows, the limbs 2- to 4-lobed; tubular florets more than ligulate ones, yellow, 4- or 5-toothed. Cypselas 1.25–1.5 (–2) mm long, obovoid to ellipsoid, weakly flattened, with scattered short hairs and glands; pappus a crown-like, cylindrical membrane, 0.5–1.3 mm long, fimbriate at tip.

Annual or perhaps occasionally perennial. Found at the edges of tanks, canals and ditches, also in rice fields but not considered to be serious weed. The seedlings are often found submerged in water. It is not exclusively found in wetlands. Pollinated by insects or autogamous. The disseminules are small cypselas, the dispersion unknown. Used for pain relief especially for stomach and menstrural disorders.

{(⊙) Hel}. Asiatic but now widespread in warm and tropical regions of the World: Ban, Bur, Cey, Ind, Nep, Pak: Ap, As, Bh, Dl, Gj, Go, Kl, Kt, Mg, Mh, Mn, Mp, Ng, Or, Pj, Rj, Tn, Up, Wb.

Sphaeranthus: 38 species: Old World tropics (32 species confined to Africa)

Much branched herbs. Stems and branches in Indian species winged. Leaves alternate, toothed decurrent along the stem. Flowering heads small, very numerous, sessile, crowded on a common receptacle into ± globose terminal compound heads with or without an involucre of empty bracts at the base, the compound heads borne terminally on the stem and its branches; involucral bracts of the flowering heads subequal, scale-like. Ligulate flowers absent; tubular florets female at the periphery of each head and bisexual or functionally male in the centre, purple; receptacle small, glabrous. Cypselas oblong, laterally compressed; pappus absent.

Highly aromatic herbs, often used locally in medicine. Ross-Craig, S. A revision of the genus *Sphaeranthus.* Hooker's Icones Plantarum, Ser. 5, 6(1): 1–90, t. 3501–3525 (1955).

1. Subtending bracts not visible or if visible then only covering the base of the flowering heads at anthesis; compound heads stalked, ± spherical not resembling conifer cones

2. Margins of the leaf blades and the wings of the stems, branches and peduncles entire or sometimes simply toothed but never spiny or spiny-dentate ***S. africanus***

2. Margins of the leaf blades and the wings of the stems, branches and peduncles spiny, spiny-dentate or spiny-double dentate

3. Leaf blades villous or hirsute, bearing stalked glands on both surfaces; margins of leaf blades usually double dentate; peduncle wings strong and running up to the compound heads ***S. indicus***

3. Leaf blades persistently white- or tawny-woolly or tomentose, not bearing stalked glands; margins of leaf blades usually simply dentate; peduncle wings tapering off and reduced to spines a short distance below the compound heads ***S. senegalensis***

1. Subtending bracts visible throughout the compound heads, closely overlapping and hiding the flowering heads at anthesis; compound heads sessile or nearly so, ± conical, resembling conifer cones

4. Stems and leaves glabrous or puberulous when young; subtending bracts glabrous or sparsely puberulous ***S. amaranthoides***

4. Stems and leaves hispidulous, villous when young; subtending bracts hispidulous and ciliate ***S. peguensis***

Sphaeranthus africanus Linnaeus [*S. microcephalus* Willdenow]: Fig. 57

Stems usually one arising from the base, erect and bearing numerous ascending or spreading branches throughout its length, up to ± 60 (–90) cm tall. Wings on stems and branches entire or with here and there a minute mucro, glabrous or pubescent. Leaves of the main stem obovate to oblanceolate, 2–10 cm long, 0.7–3.3 cm wide, pubescent or sometimes almost glabrous, the margins subentire or denticulate, the tip rounded to subacute, the base gradually narrowed and and decurrent, forming the wings of the stems; leaves of the branches generally smaller and narrower, sometimes elliptical or linear-lanceolate. Compound heads globose or somewhat ellipsoid, 7–13 mm long, 6–12 mm in diameter; peduncles up to 3 cm long, winged below and tapering off towards the tip, the wings entire. Subtending bracts just visible but only covering the base of the flowering heads at anthesis, oblong-spathulate, apiculate or rounded at the tip.

Annual. In wet places, swamps, rice fields, often in brackish water, sometimes temporarily submerged.

{⊙ Hel}. Old World tropics: Ban, Bur, Cey, Ind: An, Go, Kl, Kt, Rj, Tn, Wb.

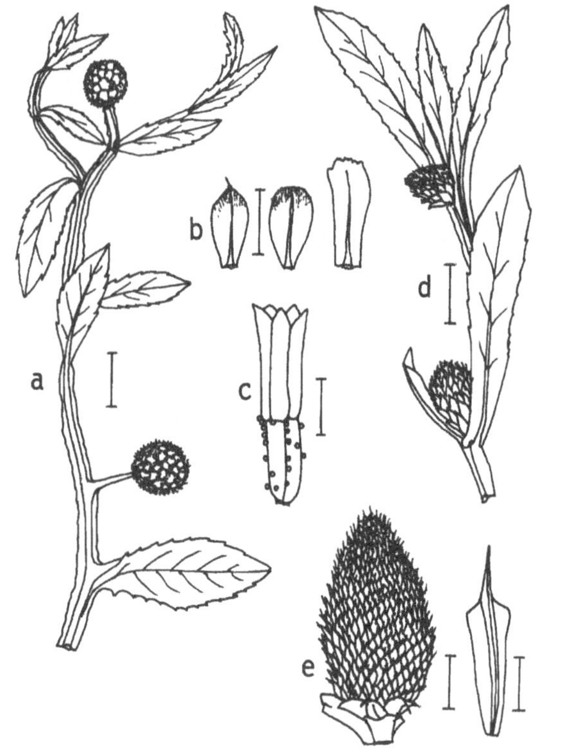

Fig. 57. *Sphaeranthus africanus*: a, upper part of flowering shoot (1 cm); b, three bracts (1 mm); c, floret (0.5 mm). *S. amaranthoides*: d, upper part of flowering shoot (1 cm); e, compound head (5 mm); f, bract (1 mm).

Sphaeranthus amaranthoides N. L. Burman [*S. zeylanicus* Heyne ex Robyns]: Fig. 57

Stems usually one arising from the base, erect and bearing numerous ascending or spreading branches throughout its length, up to ± 40 cm tall. Wings on stems and branches narrow, occasionally running out on older parts, entire or with here and there a minute mucro, glabrous or sparsely pubescent. Leaves of the main stem oblong, up 3.5–9 cm long, 0.6–1.5 cm wide, pubescent or sometimes almost glabrous, the margin distantly serrate, rounded to subacute at the tip, gradually narrowed and decurrent at the base and forming the wings of the stems. Compound heads sessile or nearly so, elongate-ovate at maturity, 1.5–2.5 cm long, 0.8–1.2 cm in diameter. All subtending bracts visible throughout the compound heads, closely overlapping and hiding the flowering heads at anthesis, the heads resembling conifer cones; subtending bracts closely imbricate, glabrous or sparsely pubescent, concave, ovate, 5.5–7 mm long, with a spine-like, spreading or reflexed tip.

Annual. In moist, probably brackish situations near the coast. {⊙ Hel}. Endemic to Sri Lanka and SE India: ♦Cey, ♦Ind: Kl, Kt, Tn.

Sphaeranthus indicus Linnaeus [*S. hirtus* Willdenow]: Fig. 58

Stems usually several arising from the base, erect, ascending or spreading and again branched above, up to ±45 cm tall. Wings on stems and branches strong, irregularly spinulose-dentate, densely hairy or villose and bearing stalked glands. Leaves obovate to oblanceolate or sometimes elliptical, 1.5–6 cm long, 0.7–2.7 cm wide, usually smaller on the ultimate branches, villose to sparsely hirsute, never woolly or tomentose, the margins usually double dentate, irregularly spinulose-denticulate to coarsely dentate, the tip rounded to subacute and aristate, the base gradually rounded and decurrent, forming the wings of the stems. Compound heads globose to somewhat ellipsoidal, 10–16 mm long, 10–16 mm in diameter; peduncles 1–4.5 cm long, strongly winged from base to just below the compound head. Subtending bracts of the flowering heads not visible at anthesis, linear to linear-oblanceolate, apiculate at the tip.

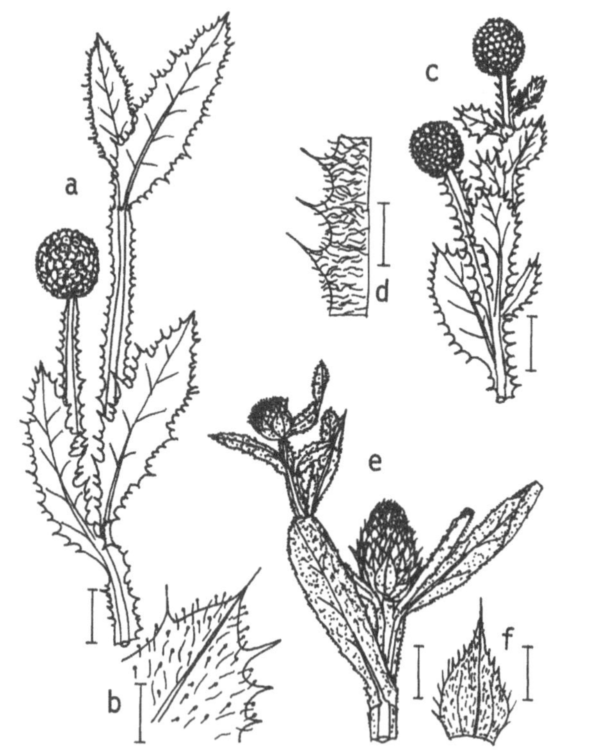

Fig. 58. *Sphaeranthus indicus*: a, upper part of flowering shoot (1 cm); b, part of a leaf blade (2 mm). *S. peguensis*: e, upper part of flowering shoot (1 cm); f, bract (5 mm). *S. senegalensis*: c, upper part of flowering shoot (1 cm); d, part of a wing (2 mm).

Annual. Sometimes found in wet areas, occasionally temporarily submerged. Common in and around irrigation ditches and rice fields but not considered to be a serious weed. The plants are highly aromatic and a range of medicinal uses has been ascribed to this plant: anthelminthic, tonic, aphrodisiac, cure for toothache, diuretic, laxative and fish poison.

{⊙ Hel}. Tropical S and SE Asia and Australia: Ban, Bhu, Bur, Cey, Ind, Nep: An, Ap, Ar, As, Bh, Go, Kl, Kt, Mh, Mp, Or, Rj, Sk, Tn, Up, Wb. 0–800 m.

Sphaeranthus peguensis C. B. Clarke: Fig. 58

Like *S. amaranthoides* but stems and leaves hispidulous, villous when young; all bracts hispidulous and ciliate, up to 10 mm long including the elongate tip.

Annual. Ecology unknown.

{⊙ Hel?}. Endemic to lower Myanmar: ♦Bur.

Sphaeranthus senegalensis A. P. de Candolle: Fig. 58

Like *S. indicus* but usually tall and more robust, stems branching above, up to 65 cm tall; leaf blades persistently white- or tawny-woolly or tomentose, not bearing stalked glands, the margin usually simply dentate; peduncle wings tapering off and reduced to spines a short distance below the compound heads.

Annual. Wet places.

{⊙ Hel}. Tropical Africa and Asia: Ban, Bur, Ind, Nep: As, Bh, Pj, Up, Wb. 500?–1000 m.

Spilanthes: 42 species: tropics

Stems weak, erect or decumbent, branching, leafy, often rooting at lower nodes. Leaves simple, in opposite pairs, ovate or elliptical. Heads solitary on terminal or axillary peduncles, with ligulate and tubular or tubular flowers only. Involucral bracts in 1 or 2 rows. Anthers rounded at the apex, minutely sagittate at the base. Receptacle convex at first, becoming ovoid-conical at maturity, with scales enfolding the tubular flowers. Cypselas of the ligulate flowers 3-angled; pappus of 2 weak bristles or absent.

Annuals. *Spilanthes* is sometimes included in the genus *Acmella*; there has also been confusion with the genus *Blainvillea*. Sivarajan, V. V. and Remesan, C. The genus *Spilanthes* in India. Journal of Economic and Taxonomic Botany 10(1): 141–147 (dated 1987, published 1988).

Excluded: ***Spilanthes oleracea*** Linnaeus and ***S. radicans*** Jaquin; not found in wetlands.

1. Heads without ligulate flowers
2. Cypselas without ciliate margins and without a pappus; involucre 1-seriate ***S. calva***
2. Cypselas with cilia on one or both margins and with a pappus of weak fragile bristles; involucre 2-seriate ***S. paniculata***
1. Heads with ligulate flowers
3. Heads 8–10 mm in diameter; ray florets 8–12, the limb ± 2.5 mm long, ± 2 mm wide ***S. ciliata***
3. Heads 3–5 mm in diameter; ray florets 4–6, the limb 1–2 mm long, 1–1.3 mm wide ***S. uliginosa***

Spilanthes calva A. P. de Candolle [*S. acmella* auct. Ind., non (Linnaeus) Murray var. *calva* (A. P. de Candolle) C. B. Clarke, *S. pseudo-acmella* Linnaeus, *Acmella calva* (A. P. de Candolle) R. K. Jansen, *Cotula conica* Wallich]: Fig. 59

Stems trailing or ascending, 15–40 cm tall, scarcely branched. Petioles up to 1.5 cm long, often pubescent near the base; leaf blades ovate, 2–6 (–9) cm long, 1–2.5 (–4) cm wide, irregularly crenate or sometimes entire, glabrous or nearly so, acute or obtuse at tip. Flowering heads long-pedunculate, solitary, at first ± 5 mm in diameter, later reaching 7–8 (–10) mm in diameter and 8–9 mm long; peduncles 4–10 cm long; involucral bracts 1-seriate, ovate, 3.5–4.5 mm long. Ligulate flowers absent; tubular flowers yellow, 1.7–2 (–3) mm long. Cypselas oblong or slightly obovoid,

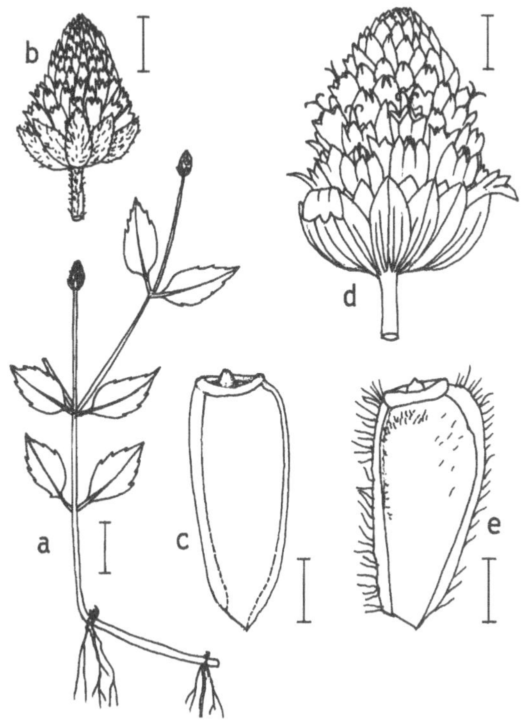

Fig. 59. *Spilanthes calva*: a, plant (2 cm); b, flowering head (3 mm); c, cypsela (0.5 mm). *S. ciliata*: d, flowering head (3 mm); e, cypsela (0.5 mm).

laterally flattened, ± 2 mm long, almost black, nearly glabrous; pappus absent.

Annual. It is not confined to wetlands but is frequently found in boggy ground, margins of pools, marshes, along water courses and in rice fields. In the south it is mostly found at higher elevations. It is used locally to cure toothache.

{⊙ Hel}. S and SE Asia: Ban, Bur, Cey, Ind, Nep: Ap, As, Kl, Kt, Mg, Mp, Rj, Tn. 0–2300 m.

Spilanthes ciliata Humboldt, Bonpland et Kunth: Fig. 59

Like *S. uliginosa* but stems usually prostrate; leaf blades ovate, up to 7 cm long, up to 4 cm wide, with margin serrate; flowering heads 1–3, 8–10 mm in diameter; peduncles 5–8 cm long; limb of ligulate flowers ± 2.5 mm long, ± 2 mm wide. Cypselas ± 2 mm long, distinctly ciliate along the margins; pappus from ligulate flowers 3-awned, from tubular flowers with 2 unequal bristles.

Annual. It is not confined to wetlands but is frequently found in wet or marshy places.

{⊙ Hel}. Tropical America, naturalized in India: <Ind: <Kl, <Mp.

Spilanthes paniculata Wallich ex A. P. de Candolle [*S. acmella* auct. Ind., non (Linnaeus) Murray var. *paniculata* (A. P. de Candolle) C. B. Clarke]: Fig. 60

Like *S. calva* but usually larger, stems erect, 20–50 (–75) cm tall and more branched; flowering heads 1 or 2 in leaf axils, ultimately appearing as terminal panicles, at first 7–9 mm in diameter, becoming conical, up to 1 cm in diameter and 1.5 cm long; involucral bracts 2-seriate; cypselas dimorphic, the marginal ones 3-sided, the central ones flat, up to 3 mm long, the margins densely ciliate, with long hairs; pappus of 1–3, 1–1.3 mm long, stiff bristles.

Annual. Not confined to wetlands but often found in wet places, moist grassy areas, ditches and rice fields. It is often cultivated and may escape. The leaves and flowers have a hot and pungent taste and are used for the treatment of toothache and sore throats.

{⊙ Hel}. India and Malaysia: Bur, Cey, Ind: As, Go, Kl, Kt, Mp, Rj, Sk, Up.

Spilanthes uliginosa Swarts [*S. iabadicensis* A. H. Moore]: Fig. 60

Stems erect or diffuse, up to 30–40 cm tall, branched above, glabrous or sparsely pubescent when young. Petioles up to 1.5 cm long, the upper leaves usually sessile or nearly so; leaf blades narrowly ovate or elliptical, 1.5–5 (–7) cm long, 0.5–2 (–3) cm wide, margin subentire or undulate, subglabrous or sparsely pubescent, subacute or obtuse at tip. Flowering heads pedunculate, mostly solitary below becoming paniculate above, ± 3 mm in diameter at first, later becoming ± 5 mm in diameter and ± 8 mm long; peduncles 1.5–5 cm long; involucral bracts 1-seriate ovate-lanceolate, 2.5–3 mm long, 1 mm wide. Ligulate flowers 4–6, yellow, petal tube ± 2.5 mm long, the limb 1–2 mm long, 1–1.3 mm wide, 3-lobed; tubular florets yellow, 1–1.3 (–2) mm long, 3- to 5-lobed. Cypselas oblong, laterally flattened, 1.2–1.5 mm long, 0.5–0.6 mm wide, weakly ciliate along one or both margins, almost black with pale margins; pappus bristles weak, fragile, ± 0.25 mm long.

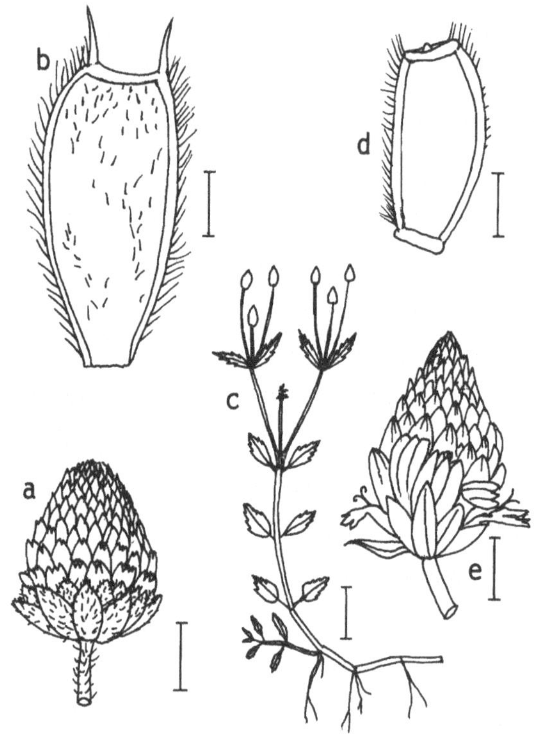

Fig. 60. *Spilanthes paniculata*: a, flowering head (3 mm); b, cypsela (0.5 mm). *S. uliginosa*: c, plant (2 cm); d, cypsela (0.5 mm); e, flowering head (2 mm).

Annual. Often abundant, mostly found in wet places, ditches and rice fields.

{⊙ Hel}. Tropics, probably native in the New World, naturalized in the Old World: <Cey, <Ind: <An, <Kt, <Sk, <Tn.

BALSAMINACEAE: 2 genera

Excluded: ***Impatiens*** has many species in India of which some are found in wet places by rivers and waterfalls in montane forests; none are considered to be true wetland plants.

Hydrocera: 1 species

Hydrocera triflora (Linnaeus) Wight et Arnott [*Impatiens natans* Willdenow, *I. angustifolia* Blume]: Fig. 61

Glabrous. Stems ascending or floating, up to 1 m long, usually branched below and simple above, 5-angled; internodes swollen and spongy when submerged or floating, bearing numerous roots. Leaves alternate, sessile or with petiole up to 1 cm long; stipules absent; leaf blades linear to elliptical, up to 10–28 cm long, 6–37 mm wide, the bases bearing paired glands; leaf margin toothed. Inflorescences axillary, cymose, (1–) 3- to 5-flowered. Flowers bisexual, bilaterally symmetrical, mostly borne in pairs, pinkish to purple, the throat yellowish. Sepals 5, free, petaloid; the 2 outer laterals flat becoming adaxial due to twisting; the abaxial one prolonged backwards into a short, hollow spur. Petals 5, free, the 2 outer abaxial ones longer, concave, reddish. Stamens 5; filaments broad, united towards the connective; anthers whitish, united around the style. Carpels 5, united, superior; ovules 2 or 3 in each carpel, axile. Fruit a purplish-red, subglobose berry, containing up to 5 stony seeds.

Annual or perennial. Found in shallow water at the edge of pools, in marshes, rice fields and other wet places. Flowers pollinated by insects. The disseminules are 5-seeded berries, dispersed by animals or the seeds dispersed in water. Sometimes planted for ornament. The berries are used to dye fingernails in parts of India. Ramadevi D. and Narayana L. L. Morphology of the flower and fruit of *Hydrocera triflora* Wight and Arn. emend [sic] Venkat. and Dutt — an elucidation. Proceedings of the Indian Academy of Science (Plant Science) 100(1): 43–49 (1990).

{(⊙) Hel}. S and SE Asia: Ban, Bur, Cey, Ind: Ap, Kl, Tn, Up, Wb.

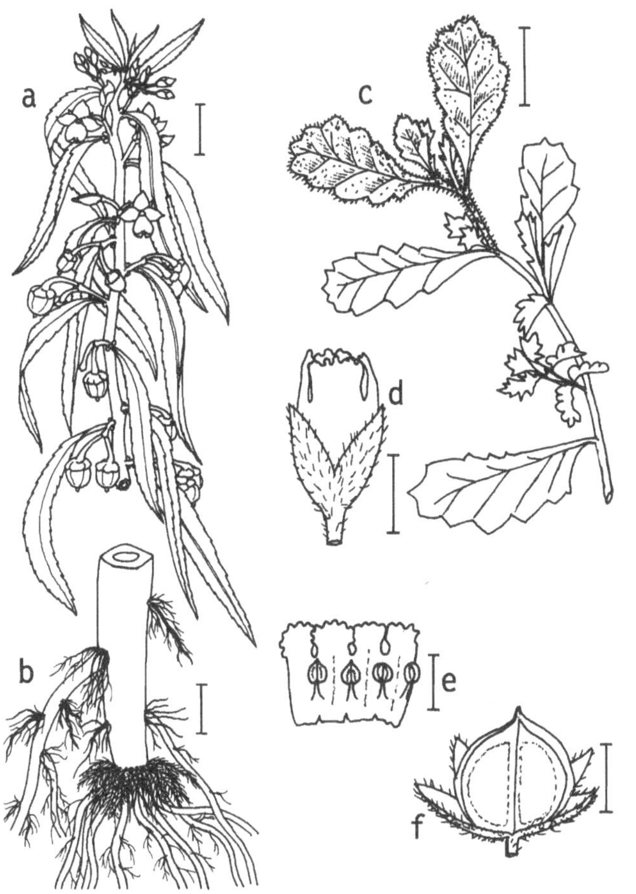

Fig. 61. *Hydrocera triflora*: a, inflorescence (2 cm); b, submerged stem base (2 cm). *Coldenia procumbens*: c, shoot with hairs omitted from lower part (1 cm); d, flower (2 mm); e, petal tube opened (1 mm); f, fruit (2 mm).

BARCLAYACEAE [included in NYMPHAEACEAE]

BARRINGTONIACEAE [LECYTHIDIACEAE] excluded see p. 12

BORAGINACEAE [CORDIACEAE, EHRETIACEAE]: 156 genera

Leaves without stipules, usually alternate, simple, pinnately nerved. Inflorescence cymose, usually a coiled cincinnus, with or without bracts. Flowers bisexual, radially symmetrical. Sepals (4) 5, united below. Petals (4) 5, united below. Stamens (4) 5, epipetalous. Ovary superior, entire or 2- or 4-lobed; style terminal or basal, simple or branched. Fruit of 4 mericarps or drupaceous with 1–4 pyrenes.

Excluded: ***Bothriospermum tenellum*** Fischer et Meyer; found in damp or moist places but not in wetlands.

1. Woody ± prostrate shrubs; fruits orange or red drupes ***Rotula***

1. Herbs, sometimes woody at base; fruits dry, breaking into 1-seeded mericarps

  2. Leaves crisped, crenate to serrate, oblique; flowers solitary or in very short, 1-sided, leafy spikes ***Coldenia***

  2. Leaves flat, entire, bilaterally symmetrical; flowers numerous, in long scorpioid cymes ***Heliotropium***

Coldenia: 1 or ± 40 species (depending on generic delimitation)

Coldenia procumbens Linnaeus: Fig. 61

Stems prostrate, trailing, white-hairy. Petioles 3–9 mm long; blades obovate-oblong, oblique, 1–2.5 cm long, 0.6–1 cm wide, white-hairy, crisped and plicate, base tapering, margins crenate to coarsely serrate, tips rounded. Flowers minute, usually extra axillary, sessile or nearly so, the upper ones sometimes in short, 1-sided leafy spikes. Sepals 4 (5), ±2.5 mm long, divided almost to the base; lobes ovate, very hairy, ciliate at margin, persistent, Petals 4 (5), the tube ±2.5 mm long; lobes, white or pale yellow, spreading, usually more than 1 mm long. Ovary shallowly 4-lobed; style terminal, bifid. Fruit dry, at first breaking into 2, 2-seeded halves, these finally breaking into 1-seeded mericarps.

Annual. In shallow water, on moist ground in seasonally flooded areas, beds and margins of drying tanks and harvested rice fields. Used for rheumatic swellings and as a suppurative for boils.

{⊙ Hel}. Tropical and warm regions of the Old World: Cey, Ind, Pak: An, Ap, As, Dl, Gj, Go, Kl, Kt, Mp, Rj, Tn, Up, Wb.

Heliotropium: ± 280 species: almost cosmopolitan

Stems herbaceous. Leaves alternate but sometimes opposite towards the stem base, margins entire. Flowers in long scorpioid cymes. Sepal lobes usually unequal. Petal tube constricted near the throat; lobes spreading. Anthers often apiculate and sometimes joined at the tips. Ovary 4-locular; stigma conical, sometimes on a fleshy annular ring. Fruit dry, splitting into 4, 1-seeded mericarps or into 2, 2-seeded portions.

A large genus with most species characteristically found in dry sandy or rocky conditions; a few species are found in wetlands.

1. Petals yellow at the throat; stems and leaves glaucous, somewhat succulent, without hairs ***H. curassavicum***

1. Petals white to blue or violet, never yellow; stems and leaves with hairs, not glaucous, not succulent

  2. Petal tube bearing bulbous-based, bristle-like hairs ***H. keralense***

  2. Petal tube glabrous or bearing minute soft hairs

    3. Stems woody at the base; leaf blades more than 5 cm long, sparsely hairy; mericarps beaked ***H. indicum***

    3. Stems not woody at the base; leaf blades 5 cm or less long, clothed in white hairs; mericarps obtuse, not beaked

      4. Mericarps with hairs, not corky; stigmas sessile or nearly so; fruiting sepals spreading ***H. ovalifolium***

      4. Mericarps glabrous, corky; stigmas borne on a style; fruiting sepals enclosing the fruit ***H. supinum***

Heliotropium curassavicum Linnaeus: Fig. 62

Stems prostrate, ascending or erect, branched, succulent, glaucous; branches 8–50 cm long. Leaves alternate or subopposite, subsessile to shortly petiolate; blades obovate to oblanceolate or linear-spathulate, (0.5–) 1–4.5 cm long, 0.2–1 cm wide, fleshy, glabrous, glaucous, tips rounded. Flowers in 1.5–5.5 cm long scorpioid spikes. Sepals ± 1.5 mm long; lobes subequal. Petals yellow at the throat with white lobes; lobes obtuse. Stamens not joined at their tips. Stigmas sessile. Mericarps 1–2 mm long, glabrous, rounded on the back, with corky margins, rugulose, pale brown.

Perennial or sometimes annual. Mostly in coastal areas by brackish backwaters but also in and around drying pools, rice fields, wastelands and fallow fields.

{⊙? Hel}. Tropical America, naturalized in Africa and Asia: <Ind, <Pak: <Dl, <Kt, <Pj, <Rj.

Heliotropium indicum Linnaeus: Fig. 62

Stems hispid, erect, woody at base, up to 50 (–75) cm tall. Leaves alternate above; petioles partially winged, up to 8 cm long; blades ovate to ovate-orbicular, 5–12 cm long, 5–8 cm wide, hispid to pubescent along the nerves, base cordate or narrowed, decurrent into the winged petiole, tips acute. Flowers 2-ranked in elongated, up to 22 cm long, terminal or leaf-opposed,

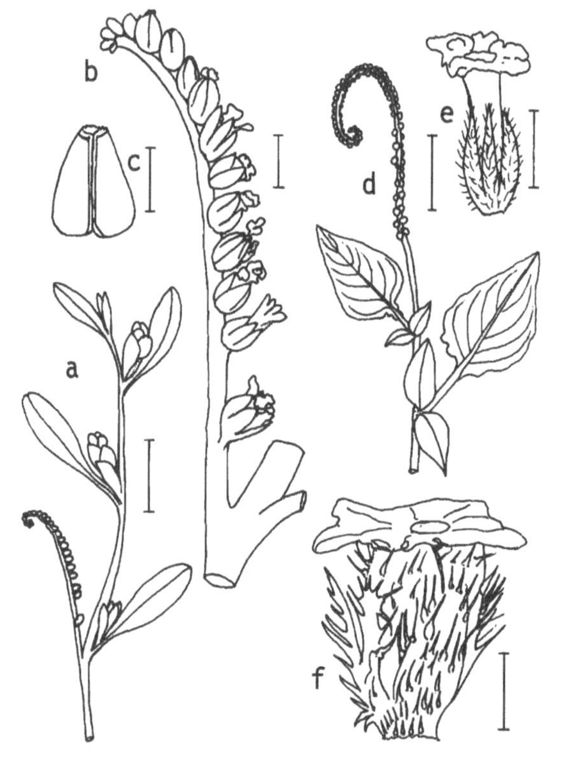

Fig. 62. *Heliotropium curassavicum*: a, flowering shoot (1 cm); b, inflorescence (2 mm); c, two mericarps (1 mm). *H. indicum*: d, flowering shoot with hairs omitted (4 cm); e, sepals and petals (2 mm). *H. keralense*: f, sepals and petals (1 mm).

scorpioid spikes. Sepals 2–3 mm long, divided almost to the base, sparsely hairy. Petal tubes 3–4 mm long, cylindrical, longer than the sepals; petal lobes broadly ovate, ± 1.5 mm long, blue-white to pale violet. Stamens not joined at the tips. Styles 0.5–0.7 mm long. Fruits deeply bifid, each lobe ± 0.7 mm long and beaked, sometimes remaining in pairs or finally dividing into 1-seeded mericarps.

Annual. Found in seasonally flooded places, in rice fields, on sandy river banks and along steams. Not confined to wetlands, also found on roadsides and in waste places. It is used to cure skin and eye complaints.

{⊙ Hel}. Pantropics: Bur, Ind: <An, Ap, As, Gj, Go, Kl, Kt, Mp, Tn, Up, Wb.

Heliotropium keralense V. V. Sivarajan et K. S. Manilal: Fig. 62

Like *H. indicum* but sepals and petal tube clothed in bulbous-based, bristle-like hairs; petal tube ± 2.5 mm long, ± as long as the sepals; petals uniformly white.

Annual. Found in flooded and moist fields during the rainy season.

{⊙ Hel}. Endemic to S India: ♦Ind: Kl.

Heliotropium ovalifolium Forsskål [*H. coromandelianum* J. G. König ex Retzius]: Fig. 63

Stems erect or decumbent, herbaceous from a frutescent base, branching from the base, 10–40 cm tall, with silky white hairs. Petioles channelled, up to 1.2 cm long, shorter towards the tips of the shoots; blades obovate to elliptical, (0.25–) 1–2.5 cm long, (0.24–) 0.5–1.3 cm wide, with long silky hairs, at the base rounded, margins narrowly revolute, tips rounded with a mucro. Flowers 2-ranked, in 2–5 cm long, scorpioid spikes. Sepals divided almost to the base, 1–2.5 mm long at anthesis, persistent, enlarging and spreading in fruit, with appressed hairs; sepal lobes densely ciliate, very unequal, 1 almost twice as wide as the others. Petal tubes ± 1.5 mm long, cylindrical; petal lobes white, hairy, 0.5–1 mm long, ovate, oblong, acute. Stigmas sessile. Mericarps exceeding 1 mm long, densely appressed-hairy with white silky hairs on the convex back, obtuse.

Perennial perhaps occasionally annual. Drying ponds, ditches, common in fallow rice fields and along roadsides.

{⊙? Hel}. Africa, Arabia, India, Indo-China and Australia: Ind, Nep, Pak: Gj, Go, Kl, Kt, Mh, Mp, Rj, Tn, Up, Wb.

Heliotropium supinum Linnaeus [*H. malabaricum* Retzius]: Fig. 63

Stems herbaceous, up to 50 cm long, much branched, prostrate or ascending, thickly clothed with soft, white hairs. Leaves usually alternate, rarely opposite; petioles channelled, 5–10 mm long, villous; blades ovate, 1–2.5 cm long, 1–1.5 cm wide, softly villous, the base mostly attenuated but sometimes rounded, margins very slightly undulate, tips obtuse. Flowers subsessile in usually simple, 1-sided, 5–9 cm long, woolly spikes. Sepals 4–5 mm long, woolly; lobes minute, ciliate, the division between the lobes marked by a thin and tearable membrane, the tube often torn to the base. Petal tubes ± 2.5 mm long; petal lobes ± 2 mm long, white, hardly exerted beyond the sepals. Stigmatic ring conspicuous, surmounted by a conical appendage. Fruit of 2 or 4 mericarps, when 2 the margins large and conspicuous; when 4 the margins divided and rugose; mericarps up to 2 mm long, glabrous and corky.

Annual. Found in the dried out beds of tanks, edges of water holes and pools, in rice fields, particularly noticeable in the cold season.

{⊙ Hel}. S Europe, Canary Islands, N and S Africa, warm and tropical Asia: Cey, Ind, Pak: Dl, Gj, Kt, Mh, Mp, Pj, Rj, Tn, Up.

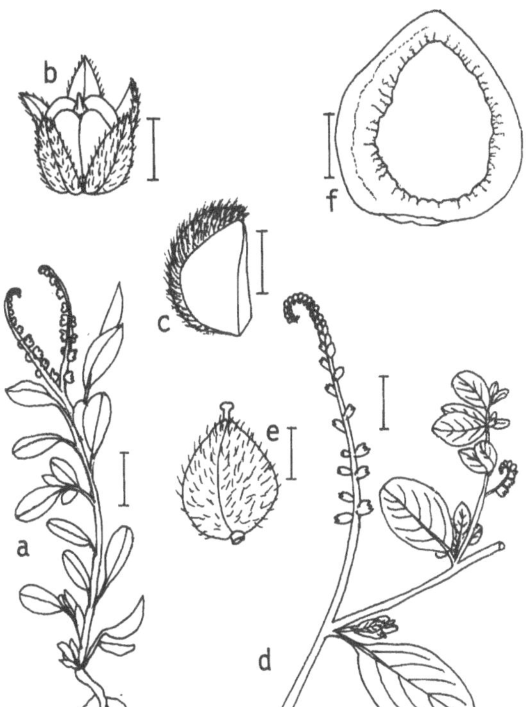

FIG. 63. *Heliotropium ovalifolium*: a, flowering shoot with hairs omitted (1 cm); b, young fruit with hairs omitted from the mericarps (1 mm); c, one of four mericarps (0.5 mm). *H. supinum*: d, flowering shoot with hairs omitted (2 cm); e, fruiting sepals (1 mm); f, one of a pair of mericarps.

Rotula [*Rhabdia* pro parte]: 1 species

Rotula aquatica Loureiro [*Rhabdia lycioides* Martius, *Rhabdia fluviatilis* Edgeworth]: Fig. 64

Stems woody, branched; branches usually villous, mostly prostrate, often rooting. Leaves simple, alternate or in clusters, narrowly spathulate, up to 2 cm long, up to 8 mm wide, usually tomentose beneath, margins entire or obscurely toothed, apiculate. Flowers small, in the axils of leaves, sometimes forming lax cymes. Sepals 3–5 mm long, shortly tubular below; lobes 5, lanceolate, acuminate, imbricate. Petals pink to purple; petal tube campanulate, ±3 mm long; lobes 5, 2.5–4 mm long, oblong, overlapping in bud. Stamens exerted. Ovary ovoid, 2-locular but each locule almost completely 2-locular; style terminal, simple, 3–5 mm long, persisting in fruit; stigma capitate. Fruit a red or orange, globose drupe, ± 3 mm in diameter, later breaking into 4, hard, 1-seeded pyrenes.

Perennial, forming small, usually flattened bushes. Found on rocky river banks and among pebbles in the beds of rivers and streams; usually submerged during the wet season.

{Hel-Rheophyte}. Tropical Old World and E. Brazil: Bur, Cey, Ind: An, As, Gj, Go, Kl, Kt, Mp, Rj, Tn, Up.

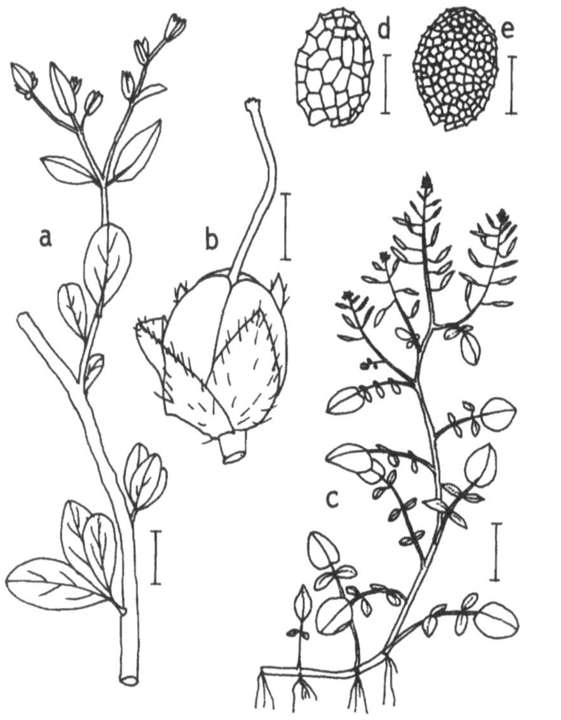

FIG. 64. *Rotula aquatica*: a, flowering shoot (1 cm); b, almost ripe fruit (1 mm). *Nasturtium officinale*: c, flowering shoot (2 cm); d, seed (0.5 mm). *N. microphylla*: e, seed (0.5 mm).

BRASSICACEAE [CRUCIFERAE]: 390 genera

Herbs. Leaves alternate, without stipules. Inflorescence a simple or branched raceme. Flowers bilaterally symmetrical, hypogynous. Sepals 4, free, in 2 decussate pairs. Petals absent or 4, free, alternating with the sepals. Stamens 6 or rarely 4. Ovary of 2 united carpels with parietal placentas, 2-locular through the formation of a membranous false septum. Fruit a dehiscent capsule opening by 2 valves.

Excluded: ***Cardamine***: — several species in the Himalayas. The following are not considered to be wetland plants but they may be found in dried pans or depressions: ***Cochlearia cochlearioides*** (Roth) Santapau et Maheshwari [*C. flava* F. Hamilton ex Roxburgh]; ***Rorippa benghalensis*** (A. P. de Candolle) H. Hara; ***R. dubia*** (Persoon) H. Hara; ***R. indica*** (Linnaeus); Hiern; ***R. micrantha*** (Roth) Jonsell [*R. madagascariensis* auct. Ind., non (A. P. de Candolle) H. Hara emend. Jonsell]; ***R. montanum*** (J. D. Hooker et Thomson) Small; ***R. palustris*** (Linnaeus) Besser; ***R. pseudoislandica*** Chowdery et R. R. Rao.

Nasturtium [*Rorippa* pro parte]: 6 species

Nasturtium officinale R. Brown [*Rorippa nasturtium-aquaticum* (Linnaeus) Hayek]: Fig. 64

Stems creeping or erect, 10–90 cm long, usually much branched, rooting below, glabrous. Leaves 1–12 cm long, petiolate; petiole with 0.1–0.3 mm long hairs on the adaxial (upper) surface; lowest basal leaves with simple suborbicular blades; upper leaf blades pinnate towards the tip becoming lobed towards the base; lateral leaflets 3–20 mm long, 2–10 mm wide; terminal leaflet ovate-cordate, 5–20 mm long, 5–15 mm long. Flowers in bractless racemes. Sepals 2–3 mm long. Petals white becoming bluish with age, 3.5–6 mm long. Pedicels in fruit forming an angle more than 60° with the inflorescence axis, 6–12 mm long. Capsules linear, ± cylindrical with prominent ridges caused by numerous seeds, 10–19 mm long, (1.8–) 2–2.7 mm in diameter. Seeds in 2 rows in each valve, with a net-like pattern on the surface, 6–12 cells showing across the broadest width.

Perennial. Grows best in fairly swiftly flowing, nutrient-rich but clear streams. It is known as watercress, widely cultivated for salad in Indian hill stations and sometimes becoming naturalized. A second species of watercress, *Nasturtium microphyllum* (Bönninghausen) ex Reichenbach, has been reported from the Himalayas; it may be more widespread; it is like *N. officinale* but has longer and narrower capsules (16–24 mm long, 1.2–2 mm thick). Seeds in 1 row in each valve, with 12–20 cells showing across the broadest width (Fig. 64e). A hybrid between these two species is sometimes cultivated but not known in India.

{Hyp/Hel}. Almost cosmopolitan through introductions, mostly in cooler regions: Cey, <Ind before 1872, <Nep, <Pak: <Go, <Mp, <Pj, <Sk, <Tn, <Up. ?–2100 m.

BUTOMACEAE [excl. from LIMNOCHARITACEAE] excluded see p. 11.

BURMANNIACEAE: 9 genera: pantropics

Burmannia: 60 species

Autotrophic or saprophytic herbs. Stems erect, simple or occasionally sparingly branched. Leaves exstipulate, alternate, green and grass-like or scale-like. Flowers solitary or in terminal cymes. Perianth tube cylindrical, urn-shaped or funnel-shaped, sometimes 3-angled or 3-winged, persisting in fruit; outer perianth lobes larger than the inner or inner lobes absent. Anthers 3, sessile, included in the perianth tube; connective with 2 apical crests. Ovary inferior; 3-angled, 3-locular with axile placentation; style filiform with 3 stigmatic branches; nectaries septal or on top of or along the sides of the ovary. Fruit a capsule, enclosed in the perianth tube, dehiscing irregularly. Seeds very small and mostly with the testa extended into a long wing at each end.

Annual or perennial. In the Indian subcontinent most species are saprophytes found in woodland. Three autotrophic species are found in wetlands. Pollinated by insects. The disseminules are seeds dispersed by wind. Sometimes used in medicine.

1. Flowers numerous in a forked cyme; robust plants up to 75 cm tall, with a persistent basal rosette of leaves up to 12 cm long; outer perianth lobes with a double margin ***B. disticha***
1. Flowers solitary or in lax clusters of 2 or 3; slender plants usually less than 12 cm tall, with or without a basal rosette of leaves, but leaves not more than 1 cm long: outer perianth lobes with a single or double margin
  2. Outer perianth lobes with a double margin; capsule 4–5 mm long ***B. caelestis***
  2. Outer perianth lobes with a single margin; capsule 1–2 mm long ***B. pusilla***

Burmannia caelestis D. Don ['*coelestis*' orth. mut., *B. triflora* Roxburgh, *B. azurea* Griffith]: Fig. 65

Stems (5–) 10–20 (–30) cm tall. Basal leaves in a rosette, linear-lanceolate to subulate 4–10 mm long, 0.5–2 (–3) mm wide, 1- or 3-nerved; stem leaves subulate, few, distant, appressed, 0.4–1 (–2) cm long, 0.5–2 mm wide, sheathing at the base. Flowers solitary or in lax clusters of 2 or 3, rarely 4. Bracts linear-lanceolate, 4–8 mm long. Perianth tube up to 6 mm long, blue or purplish with yellow lobes; wings semi-elliptical or semi-obovate, up to 2.5 mm broad; outer perianth lobes broadly ovate, ± 1.5 mm long and wide, with a double margin; inner lobes membranous, oblong-lanceolate, ± 1 mm broad. Connective crests hairy; basal spur funnel-shaped. Style 4–5 mm long. Capsule obovoid, 4–5 mm long, truncate at the top. Seeds elongate to wedge-shaped, ± 0.5 mm long

Annual. Found in swampy places, springs, wet fields and sometimes in shallow water at the edges of pools.

{⊙ Hel}. S and SE Asia: Ban, Bur, Cey, Ind, Nep, Pak: As, Bh, Mg, Mp, Or, Sk, Tn, Wb.

Burmannia disticha Linnaeus ['*distachya*' orth. mut., *B. sumatrana* Miquel]: Fig. 65

Stems robust, 14–50 (–75) cm. Basal leaves in a rosette, linear to lanceolate, acuminate, 5–15 cm long, 0.5–1.3 cm wide, 1- to 5-nerved; stem leaves overlapping below, distant above, lanceolate, appressed, 1.5–7 cm long, 4–7 mm wide. Flowers numerous, subsessile, borne on a forked cyme; the branches 3–8 cm long. Bracts linear-lanceolate, 4–5 (–12) mm long. Perianth tube up to 9 mm long, blue or purple often with yellow-tipped greenish lobes or entirely yellow; wings elliptical, 1–1.8 cm long, 1.5–3 mm broad; outer

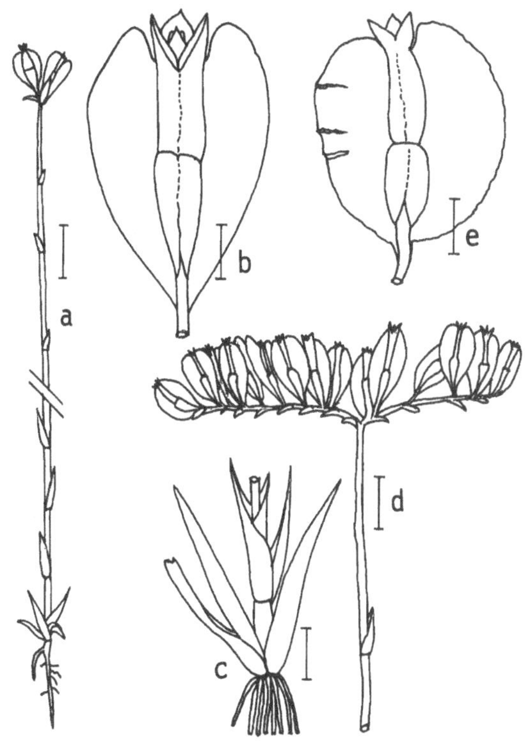

FIG. 65. *Burmannia caelestis*: a, whole plant with stem shortened (1 cm); b, flower (2 mm). *B. disticha*: c, base of shoot (3 cm); d, inflorescence (1 cm). *B. pusilla*: e, flower (1 mm).

perianth lobes broadly ovate, up to 2.5 mm long, ± 1.5 mm wide, with a double margin; inner lobes linear-spathulate, 1–1.5 mm long, yellow. Connective crests fan-like; basal spur obtuse or rounded. Style 3–3.5 mm long. Capsules obovoid, 4–5 mm long. Seeds fusiform, ± 0.7 mm long

Annual. Sometimes found in shallow water, but usually in swamps, wet flushes, boggy ground or on wet rocks.

{⊙ Hel}. S and SE Asia to tropical Australia: Ban, Bhu, Bur, Cey, Ind, Nep: Ap, As, Mg, Wb. 1500–2600 m.

Burmannia pusilla (Wallich ex Miers) Thwaites: Fig. 65

Stems filiform, 2–10 (–20) cm tall. Basal leaves in a rosette or not, linear to subulate up to 5.5 mm long, ± 1.2 mm wide, 1- or 3-nerved; stem leaves scale-like, few, distant, appressed, 1–2.5 (–4) mm long. Flowers solitary or in lax clusters of 2 or 3, the first one sessile. Bracts subulate 1–3.5 mm long. Perianth tube (2–) 4–5 mm long, bright blue to purple-blue; wings crescent-shaped or semi-obovate, up to 2 mm broad; outer perianth lobes oval-oblong to triangular-ovate, 0.5–1 mm long, with a thick, single margin; inner lobes membranous, linear-subulate, 0.25–0.5 mm long. Connective crests hairy; basal spur obtuse. Style 1.5–2.5 mm long. Capsule obovoid, 1–2 mm long. Seeds fusiform, ± 0.24 mm long

Annual. In damp and wet places, on wet rock ledges and around drying pools, usually with other dwarf herbs. In many Indian Floras the name *B. caelestis* is used indiscriminately and includes this species. It is much smaller than *B. caelestis* in all its parts and the basal rosette is often hardly distinguishable.

{⊙ Hel}. India and Indochina: Ban, Bur, Cey, Ind: As, Go, Kl, Kt, Mg, Mh, Tn.

CABOMBACEAE: 2 genera

Stems rhizomatous or erect and leafy. Leaves floating or submerged, palmately veined, either all peltate, undivided, floating and alternate or palmately dissected to the base, submerged and whorled or opposite. Flowers bisexual, radially symmetrical, solitary on long axillary pedicels, all parts hypogynous and free. Sepals (2) 3 (4). Petals (2) 3 (4), white, yellow or purple. Stamens 3–36; pollen 1-sulcate. Carpels 1–18; ovules 1–4; stigmas terminal. Fruits indehiscent follicles.

Perennial or rarely annual. Obligately aquatic plants, the leaves submerged or floating, never becoming emergent. Kasselmann, Christel (ed.). Die Familie Cabombaceae. Aqua-Planta, Sonderheft 3: 1–44 (1992).

1. Dissected leaves absent (all leaves peltate); submerged parts covered with mucilaginous jelly ***Brasenia***

1. Dissected leaves present (peltate leaves terminal, associated with flowers); submerged parts not covered with mucilaginous jelly ***Cabomba***

Brasenia: 1 species

Brasenia schreberi J. F. Gmelin: Fig. 66

Submerged parts heavily covered with mucilaginous jelly. Stem rhizomatous with floating or creeping, leaf-bearing runners. Leaves alternate; petioles up to 1 m long; blades peltate, broadly elliptical to suborbicular, (2–) 4–10 (–14) mm across. Petals linear, 1–2 cm long, purple, reflexed at anthesis, becoming erect in fruit. Stamens 18–36; filaments 8–12 mm long. Carpels 4–18, oblong-ovate. Seeds 2 (–4), ellipsoid, 2–5 mm long, dark brown.

Perennial. Found in still or slowly flowing water in a wide variety of aquatic habitats. The juvenile plants are submerged; the adults have floating leaves. The flowers are strongly protoanderous and pollinated by wind. The disseminules are indehiscent follicles; dispersed by water and perhaps also by animals. Specialized buds (hibernacules or turions) are sometimes produced. Young leaves are eaten in Japan; it is a valuable food plant for waterfowl.

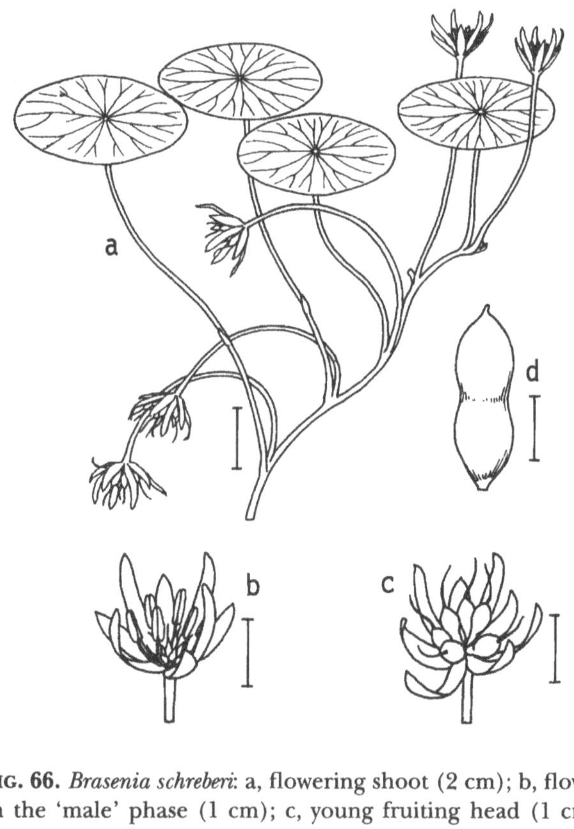

FIG. 66. *Brasenia schreberi*: a, flowering shoot (2 cm); b, flower in the 'male' phase (1 cm); c, young fruiting head (1 cm); d, fruit (5 mm).

{Eph}. Warm to subtropical N and C America, Africa, E Asia and Australia: Bhu, Ind: Mg. Mostly above 1400 m.

Cabomba: 5 species: New World, introduced in the Old World

Stems rhizomatous or erect and leafy. Leaves with divided or entire blades; submerged leaves whorled or opposite, palmately dissected into many capillary segments; floating leaves alternate, associated with flowering but sometimes absent, peltate with elliptical to broadly ovate or sagittate blades. Flowers bisexual, radially symmetrical, solitary on long axillary pedicels, all parts hypogynous and free. Sepals 3, petaloid. Petals 3, each with paired nectar-bearing auricles at the base, white, yellow or purple. Stamens 3 or 6; pollen anasulcate. Carpels 1–4, elongate, with (1–) 3 (–5) seeds. Follicles indehiscent, the wall disintegrating at maturity.

Perennial or rarely annual. Submerged, usually developing some floating leaves when flowering. Pollinated by insects, some species self-incompatible and rarely setting seed outside the native range. The disseminules are indehiscent follicles or vegetative fragments dispersed by water or mud and, perhaps, also by waterfowl. They are popular aquarium plants. Ørgaard, Marian The genus *Cabomba* — a taxonomic study. Nordic Journal of Botany 11(2): 179–203 (1991).

1. Sepals and petals bright yellow; blades of entire leaves broadly elliptical, 3–35 mm wide; blades of divided leaves with segments spreading in three dimensions
C. aquatica

1. Sepals and petals milk white; blades of entire leaves narrowly elliptical, 1–3 mm wide; blades of divided leaves with segments spreading in one plane
C. caroliniana var. ***caroliniana***

Cabomba aquatica Aublet: ({Vit/Eph}. Mexico to Brazil) is given in Gopal (1990)

I have not been able to confirm this record, perhaps it is a second incorrectly determined record of *C. caroliniana* for India.

Cabomba caroliniana A. Gray var. ***caroliniana*** [*C. australis* Spegazzini, *C. aquatica* auct. Ind., non Aublet]: Fig. 67

Stems up to 2 m long. Divided leaves submerged, in opposite pairs or very rarely in whorls of 3, reddish-brown, grass green to olive green; leaf segments lying in one plane. Laminate leaves floating, with narrowly elliptical or sagittate, 5–20 mm long, 1–3 mm wide blades. Sepals and petals 4–12 mm long, milk white except for yellow claws and nectaries of petals. Stamens (3–) 6. Carpels (2) 3 (4). Seeds ovate to ellipsoid-oblong, 1.5–3 mm long, 1–1.5 mm wide.

{Vit/Eph}. Temperate and subtropical America, naturalized in India, Malaysia, New Guinea and Japan: <Ind: <Kl 1977.

CALLITRICHACEAE: 1 genus

Callitriche: 17 species: almost cosmopolitan most in cooler regions

Callitriche stagnalis Scopoli [*C. wightiana* Wallich ex Wight et Arnott]: Fig. 67

Stems elongate when submerged (up to 1 m long), contracted when emersed, branched. Leaves without stipules, opposite (often rosette-like at the tips of floating stems); submerged leaves without petioles, linear to linear-lanceolate or linear-spathulate, (8–) 18–30 (–40) mm long, 1–5 (–7) mm wide, entire, often with a forked or spanner-like apex; floating leaves forming a rosette of 6–8 leaves, spathulate, 6–29 mm long including the short petiole, (2.5–) 4–7.5 (–9) mm wide; blade oval to almost orbicular. Flowers minute, unisexual, solitary or rarely 1 female and 1 male in the same leaf axil. Sepals and petals absent. Male flower of 1, naked stamen, 0.5–2 mm long; filament slender,

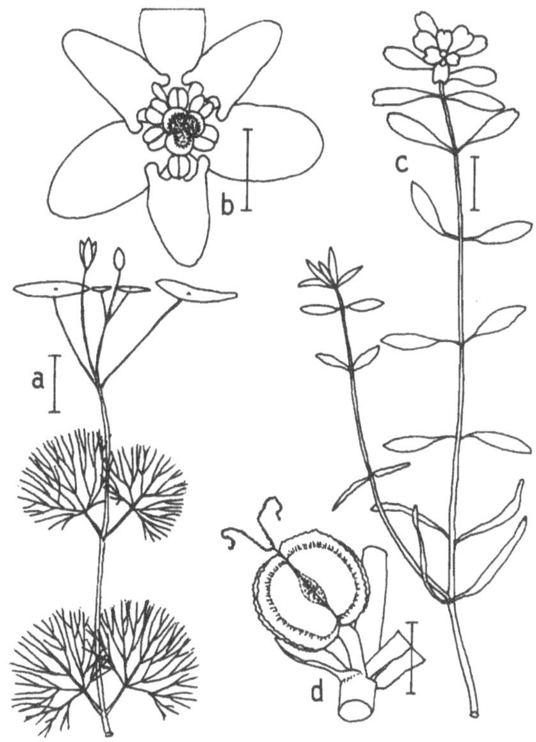

FIG. 67. *Cabomba caroliniana*: a, flowering shoot with entire and divided leaves (1 cm); b, flower (2 mm). *Callitriche stagnalis*: c, shoot (1 cm); d, fruit (1 mm).

elongate; anther reniform, with longitudinal dehiscence slits becoming continuous over the apex. Female flower of 1, naked, 4-locular ovary; ovules 1 in each locule, pendulous; styles 2, free, elongate, often papillose. Fruit 4-lobed, each lobe winged or keeled, splitting into 4 flattened mericarps, from the side almost orbicular, 1.6–1.8 mm long and wide; mericarps with sharp angles and an at least 0.1 wide wing.

Annual or perennial. Totally submerged, partly floating or temporarily terrestrial. Found in still and flowing, nutrient-rich water. The disseminules are flattened mericarps, dispersed by water or by surface adhesion to animals.

{(⊙) Vit/Ros/Eph/(Ten)}. North Hemisphere of the Old World, in the Himalayas and the mountains of SW India and Sri Lanka where it is perhaps introduced: <?Cey, Ind, Nep: Jk, <Tn. 200–3000 m

CAMPANULACEAE [including LOBELIACEAE, excluding SPHENOCLEACEAE]: 87 genera

Herbs. Leaves alternate. Flowers 5-merous. Sepals united to the ovary below. Petals united below, bell-like or 2-lipped. Stamens free from the petals. Ovary inferior or partly so, 2- or 3-locular; ovules numerous, with axile placenta.

Excluded: ***Laurentia longiflora*** (Linnaeus) Endlicher: native in America, established in Sri Lanka were it is found in damp places but not characteristically in wetlands; ***Wahlenbergia erecta*** (Roemer et Schultes) Möller et Tuyn; ***W. marginata*** (Thunberg) A. P. de Candolle [*W. gracilis* A. P. de Candolle]: both sometimes found in wet places as or after they dry out but not in wetlands.

Lobelia [*Pratia* pro parte]: 365 species: cosmopolitan

Monoecious or rarely dioecious. Stems corm-like or elongate, erect, creeping or floating, often somewhat fleshy. Leaves sessile or petiolate, cauline or in basal rosettes, the blades linear to ovate or orbicular. Flowers pedicellate or sessile, solitary or in terminal racemes, sometimes cleistogamous when submerged, sometimes unisexual. Sepals tubular below; lobes 5 subequal. Petal tubes longer than the sepals, 2-lipped, split adaxially to the base; abaxial lobes 3, spreading or reflexed; the adaxial lobes 2, erect. Stamens 5; filaments not inserted on the petal tube, free below but united above; anthers united into a tube around the style. Ovary ± inferior, 2-locular. Fruit an obconical to ovate capsule opening by 2 apical valves or a berry. Seeds small and numerous.

Annual or perennial. Pollinated by insects or autogamous. The disseminules are seeds or berries, some species dispersed by animals, others unknown. Moeliono, B. and Tuyn, P. Campanulaceae in Flora Malesiana, series 1, 6: 107–141 (1960).

1. Only the 2 anterior anthers with an apical tuft of hairs. Seeds ellipsoid; sepal lobes minutely toothed ***L. heyneana***

1. All anthers with an apical tuft of hairs. Seeds 3-sided; sepal lobes entire

  2. Stems 3-sided and 3-winged; free basal part of the 2 anterior filaments twice as broad as the other filaments; stems, leaf blades, pedicels and sepal tube glabrous or nearly so ***L. alsinoides***

  2. Stems terete or indistinctly 3-sided, not winged, at most with ridges below the leaves; free basal part of all filaments equally wide; stems, leaf blades, pedicels and sepal tube distinctly hairy ***L. zeylanica***

Lobelia alsinoides Lamarck [*L. trigona* Roxburgh, *L. griffithii* J. D. Hooker et Thomson, *L. stipularis* Roth, *L. dopatrioides* S. Kurz, *L. microcarpa* C. B. Clarke, *L. terminalis* C. B. Clarke]: Fig. 68

Stems prostrate, ascending or erect, 3–20 (–40) cm long, 3-sided and 3-winged, often rooting at lower nodes. Petioles 1–9 mm long, mostly very short; upper

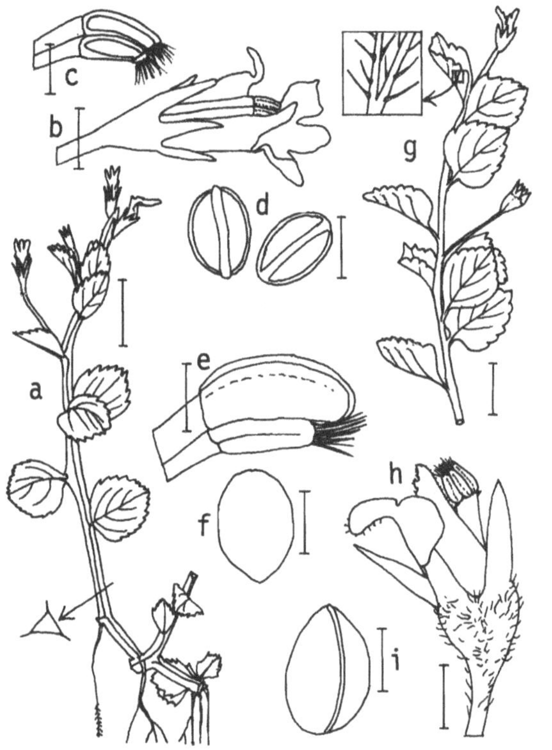

FIG. 68. *Lobelia alsinoide*: a, flowering shoot (1 cm) with transverse section of stem (inset); b, flower (2 mm); c, anthers (1 mm); d, two seeds (0.25 mm). *L. heyneana*: e, anthers (0.25 mm); f, seed (0.25 mm). *L. zeylanica*: g, flowering shoot (1 cm) with underside of leaf blade (inset); h, flower (2 mm); i, seed (0.25 mm).

stem leaves usually sessile or nearly so; leaf blades broadly ovate, 0.5–2 cm long, 0.5–1.8 cm wide, glabrous on both surfaces, base truncate to subcordate, margins toothed, tip subacute to obtuse. Flowers solitary in the axils of leaf-like bracts or on old shoots sometimes raceme-like. Bracts subulate up to 3 mm long. Pedicels 1–3.5 cm long. Sepal tubes urn-shaped, up to 3 mm long; sepal lobes triangular, subulate, 2–6 mm long, entire. Petal tubes 4–4.5 mm long, bluish-white to violet; adaxial (upper) lip erect with ovate-elliptical lobes; abaxial (lower) lip spreading, 6–7 mm across, bright blue with 2 white streaks at the base of the middle lobe. Stamens 5; free basal part of the 2 anterior filaments twice as broad as the other filaments; all anthers with an apical tuft of hairs. Capsules obovoid to hemispherical, 2–3 mm long, 1.5–3 mm wide. Seeds ovoid, 0.4–0.75 mm long, 3-angled, with thickened margins.

Annual. Often very abundant in wet rice fields, roadside ditches and in other wet or inundated places.

{⊙ Hel}. S and SE Asia: Bur, Cey, Ind, Nep: Ap, As, Gj, Go, Kl, Kt, Mg, Mh, Mp, Tn, Wb. 0–2400 m.

Lobelia heyneana Roemer & Schultes [*L. trialata* F. Hamilton ex D. Don, *L. subincisa* Wallich ex A. P. de Candolle, *L. dichotoma* Miquel]: Fig. 68

Like *L. alsinoides* but stems erect, 5–50 (–60) cm tall, not rooting at nodes; upper leaves shorter and narrower, ovate, 1.5–2 (–4) cm long, 1.3–1.8 (–3) cm wide, mostly truncate or attenuate at the base; pedicels 5–20 mm long; sepal tubes bell-shaped, ± 2 mm long; sepal lobes lanceolate-elliptical, 2–4 mm long, sparsely toothed; petal tubes ± 3 mm long, pale blue to pinkish; abaxial (lower) lip blue with a purplish streak down the base of each lobe; filaments equally wide; only the anterior pair of anthers with apical tufts of hairs; capsules 3–8 mm long, 2–3 mm wide; seeds ellipsoid, biconvex, never 3-angled, smooth and obtuse at each end.

Annual. In moist and swampy places, along streams, in marshes and rice fields; it tolerates considerable shade.

{⊙ Hel}. E Africa, S and E Asia: Bhu, Bur, Cey, Ind, Nep: Jk, Kl, Kt, Mg, Mp, Mp, Sk, Tn.

Lobelia zeylanica Linnaeus [*L. succulenta* Blume, *L. affinis* Wallich ex G. Don]: Fig. 68

Like *L. alsinoides* but lower part of stem creeping and rooting at nodes, terete to indistinctly 3-sided, without wings but sometimes ridged just below the leaves; petioles 1–20 mm long; leaf blades ovate to suborbicular, (0.7–) 1–6 cm long, (0.5–) 1–3.5 cm wide, with blunt to subcordate base, sparsely hairy or glabrous on the adaxial (upper) surface, usually distinctly hairy on the abaxial (lower) surface, particularly along the nerves; pedicels usually hairy, 1–3.6 cm long; sepal tubes usually hairy, 2–4 mm long; sepal lobes usually hairy, 3–5 mm long, entire; petal tubes pale blue or lilac; lobes of adaxial lip lanceolate, 2 mm long, each with bristly hairs beneath; abaxial lip with bristly hairs beneath; free basal part of all filaments equally wide; capsule obconical to oblong-ovate, 3–7 mm long, 2–4 mm wide. Seeds 0.4–0.6 mm long.

Annual. In wet and waterlogged places, often in shade.

{⊙ Hel}. S and SE Asia: Ban, Cey, Ind: As, Kl, Kt, Mg, Sk, Tn, Wb.

CANNACEAE: 1 genus

Canna: 25–50 species and numerous hybrids: tropical and subtropical America

The taxonomy of *Canna* is inadequately known. *Canna* × *generalis*, *C.* × *orchidioides* (both usually called *C. indica*) and *C. edulis* are widely cultivated in India, the former two for decoration the latter for the edible rhizomes. They sometimes escape cultivation. They are not wetland plants but may occur in wetlands. They are, however, very variable and hybridize with each other and with other species. The plants found in Indian wetlands are, therefore, difficult to determine and I am by no means sure that the name given below is correct.

Canna ?speciosa Roscoe: Fig. 69

Rhizomatous herbs. Stems erect, leafy, up to 2 m tall. Leaves alternate, sheathing at the base, usually without a distinct petiole; blades ovate-elliptical to oblong, up to 50 cm long, up to 15 cm wide, pinnately nerved from a prominent midrib, with parallel nerves. Inflorescence a terminal panicle, up to 20 cm or more long. Flowers bisexual, bilaterally symmetrical. Sepals 3, free, overlapping, pale red, up to 1.5 cm long. Petals 3, sepal-like, united at the base and adnate to the staminal column. Stamens forming the showy part of the flower, up to 6 cm long, bright red; the 3 outer ones, sterile, lanceolate and petaloid, the largest reflexed and rolled back on itself; 2 of the inner ones united and sterile, 1 inner one fertile with a solitary 1-locular anther on its side. Ovary inferior, 3-locular; ovules numerous. Fruit a capsule, 2–4 cm long, 1–2 cm in diameter, with warty pericarp. Seeds globose, with very hard endosperm.

Perennial. It is characteristically found in bogs, marshes, in inundated places that remain perennially wet; sometimes growing in shade. It is pollinated by insects. The disseminules are relatively large seeds, dispersal unknown; the cultivated varieties are usually propagated by rhizome cuttings.

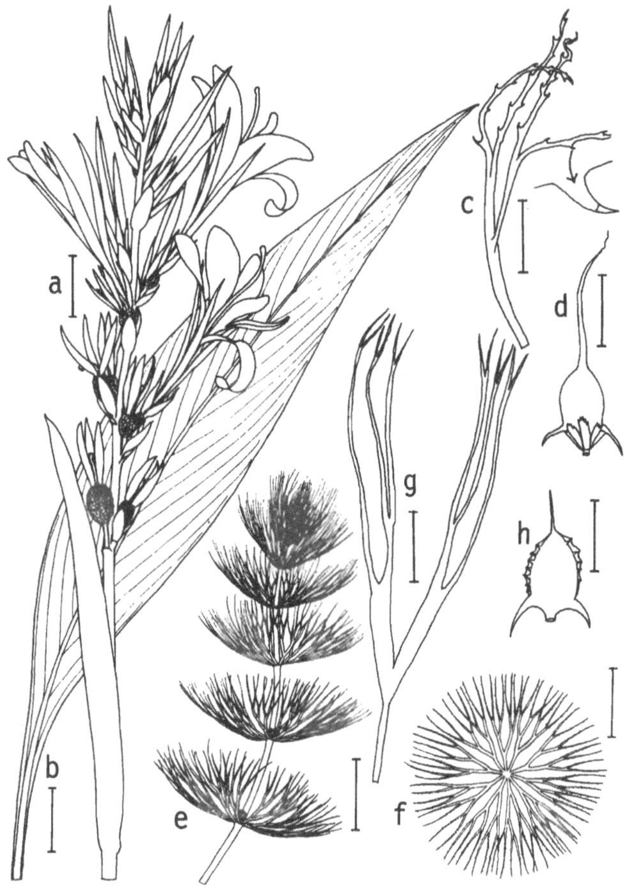

Fig. 69. *Canna* sp.: a, inflorescence (2 cm); b, leaf blade (4 cm). *Ceratophyllum demersum*: c, one leaf (4 mm) with spine (inset); d, fruit (2 mm). *C. muricatum*: e, shoot (2 cm); f, whorl of leaves (2 cm); g, one leaf (4 mm); h, fruit (2 mm).

{Hel}. Tropics and subtropics of America, naturalized in India and SE Asia: <Ind: <As, <Hp, <Mg, <Rj, <Wb.

CAPPARIDACEAE excluded see p. 11.

CARYOPHYLLACEAE excluded see p. 11.

CERATOPHYLLACEAE: 1 genus

Ceratophyllum: ± 4 polymorphic species: almost cosmopolitan

Roots absent (even in the embryo) but branches sometimes modified as 'rhizoids'. Stems branched but not more than one branch at a node. Leaves whorled, (3–) 6–8 (–12) at a node, filiform, 1–4 times forked, with 2 rows of teeth along the ultimate segments, each segment tipped by 2 bristles. Flowers unisexual, solitary in the axil of 1 leaf in a whorl. Perianth of 8–13 linear segments, united at the base. Stamens 5–30; filaments short or absent. Ovary superior, 1-locular. Fruit a 1-seeded nut, tipped by the persistent style and often with additional basal and marginal spines, sometimes also warty.

Perennial or sometimes annual, totally submerged, rootless, free-swimming. Pollinated under water (hypohydrophilous), the stamens are liberated and dehisce floating just below the water surface. The disseminules are relatively large and spiny nuts or specialized buds, dispersal by water and probably also by animals. Large populations may be troublesome, blocking or constricting water channels. Wilmot-Dear, M. *Ceratophyllum* revised, a study in leaf and fruit variation. Kew Bulletin 40: 243–271 (1985); Les, D. H. The evolution of achene morphology in *Ceratophyllum*. Systematic Botany 11: 549–558 (1986); 13: 73–86, 509–518 (1988); 14: 254–262 (1989).

Excluded: ***Ceratophyllum submersum*** Linnaeus is confined to Europe with a few localities in Africa but it is often recorded in India. *Ceratophyllum submersum* sensu stricto resembles *C. muricatum* vegetatively but it has a nut with a very short apical spine which is otherwise almost smooth.

1. The majority of the leaves 1 or 2 times forked (lower part of main axis sometimes 3 (or 4) times forked), usually robust and rigid, marginal teeth numerous, prominent and set on a broad base of tissue; nuts smooth or minutely tuberculate except for 1 apical and usually 2 basal spines; male flowers mostly with more than 10 stamens ***C. demersum***

1. The majority of the leaves 3 or 4 times forked, usually delicate and flaccid, marginal teeth few, inconspicuous and not set on a broad base of tissue; nuts with few to many small spines in addition to 1 apical and usually 2 basal spines; male flowers mostly with less than 10 stamens ***C. muricatum***

Ceratophyllum demersum Linnaeus: Fig. 69

Shoot tips often with shortened internodes giving it a 'bottle brush' appearance. Leaves usually bright or olive green, robust, rigid, brittle, feeling hard in the hand, the majority 1 or 2 times forked (lower part of main axis sometimes 3 or rarely 4, times forked); teeth along the margins numerous, conspicuous, arising from a pad of tissue, present on lower and upper parts of the leaves. Male flowers mostly with considerably more than 10 stamens; connective prolonged at the tip into 2 spines and a central projection. Nut body 3–4 (–6) mm long, 2–4 mm wide, the surface smooth or slightly tuberculate; nut margins winged or wingless but without spines; apical spine straight, 2–7 mm long; basal spines 2, up to 8 mm long or reduced to minute tubercles.

Often gregarious in eutropic water and sometimes troublsome.

{(⊙) Pla-Vit}. Almost cosmopolitan: Ban, Cey, Ind, Pak: An, As, Bh, Dl, Gj, Go, Hp, Jk, Kl, Kt, Mh, Mp, Or, Pj, Rj, Tn, Wb.

Ceratophyllum muricatum Chamisso subsp. ***muricatum*** [*C. submersum* Linnaeus subsp. *muricatum* (Chamisso) Wilmot-Dear]: Fig. 69

Shoot tips usually diffuse, stem visible between the leaf whorls. Leaves usually light green to yellowish-green, fine, flaccid, feeling soft in the hand; teeth along the margins inconspicuous, arising direct from the edge not from a pad of tissue, sometimes absent from lower part of the leaf. Male flowers mostly with less than 10 stamens; connective truncate, without spines and a central projection. Nut body 2–3 (–4) mm long, 1–2 (–2.5) mm wide, the surface rarely smooth, usually tuberculate or with small spines; nut margins winged; the wing entire, irregularly toothed or spiny, ranging from membranous to hard and thick; apical spine very variable, straight or curved, up to 8 mm long or rudimentary; basal spines 2, up to 8 mm long or rudimentary.

{(⊙) Pla-Vit}. Tropical and warm regions: NE Africa, India and SE Asia: Ban? Ind: Kl, Mp, Or, Rj, Tn.

CHENOPODIACEAE excluded see p. 11

COMPOSITAE see ASTERACEAE

COMMELINACEAE: 50 genera

Herbs. Roots fibrous or tuberous. Stems erect to creeping, often rooting at nodes; nodes usually swollen. Leaves alternate or radical, the base forming a closed sheath; blades sessile or shortly petiolate, with parallel veins. Inflorescence a panicle or cyme, terminal or axillary or both, bracteolate. Flowers bisexual, 3-merous, regular or bilaterally symmetrical, subtended by bracts; bracts scale-like, spathe-like or tubular. Sepals (2) 3, green, membranous. Petals (2) 3, free or united at base, ephemeral. Fertile stamens 2, 3 or 6; staminodes 1–3, often elaborate; filaments adherent to petals or not, glabrous or with hairs; anthers basifixed or versatile, often with expanded connective. Ovary superior, (2-) or 3-locular; ovules 1 to many, axile; style 1; stigma 1, capitate or not. Fruit a loculicidal capsule with persistent valves or rarely indehiscent. Seeds few to many.

Mostly perennial, some annual. Many species grow in damp and shady places but relatively few are wetland plants. Pollination by insects, usually offering pollen as a reward but some are autogamous or cleistogamous. The disseminules are seeds, fruits or vegetative fragments. Some used locally in medicine. Ms Ursula M. Kradolfer helped with the preparation of this account.

1. Flower buds and fruits borne within and hidden by inflated leaf sheaths ***Cyanotis***
1. Flower buds and fruits exposed or borne within folded spathe-like bracts
  2. Flower buds and fruits borne within folded spathe-like bracts; one petal very much smaller than the other two ***Commelina***
  2. Flower buds and fruits exposed, the bracts scale-like and minute or absent; all three petals nearly alike
    3. Peduncles, pedicels and sepals densely clothed in brown hairs; fertile stamens 6 ***Floscopa***
    3. Peduncles, pedicels and sepals glabrous or nearly so; fertile stamens 2–3; staminodes 0–4, apex usually 3-lobed ***Murdannia***

Commelina: ± 170 species; pantropics and subtropics

Stems erect or prostrate, often subsucculent, often rooting at the nodes. Leaf blades sessile or with a short 'petiole' between sheath and blade. Inflorescence an axillary or terminal cyme, enclosed in a folded spathe-like bract. Sepals 3, free or united at the base, the adaxial (upper) one often smaller. Petals 3, the abaxial (lower) one often reduced to a scale, the adaxial pair long-clawed, usually blue. Fertile stamens 3; staminodes (0–) 2 or 3, with showy pseudoanthers; filaments naked. Ovary 3-locular, the upper median one smaller, 1-ovuled or abortive and empty or obsolete, the lower 2 locules with 1 or 2 ovules each. Fruit a capsule with 3, unequal valves, hidden within the bract.

Annual or perennial. Most species are found in damp shady places, only four seem to be regularly found in wetlands. The flowers are very delicate; they open early in the morning and wither in the afternoon.

Excluded: found in damp or moist places but not wetland species: ***Commelina albescens*** Hasskarl; ***C. attenuata*** J. G. König; ***C. erecta*** Linnaeus [*C. undulata* R. Brown, *C. kurzii* C. B. Clarke]; ***C. clavata*** C. B. Clarke; ***C. longifolia*** Lamarck [*C. salicifolia* Roxburgh]; ***C. paludosa*** Blume [*C. obliqua* F. Hamilton ex D. Don].

1. Spathe-like bracts sessile or nearly so
  2. Spathe-like bracts erect, funnel-shaped, with the margins united below; leaf blades ovate to broadly lanceolate; petals blue ***C. benghalensis***
  2. Spathe-like bracts deflexed, sickle-shaped, folded with the margins free; leaf blades linear to linear-lanceolate; petals yellow ***C. subulata***
1. Spathe-like bracts with peduncles at least 1 cm long
  3. Seeds ± cylindrical or subquadrate, ± 2 mm long, reticulately pitted on the side opposite the hilum ***C. diffusa***
  3. Seeds globose to ovoid, at least 3 mm long, smooth on both sides ***C. hasskarlii***

Commelina benghalensis Linnaeus: Fig. 70

Stems erect at first, later branching and becoming decumbent, 20–40 (–90) cm long, rooting at the nodes, with secondary branches ascending, 10–30 cm tall, some creeping stems grow underground and develop white, cleistogamous flowers. Leaf sheaths 0.5–1.5 cm long, hairy, with rust-coloured cilia on the margin; blades ovate to ovate-elliptical, 2–7 cm long, 1–2.5 cm wide, often asymmetrical at base, the base suddenly contracted or rounded and subpetiolate, the tip ± acute. Inflorescence axillary. Bracts erect, sessile or nearly so, 1–3 clustered together, funnel-shaped, 1–2 cm long. Sepals oblong, ± 3 mm long. Adaxial pair of petals of aerial flowers 3–7 mm long, long-clawed at base, orbicular above, blue. Fertile stamens 3; staminodes 2–3 with a butterfly-like pseudoanther. Capsules ellipsoid, ± 5 mm long, ± 3 mm wide, functionally 2-locular, 5-seeded. Seeds somewhat kidney-shaped but flattened at one end, 1.6–2.3 (–3) mm long, rough, with shallow furrows between hilum and operculum and a superposed reticulate patten.

Perennial or sometimes annual. A widespread weed not confined to wetlands but often found in ditches, wet fields and places subject to flooding. The cleistogamous flowers set good seed, the aerial chasmogamous flowers are often without seed. Used locally in medicine as a refrigerant for skin inflammations and as a cure for leprosy and constipation; also used as fodder and a pot herb.

{⊙/(⊙) Hel}. Old World tropics: Bhu, Ind, Nep, Pak: An, Ap, As, Bh, Dl, Gj, Go, Kl, Kt, Mh, Rj, Sk, Tn, Tp, Up, Wb. 0–1800 m.

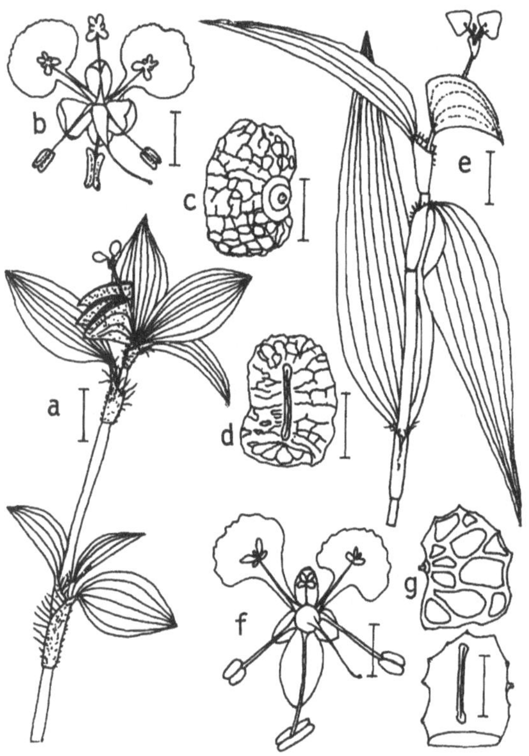

Fig. 70. *Commelina benghalensis*: a, flowering shoot (1 cm); b, flower (2 mm); c, seed, operculum view (1 mm); d, seed, hilum view to left (1 mm). *C. diffusa*: e, flowering shoot (1 cm); f, flower (2 mm); g, seed, hilum view below (1 mm).

Commelina diffusa N. L. Burman [*C. nudiflora* auct. Ind., non Linnaeus, *C. longicaulis* Jacquin, *C. aquatica* Morton]: Fig. 70

Stems erect at first, later branching and becoming decumbent, up to 1 m long, rooting at the lower nodes, with secondary branches ascending, forming diffuse bushes up to 1.2 m high. Leaf sheaths (1–) 2–3 cm long, glabrous or nearly so but with white hairs on the margin; blades linear-lanceolate to ovate, (1.5–) 3–7 (–10) cm long, 0.5–3 cm wide, the base tapering and sometimes subpetiolate, the tip ± acute. Inflorescence subterminal. Bracts erect, folded in the middle with free margins, ovate-lanceolate, 1.2–3.5 cm long, usually solitary, born on (1–) 2–4 (–5) cm long peduncles. Underground flowers not developing. Adaxial pair of petals up to 1 cm, long-clawed at base, reniform above, blue to violet or lilac. Fertile stamens 3; staminodes 2–3 with club-like or butterfly-like pseudoanthers. Capsules 4–5 (–6) mm long, ± 3 mm wide, unequally 3-locular, the basal loculus dehiscent, 3- to 5-seeded. Seeds ± cylindrical or subquadrate, 2–2.5 (–3) mm long, reticulately pitted on the side opposite the hilum, smooth on the hilum side.

Perennial or annual. A widespread and common weed usually found in damp shady places near water, but also often in open swamps and marshes and sometimes found floating in mats. It is very variable in form: from creeping to erect, with stems a few centimetres long to well over 1 m tall. Some subspecies and varieties have been described. The bruised plant is used locally against burns, itches and boils.

{⊙/(⊙) Hel}. Pantropics and subtropics: Ban, Bhu, Cey, Ind, Nep: An, Bh, Gj, Go, Kl, Kt, Mz, Pj, Sk, Tn, Wb. 0–1400 m.

Commelina hasskarlii C. B. Clarke [*C. nudiflora* auct. Ind., non Linnaeus]: Fig. 71

Like *C. diffusa* but leaf sheaths not more than 1.5 cm long, glabrous, with margin sparsely ciliate; leaf blades narrowly lanceolate to lanceolate, 2.5–7.5 cm long, 1.2–1.6 cm wide, tip subacute; inflorescences mostly axillary, pubescent; spathe-like bracts 1.2–2.5 cm long; petals blue (Gamble, 1931: petals yellow drying blue); seeds globose to ovoid, 3 mm or more long, smooth on both sides.

Probably annual and perennial. In moist places, swamps and in rice fields. Stems may creep on the surface of water with short erect shoots.

{(⊙) Hel}. Probably endemic to the Indian subcontinent: ♦?Ban, ♦?Ind, ♦?Nep: An, Dl, Gj, Go, Kt, Mh, Rj, Tn, Up. 0–1400 m.

Commelina subulata Roth: Fig. 71

Slender herbs. Stems erect, simple or branching from the base, 10–40 (–60) cm tall, rooting at lower nodes. Leaf sheaths 0.5–1.5 cm long, with or without hairs; blades linear to linear-lanceolate, 3–16 cm long, 1–5 (–7) mm wide, almost glabrous, tip acute. Inflorescences axillary. Bracts deflexed, folded in the middle with free margins, ovate-lanceolate, 0.5–1 (–1.5) cm long, usually solitary or 2 or 3 together, sessile or nearly so. Petals usually chrome yellow but sometimes apricot to orange and sometimes drying blue or violet, ± 4 mm long. Capsules 3-locular, 3–4 mm long, 3- to 5-seeded. Seeds 2–2.5 mm long, with 2–5, coarse, irregular furrows radiating out on the side opposite the hilum.

Annual. In open marshy soil and irregularly flooded places, by no means confined to wetlands. Rao S. R. and Kammathy, R. V. Notes on Indian Commelinaceae. Journal of the Bombay Natural History Society 59: 58–70 (1962).

{⊙ Hel}. Tropical Africa and India (central Deccan): Ind: Ap, Kt, Mh.

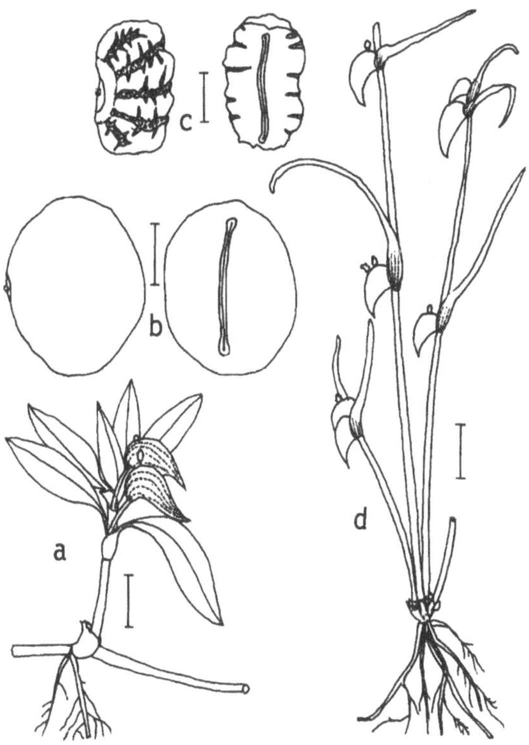

Fig. 71. *Commelina hasskarlii*: a, flowering shoot (1 cm); b, seed, hilum view on the right (1 mm). *C. subulata*: c, seed, hilum view on the right (1 mm); d, flowering shoot (1 cm).

Cyanotis [*Amischophacelus, Tonningia, Zygomenes*]: ± 50 species; pantropics

Flowers of wetland species borne in inflated leaf sheaths, just emerging in the morning at anthesis. Stamens 6, equal, all fertile; filaments densely hairy above the middle. Ovary 3-locular. Capsules 3-valved, with 2 seeds in each loculus.

Annual or perennial. Fruits of wetland species developing within the inflated leaf bases. Disseminules relatively large seeds, dispersion unknown.

Excluded: found in moist places but not in wetlands: ***Cyanotis cristata*** (Linnaeus) D. Don; ***C. fasciculata*** (Heyne ex Roth) J. A. et J. H. Schultes fil.: ♦Bhu, ♦Cey, ♦Ind.

1. Capsule valves notched at apex; capsule before dehiscence with a single apical projection; seeds deeply pitted; fertile leaf sheaths greatly swollen in fruit; filaments densely hairy ***C. axillaris***

1. Capsule valves each with 2, pointed, horn-like projections at the tip; capsule before dehiscence depressed in the centre with 3 horn-like projections; seeds shallowly pitted; fertile leaf sheaths greatly swollen in fruit ***C. cucullata***

Cyanotis axillaris (Linnaeus) Sweet: Fig. 72

Stems creeping, ascending or erect, 15–45 cm long. Leaf sheaths inflated, glabrous or somewhat hairy, margin ciliate; blades sessile, slightly succulent, linear-lanceolate, 4–15 (–20) cm long, 0.3–1.2 cm wide, tip acuminate. Inflorescence a dense cyme with 3–6 flowers, embedded in the hollow axil of a swollen leaf sheath. Flowers exerted only at anthesis. Petals 3, all alike, long-clawed, blue, pink or purple. Fertile stamens 6; filaments densely hairy above the middle. Ovary 3-locular, glabrous; style filiform, glabrous, swollen at the tip. Capsules ± 5 mm long, acute, before dehiscence with a single apical projection; valves each with a notched tip. Seeds subcylindrical, 1.4–2.4 mm long, deeply pitted.

Annual or occasionally perennial. Found in swamps and marshes, often abundant in rice fields and irrigation ditches. Apparently more common in coastal regions. Used locally in medicine for tympanites, ascites and abortions.

{(⊙) Hel}. S and SE Asia and Australia: Ban, Cey, Ind, Nep: Ap, As, Bh, Dl, Gj, Hr, Kl, Kt, Mh, Mp, Rj, Up, Wb.

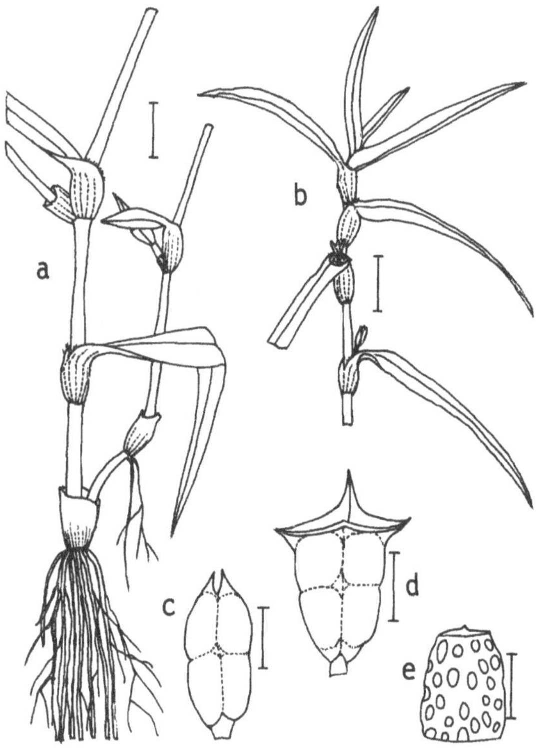

Fig. 72. *Cyanotis axillaris*: a, base of shoot (1 cm); b, flowering shoot (1 cm); c, capsule valve (2 mm). *C. cucullata*: d, unopened capsule (2 mm); e, seed (1 mm).

Cyanotis cucullata (Roth) Kunth: Fig. 72

Like *T. axillaris* but capsule before dehiscence depressed in the centre with 3 horn-like projections; capsule valves with tips entire or very slightly notched; filaments almost glabrous to distinctly hairy but when hairy then less so than in *C. axillaris*; leaf sheaths in fruit less swollen; leaf blades usually oblong-lanceolate, obtuse; petals usually pale lilac; seeds subquadrate, shallowly pitted.

Annual or occasionally perennial. Found in swampy areas, wet fields and along the banks of ponds and ditches. It is also found in irrigated fields. It is rather more common in inland districts. It is often misidentified as *C. axillaris*.

{(⊙) Hel}. Probably endemic to Peninsular India: ♦?Ind: Ap, Kt, Mh, Mp, Rj.

Floscopa: 20 species; pantropics to warm regions

Floscopa scandens Loureiro [*Tradescantia paniculata* Roxburgh, *Aneilema hispida* D. Don, *Floscopa hamiltonii* (Spengel) Hasskarl]: Fig. 73

Stems creeping and rooting below, ascending to erect above, up to 60 cm long, often rooting at nodes of creeping stems. Leaf sheaths loose, 1–2 cm long, often pubescent, margin and mouth fringed with long hairs; blades elliptical-lanceolate, 4–10 cm long, 1–2 cm wide, glabrous or hairy, margin often minutely serrate, gradually narrowed below and sometimes subpetiolate, tip acuminate. Inflorescences terminal and subterminal, hairy panicles. Bracts minute and scale-like or absent. Flowers crowded, shortly pedicelled. Sepals 3, all alike, oblong, 2–4 mm long, free, the outer surface clothed with simple and gland-tipped hairs. Petals 3, subequal, obovate, slightly longer than the sepals, purple to pink or white. Stamens 6, all fertile, subequal; filaments naked, ± as long as the petals. Capsules 2-locular, 2-seeded, suborbicular, up to 2 mm long. Seeds globose to ovoid, somewhat laterally flattened, ± 1.5 mm long, often bluish, with numerous furrows radiating out from the hilum.

Perennial or occasionally annual. Mostly found in muddy soil along streams and in swamps, in shade, in and at the edges of forests, but it also grows in the open in marshes, along ditches and canals and in and around drying pools.

{(⊙) Hel}. S and SE Asia and Australia: Ban, Cey, Ind, Nep: As, Kl, Kt, Mg, Mh, Mp, Mz, Sk, Up. 0–1800 m

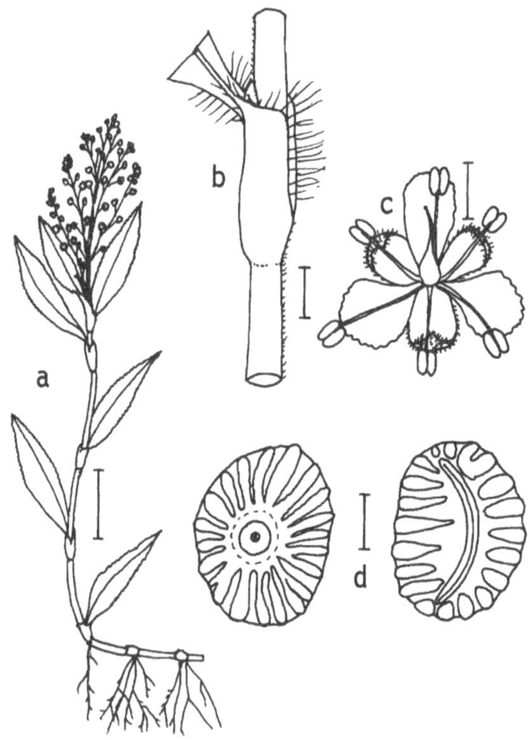

Fig. 73. *Floscopa scandens*: a, flowering shoot (2 cm); b, leaf sheath (3 mm); c, flower (3 mm); d, seed, hilum view on the right (0.5 mm).

Murdannia [*Aneilema* pro parte, *Dichoespermum*, '*Dichaespermum*' orth. mut.]: 50 species; tropical and warm temperate regions, mainly Asiatic

Herbs. Stems floating, creeping or erect. Leaves alternate. Flowers 1–3 in leaf axils and/or in terminal and axillary inflorescences, sometimes cleistogamous. Sepals 3, subequal, green. Petals 3, free, subequal, blue, pink or white. Stamens usually 6, (2) 3 fertile and (1–) 3 (4) staminodal; staminodes alternating with fertile stamens or sometimes reduced to 1 or absent, apex of the staminodes bright yellow, 3-lobed (often like a pawnbroker's sign) or the lateral lobes sometimes so near to the filament as to make the apex hastate. Ovary 3-locular; style simple with a terminal, capitate or 3-lobed stigma. Fruit a 3-valved capsule. Seeds 1–30 or more, in 1 or 2 rows in each locule.

Flowers open before midday and fade in the afternoon.

Excluded: temperate wetland species found in the Himalayas: ***Murdannia keisak*** (Hasskarl) Handel-Mazzetti.

Species found in damp or moist places but not in wetlands: ***M. elata*** (Vahl) Brückner [*Aneilema lineolatum* Kunth]; ***M. esculenta*** (Wallich ex C. B. Clarke) Rolla Rao et Kammathy; ***M. ochracea*** (Dalzell) Brückner [*Dichoespermum repens* Wight; *M. crocea* subsp. *ochracea* (Dalzell) Faden]; ***M. simplex*** (Vahl) Brenan [*Aneilema sinicum* Lindley].

1. Upper flowers subtended by bracts shorter than the sepals; inflorescence terminal or nearly so and appearing almost naked
  2. Bracts funnel-shaped; leaf blades folded, grass-like, not more than 2 mm wide; capsules not more than 3 mm long; seeds ± 0.25 mm long ***M. semiteres***
  2. Bracts not funnel-shaped, scale-like or the lowermost may be leaf-like; leaf blades flat, not grass-like, more than 3 mm wide; capsules 4 mm or more long; seeds 0.5 mm or more long
    3. Blades of lower leaves linear; seeds with furrows and pits, 2 in each locule, one borne above the other ***M. nudiflora***
    3. Blades of lower leaves ovate to oblong; seeds almost smooth, 4–6 in each locule, borne in 2 rows ***M. spirata***
1. Upper flowers subtended by bracts longer than the sepals; solitary flowers or inflorescences mostly axillary and appearing leafy
  4. Seeds 1 in each capsule locule; bracts sheath-like, differing from normal foliage leaves ***M. vaginata***
  4. Seeds more than 4 in each capsule locule; bracts resembling normal foliage leaves
    5. Flowers and fruits remaining in axils of the leaf-like bracts; petals pink or violet or white; leaf blades linear oblong; seeds 7–10 in each capsule locule ***M. blumei***
    5. Flowers and fruits borne remote from the axils of the leaf-like bracts; petals yellow; leaf blades ovate; seeds 4–6 in each capsule locule ***M. pauciflora***

Murdannia blumei (Hasskarl) Brenan [*Aneilema hamiltonianum* Wallich ex C. B. Clarke]: Fig. 74

Stems erect or creeping with ascending or erect branches, rooting at nodes of creeping stems, 15–30 cm long. Leaf sheaths glabrous, with long hairs along the margin; blades linear-oblong, 1–5 cm long, 2–7 mm wide. Flowers in axils of consecutive leaves, 2 or 3 in clusters, shortly pedicelled; pedicels 4–6 mm long. Sepals oblong, 3–4 mm long, obtuse. Petals elliptical, 4–5 mm long, acute to shortly acuminate, pink to violet or white. Filaments naked or slightly hairy. Capsules linear-oblong, acute, 6–8 mm long, ±2 mm wide. Seeds in 2 rows of 7–10 in each locule, rough.

Annual or sometimes perennial. In marshy areas, rice fields, ditches and other wet places. Stems sometimes floating. The staminodes are reported in Flora Bangladesh 4: 29 (1977) to be absent.

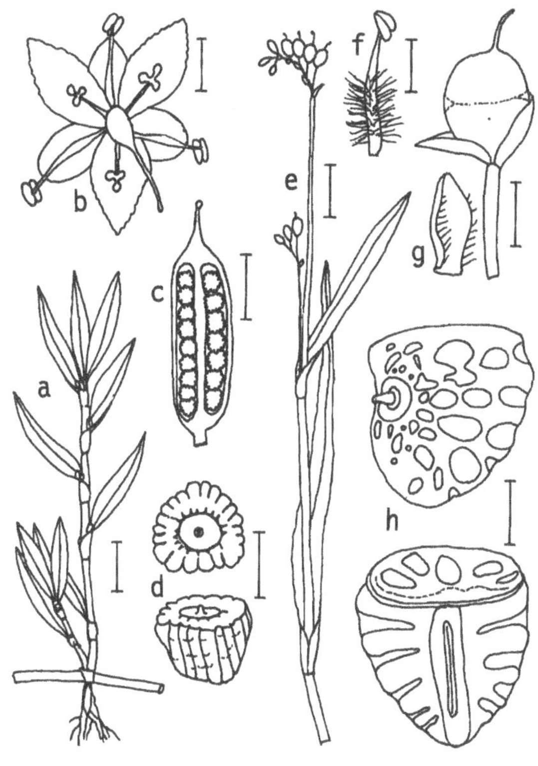

FIG. 74. *Murdannia blumei*: a, flowering shoot (2 cm); b, flower (2 mm); c, capsule valve with seeds in place (2 mm); d, seed, operculum view above (0.3 mm). *M. nudiflora*: e, flowering shoot (1 cm); f, stamen (1 mm); g, fruit, with bract detached (2 mm); h, seed, hilum view below (0.5 mm).

{Hel}. Himalayas, India, Burma and Malaysia: Ban, Bur, Ind, Nep: As, Bh? Mg, Mp?

Murdannia nudiflora (Linnaeus) Brenan [*M. malabarica* (Linnaeus) Brückner, *M. loriformis* (Hasskarl) Rolla Rao et Kammathy, *Aneilema radicans* D. Don, *A. nudiflorum* (Linnaeus) R. Brown var. *terminale* (Wight) C. B. Clarke]: Fig. 74

Stems erect and branching from below or decumbent below and ascending above; branchlets usually reddish with white nodes. Leaf sheaths 5–10 mm long, glabrous or nearly so but with long hairs along the margin; blades linear near the stem base, becoming lanceolate above, (3–) 5–8 (–15) cm long, 0.3–1.5 cm wide, slightly hairy at base. Flowers clustered in terminal and axillary cymes; pedicels up to ± 5 mm long, becoming shorter towards the tip of the cyme; bracts scale-like, up to ± 3 mm long, becoming shorter above, caducous. Petals blue or pinkish. Filaments with long blue hairs. Capsules subglobose, 3-sided with rather flat sides, ± 4 mm long. Seeds 2 in 1 row in each capsule locule, rounded at one end, flat at the other, 1.5–2 mm long, furrows arise along the hilum which become pits towards the operculum.

Annual or perhaps also perennial. In wet places, paddy fields, marshes, along ditches and sometimes in or floating on water. It is not confined to wetlands and is also found in moist shady and grassy places and as a weed in gardens. It has become an aggressive weed in parts of the West Indies. Used locally in the treatment of asthma, leprosy, piles and other diseases. *Murdannia loriformis* is considered by some botanists to be a distinct species; it is more robust with leaves in a basal rosette.

{(⊙) Hel(Eph)}. Old World tropics, naturalized in the West Indies: Ban, Bhu, Cey, Ind, Nep, Pak: An, Ap, As, Bh, Gj, Go, Kl, Kt, Mg, Mh, Mp, Mz, Pj, Rj, Sk, Tn, Tp, Up, Wb. 0–1500 m.

Murdannia pauciflora (Linnaeus) Brückner [*M. wightii* Rolla Rao et Kammathy]: Fig. 75

Stems creeping and ascending, much branched, rooting at lower nodes, up to 15 cm or more long. Leaf sheaths ± 0.5 cm, glabrous except for a line of hairs along the margin; blades ovate, 2–5 cm long, 0.5–2 cm wide, glabrous or sparsely hairy, usually cordate at base, acute or subacute at tip. Flowers 1–3 (–5), in axillary cymes; pedicels usually recurved in fruit. Petals yellow (Gamble, brownish-yellow drying blue). Filaments glabrous. Capsules linear-oblong, 6–8 mm long, somewhat 3-sided. Seeds in 2 rows of 4–6 in each locule, almost cubical, nearly smooth, black.

Perennial or annual. Found on wet muddy soil in rice fields and marshes and sometimes floating on water. It is not confined to wetlands.

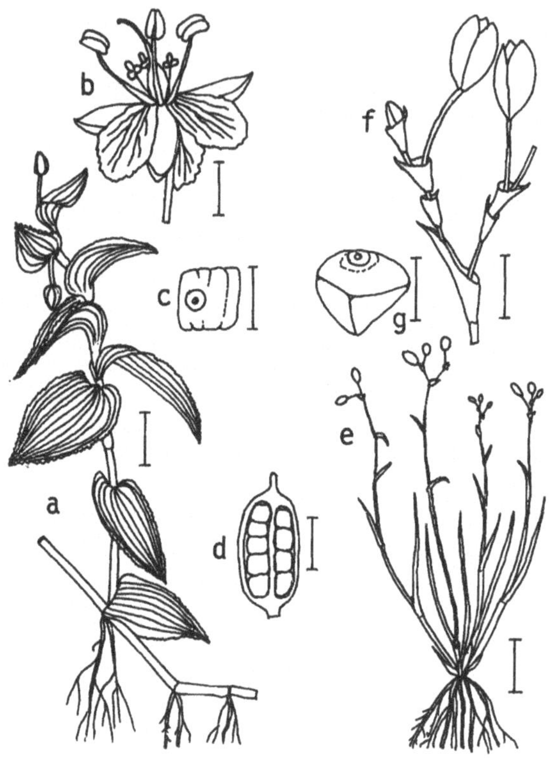

Fig. 75. *Murdannia pauciflora*: a, flowering shoot (1 cm); b, flower (2 mm); c, seed (1 mm); d, capsule valve with seeds in place (3 mm). *M. semiteres*: e, whole plant (1 cm); f, inflorescence (2 mm); g, seed (0.25 mm).

{(⊙) Hel}. Perhaps endemic to SW India, perhaps also found in Malaya: ♦?Ind: Kl, Kt, Mh, Tn?

Murdannia semiteres (Dalzell) Santapau [*Aneilema paniculatum* Wallich ex C. B. Clarke, *Dichoespermum juncoides* Wight]: Fig. 75

Stems tufted, branched from below, 2–12 cm or rarely more long; branchlets reddish. Leaf sheaths glabrous; blades grass-like, filiform to narrowly linear, succulent, often longitudinally folded, up to 10 (–16) cm long, 1–2 mm wide, glabrous, tinged with red. Flowers in terminal, paniculate cymes; pedicels up to 1 cm long; bracts funnel-shaped and enclosing the base of the pedicel, up to 3 mm long, toothed at apex, persistent. Sepals elliptical, 2–3 mm long, obtuse. Petals 3–4 mm long, blue or rose. Filaments glabrous. Capsules oblong or subglobose, 2.5–3 mm long. Seeds in 2 rows of 3–4 in each capsule locule, almost pyramidal, ± 0.25 mm long, nearly smooth, brown.

Annual. Usually gregarious around drying pools, along streams and on wet rocks.

{⊙ Hel}. Africa and S India: Ind: Ap, Go, Kl, Kt, Mh, Tn.

Murdannia spirata (Linnaeus) Brückner [*Aneilema nanum* (Roxburgh) Kunth]: Fig. 76

Stems branched, creeping with erect branches or decumbent, 5–25 cm long. Leaf sheaths glabrous except for a line of hairs along the margin; blades oblong to ovate, glabrous, 1–2 (–4) cm long, 0.5–1.3 cm wide, margin cartilaginous and minutely serrate, base cordate and amplexicaul, tip almost rounded. Flowers in terminal panicles or cymes and sometimes also from the upper axils; lowest bract leaf-like, the upper ones scale-like, ovate, to 2 mm long, persistent. Petals 3–3.5 mm long, blue to violet or pink. Filaments bearing long hairs or sometimes naked. Capsules oblong, 3-sided, acute, 4–6 (–8) mm long, bearded. Seeds in 1 row of 4–6 in each capsule locule, ± cubical or angular, ± 0.5 mm long, nearly smooth, straw-coloured or bluish-glaucous.

Annual or perhaps also perennial in permanently wet places. Found in a wide variety of wet habitats: rice fields, irrigation ditches, river beds and banks and in marshes. It is often gregarious and forms distinct clumps or zones around drying pools.

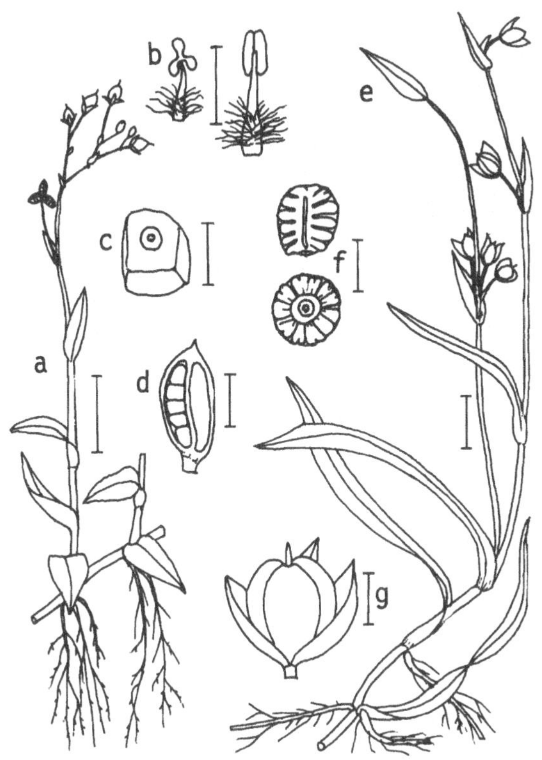

Fig. 76. *Murdannia spirata*: a, flowering shoot (1 cm); b, staminode and stamen (2 mm); c, seed (0.5 mm); d, capsule valve with seeds in place (2 mm). *M. vaginata*: e, flowering shoot (1 cm); f, seed, hilum view above (1 mm); g, capsule with sepals (1 mm).

{⊙ (⊙)? Hel}. Indomalaysia: Ban, Bhu, Cey, Ind, Nep: An, Ap, Bh, Gj, Go, Kl, Kt, Mh, Mp, Sk, Tn, Up, Wb. 0–1800 m.

Murdannia vaginata (Linnaeus) Brückner: Fig. 76

Like *M. nudiflora* but flowers 1–3 in clusters; bracts persistent, narrowly oblong, sheathing, strongly ribbed, 1–2 cm long; capsules 3-seeded, the outer walls rounded; seeds solitary in each capsule locule, ± globose, ± 1.5 mm across, with furrows radiating between the hilum and the operculum.

Annual or perennial. Common in rice fields and wet waste places, sometimes in water.

{(⊙) Hel}. Sri Lanka, tropical India and China: Ban, Ind: Bh, Go, Kl, Kt? Mh, Mn, Mp, Tn, Wb.

CONVOLVULACEAE: 55 genera

Herbs or shrubs, often twining; sometimes parasitic. Leaves alternate, usually petiolate; blades variable, cordate or sagittate to triangular or linear. Inflorescence axillary, 1- to many-flowered. Flowers bisexual, usually radially symmetrical. Sepals 5, free, persisting in fruit. Petals 5, united, usually trumpet- or funnel-shaped. Stamens 5, inserted at the base of the petals, alternating with the lobes. Ovary superior, of 2–4, united carpels; style 1, simple. Fruits 1- to 4-locular, either a capsule or nut-like. Seeds 1–4.

1. Outermost 2 or 3 sepals enlarged, much broader and longer than the inner ones; leaf blades narrowly lanceolate, entire ***Aniseia martinicensis***

1. Outermost 2 or 3 sepals not enlarged, not broader and longer than the inner ones; leaf blades ovate or reniform to deltate or hastate, entire or lobed

2. Stems delicate, creeping and firmly rooted; petals yellow, not more than 7 mm in diameter; sepals bearing long white hairs; pollen smooth, colpate ***Merremia gangetica***

2. Stems robust, floating or erect; petals pink to purple or white, at least 4 cm in diameter; sepals without hairs; pollen spinose, porate

3. Stems erect, never twining, more than 1 m tall, somewhat woody ***Ipomoea fistulosa***

3. Stems prostrate or if erect then not more than 25 cm tall or twining, never woody

4. At least some leaf blades sagittate or hastate; sepal lobes subequal, ovate, ± 6 mm long, ± 4 mm wide, acute, bearing nectaries on the outside ***Ipomoea aquatica***

4. Leaf blades rounded-cordate to subreniform; sepal lobes unequal, the outer 2 crested, 5–6 mm long, the inner 10–12 mm long, elliptical to ovate, leathery, warty, rounded and mucronate at the tip ***Ipomoea asarifolia***

Aniseia: ± 5 species: New World, 1 naturalized in Old World tropics

Aniseia martinicensis (Jacquin) Choisy [*Convolvulus uniflorus* N. L. Burman]: Fig. 77

Stems twining, glabrous or nearly so. Leaf blades narrowly lanceolate, 4–8 cm long, entire, base obtuse or acute, tip obtuse with a small mucro. Flowers usually solitary in axils but occasionally in cymes. Peduncles up to 5 cm long. Sepals broadly ovate, 1.2–2 cm long, acuminate, the outer 2 (3) broader and longer than the inner (2) 3. Petals white, campanulate, 2.5–3 cm long. Stamens included; pollen pantocolpate. Capsules 4-valved, ovoid, ± 2 cm long, subtended by the enlarged sepals. Seeds 4, black, glabrous.

Annual or perennial. In moist and swampy ground, often abundant in rice fields. Mostly near to the coast. The leaves are sometimes gathered and eaten.

{⊙? Hel}. Usually considered to be native in New World tropics: <Cey, <Ind by 1688: <Kl, <Kt.

Ipomoea: 500 species: tropical to warm regions of the world

Stems erect, twining or floating. Leaves petiolate; blades variable, cordate or sagittate to triangular or rarely linear. Inflorescence axillary, 1- to few-flowered, cymose. Sepals persisting but not enclosing the fruit. Petals funnel-shaped, glabrous outside. Stamens usually unequal in length; anthers straight; pollen spinulose, pantoporate. Fruit a thin-walled capsule, either 4-valved or splitting irregularly; seeds 4 or less.

Shrubs or herbs. The flowers are pollinated by insects. The disseminules are relatively large and sometimes hairy seeds, dispersal unknown.

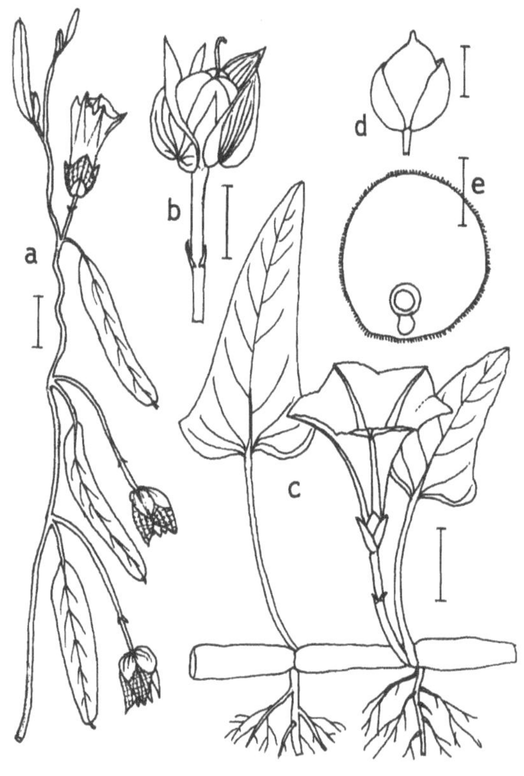

Fig. 77. *Aniseia martinicencis*: a, shoot with flower and fruits (1 cm); b, capsule with sepals (1 cm). *Ipomoea aquatica*: c, flowering shoot (2 cm); d, capsule with sepals (1 cm); e, seed (2 mm).

Ipomoea aquatica Forsskål [*I. reptans* auct. Ind., non Linnaeus]: Fig. 77

Stems usually floating and somewhat swollen. Leaves simple; petioles up to 10 cm long; leaf blades broadly lanceolate to sagittate or hastate, rarely subcordate, 4–9 (–12) cm long, 2–5 cm wide, base broadly hastate, basal lobes triangular-acute or obtuse, tips acuminate. Flowers solitary or occasionally in 2- or 3-flowered cymes; pedicel 2–7 cm long. Sepal lobes subequal, ovate, ± 6 mm long, ± 4 mm wide, acute, bearing nectaries on the outside. Petals usually pink with or without a purplish throat or white, funnel-shaped, 4–5 cm long, up to ± 6 cm in diameter. Stamen filaments villous at the base, 0.7–1 cm long; anthers ±4 mm long. Capsules subglobose, 8–10 mm in diameter, persistent sepals embrace but do not enclose the fruit. Seeds subglobose, 5–6 mm in diameter, dark brown, with fine velvet-like hairs or glabrous.

Perennial. Usually floating on stagnant water but sometimes found on the banks of pools, canals and rivers. It is often cultivated for its edible shoots and as a medicine. It can be a serious weed if left to grow unchecked. Flowers open after 9 a.m.

{Eph/Hyp/Ten}. Old World tropics and subtropics, naturalized in the New World: Ban, <?Cey, Ind, Nep: An, As, Bh, Gj, Go, Kl, Kt, Mh, Mp, Pj, Rj, Tn, Up, Wb.

Ipomoea asarifolia (Desrousseaux) Roemer et Schultes [*I. repens* Lamarck; *I. rugosa* (Rottler) Choisy; *I. beladamboe* Roemer et Schultes]: Fig. 78

Stems decumbent, the tips twining, rooting at nodes. Petioles 4–10 (–20) cm long; leaf blades rounded-cordate to subreniform, 4–8 (–12) cm long, 3–10 (–14) cm wide, base cordate, tips rounded. Flowers solitary in cymes; pedicel 1–2.4 cm long. Sepal lobes unequal, the outer 2 crested, 5–6 mm long, the inner 10–12 mm long, elliptical to ovate, leathery, warty, rounded and mucronate at the tip. Petals lavender to purple or rarely white, funnel-shaped, up to ± 6 cm in diameter, 6–8 cm long. Stamen filaments pilose at the base. Capsules subglobose, 10–12 mm long, brown glabrous. Seeds 6–7 mm long, brown to dark brown, minutely pubescent.

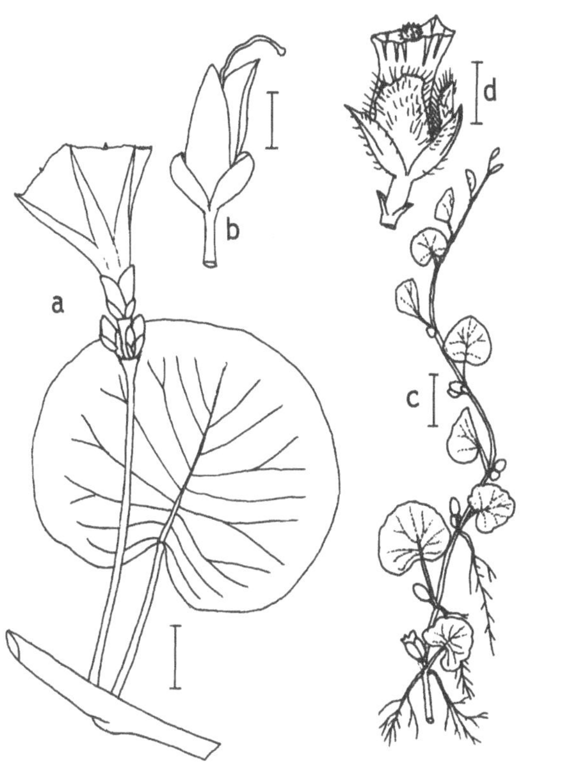

Fig. 78. *Ipomoea asarifolia*: a, shoot with inflorescence (2 cm); b, sepals (5 mm). *Merremia gangetica*: c, flowering shoot (1 cm); d, flower (2 mm).

Perennial. Usually associated with wetlands, particularly on the banks of streams, canals, lagoons and in rice fields. Flowers open before dawn.

{Hel}. Tropical S America, naturalized in the Old World: <Cey, <Ind: <Tn.

Ipomoea fistulosa Martius ex Choisy [*I. carnea* Jaquin subsp. *fistulosa* (Martius ex Choisy) D. Austin, *I. crassicaulis* (Bentham) Robinson]: Fig. 79

Usually shrubby, 2.5 m or more tall, stems swollen when in water. Leaves simple; petioles up to 8 cm long, with nectaries above; leaf blades broadly ovate to deltate or hastate, 8–15 (–25) cm long, 4–7 (–10) cm wide, base cordate to truncate, basal lobes rounded, tips gradually acuminate. Flowers in axillary cymes or panicles, rarely subterminal; peduncles up to 10 cm long; bracts deciduous; pedicels up to 1.5 cm long. Sepal lobes 5, subequal, ovate, ± 8 mm long, ± 5 mm wide, puberulous at first, obtuse, with nectaries outside (active while in bud). Petals usually pink to rose, funnel-shaped, up to ± 7 (–10) cm in diameter; tube ±4.5 cm long. Stamen filaments villous at the base, 1–2 cm long; anthers ± 7 mm long. Capsules globose, ±1.5 cm in diameter, persistent sepals embrace the fruit. Seeds ovoid, up to ± 1 cm long, densely covered with long, brownish, woolly hairs.

Perennial. It shows an exceptional ecological tolerance; it may be found on dry banks among rocks, in places subject to flooding, in permanent swamps or in water up to ± 2 m deep. It grows in fresh and brackish water. It is often planted for hedging or decoration but it has become a serious weed in some irrigation and drainage channels. It flowers at any time of year, the flowers open before dawn and wilt by noon. It can be controlled by 2,4D at 3–4 kg/ha, a few weeks after treatment the plants should be burnt to avoid regeneration. It is used locally as fuel. Cook, C. D. K. *Ipomoea fistulosa*: a new problem for India? Aquaphyte 7: 12 (1987).

{Hyp/Ten/Hel}. Native in S America, naturalized in Africa and Asia: <Ban ±1890, <Cey, <Ind, <Nep, <Pak: <An, <As, <Bh, <Dl, <Gj, <Hp, <Kl, <Kt, <Mh, <Mp, <Or, <Pj, <Rj, <Tn, <Up, <Wb ±1890.

Merremia ± 80 species: pantropical

Merremia gangetica (Linnaeus) Cufodontis [*M. emarginata* (N. L. Burman) H. Hallier, *Convolvulus reniformis* Roxburgh]: Fig. 78

Stems creeping, rooting at nodes. Leaves simple; petioles up to 2 cm long; leaf blades reniform to

Fig. 79. *Ipomoea fistulosa*: a, habit (50 cm); b, flowering shoot (2 cm).

broadly ovate, 1–2.5 cm long, 1–3 cm wide, base cordate, margins crenate and wavy, tips obtuse. Flowers solitary or occasionally in 2- or 3-flowered cymes; peduncles and pedicels very short. Sepal lobes unequal; outer 3 broadly ovate to orbicular, folded, 2.5–3 mm long, retuse, mucronate; inner 2 longer, ovate, 3–4 (–6) mm long, bearing long white hairs. Petals yellow, up to 7 mm in diameter. Stamen filaments ± 4.5 mm long; anthers spiral, ± 0.7 mm long, pollen grains smooth, colpate. Capsules globose, 4–6 mm in diameter, persistent sepals embrace the fruit. Seeds subglobose, 2–2.5 mm in diameter, glabrous.

Perennial. By no means dependant on wetland conditions but often found in marshes, on the banks of ditches, canals and pools where it is frequently inundated. Flowers open after 9 a.m.

{Hel}. Old World tropics: Cey, Ind, Nep: Bh, Gj, Kl, Kt, Mh, Mp, Rj, Tn.

CRUCIFERAE see BRASSICACEAE

CYPERACEAE: ± 110 genera

Perennial or annual. Culms (stems bearing inflorescences) simple, usually solid and triangular in transverse section, rarely hollow, terete or jointed. Leaves usually in 3 rows, mostly basal, either scale- or sheath-like or linear and grass-like, with or without a ligule. Inflorescence variable: from a terminal few-flowered spikelet to a compound panicle with many spikelets, either naked or with leaf- or scale-like involucral bracts. Spikelets unisexual or bisexual, sessile on the rachis, few- to many-flowered, often with many sterile flowers. Flowers unisexual or bisexual, sessile on the rachilla, solitary within a glume. Perianth reduced to hairs, bristles, scales or absent, sometimes forming a sac (utricle) surrounding the nut. Stamens usually 3, rarely 1–12 or more. Ovary superior, with 1 erect ovule; style 1; stigmas 2 or 3. Fruit a nut, either 3-sided, terete or flattened, often with a persistent beak.

There are many different interpretations of the cyperoid flower. In this account the flowers are considered to be basically bisexual with a perianth reduced to scales or bristles; the number of stamens in the flower is considered to be 3, in some cases reduced to 1; 'flowers' with many stamens arise by the grouping of a number of 1-staminate flowers. It is sometimes difficult to decide what constitutes a single spikelet, the generic key has been written to incorporate different views. The following works have proved particularly useful while preparing this account: Haines, R. W. and Lye, K. A. The sedges and rushes of East Africa. East African Natural History Society, Nairobi 1–404 (1983); Kern, J. H., in Steenis, C. G. G. J. van (ed.) Flora Malesiana 7(3): 435–753 (1974); Rao, A. S. and Verma, D. M. Cyperaceae of north east India. Botanical Survey of India i–vi, 1–100 (1982); Koyama, T., in Dassanayake, M. D. (ed.) A revised handbook to the Flora of Ceylon 5: 125–405 (1985). Kral, R. A treatment of *Abildgaardia, Bulbostylis* and *Fimbristylis* (Cyperaceae) for North America. Sida 4(2): 57–227 (1971). I am very indebted to Professor E. Govindarajalu for his help with this family.

Excluded: the following genera have wetland species found in the Himalayas but which do not extend further south into India: ***Cladium jamaicense*** Crantz subsp. ***chinense*** Nees ex Hooker et Arnott [*C. mariscus* (Linnaeus) Pohl subsp. *jamaicensis* (Crantz) Kükenthal]; ***Eriophorum*** several species; ***Scirpus ternatanus*** Reinwardt ex Miquel [*S. chinensis* Munro]; ***S. wichurai*** Boeckeler [*S. asiaticus* Beetle].

The following may be found in moist or damp places but it is not considered to be a wetland plant: ***Hypolytrum nemorum*** (Vahl) Sprengel [*H. latifolium* L. C. Richard, *H. wrightianum* Boeckeler].

1. Spikelets compound, with a terminal (central) and naked female flower (ovary) surrounded by 2 or more 1-staminate male flowers, the whole in turn surrounded by numerous glume-like scales
  2. Culms with transverse septa; leaves reduced to bladeless sheaths at the base of the culms ***Lepironia***
  2. Culms without transverse septa; leaves with blades, radical and cauline
    3. Inflorescence of several to ± 20, subsessile or shortly pedunculate, globose, head-like clusters of spikelets, 3–5 mm in diameter, borne singly or in pairs along the culm in the axils of leaf-like bracts ***Diplacrum***
    3. Inflorescence terminal
      4. Inflorescence a globose to globose-ellipsoidal, head-like spike, with leafy bracts; leaves many times longer than the culms ***Mapania***
      4. Inflorescence a panicle, with compound branches bearing numerous spikelets, without leafy bracts; leaves shorter or slightly exceeding the culms ***Scleria***
1. Spikelets simple, glumes either bearing an axillary flower or empty or reduced to a single flower
  5. All glumes arranged in two rows on the rachilla, each subtending a flower or the lowest 1 or 2 sometimes empty; flowers entirely bisexual or bisexual with some unisexual; spikelets usually flattened
    6. Spikelets remaining attached to the rachis whilst the lower mature glumes and nuts are shed, exposing the ± deeply pitted rachillas, generally composed of a large number of glumes
      7. Style constricted and swollen at the base ***Fimbristylis***

7. Style neither constricted nor swollen at the base

8. Nuts usually 3-sided or sometimes flattened but in either case with the flat side facing the rachilla; stigmas 3 (very rarely 1 or 2) ***Cyperus***

8. Nuts flattened with one angle (sharp edge) facing the rachilla; stigmas 2 (rarely 1) ***Pycreus***

6. Spikelets falling entire from the rachis at maturity, leaving the subtending bract and the bract (prophyll) surrounding the rachilla scar, often very numerous, sessile and crowded on short rachis, generally composed of a small number of glumes

9. Spikelets mostly with 2 or 3 glumes, each folded one within another and maturing only 1 or 2 nuts; stigmas 2; inflorescence of clustered spikelets forming a head ***Kyllinga***

9. Spikelets generally with several to many glumes maturing 2 to several nuts (if 1 then this not sheathed in a thickened rachilla); stigmas 3; inflorescence open or head-like ***Mariscus***

5. All or the upper glumes arranged spirally on the rachilla, at least the lowest 2 glumes empty; flowers unisexual or bisexual; spikelets rounded, terete, not flattened

10. Spikelets with bisexual flowers

11. Leaves in two ranks, equitant, the blades unifacial (vertically flattened); nuts with a whitish cap on the apex; pollen polyphorate ***Machaerina***

11. Leaves in three ranks or appearing spiral, not equitant, the blades absent or bifacial (horizontally flattened); nuts without a whitish cap on the apex; pollen with 0–4 pores

12. Spikelets with 1 or 2 bisexual flowers, with several empty glumes below and male or empty glumes above; perianth scales present ***Rhynchospora***

12. Spikelets with several bisexual flowers (if 1- or 2-flowered then without perianth segments), with rarely more than 2 empty glumes below; perianth present or absent

13. Stems with branches arising in whorl-like clusters, usually subtended by 2 bracts or scale-like leaf sheaths (prophylls); internodes between successive whorls progressively shorter except for the last whorls; the terminal branches longer than the lower internodes, 2–6 cm long, less than 0.1 mm thick, simple, hair-like and often very numerous ***Websteria***

13. Stems unbranched or branches not in whorl-like clusters

14. Style constricted and swollen at the base

15. Perianth bristles persisting in fruit, filiform, usually with backwardly directed barbs ***Eleocharis***

15. Perianth bristles absent

16. Style base falling off with style, leaving a dark button-like scar or tumour on the fruit; style smooth ***Bulbostylis***

16. Style base usually persistent on the nut as a short knob or if falling then not leaving a scar or tumour on the fruit; style usually hairy ***Fimbristylis***

14. Styles neither constricted nor swollen at the base

17. Leaf blades absent or very reduced; inflorescence with a culm-like bract continuing in the direction of the culm ***Schoenoplectus***

17. Leaf blades present; inflorescence with scale-like or leaf-like, never culm-like involucral bracts

18. Leaf blades and leaf sheaths hairy, but sometimes only on the leaf tip and near the throat of the leaf sheath ***Fuirena***

18. Leaf blades and leaf sheaths glabrous

19. Glumes 5–6 mm long with an emarginate or frayed apex and an excurrent midrib elongated into an awn; culms bearing leaves above the base ***Bolboschoenus***

19. Glumes less than 4 mm long, entire at the apex, with or without an excurrent midrib elongated into an awn; culms bearing leaves only in the lower third

20. Inflorescence paniculate with stalked terminal spikelets; involucral bracts numerous and leaf-like ***Schoenoplectus***

20. Inflorescence terminal and head-like; involucral bracts not leaf-like

21. Inflorescence of 1, terminal spikelet, subtended by 1 bract, with (3–) 5–10 flowers ***Eleogiton***

21. Inflorescence with 1 to many spikes, surrounded by 1–few, spike-bearing or empty bracts, the lowermost elongate, the uppermost scale-like; the spikes composed of a cylindrical axis with many well-developed spikelet bracts in a dense spiral, the lateral spikes each with an empty spike prophyll; each spikelet bract subtends a 1-flowered spikelet with 2 small hyaline scales ***Lipocarpha***

10. All flowers unisexual or sterile

22. Nut enclosed in a sac- or bottle-like sheath (prophyll or utricle) ***Carex***

22. Nut exposed, not enclosed in a sac- or bottle-like sheath (prophyll or utricle)

23. Culms with transverse septa; leaves reduced to bladeless sheaths at the base of the culms ***Lepironia***

23. Culms without transverse septa; leaves with blades, radical and a few cauline

24. Inflorescence of several to ± 20, subsessile or shortly pedunculate, globose, head-like clusters of spikelets, 3–5 mm in diameter, borne singly or in pairs along the culm in the axils of leaf-like bracts ***Diplacrum***

24. Inflorescence terminal

25. Inflorescence a globose to globose-ellipsoidal, head-like spike, with leafy bracts; leaves many times longer than the culms ***Mapania***

25. Inflorescence a panicle, with compound branches bearing numerous spikelets, without leafy bracts; leaves shorter or slightly exceeding the culms ***Scleria***

Bolboschoenus [excluded from *Schoenoplectus* and *Scirpus*]: ± 4 or 1 polymorphic species: cosmopolitan

Bolboschoenus maritimus (Linnaeus) Palla [*Scirpus maritimus* Linnaeus, *Schoenoplectus maritimus* (Linnaeus) K. Lye, *Scirpus macrostachys* Willdenow, *Scirpus compactus* Hoffmann, *Scirpus tuberosus* Desfontaines, *Scirpus affinis* Roth]: Fig. 80

Rhizomes hard, creeping often bearing subglobose corms. Culms solitary, sharply 3-angled, leafy, the lower part swollen and covered by leaf sheaths, (15–) 20–100 (–180) cm tall, 2–5 mm thick above, smooth, often somewhat nodding at the tip. Lower leaf sheaths brown to almost black; leaf blades flat, with revolute margins, the upper longer than the culm, 25–40 cm long, (1–) 3–8 (–12) mm wide, midrib distinct and strongly keeled below, tapering to a 3-angled tip, margin and keel with densely set minute teeth. Involucral bracts leaf-like, the lower up to 15 cm long and overtopping the inflorescence. Inflorescence terminal, compound or simple, compact or loose and umbel-like, with (1–) 20–40 spikelets. Spikelets ovoid, single or in clusters, many-flowered, ovoid, (5–) 10–20 (–40) mm long, 3–8 mm in diameter. Glumes ovate, 4–7 mm long, up to 4 mm wide, pubescent on the back, often with emarginate or frayed apex and excurrent midrib elongated into a 2–3 mm long awn; keel absent. Perianth of 6 or fewer, retrorsely scabrid bristles or rarely absent. Stamens 3, with flattened, white filaments; anthers 2–5 mm long. Styles (2-) 3-cleft, smooth. Nuts 3-sided or biconvex with one side flattened, obovoid with an acuminate apex, 2–3 (–4) mm long, 1.4–1.8 mm wide, dark brown to black, shiny; style base not thickened.

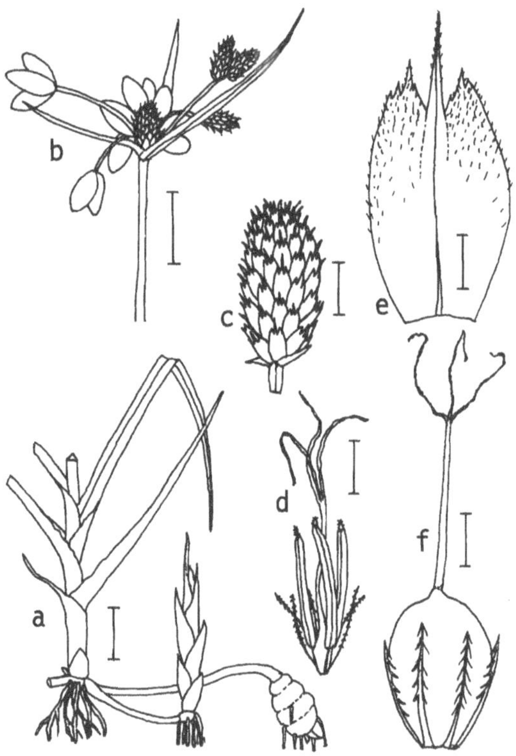

FIG. 80. *Bolboschoenus maritimus*: a, base of shoot (2 cm); b, inflorescence (2 cm); c, spikelet (5 mm); d, flower (2 mm); e, glume (1 mm); f, fruit (1 mm).

Perennial. Often locally dominant in shallow permanent or semi-permanent water. It can tolerate saline conditions. Large stands are undoubtedly important for wildlife. It is delivers poor quality forage but can be used as hay if cut early. In arid regions it is of importance and can yield up to 50 centners ha^{-1} of hay or up to 150 centners ha^{-1} as silage. Tubers are edible but apparently not always developing. The leaves and culms are used for weaving. In rice fields it can become a serious pest. Morphologically very variable but not clearly divisible into decreet taxa. It is sometimes placed in *Scirpus* or *Schoenoplectus*.

{Hyp}. Almost cosmopolitan: Cey? Ind, Pak: Ap, Bh, Gj, Kt, Mg, Mh, Mp, Pj, Rj, Up, Wb.

Bulbostylis [excluding *Abildgaardia*]: ± 100 species: pantropics

Like *Fimbristylis* but the styles smooth, articulate with the ovary; style base bulb-like, persisting as a dark-coloured button on the apex of the nut.

There are good grounds for placing *Bulbostylis* in the genus *Abildgaardia* but pending a complete revision of Indian Cyperaceae it is felt better to retain the genera *Fimbristylis* and *Bulbostylis* in the traditional sense.

Most species grow in dry sandy places but in India one grows in wetlands.

Bulbostylis densa (Wallich) Handel-Mazzetti [*Isolepis trifida* Nees, *Bulbostylis capillaris* (auct. Ind., non Linnaeus) Boeckeler var. *trifida* (Nees) C. B. Clarke, *Fimbristylis densa* (Wallich) T. Koyama et Chuang, *Abildgaardia densa* (Wallich) K. Lye]: Fig. 81

Culms tufted, bristle-like, (1–) 5–40 cm long, 0.2–0.5 mm thick, glabrous, smooth. Leaf sheaths membranous, pale; ligule of 1–2 mm long white hairs; blades capillary, very much shorter than the culms, 2–10 (–18) cm long, 0.2–0.5 mm wide, soft, light green, gradually acute at apex, glabrous or sparingly hairy on lower surface. Involucral bracts usually very short and glume-like, occasionally 1 is bristle-like and up to 7 cm long. Inflorescence very variable, simple or compound, spreading or contracted and nearly head-like, sometimes reduced to single spikelets. Spikelets solitary or clustered, oblong-ovoid to oblong, 3–6 mm long, 1–2 mm in diameter, 5- to 18-flowered (1 or 2 lower glumes empty). Glumes ovate to broadly ovate, 1.5–2 mm long, 1–1.5 mm wide, brown to reddish-brown, rounded or subacute, usually a distinct but short mucro, chestnut brown, sides without nerves, glabrous but the margins ciliolate; keel strong. Stamens 2; anthers 0.2–0.4 mm long. Styles 0.7–1 mm long, 3-cleft, glabrous; stigmas shorter than the style. Nuts 3-angled, obovoid to broadly obovoid, 0.7–0.9 mm long, 0.5–0.8 mm broad, straw-coloured to brownish, densely covered in small tubercles.

Annual. Grows in seasonally flooded places; flowering as the water recedes. It is also found in open wet places and wet grasslands. It is not an exclusively wetland plant and is also found in dry sandy places and in dry forests. It is very near to the American species *B. capillaris* (Linnaeus) C. B. Clarke but differs by the more acute, glabrous or slightly hairy glumes and the granular, not transversely rugulose nuts.

{⊙ Ten/Hel}. Tropics of the Old World: Bhu, Cey, Ind, Nep: As, Bh, Go, Hp, Kl, Kt, Mg, Mh, Mp, Ng, Sk, Tn, Up, Wb. 0–3500 m.

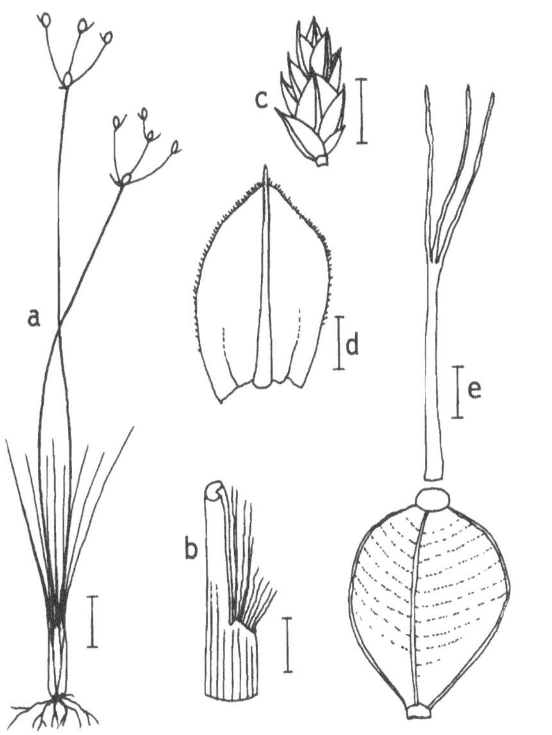

Fig. 81. *Bulbostylis densa*: a, whole plant (2 cm); b, hairy ligule (1 mm); c, spikelet (2 mm); d, glume (0.5 mm); e, nut with detached style (0.2 mm).

Carex: ± 1500 species: cosmopolitan but mainly in temperate climates

Monoecious or dioecious, rhizomatous or tufted perennials. Culms mostly leafy. Leaf blades flat, keeled or involute. Inflorescence a simple spike to a much-branched panicle, with or without bracts. Spikes unisexual, the upper usually male or if bisexual then the uppermost flowers usually male. Spikelets 1-flowered. Flowers unisexual, each subtended by 1 glume; male flowers with 2 or 3 stamens, perianth absent; female flowers with 1 nut surrounded by a globular, 3-sided or compressed utricle sometimes inflated and beaked in fruit. Stigmas 2 or 3. Nuts biconvex or 3-sided.

Many species of *Carex* are found in wet soil or in damp places particularly at high altitudes but in Peninsular India only three species grow in wetlands.

1. Midrib of female glumes ending below the apex, not developing into an awn ***C. maculata***
1. Midrib of female glumes excurrent into a recurved, scabrous-margined, 0.2–0.5 mm long awn
  2. Rhizomes stoloniferous; lateral spikelets erect, 1–2 cm long, rarely bearing male flowers; stigmas 3; utricle with a 0.2–0.5 mm long beak ***C. heterostachya***
  2. Rhizomes short and creeping; lateral spikelets nodding, 3–9 cm long, usually bearing male flowers towards the tips; stigmas 2; utricle without a beak ***C. phacota***

Carex heterostachya Bunge: Fig. 82

Rhizomes stoloniferous. Culms tufted, obtusely 3-angled, 1- or few-leafed, 30–45 cm tall, 1–1.5 mm wide. Leaf sheaths brown to purple; leaf blades folded, most shorter or almost equalling the culms, 2–3 mm wide, soft, gradually tapering to an acute tip. Bracts not sheathing, the lowest leaf-like, usually exceeding the inflorescence. Inflorescence of 4–6 spikelets. Spikelets cylindrical or oblong; the uppermost 1–3 spikelets male; the terminal spikelet male, linear, 2–3 cm long, light orange brown; the lower 2–4 spikelets female, shortly peduncled to subsessile and erect, oblong to oblong-cylindrical, 1–2 cm long, ± 5 mm in diameter, dark brown when dry. Female glumes ovate, 2–2.5 long, ± 1.5 mm wide, pale brown, with hyaline margins; midrib excurrent into a recurved, scabrous-margined, 0.2–0.5 mm long awn. Utricles ovoid, 3–4 mm long,

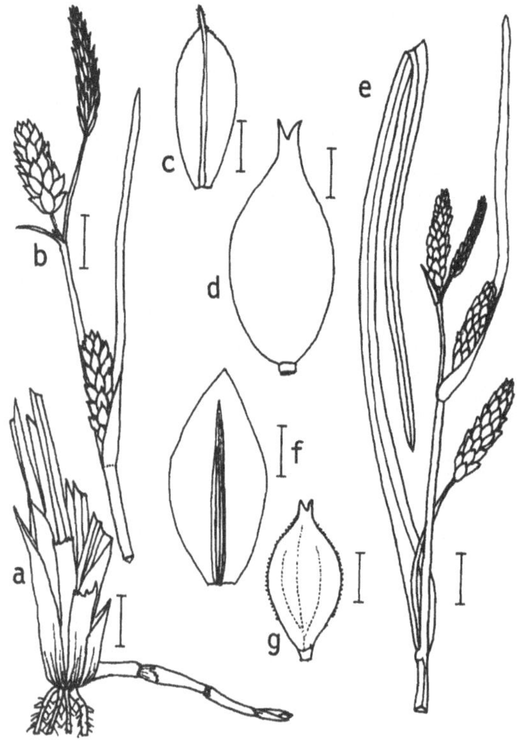

Fig. 82. *Carex heterostachya*: a, base of shoot (1 cm); b, inflorescence (1 cm); c, female glume (1 mm); d, utricle (1 mm). *C. maculata*: e, inflorescence (1 cm); f, female glume (0.5 mm); g, utricle (0.5 mm).

±1.5 mm wide, yellowish-brown below, dark brown above when dry, with a 0.2–0.5 mm long beak. Styles 3-cleft. Nuts obovoid to ellipsoid, 1.5–1.8 mm long, 1–1.2 mm wide.

Perennial. Margins of ponds, along streams and rivers, wet grassland, but also found in dryer places.

{Hyp/(Hel)}. C and E Asia extending southwards to N India: Ind: Jk, Rj.

Carex maculata Boott: Fig. 82

Rhizomes short, ascending. Culms loosely tufted, obtusely 3-angled, 1- or few-leafed, 20–40 (–60) cm tall, 1.5–2 mm thick. Leaf sheaths orange-brown; leaf blades folded, with slightly recurved margins, shorter or ± equalling the culms, 2–5 (–8) mm wide, soft, gradually tapering to an acute tip. Bracts not sheathing, the lowest 2 leaf-like, exceeding the inflorescence. Inflorescence of 3–4 (–6) spikelets. Spikelets cylindrical; the terminal spikelet male, linear, 1–4.5 cm long, orangey- to reddish-brown; lateral spikelets female or rarely with some male flowers towards the tip, peduncled and erect, 1–4 cm long, 2.5–4 mm in diameter, green and white powdery when fresh becoming dark brown when dry. Female glumes oblong-ovate, 1.8–2 mm long, 0.7–1 mm wide, semitranslucent, brown, obtuse to acutish; midrib 5-nerved, ending below the apex. Utricles ellipsoid to broad ellipsoid, 2–3 mm long, 1–1.8 mm wide, pale green becoming dark brown when dry, with a 0.2–0.5 mm long beak. Styles 3-cleft. Nuts obovoid to ellipsoid, 1.5–1.8 mm long, 1–1.2 mm wide.

Perennial. Margins of ponds, along streams and wet places in highland forests.

{Hyp/(Hel)}. India and Sri Lanka through China to Japan: Cey, Ind: Mg, Tn. 1600–2300 m.

Carex phacota Sprengel [*Carex lenticularis* D. Don]: Fig. 83

Rhizomes short, creeping. Culms loosely tufted, sharply 3-angled, 1- or 2-leafed, 20–90 (–200) cm tall, 1–3 mm thick. Leaf sheaths rusty to yellow-brown; leaf blades folded, with recurved margins, 15–100 cm long, 3–8 mm wide, scabrous, gradually tapering to an acute tip. Bracts not sheathing, the lowest leaf-like, much exceeding the inflorescence. Inflorescence of 4–6 (–11) spikelets. Spikelets cylindrical; the terminal

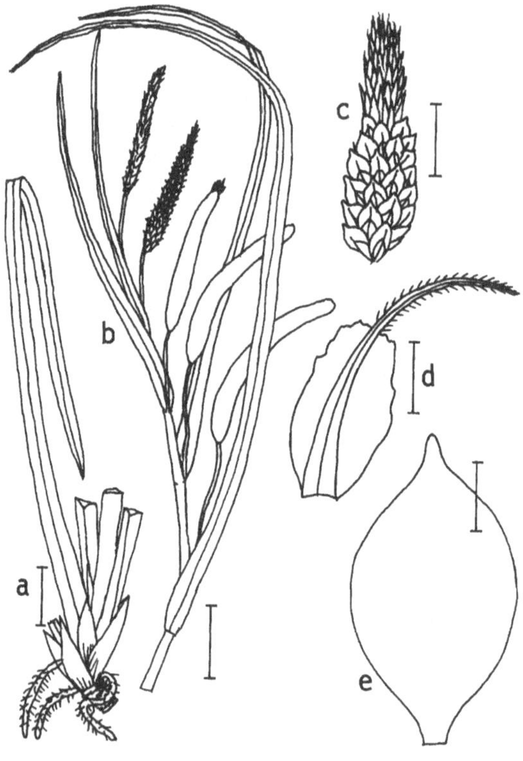

Fig. 83. *Carex phacota*: a, base of shoot (1 cm); b, inflorescence, in life female spikelets hanging (2 cm); c, spikelet showing transition from female to male at the tip (4 mm); d, female glume (0.5 mm); e, utricle (0.5 mm).

spikelet male, linear, 2–7 cm long, light brown; lateral spikelets female with some male flowers towards the tip, peduncled and nodding, 3–9 cm long, 4–6 mm in diameter, green and white powdery when fresh becoming brownish-green when dry. Female glumes ovate-oblong or ovate-oval, 1.2–2.5 mm long, 0.7–1.5 mm wide, rust-coloured or pale with rusty-coloured dots, pale along the margins, obtuse to shallowly emarginate; midrib green, 3-nerved, excurrent into a recurved, scabrous-margined, 0.2–0.5 mm long awn. Utricles broadly obovate to suborbicular, 2.5–3.2 mm long, 1.5–2.2 mm wide, green with red-brown dorsal side, without a beak. Styles 2-cleft. Nuts flattened, suborbicular, 1.8–2 mm long, 1.2–1.8 mm wide.

Perennial. In swamps, marshes and seasonally flooded places, at high altitudes.

{Hyp/(Hel)}. S and E Asia: Bhu, Bur, Cey, Ind, Nep: As, Mg, Sk, Tn. 300–1300 m.

Cyperus [including *Dichostylis, Juncellus, Sorostachys*; excluding *Kyllinga, Mariscus, Pycreus*]: ± 380 species: cosmopolitan, mostly tropical

Annual or perennial, some with creeping or tuberous rhizomes. Culms usually erect, from minute annuals to 5 m tall reeds. Leaves mostly radical, usually shorter than the culm, leaf-like or reduced to scale-like sheaths. Involucral bracts long and leaf-like or reduced and scale-like. Inflorescence terminal, rarely pseudolateral, simple or compound panicles, usually with sessile and stalked clusters of spikelets (spikes), sessile spikelets or compact and head-like; prophylls keeled, tubular at base. Spikes oblong to ovate, spicate or digitate. Spikelets usually flattened, 1- to many-flowered. Glumes distichous, clearly in 2 opposite rows, often imbricating, margins usually hyaline, with a distinct keel, caducous (glumes and nuts are shed leaving the ± deeply pitted rachillas). Flowers bisexual, all fertile except sometimes the very lowest ones sterile. Perianth bristles absent. Stamens (1) 2 or 3; connectives often crested but usually smooth. Stigmas 3 (or rarely 2); style not swollen at the base. Nuts usually 3-sided or sometimes flattened (biconvex or plano-convex), if flattened then one side (not the edge) against the rachilla.

Many species are of considerable economic importance, some cultivated for food (*C. esculentus*), some for fibres or perfume (*C. corymbosus, C. exaltatus, C. pangorei*), others are serious weeds (*C. rotundus*) while others are dominant over large areas. Clarke, C. B. On the Indian species of *Cyperus*; with remarks on some others that specially illustrate subdivisions of the genus. Journal of the Linnean Society (Botany) 21: 1–202 (1884) — very old but still of value; Kükenthal, G. *Cyperus*; in Engler, A. (ed.), Das Pflanzenreich 101(IV.20): 1–671 (1935, 1936) — a standard work treating *Cyperus* in the wide sense; many new species have been described since this work was published.

Excluded: the following wetland species of ***Cyperus*** are found in the Himalayas but do not extend south into Peninsular India: ***C. atkinsonii*** C. B. Clarke; ***C. glaber*** Linnaeus; ***C. glomeratus*** Linnaeus; ***C. fuscus*** Linnaeus.

The following species may grow in damp or moist places but they are not considered to be wetland plants: ***C. atkinsonii*** C. B. Clarke; ***C. bulbosus*** Vahl; ***C. circinatus*** Reidl; ***C. conglomeratus*** Rottbøll [*C. pachyrrhizus* Nees ex Boeckeler]; ***C. diffusus*** Vahl [*C. nigro-viridis* Thwaites]; ***C. leucocephalus*** Retzius [*Scirpus coronarius* Vahl]; ***C. malabaricus*** (C. B. Clarke) T. Cooke: ♦?Ind; ***C. niveus*** Retzius; ***C. pentabracteatus*** E. Govindarajalu: ♦Ind: Mh; ***C. polyanthelus*** E. Govindarajalu: ♦Ind: Tn; ***C. pubisquama*** Steudel [*C. diffusus* subsp. *macrostachyus* (Boeckeler) T. Koyama]; ***C. pulcherrimus*** Willdenow ex Kunth; ***C. serotinus*** Rottbøll [*Juncellus serotinus* (Rottbøll) C. B. Clarke, *J. serotinus* C. B. Clarke, *C. monti* Linnaeus]; ***C. radians*** Nees et Meyen ex Nees; ***C. rubicundus*** Vahl [*C. teneriffae* Poiret]; ***C. sphacelatus*** Rottbøll [*C. locuples* C. B. Clarke] <Cey before 1900; ***C. stoloniferus*** Retzius; ***C. zollingeri*** Steudel [*C. ramosii* Kükenthal].

1. Spikelets in spikes (each at some distance from one another upon a more or less elongate rachis); inflorescence of 1 or more spikes; at least some of the spikes stalked, but sometimes the peduncles are only 2–5 mm long

2. Culms without bladed leaves (but involucral bracts may be leafy) or some upper sheaths may occasionally bear short blades up to 3 cm long; plants large perennials

3. Involucral bracts entirely scale-like, not more than 1.5 cm long; culms transversely septate, terete just below the inflorescence ***C. articulatus***

3. Involucral bracts leaf-like below, at least some more than 2 cm long; culms not transversely septate, 3-sided just below the inflorescence

4. Culms just below the inflorescence with rounded angles

5. Involucral bracts brown (never green), numerous (up to 12); primary rays of the inflorescence 30–100; gigantic herbs with culms (1–) 2–5 m tall ***C. papyrus***

5. Involucral bracts green, 2–4; primary rays of the inflorescence much less than 30; medium to large herbs with culms much less than 2 m tall

6. Rhizome stolon-like, elongate, 2–5 mm thick; spikelets narrow-oblong, 5–18 mm long, 1–1.5 mm wide; glumes broadly lanceolate to oblong, 1.8–2.2 (–3) mm long, 0.8–1 mm wide; nuts 1–1.2 mm long, 0.2–0.3 mm wide *C. corymbosus*

6. Rhizomes rhizomatous, up to 5 cm long, usually more than 5 mm thick; spikelets linear, 10–20 (–25) mm long, 2–2.5 mm wide; glumes ovate, ± 2.8 mm long, 1.5–2 mm wide; nuts 1.2–1.5 mm long, ± 0.5 mm wide *C. scariosus*

4. Culms just below the inflorescence with 3 sharp angles or 3-winged

7. Culms with concave sides, 3-winged just below the inflorescence; glumes rarely more than 2.2 mm long, the sides 5- or 7-nerved, crispidly incurved when dry; nuts 1.7–2 mm long *C. malaccensis*

7. Culms with flat or convex sides, 3-angled just below the inflorescence; glumes more than 2.2 mm long, the sides without nerves, not incurved when dry; nuts 1–1.5 mm long *C. pangorei*

2. Culms with bladed leaves more than 3 cm long or if shorter then plants annual; plants annual or perennial

8. Plants annual with minute root systems

9. Glumes nearly as wide as long, obtuse at the apex, hardly imbricating at maturity; nuts almost as long as the glumes

10. Glumes 7- or 9-nerved, with a scabrid and winged keel; rachis hispid on the angles; spikelets spreading at right angles to the rachilla or reflexed, 2–2.5 mm wide *C. alulatus*

10. Glumes 3- or 5-nerved, with a flattened keel; rachis glabrous; spikelets erect to spreading, rarely reflexed, 1.5–2 mm wide *C. iria*

9. Glumes much longer than wide, acute at the apex, imbricating at maturity; nuts much shorter than glumes

11. Glumes 2.5–5 mm long, with a straight, ± 0.8 mm long mucro; spikelet not squarrose *C. compressus*

11. Glumes 1–2 mm long, with a recurved, 0.4–1.3 mm long awn; spikelet squarrose *Mariscus squarrosus*

8. Plants perennial with woody rhizomes, slender stolons or stem-bases swollen and corm-like

12. Style with 2 branches; nuts flattened; spikelets arranged very densely and obscuring the rachis *C. alopecuroides*

12. Style with 3 branches; nuts rounded or 3-sided

13. Plants with slender stolons

14. Culms terete or obtusely 3-angled, usually with some indistinct articulations *C. corymbosus*

14. Culms sharply 3-angled, with sides flat or concave, without articulations

15. Glumes mostly 4 mm or more long; stolons usually ending in tubers

16. Lower involucral bracts 2–3 times as long as the inflorescence; plants freely fruiting *C. bifax*

16. Lower involucral bracts shorter than or scarcely longer than the inflorescence; plants mostly not developing fruits *C. rotundus*

15. Glumes less than 4 mm long; stolons with or without tubers

17. Rachis with hairs or scabrid edges

18. Glumes obtuse, without apical mucro, 2.5–4 mm long, 2–2.2 mm wide, with a broad hyaline margin, hardly keeled *C. procerus*

18. Glumes acute, with usually mucronulate apex, 1.8–2.5 mm long, 1.2–1.5 mm wide, without a broad hyaline margin, distinctly keeled *C. pilosus*

17. Rachis glabrous

19. Plants with knotty corm-like rhizomes and elongate stolons *C. tenuiculmis*

19. Plants without knotty, corm-like rhizomes

20. Stolons stout, almost rhizome-like, never ending in tubers; culm base not much swollen *C. longus*

20. Stolons slender, frequently ending in tubers; culm base swollen and corm-like

21. Spikelets yellowish or rusty brown; glumes not more than 2.6 mm long, with raised nerves almost to the margin *C. esculentus*

21. Spikelets dark reddish-brown; glumes 3 or more mm long, with a wide marginal border without nerves *C. rotundus*

13. Plants without stolons; culm base swollen or hardened or culms arising from woody rhizomes

22. Spikelets cylindrical, terete or subterete

23. Spikes stalked; spikelets spreading; leaves and involucral bracts mostly 4–15 mm wide; culms smooth along the angles *C. digitatus*

23. Spikes subsessile forming secondary corymbs; spikelets erect to suberect; leaves and involucral bracts mostly 15–30 mm wide; culms scabrous along the angles *C. platyphyllus*

22. Spikelets distinctly flattened

24. Spikelets very numerous and crowded together so as to obscure the axis of the spike

25. Nuts dorsiventrally flattened

26. Glumes 1.4–2.2 mm long; nuts 0.7–1 mm long *C. alopecuroides*

26. Glumes 0.9–1.2 mm long; nuts 0.5–0.6 mm long *C. imbricatus*

25. Nuts 3-sided, not flattened
27. Glumes less than 2 mm long; spikelets ± spreading
28. Prophylls at base of the rays with scabrous keels; nuts 0.8–0.9 mm long; connective appendage ± one-third as long as the anther *C. elatus*
28. Prophylls at base of the rays with smooth keels; nuts ± 1.2 mm long; connective appendage ± half as long as the anther *C. ohwii*
27. Glumes more than 2 mm long; spikelets ± erect
29. Primary rays suberect; spikelets 1.5–2.5 mm wide (4–8 mm long); nuts 1–1.3 mm long *C. nutans* var. *eleusinoides*
29. Primary rays oblique, spreading; spikelets 1–1.5 mm wide (6–15 mm long); nuts 1.3–1.5 mm long *C. nutans* var. *nutans*
24. Spikelets not very numerous, mostly distant, the axis of the spike visible between them
30. Rachis distinctly hairy along the margins *C. pilosus*
30. Rachis glabrous
31. Glumes 3 mm or more long *C. tenuiculmis*
31. Glumes less than 3 mm long
32. Spikelets not more than 1 mm wide; glumes rather remote on a zigzag rachilla *C. distans*
32. Spikelets more than 1 mm wide; glumes close together on a straight rachilla
33. Glumes not more than 1.5 mm long, almost as wide as long, about as long as the nut *C. iria*
33. Glumes more than 1.5 mm long, distinctly longer than wide, shorter than the nut
34. Blades of leaves scarcely half as long as the culm
35. Culms with concave sides, 3-winged just below the inflorescence; glumes rarely more than 2.2 mm long, the sides 5- or 7-nerved, crispidly incurved when dry; nuts 1.7–2 mm long *C. malaccensis*
35. Culms with flat or convex sides, 3-angled just below the inflorescence; glumes more than 2.2 mm long, the sides without nerves, not incurved when dry; nuts 1–1.5 mm long *C. pangorei*
34. Blades of leaves more than half as long as the culm
36. Nuts 0.6–1 mm long; keel of glume excurrent to a short mucro; spikes ± cylindrical, distinctly longer than wide *C. exaltatus*
36. Nuts 1.4–1.6 mm long; keel of glume at most apiculate, not excurrent to a mucro; spikes ovoid to broadly ovoid *C. longus*

1. Spikelets arranged digitately or stellately in clusters (spikelets at practically the same level upon a much shortened rachis); inflorescence of 1 (usually dense) head (very rarely 1–4 additional heads) or inflorescence more open but then at least some spikelets arranged in stalked clusters
37. Inflorescence open, with at least some of the spikelet clusters borne on elongate rays
38. Culm leaves bladeless, reduced to sheaths or scales or the upper sheath with a rudimentary blade
39. Involucral bracts few (up to 4), unequal in length, arising close together *C. haspan* subsp. *juncoides*
39. Involucral bracts numerous (up to 20), almost equal in length, arising distant from each other *C. involucratus*
38. Culm leaves with blades
40. Plants annual with a minute root system
41. Glumes 2.5–5 mm long, with silvery margins *C. compressus*
41. Glumes 0.5–2.5 mm long, without silvery margins
42. Glumes with long recurved mucros
43. Spikelets at anthesis 1.7–3 mm wide (in fruit usually spreading and wider), rusty brown; mucro 0.7–1 mm long, strongly recurved; nuts obovoid, (0.5–) 0.6–0.8 mm long, 0.3–0.4 mm wide *C. cuspidatus*
43. Spikelets at anthesis not more than 1.5 mm wide (in fruit hardly spreading), chestnut brown; mucro 0.3–0.8, slightly recurved; nuts cylindrical, parallel-sided, 0.6–0.9 mm long, 0.2–0.3 mm wide *C. castaneus*
42. Glumes either without mucros or with short and straight mucros
44. Spikelets in almost spherical clusters; nut about as long as the subtending glume *C. difformis*
44. Spikelets never in spherical clusters; nut much shorter than the subtending glume
45. Nuts white and visible between the glumes in fruiting spikelets; glumes 0.7–1.3 mm long *C. tenuispica*
45. Nuts brown, not visible between the glumes in fruiting spikelets; glumes 1–1.5 mm long
46. Keel of glume extending beyond the apex forming an erect mucro; spikelets orange-brown to golden, in tight clusters *C. amabilis*

46. Keel of glume ending below the apex, at most apiculate; spikelets reddish-brown, in loose clusters *C. haspan*

40. Plants perennial with a woody rhizome or slender stolons or stem-base swollen

47. Nuts unequally 3-sided, dorsoventrally flattened

48. Nuts 0.5–0.6 mm long, ± 0.4 mm wide, without thickened angles, uniformly reddish-brown; spikelets with 6–16 flowers *C. imbricatus*

48. Nuts 1.8–2 mm long, 0.8–1 mm wide, with corky or spongy thickened angles, the sides shining greyish-brown to blackish with pale whitish to yellowish angles; spikelets with 16–60 flowers *C. platystylis*

47. Nuts ± equally 3-sided, not flattened;

49. Prophylls at base of the rays with scabrous keels; nuts 0.8–0.9 mm long; connective appendage ± one-third as long as the anther *C. elatus*

49. Prophylls at base of the rays with smooth keels; nuts ± 1.2 mm long; connective appendage ± half as long as the anther *C. ohwii*

37. Inflorescence closed, of 1, sessile, usually dense head; the head consisting of 1 or more sessile clusters of spikelets

50. Inflorescence appearing lateral; 1 involucral bract erect, appearing to be an extension of the culm *C. laevigatus*

50. Inflorescence terminal; the involucral bracts leaf-like or scale-like, spreading or reflexed, not appearing to be an extension of the culm

51. Plants annual with a minute root system

52. Glumes very small, 0.5–0.8 (–1) mm long, nearly orbicular, without mucro or awn; nuts almost as long as the subtending glume *C. difformis*

52. Glumes at least 1 mm long, distinctly longer than wide, with mucro or awn; nuts much shorter than the subtending glume

53. Glumes 3–4 mm long, conspicuously multinerved; the nuts at least 1 mm long, some remaining attached to the rachilla at maturity; whole plant grey to silvery-green *C. compressus*

53. Glumes not more than 2 mm long, never more than 7-nerved; the nuts less than 1 mm long, released as the glumes fall; whole plant green to brown

54. Heads whitish-green; involucral bracts many, leaf-like, spreading, nearly all of them overtopping the inflorescence *C. pygmaeus*

54. Heads golden to brown or reddish-brown; involucral bracts few, rarely more than 3 overtopping the inflorescence

55. Spikelets at anthesis not more than 1.5 mm wide (in fruit hardly spreading); mucro 0.3–0.8 mm long, slightly recurved *C. castaneus*

55. Spikelets at anthesis 1.7–4 mm wide, in fruit usually spreading and wider; mucro 0.7–1.3 mm long, strongly recurved

56. Each head with 20–60 spikelets *Mariscus squarrosus*

56. Each head with 3–15 spikelets

57. Glumes 1.4–2 mm long *C. cuspidatus*

57. Glumes ± 2.5 mm long *Mariscus maderaspatanus*

51. Plants perennial with rhizomes, stolons or swollen culm bases

58. Plants with long, thin creeping stolons

59. Heads whitish; nuts much dorsoventrally flattened and not obviously 3-sided, 1.5–1.8 mm long, 1.1–1.3 mm wide, maturing black-brown, stipe minute *C. arenarius*

59. Heads yellowish-brown; nuts dorsoventrally flattened but clearly 3-sided, 1–1.5 mm long, 0.7–1 mm wide, stipe 1–1.2 mm long winged at anthesis, the wings developing into yellowish corky ridges covering the lower half of the nut *C. cephalotes*

58. Plants lacking stolons, either with swollen culm bases or rhizomes

60. Lowest involucral bract erect, pushing the inflorescence somewhat to the side; nuts 1.2–1.5 mm long *C. scariosus*

60. Lowest involucral bract spreading or reflexed, not pushing the inflorescence to the side; nuts 0.5–0.8 mm long

61. Heads yellowish or brown; nuts pear-shaped *C. meeboldii*

61. Heads dirty white; nuts cylindrical to obovoid, obscurely 3-sided *C. pulchellus*

Cyperus alopecuroides Rottbøll [*Juncellus alopecuroides* (Rottbøll) C. B. Clarke]: Fig. 84

Rhizomes very short, woody. Culms robust, tufted, 3-sided above, smooth, 50–150 cm tall, 7–12 mm thick, inflated when in water. Leaves 3–5; sheaths somewhat spongy, yellowish to blackish-brown, up to 25 cm long; blades inverted W-shaped, leathery, 40–70 cm long, 5–17 mm wide, scabrous on nerves and margins. Involucral bracts 4–7 (–12), unequal, the longer ones much overtopping the inflorescence, up to 65 cm long. Inflorescence compound paniculate, consisting of 1 sessile and few- to many-stalked clusters of spikes; rays 5–10, unequal, up to 20 cm long, the larger ones sometimes with tertiary clusters of spikes. Spikes digitate, upright, cylindrical to

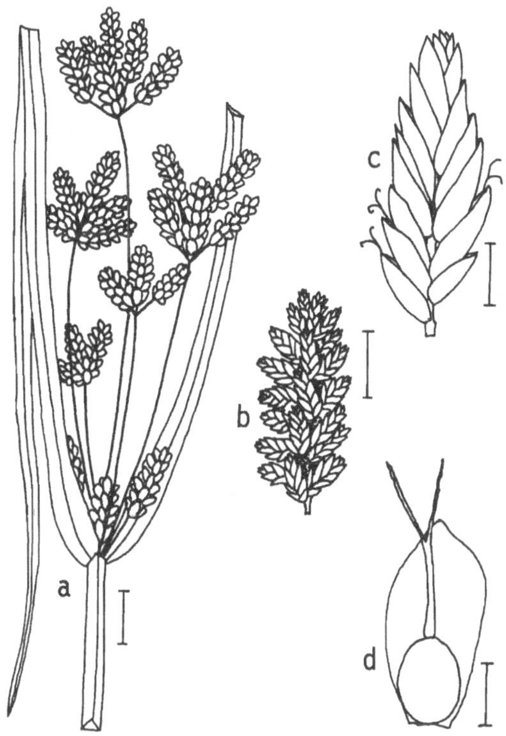

FIG. 84. *Cyperus alopecuroides*: a, inflorescence, outer bracts removed (4 cm); b, spike (1 cm); c, spikelet (1 mm); d, nut with glume (1 mm).

oblong-cylindrical, 1–4 cm long, 4–15 mm in diameter. Spikelets spicate, eventually spreading, ovate to oblong-lanceolate, 2–6 (–8) mm long, 1.5–2.5 (–3) mm wide, 10- to 30-flowered, pale to golden or reddish-brown; rachilla persistent, strongly flattened, winged; wings whitish, very narrow, long-persistent. Glumes ovate to elliptical-ovate, obtuse, 1.4–2.2 mm long, 1–1.3 mm wide, apiculate or minutely mucronate, 5- or 7-nerved, yellowish or golden with purple stripes, densely imbricate; keel absent, midrib excurrent in a short mucro. Stamens 2 (3). Style 1–1.3 mm long, 2 (3) cleft; stigmas 2 or 3, one-quarter to one-third longer than the style. Nuts dorsoventrally flattened, plano-convex or ellipsoid, 0.7–1 mm long, 0.5–0.6 mm wide, yellow to golden yellow, smooth or minutely reticulate.

Perennial. Usually in standing water in tanks, pools and lakes but also in swamps and other wet places, often found in rice fields. Like *C. imbricatus* but the styles usually 3-cleft; also has longer glumes and larger nuts.

{Hyp/Ten/(Hel)}. Warm and tropical Old World and Australia. Ban? Cey, Ind, Pak: Ap, Bh, Dl, Kl, Kt, Mh, Mp, Pj, Rj, Tn, Up, Wb. 0–1500 m.

Cyperus alulatus Kern [*C. iria* Linnaeus var. *rectangularis* Kükenthal]: Fig. 85

Culms erect, slender, smooth, 10–70 cm tall, 1–2 mm thick, 3-sided. Leaves 2 or 3, as long as or shorter than the culm; blades flat, herbaceous, 1–5 mm wide, long-acuminate, smooth below, scabrous above. Involucral bracts 2–4, unequal, the longer ones overtopping the inflorescence. Inflorescence simple or compound, 1 spike ± sessile, the others stalked; rays 3–9, unequal, up to 16 cm long. Spikes ovate to oblong-ovate, lax or in dense clusters, 1–3 cm long, 8–25 (–30) mm wide, bearing 5–20 spikelets; rachis hispid on the angles. Spikelets loose, ovate to linear-oblong, spreading at right angles to the rachilla or reflexed, 3–12 mm long, 2–2.5 mm wide, yellowish, 4- to 18-flowered; rachilla persistent, unwinged. Glumes broadly obovate to suborbicular, obtuse, 1.7–2 mm long and wide, 7- or 9-nerved, with a green, scabrid and winged keel, the sides brown with purplish flecks, not or hardly mucronate, hardly imbricating at maturity. Stamens 2; connective appendage very short. Style 3-cleft, less than 0.2 mm long. Nuts 3-sided, ellipsoid to ovoid-oblong, ± 1.5 mm long, 0.8–0.9 mm wide, shining, dark brown.

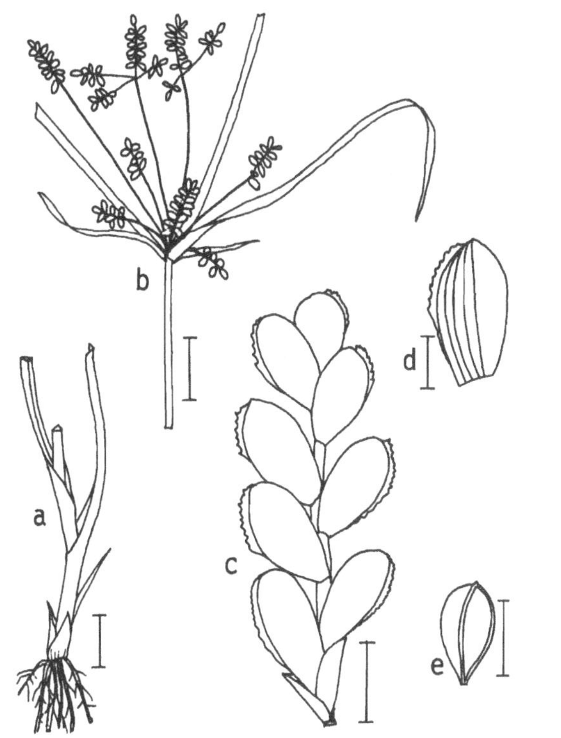

FIG. 85. *Cyperus alulatus*: a, base of shoot (1 cm); b, inflorescence (2 cm); c, spikelet (1 cm); d, glume (0.5 mm); e, nut (1 mm).

Annual. Found in marshes, ditches and in rice fields. It is mostly found on hills and is not confined to wetlands but also in dryer places. Like *C. iria* but spikelets are not erect but spread at right angles to the rachis or are reflexed; rachis hispid on the angles, glumes 7- or 9-nerved, with a scabrid and winged keel.

{⊙ Ten/Hel}. Endemic to northern India, Nepal and Pakistan: Ind, Nep, Pak: Ap, Gj, Hp, Jk, Kt, Mh, Mp, Or, Pj, Rj, Up, Wb. 0–1100 m.

Cyperus amabilis Vahl: Fig. 86

Culms solitary or tufted, 5–25 cm tall, 0.3–1.3 mm thick, 3-sided, glabrous. Leaves mostly shorter than the culm; sheaths membranous, striate, reddish-brown to purple, rather short; blades flat or enrolled, 2–10 cm long, 0.5–1 (–2.5) mm wide, the margins scabrous towards the tip. Involucral bracts 3–6, unequal, mostly shorter than the inflorescence, 1–10 cm long, 0.5–2.5 mm wide. Inflorescence simple or compound, of 1–3 sessile and 1–10 stalked clusters of digitately arranged spikelets; rays 1–7 cm long. Spikelets linear, strongly flattened, 3–18 mm long, 1–1.5 mm wide (or up to 3.5 mm wide with glumes spreading at right angles), reddish-brown to golden; rachilla slightly flexuous, persistent, without wings. Glumes ovate, spreading, acute, mucronate, 1.3–1.5 mm long, ±0.6 mm wide, orange to red-brown with a greenish midrib; keel 3-nerved, yellowish-green, sometimes with purple stripes, elongated into a short (± 0.1 mm long), erect or slightly recurved mucro. Stamen 1 (2). Style as long as the nut; stigmas shorter than the style. Nuts 3-sided, obovoid, half as long as the subtending glume, 0.5–0.6 mm long, 0.3–0.4 mm wide, shortly apiculate, brownish, minutely papillose.

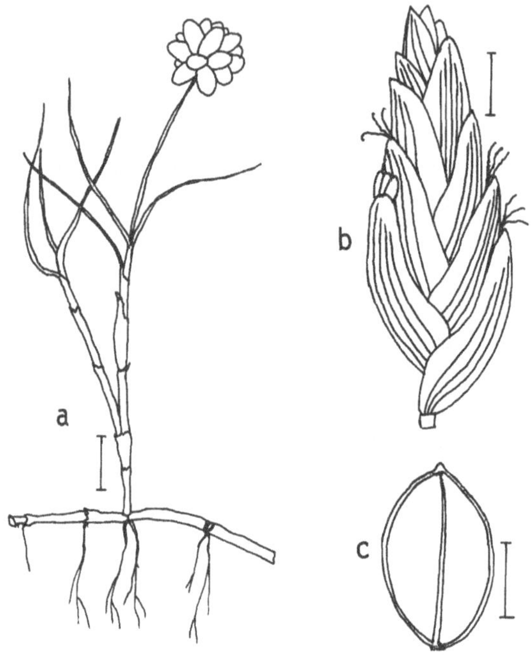

FIG. 87. *Cyperus arenarius*: a, whole plant (1 cm); b, spikelet (1 cm); c, nut (0.3 mm).

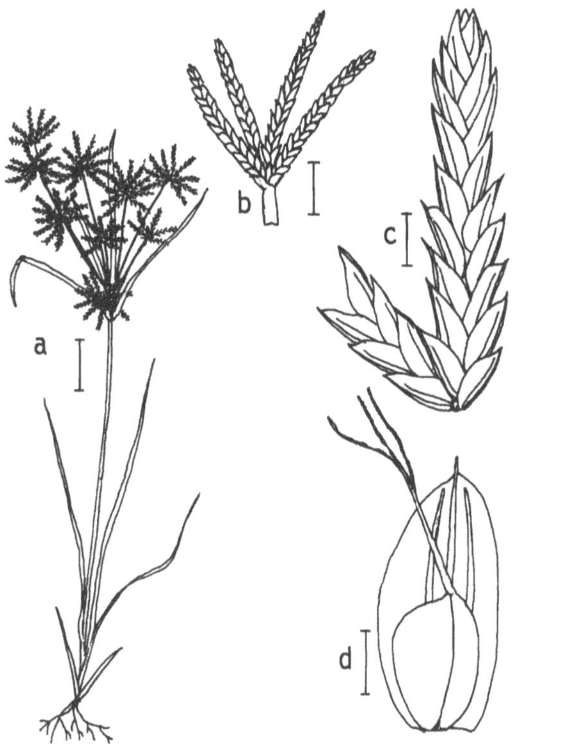

FIG. 86. *Cyperus amabilis*: a, whole plant (1 cm); b, spikelet cluster (3.5 mm); c, one and a half spikelets (1 mm); d, nut with glume (0.3 mm).

Annual. In seasonally flooded places, swamps, edges of lakes, usually on sandy soil. Flowering as the water recedes. Like *C. castaneus* but with numerous stalked spikelet clusters, a short and erect mucro on the glumes and dark orange-brown spikelets. It is so similar to *C. castaneus* that it should perhaps be recognized at the rank of subspecies or variety.

{⊙ Hel}. Tropical America, Africa and S India: Ind: As, Bh, Kl, Or, Up. 0–900 m.

Cyperus arenarius Retzius [*Bobartia indica* Linnaeus, *C. conglomeratus* var. *arenarius* (Retzius) Cosson]: Fig. 87

Rhizomes stolon-like, horizontally long-creeping, up to 1 m long, 0.8–2 mm thick, laxly branched, dark brown, covered with membranous scales. Culms solitary or few together arising from the apices of rhizome branches, obtusely 3-sided, smooth, 6–20 cm tall, 0.8–1.5 mm thick, glaucous green. Leaves 4–8, shorter than or slightly longer than the culm; sheaths membranous, straw-coloured to light brown, 1.5–7 cm long, the lower ones without blades; blades channelled, recurved, (2.5–) 5–20 cm long, 1–3 mm wide, smooth, glaucous green, obtuse at the apex. Involucral bracts 1–3, very unequal, the lowest 3–10 cm long, the remaining much shorter but still overtopping the inflorescence. Inflorescence head-like (rarely ± lobed or bearing up to 3 heads with rays up to 1 cm long), globose, 1.5–2.5 (–3) cm in diameter, with 5–30 spikelets. Spikelets ovate-lanceolate, 6–13 mm long, 2–5 mm wide, 8- to 17-flowered, pale straw-coloured; rachilla straight, not winged. Glumes ovate to oval, 4–5 mm long, 2–3 mm wide, obtuse or shallowly emarginate, minutely mucronate, 9- or 11- (13)-nerved, straw-coloured and tinged with brown with red-brown along the nerves, densely imbricate; keel weak. Stamens 3. Style 5–6 mm long, 3-cleft; stigmas up to half as long as the style. Nuts dorsally flattened, broadly obovate, 1.5–1.8 mm long, 1.1–1.3 mm wide, maturing black-brown.

Perennial. In shallow water in sandy soil, on river banks and also found in dry sandy situations, mostly near the sea. Often forming large single-species stands. Easily recognizable by the whitish head with relatively broad spikelets and the extensive rhizome system.

{Hyp/Hel}. Asia, from S Iran to Indochina: Cey, Ind, Pak: Gj, Kl, Kt, Mh, Or, Pj, Tn.

Cyperus articulatus Linnaeus: Fig. 88

Like *C. corymbosus* but culms terete, transversely septate, the intersepta 5–20 mm long; leaf blades completely absent; involucral bracts scale-like, not more than 15 mm long; floral glumes 2.25–3.5 mm long; anthers 1–1.5 mm long.

Perennial. In or by water, in swamps, lake shores, streams and rivers.

{Hyp/Ten?}. Pantropics and subtropics: Ban, Cey, Ind: Ap, Dl, Gj, Kl, Kt, Mh, Tn. 0–1300 m.

Cyperus bifax C. B. Clarke [*C. retzii* Nees, *C. rotundus* subsp. *retzii* (Nees non Poiret) Kükenthal, *C. disruptus* C. B. Clarke, *C. tuberosus* auct. Ind., non Rottbøll]: Fig. 89

Like *C. rotundus* but larger in every respect: culms up to 80 cm tall; scales of stolons persisting into the second year; lowest involucral bract 2–3 times longer than the inflorescence; broader spikelets (up to 5 mm wide when ripe); glumes thinner, membranous, elliptic-oblong, 3.5–5 mm long, paler (ferruginous); straight keel of the glumes; freely fruiting. Nuts obovoid, up to 2 mm long, smooth, dark brown.

Perennial. Found in shallow water at the edges of lakes, on open wet ground, along ditches and the margins of rice fields.

{Hel}. S and SE Asia and Australia: Cey, Ind: Ap, Dl? Kl, Lk, Tn.

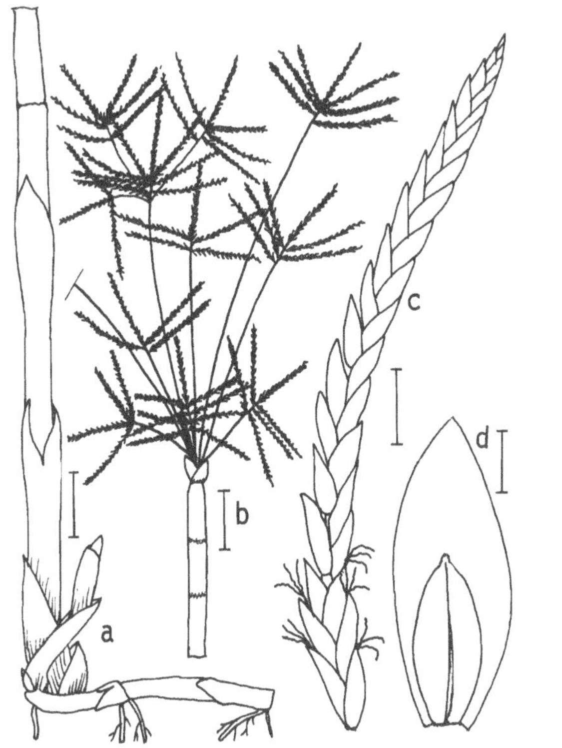

FIG. 88. *Cyperus articulatus*: a, base of shoot (2 cm); b, part of inflorescence (2 cm); c, spikelet (3 mm); d, nut and glume (0.5 mm).

Cyperus castaneus Willdenow: Fig. 90

Like *C. cuspidatus* but inflorescence often with several stalked clusters of spikelets; spikelets narrower, ± 1.5 mm wide (with mucros included ± 2 mm), often many-flowered (up to 50 (–80) flowers); glumes less spreading, chestnut brown; mucros less recurved, shorter 0.3–0.8 mm long; stamen 1; nuts oblong, with exactly parallel sides, 0.6–0.9 mm long, 0.2–0.3 mm wide, rarely smaller and subovoid.

Annual. Found in seasonally flooded places; flowering as the water recedes.

{⊙ (Ten)/Hel}. Tropical Asia and Australia: Ban, Bur, Cey, Ind, Nep: Kl, Kt, Mh, Mp, Sk, Tn, Up.

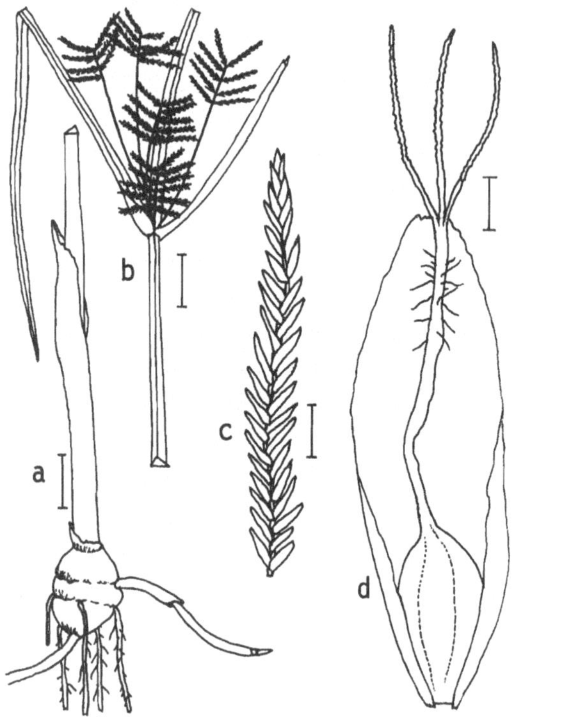

Fig. 89. *Cyperus bifax*: a, base of shoot (2 cm); b, part of inflorescence (2 cm); c, spikelet (5 mm); d, nut and glume (0.5 mm).

Cyperus cephalotes Vahl [*C. hookerianus* Thwaites]: Fig. 91

Rhizomes short but developing stolons; stolons up to 50 cm long, 0.5–1.2 mm thick, rooting at nodes. Culms solitary or few together, 3-sided, becoming 3-angled below the inflorescence, (10–) 15–40 (–55) cm tall, (0.5–) 1–2 mm wide, rather soft, smooth. Leaves half as long as or overtopping the culm; sheaths 1–11 cm long, scale-like below, becoming brown; blades flat, stiffish, 6–55 cm long, (0.5–) 2–4 mm wide, gradually narrowing to a 3-angled tip, scabrous on nerves and margins. Involucral bracts 3–5, unequal, the longer ones much overtopping the inflorescence, up to 30 cm long. Inflorescence an ovoid to conical, ± lobed head; heads 0.5–1.5 cm across, bearing 1–4 glomerules of 3–8 spikelets. Spikelets ovate to lanceolate-ovate, 4–8 (–10) mm long, 3–4 mm wide, densely 10- to 25-flowered, reddish-brown with yellowish margins; rachilla straight, not winged. Glumes leathery, ovate, 3–4 mm long, 1.8–2.5 mm wide, mucronate at acute apex, obscurely 3- to 5-nerved, yellow-brown and stained with chestnut brown; keel 3-nerved, projecting into a straight or slightly curved ± 0.3 mm long mucro. Stamens 3. Style 1–3 (–4) mm long, 3-cleft or sometimes entire. Nuts 3-sided and dorsally flattened, 1–1.5 mm long, 0.7–1 mm wide, dark brown, minutely punctuate, apiculate at the apex with a ± 0.6 mm long style base; stipe 1–1.2 mm long, 3-winged at anthesis, the stipe and wings developing into yellowish corky ridges covering the lower half of the nut.

Perennial. An emergent aquatic often forming floating mats in tanks, pools, ditches and canals. Also found in marshy places and rice fields. Used locally as fodder. The nuts have corky ridges and float; they are presumably dispersed by water.

{Hyp/Ple}. Tropical Asia and Australia: Ban, Bur, Cey, Ind: An, Ap, As, Bh, Go, Kl, Kt, Mn, Mp, Tn, Wb. 0–100 m.

Cyperus compressus Linnaeus [including subsp. *micranthus* T. Koyama, var. *capillaceus* C. B. Clarke apud Urban, *C. brachiatus* Poiret, *C. pectiniformis* Roemer et Schultes, *C. meyenii* Nees]: Fig. 92

Culms tufted, somewhat spreading, (8–) 10–40 (–50) cm tall, 1–2.5 mm thick, smooth, 3-sided. Leaves shorter to somewhat longer than the culm; sheaths membranous, reddish-brown; blades flat or folded, up to 15 cm long, (0.5–) 2–4 mm wide, gradually acuminate, the margins scabrous towards the tip. Involucral bracts 2–5, unequal, the lowest overtopping the

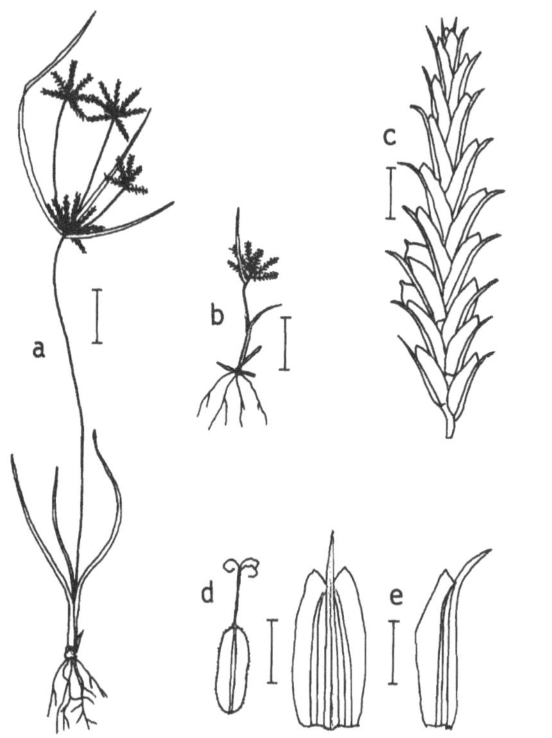

Fig. 90. *Cyperus castaneus*: a, whole plant (1 cm); b, dwarf plant (1 cm); c, spikelet (1 mm); d, nut (0.5 mm); e, glume viewed from back and side (0.5 mm).

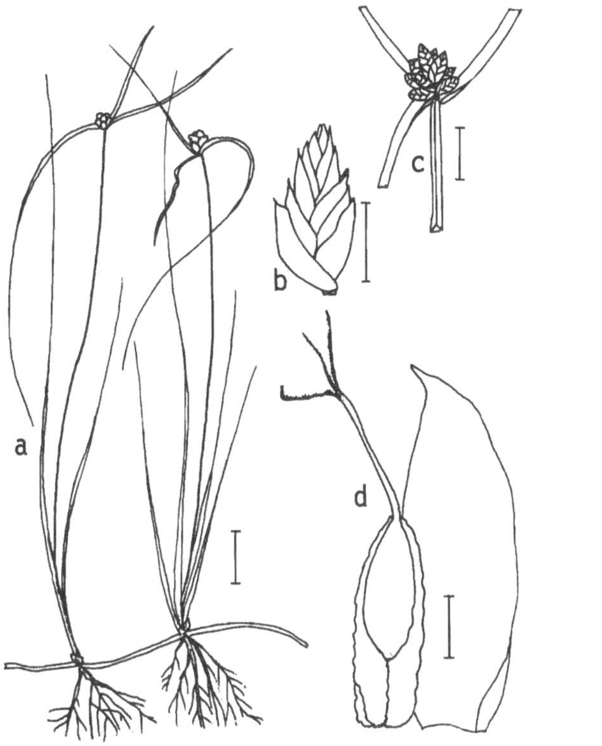

Fig. 91. *Cyperus cephalotes*: a, whole plant (3 cm); b, spikelet (3 mm); c, inflorescence (1 cm); d, nut and glume (0.5 mm).

inflorescence, up to 30 cm long. Inflorescence simple, open, sometimes reduced to a single sessile cluster; rays 0–8, unequal, up to 15 cm long. Spikes of 3–7 (–10) digitately or racemosely arranged spikelets. Spikelets linear to oblong, 10–35 mm long, 3–5 mm wide, 15- to 40-flowered, flattened, greenish turning straw-coloured at maturity; rachilla flexuous, persistent, with very thin caducous wings. Glumes ovate, acute, 3–5 mm long, conspicuously multinerved, with green keel, pale green to yellowish-brown sides and silvery hyaline margins, mucro ± 0.8 mm long, straight to slightly curved. Stamens 3. Style 3-cleft. Nuts 3-sided, broadly obovoid, (1–) 1.3–1.7 mm long, ± 1 mm wide, stipitate, apiculate, shiny, minutely punctate, dark brown to blackish.

Annual. Widespread and common on sandy soil, a weed of waste places, roadsides and cultivated land but also found in wet places such as rice fields, along water courses and at the edge of drying pools. The nuts remain attached to the rachilla and germinate together thus it often grows in tufts, it is without stolons or runners. The whole plant is grey or silvery-green. It is very variable in size and some intraspecific taxa have been recognized. It is distinguishable from most other similar looking sedges by its large glumes (3–4 mm long) with silvery margins.

{⊙ Ten/Hel}. Almost cosmopolitan: Ban, Bhu, Bur, Cey, Ind, Nep, Pak: <An, Ap, Ar, As, Dl, Gj, Go, Hp, Jk, Kl, Kt, Mg, Mh, Mp, Pj, Rj, Sk, Tn, Up, Wb. 0–3300 m.

Cyperus corymbosus Rottbøll [*C. tegetiformis* Roxburgh, *C. corymbosus* var. *longispiculatus* (O. Kuntze) Kükenthal]: Fig. 93

Rhizomes long-creeping, stoloniferous, 2–5 mm thick. Culms arranged in a row along the rhizome, spaced or 2 or 3 close together, stiffly erect, 60–150 (–200) cm tall, 4–10 mm thick, subterete to weakly 3-sided, rarely septate, glaucous green, smooth. Leaves mostly reduced to bladeless sheaths; sheaths scale-like below, the uppermost one 20–35 cm long, reddish-brown to dark brown; blades rarely present, up to 3 (–18) cm long. Involucral bracts 2–4, unequal, leafy, shorter than the inflorescence, the longest 2–3.5 (–20) cm long, up to 7 mm wide. Inflorescence compound paniculate, consisting of 1 sessile spike and several stalked spikes with or without secondary spikes; rays 6–15. Spikes corymbose, ovoid,

Fig. 92. *Cyperus compressus*: a, base of shoot (5 mm); b, inflorescence (1 cm); c, spikelet (3 mm); d, nut and glume (0.5 mm).

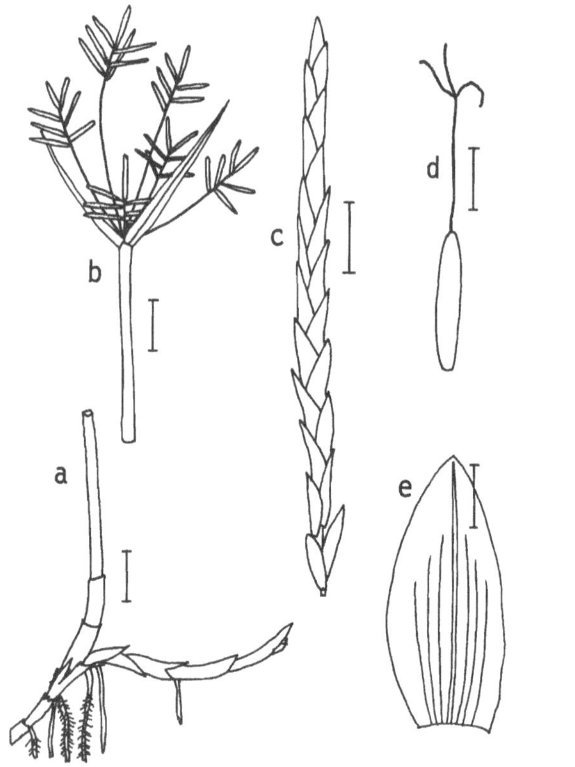

FIG. 93. *Cyperus corymbosus*: a, base of shoot (2 cm); b, inflorescence (2 cm); c, spikelet (3 mm); d, young fruit (1 mm); e, glume (1 mm).

10–25 mm long and wide, with a 2–10 mm long rachis, bearing 10–15 spikelets. Spikelets spicate, narrow-oblong, 5–18 mm long, 1–1.5 mm wide; rachilla straight, persistent, winged; wings 0.2 mm wide, reddish-brown, persistent. Glumes membranous, broadly lanceolate to ovate-oblong, 1.8–2.2 (–3) mm long, 0.8–1 mm wide, obtuse, mucronulate at tip; sides with 2 or 3 veins, grey to reddish-brown on the back, with scarious, weakly incurved margins; keel green. Stamens 3; anthers 0.7–1 mm long. Styles deeply 3-cleft. Nuts 3-sided, oblong, 1–1.2 mm long, 0.2–0.3 mm wide, maturing brown or grey.

Perennial often growing in large colonies. Marshy places, pools, ditches, streams, river beds and seasonally flooded areas. Sometimes found in brackish conditions. The culms are used for making fine mats and is sometimes cultivated, particularly in Tamilnadu where it is called 'pangorai'. It is like *C. articulatus* but culms not so clearly septate, involucral bracts leaf-like.

{Ten/Hel}. Pantropics and subtropics: Ban, Bur, Cey, Ind, Nep: As, Bh, Dl, Gj, Go, Kl, Kt, Mh, Mn, Mp, Or, Pj, Rj, Tn, Up, Wb. 0–600 m.

Cyperus cuspidatus Kunth [*C. angustifolius* Nees, *C. uncinatus* auct. Ind., non Poiret]: Fig. 94

Culms solitary or tufted, smooth, 2–15 cm tall, 0.2–0.7 (–1) mm thick. Leaves shorter to about as long as the culm; sheaths membranous, striate, reddish-brown to purple; blades very narrow, almost wire-like, 1–8 cm long, 0.2–1 mm wide, gradually acuminate, the margins scaberulous towards the tip. Involucral bracts 3–6, unequal, the lowest overtopping the inflorescence, 2–8 cm long. Inflorescence with 1 sessile and 1–3 stalked spikelet clusters or of 1 sessile cluster; rays 0–3, filiform, unequal, up to 3 cm long. Spikes of digitately arranged, stellately spreading clusters of (1–) 4–20 spikelets. Spikelets linear, strongly flattened, 4–15 (–20) mm long, 1.7–3 mm wide, 8- to 40- (–60)-flowered; rachilla slightly flexuous, persistent, without wings. Glumes narrowly elliptical to oblong-spathulate, spreading, acute, awned, 1.4–2 mm long (including the arista), 0.6–1 mm wide, emarginate at apex, the sides rust to red-brown; keel 3-nerved, yellowish-green, sometimes with purple stripes, elongated into 0.7–1 mm long, spreading or recurved awn. Stamens (2) 3. Style ± 0.5 mm long; stigmas shorter than the style. Nuts 3-sided, obovoid, half as long as the subtending glume, (0.5–) 0.6–0.8 mm long, 0.3–0.4 mm wide, shortly apiculate, brown to chestnut brown.

FIG. 94. *Cyperus cuspidatus*: a, whole plant (1 cm); b, spikelet (1 mm); d, nut (0.3 mm).

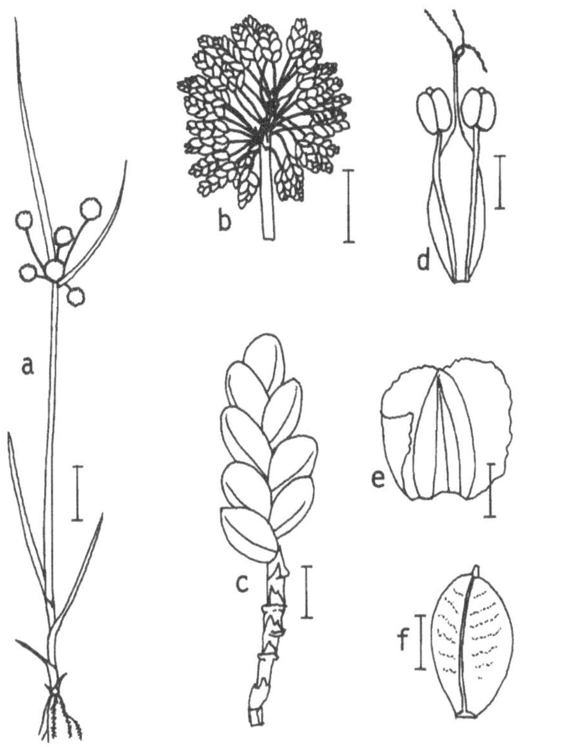

Fig. 95. *Cyperus difformis*: a, whole plant (2 cm); b, cluster of spikelets in longitudinal section (4 mm); c, spikelet (1 mm); d, flower (0.2 mm); e, glume (0.3 mm); f, nut (0.3 mm).

Annual. Seasonally wet places, margins of ponds, rice fields, usually on sandy soil. It is rather common but it superficially resembles *Lipocarpha kernii, L. squarrosa* and *Mariscus squarrosa* which differ in having true spikelets (not digitate clusters). It is most similar to *C. castaneus*, which has narrower spikelets (± 1.5 mm wide), less curved awns and parallel-sided nuts, and *Mariscus maderaspatanus*, which has larger glumes (± 2.5 mm long) and the bracts at base of rays (prophylls) are keeled and tubular at base.

{⊙ Hel}. Pantropics: Ban, Bhu, Bur, Cey, Ind, Nep, Pak: An, Ap, Ar, As, Jk, Kl, Kt, Mg, Mh, Ng, Pj, Rj, Sk, Tn, Up, Wb. 0–1700 m.

Cyperus difformis Linnaeus: Fig. 95

Culms tufted, smooth, 8–40 (–75) cm tall. Leaves usually slightly shorter than the culm; sheaths yellowish to brown; blades folded, up to ± 20 cm long, 2–5 (–8) mm wide, acuminate, the margins scabrous towards the tip. Involucral bracts 2–4, unequal, the lower two overtopping the inflorescence, up to 25 cm long. Inflorescence simple or in part compound, rather lax but sometimes reduced to a single, globose cluster of spikelets; rays 3–9, unequal, 0.2–6 cm long. Spikelet clusters aggregated into spherical heads, (3–) 8–15 mm in diameter. Spikelets sessile, linear, 2–5 (–8) mm long, 1–1.3 mm wide, 5- to 30-(–40)-flowered; rachilla straight, not winged. Glumes very small, nearly orbicular, 0.5–0.8 (–1) mm long, 0.5–1 mm wide, closely imbricate, at first green, becoming reddish; keel green becoming yellowish, faintly 3-nerved, not prolonged. Stamens 1 (2 or 3). Style ± half as long as the nut. Nuts 3-sided, ellipsoid, 0.5–0.7 mm long, ± 0.3 mm wide, yellowish to pale brown.

Annual. Very common, often found in shallow water in seasonally flooded areas, ponds, ditches, frequently in rice fields. Usually forming clumps. In certain situations it can be a very undesirable weed. Easily identified by its head-like clusters of spikelets with orbicular glumes less than 1 mm long.

{⊙ Hyp/Ten/Hel}. Warm and tropical Old World, naturalized in America: Ban, Bhu, Bur, Cey, Ind, Nep, Pak: An, Ap, As, Dl, Gj, Go, Hp, Jk, Kl, Kt, Mh, Mp, Pj, Rj, Sk, Tn, Up, Wb. 0–2700 m.

Cyperus digitatus Roxburgh [*C. hookeri* Boeckeler]: Fig. 96

Rhizomes short, suberect, woody, 8–12 mm thick. Culms tufted or solitary, smooth, 50–150 (–200) cm tall, (3–) 5–10 (–20) mm thick, 3-sided below, 3-angled to 3-winged above. Leaves few, shorter than to as long as the culm; sheaths purplish; blades flat or the larger ones somewhat plicate with 1 or 3 prominent nerves, herbaceous to leathery, up to 50 cm long, 4–10 (–15) mm wide, glaucous, scabrous along the margins and midrib. Involucral bracts 3–8, the lower ones much overtopping the inflorescence, up to 60 cm long, 4–12 mm wide. Inflorescence of 1 sessile and 2–9 stalked groups of spikes, rather dense, rarely congested; rays 4–9, spreading, unequal, up to 30 cm long. Spikes 4–10, cylindrical, 2–6 cm long, 1–3 cm in diameter, rather loose, with (4–) 15–60 spikelets. Spikelets sessile, finally at right angles to the rachis, subterete, linear, 4–20 mm long, 1–1.5 mm in diameter, 5- to 30-(–45)-flowered, straw-coloured to golden or reddish-brown; rachilla straight, persistent, with caducous wings up to 1 mm wide. Glumes elliptic-oblong, acute, apiculate, 1.6–2.3 mm long, ± 1 mm wide; sides nerveless, yellow to brown; keel 1-, 3- or (5)-nerved, green to reddish-brown, ending at the apiculate apex, not forming a mucro. Stamens 3; anthers with a very short appendage. Style up to 3 times longer than the nut, 3-cleft. Nuts 3-sided, ellipsoid to oblong-obovoid, 0.9–1.4 mm long, 0.4–0.5 mm wide, yellowish brown, minutely papillose.

Perennial. Found in or near water in swamps, or seasonally flooded areas, wet rice fields, ditches and river-

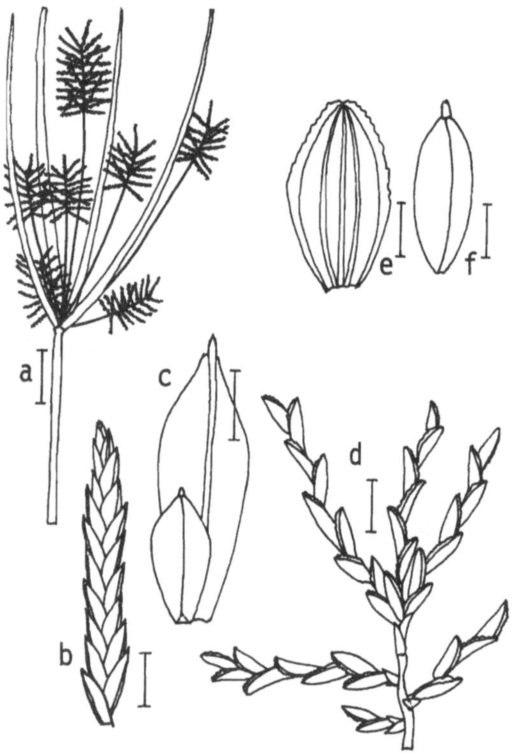

FIG. 96. *Cyperus digitatus*: a, part of inflorescence somewhat simplified (2 cm); b, spikelet (2 mm); c, glume and nut (0.5 mm). *C. distans*: d, upper part of spike (2 mm); e, glume (0.5 mm); f, nut (0.5 mm).

banks. It is a very variable species: *Cyperus hookeri* is sometimes recognized at varietal rank on the basis of dense and rigid spikelets and brown glumes.

{Ten/Hel}. Pantropics and subtropics: Ban, Bhu, Bur, Cey, Ind, Nep, Pak: Ar, As, Jk, Kt, Mg, Mh, Mn, Mz, Ng, Pj, Rj, Sk, Tn, Up, Wb. 0–2000 m.

Cyperus distans Linnaeus fil. [including var. *pseudonutans* Kükenthal]: Fig. 96

Rhizomes short and knotty. Culms solitary or few together in a row, rather slender, 3-sided, 20–60 (–100) cm tall, 2–3 (–5) mm wide. Leaves shorter than or about as long as the culm; sheaths 5–20 cm long, light brown to purplish; blades weakly plicate, 5–30 cm long, 4–6 (–10) mm wide, acuminate; margins scabrous towards tip. Involucral bracts 4–6, unequal, the lower 2 or 3 overtopping the inflorescence, 20–40 (–60) cm long. Inflorescence paniculate, corymbose; primary rays 5–12, very unequal, up to 18 cm long, secondary or tertiary rays bearing spikes of 4–8 subsessile spikelets. Spikes broadly ovate-pyramidal in outline, 2–4 cm long, loose, with 8–20 spikelets. Spikelets arranged in spikes, divergent, linear, subterete, 8–40 mm long, 0.5–1 mm wide, 6- to 32-flowered; rachilla slightly flexuous, zigzag, persistent, winged. Glumes distant on a weakly zigzag rachilla, membranous, elliptical or ovate-oblong, 1.8–2 mm long, 1–1.2 mm wide, obtuse at apex, 3- to 5-nerved near the keel, light to dark reddish-brown; keel greenish, hardly reaching the glume apex; margins of the upper half white hyaline. Stamens 3. Nuts 3-sided, oblong to oblong-ellipsoid, 0.5–1 mm long, 0.2–0.5 mm wide, apiculate, smooth with a metallic shine, yellowish maturing dark brown.

Perennial. Wet places, edges of pools, ditches and streams, in rice fields, often in shallow water, sometimes in mats of floating vegetation, also found in wet grassland and as a weed of roadside banks and cultivated fields. A rather variable species and some infraspecific taxa have been recognized. The spikelets break up rather irregularly and within the same plant some may even fall off whole but this is not grounds enough to place it in *Mariscus*.

{Ten/Hel}. Pantropics and subtropics: Ban, Bhu, Bur, Cey, Ind, Nep: An, Ap, As, Jk, Kl, Kt, Mg, Mh, Mp, Pj, Sk, Tn, Up, Wb. 0–2600 m.

Cyperus elatus Linnaeus [*C. racemosus* Retzius]: Fig. 97

Rhizomes very short. Culms tufted, smooth, 50–150 (–200) cm tall, (3–) 4–7 (–10) mm thick, 3-sided below, 3-angled above. Leaves few, shorter than or as long as the culm; sheaths purplish to almost black; blades flat or the larger ones somewhat plicate with 1 or 3 prominent, somewhat winged nerves, (4–) 6–15 mm wide, glaucous, scabrous along the margins and midrib. Involucral bracts 4–8, the lower ones much overtopping the inflorescence, up to 75 cm long, up to ± 10 mm wide. Inflorescence compound, of sessile and stalked, digitately arranged groups of spikes; rays up to 10, spreading, unequal, up to 20 cm long; bracts (prophylls) at the base of the rays with a scabrous keel. Spikes 4–10, cylindrical, (1–) 2–5 (–7) cm long, 3–5 mm in diameter, dense (rachis not visible at anthesis), with up to 40 or more spikelets. Spikelets sessile, suberect, flattened, at anthesis lanceolate, 3–6 mm long, 1–2 mm wide, 6- to 16-flowered; rachilla slightly flexuous, persistent, with yellow caducous wings. Glumes ovate, acute, 1.2–1.8 mm long, ± 1 mm wide; sides nerveless, golden to reddish-brown; keel 3- to 5-nerved, green to reddish-brown. Stamens 3; connective appendage bristle-like, ± one-third as long as the anther. Style ± 1 mm long, 3-cleft; stigmas slightly shorter than the style. Nuts 3-sided and 3-angled, ellipsoid to oblong-obovoid, 0.8–0.9 mm long, 0.3–0.4 mm wide, shiny, yellowish-brown to grey.

Perennial. In swamps, seasonally flooded places, in ditches, along rivers and in wet rice fields.

{Hyp/Ten}. S and SE Asia: Ban Ind: Tn, Up.

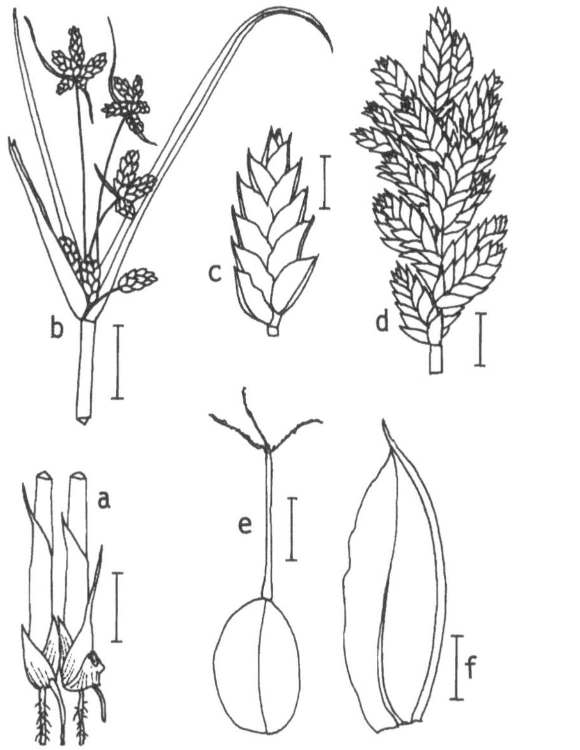

Fig. 97. *Cyperus elatus*: a, base of shoot (4 cm); b, part of inflorescence (4 cm); c, spikelet (1 mm); d, small spike (2 mm); e, nut (0.5 mm); f, glume (0.5 mm).

Cyperus esculentus Linnaeus: Fig. 98

Like *C. rotundus* but tubers covered with a grey tomentosum; spikes shorter (1–3 cm long); spikelets yellowish to rusty brown, obtuse, 5–20 mm long; glumes yellow to light brown, smaller (2.2–2.6 mm long), with more nerves (7 or 9), the nerves reaching the margin more closely.

Perennial. By no means a wetland plant but often found in seasonally flooded areas and in swamps, but usually found in cultivated land. Rarely cultivated for its edible tubers.

{(Hel)}. Almost cosmopolitan in warmer regions: Ban, Ind, Nep: Dl, Jk, Kl, Kt, Mh, Pj, Tn. 0–2300 m.

Cyperus exaltatus Retzius [including var. *serpens* Kükenthal, *C. venustus* R. Brown, *C. altus* Nees]: Fig. 99

Rhizomes very short, ± 1 cm thick, woody. Culms densely tufted, smooth, (60–) 90–180 cm tall, 6–10 mm thick, 3-sided, smooth. Leaves nearly equalling the culm; sheaths somewhat spongy, reddish to blackish-brown; blades flat or plicate, herbaceous, up to ± 80 cm long, 6–15 mm wide, 3-nerved, scabrous margins. Involucral bracts 4–6, unequal, the lower ones overtopping the inflorescence, 30–80 cm long, 8–12 mm wide, erect or spreading. Inflorescence compound, of groups of subsessile and stalked spikes; rays 5–15, unequal, up to 20 cm long. Spikes 2–5 cm long, 1.5–2.5 cm wide when mature, with 15–60 spreading and rather distant spikelets. Spikelets loose or dense, semi-erect, linear to oblong, flattened, reddish-brown, 4–12 mm long, ± 1.5 mm wide, (6–) 12-to 20-flowered; rachilla persistent, winged; wings whitish or yellowish, caducous. Glumes ovate to broadly ovate, 1.5–2.4 mm long, 1.2–1.5 mm wide, 3- to 5- (7)-nerved, shining brown or greenish, with hyaline margins; keel green, excurrent to a short mucro. Stamens 3. Style slender, 3-cleft. Nuts 3-sided, ellipsoid to obovoid-ellipsoid, 0.6–1 mm long, 0.3–0.6 mm thick, yellow, yellowish-brown or grey, almost smooth.

Perennial. Mostly gregarious in swamps or in open water, in wet or marshy places, also found in rice fields. Cultivated for making course mats, called 'padupai korai' in Tamilnadu. Differs from *C. digitatus* in having flattened and wider spikelets which are semi-erect; it also has a characteristic reddish-brown colour.

{Ten/Hel}. Old World tropics and subtropics, doubtfully recorded from S America: Ban, Cey, Ind: Ap, As,

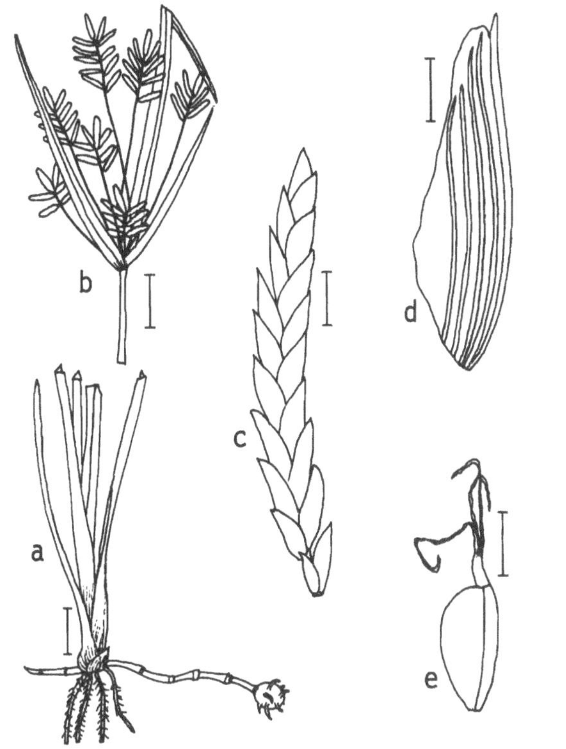

Fig. 98. *Cyperus esculentus*: a, base of shoot (1 cm); b, inflorescence (2 cm); c, preanthesis spikelet (2 mm); d, glume (0.5); e, nut (0.5 mm).

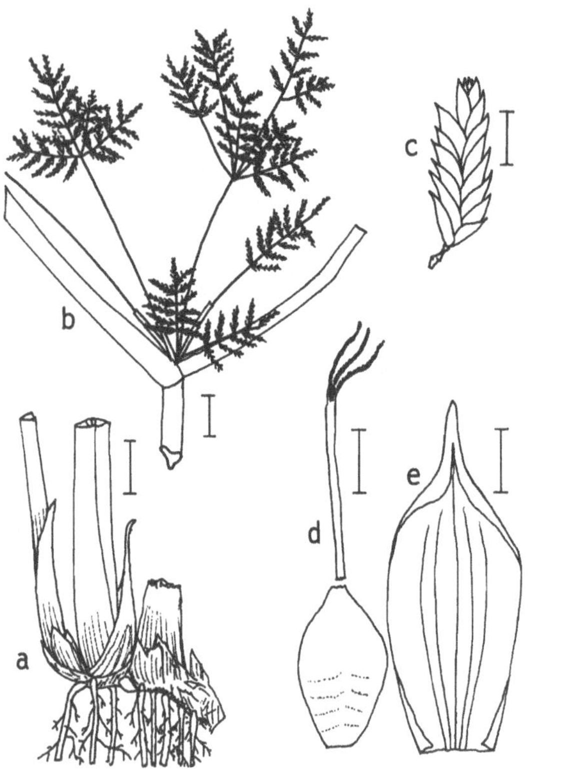

Fig. 99. *Cyperus exaltatus*: a, base of shoot (1 cm); b, part of inflorescence (2 cm); c, spikelet (2 mm); d, nut (0.25 mm); e, glume (0.25 mm).

Bh, Gj, Go, Kl, Kt, Mg, Mh, Mp, Or, Pj, Rj, Tn, Up, Wb. 0–2100 m.

Cyperus haspan Linnaeus subsp. ***haspan*** [*'halpan'* orth. mut.]: Fig. 100

Rhizomes elongate, short or absent. Culms slender, weak, tufted or sometimes solitary, smooth, (5–) 10–40 cm tall, 0.5–2.5 mm thick, acutely 3-angled. Leaves 2–4, as long as or shorter than the culm; sheaths bladeless below, bladed or rarely bladeless above, up to 6 cm long, yellowish to red-purple or brown; blades flat or slightly plicate, flaccid, 2–20 cm long, 2–5 mm wide, gradually acuminate, smooth or slightly scabrous above. Involucral bracts 2 or 3, unequal, the lower one usually overtopping the inflorescence, up to 8 cm long. Inflorescence compound, lax, relatively large; primary rays 10–22, up to 10 (–20) cm long; secondary rays, 5–7, 1–2.5 mm long, tertiary rays when present very short. Spikelets digitately arranged in clusters of 3–10 spikelets, stellately spreading, oblong to linear, acute, flattened, 3–12 mm long, 1–2 mm wide, with 10–28 (–40) flowers; rachilla slightly flexuous, hidden by the glumes, persistent, without wings. Glumes ovate, slightly keeled towards the tip, obtuse, soon excurved, 1–1.5 mm long, 0.5–0.7 mm wide, reddish-brown often with purple stripes or a purple spot on one or each side, the sides nerveless, hyaline; keel green, 3-nerved, the midrib excurrent as a short and straight mucro just below the tip. Stamens (1–) 3; connective appendage hairy. Style 1.5 times as long as the nut, 3-cleft; stigmas ± as long as the style. Nuts obtusely 3-sided with convex sides, subglobose to ellipsoid, 0.4–0.7 mm long, 0.3–0.4 mm wide, broadly stipitate, hardly apiculate, densely tuberculate, marbly white to cream-yellow.

Short-lived perennial or sometimes annual. In very wet places, often standing in water or in mats of floating vegetation. A common weed in drainage and irrigation ditches and rice fields. *Cyperus haspan* is very variable and may not be distinct from subspecies *juncoides* in India; further investigations are needed. It is like *C. tenuispica* but usually perennial; glumes straight, obliquely erect, glumes larger (more than 1 mm long), not excurved at top; white nuts not visible between the opened glumes in fruiting spikelets; connective appendage with bristles.

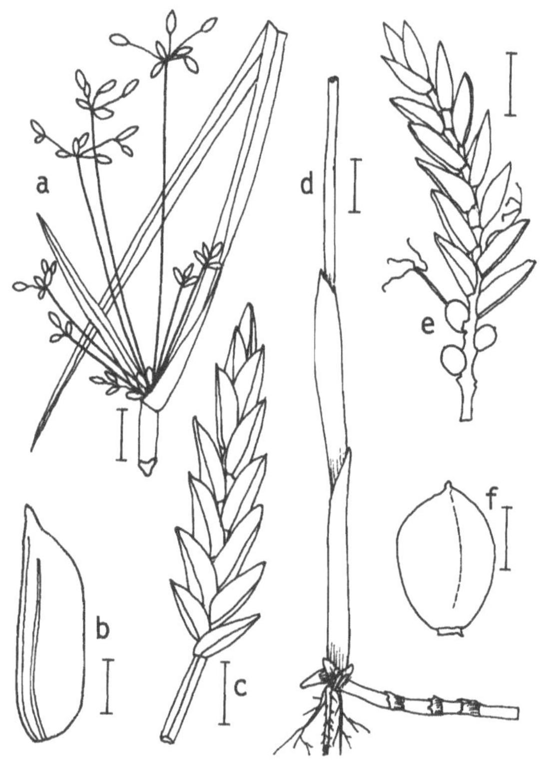

Fig. 100. *Cyperus halpan* subsp. *halpan*: a, inflorescence (1 cm); b, glume (0.25 mm); c, spikelet (1 mm). *C. halpan* subsp. *juncoides*: d, base of shoot (1 cm); e, spikelet (1 mm); f, nut (0.25 mm).

{(⊙) Hyp/Ten}. Tropical and subtropical regions of the world: Ban, Bhu, Bur, Cey, Ind, Nep: An, Ap, Ar, As, Go, Kl, Kt, Mg, Mn, Mp, Ng, Sk, Tn, Up, Wb.

Cyperus haspan Linnaeus subsp. ***juncoides*** (Lamarck) Kükenthal [*C. juncoides* Lamarck]: Fig.100

Differs from subsp. *haspan* by having a long, creeping rhizome; longer culms, up to 80 (–100) cm long, at the base surrounded by bladeless, reddish sheaths; blades rarely developed; glumes not completely hiding the rachilla, mucronulate; nuts larger, 0.5–0.8 mm long.

Perennial. In seasonally flooded places, waterlogged soils, damp open situations and along the banks of streams. Subspecies *juncoides* is essentially American. The plants called subsp. *juncoides* in India are not always not clearly separated from subspecies *haspan*. In many ways *C. haspan* subsp. *juncoides* is like the African species *Cyperus denudatus* Linnaeus fil.; further investigations are needed.

{Ten/Hel}. American and Asian tropics and subtropics: Ind: Ap, Kl, Kt, Up, Wb.

Cyperus imbricatus Retzius [*C. radiatus* Vahl]: Fig. 101

Rhizomes short, woody. Culms robust, tufted, 3-sided, smooth, 40–120 cm tall, 2–5 mm thick. Leaves few; sheaths light brown and purple; blades flat, the largest 20–40 cm long, 4–8 mm wide, scabrous on nerves and margins. Involucral bracts 4–7, unequal, the longer ones much overtopping the inflorescence, 25–50 cm long, 5–8 mm wide. Inflorescence of a few sessile spikes and 3–8 groups of spikes on 1–12 cm long peduncles, each group containing 2–12 densely clustered spikes. Spikes digitate, spreading, ± cylindrical, (0.5–) 2–3.5 cm long, 3–8 mm in diameter, with (10–) 30–80 spikelets. Spikelets spicate, ovate to oblong-lanceolate, 2–5 mm long, 0.8–1.5 mm wide, 6- to 16-flowered, pale brown to golden; rachilla persistent, strongly flattened, winged; wings whitish, very narrow, long-persistent. Glumes ovate to orbicular, obtuse, 0.9–1.2 mm long, 1–1.3 mm wide, 5- or 7-nerved, light brown or golden with or without purple stripes, densely imbricate; keel obscure, midrib green and excurrent in a short mucro. Stamens 3. Style ± as long as the nut; stigmas as long as or slightly longer than the style. Nuts dorsoventrally flattened, planoconvex or ellipsoid, 0.5–0.6 mm long, ± 0.4 mm wide, reddish-brown, smooth or minutely pitted.

Perennial. Seasonally flooded areas, margins of ponds and ditches, swamps, river banks, mostly on sandy soil. Used as fodder in Bengal. Like *C. alopecuroides* but has shorter glumes, smaller nuts and styles 3-cleft.

{Ten?/Hel}. Almost pantropical: Ban, Ind, Nep: As, Kl, Mp, Rj, Tn, Wb.

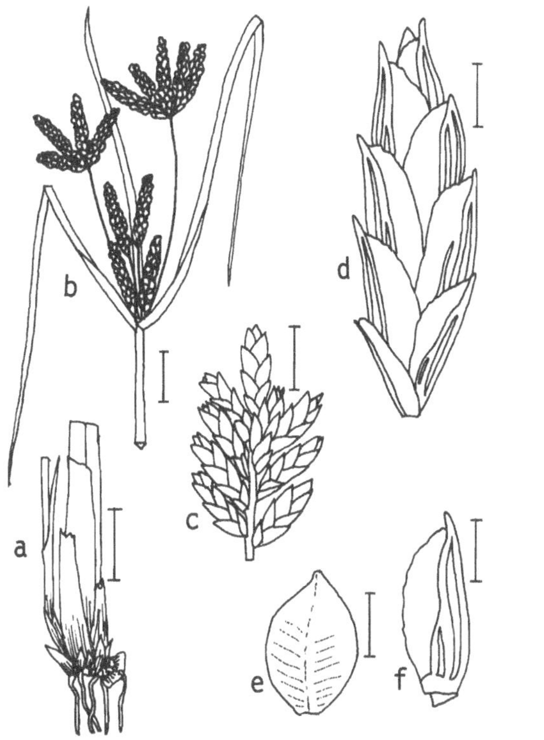

FIG. 101. *Cyperus imbricatus*: a, base of shoot (2 cm); b, part of inflorescence (2 cm); c, upper part of spike (2 mm); d, spikelet (0.5 mm); e, nut (1 mm); f, glume (0.5 mm).

Cyperus involucratus Rottbøll [*C. flabelliformis* Rottbøll, *C. alternifolius* Linnaeus subsp. *flabelliformis* (Rottbøll) Kükenthal]: Fig. 102

Rhizomes creeping, woody, 2–10 mm thick. Culms usually in a straight row, obtusely 3-angled to almost terete, 25–170 cm tall, 5–10 mm in diameter, scabrous just below the inflorescence. Bladed leaves confined to first-year sterile shoots; culm leaves bladeless; lower sheaths scale-like, yellow-brown; upper sheaths 10–20 cm long, obliquely truncate at the orifice, pale green or straw-coloured above, dark purple to black below. Involucral bracts numerous (15–25), rather equal, (10–) 20–50 cm long, arising distant from one another, longer ones overtopping the inflorescence. Inflorescence compound, subdense, 15–30 cm in diameter; primary rays numerous, rather equal, each bearing 4–10 raylets, up to 10 cm long; secondary rays spreading, 1–1.5 cm long. Spikelets in digitate clusters of 3–20, lanceolate-oblong to elliptic, flattened, 2–10 mm long, (1–) 1.7–3 mm wide, densely (6–)- 15- to 40-flowered, light green; rachilla without wings. Glumes ovate, acute or mucronate at apex, 1–2 mm long, 1.3–1.5 mm wide; membranous, pale green and spotted with rusty brown, 3- or 5-nerved, the back keeled. Stamens 3; connective

Fig. 102. *Cyperus involucratus*: a, base of shoot (4 cm); b, inflorescence, somewhat simplified (3 cm); c, spikelet (2 mm); d, nut (0.25 mm); e, glume (0.25 mm).

appendage bristly. Style nearly as long as the nut, 3-cleft. Nuts 3-sided, obovoid-ellipsoid, 0.6–0.9 mm long, 0.4–0.5 mm wide, maturing brown, minutely papillose.

Perennial. Widely cultivated as an ornamental, frequently escaping and becoming established. In wet places or standing in still or flowing shallow water.

{Hyp/Hel}. Tropical and subtropical Africa and Arabia, introduced Asia and America: <Cey, <Ind: <Dl, <Kt, <Sk, <Wb.

Cyperus iria Linnaeus: Fig. 103

Culms sparsely tufted or solitary, smooth, (3–) 15–50 (–80) cm tall, 2–3 (–5) mm thick, 3-sided. Leaves 2 or 3, as long as or shorter than the culm; sheaths purplish; blades flat or weakly folded, herbaceous, up to ± 15 cm long, 2–6 mm wide, acuminate, smooth below, scabrous above. Involucral bracts 3–6, unequal, the longer ones overtopping the inflorescence, up to 40 cm long. Inflorescence simple or compound, 1 spike ± sessile, the rest stalked; rays (3–) 5–8, unequal, 0.3–8 (–18) cm long. Spikes 5–10 in clusters, 1–5 cm long, bearing 4–30 spikelets; rachis glabrous. Spikelets erect or spreading, loose, linear-oblong, 4–8 (–15) mm long, 1.5–2 mm wide, yellowish, 5- to 22-flowered; rachilla persistent, unwinged. Glumes broadly obovate to suborbicular, obtuse, 1.2–1.5 mm long, 1–1.5 mm wide, 3- or 5-nerved, yellowish or brown, with white membranous margins, terminating in a short green mucro, hardly imbricating at maturity. Stamens (2) 3; connective appendage reddish. Style 3-cleft, ± one-third as long as the nut. Nuts 3-sided, ellipsoid to ovoid-oblong, 1–1.3 (–1.5) mm long, 0.5–0.8 mm wide, shining, dark brown to grey or black.

Annual or sometimes perennial under favourable conditions. A common weed of cultivated ground and rice fields. During the monsoon it is found in water in marshes and ditches. Fresh plants, particularly the roots, are aromatic and used locally in medicine; also it is used for making mats. It has distinct broad, short, scarcely imbricating glumes, with 3 or 5 nerves.

{(⊙) Ten/Hel}. Warm and tropical Eurasia and Australia, naturalized in America: Bur, Cey, Ind, Nep: <An, Ap, Ar, As, Bh, Dl, Gj, Go, Hp, Jk, Kl, Kt, Mg, Mh, Mn, Mp, Ng, Pj, Rj, Sk, Tn, Tp, Up, Wb. 0–1800 m.

Cyperus laevigatus Linnaeus [*Juncellus laevigatus* (Linnaeus) C. B. Clarke]: Fig. 104

Rhizomes short or long and creeping, 30 cm or more long, 1–5 mm thick, light brown to purple-black,

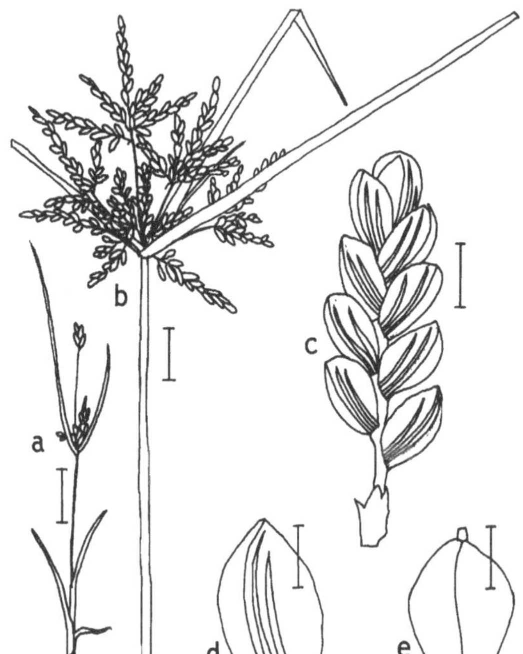

Fig. 103. *Cyperus iria*: a, dwarf plant (1 cm); b, inflorescence (1 cm); c, spikelet (1 mm); d, glume (0.5 mm); e, nut (0.5 mm).

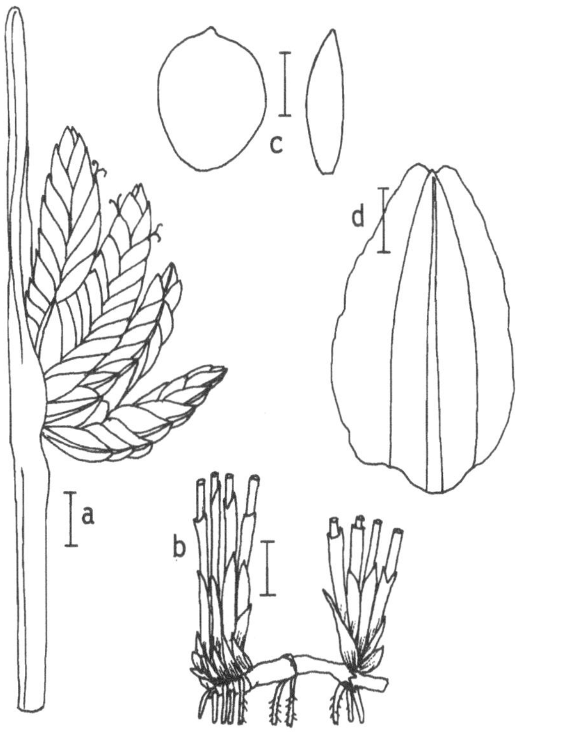

FIG. 104. *Cyperus laevigatus*: a, inflorescence (2 mm); b, base of shoot (1 cm); c, nut viewed from both sides (0.5 mm); d, glume (0.5 mm).

becoming woody. Culms close together in rows or solitary along the creeping rhizome, (5–) 10–35 (–60) cm tall, 0.5–2 mm wide, smooth, terete to 3-sided but flattened. Leaves often reduced to sheaths; sheaths grey or purplish-brown; blades flat or inrolled and culm-like, up to 2.5 (–4) cm long, 0.5–2 mm wide. Involucral bracts 1 or 2, the longer erect and culm-like, overtopping the inflorescence, 1–3 cm long. Inflorescence pseudolateral, a single cluster of spikelets, 0.6–1.5 cm in diameter. Spikelets sessile, 1–12 (–30), linear to lanceolate, sometimes curved, 2–10 (–20) mm long, 1.5–3.5 mm wide, 10- to 30-flowered; rachilla straight, persistent, 4-sided, hardly winged. Glumes broadly elliptic to ovate, subapiculate or rarely mucronate, 2–3.5 mm long, obscurely nerved, whitish to straw-coloured or brownish with or without reddish dots; sides faintly striped, margin thin; keel indistinct. Stamens 3. Style 2-cleft. Nuts plano- or concave-convex, obovoid, 1.5–2 mm long, 0.8–1 mm wide, stipitate, apiculate, grey to brown, surface smooth but densely punctate, with isodiametric surface cells.

Perennial. Wet places, particularly on the shores of lakes and tanks, mostly on sandy substrates, sometimes standing in brackish water. It often forms dense swards. It is sometimes confused with *Pycreus* because of the 2-cleft style and flattened nut but the flat side faces the rachis and not the angle as in *Pycreus*.

{Ten/Hel}. Widespread in warm and tropical regions: Ban, Ind: Dl, Kt, Mp, Pj, Rj, Tn, Up.

Cyperus longus Linnaeus: Fig. 105

Like *C. rotundus* but stolons stout and almost rhizome-like, never ending in tubers; culm base not or only slightly swollen; glumes shorter (3–3.2 mm long); nuts maturing, ellipsoid, 1.4–1.6 mm long, 0.5–0.6 mm wide, 3-sided, brown, almost smooth.

Perennial. In seasonally flooded areas, wet grassland, river and pond margins.

{Hel}. Europe, Mediterranean, Africa, Asia minor and W and S India: Ban, Ind, Nep, Pak: Kl, Kt, Rj, Tn. 0–700 m.

Cyperus malaccensis Lamarck: Fig. 106

Rhizomes stoutly stoloniferous, clothed with dark brown scales, becoming woody later. Culms usually close together, 60–175 cm tall, up to 12 (–15) mm thick, smooth, with concave sides, 3-angled below to 3-winged near the inflorescence. Leaves bladeless or

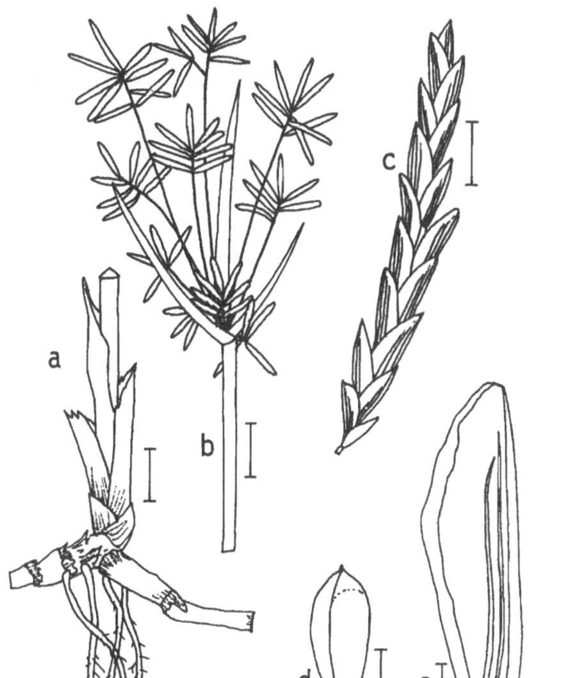

FIG. 105. *Cyperus longus*: a, base of shoot (1 cm); b, relatively small inflorescence (1 cm); c, spikelet (0.5 mm); d, nut (0.5 mm); e, glume (0.5 mm).

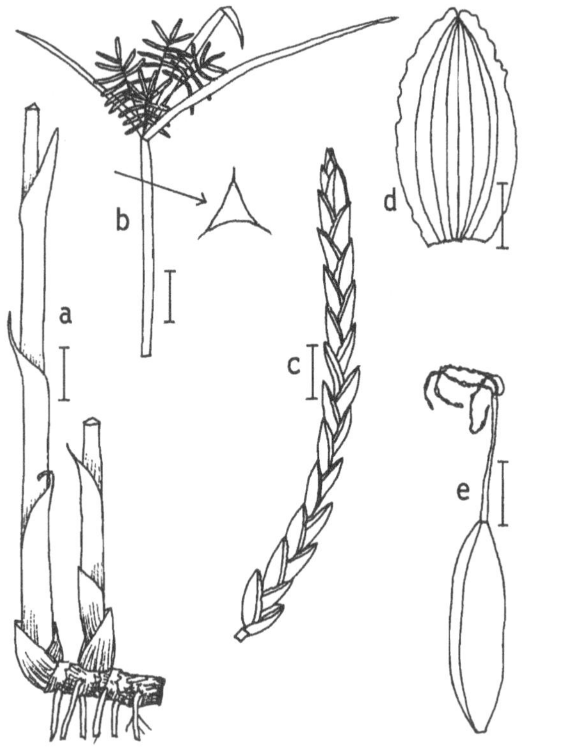

Fig. 106. *Cyperus malaccensis*: a, base of shoot (1 cm); b, inflorescence (2 cm) with transverse section of the culm (inset); c, spikelet (2 mm); d, glume (0.5 mm); e, nut (0.5 mm).

with short blades; sheaths spongy, greyish-brown to purplish, the upper ones very long (up to 20 cm long); blades absent or the upper one reaching halfway up the culm; blades on sterile shoots abruptly acuminate, 5–10 (–18) mm wide. Involucral bracts 3 or 4, unequal, the lower ones far overtopping the inflorescence, erect, up to 30 cm long. Inflorescences compound, broader than long, sometimes congested; primary rays 3–6 (–10), very unequal, 3–10 cm long; secondary rays bristle-like. Spikes broadly ovoid; rachis glabrous. Spikelets 6–12, in terminal spikes, linear, often somewhat curved, subterete, 10–30 mm long, 12–18 mm in diameter, 16- to 40-flowered; rachilla straight, persistent, very narrowly winged. Glumes ovate to elliptic, obtuse, 2–2.3 mm long, 1–1.5 mm wide, indistinctly 5- or 7-nerved, pale brown with yellowish margins, when dry crispidly incurved, hardly keeled (rounded on the back), muticous. Stamens 3; connective appendage pyramidical, reddish. Styles ± half as long as the nut, 3-cleft; stigmas 3, longer than the style. Nuts 3-sided, narrowly oblong, slightly flattened dorsally, hardly apiculate, 1.7–2 mm long, ± 0.5 mm wide, dark brown to blackish.

Perennial. In wet places and seasonally flooded areas, river banks, roadside ditches and marshes. Gregarious and often abundant. Often in brackish conditions on mud flats and sandy shores covered by spring tides. The rhizomes are used in medicine; the culms are used for making ropes, baskets, hats, slippers, etc., in Java it is used to make special traps for catching fish fry.

{Hyp/Ten/Hel}. From Mediterranean through India to Malaysia: Ban, Bur, Ind, Nep, Pak: Go, Kl, Kt, Mh, Wb. 0–700 m.

Cyperus meeboldii Kükenthal: Fig. 107

Rhizomes corm-like. Culms swollen at the base, tufted, erect, 4–20 cm tall, 0.4–1 mm thick, 3-sided, glabrous or slightly scabrid. Leaf sheaths grey to light brown, the oldest disintegrating into blackish fibres; blades flat or inrolled, 3–12 cm long, 0.5–2 mm wide, slightly scabrid on margin. Involucral bracts leaf-like, unequal, 4–10 cm long, 1–4 mm wide, all much overtopping the inflorescence. Inflorescence a brownish, congested, 0.7–2 cm in diameter head of 15–60, sessile spikelets. Spikelets flattened, linear, 4–10 mm long, 1–2 mm wide (or to 2.5 mm wide with glumes spreading), light to dark brown, 15- to 30-flowered. Glumes

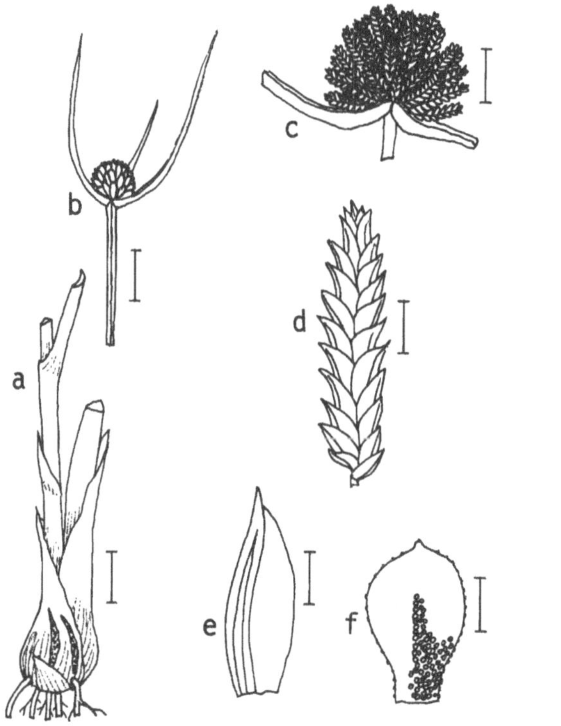

Fig. 107. *Cyperus meeboldii*: a, base of shoot (2 mm); b, inflorescence (1 cm); c, inflorescence (3 mm); d, spikelet (1 mm); e, glume (0.3 mm); f, nut (0.1 mm).

ovate, 1.3–1.5 mm long, ± 0.5 mm wide, reddish-brown, mucronate; keel pale, 3-nerved, ending in a slightly recurved apex. Stamen usually 1. Style unbranched or 3-cleft; stigmas ± half as long as the style. Nuts pear-shaped, 0.5–0.6 mm long, 0.3–0.4 mm wide, grey, minutely punctate.

Perennial and perhaps also sometimes annual. In seasonally flooded areas, by pools and in swamps. Superficially like *C. pulchellus* but it has brownish (not whitish) heads and pear-shaped nuts.

{(⊙)? Ten/Hel}. India and E Africa. Ind: Kt, Rj, Up. ?–800 m.

Cyperus nutans Vahl var. ***nutans*** [*C. distans* var. *major* Thwaites]: Fig. 108

Rhizomes corm-like, very short. Culms solitary or few, 3-sided, becoming 3-angled below the inflorescence, 40–120 (–150) cm tall, 4–10 mm in diameter, smooth. Leaves few; sheaths purplish, pale green above, 10–30 cm long; blades flat or the larger ones plicate, 20–50 cm long, 4–15 mm wide, rather leathery, dark green or glaucous above, whitish striate below; margins scabrous towards tip. Involucral bracts 4–6, unequal, the longer ones overtopping the inflorescence, 10–75 cm long, 4–8 mm wide. Inflorescence compound, paniculate, lax; primary rays 6–12, oblique, very unequal, up to 30 cm long, the longer ones bearing 3–7 secondary rays; secondary rays up to 6 cm long, bearing spikelets in corymbs. Spikes clustered, almost sessile, narrow, longer than broad, 1.5–4 cm long, 5–10 mm wide, loose, with (7–) 10–25 spikelets; rachis smooth. Spikelets arranged in spikes, suberect, linear-oblong, 6–15 mm long, 1–1.5 mm wide, 8- to 14-flowered, light brown; rachilla slightly flexuous, persistent, winged; wings whitish, hyaline, persistent, ±0.2 mm wide. Glumes closely appressed to the rachilla, oblong, obtuse, 2–2.5 mm long, ± 1 mm wide, 3- to 7-nerved, brown to yellow-brown, incised at tip (bicuspidate) and mucronulate by the excurrent midrib, mucro up to 0.5 mm long; keel greyish-green; margins of the upper half, white broad and hyaline. Stamens 3. Nuts 3-sided, oblong to oblong-ovoid, 1.3–1.5 mm long, ± 0.5 mm wide, apiculate, brown.

Perennial. Swamps, wet rice fields, margins of pools, river banks and seasonally wet habitats.

{Ten/Hel}. Tropical India and SE Asia: Ban, Bhu, Cey, Ind, Nep, Pak: Ap, As, Bh, Dl, Gj, Kl, Kt, Mh, Or, Pj, Sk, Tn, Up, Wb. ?–2000 m.

Cyperus nutans Vahl var. ***eleusinoides*** (Kunth) Haines [*C. eleusinoides* Kunth, *C. mysurensis* Heyne ex Wallich]: Fig. 108

Like *C. nutans* var. *nutans* but inflorescence less compound; primary rays suberect, the secondary ones often clothed in spikelets almost to the base; spikes shorter (1.5–3 cm long), denser. Spikelets much denser, 4–8 mm long, 1.5–2.5 mm wide, fewer flowered (4–10 flowers); glumes less remote, distinctly mucronate; mucro 0.2–0.5 mm long; nuts obovoid, 1.1–1.3 mm long.

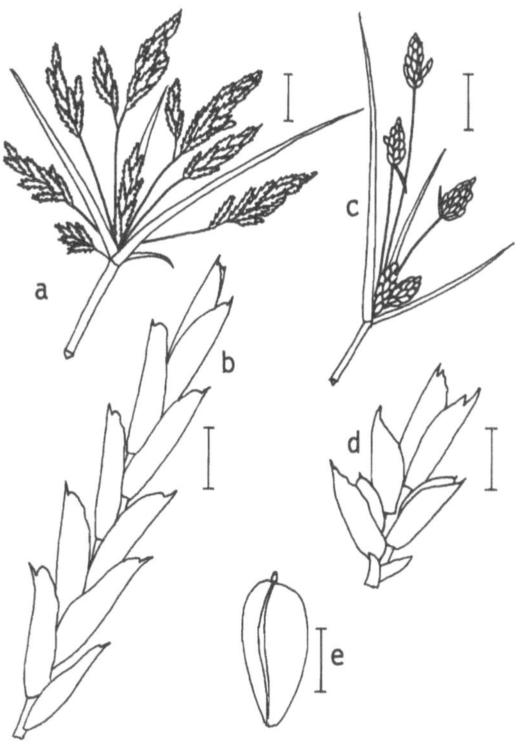

Fig. 108. *Cyperus nutans* var. *nutans*: a, inflorescence (1 cm); b, spikelet (1 cm). *C. nutans* var. *eleusinoides*: c, inflorescence (1 cm); d, spikelet (1 mm); e, nut (0.5 mm).

Perennial. In or near pools, streams, swamps and in seasonally flooded areas. The leaves and culms appear to be highly polished. In older Floras it is often recognized as a distinct species.

{Ten/Hel}. Tropical and subtropical Old World and Australia: Cey, Ind, Nep: Ap, As, Go, Jk, Kl, Kt, Mg, Mh, Mp, Ng, Rj, Sk, Tn, Up. 0–2000 m.

Cyperus ohwii Kükenthal: Fig. 109

Like *C. elatus* but usually more robust, leaf blades up to 18 mm wide; bracts (prophylls) at the base of the rays with a smooth keel; spikes very dense (with ± 70 spikelets), 6–8 mm (ultimately 10 mm) wide; glumes ovate, (1.2–) 1.7–2 mm long, 1–1.3 mm wide; connective appendage ± half as long as the anther; Nuts oblong-ellipsoid, ± 1.2 long; ± 0.4 mm wide.

Perennial. In swamps and bogs, sometimes in floating vegetation.

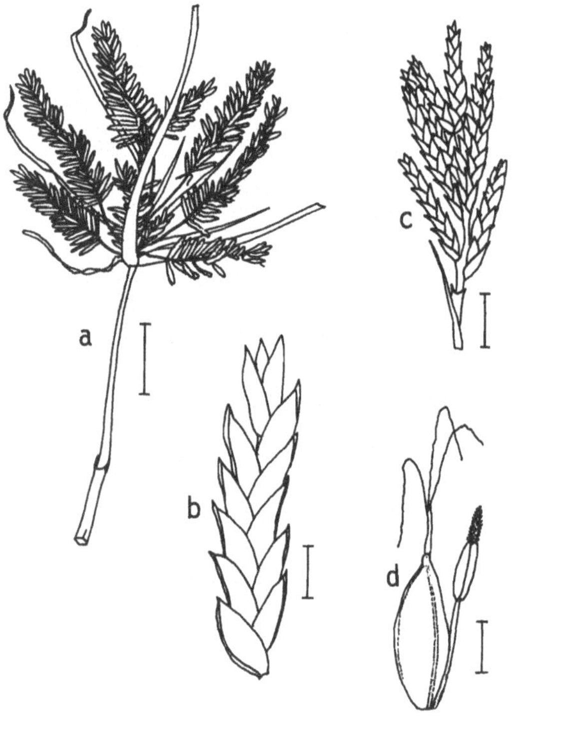

Fig. 109. *Cyperus ohwii*: a, inflorescence (1 cm); b, spikelet (1 mm); c, small spike (3 mm); d, ovary and stamen (0.5 mm).

{Hyp}. Temperate and warm E and SE Asia, probably introduced in Japan: Ban, Ind: Up.

Cyperus pangorei Rottbøll [*C. tegetum* Roxburgh, *C. dehiscens* Nees]: Fig. 110

Rhizomes decumbent, 3–8 mm thick. Culms laxly tufted or arranged in a row along the rhizome, stiffly erect, 50–150 (–200) cm tall, 2–7 mm thick, 3-angled above, the sides flat or convex, light green, smooth. Leaves usually reduced to bladeless sheaths; sheaths scale-like below, the uppermost 20–32 cm long, cinnamon-coloured to purplish or grey; blades rarely present, erect, up to 10 cm long and 5 mm wide. Involucral bracts 3–5, unequal, overtopping the inflorescence, the lowest 17–30 cm long. Inflorescence compound, lax; rays 4–12, up to 16 cm long. Spikes up to 4 cm long, arranged in corymbs, with 2–14, subremote, spreading to obliquely erect spikelets. Spikelets spicate, linear, laterally flattened, 10–30 mm long, 1–2 mm wide, straw-coloured with a reddish tinge, 10- to 36-flowered; rachilla flexuous, persistent, winged; wings 1.5 mm wide, reddish-brown, persistent. Glumes oblong-elliptical, 2.2–3.5 mm long, subobtuse, mucronulate at tip; sides nerveless or striped, cinnamon to red-brown, with narrow hyaline margins; keel 3-nerved, green, ending below the apex. Stamens 3. Styles ±3 mm long, 3-cleft. Nuts 3-sided, oblong-obovoid to ellipsoid, 1–1.5 mm long, rounded to apiculate, straw-coloured maturing to dark brown.

Perennial. Wet places, marshes, river banks, stream beds, often found standing in water, frequent in rice fields. The culms are very flexible and can withstand strong floods. It is widely used for making fine mats, sometimes cultivated in India. In Tamilnadu it is called 'pangorai'. It differs from *C. corymbosus* in having a short, thick rhizome and remote oblong-elliptic glumes.

{Hyp/Ten}. Endemic to the Indian subcontinent, cultivated in Mauritius and Sulawesi: ♦Ban, ♦Bur, ♦Cey, ♦Ind, ♦Nep, ♦Pak: As, Bh, Jk, Kl, Kt, Mg, Mh, Mn, Mp, Or, Rj, Sk, Tn, Up, Wb. 0–1800 m.

Cyperus papyrus Linnaeus: Fig. 111

Rhizomes creeping, 3–6 cm thick, densely covered by 5–10 cm long and wide blackish scales. Culms erect, (1–) 2–5 m tall, 3-sided with rounded angles, green. Leaves bladeless on culms; sheaths blackish or reddish-brown, becoming woody below; sterile shoots with leaf blades up to 8 mm wide. Involucral bracts up to 12, much shorter than the inflorescences; 5–10 cm

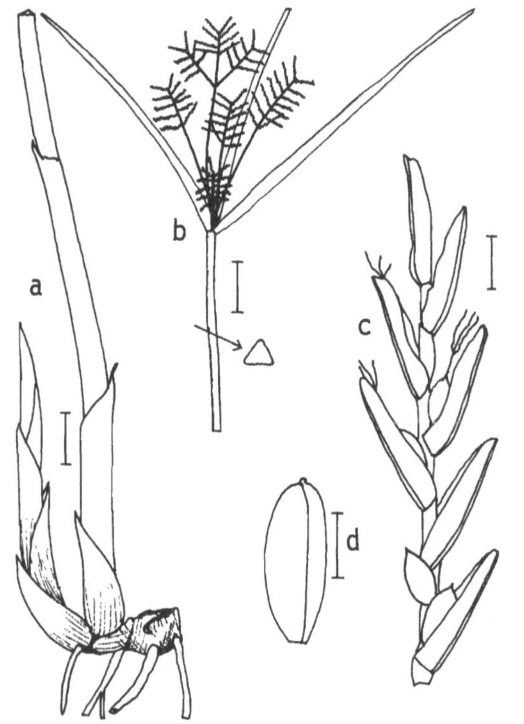

Fig. 110. *Cyperus pangorei*: a, base of shoot (1 cm); b, part of inflorescence (4 cm) with transverse section of culm (inset); c, apical part of spikelet (1 mm); d, nut (0.5 mm).

FIG. 111. *Cyperus papyrus*: a, distal part of spikelet (2 mm); b, inflorescence (10 cm); c, whole plant (20 cm).

long, 1–3 cm wide, light brown, never green. Inflorescences compound, spreading, umbel-like with 30–100 subequal rays 10–60 (–80) cm long; secondary rays 3–5 cm long. Spikes cylindrical, 1–3 cm long, 6–12 mm wide, bearing many spikelets. Spikelets cylindrical or slightly flattened, 3–12 mm long, 0.7–1 mm in diameter, 5- to 17-flowered. Glumes elliptical to ovate-elliptical, 2–2.5 mm long, obtuse, light brown to golden; keel green 3-nerved. Stamens 3. Styles ± 0.8 mm long, 3-cleft. Nuts oblong, 3-sided, 0.9–1.1 mm long, 0.4–0.5 mm wide, smooth, dark brown to dark grey.

Perennial. Rooted in water in lakes and along rivers or forming floating mats or islands. Often dominant and of considerable importance in C and E Africa, a potentially dangerous weed in India. It is not clear if it has become naturalized.

{Hyp/Ple}. N and C Africa and Palestine: <Cey in gardens, <Ind: <Gj, <Rj.

Cyperus pilosus Vahl [*C. paniculatus* D. Don, *C. donianus* A. Dietrich, *C. obliquus* Nees, *C. marginellus* Nees]: Fig. 112

Rhizomes stoloniferous. Culms usually solitary, 25–85 (–120) cm tall, 5–8 mm thick, acutely 3-angled, smooth or scabrid. Leaves somewhat shorter than or longer than the culms; sheaths spongy, pale green to brown or purplish; blades channelled below, flat towards the tip, leathery, gradually acuminate, 7–9 (–15) mm wide, scabrid towards tip. Involucral bracts 3–5, unequal, the lower ones far overtopping the inflorescence, up to 50 cm long. Inflorescences simple or compound, loose to subdense; primary rays 3–7, very unequal, 6–12 (–20) cm long, sparsely scabrid to glabrous; secondary rays when present very short. Spikes subpyramidal, subdense, 2-3 cm long, 1–2 cm wide; the rachis densely hirsute with yellow-brown short hairs. Spikelets 5–25 in terminal spikes, linear-lanceolate to elliptical-oblong, 5–20 mm long, 1.5–3 mm wide, 8- to 25- (–50)-flowered, straw-coloured and tinged with red-brown; rachilla straight, persistent, not winged. Glumes ovate to broadly deltoid-ovate, acute with usually mucronulate apex, 1.8–2.5 mm long, 1.2–1.5 mm wide, 5- or 7-nerved, pale

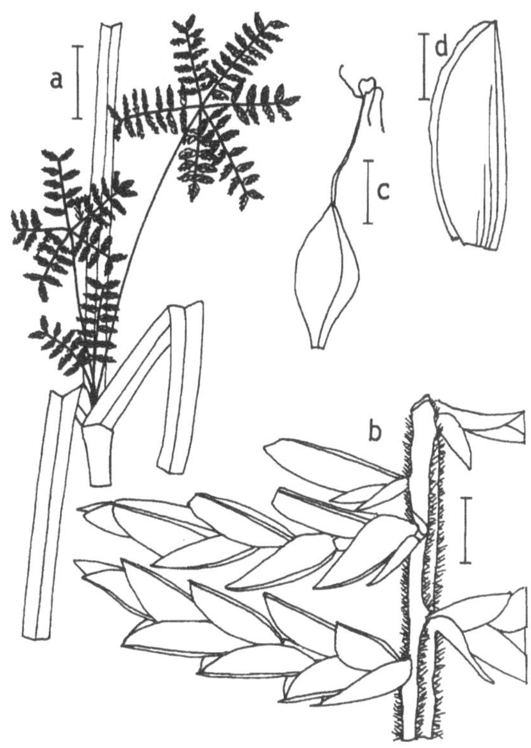

FIG. 112. *Cyperus pilosus*: a, part of inflorescence (2 cm); b, spikelets and rachis (1 mm); c, nut (0.5 mm); d, glume (0.5 mm).

straw-coloured and tinged or flecked with reddish-brown, distinctly keeled, with green midrib, with whitish hyaline margins towards the tip. Stamens 3. Styles rather short, 3-cleft. Nuts 3-sided, obovoid to ellipsoid, stipitate, minutely apiculate, 1–1.3 mm long, 0.5–0.8 mm wide, dark brown.

Perennial. Seasonally flooded places, wet grassland, rice fields and drains. Like *C. procerus* but the rachis densely hirsute with yellow-brown short hairs; glumes acute at apex without a broad hyaline margin at the apex, distinctly keeled.

{Ten?/Hel?}. Old World tropics: Ban, Bhu, Bur, Cey, Ind, Nep: An, Ar, As, Bh, Hp, Hr? Jk, Kl, Kt, Mg, Mh, Mn, Mp, Ng, Or, Rj, Sk, Tn, Up, Wb. 0–1500 m.

Cyperus platyphyllus Roemer et Schultes [*Papyrus latifolia* Willdenow, *C. roxburghii* Nees]: Fig. 113

Rhizomes up to 8 cm long, 2–3 cm thick, obliquely ascending. Culms robust, 70–240 cm tall, 7–12 mm thick, 3-sided below, sharply 3-angled with concave sides above, scabrous on the angles. Leaves up to two-thirds as long as the culm; sheaths straw-coloured; blades plicate (like an inverted 'W' in transverse section), with 3 prominent nerves, leathery, 50–140 cm long, 15–30 mm wide, deep green, scabrous along the margins. Involucral bracts 4–6, the two lower ones much overtopping the inflorescence, up to 100 cm long and 30 mm wide. Inflorescences usually of 1 sessile spike and up to 12 stalked spikes; rays 6–12, spreading, unequal, up to 20 cm long. Spikes linear-cylindrical, subsessile, forming digitate secondary corymbs, 3–6 cm long, 3–6 mm in diameter, rather loose, yellowish-brown. Spikelets sessile, erect to suberect or slightly spreading, subterete, linear, 5–8 mm long, 1–1.8 mm in diameter, 8- to 14-flowered; rachilla straight, flattened, with caducous wings. Glumes obovate-elliptical, 1.8–2 mm long, 1–1.3 mm wide, incurved margins, subacute at mucronate apex, 3- or 5-nerved; midrib green; sides nerveless, straw-coloured tinged with gold or brown. Stamens 3; connective appendage subulate, 0.6 mm long. Style 1.7–2 mm long; stigmas 3, longer than the style. Nuts 3-sided, ellipsoid, 1–1.2 mm long, ±0.5 mm wide, greyish-brown.

Perennial. Found in marshy places at low altitudes. The linear-cylindrical, subsessile spikes which form digitate secondary corymbs is diagnostic.

{Hel}. Endemic to Sri Lanka and S India: ♦Cey, ♦Ind; Tn.

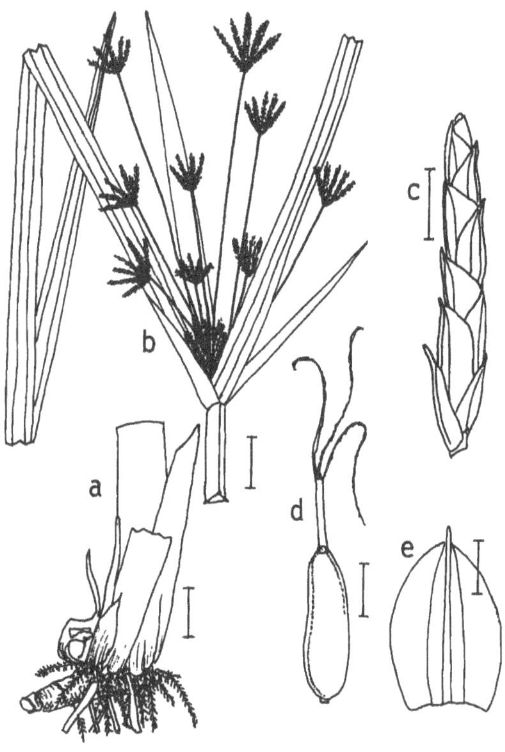

FIG. 113. *Cyperus platyphyllus*: a, base of shoot (2 cm); b, inflorescence (3 cm); c, spikelet (1 mm); d, nut (0.3 mm); e, glume (0.3 mm).

Cyperus platystylis R. Brown [*C. pallidus* Nees]: Fig. 114

Rhizomes either very short or up to 10 cm long. Culms solitary or 2 or 3 together, smooth or scabrous along the angles towards the tip, 30–80 (–120) cm tall, 3–5 (–8) mm thick. Leaves as long as or longer than the culm; sheaths bladeless below, strongly keeled, brownish to purple; blades flat or slightly plicate, septate, nodulose, leathery, gradually acuminate, 5–12 (–20) mm wide, greyish-green or glaucous above; margins and midrib very scabrous (cutting). Involucral bracts 4–12, unequal, the longer ones overtopping the inflorescence, up to 80 cm long. Inflorescence compound, depressed-corymbose or semiglobose, often with very numerous spikelets; primary rays 10–17, widely spreading, often slightly upcurved, up to 10 cm long; secondary rays spreading, up to 4 cm long; tertiary rays when present very short. Spikelets digitately arranged in clusters of 2–8, widely spreading, ovate to linear-lanceolate, flattened, 5–20 mm long, 2.5–3 mm wide, densely 15- to 60-flowered, straw-coloured and stained with rusty brown; rachilla straight, persistent, without wings. Glumes closely appressed to the rachilla, ovate to broadly ovate, obtuse, mucronulate, 2–2.5 mm long, 1.5–2 mm wide; the back rounded, strongly 3-nerved; the sides nerveless, cellular-reticulate, yellowish or brownish with narrow, whitish hyaline

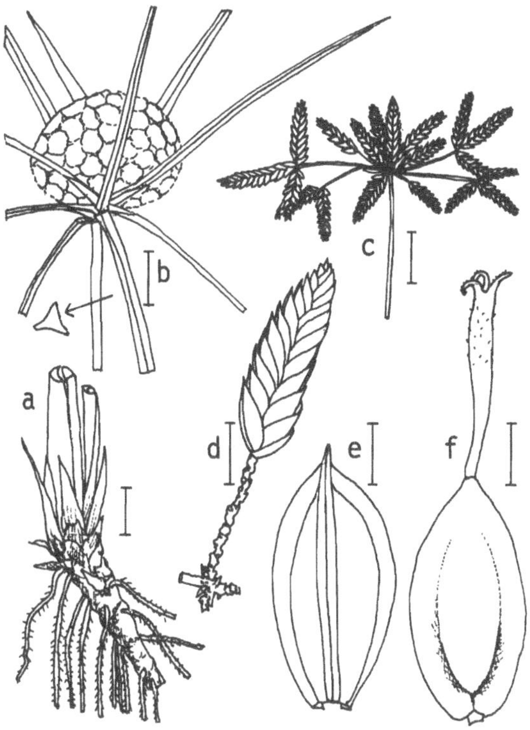

Fig. 114. *Cyperus platystylis*: a, base of shoot (1 cm); b, schematic illustration of an inflorescence (1 cm) with transverse section of the culm (inset); c, part of inflorescence (1 cm); d, mature spikelet (2 mm); e, glume (0.5 mm); f, nut (0.5 mm).

margins. Stamens 3; connective appendage bristly. Style flattened, slightly shorter than the nut, ciliate, 3-cleft. Nuts unequally 3-sided, dorsoventrally flattened, with corky or spongy thickened angles, ellipsoid, 1.75–2 mm long, 0.8–1 mm wide, apiculate, the sides shining greyish-brown to blackish with pale whitish to yellowish angles.

Perennial. Grows standing in shallow water or in floating mats of vegetation with plants such as *Pistia* and *Salvinia*. Also found in marshes, ponds, swamps, swinging bogs and wet rice fields. Usually gregarious and forming large communities.

{Hyp/Ple}. Tropical Asia and Australia: Ban, Bur, Cey rare, Ind: An, Ap, As, Kl, Mp, Rj, Tn, Up, Wb.

Cyperus procerus Rottbøll [*C. carnosus* Heyne ex Nees]: Fig. 115

Rhizomes developing stolons. Culms usually solitary, 50–125 (–175) cm tall, 4–7 (–10) mm thick, 3-angled, smooth. Leaves somewhat shorter than or longer than the culms; sheaths spongy, greyish-brown to purplish; blades channelled below, flat towards the tip, spongy or leathery, gradually acuminate, 7–9 (–15) mm wide, scabrid towards tip. Involucral bracts 3 or 4, unequal, the lower ones far overtopping the inflorescence, 20–50 (–70) cm long. Inflorescences simple or subcompound, loose, relatively small; primary rays 3–7, very unequal, 6–12 (–20) cm long, sparsely scabrid to glabrous; secondary rays when present very short. Spikes broadly ovoid, loose, 2–4 cm long; rachis weakly hispid or scabrous on the wings. Spikelets 4–10 (–18) in terminal spikes or 2–4 in lateral spikes, the lower ones ± at right angles to the rachis, oblong, 1–3.5 cm long, 10- to 40-flowered, pale and tinged with red-brown; rachilla straight, persistent, not or hardly winged. Glumes ovate to elliptic, obtuse, 2.5–4 mm long, 2–2.5 (–3) mm wide, 5- or 7-nerved, reddish-brown, smooth, hardly keeled, with broad hyaline margins towards the tip. Stamens 3; connective appendage red. Styles as long as the nuts, 3-cleft. Nuts 3-sided, obovoid to ellipsoid, smooth, stipitate, minutely apiculate, 1–1.5 mm long, 0.6–0.8 mm wide, blackish-brown.

Perennial. In swamps, marshes, pools, wet rice fields and in seasonally flooded areas. Often abundant. The

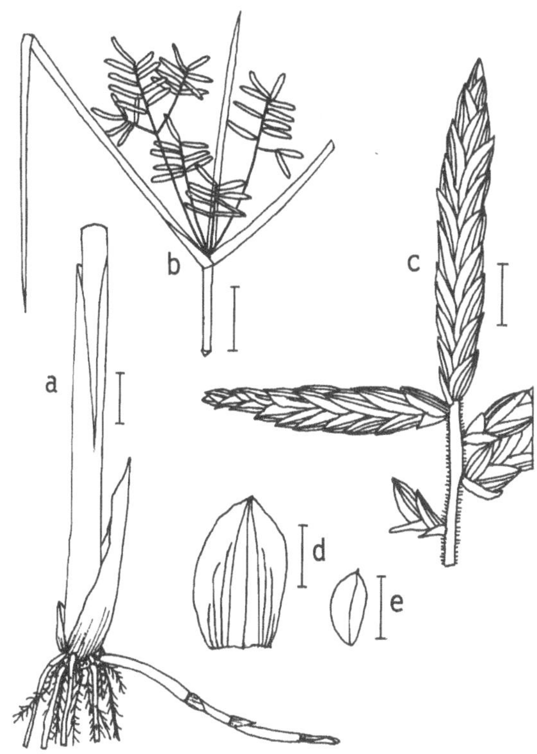

Fig. 115. *Cyperus procerus*: a, base of shoot (1 cm); b, inflorescence (2 cm); c, terminal spikelets on rachis (2 mm); d, glume (0.5 mm); e, nut (0.5 mm).

stems are sometimes collected then split and used as string or twine.

{Ten/Hel}. Tropical Old World and N Australia: Ban, Bur, Cey, Ind: As, Kl, Kt, Mp, Tn, Wb. Low altitudes

Cyperus pulchellus R. Brown [*Sorostachys kyllingioides* Steudel]: Fig. 116

Rhizomes corm-like. Culms solitary or loosely tufted, swollen at base and sometimes appearing rhizome-like, 6–31 cm tall, 0.5–1.5 mm wide, terete to 3-sided, glabrous or slightly scabrid. Leaf sheaths up to 4 cm long, green to light reddish-brown, the oldest darker; blades flat or V-shaped, 2–25 cm long, 0.5–2.3 mm wide, gradually narrowed to an acuminate tip, midrib and margins scabrid. Involucral bracts 2–4, leaf-like, unequal, deflexed or spreading, the longest (3–) 5–19 cm long, (0.8–) 0.9–1.8 mm wide, all much overtopping the inflorescence. Inflorescence of 1, sessile, congested, globose cluster of up to 60 spikelets; the cluster dirty white, 0.7–1.5 cm in diameter. Spikelets flattened, lanceolate to ovate-lanceolate, 4–8 mm long, 1–2 mm wide, greyish-white with a light pinkish-brown tinge, 10- to 20-flowered. Glumes lanceolate, 1.2–1.5 mm long, 0.3–0.4 mm wide, obtuse, whitish or light brown; keel and midrib indistinct. Stamen usually 1. Style up to 0.7 mm long, 3-cleft; stigmas shorter than the style. Nuts cylindrical to obovoid, obscurely 3-sided, 0.5–0.8 mm long, 0.2–0.3 mm wide, grey to pale brown, minutely papillose.

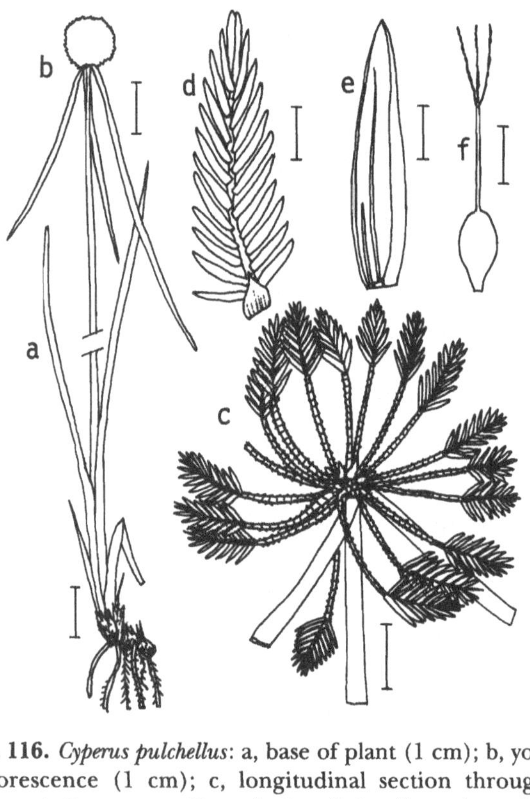

Fig. 116. *Cyperus pulchellus*: a, base of plant (1 cm); b, young inflorescence (1 cm); c, longitudinal section through a mature inflorescence (2 mm); d, spikelet (1 mm); e, glume (0.25 mm); f, nut (0.5 mm).

Perennial. Seasonally wet places, temporary swamps and grasslands. Superficially like *Kyllinga* and *Lipocarpha* but differs in having flattened spikelets, strictly distichous glumes and different structure of the flower and its surrounding scales. It differs from other single-headed *Cyperus* species by the dirty white colour of the heads. It is sometimes treated as a synonym of *C. leucocephalus*, a species found in NE India usually growing in sandy places and not in wetlands. Simpson, D. A. A revision of *Cyperus* sect. *Leucocephali*. Kew Bulletin 45(3): 485–501 (1990).

{Hel}. Old World tropics: Ind: Ap, Bh, Gj, Kl, Kt, Mh, Or, Rj, Tn.

Cyperus pygmaeus Rottbøll [*Juncellus pygmaeus* (Rottbøll) C. B. Clarke, *C. michelianus* (Linnaeus) Link subsp. *pygmaeus* (Rottbøll) Ascherson et Graebner, *Scirpus occultus* C. B. Clarke, ?*Pycreus lancelotii* S. M. Almeida]: Fig. 117

Rhizomes absent, roots fibrous. Culms densely tufted, erect or spreading, (0–) 1–10 (–25) cm tall, 0.5–1.5 mm thick, smooth, rather soft. Leaves 1–3 to a culm, as long as or shorter than the culm; sheaths brown or reddish; blades flat or grooved (becoming folded and twisted as they dry), 1–6 (–20) cm long, 1–2 (–3) mm wide, gradually acuminate, smooth below, scabrous above. Involucral bracts (2–) 4–6 (–8), leaf-like, unequal, spreading to slightly reflexed, widened at the base, all overtopping the inflorescence, the longest 4–15 cm long. Inflorescence consisting of 1 to several very dense glomerules, contracted into a triangular-ovoid to subglobose, lobed head (0.4–) 0.7–1.5 cm in diameter. Spikelets strongly flattened, sometimes incurved and contorted, ovate to lanceolate, 2.5–5 mm long, 1–2.2 mm wide, whitish-green, 8- to 20-flowered; rachilla persistent, unwinged. Glumes sometimes appearing spiral due to torsion of the rachilla, ovate-lanceolate, 1.3–2.2 mm long, ±0.5 mm wide, 3- to 7-nerved, thinly membranous, pale white to straw-coloured, subtranslucent, acute or mucronate (mucro up to 0.5 mm long and curved); keel thick, pale green. Stamens 1 or 2. Style 0.5–0.7 mm long, 2- or 3-cleft; stigmas ± 1 mm long. Nuts 3-sided or plano-convex, oblong, 0.8–1.2 mm long, 0.2–0.4 mm wide, yellow to pale brown, minutely punctate.

Annual. In seasonally flooded places, flowering as the water recedes but also in and around pools, ponds, watercourses and in rice fields. It is sometimes domi-

Fig. 117. *Cyperus pygmaeus*: a, whole plant (1 cm); b, inflorescence (2 mm); c, spikelet (0.5 mm); d, ovary (0.5 mm).

nant and often forms an almost closed turf. It is very different from all other *Cyperus* species with its whitish-green, congested heads with many, spreading, leaf-like bracts. The generic position has been disputed and it has appeared in *Dichostylis, Eleocharis, Fimbristylis, Isolepis, Juncellus* and *Scirpus.*

{⊙ Ten/Hel}. Warm and tropical Old World: Bur, Cey, Ind, Nep, Pak: Dl, Gj, Go, Jk, Kt, Mg, Mh, Mp, Pj, Rj, Tn, Up, Wb. Low altitudes.

Cyperus rotundus Linnaeus [*C. tuberosus* Rottbøll, ?*Schoenus tuberosus* N. L. Burman]: Fig. 118

Stolons slender, wiry, 5–20 cm long, ending in globose or ellipsoidal tubers; tubers 3–20 mm in diameter, rust-coloured becoming black. Culms cormlike at the base, usually solitary, (7–) 15–40 (–90) cm tall, 1–4 mm thick, 3-angled, smooth. Leaves crowded at the base; sheaths green to reddish brown, soon disintegrating into parallel fibres; blades flat, gradually acuminate, usually shorter than the culm, 15–30 cm long, 2–6 mm wide, smooth, scabrous above and on the margins. Involucral bracts 2–4 (–6), very unequal, usually short, the lowest sometimes just overtopping the inflorescence, up to 30 cm long but usually much shorter. Inflorescence simple or compound, of 1 sessile and 1–10 stalked spikes, with or without secondary spikes; rays 2–10, very unequal, up to 10 cm long. Spikes loose to dense, ovoid, 1–5 cm long, 1.5–7 cm wide, with (2–) 4–15 erect or spreading spikelets. Spikelets linear to oblong, flattened with acute edges, acute at the tip, 1–3.5 (–7) cm long, 1.5–3 mm wide, with 8–40 (–100) flowers, red-brown; rachilla flexuous, persistent, winged; wings 0.5 mm wide, persistent. Glumes membranous, ovate, 3–4.5 mm long, ± 2 mm wide, subobtuse, apiculate at tip, 5- to 7-nerved, red to deep reddish-brown, the sides without nerves and with narrow hyaline margins; keel green, ending in or below the apex of the glume. Stamens 3; connective appendage reddish. Styles 3-cleft. Nuts very rarely maturing, 3-sided, oblong-obovoid to ellipsoid, 1.3–2 mm long, 0.5–1 mm in diameter, apiculate, shining, brownish to black.

Perennial. Found on river banks, in dried up pools and ditches and in rice fields. It is not confined to wetlands and is a common and sometimes serious weed in cultivated land. It is difficult to eradicate. The seeds rarely mature, the effective spread is by tubers. The

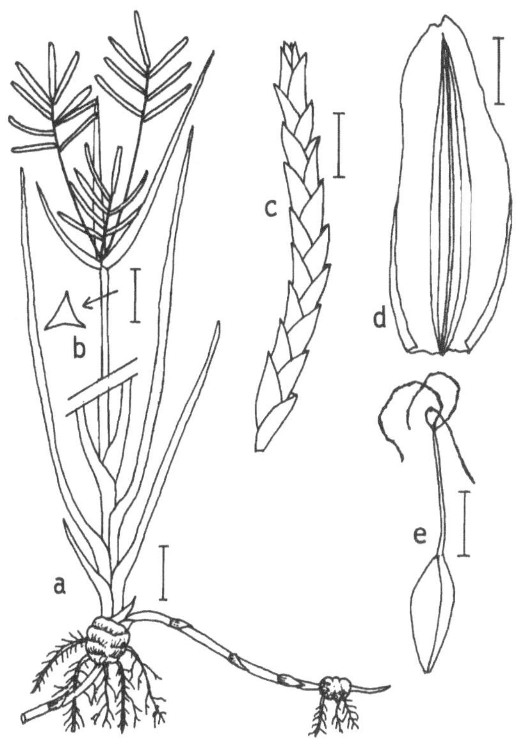

Fig. 118. *Cyperus rotundus*: a, base of shoot (1 cm); b, part of inflorescence (1 cm) with transverse section of the culm (inset); c, spikelet (2 mm); d, glume (0.5 mm); e, ovary (0.5 mm).

leaves remain green throughout the year. Tubers are collected, roasted and eaten. Powder from tubers used in fragrant incense sticks (agarbotti). It is also a medicinal plant; the tubers contain cyperene, cyperone, cyperol and 1-pinene which are used for spasms or as an emmenagogue, it is also used in cooling indigenous medicines.

{Hel?}. Almost cosmopolitan: Ban, Bhu, Bur, Cey, Ind, Nep, Pak: An, Ap, Ar, As, Bh, Dl, Gj, Go, Hp, Jk, Kl, Kt, Mg, Mh, Mn, Mp, Or, Pj, Rj, Sk, Tn, Up, Wb. 0–2400 m.

It is very variable and many taxa of different rank have been described. One of the more distinct ones is *Cyperus mitis*.

Cyperus mitis Steudel [*C. stenostachys* Bentham var. *indica* C. B. Clarke, *C. subcapitatus* C. B. Clarke]

Like *C. bifax* and *C. rotundus* but the spikelets are filiform, obtuse-tipped, 1.5–2 mm in diameter; inflorescences tending to become pseudolateral and subtended by the lowest involucral bract.

Perennial. I have no ecological information, it is perhaps not even a wetland plant.

{Hel?}. Madagascar, S and SE Asia and Australia: Ban, Bur, Cey, Ind: Go, Mh, Mp, Pj, Tn.

Cyperus scariosus R. Brown [*C. corymbosus* var. *scariosus* (R. Brown) Kükenthal: Fig. 119

Rhizomes creeping, up to 5 cm long, forming small tubers. Culms slender but rigid, 50–80 cm tall, 1.5–2 mm in diameter, swollen and almost terete below becoming 3-sided above, not septate, smooth. Leaves up to 3, short, rarely half as long as the culm; sheaths yellowish to purplish; blades poorly developed, weak, 0.5–12 cm long, up to ± 2 mm wide, gradually acuminate, 3-angled at the tip. Involucral bracts 2–4, unequal, very variable (when blades develop on the culms then they are long, when blades do not develop they are short), leafy, shorter than or much longer than the inflorescence, the longest up to 6 cm long, embracing the culm with dark brown auricles. Inflorescence simple with several stalked spikelet clusters or rarely reduced to 1 sessile cluster, spikelet clusters usually pushed to one side by the erect lowest involucral bract; rays filiform, not more than 3 cm long. Spikes subglobose, 2–5 cm long and wide. Spikelets spicate, obliquely erect to spreading, linear, strongly flattened, 1–2 (–2.5) cm long, 2–2.5 mm wide, up to 26-flowered; rachilla flexuous, persistent, broadly winged. Glumes membranous, ovate, ± 2.8 mm long, 1.5–2 mm wide, obtuse, mucronulate at tip; sides with 3 or 4 faint veins, straw-coloured, tinged with red, with a green midrib and with margins banded with brown. Stamens 3; connective appendage short. Styles 3-cleft. Nuts 3-sided, oblong to ellipsoid, 1.2–1.5 mm long, ±0.5 mm wide, maturing brown.

Perennial. Coastal regions, almost restricted to swampy, brackish or near-brackish localities. Near to *C. corymbosus* but: rhizome rhizomatous, up to 5 cm long, usually more than 5 mm thick; spikelets linear, 10–20 (–25) mm long, 2–2.5 mm wide; glumes ovate, ±2.8 mm long, 1.5–2 mm wide; nuts 1.2–1.5 mm long, ± 0.5 mm wide.

{Ten/Hyp/Hel}. India, SE Asia and tropical Australia: Ban, Bur, Ind?: Wb?

Cyperus tenuiculmis Boeckeler [*C. zollingeri* auct. Ind., non Steudel]: Fig. 119

Rhizomes corm-like or knotty, often with creeping stolons. Culms solitary or few together, stiffly erect, smooth, 35–70 (190) cm tall, 1–3 mm thick, 3-sided, smooth. Leaves few, much shorter than the culm; sheaths green to light brown; blades plicate, 10–40 cm long, 2–6 mm wide, gradually narrowed to a very acute apex. Involucral bracts 2–5 (–7), unequal, erect-spreading, the lower one equalling or overtopping the

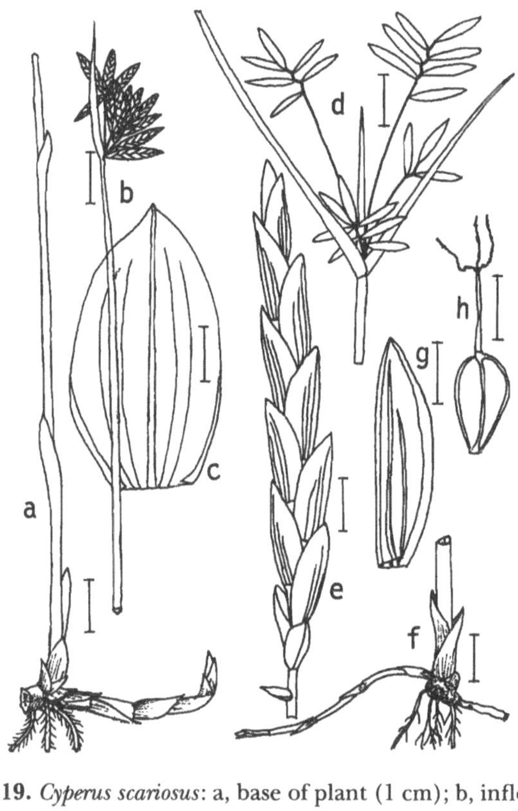

FIG. 119. *Cyperus scariosus*: a, base of plant (1 cm); b, inflorescence (1 cm); c, glume (0.5 mm). *C. tenuiculmis*: d, inflorescence (1 cm); e, spikelet (2 mm); f, base of shoot (1 cm); g, glume (0.5 mm); h, nut (0.5 mm).

inflorescence, 8–20 cm long. Inflorescence simple or subcompound, of 1 sessile and 2–8 stalked spikes (rarely with secondary spikes); rays 2–8, unequal, 1–16 cm long. Spikes ovoid, subdense, 1.5–4 (–5) cm long, bearing 3–14 spikelets. Spikelets linear to linear-lanceolate, 8–30 mm long, 1.3–2 mm wide (or to 4 mm with spreading glumes), slightly flattened, nearly 4-sided but with 2 acute edges, loosely 6- to 20-flowered; rachilla flexuose, widely winged; wings pale, hyaline, caducous. Glumes suberect, ovate-elliptical or elliptical, boat-shaped, 3–4 mm long, 2–2.5 mm wide, 7- or 9-nerved, straw-coloured to golden brown, obtuse or acutish at the apex; keel obtuse, green, not excurrent. Stamens 3. Style slender; stigmas 3, about as long as the style. Nuts 3-sided with concave sides, ellipsoid to obovoid, 1.6–2 mm long, 0.8–1.2 mm wide, maturing blackish, minutely punctate.

Perennial or occasionally annual. Seasonally flooded areas, roadside ditches, swamps, usually on sandy soil; it is not confined to wetlands. The creeping stolons and straw- to gold-coloured inflorescence are characteristic.

{(⊙) Hel}. Old World tropics and subtropics: Bhu, Cey, Ind, Nep: Ap, As, Kt, Mh, Ng, Sk, Tn, Wb. 0–1700 m.

Cyperus tenuispica Steudel [*C. flavidus* auct. Ind., non Retzius]: Fig. 120

Culms slender, weak, solitary or tufted, smooth, (2–) 5–40 cm tall, 1–2.5 mm thick. Leaves as long as or shorter than the culm; sheaths scarious, yellowish to brownish or grey, the lower ones bladeless or shortly bladed; blades flat, flaccid, up to 20 cm long but usually less, 1–2 (–4) mm wide, gradually acuminate, smooth or slightly scabrous above. Involucral bracts 2–3 (–5), unequal, the longer one overtopping the inflorescence, up to 15 (–25) cm long. Inflorescence compound, lax, relatively large; primary rays 5–10, up to 10 cm long; secondary rays, 5–7, 2–2.5 mm long, tertiary rays when present very short. Spikelets digitately arranged in clusters of 3–9 spikelets, stellately spreading, oblong to linear, acute, flattened, 3–8 (–13) mm long, 0.8–1.3 mm wide, with 8–30 (–50) flowers; rachilla slightly flexuous, soon visible between the glumes, persistent, without wings. Glumes rather loose, exposing parts of the rachilla and nuts, ovate to ovate-oblong, obtuse, 0.7–1.3 mm long, sides nerveless, hyaline, reddish-brown with purple stripes or spots; keel green, 3-nerved, excurrent and curved into a minute mucro just below the tip. Stamens (1) 2 or 3. Style 1.5 times as long as the nut. Nuts obtusely 3-sided with convex sides, subglobose to ellipsoid, 0.3–0.5 mm long, 0.2–0.3 mm wide, broadly stipitate, hardly apiculate, rectangular tuberculate surface, marbly white.

Annual or perhaps sometimes longer-lived in permanently wet habitats. In swamps, streams, ditches, seasonally flooded areas and rice fields. Like *C. haspan* but annual; glumes small, obtuse, spreading at maturity; nuts white and visible between the opened glumes in fruiting spikelets; like *C. difformis* but nuts shorter than the glumes.

{(⊙) Ten/Hyp/Hel}. Old World tropics: Bhu, Bur, Cey, Ind, Nep: Ap, Ar, As, Bh, Go, Hp, Jk, Kl, Kt, Mg, Mp, Ng, Rj, Tn, Up, Wb. 0–1400 m.

Diplacrum [excluded from *Scleria*]: 7 species: tropical Africa, E Asia and America

Diplacrum caricinum R. Brown [*D. tridentatum* Brongniart, *D. zeylanicum* Nees, *Scleria caricina* (R. Brown) Bentham]: Fig. 120

Culms slender, loosely tufted, erect or procumbent, sharply 3-angled, (3–) 6–30 cm tall, ± 1 mm wide, glabrous. Leaves basal and on the culm, glabrous; sheaths 3–10 mm long, acutely 3-angled without wings; blades 1–5 cm long, 2–5 mm wide, flattish, 3-nerved, abruptly acute at the apex. Inflorescence of several to

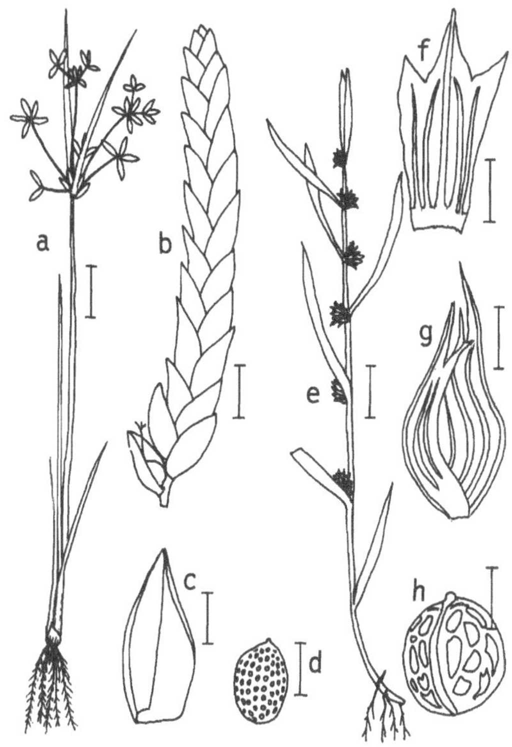

FIG. 120. *Cyperus tenuispicata*: a, whole plant (1 cm); b, spikelet (0.5 mm); c, glume (0.3 mm); d, nut (0.3 mm). *Diplacrum caricinum*: e, whole plant (1 cm); f, female glume (0.5 mm); g, fruit (0.5 mm); h, nut (0.5 mm).

± 20, subsessile or shortly pedunculate, ± globose, head-like clusters of spikelets, 3–5 mm in diameter, born singly or in pairs in the axils of leaf-like bracts at ± regular intervals along the culm. Spikelets unisexual but male spikelets sometimes appearing bisexual, with 1 terminal female flower and 1 or 2 male flowers; female spikelet without rudiments of upper flowers, thus appearing terminal, ± 2 mm long; male spikelets 3 or 4 at base, 1- or 2-flowered, with 3 or 4 glumes. Flowers unisexual. Glumes 3 or 4; lower glumes bristle-like, thinly membranous; male glumes broadly lanceolate; the 2 glumes subtending the female flower ovate-lanceolate or elliptical, 2–2.5 mm long, 5- to 8-nerved, strongly folded, membranous, greenish, 3-lobed and falling with the nut. Perianth bristles absent. Stamens 3. Stigmas 3. Nuts globose, 0.7–1 mm in diameter, bony, white, irregularly ribbed or reticulate, disk ± 0.7 mm in diameter and completely joined to the base of the nut.

Annual. In marshes and wet open places.

{⊙ Ten/Hel}. Tropical Asia and Australia: Cey, Ind: Kl, Kt, Tn. 0–1000 m.

Eleocharis [*Limnochloa, Scirpus* pro parte; excluding *Websteria*]: ± 180 species: cosmopolitan

Annual or perennial, often with creeping rhizomes or stolons. Culms tufted, terete or sometimes 4-angled, sometimes septate. Leaf blades where present, similar to the culms, at least the upper leaves reduced to membranous sheaths. Inflorescence a terminal, solitary spikelet, not subtended by bracts. Spikelets few- to many-flowered, terete, globose or ovoid, rachilla persistent. Glumes all spirally arranged or sometimes in 2 rows at the base of the spikelet, caducous, the lower ones usually sterile or at least different in shape from the others. Flowers bisexual. Perianth bristles 4–10 (–12) or rarely absent, shorter or not much longer than the nut, with backwardly directed barbs, persistent in fruit. Stamens 1–3; connectives appendaged. Stigmas 3 or 2; style usually with a persistent and swollen base. Nuts 3-sided or biconvex.

Most occur in shallow water, some persist submerged in deeper water where they sometimes remain sterile. Some perennial species are often locally dominant over relatively large areas. Some annual species are found in seasonally wet areas and pass the dry period as seeds. Disseminules have barbed perianth segments and are dispersed on the surface of mammals and birds. Many are reported as weeds; some are cultivated for matting, as decorative aquarium plants or as food (*E. dulcis* the Chinese water-chestnut).

Excluded: the following are essentially temperate species of ***Eleocharis*** which are found in the Himalayas but do not extend south in the Indian subcontinent: ***E. acicularis*** (Linnaeus) Roemer et Schultes; ***E. mitracarpa*** Steudel; ***E. obscura*** T. Koyama; ***E. palustris*** (Linnaeus) R. Brown (frequently but erroneously recorded from Peninsular India); ***E. parishii*** Britton; ***E. quinqueflora*** (F. X. Hartman) O. Schwarz [*E. pauciflora* (Lightfoot) Link]; ***E. thomsonii*** Boeckler; ***E. tuberculosa*** R. Brown; ***E. uniglumis*** (Link) Schultes.

1. Spikelets cylindrical tapered above, as wide or slightly narrower than the culms; glumes thick and leathery; perennials
  2. Culms hollow, septate; glumes with distinct keels
    3. Rhizomes stoloniferous; culms soft; floral glumes truncate at the apex ***E. dulcis***
    3. Rhizomes not stoloniferous; culms subrigid; floral glumes subacute at the apex (Sri Lanka) ***E. lankana***
  2. Culms solid, continuous within, not septate; glumes without keels or keels obscure
    4. Culms terete or nearly so; hyaline margin of glumes widest at the tip
      5. Sheaths pale, tinged with red or brown; floral glumes (excluding hyaline margin) 4–5 mm long, 2–2.2 mm wide; nut with (15–) 25–30 rows of transversely elongated cells on each face ***E. ochrostachys***
      5. Sheaths dirty black; floral glumes (excluding hyaline margin) 3.5–4 mm long, ± 2 mm wide; nut with 20–25 rows of transversely elongated cells on each face ***E. swamyii***
    4. Culms distinctly 3-angled above; hyaline margin of glumes widest at the sides
      6. Glumes ovate; perianth bristles unequal, most as long as or longer than the nut; nuts constricted apically into a neck with an expanded tip; sheaths without appendages ***E. acutangula***
      6. Glumes cuneate; perianth bristles almost equal, as long as or shorter than the nut; nuts without a neck, apically annular; sheaths with bristle-like appendages ***E. spiralis***
1. Spikelets ovoid to globose, much wider than the culms; glumes membranous; annuals
  7. Styles 2-cleft; nuts biconvex
    8. All glumes fertile; culms rarely more than 10 cm tall; spikelets fuscous brown; perianth bristles white, ± as long as or shorter than the nut; style base discoid ***E. atropurpurea***
    8. Lower two glumes sterile and covering the fertile glumes in bud; culms usually at least 15 cm tall; spikelets straw-coloured; perianth bristles purplish-grey, longer than the nut; style base conical ***E. geniculata***
  7. Styles 3-cleft; nuts 3-sided

9. Culms acutely 4- or sometimes 5-angled; perennial with stolons; spikelets oblique ***E. tetraquetra***

9. Culms terete or obscurely angled; annual or short-lived perennial with fibrous roots only; spikelets upright

10. Glumes spiral, numerous (many more than 10), compact, without nerves; culms 5–30 (–60) cm tall, longitudinally striate, not ribbed nor punctulate; perianth bristles longer than the nut; nuts smooth ***E. congesta***

10. Glumes in 2 rows below, less than 10, loose, several-nerved; culms not more than 15 cm tall, deeply 5-ribbed, punctulate; perianth bristles ± as long as the nut; nuts pitted ***E. retroflexa*** subsp. ***chaetaria***

Eleocharis acutangula (Roxburgh) Schultes [*'actangula'* orth. mut., *E. fistulosa* (Poiret) Schultes, *Limnochloa media* (Roxburgh) Nees]: Fig. 121

Stoloniferous, the stolons often terminated by a small tuber 3–5 mm in diameter. Culms tufted, erect, 30–90 cm tall, 2–5 mm in diameter, sharply 3-angled with flat or shallowly concave sides, pale green, spongy inside. Sheaths 3 or 4, the lower ones scale-like brownish, the uppermost ones 5–10 cm long, pale green, obliquely truncate above. Spikelet cylindrical to oblong-elongate, 10–60 mm long, 3–5 mm in diameter, the glumes often projecting stiffly to give a saw-like appearance. Glumes firm, ovate to ovate-elliptical, 4–5 (–7) mm long, 2–4 mm wide, dirty straw-coloured, several nerved on the sides, narrowly hyaline margins, rounded at apex, obscurely keeled. Perianth bristles with barbs or smooth or nearly so, 6 or 7, unequal, one long up to 10 mm long, one very short, the others about as long as the nut. Styles 6–7 mm long, (2) 3-cleft; style base articulate, ovoid-deltoid, 0.5–0.7 mm long, about half as wide as the nut. Nuts unequally obovoid, flattened-triangular to biconvex, 1.4–2 mm long, 1.2–1.6 mm wide, glossy, yellowish-brown, longitudinally striate with 13–15 rows of cells, apically constricted to an annular neck almost half as wide as the nut.

Perennial. Found in shallow water, in seasonal or permanent pools, along the marshy banks of pools, in swamps and in rice fields; often gregarious and forming tussocks.

{Hyp/(Ten)}. Pantropics: Bur, Cey, Ind: As, Bh, Kl, Kt, Mg, Mp, Pj, Tn, Up. 0–1000 m.

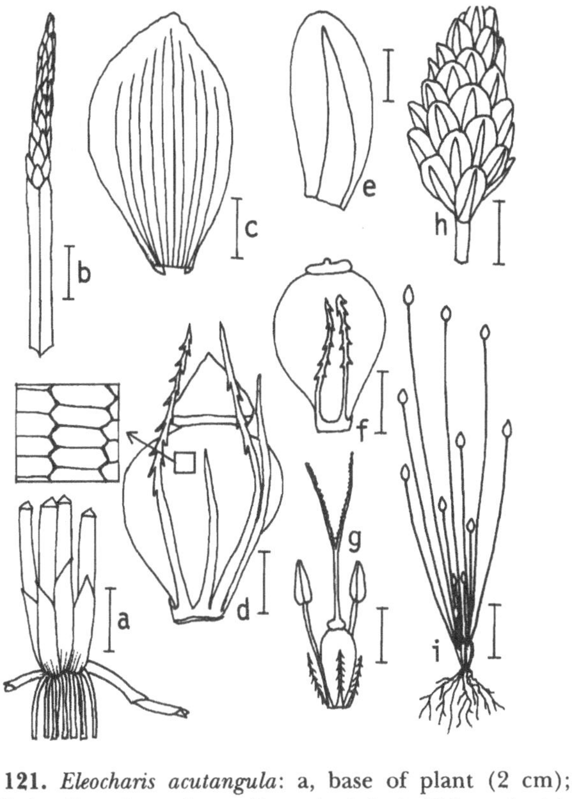

Fig. 121. *Eleocharis acutangula*: a, base of plant (2 cm); b, spikelet (1 cm); c, glume (1 mm); d, fruit (0.5 mm) with surface pattern (inset). *E. atropurpurea*: e, glume (0.3 mm); f, fruit (0.25 mm); g, flower (0.25 mm); h, spikelet (1 mm); i, whole plant (1 cm).

Eleocharis atropurpurea (Retzius) Presl: Fig. 121

Culms 1–10 (–15) cm tall, 0.2–0.4 mm in diameter. Sheaths 2, mucronate, purplish below, pale above. Spikelet ovoid to broadly lanceolate, 2–4 mm long (finally elongating to 8 mm), 1.5–2 mm in diameter, wider than the culm, brown turning fuscous-dark in fruit. Glumes all fertile, ovate, 0.9–1.2 mm long, 0.5–0.7 mm wide, the sides purplish or hyaline, nerveless; keel green. Perianth bristles white, 3–4 (–6) or rarely vestigial or absent, equal, ± 1 mm long, ± as long as or shorter than the nut. Styles 2-cleft; style base discoid, 0.3–0.6 (–1) times as wide as the nut. Nuts obovoid, biconvex, up to 1 mm long, 0.3–0.5 mm wide, very shiny, black, stipitate at base, truncate at apex with a greyish and flattened appendage.

Slender annual. Marshes and drying margins of tanks. Frequently gregarious in shallow water and in seasonally flooded places, often grows in rice fields. It can withstand considerable trampling. Diagnostic features: styles 2-cleft; bristles white and mostly shorter than the nut; nuts never more than 1 mm long and black.

{⊙ Hyp/Ten?}. Pantropics and subtropics, extending to Europe and USA: Ind, Nep: As, Dl, Gj, Hp, Jk, Kl, Kt, Mh, Mp, Pj, Rj, Tn, Up, Wb.

Eleocharis congesta D. Don [*E. japonica* Miquel, *E. pellucida* J. et C. Presl]: Fig. 122

Rhizomes absent or very short. Culms densely tufted, erector oblique, 5–30 (–60) cm tall, 0.2–1 mm in diam-

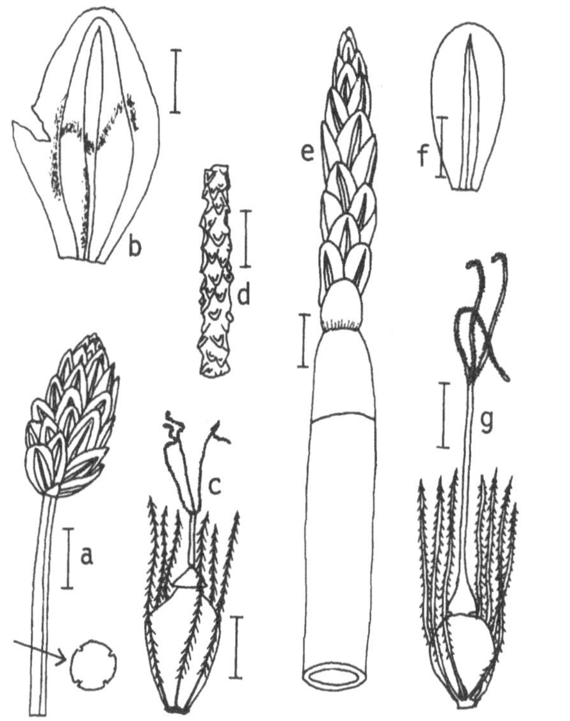

Fig. 122. *Eleocharis congesta*: a, spikelet (2 mm) with transverse section of culm (inset); b, glume (0.5 mm); c, fruit (0.5 mm); d, rachilla (1 mm). *E. dulcis*: e, spikelet (3 mm); f, glume (2 mm); g, fruit (1 mm).

eter, terete or nearly so, longitudinally striate, bright green. Sheaths 2 or 3, membranous, purplish below becoming pale above, orifice truncate with a mucro 0.3–1 mm long. Spikelet ovoid, 5–7 (–11) mm long, 1.5–3 (–4) mm in diameter (occasionally showing proliferations at the base). Glumes spiral, numerous (many more than 10), thin and membranous, ovate to broadly ovate-elliptical, 1.5–2 mm long, 0.7–1 mm wide, reddish-brown, with whitish hyaline margins, without nerves; keel obscure and green, the nerve ending below the apex. Perianth bristles whitish becoming rusty-coloured, 6–8, subequal, ± 1.5 mm long, mostly longer than the nut. Styles 1–1.2 mm long, 3-cleft; style base conical, basally constricted and articulate, as wide as or wider than the nut. Nuts obovoid, 3-angled, 0.8–1.2 mm long, 0.5–0.6 mm in diameter, contracted at the apex, greenish-yellow to olive-coloured, glossy, smooth.

Annual or sometimes a short-lived perennial. Often forming discrete circular clumps in shallow water or on marshy land. Sometimes found in rice fields. Often wrongly called *E. palustris.* Small and slender plants with floral glumes not exceeding 1.2 mm long and nuts not more than 1 mm long are sometimes recognized as *E. pellucida* or *E. congesta* var. *japonica* (Miquel) T. Koyama.

{(⊙) Hyp/(Hel)}. Nepal, India, Indo-China, China and Japan: Bhu, Cey, Ind, Nep: As, Hp, Jk, Kt, Mg, Mh, Mn, Mp, Pj, Rj, Sk, Tn, Up. 0–2800 m.

Eleocharis dulcis (N. L. Burman) Trinius ex Henschel [*E. plantaginea* (Retzius) Roemer et Schultes, *E. equisetina* J. Presl et C. Presl, *E. tuberosa* (Schultes) Schultes, *E. andamanensis* E. Govindarajalu, *Scirpus tumidus* Roxburgh, *Scirpus plantaginoides* Rottbøll]: Fig. 122

Stolons present, often terminating in tubers; tubers brown to blackish, ± 1 cm in diameter (in cultivated races up to 5 cm in diameter). Culms tufted, erect, 40–100 (–200) cm tall, (1–) 3–7 (–10) mm in diameter, terete, hollow but with transverse septa (seen by squashing the stem), deep green, somewhat shiny. Sheaths obliquely obtuse, purplish. Spikelet cylindrical to oblong-elongate, 15–60 mm long, 2–6 mm in diameter, not wider than the culm, whitish to pale straw-coloured. Glumes thick and slightly leathery, elliptical to oblong-obovate, (2–) 5–7 mm long, 1.7–4 mm wide, greyish-green to straw-coloured, many nerved, margins narrowly hyaline, rounded at apex, distinctly keeled; keel green. Perianth bristles 6–8, subequal, as long as or up to twice as long as the nut. Styles 7–8 mm long, 2- or 3-cleft; style base elongate-conical, flattened, 2–2.5 mm long, almost as wide as the nut. Nuts obovoid, biconvex with obtuse edges, 1.5–2.2 mm long, 1.3–1.8 mm wide, smooth, yellow or greyish-brown, shiny.

Perennial. Gregarious in shallow water in ponds, rice fields and along irrigation canals. Sometimes it grows in salt or brackish water. Often locally dominant. It is a widespread and very variable species but immediately recognizable by the culms with transverse septa and spikelets narrower than the culms. A cultivar *'tuberosa'* is sometimes cultivated for its edible tubers. These tubers may be up to 5 cm across; in Western markets they are often sold under the name 'Chinese water-chestnut'. The above-ground parts are high in protein and low in fibre; it has been recommended to be cultivated for its leaf proteins — Pandey, V. N. and Srivastava, A. K. Aquatic Botany 41 (4): 369–374 (1991). The culms are sometimes used for weaving and mat making. Plants with compact horizontal rhizomes, with broader spikelets and longer glumes collected in N Kerala have been identified as *E. sphacelata* R. Brown, a species from temperate Australasia.

{Hyp/Ten}. Old World tropics: Ban, Cey, Ind, Nep: An, As, Bh, Gj, Kl, Kt, Mh, Mp, Pj, Rj, Tn, Up, Wb. 0–2000 m.

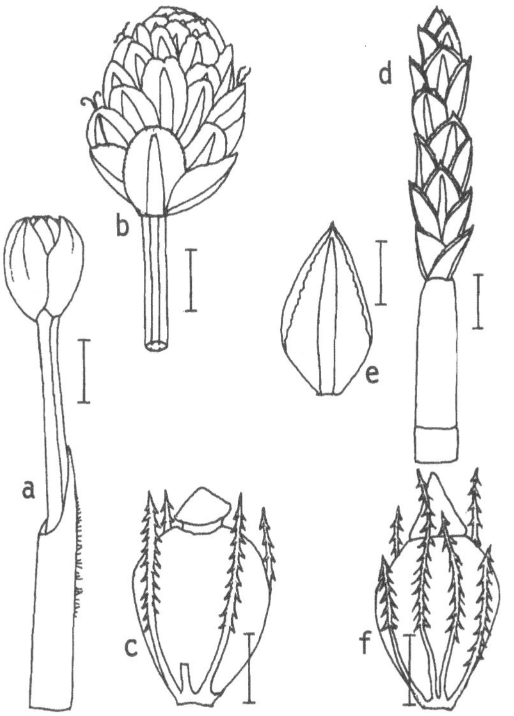

Fig. 123. *Eleocharis geniculata*: a, young spikelet (1 mm); b, mature spikelet (1 mm); c, fruit (0.25 mm). *E. lankana*: d, spikelet (5 mm); e, glume (2 mm); f, fruit (1 mm).

Eleocharis geniculata (Linnaeus) Roemer et Schultes [*E. capitata* R. Brown]: Fig. 123

Culms densely tufted, slender but rather rigid, (7–) 15–25 (–40) cm tall, 0.2–0.4 (–1) mm in diameter. Sheaths herbaceous, obliquely acute or attenuate, green, often tinged with red-brown below. Spikelet globose to ovoid, very obtuse, (2–) 3–4 (–8) mm long, 2–4 mm in diameter, much wider than the culm, usually rusty brown. Glumes membranous, the lower ones sterile and covering the fertile glumes in young buds, ovoid to suborbicular, 1.8–2 mm long, 1.3–1.6 mm wide, greyish below, pale brown to straw-coloured above, the sides sometimes tinged with purple, nerveless, very obtuse at apex, indistinctly keeled; keel green. Perianth bristles smooth, rusty to purplish grey, 6–10, equal, up to 1.5 mm long, slightly longer than the nut. Styles 1.5–1.7 mm long, 2-cleft; style base depressed-conical, whitish, spongy, not as wide as the nut. Nuts broadly obovoid, biconvex or sometimes 3-sided, 0.8–1 mm long, 0.6–0.8 mm in diameter, suddenly contracted to a rounded apex, maturing shiny blackish-purple.

Annual. Often growing in clumps, gregarious in shallow water, particularly common in fallow rice fields where it may sometimes look like a crop. Sometimes in brackish water by the sea. Diagnostic features: almost spherical heads and dark, smooth nut.

{⊙ Hyp/Ten}. Almost cosmopolitan in tropical to warm regions: Cey, Ind: An, Ap, Gj, Go, Kl, Kt, Mp, Rj, Tn, Up, Wb. 0–900 m.

Eleocharis lankana T. Koyama: Fig. 123

Like *E. dulcis* but without stolons and with subacute glumes.

Known from a single locality in Sri Lanka, Colombo District. ♦Cey.

Eleocharis ochrostachys Steudel [*E. variegata* Presl var. *laxiflora* (Thwaites) C. B. Clarke, *Scirpus laxiflorus* Thwaites]: Fig. 124

Rhizomes short with long stolons. Culms tufted, erect, slender, rigid, 35–80 cm tall, (0.8–) 2–3 (–5) mm in diameter, terete or weakly angular, green, smooth, slightly shiny. Sheaths 2 or 3, membranous, pale or tinged with red or brown, obliquely attenuate with a small mucro. Spikelet cylindrical to oblong-elongate, (5–) 10–25 mm long, 2–4 mm in diameter. Glumes firm, ovate to ovate-oblong, 4–5 mm long, 2–3.5 mm

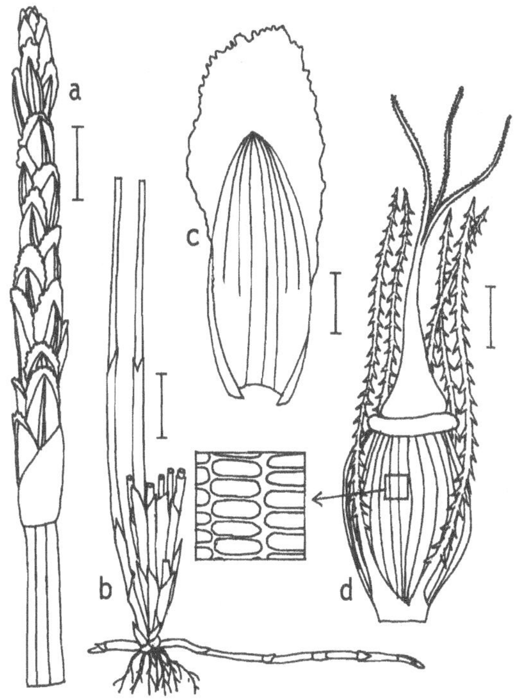

Fig. 124. *Eleocharis ochrostachys*: a, spikelet (4 mm); b, base of plant (2 cm); c, glume (1 mm); d, fruit (0.5 mm) with surface pattern (inset).

wide, pale green with brown-tinged hyaline margins up to 2 mm wide at the tip, weakly many-nerved, obtuse at apex, obscurely keeled. Perianth bristles 4–7, subequal, 2–3 times as long as the nut, yellowish or light brown. Styles (2-) 3-cleft; style base pyramidal, 1.5–3 mm long, 1–2 mm wide at base. Nuts flattened 3-sided to unequally biconvex, broadly obovoid, 1.5–2 mm long, 1–1.6 mm in diameter, glossy, straw-brown to greyish, longitudinally striate with 15–30 rows of transversely elongated cells on each face, contracted to an annulate apex.

Perennial. Found in shallow water, in marshes and along irrigation canals, sometimes found in floating islands. Occasionally used in weaving.

{Hyp/Ten}. India, Malaysia, Japan and Pacific: Cey, Ind: Ap, Kl, Tn.

Eleocharis retroflexa (Poiret) Urban subsp. ***chaetaria*** (Roemer et Schultes) T. Koyama [*E. chaetaria* Roemer et Schultes, *Cyperus setaceus* Retzius]: Fig. 125

Culms tufted, filiform, mostly curved, sometimes erect, 3–15 (–25) cm long, 0.2–0.5 mm in diameter, light green, 4- or 5-ribbed, punctulate. Sheaths 1 or 2, thinly membranous, obliquely attenuate, pale, at the base purplish. Spikelet ovoid, ± flattened, 2–6 mm long, 1–2.5 mm wide, loose, 3- to 10-flowered. Glumes basally distichous, not more than 10, thin and membranous, ovate to oblong-lanceolate, 2–3.3 mm long, 1.5–2 mm wide, sides purplish, nerveless or indistinctly striate, margin hyaline, obtuse at apex, keeled with a greenish midnerve. Perianth bristles rather delicate, flattened, whitish to yellowish, 6–8, unequal, the longest equalling to just surpassing the style base, scabrid towards the apex. Styles (2-) or 3-cleft; style base depressed conical, 3-lobed, as wide as and decurrent on the nut angles. Nuts obovoid, sharply 3-angled, with ribbed edges, 0.7–0.8 (–1) mm long, 0.6–0.9 mm wide, straw-coloured becoming almost black, pitted, longitudinally 6–10 striate, ribbed on the angles, apically truncate.

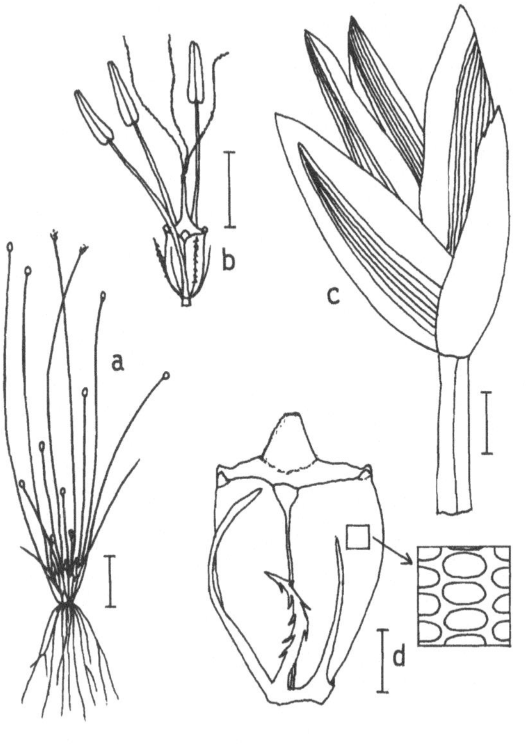

FIG. 125. *Eleocharis retroflexa* subsp. *chaeteria*: a, whole plant (1 cm); b, flower (1 mm); c, spikelet (0.5 mm); d, fruit (0.25 mm) with surface pattern (inset).

Annual. Gregarious in shallow water and along the marshy banks of pools or in ditches and rice fields. In deep or flowing water it sometimes develops hair-like culms like *Websteria*. It occasionally propagates by curving over and developing young plants in the axils of glumes. It is not confined to wetlands. *Eleocharis retroflexa* in the strict sense is confined to the New World.

{⊙ Hyp/Ten/(Hel)}. Tropics of the Old World: Ban? Bhu, Cey, Ind, Nep: An, As, Hp, Kl, Kt, Mp, Tn, Wb. 0–1600 m.

Eleocharis spiralis (Rottbøll) Roemer et Schultes: Fig. 126

Rhizomes short with creeping stolons. Culms densely tufted, erect, rather robust, 25–60 (–70) cm tall, 2–4 mm in diameter, deep green, smooth, shiny, apically 3-sided. Sheaths 2 or 3, pale green becoming purple-red above, obliquely attenuate with a bristle-like appendage 2–4 mm long. Spikelet cylindrical to oblong-elongate, 15–40 mm long, 3–6 mm in diameter. Glumes rather firm, suborbicular to broadly obovate, very closely packed (the exposed portion broader than long), 2.8–4 mm long, 2–3.8 mm wide, purplish becoming yellowish below apex, weakly many-nerved, margin narrow and hyaline, obtuse at apex, obscurely keeled. Perianth bristles delicate, 4–6, subequal, from much shorter than to about as long as the nut. Styles ± 1.7 mm long, (2-) or 3-cleft; style base pyramidal, confluent, not as long and wide as the nut. Nuts biconvex (the margins not ribbed), obovoid to oval, 1.5–1.8 mm long, 1.2–1.4 mm in diameter, glossy, straw-coloured but becoming deep brown later, longitudinally striate with 17–20 rows of cells, without neck and apical annulus or annulus indistinct.

Perennial. Gregarious in shallow water, particularly common in rice fields and along irrigation canals. It often forms extensive stands in brackish water and brackish marshes.

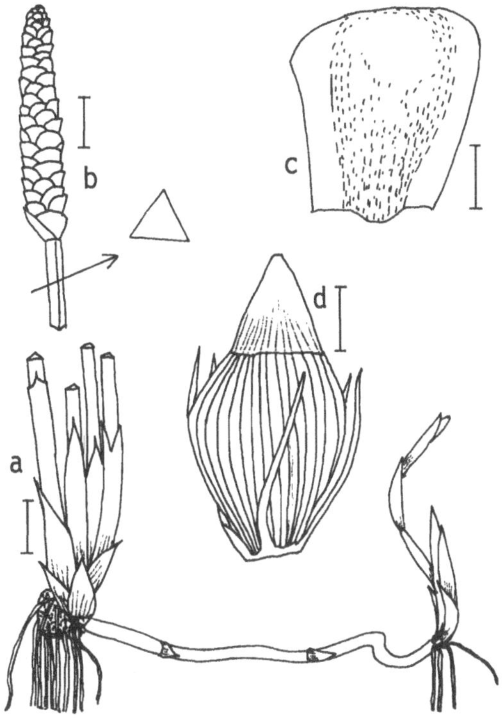

Fig. 126. *Eleocharis spiralis*: a, base of plant (1 cm); b, spikelet (5 mm) with transverse section of culm (inset); c, glume (1 mm); d, fruit (0.5 mm).

{Hyp}. Tropics of the Old World: Cey, Ind: Kl, Kt, Mh, Tn, Wb.

Eleocharis swamyii E. Govindarajalu: refer to Fig. 124

Like *E. ochrostachys* but culms always terete just below the inflorescence; sheaths dirty black; floral glumes loosely ± 4-ranked, 3.5–4 mm long, ± 2 mm wide, the midrib distinct, the upper half hyaline with a lacerate margin; style 2–2.5 mm long; nuts 1.7–2 mm long, ±1.5 mm wide, with 20–25 rows of transversely elongated cells on each face.

Perennial. Ecology unknown. It is doubtfully distinct from *E. ochrostachys*; the whole *Eleocharis variagata* group is in need of taxonomic revision.

{Hyp?/Ten?}. Endemic in Peninsular India: ♦Ind: Ap, Tn. 0–600 m.

Eleocharis tetraquetra Nees: Fig. 127

Rhizomes short, sometimes with long and slender stolons. Culms tufted, erect, slender, rigid, (15–) 30–90 cm tall, 1–2.5 mm in diameter, acutely 4- or rarely 5-angled, deep green. Sheaths membranous to herbaceous, pale brown tinged with purple below, orifice thickened with an obtuse mucronulate tip. Spikelet often somewhat inclined, ovoid to elliptical-ovoid, 9–20 mm long, 3–5 mm, densely many-flowered. Glumes spiral, chartaceous, oblong-elliptical or lanceolate, 3–4 mm long, ± 2 mm wide, rusty brown above, with narrow whitish hyaline margins, without nerves; keel obscure, green, the nerve ending below the apex. Perianth bristles whitish or reddish, 6 (–8), subequal, slightly longer than to twice as long as the nut, very densely retrorsely scabrous or subplumose. Styles 1–2 (–3) mm long, 3-cleft; style base conical, basally constricted and articulate, narrower than the nut. Nuts obovoid, obtusely 3-sided, flattened, 1.5–2 mm long, 1–1.2 mm in diameter, greenish-brown, glossy, smooth.

Perennial. Found in wet soil and rice fields. Mostly at higher altitudes.

{Hyp/Ten?}. India to Japan, south to Malaysia and N Australia: Bhu, Cey, Ind, Nep: Ar, Hp, Kl, Mg, Sk, Tn, Up. ?–3500 m.

Eleogiton [*Scirpidiella*, *Isolepis* pro parte, *Scirpus* pro parte]: 5 species

Eleogiton fluitans (Linnaeus) Link [*E. curvulus* Nees]: Fig. 128

Submerged or floating. Stems often rooting at nodes, leafy, usually much branched, 8–80 cm long,

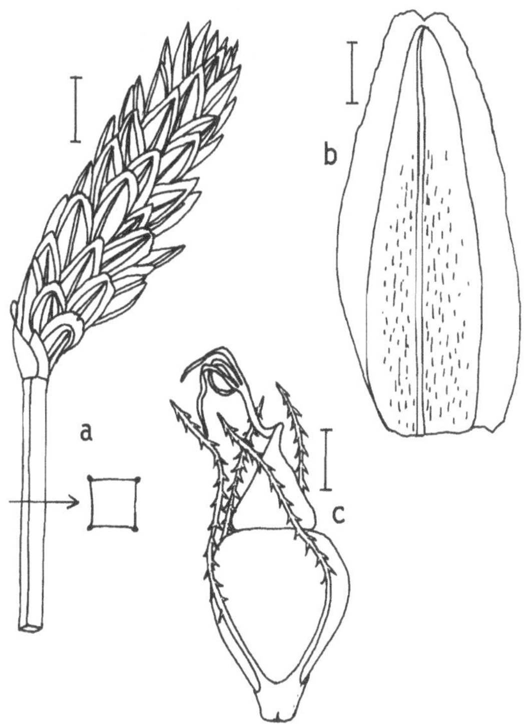

Fig. 127. *Eleocharis tetraquetra*: a, spikelet (2 mm) with transverse section of culm (inset); b, glume (0.5 mm); c, fruit (0.5 mm).

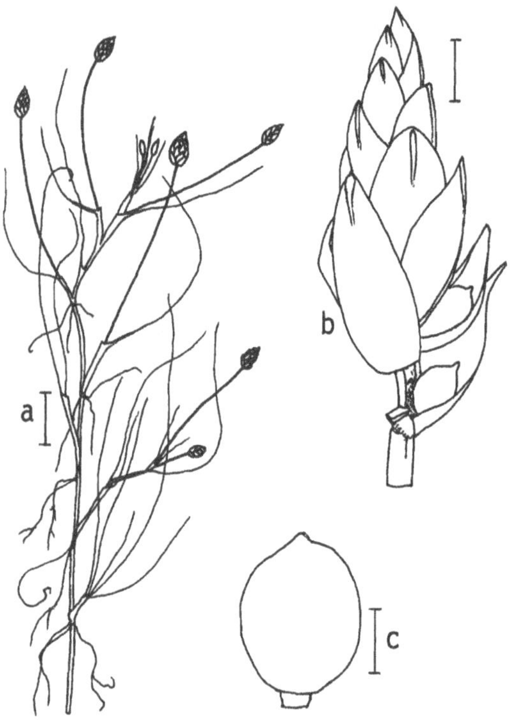

Fig. 128. *Eleogiton fluitans*: a, whole plant (1 cm); b, spikelet (1 mm); c, nut (0.5 mm).

0.3–1.2 mm in diameter, terete. Leaves cauline, sheathing at the base with capillary blades; bladeless sheaths subtend the peduncle; blades 1.5–7 cm long, 0.3–1 mm wide, flattish or with incurved margins, light green. Inflorescence a solitary terminal spikelet. Spikelet erect, ovoid, 3–6 (–8) mm long, 1.7–3.5 mm in diameter, (3-) 5- to 10-flowered, pedunculate; peduncles 1.5–6 (–12) cm long; bract 1, scale-like, up to three-quarters as long as the spikelet. Flowers bisexual, without perianth scales or bristles. Glumes spiral, elliptical to ovate, 1.7–3 (–4) mm long, 0.8–1.5 mm wide. Stamens 2 or 3. Style ± 0.3 mm long, 2- or 3-cleft. Nuts 3-sided or biconvex and flattened, obovoid, 1.3–1.5 mm long, 0.7–1 mm wide, yellowish-brown, smooth, without a beak.

Annual or perennial. Mostly submerged with emergent spikelets but can persist for a while as a terrestrial plant as the water recedes. In cool, running water or forming mats at the margins of mountain streams. There is little agreement as to which genus this species belongs; in recent works it has appeared in *Isolepis*, *Scirpidiella* and *Scirpus*.

{(⊙) Vit/(Pla/Ple/Hyp)}. Old World from temperate lowlands to tropical mountains: Cey, Ind: As, Mg, Tn. 1400–2000 m.

Fimbristylis [*Scirpus* pro parte, including *Abildgaardia* and *Trichelostylis*, excluding *Bulbostylis*]: ± 300 species: warm to tropical regions of the world

Annual or perennial, sometimes rhizomatous. Culms mostly tufted, up to 1 m tall but usually much less, sometimes hairy. Leaves usually basal, usually shorter than the culms; sheaths loose, open or tubular; ligules membranous, hairy or absent; blades flat, sometimes grooved or inrolled or needle-like or reduced to sheaths. Involucral bracts often leafy below and setaceous above or all setaceous, 0 or several. Inflorescence usually terminal, rarely pseudolateral, loose or compact, simple or compound, corymbose, panicle-like, umbel-like or a head-like cluster of spikelets or sometimes reduced to a solitary spikelet. Spikelets usually many-flowered but sometimes reduced to 1, terete, angular or laterally flattened; rachilla persistent, usually winged. Glumes usually spirally arranged but sometimes distichous and in 2 rows at least at the base of the spikelet, caducous, the lower 1–4 empty. Perianth bristles absent. Stamens 1–3; connectives crested. Styles 2- or 3-cleft, fimbriate, pubescent or glabrous, their bases swollen, usually caducous with the style; stigmas 2 or 3. Nut terete, 3-sided, biconvex or plano-convex.

There are botanical grounds for recognizing the genus *Abildgaardia* but pending a complete revision of Indian Cyperaceae it is felt better to retain the genera *Fimbristylis* and *Bulbostylis* in their traditional sense. There are some taxonomic difficulties in this genus, I have tended to take the side of the splitters rather than the lumpers.

Excluded: the following may be found in damp or wet places but they are not considered to be aquatic or wetland species: ***Fimbristylis aggregata*** Fischer: ♦Ind: Tn; ***F. amplocarpa*** E. Govindarajalu: ♦Ind; Tn; ***F. carpopoda*** E. Govindarajalu: ♦Ind: Ng; ***F. circumciliata*** E. Govindarajalu: ♦Ind: As; ***F. contorta*** Fischer: ♦Ind: Tn; ***F. dura*** (Zollinger et Moritz) Merrill [*F. asperrima* Boeckeler]; ***F. eligulata*** E. Govindarajalu: ♦Ind: Tn; ***F. eragrostis*** (Nees) Hance [*Abildgaardia eragrostis* Nees et Meyen ex Nees, *F. nigro-brunnea* Thwaites]; ***F. falcata*** (Vahl) Knuth [*Trichelostylis torta* Nees, *T. chaetorrhiza* Nees, *T. latifolia* Nees, *F. junciformis* (Nees) Kunth]; ***F. fimbristyloides*** (F. von Mueller) Druce; ***F. fusca*** (Nees) C. B. Clarke subsp. ***fusca***; ***F. fusca*** (Nees) C. B. Clarke subsp. ***fulvescens*** (Thwaites) T. Koyama; ***F. insignis*** Thwaites; ***F. intonsa*** S. T. Blake; ***F. longistigmata*** E. Govindarajalu: ♦Ind: Tn; ***F. merguensis*** C. B. Clarke; ***F. monospicula*** E. Govindarajalu: ♦Ind: Kt, Tn; ***F. monticola*** Hochstetter ex Steudel: ♦Cey: ♦Ind; ***F. multinervia*** E. Govindarajalu: ♦Ind: Ng; ***F. nayaranii*** Fischer: ♦Ind: Kt, Up; ***F. pustulosa*** E. Govindarajalu: ♦Ind: Tn; ***F. rectifolia*** E. Govindarajalu [*F. dichotoma* (Linnaeus) Vahl var. *nilgaririca* C. B. Clarke]: ♦Ind: Tn; ***F. rigidiuscula*** E. Govindarajalu: ♦Ind: Tn; ***F. rugosa*** E. Govindarajalu: ♦Ind: Tn; ***F. scabrisquama*** E. Govindarajalu: ♦Ind: Tn; ***F. semi-disticha*** E. Govindarajalu: ♦Ind: Kt, Tn; ***F. stolonifera*** C. B. Clarke; ***F. strigosa*** E. Govindarajalu: ♦Ind: Tn; ***F. swamyii*** E. Govindarajalu: ♦Ind: Kt, Tn; ***F. thomsonii*** Boeckeler; ***F. tortifolia*** E. Govindarajalu: ♦Ind: Tn; ***F. tristachya*** R. Brown; ***F. uliginosa*** Steudel: ♦Ind; ***F. unispicularis*** E. Govindarajalu et Hemadri: ♦Ind: Mh; ***F. velata*** R. Brown [*F. propinqua* R. Brown, *F. squarrosa* Vahl var. *esquarrosa* Makino, *F. dichotoma* (Linnaeus) Vahl var. *villosa* Fischer apud Gamble, *F. bisumbellata* (Forsskål) Bubani var. *hirtistyla* Fischer, *F. aestivalis* (Retzius) Vahl var. *esquarrosa* (Makino) Koyama].

1. Nuts oblong-cylindrical (the greater part parallel-sided)
- **2.** Inflorescence corymbose with few to several, 3–6 mm long spikelets; leaves with linear blades ***F. dipsacea***
- **2.** Inflorescence of a single terminal spikelet; spikelet 5–12 mm long; leaves reduced to bladeless sheaths ***F. tetragona***

1. Nuts ellipsoidal to suborbicular (not parallel-sided)
- **3.** Stigmas 2, style flat, usually hairy; nut biconvex or plano-convex; glumes spiral never in 2 rows
  - **4.** Nut with transverse ridges; cauline leaves reduced to bladeless sheaths; inflorescences of single spikelets
    - **5.** Spikelet erect, elongate-ovoid to ovoid, 2–3.5 mm wide; nut with 5–8 transverse ridges ***F. acuminata***
    - **5.** Spikelet ± nodding, ovoid to broadly-ovoid, 3–5 mm wide; nut with 3–5 (–6) transverse ridges ***F. nutans***
  - **4.** Nut without transverse ridges; cauline leaves with or without blades; inflorescences paniculate or reduced to single spikelets
    - **6.** Culms strongly flattened, 2-edged ***F. scaberrima***
    - **6.** Culms not strongly flattened, not 2-edged
      - **7.** Glumes oblong-ovate to oblong-lanceolate, twice or more than twice as long as wide (inflorescences usually of single spikelets)
        - **8.** Upper flower bearing glumes without a mucro; nuts with ± 30 longitudinal stripes on either face; ligule absent; stigmas as long as or longer than the style ***F. polytrichoides***
        - **8.** Upper flower bearing glumes with a mucro; nuts with ± 15 longitudinal stripes on either face; ligule a fringe of hairs; stigmas shorter the style ***F. tenuicula***
      - **7.** Glumes ovate to orbicular, less than twice as long as wide (inflorescences usually of many spikelets)
        - **9.** Ligule a fringe of hairs or membranous
          - **10.** Nut smooth throughout
            - **11.** Glumes glabrous; inflorescence head-like, with 1 or 2 (3) spikelets ***F. schoenoides***
            - **11.** Glumes hairy, at least in the apical part; inflorescence corymbose, with 3 or many more spikelets
              - **12.** Lower sheaths leathery, shining brown or reddish-brown; blades of cauline leaves rather short, up to 10 cm long, 0.5–1.5 mm wide; involucral bracts usually shorter than the inflorescence; glumes pubescent ***F. ferruginea***
              - **12.** Lower sheaths herbaceous, straw-like or brown; blades of cauline leaves long, up to 30 cm long, 1.5–2 mm wide; lowest involucral bract usually overtopping the inflorescence; glumes densely hairy ***F. sieberiana***
          - **10.** Nut with longitudinal stripes on either face
            - **13.** Glumes wholly pubescent, clothed with short hairs ***F. pubisquama***
            - **13.** Glumes glabrous
              - **14.** Glumes white at base and along the margins; nuts bearing warts ***F. alboviridis***
              - **14.** Glumes brown with green keel and hyaline margins; nuts without warts
                - **15.** Spikelets 1–1.8 (–2) mm wide, somewhat angular; glumes membranous, 1.5–2 mm long, sharply keeled; style 0.7–1 mm long; nut 0.5–1 mm long, 0.5–0.6 mm broad, straw-coloured
                  - **16.** Leaf sheaths and blades glabrous; nuts with 5–9 longitudinal stripes on either face; culms 3-sided ***F. bisumbellata***
                  - **16.** Leaf sheaths and blades clothed with bristly hairs; nuts with 15–30 longitudinal stripes on either face; culms 4-sided ***F. hirsutifolia***
                - **15.** Spikelets 2–4 mm wide, ± terete; glumes chartaceous, 2–3 (–4.5) mm long, obscurely keeled; nut 1–1.5 mm long, 0.7–1 mm broad, whitish or straw-coloured, rarely brown
                  - **17.** Each face of nut with 12 or less longitudinal rows of horizontally arranged cells (appearing as pits); usually perennial; stipe of nut minute, less than 0.2 mm long ***F. dichotoma***
                  - **17.** Each face of nut with 15 or more longitudinal rows of horizontally arranged cells (appearing as pits); always annual; stipe of nut distinct 0.3 mm or more long ***F. tomentosa***

9. Ligule absent (sheaths on the inner side gradually passing into blades)

18. Edges of the nut ornamented with a row of white, club-like appendages (dwarf annual with filiform leaves, squarrose spikelets and long-awned glumes) ***F. dipsacea***

18. Edges of the nut without appendages

19. Spikelets 2–4 mm wide; rhizomatous perennials

20. Nuts blackish; spikelets ovoid to cylindrical (longer than wide); culm leaves with blades ***F. cymosa***

20. Nuts milky yellow; spikelets globose or nearly so (about as long as wide); culm leaves without blades ***F. umbellaris***

19. Spikelets 1–1.5 mm wide; small annuals

21. Inflorescence a head-like cluster of sessile spikelets; nut broadly obovoid or suborbicular, ± 0.5 mm long and wide ***F. argentea***

21. Inflorescence of numerous stalked spikelets; nut ellipsoid to obovoid, usually more than 0.5 mm long

22. Base of style fringed with long, white, pendent hairs covering at least the upper half of the nut ***F. squarrosa***

22. Base of style glabrous or microscopically hairy

23. More or less hairy; stem leaves with distinct blades; style hairy at least at the top ***F. aestivalis***

23. Glabrous; stem base clothed with 1 or 2 bladeless or very shortly laminate sheaths; style glabrous ***F. griffithii***

3. Stigmas 3, rarely in some flowers 2, style 3-angled, usually glabrous; nut 3-sided or 3-angled; glumes in 2 rows or spiral

24. Glumes, at least in young spikelets, distichous (in 2 opposite rows), the distichous arrangement in mature spikelets sometimes less pronounced by torsion of the rachilla; spikelets strongly laterally flattened

25. Glumes densely reddish gland-dotted; leaf blades very narrow, needle-like or almost so, 0.3–0.5 mm wide; spikelets up to 8-flowered but usually less ***F. cinnamometorum***

25. Glumes not gland-dotted; leaf blades flattened, usually more than 1 mm wide; spikelets usually more than 8-flowered

26. Spikelets 3–7 mm wide; nuts 2–3.2 mm long; (inflorescence with 1–4 (–6) large spikelets)

27. Inflorescence usually of a single terminal spikelet, rarely 1 or 2 peduncled spikelets added; spikelets glaucous yellowish to yellowish-green; low slender plant with culms 5–40 tall, 0.3–0.8 mm in diameter ***F. ovata***

27. Inflorescence of (2–) 3–4 (–6) spikelets; spikelets straw-coloured to brown; tall robust plant with culms 30–80 tall, 1–3 mm in diameter ***F. triflora***

26. Spikelets 1.5–3 mm wide; nuts 0.6–1 mm long; (inflorescence with either 1 or 2 or with several to numerous spikelets)

28. Leaves 1–3 mm wide; inflorescence with several to numerous spikelets; culms 10–35 cm tall ***F. disticha***

28. Leaves 0.5–0.8 mm wide; inflorescences with 1 or 2 spikelets; culms 3–10 cm tall ***F. zeylanica***

24. Glumes spirally arranged; spikelets terete or angular, not strongly flattened

29. Leaves with a ligule; ligule a row of short hairs or at least a distinct line between the sheath and blade

30. Spikelets very small, 2–4 mm long, ± 1 mm wide; glumes ± 1 mm long; style 0.5–0.75 mm long ***F. microcarya***

30. Spikelets larger, at least 5 mm long, 1.5–2.3 mm wide; glumes 2–3 mm long; style 1–2 mm long

31. Inflorescence compound, with many spikelets; culms 2–4 mm wide; glumes 2–2.5 mm long; nut warty, 0.7–0.8 mm long ***F. complanata***

31. Inflorescence simple or nearly so, with few spikelets; culms 1–1.5 mm wide; glumes 2.5–3 mm long; nut smooth, 1–1.2 mm long ***F. consanguinea***

29. Leaves without a ligule (sheaths on the inner side gradually passing into blades)

32. Culm leaves bearing well-developed blades or all the leaves basal (culm without bladeless sheaths)

33. Inflorescence a cluster of 3, free spikelets ***F. woodrowii***

33. Inflorescence corymbose, of many more than 3 spikelets

34. Nuts dark brown to black; at least some spikelets in clusters; leaf blades and glumes obtuse ***F. cymosa***

34. Nuts pale yellow, straw-coloured or whitish; spikelets solitary or paired but not clustered; leaf blades and glumes acute

35. Leaf blades 1–1.5 mm wide; some spikelets paired; glumes 1.8–2 mm long; culms fine, 10–20 cm tall ***F. crystallina***

35. Leaf blades 1.7–2 (–2.7) mm wide; all spikelets solitary; glumes 1.6–1.7 mm long; culms robust, 15–40 cm tall ***F. dauciformis***

32. All culm leaves reduced to bladeless sheaths or, at least, the upper culm leaf reduced to bladeless sheath

36. All culm leaves bladeless

37. Culms obtusely 3-sided or subterete; nuts with 12–15 rows of cells on each face *F. umbellaris*
37. Culms (4-) 5-angled, winged or grooved; nuts with 4–13 rows of cells on each face
38. Keels on the glumes obscure thus the spikelet appearing smooth and terete; nuts 0.6–0.7 mm long, densely warty, with 4–6 vertical rows of cells on each face; culms densely tufted *F. aphylla*
38. Keels on the glumes prominent giving the spikelet an angular appearance; nuts 0.8–1.2 mm long, smooth or slightly warty, with 9–13 vertical rows of cells on each face; culms usually in a single row along the rhizome
39. Culms usually hairy; spikelets 4.5–6.2 mm long; glumes 2.3–3 mm long, with keel projecting as a mucro; anthers ± 1.5 mm long *F. salbundia* subsp. *pentaptera*
39. Culms glabrous; spikelets 3.5–4.5 mm long; glumes 2–2.5 mm long, with keel ending somewhat below the apex; anthers 0.7–1 mm long *F. salbundia* subsp. *salbundia*
36. Lower culm leaves with well-developed blades
40. Spikelets ovoid-globular to almost globular, 1.5–3 mm wide; culms 4-sided, below the inflorescence flattened with 2 sharp edges; leaf sheaths bilaterally flattened with a sharp dorsal edge; keel on fertile glumes ending just below the apex *F. littoralis*
40. Spikelets ovoid, 1.2–1.7 mm wide; culms acutely 5-angled, below the inflorescence becoming almost 5-winged; leaf sheaths cylindrical with rounded dorsal side; keel on all glumes extending beyond the apex as a small mucro *F. miliacea*

Fimbristylis acuminata Vahl: Fig. 129

Shortly rhizomatous. Culms densely tufted, 5–15 (-30) cm long, 0.5–1 mm thick, erect, ridged, obtusely 3- or 4-angled. Leaves usually 2, reduced to obliquely truncate sheaths with minute triangular blade; sheaths inflated at apex; ligule absent. Involucral bracts absent. Inflorescence of 1, terminal spikelet. Spikelets solitary, erect, ovoid to elongate-ovoid, terete, 6–10 (–12) mm long, 2–3.5 mm wide. Glumes ovate-elliptical, obtuse to acute, scarcely mucronulate, 3.5–4.5 mm long, 2.2–4 mm broad, chestnut brown above, sides usually without nerves at base of spikelet becoming many-nerved above; keel green, with 3–5 (–7) nerves. Stamens 2 or 3. Styles 2-cleft, flat, less than 0.5 mm wide, scarcely expanded at base, hairy above. Nuts biconvex, with obtuse edges, broadly obovate or orbicular, 1.2–1.8 mm long and wide, sometimes broader than long, with 5–8 strong, transverse, wavy ridges, yellowish to brown, shortly stipitate.

Perennial. Open wet and muddy places, marshes, rice fields and river banks.

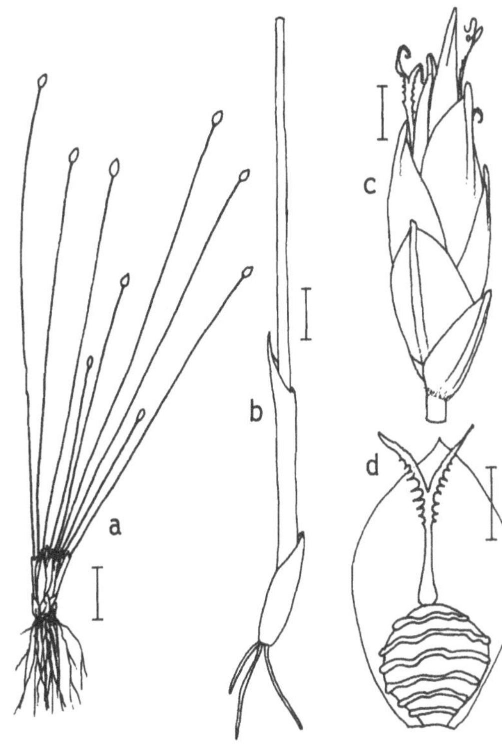

Fig. 129. *Fimbristylis acuminata*: a, whole plant (2 cm); b, base of culm (5 mm); c, spikelet (2 mm); d, fruit with glume (1 mm).

{(Ten)/Hel}. Indomalaysia extending to the Philippines: Bur, Cey, Ind: An, As, Gj, Go, Hp, Kl, Kt, Sk, Wb. 0–1200 m.

Fimbristylis aestivalis (Retzius) Vahl [including var. *trichopoda* Kern, var. *macrostachya* Bentham, subsp. *major* (Trimen) T. Koyama]: Fig. 130

Culms densely tufted, very slender, setaceous, 2–10 (–40) cm long, 0.3–0.5 mm in diameter, hairy. Leaf sheaths red-brown, membranous; ligule absent; blades filiform, up to 2.5 cm long, softly hairy, with hairy margins. Involucral bracts up to 6, some overtopping the inflorescence, the longest leaf-like, up to 4 cm long. Inflorescence compound, somewhat umbel-like. Spikelets ovoid to elongate-ovoid, 2–4 (–7) mm long, 0.8–1 (–1.5) mm wide. Glumes thinly membranous, ovate, up to 1.5 mm long, pubescent, brown to purplish-green, sides without nerves, margins hyaline, cuspidate with a short mucro; keel prominent, green, 3-nerved. Stamen 1. Styles 2-cleft, ± 0.5 mm long, hairy above, style base papillose, without hairs. Nuts biconvex, flattened, broadly elliptical, ± 0.5 mm long, yellowish, smooth and shining, semi-transparent, bordered faintly reticulate.

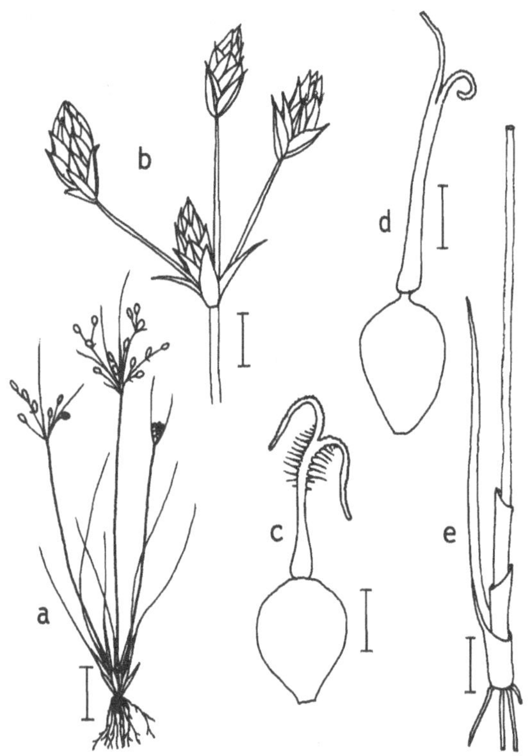

Fig. 130. *Fimbristylis aestivalis*: a, whole plant (2 cm); b, inflorescence branch bearing 4 spikelets (2 mm); c, fruit (0.2 mm). *F. griffithii*: d, fruit (0.2 mm); e, base of culm (5 mm).

Annual. Not confined to wetlands, it grows in open damp places but is also found in temporary water, swamps and as a weed of rice fields, often gregarious. Like *F. squarrosa* but without silky hairs at the style base; glumes cuspidate but not awned. It is rather variable and some infraspecific taxa have been described.

{⊙ Hel}. Warm and tropical Asia and Australia: Ban, Bhu, Bur, Cey, Ind: An, Ar, As, Gj, Go, Kl, Kt, Mg, Mh, Mp, Mz, Rj, Sk, Up, Wb. 0–1800 m.

Fimbristylis alboviridis C. B. Clarke: Fig. 131

Like *F. cymosa* but annual; the ligule a fringe of white hairs; lowest involucral bract usually overtopping the inflorescence; primary rays flattened; spikelets 5–7 (–10) mm long, ± 2.5 mm wide; glumes ovate to suborbiculate, with basal part and margins white; stamen 1; style flat, ± 1 mm long, 2-cleft, hairy above; nut biconvex, flattened, straw-coloured; somewhat warty toward the margin, with 6–10 vertical rows on each face.

Annual. At the edges of pools but also in moist and shady places, in dry grassland and along roadsides.

{⊙ Ten/Hel}. Old World tropics: Ind: An, As, Kt, Mg, Mp, Up. 100–1500 m.

Fimbristylis aphylla Steudel [*F. quinquangularis* (Vahl) Kunth var. *crassa* C. B. Clarke]: Fig. 132

Rhizomes short. Culms densely tufted, 4- or 5-angled or winged or grooved, 30–75 cm long, 1–3 mm in diameter, glabrous. Leaves on culms reduced to 3 or 4 bladeless sheaths; the sheaths loose, tubular, herbaceous, up to 20 cm long, brown, obliquely truncate; leaves sterile shoots with well developed blades, up to 30 cm long, 2–4 mm wide, with prominent midrib and scabrid margins; ligule absent. Involucral bracts 2–3, lanceolate, up to 2 cm long, not overtopping the inflorescence. Inflorescence compound, corymbose. Spikelets solitary, ovoid, 2.5–3.5 mm long, 1.5–2 mm wide, terete, acute, brown. Glumes membranous, ovate, obtuse or mucronate, ± 1.5 mm long, ± 1 mm broad, glabrous. Stamens 1 or 2. Styles 3-angled, 3-cleft, 0.75–1 mm long, glabrous. Nuts obscurely 3-sided, with somewhat incrassate angles, obovoid, 0.6–0.7 mm long, 0.4–0.5 mm broad, densely warty, epidermal cells in 4–6 vertical rows on each face, whitish or straw-coloured, shortly stalked.

Perennial. Open wet places and swamps. Very similar to *F. salbundia* but keels on the glumes obscure thus the spikelet appearing smooth and terete; nuts

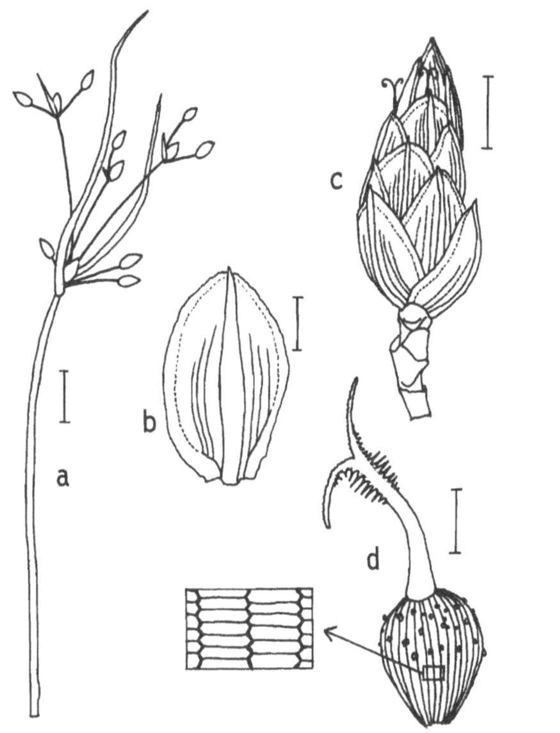

Fig. 131. *Fimbristylis alboviridis*: a, inflorescence (1 cm); b, glume (0.5 mm); c, spikelet (1 mm); d, fruit (0.5 mm) with surface pattern (inset).

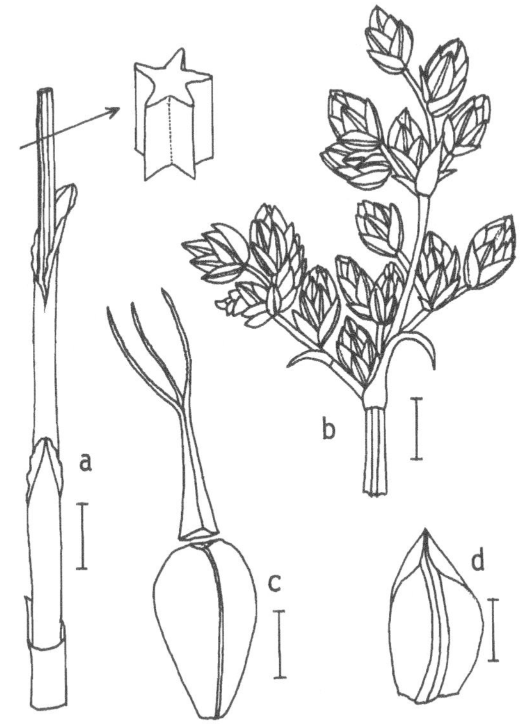

Fig. 132. *Fimbristylis aphylla*: a, base of culm (5 mm) with transverse section (inset); b, inflorescence (3 mm); c, fruit (0.3 mm); d, glume (0.5 mm).

0.6–0.7 mm long, densely warty, with 4–6 vertical rows of cells on each face; culms densely tufted.

{Ten/(Hel)}. Tropical and S Africa, India, Java and Sumatra or perhaps confined to Asia: Ind: Hp, Kt, Tn. 500–2000 m.

Fimbristylis argentea (Rottbøll) Vahl: Fig. 133

Culms densely tufted, very slender, setaceous, 3–10 (–20) cm long, 3-sided, slightly flattened, 0.5–0.7 mm wide, glabrous. Leaf sheaths loose, straw-like, red-dotted; ligule absent; blades flat or channelled, 0.5–0.7 mm wide, glaucous, hairy on the midrib, smooth or slightly scabrid at the top. Involucral bracts 2–4, the lowest much overtopping the inflorescence. Inflorescence a globose to semiglobose head, ± 1 cm in diameter, with (2–) 4–10 (–25) sessile spikelets. Spikelets sessile, oblong-ovoid to elongate-ovoid, 4–10 mm long, 1.2–1.5 mm wide. Glumes thinly membranous, triangular-ovate, obtuse, 1–1.2 mm long, ± 1 mm wide, sides without nerves and silvery-grey to brown, cuspidate with a short mucro; keel green, 3-nerved. Stamen 1. Styles 2-cleft, 0.5–0.7 mm long, smooth or minutely hairy above. Nuts biconvex, flattened, broadly obovoid to suborbicular, with acute edges, ± 0.5 mm long, 0.4–0.5 mm wide, pale yellow, smooth or slightly warty, semi-transparent, obscurely reticulate, shortly stipitate.

Annual. Grows in swampy places, at the drying edges of open water, in dried-out rice fields, often on sandy soil and in damp hollows by the sea shore. It is not exclusively a wetland plant and is found in grasslands and roadsides.

{⊙ Hel/Ten}. Indomalaysia: Ban, Cey, Ind: Gj, Kl, Kt, Tn.

Fimbristylis bisumbellata (Forsskål) Bubani [*F. pallescens* (Roxburgh) Nees, *F. dichotoma* sensu C. B. Clarke, non Vahl]: Fig. 134

Culms densely tufted, 3-sided, (1.5–) 7–10 (–25) cm long, 0.3–1.2 mm thick, smooth, glabrous. Leaf sheaths loose, membranous, 1–3 cm long, hairy, straw-coloured to brown; ligule a dense fringe of short hairs; blades flat or sometimes folded, 0.5–2 mm wide, with several equal nerves, light green, glabrous or pubescent, scabrid at the apex. Involucral bracts 1–3, hairy at the base, the lowest not or slightly overtopping the inflorescence. Inflorescence corymbose, rarely simple or compound, loose or in umbel-like clusters, 10–40 spikelets; primary rays 3–10. Spikelets solitary, oblong-ovoid to narrowly

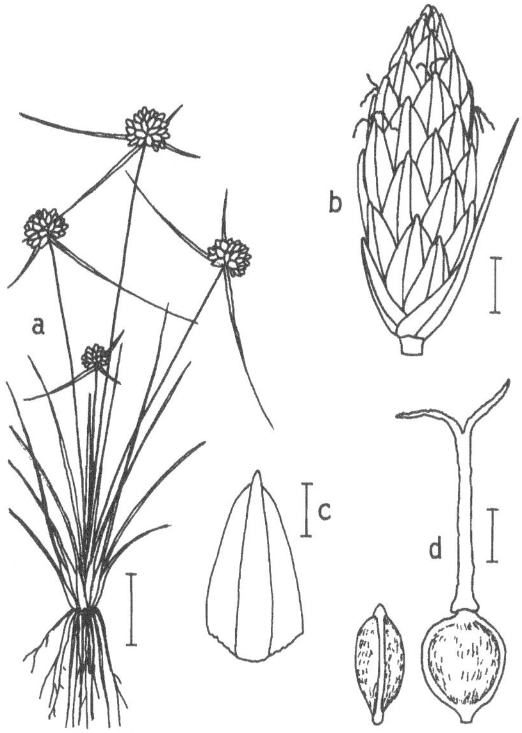

Fig. 133. *Fimbristylis argentea*: a, whole plant (2 cm); b, spikelet (0.5 mm); c, glume (0.25 mm); d, fruit viewed from side and face (0.25 mm).

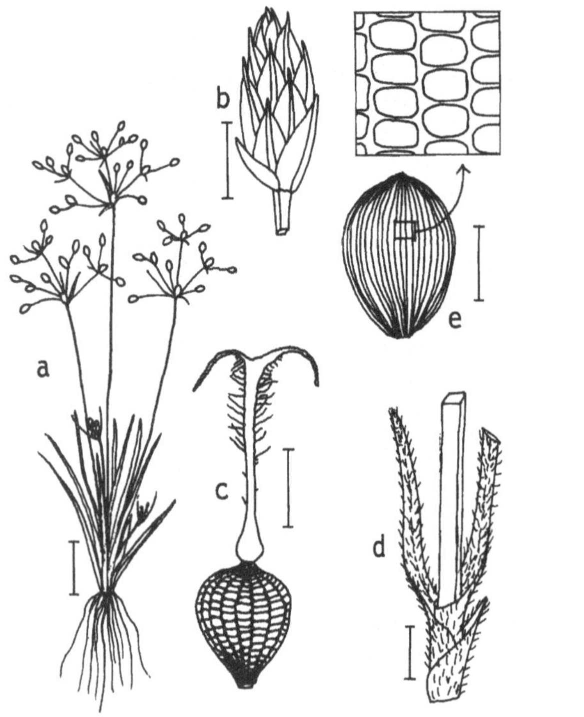

FIG. 134. *Fimbristylis bisumbellata*: a, whole plant (2 cm); b, spikelet (1 mm); c, fruit (0.5 mm). *F. hirsutifolia*: d, base of culm (3 mm); e, nut (0.5 mm) with surface pattern (inset).

oblong, densely 10- to 30-flowered, subterete to angular, acute, yellow-brown, 2–8 (–10) mm long, 1–1.5 (–2) mm wide; rachilla elongating in fruit. Glumes membranous, broadly ovate, acute, 1.2–1.8 mm long, 0.7–1.2 mm broad, glabrous, sides without nerves or obscurely nerved, red-brown, margins hyaline, cuspidate with a short mucro; keel sharp, green, 3-nerved. Stamen 1. Styles flat, 1–1.5 mm long, 2-cleft, hairy above. Nuts biconvex, obovoid to broadly-obovoid, 0.5–0.8 mm long, 0.4–0.7 mm broad, with (5–) 7 (–9) longitudinal stripes on either face, straw-coloured, outer layer transparent, shortly stipitate.

Annual. At the edges of drying pools, in wet flushes, springs, along streams and rivers on sandy river bars. It is also a common weed of rice fields. Small plants may resemble *F. aestivalis* but they have nuts with longitudinal stripes; large plants may resemble *F. dichotoma* but they have spikelets and glumes which are smaller and yellowish-brown.

{⊙ Ten/Hel}. Old World tropics and subtropics: Cey, Ind: Hp, Kl, Kt, Pj, Rj, Sk, Up, Wb. (0–900 m in Africa).

Fimbristylis cinnamometorum (Vahl) Kunth [*F. cyperoides* R. Brown var. *cinnamometorum* (Vahl) C. B. Clarke]: Fig. 135

Rhizomes creeping, woody, 1–1.5 mm thick. Culms solitary, slender, smooth, 25–50 cm long, 0.5–1 mm wide, thickened at base, flexuous, ridged, 3-sided. Leaf sheaths horny at the base of the culm; ligules absent; leaf blades needle-like, about half as long as the culms, 0.3–0.5 mm wide. Involucral bracts 1–2, needle-like, from shorter than to overtopping the inflorescence, the lowest up to 4 cm long. Inflorescence compound, loose, with many spikelets. Spikelets solitary, oblong to oblong-linear, strongly flattened, few-flowered, 4–5 mm long; ± 1 mm wide. Glumes in 2 rows, thinly membranous, ovate-lanceolate, acute, mucronate, 2.7–4 mm long, 1.5–2 mm broad, glabrous, sides with red dots, without nerves; keel sharp, 3-nerved. Stamens 3. Styles 3-cleft, glabrous except for the shortly hairy, pyramidal style base. Nuts 3-sided, with convex sides, obovoid to oblong-obovoid, 0.7–0.9 mm long, 0.4–0.7 mm wide, yellowish to brown, warty, finely transversely lined, shortly stipitate.

Perennial. In swamps and marshy areas, but also on wet flats in savannah forests, locally abundant.

{Ten/(Hel)}. N Australia, S and SE Asia: Cey, Ind: Kl, Kt, Mg, Mh, Tn. 0–1200 m.

Fimbristylis complanata (Retzius) Link [*F. autumnalis* auct. Ind., non Roemer et Schultes, *F. autumnalis* var. *complanata* (Retzius) Kükenthal or Barros]: Fig. 136

Rhizomes short. Culms densely tufted, quadrangular at the base becoming flattened and almost 2-winged in

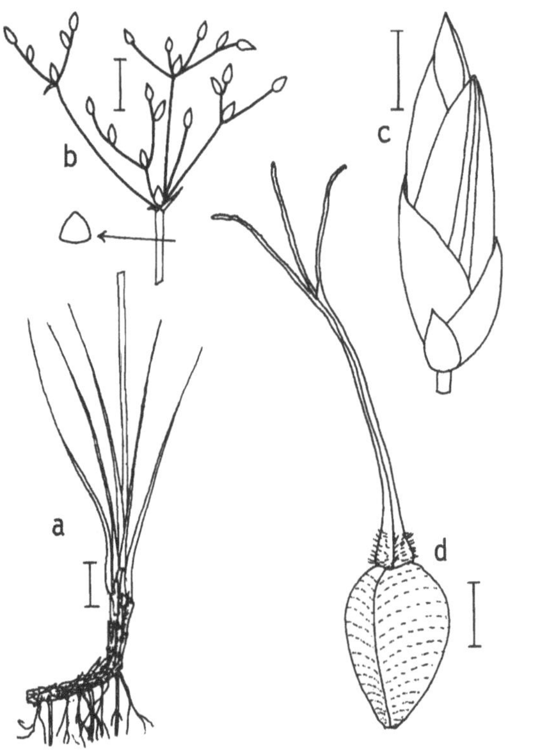

FIG. 135. *Fimbristylis cinnamometorum*: a, base of shoot (1 cm); b, inflorescence (1 cm) with transverse section of culm (inset); c, spikelet (1 mm); d, fruit (0.5 mm).

FIG. 136. *Fimbristylis complanata*: a, base of shoot (1 cm); b, ligule (4 mm); c, inflorescence (1 cm) with a transverse section of culm (inset); d, spikelet (1 mm); e, fruit (0.5 mm). *F. consanguinea*: f, inflorescence (5 mm); g, spikelet (1 mm).

the upper part, 20–90 (–120) cm long, 2–4 mm wide. Leaf sheaths flattened, keeled on the back, the lower ones bladeless; ligule a dense fringe of short hairs; blades flat below, becoming constricted above halfway up, 1–35 cm long, 3–5 mm wide, glabrous, abruptly acuminate. Involucral bracts 2–4, mostly not overtopping the inflorescence, the lowest 1–5 cm long. Inflorescence complexly compound, loose to dense; primary rays flattened, up to 7 cm long. Spikelets solitary and usually some clustered, oblong-ovoid to narrowly oblong, angular, acute, red-brown, 5–8 mm long at anthesis elongating to 12 mm in fruit, ± 2 mm wide; rachilla elongating in fruit. Glumes membranous, ovate to elliptical, acute, apiculate, 2–3 mm long, ± 1.5 mm broad, shortly hairy, the sides whitish and without nerves, margins hyaline; keel straw-coloured, slender, 3-nerved. Stamens 3. Styles 3-angled, 3-cleft, 1–1.2 mm long. Nuts 3-sided, broadly obovoid, 0.7–1 mm long, 0.5–0.7 mm broad, warty or rarely smooth, whitish or straw-coloured becoming brown later, shortly stipitate.

Perennial. Grows on muddy river banks, swampy grassland, sea shore, wet rice fields but it is not confined to wetlands and is common on open grasslands and on cultivated or disturbed ground.

{Ten/Hel}. Pantropics: Bhu, Bur, Cey, Ind, Pak: An, Ap, As, Hp, Kl, Kt, Mg, Mh, Ng, Pj, Sk, Tn, Wb. 0–1300 m.

Fimbristylis consanguinea Kunth [*F. connectens* Thwaites, *F. complanata* var. *kraussiana* C. B. Clarke]: Fig. 136

Like *F. complanata* but culms narrower, 1–1.5 mm wide at the top; leaf sheaths rounded on the back or slightly keeled; leaf blades narrow, 1–2 mm wide, the cauline ones sometimes much reduced; inflorescence simple or nearly so; glumes larger, (2.2–) 2.5–3 mm long, ± 2 mm wide; anthers ± 1.5 mm long; style 1.6–2.3 mm long; nut smooth, larger, 1–1.2 mm long, ± 0.7 mm wide.

Perennial. In wet grasslands, swamps, swinging bogs, swampy lake margins; often dominant and forming large clumps.

{Ten/Hel}. S Africa, Madagascar and tropical Asia: Cey, Ind. 1200–2300 m.

Fimbristylis crystallina E. Govindarajalu: Fig. 137

Culms densely tufted, erect, rigid, 10–20 cm long, 0.3–0.8 mm in diameter, 3–5 angled, smooth. Leaf sheaths at base of culm bladeless, rusty brown, membranous; blades flat, nearly as long as the culms, 1–1.5 mm

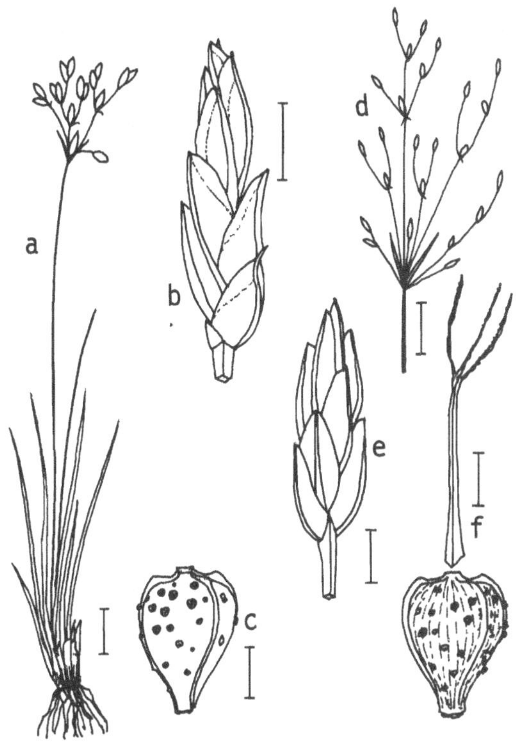

FIG. 137. *Fimbristylis crystalina*: a, whole plant (1 cm); b, spikelet (2 mm); c, nut (0.25 mm). *F. dauciformis*: d, inflorescence (1 cm); e, spikelet (1 mm); f, nut (0.25 mm).

wide, stiff, 7- to 13-nerved, acute, glabrous, margin thickened and smooth throughout. Involucral bracts 2 (3), not overtopping the inflorescence. Inflorescence compound, contracted, umbel-like, with 10–25 spikelets, stalked. Spikelets solitary or in pairs, narrowly ovoid-ellipsoid, 3–3.5 mm long at anthesis, becoming up to 8 mm long in fruit, 1–1.2 mm wide, 6- to 8-flowered, rusty brown, conspicuously angular because of the prominent keels. Glumes narrowly ovate, 1.8–2 mm long, thin, brown, glabrous, acute, sides without nerves, mucronate; keels not excurrent, very prominent, 3-nerved but appearing 1-nerved. Stamens 3. Styles 3-cleft, 3-sided, glabrous, style base pyramidal; stigmas as long as styles. Nuts 3-sided, slightly asymmetrical, obovoid, 0.6–0.7 mm long, ±0.3 wide, glossy, whitish-yellow, vertically 3-ridged, warty, shortly stipitate.

Annual. Seasonally wet places, marshes, usually on hills. It is rather like *F. dauciformis.*

{⊙ Ten/Hel}. ♦Ind: As, Tn. 1000–1800 m.

Fimbristylis cymosa R. Brown [*F. glomerata* Nees, *F. spathacea* Roth, *F. wightiana* Nees]: Fig. 138

Rhizomes short. Culms densely tufted, flattened to 3-sided or subterete, 10–50 cm long, 1–2 mm wide. Leaves numerous; leaf sheaths rather short, brown; ligule absent; blades flat or channelled, leathery, much shorter than the culms, 1–2 (–3) mm wide, scabrous along the margins, very abruptly acuminate. Involucral bracts 2–3, short, 0.5–1 cm long, mostly not overtopping the inflorescence. Inflorescence compound, loose to very dense, sometimes almost head-like; primary rays 3–8. Spikelets solitary or in clusters, ovoid, oblong-ovoid, ellipsoid or cylindrical, somewhat angular, acute, brown, 3–6 mm long, ± 2 mm wide. Glumes membranous, broadly ovate to oblong-ovate, obtuse, often notched, 1.5–2.2 mm long, 1.2–1.7 mm broad, glabrous, sides without nerves, brown with pale scarious margins; keel not prominent, 3- or 5-nerved. Stamens 1–3. Styles flat or 3-angled, 2- or 3-cleft, glabrous. Nuts biconvex or 3-sided, obovoid, 0.6–0.8 mm long, 0.5–0.7 mm broad, smooth or slightly warty, dark brown to black, shortly stipitate.

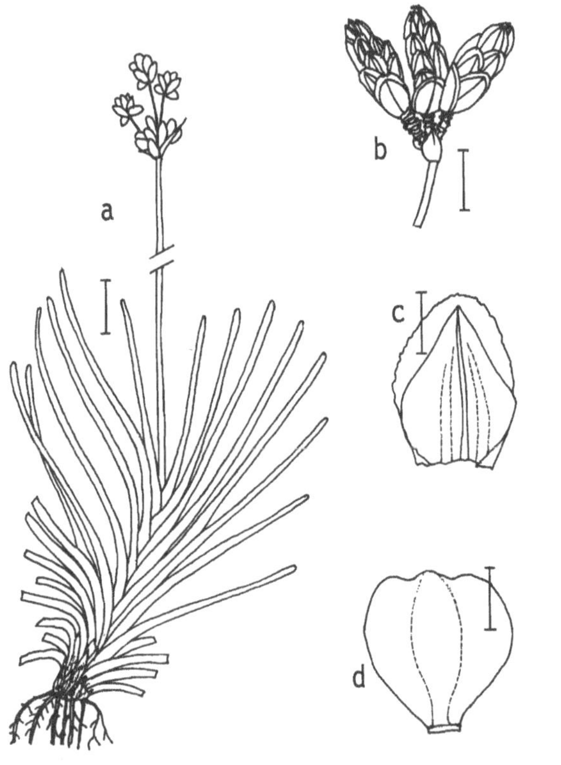

Fig. 138. *Fimbristylis cymosa*: a, whole plant with the culm shortened (1 cm); b, three spikelets (2 mm); c, glume (0.5 mm); d, nut (0.25 mm).

Perennial. Frequently found by the sea in open places, sandy beaches, wet dune hollows, river banks and muddy places in mangroves. Inland it grows along the margins and embankments of pools, ponds and watercourses. It has also been recorded near hot springs. It is characterized by having a dense rosette of rather short, thick, ± equally long, blunt-tipped leaves.

{Ten/(Hel)}. Pantropics: Cey, Ind, Pak: An, Ap, Gj, Go, Kl, Kt, Lk, Or, Tn, Wb.

Fimbristylis dauciformis E. Govindarajalu: Fig. 137

Rhizomes absent. Culms densely tufted, 5-sided, ribbed, glabrous, smooth, 15–40 cm long, 0.5–0.6 mm wide. Leaf sheaths many-nerved, obliquely truncate; blades flat, 8–16 cm long, 1.7–2 (–2.7) mm wide, glabrous, acute, scabrid at tip. Involucral bracts not overtopping the inflorescence, the lowest 1–2.5 cm long. Inflorescence complexly compound, lax, with 15–30 spikelets; primary rays flattened, up to 7 cm long. Spikelets solitary, never in pairs, narrowly ovate to oblong-cylindrical, subangular, acute, red-brown, 3.5–5 mm long at anthesis, elongating to 12 mm in fruit, 1.3–1.4 mm wide; rachilla winged, elongating in fruit. Glumes membranous, broadly ovate, rounded, minutely mucronulate, 1.6–1.7 mm long, 1.2–1.3 mm broad, sides shining and without nerves, margins hyaline; keel distinct, 3-nerved. Stamens (2) 3. Styles 3-angled, 3-cleft, up to 1.1 mm long; stigmas 3, shorter than the style. Nuts 3-angled, obconical, ± 0.8 mm long, 0.6 mm broad, smooth or warty, whitish or pale straw-coloured, shortly stipitate.

Annual. Found exclusively in marshes. Very like *F. crystallina.*

{⊙ (Ten)/Hel}. ♦Ind: Kl (Anakayam, Sholayar, Poringalkuthu).

Fimbristylis dichotoma (Linnaeus) Vahl [*F. diphylla* (Retzius) Vahl, *F. glauca* Vahl, *F. annua* var. *diphylla* (Retzius) Kükenthal, *F. sincta* Nees, *Scirpus brevifolius* Rottbøll]: Fig. 139

Rhizomes very short and woody or absent. Culms tufted, 3-angular, flattened below the inflorescence, (5–) 10–75 (–100) cm long, 1–2.5 mm broad, smooth or with few hairs. Leaf sheaths loose, membranous, brown-spotted, persisting but not splitting; ligule a dense fringe of short hairs; blades flat or sometimes folded, (3–) 10–30 (–50) cm long, (1–) 2–5 mm wide, glabrous or sparsely pubescent, scabrid at the top, abruptly acuminate at apex. Involucral bracts 2–5, mostly not overtopping the inflorescence, the longest up to 10 cm long. Inflorescence, corymbose, simple or compound, loose or dense; primary rays 3–5 (–10), secondary rays 3–5. Spikelets solitary or in clusters of 2–10, ovoid to oblong-ovoid, terete, acute, drab, brown to red-brown, (3–) 5–10 (–20) mm long, (2–) 2.5–3 (–5) mm wide. Glumes chartaceous, very broadly ovate to oblong-ovate, obtuse, (1.5–) 2–3 (–4.5) mm long, glabrous, sides without nerves or obscurely nerved, brown margins hyaline, often mucronulate; keel flattened, obscure, green, 3-nerved. Stamens 1–3. Styles flat, 2-cleft, 0.8–1.5 mm long, hairy above, base swollen. Nuts biconvex, obovoid to broadly obovoid, 0.7–1.2 mm long, 0.6–1 mm broad, with 5–13 longitudinal rows of horizontal cells (appearing as pits) on either face, glistening white to straw-coloured or rarely brown, stipe obdeltoid, 0.1–0.2 mm long and wide.

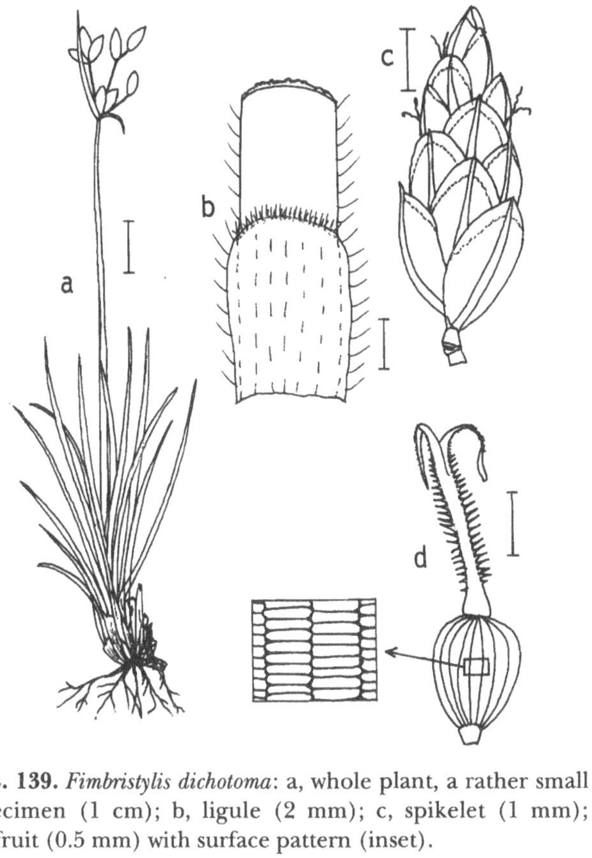

FIG. 139. *Fimbristylis dichotoma*: a, whole plant, a rather small specimen (1 cm); b, ligule (2 mm); c, spikelet (1 mm); d, fruit (0.5 mm) with surface pattern (inset).

Perennial or rarely annual. A very widespread and common species. It is by no means confined to wetlands and is a weed of cultivation but it is also grows in seasonally inundated areas, marshy localities, in and by streams, ditches and in rice fields. The roots are aromatic. It is used as forage and for making mats. It is a highly polymorphic species. Various botanists have recognized some of the cited synonyms as distinct species while others have described them as subspecies or varieties. As there are so many intermediate plants I feel it is better to lump them together. Within this species complex I have, however, kept *F. bisumbellata*, *F. hirsutifolia*, *F. pubisquama* and *F. tomentosa* as separate species because they seem to be more distinct than the others.

{(⊙) Ten/(Hel)}. Tropical and warm temperate regions of the world: Ban. Bhu, Bur, Cey, Ind, Nep, Pak: An, Ap, Ar, As, Dl, Gj, Go, Hp, Jk, Kl, Kt, Lk, Mg, Mh, Mn, Mp, Ng, Pj, Sk, Tn, Tp, Up, Wb. 0–1300 m.

Fimbristylis dipsacea (Rottbøll) C. B. Clarke: Fig. 140

Culms tufted, very slender, setaceous, ribbed, (–2) 5–15 (–25) cm long, 0.2–0.6 mm in diameter. Leaf sheaths obliquely truncate, membranous, pinkish-grey, the lower ones bladeless or with reduced blades; ligule absent; blades filiform, 2–5 (–10) cm long, 0.2–0.3 mm wide, somewhat scabrid. Involucral bracts 3–5, frequently overtopping the inflorescence, the longest up to 2 cm long. Inflorescence usually simple with one sessile and 2–10 stalked spikelets or occasionally compound, at times compact and almost head-like. Spikelets solitary, subglobose to oblong, squarrose, very densely many-flowered, pale green, 3–6 mm long, 1–3 mm in diameter. Glumes membranous, lanceolate, up to 2 mm long, glabrous, sides hyaline and obscurely 1- or 2-nerved or without nerves; keel strong, green, 1-nerved, excurrent into an awn; awn recurved, up to 1 mm long, scabrid. Stamen 1. Styles 2-cleft, fimbriate, style base pubescent. Nuts subterete, subcylindrical to elongate obovoid, sometimes curved but the sides almost parallel, 0.5–0.7 mm long, 0.15–0.4 mm in diameter, brown to greyish-black, with whitish, somewhat irregularly formed, club-shaped outgrowths on the sides.

Annual. Found on muddy or sandy lake or pond shores, river banks, wet rice fields, in marshes and in seasonally wet places.

{⊙ Hel/Ten}. Old World tropics: Ban? Bur, Cey, Ind, Pak: As, Kl, Kt, Mg, Mh, Rj, Tn, Up. At low altitudes.

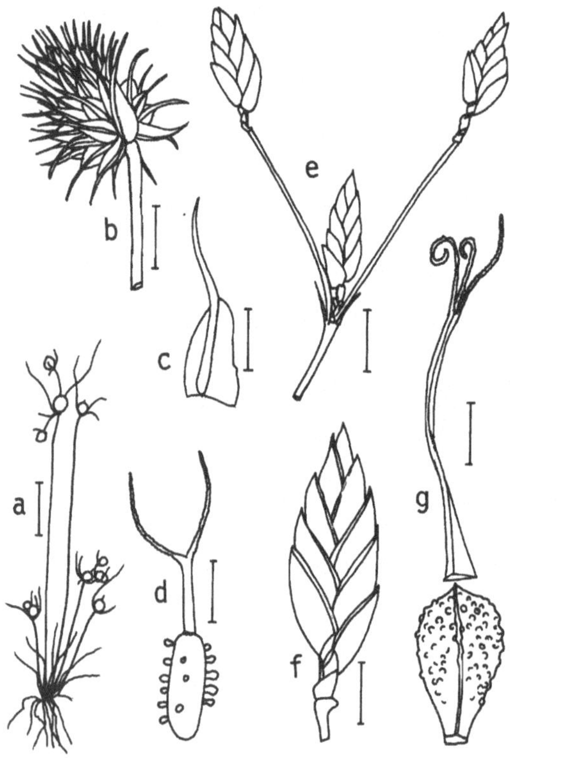

Fig. 140. *Fimbristylis dipsacea*: a, whole plant (1 cm); b, spikelet (1 mm); c, glume (0.5 mm); d, fruit (0.25 mm). *F. disticha*: e, part of inflorescence (2 mm); f, spikelet (1 mm); g, fruit (0.25 mm).

Fimbristylis disticha Boeckeler: Fig. 140

Rhizome absent, roots fibrous. Culms solitary or tufted, rigid, acutely 4- or 5-angled, 10–35 cm long, 0.3–0.5 mm wide, the base clothed with 1 or 2 short-bladed sheaths. Leaf sheaths cylindrical; ligules absent; leaf blades flat, rather rigid, curved, less than half as long as the culms, 1–3 mm wide, abruptly acuminate, margins scabrid towards the tip. Involucral bracts 3 or 4, thread-like, much shorter than the inflorescence, the lowest up to 2.5 cm long. Inflorescence compound, very loose, with several to many spikelets. Spikelets solitary, elongate-ovate, strongly flattened, 8- to 12-flowered, 4–6 mm long; ± 1.5 mm wide. Glumes in 2 rows, thinly membranous, ovate, rather obtuse, minutely apiculate just below the apex, 1.5–2 (–2.5) mm long, ± 1.5 mm broad, glabrous, with narrow hyaline margin, without nerves; keel sharp. Stamens 2 (3). Styles 3-angled, 3-cleft, glabrous, style base shortly pyramidal; stigmas much shorter than the style. Nuts 3-angled, with convex sides, obovoid to narrowly obovoid, 0.6–0.75 mm long, 0.3–0.5 mm wide, whitish or yellowish, warty, shortly stipitate.

Annual. In marshes and seasonally flooded areas and in grassy plains. Like *F. eragrostis* (not a wetland species) but annual with fibrous roots; leaf blades 1–3 mm wide; spikelets ± 1.5 mm wide; glumes exactly in 2 opposite rows, thinly membranous, ovate, 1.5–2 (–2.5) mm long, minutely apiculate just below the apex.

{⊙ Ten?/Hel}. SE Asia: Bur, Ind: An.

Fimbristylis ferruginea (Linnaeus) Vahl: Fig. 141

Rhizomes woody, shortly creeping. Culms tufted, striate, flattened, 20–80 (–100) cm long, 1.5–3 mm wide. Leaves basal and cauline; sheaths at base of culm leathery, shining brown to reddish-brown, the upper ones membranous gland-dotted, hairy; ligule a dense fringe of short hairs; blades of cauline leaves absent or much shorter than the culms, rigid, erect, folded or 3-angled, scabrid at top, up to 10 cm long, 0.5–1.5 mm wide, those of the sterile shoots much longer. Involucral bracts 2 or 3, usually not overtopping the inflorescence. Inflorescence simple or subcompound, usually contracted and sometimes head-like; primary rays flattened, up to 2.5 cm long. Spikelets mostly subsessile or sessile, ovoid to oblong-ovoid, terete, acute, dull brown, 5–20 mm long, 2.5–4 mm wide. Glumes subchartaceous, ovate to oblong, obtuse, apiculate,

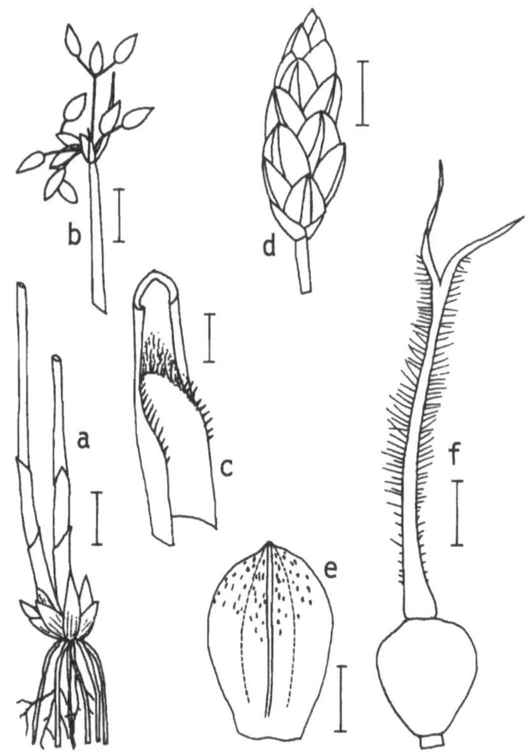

Fig. 141. *Fimbristylis ferruginea*: a, base of plant (1 cm); b, inflorescence (1 cm); c, ligule (2 mm); d, spikelet (2 mm); e, glume (1 mm); f, fruit (0.5 mm).

3–4.5 mm long, 2.5–3 mm wide, sides red-brown without nerves and shortly hairy at the upper edge, the apical part appressed-puberulous with whitish-grey hairs or rarely subglabrous; keel obscure, green, 1-nerved. Stamens 3. Styles flat, 2-cleft, ± 3 mm long; densely hairy. Nuts biconvex, strongly flattened, obovate to oblong-ovate, 1–1.3 mm long, 0.7–1 mm broad, smooth, straw-coloured to brown, shortly stipitate.

Perennial. A gregarious sedge of marshy places especially with sandy or clayish soil. It may grow standing in shallow water and is sometimes found in brackish water near the sea. The pubescent nature of the upper half of the glumes is a good diagnostic character. It is close to *F. sieberiana* Kunth but is distinguished by the reduced leaf blades.

{Hyp/Ten}. Pantropics: Cey, Ind, Pak: An, Ap, Dl, Gj, Go, Kl, Kt, Lk, Mh, Pj, Rj, Tn, Tp, Up, Wb. 0–2000 m.

Fimbristylis griffithii Boeckeler: Fig. 130

Like *F. aestivalis* but glabrous, larger (culms (3–) 10–30 cm long); leaf sheaths tubular, up to 3 cm long; cauline leaves without blades or blades minute; style glabrous.

Annual. In wet or swampy places, river banks, edges of lakes and rice fields.

{⊙ Ten/(Hel)}. Bengal, Myanmar and SE Asia: Ban? Bur, Ind: An, As. 0–700 m.

Fimbristylis hirsutifolia E. Govindarajalu: Fig. 134

Like *F. bisumbellata* but culms 4-sided and 4-ribbed; leaf sheaths and blades clothed in bristly hairs; leaves much shorter than the culms (3–5 cm long, 0.4–0.8 mm wide); involucral bracts much shorter than the inflorescence; styles hairy throughout; glumes slightly longer (± 2 mm long) and with a longer mucro (0.3–0.4 mm long); nuts slightly larger (0.7–0.8 mm long, 0.6–0.7 mm wide), yellowish-brown, with 15–20 longitudinal stripes on each face.

Annual. In seasonally flooded places.

{⊙ Ten/Hel?}. ♦Ind: Kl (Malappuram District), Kt?

Fimbristylis littoralis Gaudichaud [*F. miliacea* auct., non (Linnaeus) Vahl, see Blake, S. T. Journal of the Arnold Arboretum 35: 217 (1954)]: Fig. 142

Culms arising from the axils of bladed leaves, densely tufted, erect, 10–60 (–100) cm tall, 1–3 mm thick below, flattened, 4-sided with 2 sharp edges, smooth, clothed at base with 2 or 3 bladeless sheaths. Basal leaves (or leaves of sterile shoots) with sheaths 2–10 cm long, the ventral side splitting as the culm grows; ligule absent; blades equitant, bilaterally flattened, 10–50 cm long, 1–4 mm wide, soft, as long as or exceeding the culm, narrowed to an acute tip. Culm leaves reduced to bladeless sheaths; sheaths laterally flattened with a sharp dorsal edge, brownish, the orifice obliquely truncate, hyaline on the margin, gradually tapering above to bristle-like elongation up to 10 mm long, the uppermost sheath (3–) 6–12 (–18) cm long, the lowest scale-like. Involucral bracts 2–4, rigid, bristle-like, not overtopping the inflorescence. Inflorescence compound. Spikelets ovoid-globular to globular, (1.5–) 2–3 (–5) mm long, 1.5–3 mm in diameter, rusty brown, densely many-flowered. Glumes broadly ovate, 1–1.3 mm long, 0.6–0.8 mm wide, boat-shaped, obtuse at apex, membranous, reddish-brown, the sides with white hyaline margins and without nerves; keel 1- or 3-nerved, ending below the apex (on fertile glumes). Stamens 1. Styles 3-cleft, 0.5–0.7 mm long, glabrous; style base pyramidal. Nuts 3-sided, obovoid to broadly obovoid, 0.6–0.8 mm long, 0.4–0.5 mm wide, cream-yellow, the surface sparsely warty particularly on the ridges.

Annual, biennial or perhaps sometimes perennial. Found in wet places, at the edges of drying pools, in rice fields, along streams and at the edges of pools. Often very abundant and locally dominant. In many

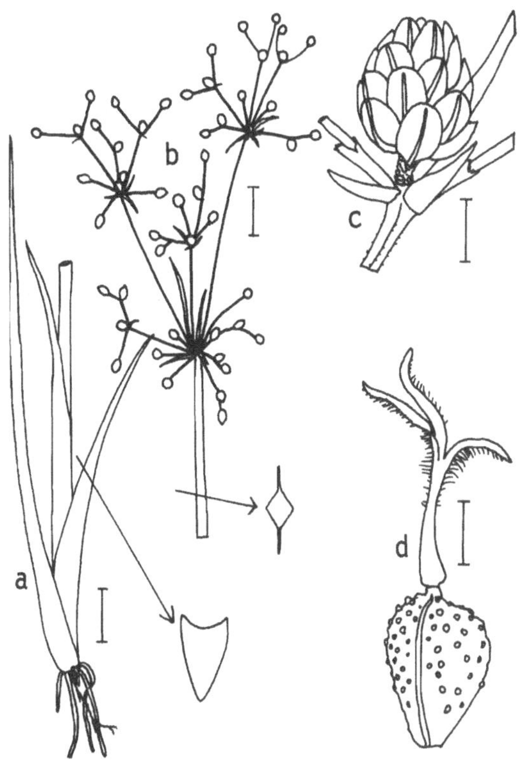

Fig. 142. *Fimbristylis littoralis*: a, base of plant (1 cm) with transverse section of leaf blade (inset); b, rather small inflorescence (1 cm) with transverse section of culm (inset); c, spikelet (1 mm); d, fruit (0.25 mm).

Indian floristic works this species bears the name *F. miliacea*; I am, therefore, not confident that all the cited distribution records are correct. My impression is that it is common in the south while genuine *F. miliacea* is common in the north. It is distinguished from *F. miliacea* by: spikelets globose; culm flattened and 2-winged; leaf sheaths flat; leaf blades flat, ensiform, without prominent midrib.

{⊙, biennial or perennial Ten/Hel}. Tropical and warm temperate regions of the World: Bhu, Bur, Cey, Ind, Nep, Pak: Ap, As, Bh, Gj, Hp, Jk, Kl, Kt, Mg, Mh, Mp, Ng, Or, Pj, Rj, Sk, Tn, Up, Wb. 0–500 m.

Fimbristylis microcarya F. von Mueller [*F. complanata* var. *microcarya* (F. von Mueller) C. B. Clarke misspelt '*microcarpa*' in Hooker 1893, *F. autumnalis* (Linnaeus) Roemer et Schultes var. *microcarya* (F. von Mueller) Kükenthal]: Fig. 145

Like *F. complanata* but annual; culms 10–30 cm long; spikelets very small (the smallest in the genus), 2–4 mm long, ± 1 mm wide; glumes ± 1 mm long excluding the cusp; style 0.5–0.75 mm long; nut minutely warty, 0.5–0.6 mm long, 0.3–0.4 mm broad.

Annual. Open savannah by the sea, along lake margins and in water holes. It is sometimes confused with *F. woodrowii* in Indian Floras. The distribution is poorly known. I have seen material only from Maharashtra.

{⊙ Hel}. Australia, S and SE Asia: Ind: Mh, Sk? Up?

Fimbristylis miliacea (Linnaeus) Vahl [*F. quinquangularis* (Vahl) Kunth]: Fig. 143

Rhizomes very short, roots fibrous. Culms arising from the axils of bladed leaves, densely tufted, erect, (5–) 15–70 cm tall, 1.5–2.7 mm thick below, acutely (4-) 5-angled, smooth, clothed at base with 2–4 sheaths. Basal leaves (or leaves of sterile shoots) with sheaths 1–7 cm long, the ventral side splitting as the culm grows; ligule absent; blades dorsoventrally flattened, 8–30 cm long, 1.5–3 mm wide, soft, as long as or exceeding the culm. Culm leaves reduced to bladeless sheaths; sheaths subterete, pale green to straw-coloured, the ventral side hyaline and pale yellowish-brown, the orifice obliquely truncate, hyaline on the margin, the uppermost sheath 3–18 cm long, the lowest scale-like. Involucral bracts 3–5, rigid, bristle-like, much shorter than the inflorescence. Inflorescence compound. Spikelets ovoid, 2–5 mm long, 1–1.7 mm in diameter, rusty brown, densely 20- to 40-flowered. Glumes ovate, 1–1.8 mm long, 0.8–1 mm wide, boat-shaped, rounded or mucronate, membranous, yellow- or rusty brown, the sides pale brown and without nerves; keel 1- or 3-nerved. Stamens 1 (2). Styles 3-sided, 0.5–0.7 mm long, glabrous, 3-cleft, style base pyramidal. Nuts 3-sided, obovoid, (0.4–) 0.6–0.8 mm long, 0.4–0.5 mm wide, cream-yellow, the surface warty.

Annual or biennial. Found in wet places, marshes, at the edges of drying pools, in rice fields, along railway tracks and roads. Often very abundant and locally dominant. Often mistaken for *F. littoralis* but: spikelets ovoid to lanceolate; culms 5- or 6-angled; leaf sheaths cylindrical; leaf blades present on upper sheaths and orientated dorsoventral; the scales left on the rachilla after the fall of the glumes are brown, lanceolate and very conspicuous.

{⊙ or biennial Ten/Hel}. Old World tropics and N Australia, naturalized in Puerto Rico: Bhu, Cey, Ind: An, Ar, As, Dl, Gj, Go, Hp, Kl, Kt, Mg, Mh, Mp, Rj, Sk, Up, Wb.

Fimbristylis nutans (Retzius) Vahl: refer to Fig. 129

Like *F. acuminata* but culms longer (15–) 25–40 (–70) cm long); spikelet at an angle to the culm or ± nodding, broadly ovoid or ovoid, 3–5 mm wide; styles strap-like, 0.5–0.7 mm wide; nuts with 3–5 (–6) transverse ridges.

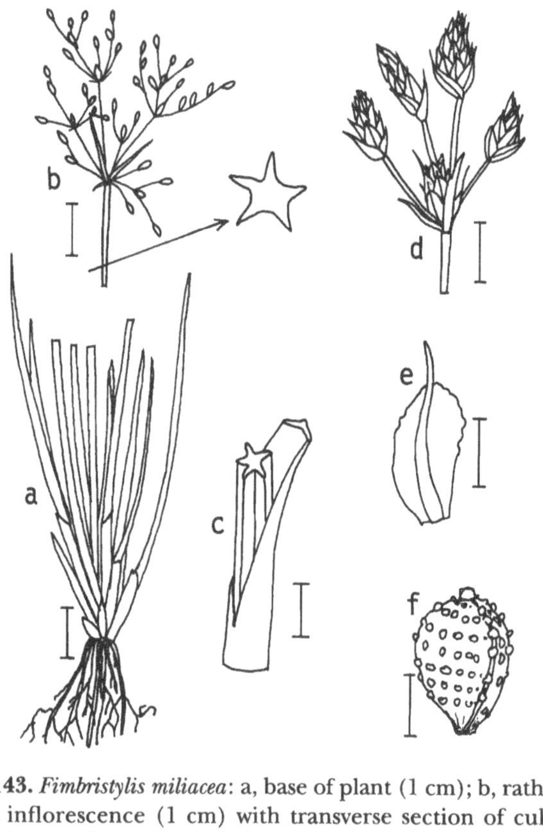

Fig. 143. *Fimbristylis miliacea*: a, base of plant (1 cm); b, rather small inflorescence (1 cm) with transverse section of culm (inset); c, culm emerging from sheath (2 mm); d, five spikelets (2 mm); e, glume (0.5 mm); f, nut (0.25 mm).

Perennial. Grows in damp sandy places, moist hollows in forests, swamps, rice fields and also saline situations.

{Ten/Hel}. Warm and tropical Asia and Australia: Cey, Ind: An, As, Mg, Tn. 0–70 m.

Fimbristylis ovata (N. L. Burman) Kern [*Carex ovata* N. L. Burman, *Abildgaardia ovata* (N. L. Burman) Kral, *F. monostachyos* (Linnaeus) Hasskarl]: Fig. 144

Culms densely tufted, 5–40 (–70) cm long, 0.5–1 mm in diameter, thickened and hardened at base, flexuous, ridged, 3-sided. Leaf sheaths persistent, pale grey to dark brown, often with 2 spinous keels; ligules absent; leaf blades flat but with usually inrolled margins, usually about half as long as the culms, up to 25 cm long but usually less, up to 1 mm wide. Involucral bracts 1–3, needle-like, not overtopping the inflorescence. Inflorescence of 1 or rarely 2, terminal spikelets. Spikelets solitary, ovoid to oblong, somewhat laterally flattened below, 5–15 mm long at anthesis, becoming up to 30 mm long in fruit; 3–6 mm wide. Glumes in 2 opposite rows at the base becoming spiral above, triangular-ovate, acute, mucronate, 4–5 mm long, 3–4.5 mm broad, glabrous, straw-like, sides pale, without nerves; keel green, excurrent, 3-nerved. Stamens 2 or 3. Styles flattened, ± 2 mm long, ciliate above, 3-cleft, style base pyramidal. Nuts obscurely 3-sided, obovoid to pyriform, 2–3 mm long and 1.5–2 mm wide, pale grey or yellowish, ridged, the surface coarsely tuberculate except on ridges; sessile or with a stipe up to 0.5 mm long.

Perennial with short rhizome. A common and widespread species not confined to wetlands but often found in marshes, seasonally inundated areas and at the edges of pools and ditches.

{Ten/(Hel)}. Pantropics: Ban, Bhu, Cey, Ind, Nep: An, Ap, As, Gj, Go, Kl, Kt, Mh, Mp, Pj, Rj, Tn, Up, Wb? 0–1200 m.

Fimbristylis polytrichoides (Retzius) Vahl: Fig. 144

Culms tufted, setaceous, 4–40 (–60) cm long, 0.3–1 in diameter, flexuous, smooth, weakly 3-sided. Leaf sheaths membranous, brown; ligule a row of short hairs; leaf blades flattened, setaceous, wiry, channelled, about half as long as the culms, 0.5–1 mm wide. Involucral bracts 1 or rarely 2, either needle-like, short or slightly overtopping the inflorescence or glume-like. Inflorescence usually of 1 terminal or pseudolateral spikelet with occasionally 1 or 2 stalked lateral spikelets. Spikelets solitary, ovoid-ellipsoid to oblong, subterete, 5–15 mm long; 2–4 mm wide, pale brown. Glumes lanceolate to oblong-ovate, obtuse, 2–3 mm long, 0.5–1.3 mm broad, thin, sides without nerves, grey with a brown tinge near the tip; keel green, 1- or obscurely 3-nerved; the lower two sterile glumes often

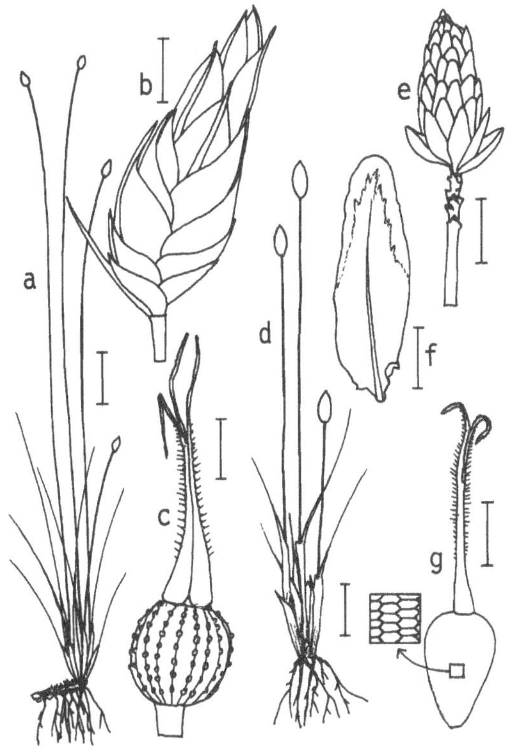

FIG. 144. *Fimbristylis ovata*: a, whole plant (3 cm); b, spikelet (2 mm); c, fruit (1 mm). *F. polytrichoides*: d, whole plant (1 cm); e, spikelet (2 mm); f, glume (0.5 mm); g, fruit (0.5 mm) with surface pattern (inset).

mucronate, all other glumes without mucro. Stamens 1 or 2. Styles 0.5–1 mm long, 2-cleft, hairy, flat stigmas as long as or longer than the styles. Nuts biconvex, acutely angled, oblong-obovoid, 0.7–1.2 mm long, 0.5–0.8 mm wide, grey or blackish-brown, the surface smooth or slightly warty, with ± 30 vertical stripes on either face; sessile or nearly so.

Perennial or perhaps sometimes annual. Gregarious and usually forming dense tufts. Found in marshes, wet hollows, ditches on saline mud or sandy soil near to the coast. Like *F. schoenoides* which has 1 or 2 spikelets per culm, glumes with several nerves and a nut with a prominent stipe.

{(⊙?) Ten/Hel}. Old World tropics: Ban? Cey, Ind: Gj, Jk, Kl, Kt, Mh, Tn, Up, Wb.

Fimbristylis pubisquama Kern [*F. compressa* sensu Boeckeler non Roemer et Schultes]: Fig. 145

Culms densely tufted, obtusely 3-sided, 8–50 cm long, 0.5–1.2 mm wide, smooth, glaucous green. Leaves 2–5, basal; sheaths 1–10 cm long, loose, membranous, with soft hairs on the margins; ligules 0.5–0.7 mm long, hyaline, ciliate; blades flattened 3-sided, half to three-

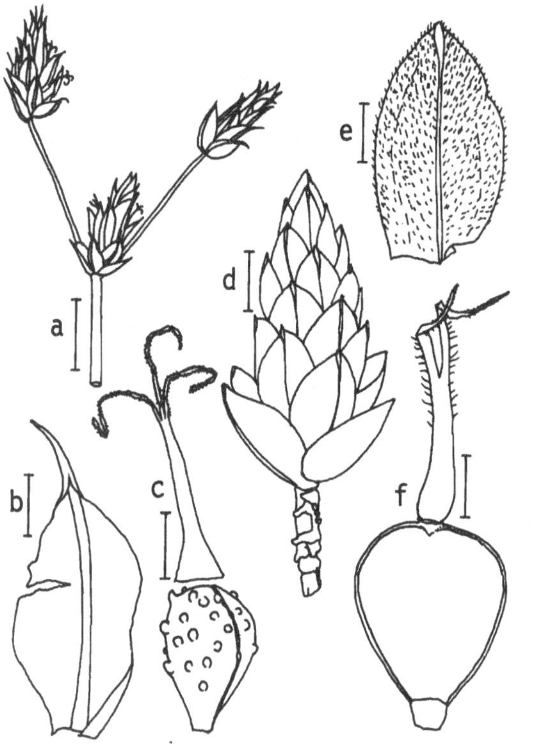

Fig. 145. *Fimbristylis microcarya*: a, three spikelets (2 mm); b, glume (0.25 mm); c, fruit (0.25 mm). *F. pubisquama*: d, spikelet with hairs omitted (1 mm); e, glume (0.5 mm); f, fruit (0. 25 mm).

quarters the length of the culms, 0.7–1.5 mm wide, glabrous, glaucous green, suddenly contracted at the subacute apex. Involucral bracts 1–3, the lowest leaf-like, overtopping the inflorescence. Inflorescence simple or compound, corymbose; primary rays 3–10. Spikelets ovoid to ovoid-ellipsoid, terete, acute, deep rusty brown, (3.7–) 4.5–8 mm long, 1.5–3 mm wide, wholly pubescent; rachilla elongating in fruit. Glumes membranous, ovate to broadly ovate, subacute to sub-obtuse, 1.5–2.7 mm long, 0.8–1.4 mm broad, wholly pubescent with short hairs, deep rusty brown, ciliate on margins hyaline; keel weak, with a sender midvein. Stamens 2. Styles flat, 0.6–1.2 mm long, 2-cleft, sparsely hairy above; stigmas shorter than the style. Nuts biconvex, obovoid, 0.8–1 mm long, 0.5–0.8 mm broad, shiny, with 26–32 rows of hexagonal cells on either face, yellowish-brown, shortly stipitate.

Annual. At the edges of drying pools, in rice fields, in wet flushes, often found in saline marshes and open grassland by the seashore. Probably near to *F. bisumbellata.*

{⊙ Ten/Hel}. Southern India and Indochina: Bur, Cey, Ind: Tn.

Fimbristylis salbundia (Nees) Kunth subsp. ***salbundia*** [*F. paludosa* Merrill, *F. haspaniformis* T. Koyama]: Fig. 146

Rhizomes short, obliquely descending, clothed with ovate-lanceolate scales. Culms close together in a single row on the rhizome, (4-) 5-angled, winged or grooved, 30–75 cm long, 1–3 mm in diameter, glabrous. Leaves on culms reduced to 2–4 bladeless sheaths; the lower sheaths scale-like, ovate to oblong-ovate; the upper sheaths tubular, herbaceous, up to 15 cm long, brown, obliquely truncate, the margin membranous; sterile shoots unknown; ligule absent. Involucral bracts 2–5, lanceolate, up to 2 cm long, not overtopping the inflorescence, the tips elongated into a recurved, setaceous points. Inflorescence compound, corymbose, loose or rather dense. Spikelets solitary or somewhat clustered, broadly ovoid to ellipsoid, 3.5–4.5 mm long, ± 2 mm wide, angular, obtuse, brown. Glumes membranous, ovate, obtuse or mucronate, 2–2.5 mm long, ± 1.5 mm broad, glabrous, the sides nerveless, yellowish with hyaline margins, the

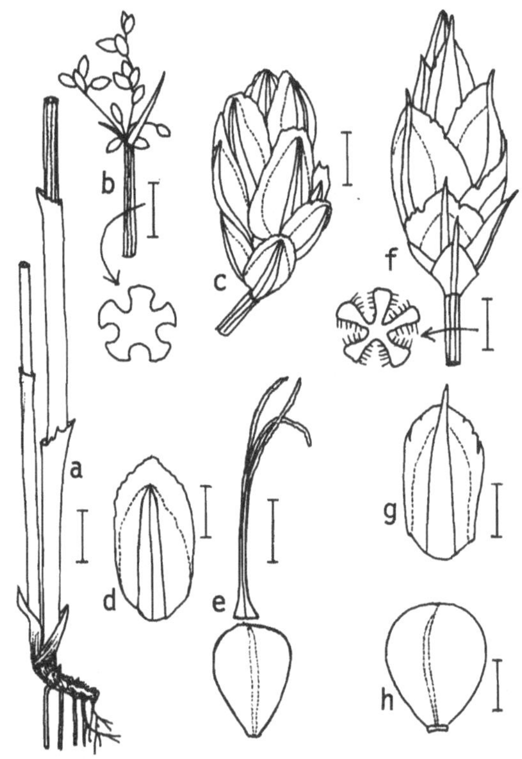

Fig. 146. *Fimbristylis salbundia* subsp. *salbundia*: a, base of shoot (1 cm); b, inflorescence (1 cm) with a transverse section of the culm (inset); c, spikelet (1 mm); d, glume (0.5 mm); e, fruit (0.5 mm). *F. salbundia* subsp. *pentaptera*: f, spikelet (1 mm) with transverse section of the culm (inset); g, glume (0.5 mm); h, nut (0.5 mm).

back reddish-brown, 3-veined; keel ending somewhat below the apex. Stamens (2) 3. Styles 3-angled, 3-cleft, ± 1mm long, glabrous. Nuts 3-sided, broadly obovoid, 0.8–1 mm long, 0.6–0.7 mm broad, smooth or slightly warty, epidermal cells in 9–13 vertical rows, whitish or brownish, shortly stalked.

Perennial. In wet open places, marshes, swamps and grasslands. Common at high altitudes. Rather similar to *F. aphylla* but is characterized by having darker spikelets, angular by reason of their keeled glumes, the slightly longer glumes, the number of stamens and the markings of the nut.

{Ten/(Hel)}. S and SE Asia: Bur, Ind: Mg. 900–2200 m.

Fimbristylis salbundia subsp. ***pentaptera*** (Nees) T. Koyama [*F. pentaptera* Kunth]: Fig. 146

Like subsp. *salbundia* but culms usually hairy; spikelets 4.5–6.2 mm long; glumes 2.3–3 mm long, with keel projecting as a mucro; anthers ± 1.5 mm long; nuts orbicular-obovoid, 1–1.2 mm long.

Perennial. Found in swamps and marshes but not confined to wetlands and also found in grasslands.

{Ten?/Hel]. Endemic to Sri Lanka and S India: ♦Cey, ♦Ind: Tn. 0–2000 m.

Fimbristylis scaberrima Nees: Fig. 147

Like *F. cymosa* but culms strongly flattened, 2-edged; leaf blades flat with sometimes incurved margins, 3–5 mm wide; lowest involucral bracts as long as or overtopping the inflorescence; spikelets 4–7 (–12) mm long, 2–2.5 mm wide; glumes acute, apiculate, 2–2.7 mm long, ±1.5 mm wide, distinctly keeled; style flat, 2-cleft, glabrous or scarcely papillose; nuts biconvex, straw-coloured to greyish-brown.

Perennial. Found in open places in swamp forest and along the shores of lakes, often locally dominant.

{Ten/(Hel)}. SE Asia: Ban/Ind (Sylhet): Mg?

Fimbristylis schoenoides (Retzius) Vahl [*Scirpus bispicatus* Roxburgh]: Fig. 147

Culms densely tufted, 5–50 cm long, 0.5–1 mm in diameter, erect, slender, often swollen at the base. Leaf sheaths membranous, yellowish to pinkish-brown, inflated above; ligule a dense fringe of short hairs; leaf blades flat or inrolled, mostly slightly shorter than the culms, 5-15 cm long, 0.6–1 (–2) mm wide, glaucous, blunt, scabrid towards the apex, often withered as the plant matures. Involucral bracts 1 or 2, glume-like or stiff and leaf-like, not overtopping the inflorescence. Inflorescence of 1 terminal spikelet, with or without 1 or 2, lateral, peduncled spikelets; the peduncles when present 1–2.5 cm long. Spikelets solitary, ovoid

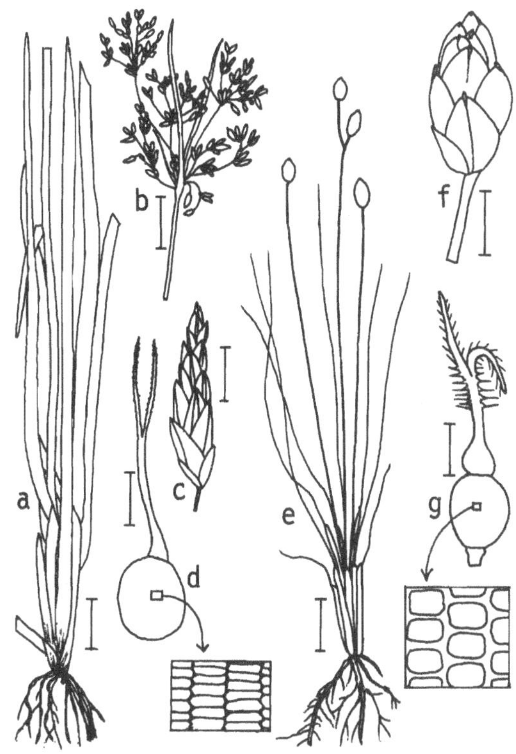

FIG. 147. *Fimbristylis scaberrima*: a, base of plant (2 cm); b, inflorescence (2 cm); c, spikelet (2 mm); d, fruit (1 mm) with surface pattern (inset). *F. schoenoides*: e, whole plant (1 cm); f, spikelet (2 mm); g, fruit (1 mm) with surface pattern (inset).

to subglobose, 4–8 mm long at anthesis, extending to 15 mm long in fruit, 3–5 mm wide. Glumes very broadly ovate to suborbicular, obtuse, 2.5–3 mm long, 3–3.5 mm broad, glabrous, pale brown, with many nerves which become less distinct towards the edges, mucronate; keel indistinct. Stamens 3. Styles flattened, 2-cleft, hairy above, flat, style base pyramidal. Nuts biconvex, obovoid, flattened, 1.3–1.8 mm long, 1–1.2 wide, whitish becoming brown at maturity, smooth and glossy, the surface finely striate, reticulate, stipe 0.3–0.5 mm long.

Annual or short-lived perennial. In wet grassy places, marshes, seasonally inundated areas, on the banks of ponds, streams and in roadside drains. Sometimes found in rice fields. It is sometimes gregarious and forms a closed turf. Like *F. polytrichoides* which has glumes with a single nerve only and a shorter stipe to the nut.

{(⊙) Ten/(Hel)}. Tropical Asia and Australia, introduced in N America: Cey, Ind, Pak: An, As, Bh, Gj, Go, Hp, Kl, Kt, Mh, Mn, Mg, Mp, Pj, Rj, Up, Wb. 0–3000 m.

Fimbristylis sieberiana Kunth [*F. ferruginea* subsp. *sieberiana* (Kunth) K. Lye]: Fig. 148

Like *F. ferruginea* but cauline leaves well-developed, less stiff, flat, 10–30 (–35) cm long, 1.5–2 mm wide, with spine-like teeth on margin and major ribs; lower leaf sheaths membranous and straw-coloured to brown, with blades; the cauline leaf sheaths pilose especially towards the top; inflorescence spreading and more open; lowest involucral bract as long as or usually overtopping the inflorescence, up to 10 cm long; spikelets somewhat obtuse; glumes very broadly ovate, 3–4.5 mm long and wide, usually dark reddish-brown, densely set with short whitish hairs in the upper half; nuts broadly obovate to orbicular, 1–2 mm long, 1–1.5 mm wide.

Perennial or annual. Grows on the drying edges of pools and tanks, along streams and rivers, in seasonally wet areas and in rice fields. It is tolerant of saline conditions. In Africa it merges into *F. ferruginea* but seems to be distinct in India.

{(⊙) Ten/(Hel)}. Warm and tropical Old World: Ind: Gj, Kt, Mh, Rj, Tn, Up. (0–2900 m in Africa).

Fimbristylis squarrosa Vahl [*F. comata* Nees, *F. aestivalis* (Retzius) Vahl var. *squarrosa* (Vahl) T. Koyama]: Fig. 149

Culms densely tufted, 2–30 cm long, up to 1 mm in diameter, suberect, hairy. Leaf sheaths membranous,

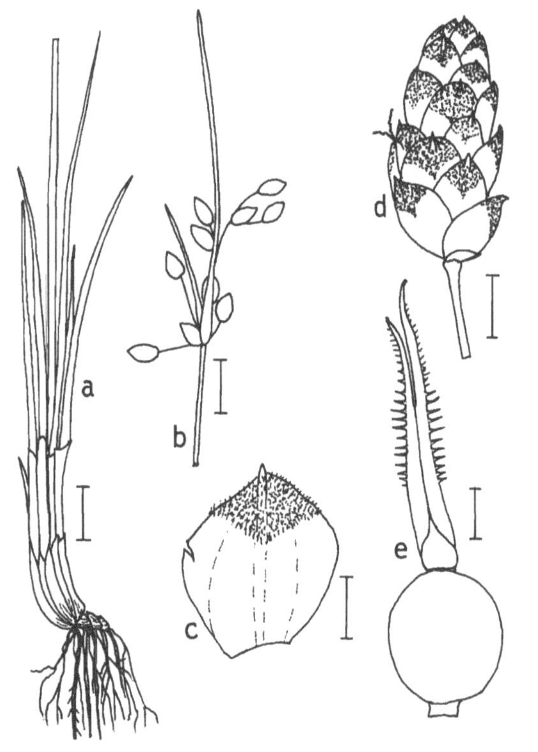

FIG. 148. *Fimbristylis sieberiana*: a, base of plant (1 cm); b, inflorescence (1 cm); c, glume (1 mm); d, spikelet (2 mm); e, fruit (0.5 mm).

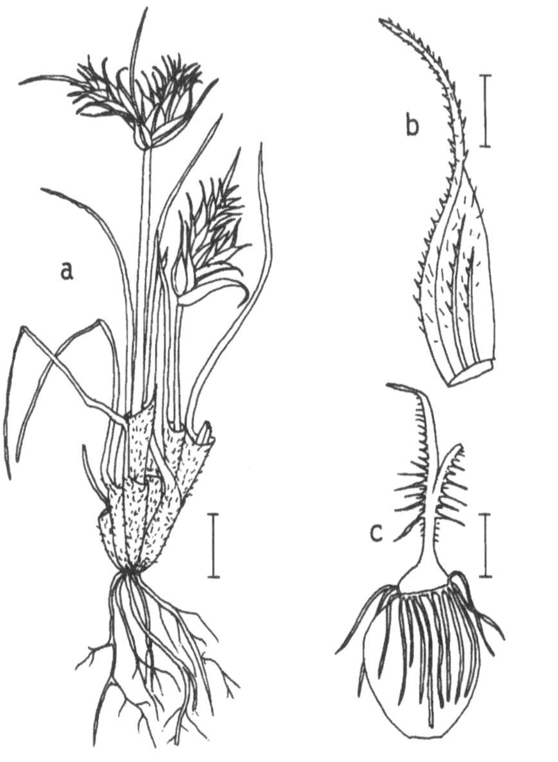

FIG. 149. *Fimbristylis squarrosa*: a, whole plant (2 mm); b, glume (0.5 cm); c, fruit (0.25 mm).

brownish, hairy; ligule absent; blades filiform, flat, up to 15 cm long, up to 1 mm wide, pubescent. Involucral bracts 3–5, overtopping the inflorescence, the longest up to 7 cm long. Inflorescence compound. Spikelets ovoid to subglobose or subcylindrical, angular, up to 7 mm long, ± 3 mm wide, greyish-green, appearing spiny. Glumes hairy, thinly membranous, ovate, 1.5–2 mm long, pubescent, purplish-brown, sides without nerves, margins hyaline, cuspidate with an up to 1.5 mm long and often recurved awn; keel excurrent, 3-nerved. Stamen 1. Styles 0.7–1 mm long, hairy, 2-cleft, style base with long, silky white, pendent hairs appressed to the nut and covering half to three-quarters of it. Nuts biconvex, obovoid, 0.5–0.7 mm long, 0.4–0.6 mm wide, shining, transparent yellowish, 7-ridged, faintly reticulate.

Annual. Wet sandy or muddy soil in seasonally inundated areas, on the shores of drying out pools and lakes, also along railways and roads and in rice fields. It often forms a closed turf. Very variable in size, hairiness and other characteristics. Often confused with *F. aestivalis* but it has longer, stronger curved mucros and long pendent hairs on the style base.

{⊙ Ten/(Hel)}. Warm and tropical regions of the world: Bur, Ind, Nep: As, Hp, Jk, Rj, Up, Wb.

Fimbristylis tenuicula Boeckeler: refer to Fig. 144

Like *F. polytrichoides* but culms always with a single terminal spikelet without involucral bracts; ligules absent; glumes shortly but distinctly mucronate; stigmas much shorter than the style; nuts with ± 15 longitudinal stripes on either face.

Perennial or perhaps also annual. Along lake shores and in swamps.

{(⊙?) Ten/Hel}. India, Tenasserim, Indochina and Malaysia: Ban, Bur, Ind: As, Up.

Fimbristylis tetragona R. Brown [*F. arnottii* Thwaites, *F. cylindrocarpa* Boeckeler]: Fig. 150

Culms tufted, 10–40 (–60) cm long, 0.5–1 mm thick, erect, ribbed, 4-angled to 4-winged. Leaves reduced to membranous, yellowish to brown or grey sheaths. Involucral bracts absent or glume-like. Inflorescence of 1, terminal spikelet. Spikelets solitary, obconical to hemispherical or subglobose, 5–12 mm long, up to 7 mm in diameter. Glumes oblong-ovate to oblong, cuneate, very obtuse, with rounded apex, 3–5 mm long, 2–3 mm broad, glabrous, thick, yellowish to pale brown, sides with many obscure nerves, obtuse; keel

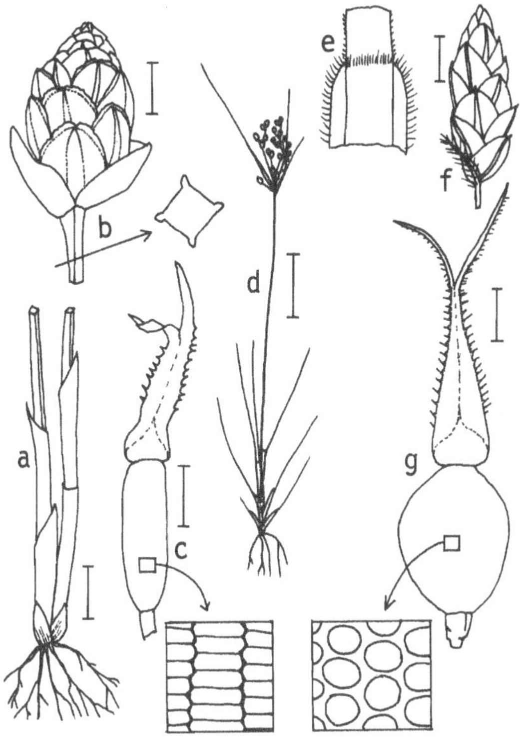

FIG. 150. *Fimbristylis tetragona*: a, base of plant (5 mm); b, spikelet (2 mm); c, fruit (0.5 mm) with surface pattern (inset). *F. tomentosa*: d, whole plant (5 cm); e, ligule; f, spikelet (1 mm); g, fruit (0.5 mm) with surface pattern (inset).

3-nerved. Stamens (1) 2 (3). Styles 2- or 3-cleft, hairy, style base pyramidal. Nuts subterete, with parallel sides, spindle-shaped to almost cylindrical, up to 3 mm long and 0.7 mm in diameter, whitish to yellow or grey, surface reticulate, with a stipe up to 1 mm long.

Annual or occasionally perennial. It is often found in shallow water but also grows in marshes, around pools, along streams, roadside drains, ditches and sometimes in rice fields. It usually germinates underwater and flowers as the water recedes.

{(⊙) Ten/(Hel)}. Tropical Asia and Australia: Bur, Cey, Ind, Nep: Bh? Gj, Go, Kl, Kt, Mh, Mg, Mp, Or, Rj, Tn, Up, Wb. 0–1000 m.

Fimbristylis tomentosa Vahl [*F. affinis* Presl, *F. podocarpa* Nees et Meyer, *F. diphylla* var. *pluristriata* C. B. Clarke, *F. annua* forma *tomentosa* (Vahl) Ohwi, *F. dichotoma* subsp. *podocarpa* (Nees et Meyen) T. Koyama]: Fig. 150

Like *F. dichotoma* but always annual; leaves consistently hairy; spikelets rust-coloured; style tongue-like, strongly flattened, gradually tapering from base to apex, swollen style base obscure; nuts thickly biconvex, orbicular-obovoid, 1.2–1.5 mm long, 1–1.2 mm broad, with 15–24 longitudinal rows of horizontal cells (appearing as pits) on either face, cream-coloured sometimes eventually turning brownish, stipe obdeltoid, ± 0.3 mm long, 0.3–0.4 mm wide.

Annual. In wet and marshy places, usually in disturbed habitats such as pond or river banks, roadside ditches, canals and irrigated arable land. It is particularly common in rice fields.

{⊙ Ten/(Hel)}. S and SE Asia, Pacific, naturalized in N America: Ban, Bhu, Ind, Pak: Ap, Ar, As, Bh, Jk, Kl, Kt, Mg, Tn, Wb.

Fimbristylis triflora (Linnaeus) Schumacher ex Engler [*Schoenus cyperoides* Retzius, *Abildgaardia triflora* (Linnaeus) Abeywickrama]: Fig. 151

Rhizomes knotty, woody, oblique or horizontal, 5–10 mm thick, clothed with remnants of leaf sheaths. Culms usually tufted, 30–80 cm long, 1–3 mm in diameter, thickened at base, stiff, subterete, smooth, glaucous. Leaf sheaths cylindrical, 2–8 cm long, brown, ventrally hyaline; ligules absent; leaf blades with usually inrolled margins, half to one-third as long as the culms, 10–30 cm long, 0.5–2.5 mm wide, leathery, glaucous green, abruptly narrowed to subacute apex. Involucral bracts 1–3, much shorter than the inflorescence. Inflorescence of (2) 3 or 4 (–6) spikelets. Spikelets solitary, elongate-ovoid, somewhat laterally flattened, 10–25 mm long at anthesis, elongating to 40 mm in fruit; 4–10 mm wide. Glumes in 2 sometimes rather irregular rows, broadly-ovate to ovate, abruptly narrowed above to a cuspidate apex, 6–8 mm long, 4–5 mm wide, glabrous, chartaceous, pale greyish-

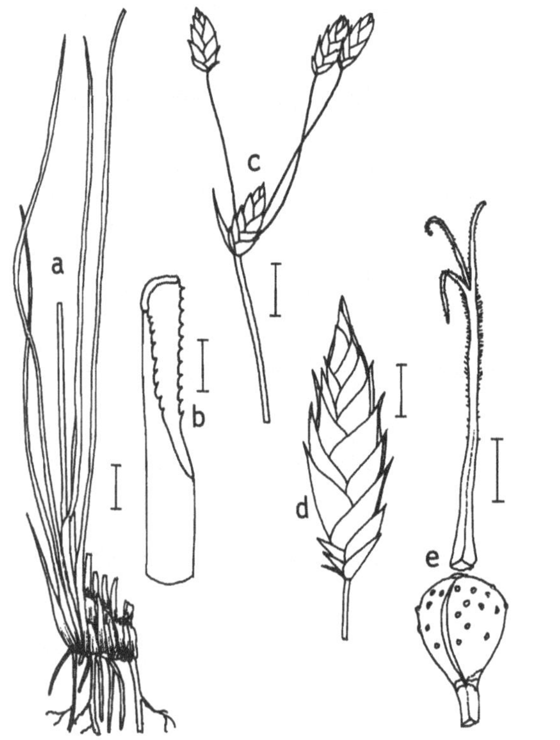

FIG. 151. *Fimbristylis triflora*: a, base of plant (1 cm); b, top of sheath (2 mm); c, inflorescence (1 cm); d, spikelet (5 mm); e, fruit (1 mm).

brown, sides paler but minutely dotted with brown or reddish spots, without nerves; keel reddish-brown, excurrent, closely 4- to 6-nerved. Stamens 3. Styles 3-cleft, fimbriate above the middle, 4.5–6 mm long; style base pyramidal. Nuts 3-sided with convex sides, obovoid, 2.4–3.2 mm long, 1.3–1.8 mm wide, brownish-grey, somewhat shiny, slightly warty; stipe up to 1 mm long.

Perennial. Grows in wet grassland, saline swamps and brackish waters, usually on sand soil. A distinct species with exceptionally large spikelets (up to 2.5 cm long at anthesis) with irregularly distichous greyish-brown glumes.

{Ten/Hel}. Old World tropics: Cey, Ind: Tn.

Fimbristylis umbellaris (Lamarck) Vahl [*F. globulosa* (Retzius) Kunth, *F. utilis* Elmer, *F. assamica* C. B. Clarke ex Guha Bakshi]: Fig. 152

Rhizomes short, horizontal, clothed with brownish scales. Culms tufted, obtusely 3-sided to subterete, 20–70 (–120) cm long, 1–3 (–5) mm wide, glabrous. Leaves on culms reduced to 2 or 3 bladeless sheaths; sheaths cylindrical; the lower sheaths scale-like, 2–4 cm long, straw-coloured with brown margins; the upper sheaths 6–10 (–20) cm long, yellowish-brown, ventral side membranous; leaves on sterile shoots usually short-bladed, ± 1.5 mm wide; ligule absent. Involucral bracts 2 or 3, narrowly lanceolate, 0.5–1 cm long, not overtopping the inflorescence. Inflorescence simple or compound, corymbose; primary rays 2–10, unequal. Spikelets solitary, globose, broadly ovoid to ellipsoid, 3–8 mm long, 3–4 mm in diameter, terete, obtuse, brown to rusty brown. Glumes membranous, elliptical to ovate-elliptical, obtuse or mucronate, 1.8–2.6 mm long, 1–1.5 mm broad, glabrous, yellow- or rusty brown, with a longitudinal dark brown stripe on either side, whitish margins; keel obscure, ending somewhat below the apex. Stamens (2) 3. Styles 3-angled or rarely flattened, 3- or rarely 2-cleft, ± 1 mm long, glabrous. Nuts 3-sided or plano-convex, obovoid to obovoid-globose, 0.75–1 mm long, 0.5–0.7 mm broad, smooth or slightly warty, epidermal cells in 12–15 vertical rows, milky yellow, sessile or nearly so.

Perennial. In wet open places, marshes, swamps and in rice fields. Cultivated in some regions for weaving hats, baskets, mats, string, etc. The cultivated race (clone) has 2 stigmas. The epithet '*umbellaris*' has priority as it was published a few months before '*globulosa*'.

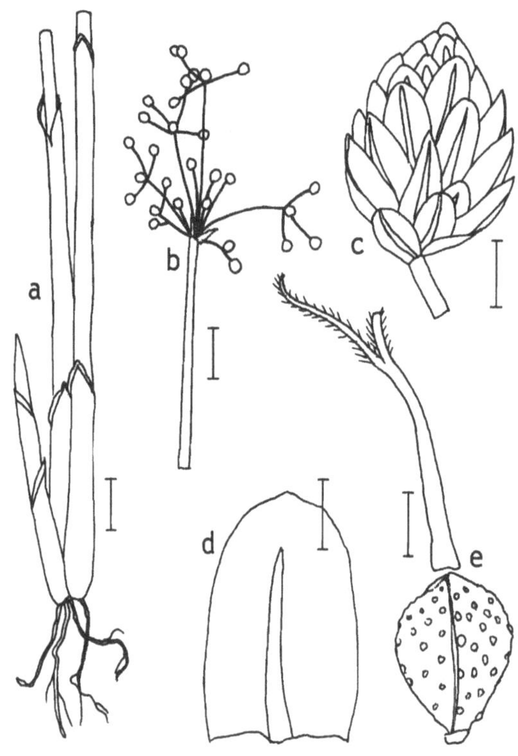

FIG. 152. *Fimbristylis umbellaris*: a, base of plant (1 cm); b, inflorescence (1 cm); c, spikelet (1 mm); d, glume (0.5 mm); e, fruit (0.25 mm).

{(Ten)/Hel}. Tropical Asia and Pacific: Bur, Cey, Ind, Nep: An, As, Kt, Mg. 0–1000 m.

Fimbristylis woodrowii C. B. Clarke: [*F. microcarya* auct. Ind., non F. von Mueller]: Fig. 153

Culms sparsely tufted, filiform, erect, rigid, 7–20 cm long, 0.5 mm in diameter, 5-angled below the inflorescence. Leaves basal and cauline; sheaths membranous; ligule absent; blades flat, up to 12 cm long, ± 0.7 mm wide, flaccid, prominently nerved, abruptly acute, margins scabrid. Involucral bracts 2, not overtopping the inflorescence, the longest up to 6 mm long. Inflorescence a cluster of 3 free spikelets, stalked. Spikelets oblong, up to 5 mm long, ± 1.5 mm wide, ± 10-flowered. Glumes ovate, up to 2 mm long, thin, brown, glabrous, acute, sides without nerves, mucronate; keel 3- or 5-nerved. Stamens 2. Styles 3-cleft, flattened, fimbriate towards the tips, style base pyramidal. Nuts 3-sided, obovoid, up to 1 mm long, glossy, white, vertically ridged, warty, shortly stipitate.

Perennial or perhaps sometimes annual. Seasonally wet places, rice fields, usually in hilly districts.

{(⊙) Ten/(Hel)}. Endemic to India: ♦Ind: Gj, Go, Kt, Mh, Rj, Tn. Hills above 1200 m in the South.

Fimbristylis zeylanica T. Koyama: Fig. 153

Rhizome absent, roots fibrous. Culms tufted, capillary, 2- or 3-angular, smooth, (3–) 5–10 cm long, 0.3–0.6 mm wide. Leaf sheaths 5–12 mm long, membranous; ligules absent; leaf blades linear, about one-third to half as long as the culms, 1.5–3 cm long, 0.5–0.8 mm wide, light green, glabrous, smooth except for the scabrous upper margin, suddenly contracted to a rounded apex. Involucral bracts 1 or 2, scale-like, 1–1.2 mm long, the lowest up to 4 cm long. Inflorescence with 1 or 2 spikelets, when 2 then the upper spikelet on a ± 5 mm long peduncle. Spikelets lanceolate-ovate or elliptical, strongly laterally flattened, 6–9 mm long; 2–2.5 mm wide, brown, 6- to 17-flowered. Glumes in 2 rows, thinly membranous, folded, ovate, 2.3–3 mm long, 1.8–2 mm broad, acuminate at apex, straw-coloured on both sides, stained with brown near the keel, sparsely hairy on upper half, ciliate on the margins; keel sharp, 1-nerved, ending in a straight short cusp ± 0.5 mm long. Stamens 2. Styles 2–2.3 mm long, 3-cleft, glabrous; stigmas ± 0.6 mm long. Nuts 3-sided, broadly obovoid, 0.7–0.8 mm long, 0.4–0.7 mm wide, whitish or yellowish-brown, with 8–10 longitudinal rows of pits, shortly stipitate.

Annual. On wet sand at edges of ponds.

{⊙ Ten/Hel}. Endemic to Sri Lanka: ♦Cey.

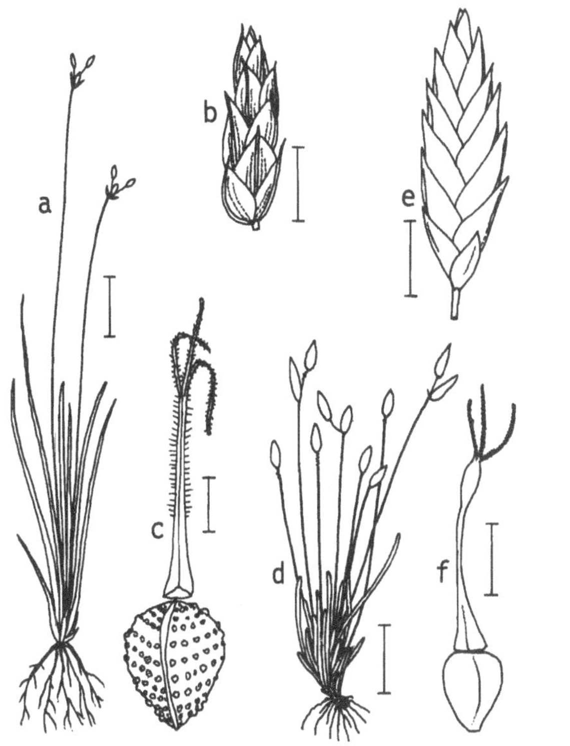

Fig. 153. *Fimbristylis woodrowii*: a, whole plant (2 cm); b, spikelet (2 mm); c, fruit (0.5 mm). *F. zeylanica*: d, whole plant (1 cm); e, spikelet (2 mm); f, fruit (0.7 mm).

Fuirena: ± 30 species: tropics and subtropics, most in Africa.

Tufted or with woody creeping rhizomes. Culms with nodes and leaves throughout their length. Leaf sheaths well-developed, closed, at base usually bladeless; ligule tubular; leaf blades cauline, flat, usually hairy. Involucral bracts mostly short, not exceeding the inflorescence. Inflorescence of 2 or more head-like clusters of spikelets or compound corymbs of spikelets. Spikelets mostly sessile, many-flowered; rachilla persistent, not winged, with vertically elongated pits. Flowers bisexual. Glumes spirally arranged, caducous, imbricate, hairy, the lower 2 sterile, the upper with bisexual flowers. Perianth of 3 bristles and 3 variable and often awned scales, of 6 bristles or absent. Stamens (2) 3. Style caducous, not swollen at the base; stigmas (2) 3. Nuts 3-sided, obovoid or ellipsoid, falling off enclosed in the perianth when present.

Annuals or perennials. Generally considered to be good fodder plants but rarely growing in large populations. Vartak, V. D. Taxonomic study of the genus *Fuirena* Rottb. from India. Journal of the University of Poona: Science Section 40: 185–203 (1971); this work is rather incomplete and not up to date. It is high time to revise *Fuirena* in India.

1. Annual, without a rhizome, with fibrous roots only
2. Awn on glume recurved, strong and thorn-like; perianth scales ± 0.5 mm wide *F. capitata*
2. Awn on glume ± straight, soft, slender, not thorn-like; perianth scales ± 0.2 mm wide
3. Perianth scales with blade subquadrate, cordate or hastate at the base, with 3 minute teeth at the distinctly thickened apex (central tooth the longest); glumes 1.5–2.5 mm long, excluding a 1–1.5 mm long awn *F. ciliaris*
3. Perianth scales with blade semicircular to lunate blade with the 2 lower lobes extending downwards, cuspidate at the apex, almost appearing 3-lobed; glumes 1.2–1.3 mm long, excluding a 1–1.3 mm long awn *F. trilobites*
1. Perennial with conspicuous, creeping rhizome
4. Culms (4-) 5-angled; perianth scales subsessile with a very short stalk *F. umbellata*
4. Culms 3-angled or rounded; perianth scales absent or with long stalks
5. Perianth absent
6. Culms 3-sided with obscure longitudinal ridges; nuts 3-sided, with 3 distinct longitudinal ridges, broadly ellipsoid, 1.2–1.7 mm long excluding the 0.2–0.5 mm long basal part of the style, 0.9–1.2 mm wide, white or rarely brownish, glossy, not ridged or furrowed *F. pubescens*
6. Culms sharply 3-angled; nuts 3-sided, with sharp angles but without distinct longitudinal ridges, ellipsoid, 0.9–1 mm long excluding the ± 0.2 mm long basal part of the style, 0.7–0.8 mm wide, chalky white to light brown, each side conspicuously marked with 5–7 longitudinal furrows bearing short transverse ridges *F. wallichiana*
5. Perianth of bristles with or without scales
7. Leaf blades with rows of hairs along the keel and adaxial grooves; spikelets greyish-black; perianth of 6 bristles, each linear, with backwardly directed barbs *F. swamyii*
7. Leaf blades glabrous along the keel and adaxial grooves; spikelets greenish-brown; perianth of 3 bristles and 3 scales; perianth bristles linear, with backwardly directed barbs; perianth scales like table tennis bats but hastate at the base, as long as or exceeding the nut *F. tuwensis*

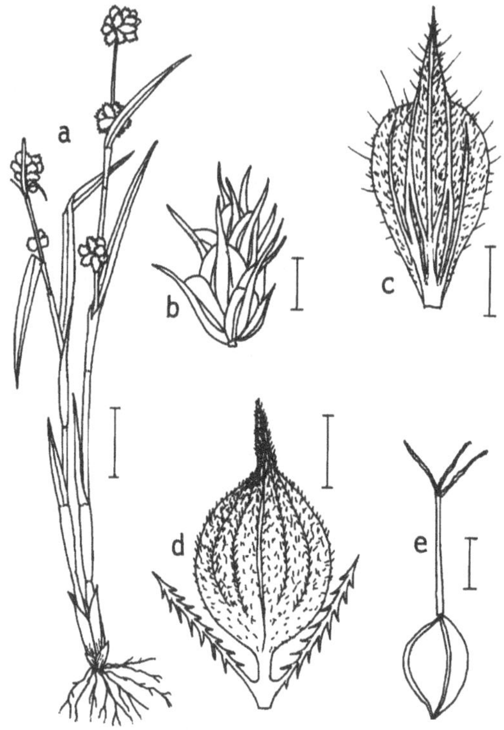

FIG. 154. *Fuirena capitata*: a, rather small plant with hairs omitted (2.5 cm); b, spikelet with hairs omitted (2 mm); c, glume (1 mm); d, one perianth scale with two perianth bristles (1 mm); e, fruit (0.5 mm).

Fuirena capitata (N. L. Burman) T. Koyama [*F. uncinata* (Willdenow) Kunth]: Fig. 154

Culms tufted, obtusely 5-sided, ribbed, hairy or rarely glabrous, (4–) 7–30 cm tall, 0.8–2 mm wide. Leaf sheaths hairy; ligules ± 1 mm long, rusty brown, membranous; leaf blades soft, flattish, 2–14 cm long, 2–6 mm wide, profusely hairy on both surfaces, with several nerves. Inflorescence of 2–4, ± globose clusters of spikelets; the 2 distal clusters proximate, the remaining 1 or 2 distant; clusters 6–17 mm in diameter, with 4–30 spikelets. Spikelets sessile, ellipsoidal, (3–) 5–7 mm long, 3–4 mm in diameter. Glumes obovate to oblong-elliptical, 2.3–2.6 mm long, excluding a 1.4–1.6 mm long recurved thorn-like awn, glaucous green, hairy between the nerves, the keel 1-nerved, the sides (2-) 3-nerved to 3-ribbed, 2 of which concurrent to the midvein, the nerves yellowish turning dark brown. Perianth bristles 3, 0.3–0.5 mm long, retrorsely scabrid; perianth scales 3, ± 0.8 mm long, ± 0.5 mm wide, the blade almost circular with a short-awned apex and a short (± 0.3 mm long) stalk. Nuts sharply 3-angled, obovoid to subglobose, 0.6–0.7 (–1) mm long, 0.4–0.6 mm wide, also smooth, light brown.

Annual or perhaps sometimes a short-lived perennial. Forming tufts in open wet habitats, often abundant in rice fields after the harvest.

{(⊙)? Hel}. Endemic to Sri Lanka and S India: ♦Cey, ♦Ind: Kl, Kt, Mh, Or, Tn. 0–600 m.

Fuirena ciliaris (Linnaeus) Roxburgh [*F. glomerata* Lamarck, *F. rottboellii* Nees]: Fig. 155

Culms slender, tufted, 3- or 5-sided just below the inflorescence, obtuse-angled, striate, hairy (at least above), (5–) 10–50 (–75) cm tall, 1–2 mm wide. Leaf sheaths pubescent; ligules short (1–2 mm long), membranous; leaf blades flaccid, flat, 2–8 (–20) cm long, 3–8 mm wide, profusely hairy or at least ciliate on the margins, with 3 or 5 (7) prominent nerves, light green. Inflorescence of 1–3, terminal clusters of spikelets, with short and pubescent peduncles in the upper axils; clusters ± globose, 1–2 cm in diameter, with 3–10 spikelets. Spikelets ovate or ellipsoidal, 4–12 mm long, 3–4 mm in diameter, greyish-green. Glumes obovate to oblong-obovate, 1.5–2.5 mm long, excluding a 1–1.5 mm long awn, hairy throughout. Perianth bristles 3, sometimes reduced, scaberulous to almost smooth, as long as or shorter than the nut; perianth scales 3, distinctly stalked, subquadrate, up to 1.2 mm long, ± 0.2 mm wide, cordate or hastate at the base, with 3 minute teeth at the distinctly thickened apex (central tooth the longest), glabrous or minutely hairy at the apex. Nuts

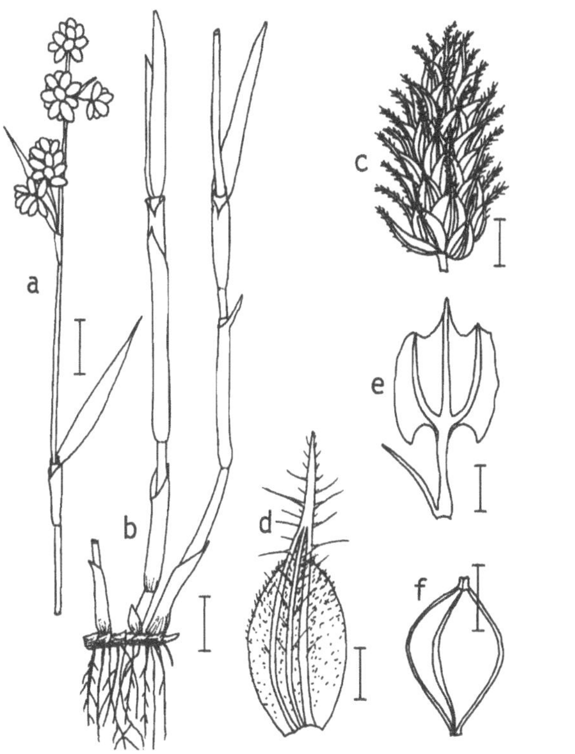

Fig. 155. *Fuirena ciliaris*: a, inflorescence with hairs omitted (2 cm); b, base of plant with hairs omitted (2 cm); c, spikelet (2 mm); d, glume (0.5 mm); e, one perianth scale with one perianth bristle (0.25 mm); f, nut (0.5 mm).

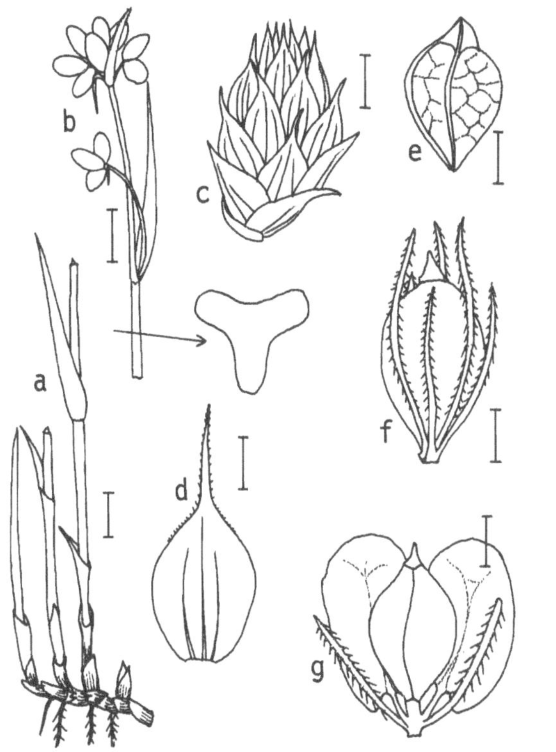

Fig. 156. *Fuirena pubescens*: a, base of plant with hairs omitted (1 cm); b. inflorescence (1 cm) with hairs omitted and a transverse section of culm (inset); c, spikelet with hairs omitted (3 mm); d, glume (1 mm); e, nut (0.5 mm). *F. swamyii*: f, nut with perianth bristles (0.4 mm). *F. tuwensis*: g, nut with two perianth bristles and two perianth scales (0.5 mm).

subsharply 3-angled, obovoid or ellipsoid, 0.7–1 mm long, 0.5–0.8 mm wide, the sides shiny but wrinkled, yellow to dark brown.

Annual. Forming small clumps or tufts in open wet or swampy places, river banks, roadside ditches, margins of ponds and in rice fields.

{⊙ Ten/(Hel)}. Old World tropics and subtropics: Cey, Ind: An, Ap, As, Bh, Kl, Kt, Mh, Mp, Or, Rj, Tn, Up, Wb. 0–2000 m.

Fuirena pubescens Kunth: Fig. 156

Rhizomes long creeping, 2–5 mm thick, densely set with scales. Culms robust, erect, tufted or arising at ± 1 cm intervals along the rhizome, 30–50 (–100) cm tall, (1–) 6–9 mm thick, 3-sided with obscure longitudinal ridges, glabrous (but sometimes hairy just below the inflorescence). Leaf sheaths usually glabrous or sometimes the upper ones hairy; ligule reddish-brown, hairy; leaf blades 6–25 cm long, 4–9 mm wide, glabrous or hairy below and at the tip. Inflorescence bracts like upper leaves but usually more hairy. Inflorescence of stalked, clusters of spikelets; each cluster with 2–6 spikelets. Spikelets sessile or 1 spikelet of a cluster stalked, ovoid, 6–10 mm long, 3–4 mm in diameter, brownish-green. Glumes broadly ovate-elliptical, 2.5–4 mm long excluding the 0.8–1.7 mm long awn, hairy throughout. Perianth bristles and scales absent. Nuts 3-sided, with 3 distinct longitudinal ridges, broadly ellipsoid, 1.2–1.7 mm long excluding the 0.2–0.5 mm long basal part of the style, 0.9–1.2 mm wide, glossy, white or rarely brownish.

Perennial. Found in seasonally flooded areas, open wet or swampy places, waterlogged marshes and rice fields. A rather variable species, several varieties have been described. I have seen no material; these records are based on Vartak (1971).

{Ten/Hel}. Warm regions of Europe, Africa and Asia Minor, extending to the Himalayas and W Ghats: Ind: Hp, Kt, Pj, Tn.

Fuirena swamyii E. Govindarajalu: Fig. 156

Like *F. pubescens* but: leaf blades with rows of hairs along the keel and adaxial grooves; spikelets greyish-black; perianth of 6 bristles, each linear, with backwardly directed barbs, the lower 3 bristles as long as the nut, the upper 3 bristles longer than the nut; nuts with a stalk ± a quarter the length of the nut.

Perennial. Marshy places in grassland.

{Hel}. Endemic to India: ♦Ind: Tn (Madurai District).

Fuirena trilobites C. B. Clarke: Fig. 157

Like *F. ciliaris* but: generally more slender and less hairy; glumes obovate, 1.2–1.3 mm long, excluding a 1–1.3 mm long awn; perianth bristles linear, minutely scabrid, shorter than the nut; perianth scales distinctly stalked, with a semicircular to lunate blade with the 2 lower lobes extending downwards, cuspidate at the apex, almost appearing 3-lobed, up to 1 mm long, glabrous. Nuts ellipsoid, 0.9–1.1 mm long, 0.3–0.4 mm wide, minutely tuberculate with very faint horizontal striations, light yellow to pale brown.

Annual. Waterlogged grassland, rice fields, pools and along streams and rivers.

{⊙ Hel}. Endemic to Peninsular India: ♦Ind: Ap, Gj, Kt, Mh, Tn.

Fuirena tuwensis M. P. Deshpande et G. L. Shah: Fig. 156

Like *F. pubescens* but: perianth with 3 bristles and 3 scales; perianth bristles linear, with backwardly directed barbs, ± as long as the nut; perianth scales like table tennis bats but hastate at the base, ± 1.5 mm long, as long as or exceeding the nut.

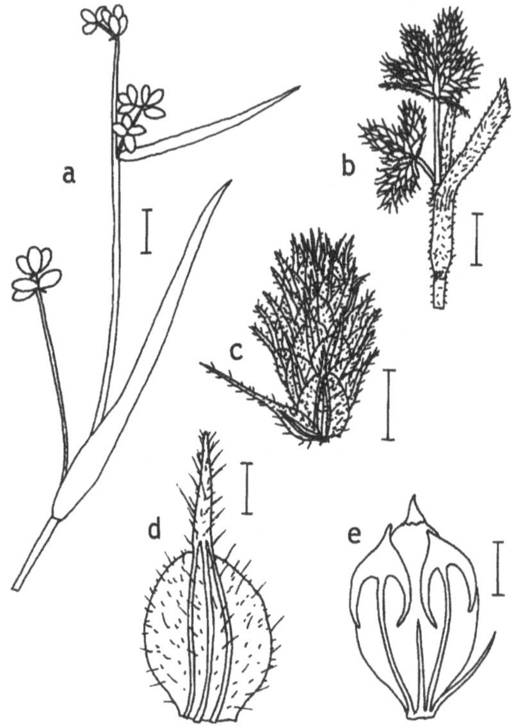

FIG. 157. *Fuirena trilobites*: a, inflorescence with hairs omitted (1 cm); b, two spikelet clusters (1 mm); c, spikelet (2 mm); d, glume (0.5 mm), e, nut surrounded by two perianth bristles and two perianth scales (0.3 mm).

Perennial. Moist ground, margins of ditches. I have seen no authentic material of this species and suspect it is a perennial race of *F. ciliaris*.

{Hel}. Endemic to India: ♦Ind: Gj, Kt, Mp.

Fuirena umbellata Rottbøll [*Fimbristylis pentagona* Wight et Arnott ex Nees]: Fig. 158

Rhizomes creeping, woody, ± 5 mm thick. Culms erect, solitary or few together, (3–) 50–90 (–150) cm tall, 3–6 mm thick, swollen and bulb-like at the base, sharply (4-) 5-angled, glabrous except just below the inflorescence. Leaf sheaths 5-angled like the culm, usually glabrous; ligule reddish-brown, densely hairy; leaf blades much shorter than the culms, 10–20 (–30) cm long, 5–15 (–25) mm wide, glabrous except at base and on the margins, glaucous green. Inflorescence bracts like leaves. Inflorescence of 2–20, stalked, compound corymbs, borne along the upper part of the culm; corymbs 15–20 mm in diameter, with 2–20 clusters of spikelets. Spikelets ovoid to ovoid-ellipsoid, 4–10 mm long, 2.5–3 mm in diameter, brownish-green. Glumes obovate to ovate-elliptical, 2–2.5 mm long excluding a 0.5–1 mm long awn, both short and long hairs throughout. Perianth bristles usually absent; perianth scales 3, scale–like, not or very shortly stalked, obovate or oblong, ± as long as the nut, 3-nerved, apex square or rounded with the midrib frequently excurrent in a short and slender mucro. Nuts 3-sided, ellipsoid to obovoid, obscurely wrinkled, 0.8–1.2 mm long, excluding a 0.2 mm long persistent style base, 0.6–0.7 mm wide, whitish when young becoming brown at maturity.

Perennial. Found in seasonally flooded places, also in open wet or swampy localities, waterlogged marshes, along lake edges and in rice fields. This species is easily recognized by its large size and 5-angled culms and leaf sheaths and scale-like (not stalked) perianth segments.

{Ten/Hel}. Pantropics and subtropics: Ind: An, Ar, As, Kl, Kt, Mg, Mh, Mn, Ng, Pj, Tn, Wb. 0–900 m.

Fuirena wallichiana Knuth: Fig. 158

Like *F. pubescens* but: culms sharply 3-sided, hollow, smooth, up to 60 cm tall, up to 3 mm in diameter; leaf sheaths 3-sided; inflorescence of 1–3 (–5), terminal

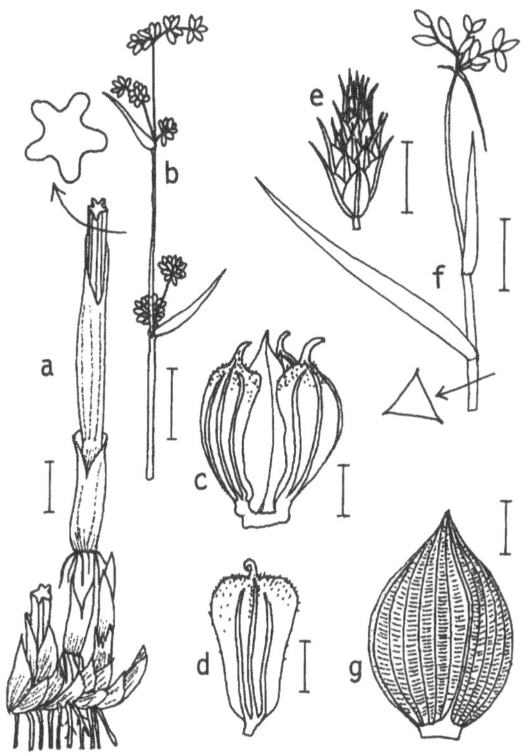

Fig. 158. *Fuirena umbellata*: a, base of culm (1 cm) with a transverse section of culm (inset); b. inflorescence (1.5 cm); c, nut surrounded by perianth scales (0.4 mm); d, perianth scale (0.4 mm). *F. wallichiana*: e, spikelet (2 mm); f, inflorescence (2 cm); g, nut (0.25 mm).

clusters of spikelets on short peduncles; spikelets olive green; nuts 3-sided, with sharp angles but without distinct longitudinal ridges, ellipsoid, 0.9–1 mm long excluding the ± 0.2 mm long basal part of the style, 0.7–0.8 mm wide, chalky white to light brown, each side conspicuously marked with 5–7 longitudinal furrows bearing short transverse ridges.

Perennial. Banks of streams and tanks, seasonally flooded areas and moist pasture land.

{Hel/Ten?}. Endemic to India: ♦Ind: Ap, Kt, Mh, Rj, Tn Up.

Kyllinga [*Queenslandiella*]: ± 60 species: almost cosmopolitan.

Annual or perennial. Culms 3-sided. Leaves mostly radical, leaf-like. Involucral bracts whorled beneath the head, usually 3, unequal, long and leaf-like. Inflorescence terminal, head-like, of 1–3 (–4), sessile, clusters of spikelets; each densely bearing many spikelets. Spikelets spirally arranged, numerous, laterally flattened, 1 or 2 or rarely more bearing flowers; rachilla caducous as a whole. Glumes distichous, clearly in 2 opposite rows, mostly pale whitish to yellow or green, folded with an acute keel. Lower 1 or 2 glumes sterile, the next subtending bisexual flowers, the uppermost male or sterile. Perianth bristles absent. Stamens (1–) 3. Stigmas 2; style not swollen at the base. Nuts strongly flattened, ellipsoid, sometimes apiculate, usually containing oily endosperm.

Excluded: the following species are sometimes found on wet soil but they are not considered to be hydrophytic: ***Kyllingia brevifolia*** Rottbøll var. ***stellulata*** (Valckenier-Suringar) S. Hooper; ***K. hyalina*** (Vahl) T. Koyama [*Cyperus hyalinus* Vahl, *Pycreus hyalinus* (Vahl) Domin, *Queenslandiella hyalina* (Vahl) Ballard, *Cyperus pumilus* auct. Ind., non Linnaeus, *Mariscopsis suaveolens* Chermezon]; ***K. nemoralis*** (J. R. et G. Forster) Dandy ex Hutchinson et Dalziel [*Thyrocephalon nemoralis* J. R. et G. Forster, *K. monocephala* Rottbøll, *Cyperus kyllingia* Endlicher]; ***K. metzii*** Hochstetter ex Steudel [*K. squamulata* Vahl]; ***K. odorata*** Vahl subsp. ***cylindrica*** (Nees ex Wight) T. Koyama; ***K. polyphylla*** Willdenow ex Kunth [*K. aromatica* Ridley].

1. Rhizomes corm-like, erect; culms densely tufted; inflorescence usually of 3 heads; spikelets not more than 2 mm long; third and fourth glumes not more than 2 mm long, without a cusp ***K. tenuifolia***
1. Rhizomes elongate, creeping; culms solitary or in a single row along the rhizome; inflorescence usually of 1 head; spikelets 3 mm or more long; third and fourth glumes 2.5 mm or more long, with a straight or curved cusp

2. Rhizomes stoloniferous (long and creeping near the surface); lowest involucral bract erect or spreading at maturity; nuts 1–1.5 mm long, 0.5–0.7 mm wide
K. brevifolia

2. Rhizomes decumbent and woody, never stoloniferous; lowest involucral bract usually reflexed at maturity; nuts 1.5–2 mm long, 0.6–0.8 mm wide
K. melanosperma

Kyllinga brevifolia Rottbøll [*Cyperus brevifolius* (Rottbøll) Hasskarl]: Fig. 159

Rhizomes stoloniferous (long, creeping and near the surface), 1–3 mm in diameter, clothed with brown scales. Culms solitary or in a single row, 3–50 (–90) cm tall, (0.5–) 1–1.5 mm thick. Leaf sheaths membranous, brown to purple, the lower ones almost bladeless; blades well-developed on upper leaves, flat, shorter than or as long as the culms, 1–3 (–5) mm wide. Involucral bracts 3 or 4 (–6), the lowest usually erect or spreading, the remainder usually spreading or some reflexed, 3–10 (–20) cm long. Inflorescence a single head (rarely of 2 or 3 heads); head globose or broadly ovoid-globose, 5–10 mm in diameter, pale green becoming straw-coloured at maturity. Spikelets lanceolate-oblong to elliptical, 2–4 mm long, ± 1 mm wide. Glumes ovate-elliptical, membranous, 5- or 7-nerved, cuspidate at apex; first and second glumes up to 1 mm long, remaining glumes 2.5–4 mm long; keel projecting beyond the glume apex into a straight or curved cusp, smooth or sparsely spiny towards the tip. Stamens 1 or 2 (3). Nuts obovate to elliptical, 1–1.5 mm long, 0.5–0.8 mm wide, brownish.

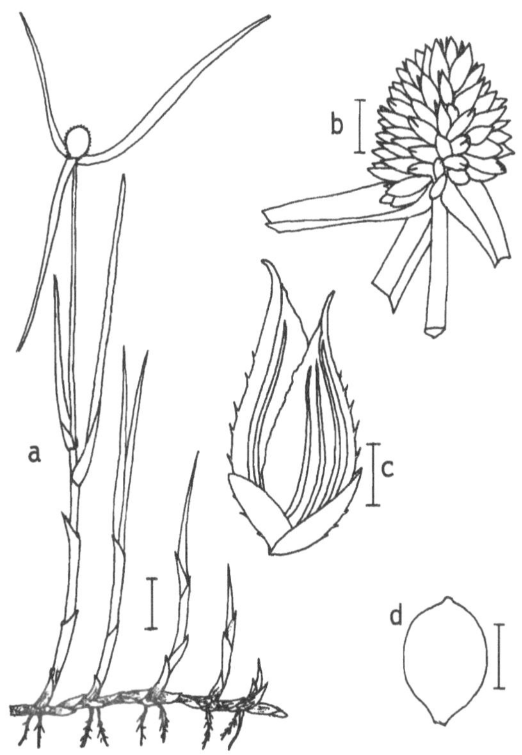

Fig. 159. *Kyllinga brevifolia*: a, whole plant (1 cm); b, inflorescence (2 mm); c, spikelet (0.5 mm); d, nut (0.3 mm).

Perennial. A weedy plant not confined to wetlands but often found along streams and rivers, at the edge of pools, in marshy irrigated land and frequent in and around rice fields. It often forms an almost single species turf. Used locally in medicine

{Hyp/Hel}. Warm and tropical regions of the whole world: Ban, Bhu, Bur, Cey, Ind, Nep, Pak: An, Ap, As, Bh, Dl, Gj, Go, Hp, Kl, Kt, Lk, Mg, Mh, Mp, Mz, Ng, Or, Pj, Rj, Sk, Tn, Up, Wb. 0–2000 m.

Kyllinga melanosperma Nees [*Cyperus melanospermus* (Nees) Valckenier-Suringar, *K. bifolia* Miquel]: Fig. 160

Rhizomes decumbent, woody, 2–4 mm in diameter, the surface, clothed with brown scales, very aromatic. Culms close together in a single row, (30–) 50–170 cm tall, 2–5 mm thick, sharply 3-angled just below the inflorescence. Leaf sheaths membranous, brown to purple or almost black, the lower ones bladeless; blades of upper sheaths if present, flat, rarely more than 3 cm long, much shorter than the culms. Involucral bracts 3 (–5), spreading to reflexed, the lowest up to 20 cm long, up to 6 mm wide. Inflorescence a single head (rarely of 2 or 3 heads); head globose, (6–) 8–14 (–16) mm in diameter, pale green becoming straw-coloured or golden-brown at maturity. Spikelets lanceolate-oblong to elliptical, 3–4.5 mm long, 1–1.5 mm wide. Glumes ovate-elliptical, membranous; first and second glumes, 7- or 9-nerved, 2.5–3 mm long; remaining glumes 5-nerved, 3–4.2 mm long, cuspidate at apex; keel projecting beyond the glume apex into a cusp, smooth or sparsely spiny. Stamens (1, 2) or 3. Nuts oblong to elliptical-oblong, 1.3–2 mm long, 0.6–0.8 mm wide, brown to black.

Perennial. In or near water, by streams, in marshes, swamps, seepage hollows, mats of floating vegetation and wet grassy places. The plants are strongly aromatic. Plants growing in water tend to have well-developed leaf blades; they are sometimes recognized as subsp. *bifolia* (Miquel) Kern.

{Ten/Hel}. Old World tropics and subtropics: Cey, Ind: Kl, Kt, Tn, Up. 0–2000 m.

Kyllinga tenuifolia Steudel [*K. triceps* Rottbøll nom. illeg., *Cyperus triceps* (Rottbøll) Endlicher, *K. bulbosa* auct. Ind., non Palisot de Beauverd]: Fig. 160

Rhizomes short, erect, without stolons. Culms densely tufted, slender but bulb- or corm-like at base, (3–) 5–15

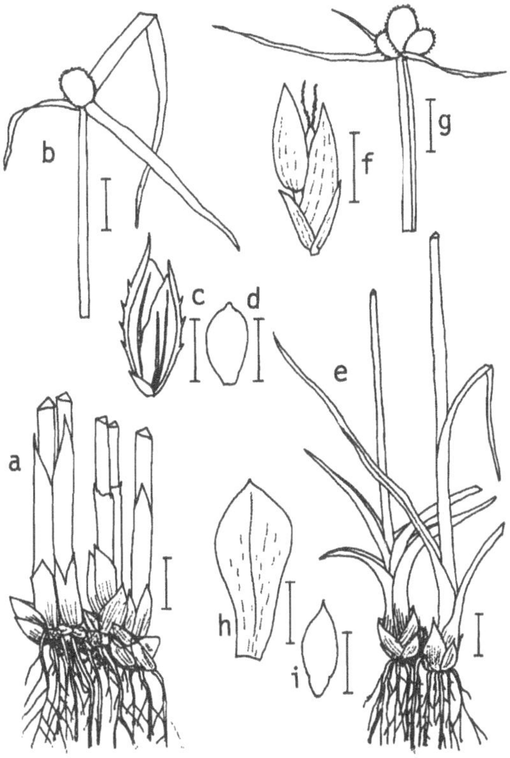

FIG. 160. *Kyllinga melanosperma*: a, base of shoot (1 cm); b, inflorescence (1 cm); c, spikelet (1 mm); d, nut (1 mm); *K. tenuifolia*: e, base of shoot (1 cm); f, spikelet (1 mm); g, inflorescence (1 cm); h, fertile glume (1 mm); i, nut (1 mm).

(–45) cm tall, 0.5–1 mm thick. Leaf sheaths stout, light brown; blades well-developed, linear, flat to plicate, shorter than the culms, 1–2 (–3) mm wide. Involucral bracts spreading, becoming reflexed at maturity, the lowest up to 10 cm long. Inflorescence of usually 3 (rarely 1, 4 or 5), sessile spikes; central spike globose to ovoid-globose, 4–9 mm long, ± 5 mm in diameter (the lateral ones somewhat smaller), pale green to greyish-white. Spikelets oblong to lanceolate-oblong, 1.7–2 mm long, ± 0.6 mm wide. Glumes ovate-oblong to lanceolate-oblong, hyaline; first and second glumes, 1- or 3-nerved, 0.3–0.5 mm long; remaining glumes 5- or 7-nerved, 1–2 mm long, cuspidate at apex; keel not or hardly extending beyond the glume apex, without a cusp at most submucronulate, smooth. Stamens 2. Nuts oblong, 1–1.5 mm long, ± 0.5 mm wide, brownish.

Perennial. Common in marshy places particularly at the margins of tanks and streams but it is also found in wet or damp grassy places.

{Ten/Hel}. Old World tropics and subtropics: Bhu, Bur, Cey, Ind, Pak: Bh, Dl, Hp, Kl, Kt, Mh, Mp, Or, Pj, Rj, Tn, Up, Wb. 0–1000 m.

Lepironia: 1 species

Lepironia articulata (Retzius) Domin [*L. mucronata* L. C. Richard, *L. compressa* Boeckeler, *L. conifera* (Poiret) Druce, *Coricarpha aphylla* Boeckeler]: Fig. 161

Rhizomes creeping, woody, up to 15 cm long, 5–8 mm in diameter, clothed with scales. Culms close together, arranged in a row along the rhizome, leafless, erect, (5–) 50–100 (–120) cm tall, 2–7 mm in diameter, terete, glaucous green, smooth, hollow and transversely septate. Leaves reduced to a few scales and sheaths at the base of the stem. Involucral bract erect, exceeding the inflorescence, 2.2–6 cm long, appearing as a continuation of the stem. Inflorescence a solitary, pseudolateral spike; spike ovoid to oblong-ellipsoid, 10–35 mm long, brown to black at maturity. Spikelets many, spirally arranged, subtended by ovate, obtuse, hard, stiff glume-like bracts or bracteoles, (or the spikelet representing a secondary spike bearing 1-flowered spikelets); bract-like glume oval to suborbic-

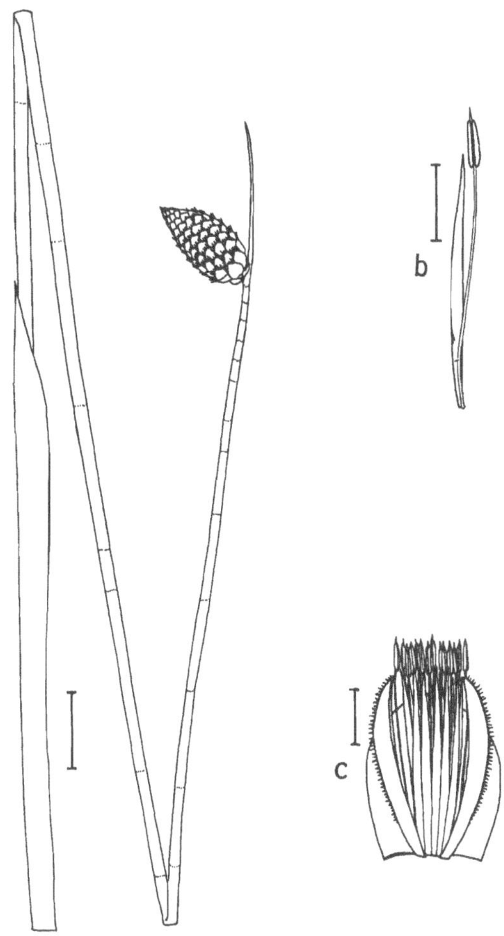

FIG. 161. *Lepironia articulata*: a, flowering shoot (2 cm); b, male flower (2 mm); c, spikelet (2 mm).

ular, leathery, 3.2–6.7 mm long, 3–6.2 mm wide, chestnut to purple-brown, not nerved, with hyaline margin. Flowers unisexual. Male flowers numerous, each with 1 stamen, anther 2.2–3 mm long. Female flower terminal (central), with up to 8 scales (glumes) and 1 ovary. Style 2–3 mm long, 2-cleft; stigmas 2. Nuts biconvex, flattened, oval to broadly obovate, 3–4 mm long with a ± 0.5 mm long beak, 2–2.8 mm wide, 7- to 9-nerved on each side, brown.

Perennial. Grows in shallow lakes, in swamps and in saline backwaters and marshes. Cultivated for fibre which is used for mats and as packing material, in southern China for junk sails.

{Ten/Hyp}. From Madagascar through S Asia to New Caledonia: Cey, Ind: <?Tn.

Lipocarpha [*Hemicarpha, Isolepis* pro parte, *Rikliella*]: 35 species: pantropics and subtropics, most in Africa.

Annual or perennial. Culms erect, ± cylindrical, glabrous and smooth. Leaves basal; sheaths rather short, closed; ligule absent; blades linear, flat or rarely inrolled, glabrous and smooth. Inflorescence terminal, head-like, with 1 to many spikes, surrounded by 1 to few, spike-bearing or empty bracts, the lowermost elongate, the uppermost scale-like. Spikes composed of a cylindrical axis with many well-developed spikelet bracts in a dense spiral, the lateral spikes each with an empty spike prophyll. Each spikelet bract subtends a 1-flowered spikelet with 2 small hyaline scales, 1 empty prophyll and 1 flower-bearing glume (in some species the prophyll and glume are reduced). Glumes obovate to spathulate, caducous, with a short to long apical part. Flowers bisexual. Stamens 1–3. Stigmas 2 or 3; styles not swollen at the base, totally caducous. Nuts broadly obovoid to ellipsoid.

I have followed the treatment of Goetghebeur, P. and Van den Borre, A. Studies in Cyperaceae 8. A revision of *Lipocarpha*, including *Hemicarpha* and *Rikliella*. Wageningen Agriculture University Papers 89(1): 1–87 (1989); Datar Rekha and Vartak, V. D. Taxonomic study of the genera *Lipocarpha* R. Br. and *Bulbostylis* Kunth. [sic] from India. Journal of the University of Poona, Science and Technology 48: 19–32 (1976): these authors consider *Lipocarpha* in a restricted sense with only 2 species.

Excluded: the following do not grow in wetlands: ***Lipocarpha hemisphaerica*** (Roth) Goetghebeur [*Hemicarpha isolepis* Nees, *Isolepis bellula* Steudel]; ***L. pygmaea*** Kern.

1. Lower involucral bract always strictly erect (awn of spikelet bract 0.4–0.9 mm long) ***L. squarrosa***
1. Lower involucral bract obliquely patent to reflexed in fruiting heads
2. Spikelet prophyll and glume absent; awn of the spikelet bract (0.7–1.5 mm long) as long as or longer than the lower expanded part ***L. kernii***
2. Spikelet prophyll and glume present; awn of the spikelet bract shorter than the lower expanded part
3. Nuts without a beak, remnant of the style base less than 0.1 mm long; midrib of the spikelet bract decurrent beyond the bract into a cusp or short awn ***L. reddyi***
3. Nuts with a conspicuous beak, remnant of the style base 0.2–0.3 mm long; midrib of the spikelet bract not decurrent beyond the bract
4. Spikelet bracts almost as long as the nut; spikes dark purplish-brown to almost black ***L. sphacelata***
4. Spikelet bracts ± twice as long as the nut; spikes whitish, silver or pale brown
5. Leaves not more than half as long as the culm; spikelets whitish to silver; nuts narrowly obovoid to oblong, 0.8–1.2 mm long, 0.2–0.5 mm wide ***L. chinensis***
5. Leaves as long as or longer than the culm; spikelets pale brown; nuts oblong, 1.3–1.6 mm long, 0.3–0.5 mm wide ***L. raynalinana***

Lipocarpha chinensis (Osbeck) Kern [*Hypaelyptum argenteum* Vahl, *Tungia laevigata* Roxburgh]: Fig. 162

Culms tufted, 15–80 cm tall, 0.5–2 mm thick. Leaf blades flat, one-quarter to half as long as the culms, up to 40 cm long, 2–4 mm wide. Involucral bracts 2–3 (–5), the largest up to 13 cm long. Inflorescence terminal, with (1–) 2–10 (–12), subequal spikes. Spikes ovoid to conical, 3–13 mm long, (1–) 4–5 mm in diameter, whitish to almost silver. Spikelet bract oblong-obovate to spathulate, 1.5–2.7 mm long, 0.5–1 (–1.3) mm wide, acute to apiculate at the tip, white to pale yellowish-brown, often with red flecks and a green midnerve. Spikelet prophyll and glume 1.2–2 mm long, often with reddish stripes. Stamens 1 (2). Style 0.2–0.8 mm long, 3-cleft; stigmas (2) 3 about as long as the style. Nuts weakly 3-sided, narrowly obovoid to oblong, 0.8–1.2 mm long, 0.2–0.5 mm wide, brown, with small style base remnant.

Annual or sometimes perennial. In marshes, seasonally flooded places, ditches and rice fields. Common and often abundant. Superficially it looks like *Kyllinga* but differs in the spikelet structure.

{(⊙) Ten/Hel}. Old World tropics and subtropics: Bhu, Cey, Ind, Nep: As, Bh, Hp, Kl, Kt, Mg, Mh, Mn, Mp, Mz, Sk, Tn, Up, Wb.

Lipocarpha kernii (Raymond) Goetghebeur [*Rikliella kernii* (Raymond) J. Raynal, *Isolepis kernii* (Raymond) K. Lye]: Fig. 163

Like *L. squarrosa* but involucral bracts obliquely patent to reflexed; most parts generally larger: culms

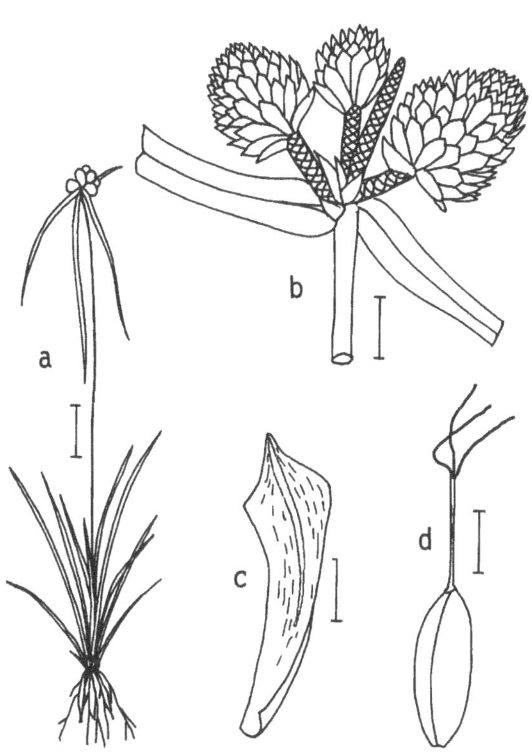

Fig. 162. *Lipocarpha chinensis*: a, whole plant (2 cm); b, inflorescence (2 mm); c, spikelet bract (0.5 mm); d, nut (0.5 mm).

to 40 cm tall; spikes sessile, up to 8; midrib of the spikelet bract elongated into a 0.7–1.5 mm long awn, longer than the expanded part; nuts 0.3–0.4 mm wide.

Annual. On sand at the edges of pools and tanks, flowering as the water recedes.

{⊙ Ten?/Hel?}. Tropical Africa and India. Ind: Kt, Mh? Mp.

Lipocarpha raynaliana E. Govindarajalu ['*raynaleana*' orth. mut.]: Fig. 164

Like *L. sphacelata* but with recurved or obliquely erect culms with curved leaves longer than the culms; spikelet bract 2–2.5 mm long, at apex rounded, without an awn; spikelet prophyll and glume 1.8–2.1 mm long; nuts oblong, 1.3–1.6 mm long, 0.3–0.4 mm wide.

Annual. On the edges of rice fields.

{⊙ Ten?/Hel?}. Endemic to S India: ♦Ind: Kl, Tn.

Lipocarpha reddyi Hooper: Fig. 164

Like *L. sphacelata* but spikelet bract abruptly narrowed with a 0.6–0.8 mm long awn; spikelet prophyll and glume 0.7–0.9 mm long; style less than 0.1 mm long; nuts 0.7–0.8 mm long and ± 0.3 mm wide.

Annual. Ecology unknown but probably in wet places. {⊙ Ten?/Hel?}. Endemic to S India: ♦Ind: Ap, Tn.

Lipocarpha sphacelata (Vahl) Kunth [*L. triceps* (Roxburgh) Nees]: Fig. 164

Culms tufted, erect, 5–40 cm tall, 0.5–1 mm thick. Leaf blades flat, straight, one-quarter to three-quarters as long as the culm, up to 25 cm long but usually much less, (0.5–) 1–2.2 mm wide. Involucral bracts 2–3, the largest up to 15 cm long. Inflorescence terminal, with (1–) 2–3 (–5), spikes. Spikes ovoid to conical, 2–8 (–15) long, 1.5–4 mm in diameter, usually dark purple-brown. Spikelet bract obovate, 1.3–1.7 mm long, 0.6–0.9 mm wide, shortly acuminate, apical part 0.2–0.4 mm long, pale to yellowish-brown, reddish-brown above. Spikelet prophyll and glume 1.1–1.6 mm long, 0.8–1 mm wide. Stamens 1 or 2. Style 0.1–0.3 mm long, 3-cleft; stigmas 3, ± 0.25 mm long. Nuts 3-sided, obovoid to subellipsoid, 1–1.4 mm long, 0.3–0.5 mm wide, dark brown to black, conspicuously beaked at the top; beak ± 0.3 mm long.

Annual. Swamps, marshes, rice fields and wet grassy places. The dark-coloured heads are a useful diagnostic character in the field.

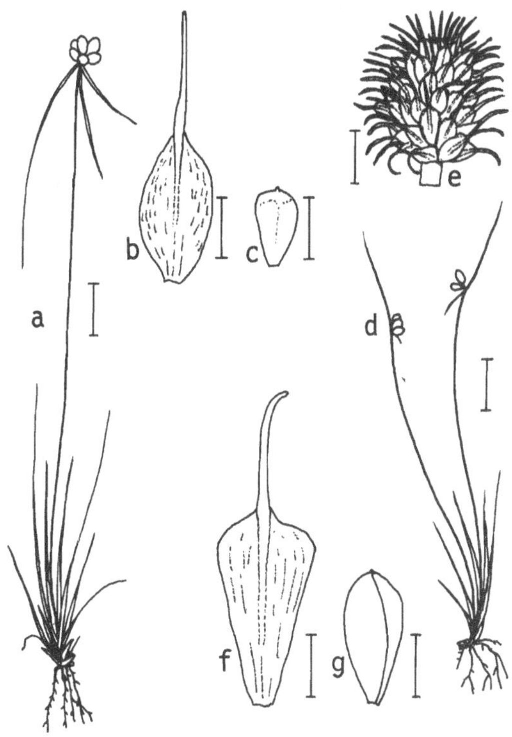

Fig. 163. *Lipocarpha kernii*: a, whole plant (1 cm); b, spikelet bract (0.5 mm); c, nut (0.5 mm); *L. squarrosa*: d, whole plant (1 cm); e, spike (1 mm); f, spikelet bract (0.3 mm); g, nut (0.3 mm).

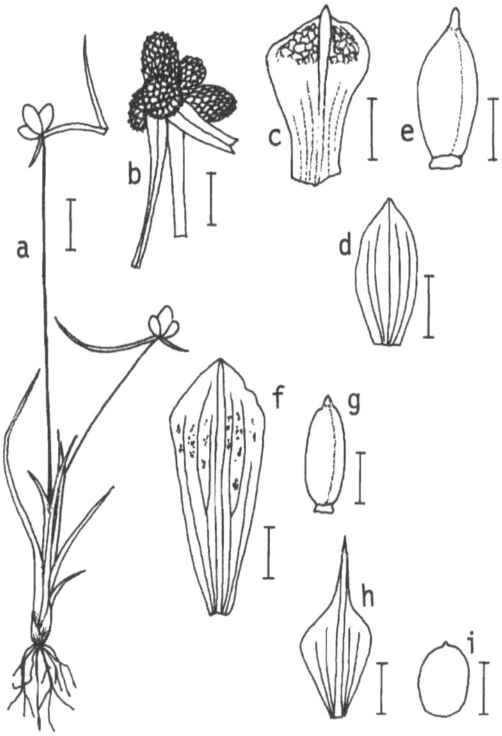

Fig. 164. *Lipocarpha raynaliana*: f, spikelet bract (0.5 mm); g, nut (0.5 mm). *L. reddyi*: h, spikelet bract (0.5 mm); i, nut (0.5 mm). *L. sphacelata*: a, whole plant (1 cm); b, inflorescence (5 mm); c, spikelet bract (0.5 mm); d, glume (0.5 mm); e, nut (0.5 mm).

{⊙ Ten/Hel}. Thailand to S Asia, tropical Africa and America: Bur, Cey, Ind, Nep: As, Bh, Kl, Kt, Or, Tn, Wb. 0–1500 m.

Lipocarpha squarrosa (Linnaeus) Goetghebeur [*Isolepis squarrosa* (Linnaeus) Roemer et Schultes, *Rikliella squarrosa* (Linnaeus) J. Raynal, *Ascolepis tenuior* Steudel]: Fig. 163

Culms tufted, 2–10 (–20) cm tall, 0.3–0.5 mm in diameter. Leaf blades 1–6 (–10) cm long, 0.6–1 mm wide, often withered at maturity. Involucral bracts (1–) 2–3 (–5), the largest erect, overtopping the inflorescence, up to 5 cm long. Inflorescence pseudolateral, with 1–3 spikes. Spikes sessile and stalked, ovoid- to oblong-ellipsoid, appearing spiny, 2–10 mm long, 1.5–4 mm in diameter. Spikelet bract obovate, pale to red-brown below, yellowish at top, 0.6–0.9 mm long (excluding awn), 0.4–0.5 mm wide; midrib elongated into a 0.4–0.9 mm long, green and usually recurved awn. Spikelet prophyll and glume absent. Stamens 1 or 2. Style up to 0.1 mm long, 3-cleft; stigmas 3, ± 0.1 mm long. Nuts obovate, obscurely 3-sided, somewhat flattened, 0.4–0.6 mm long, 0.2–0.3 mm wide, minutely apiculate.

Annual. Usually in or near water, fruiting as the water recedes, on drying mud or sand. Often forming a closed turf. Also found along coastal lagoons.

{⊙ Ten/Hel}. India and Sri Lanka to China, naturalized in N America: Bur, Cey, Ind, Nep: As, Kl, Kt, Mp, Or, Pj, Rj, Up, Wb. 0–1800 m.

Machaerina [including *Baumea*]: 48 species: widely distributed in the tropics: one hydrophytic species in India.

Machaerina rubiginosa (Sprengel) T. Koyama [*Fuirena rubiginosa* Sprengel, *Cladium glomeratum* R. Brown, *Baumia crassa* Thwaites, *Cladium riparium* Bentham, *Machaerina crassa* (Thwaites) T. Koyama]: Fig. 165

Rhizomes short, producing stolons; stolons up to 40 cm long, 2–4 mm in diameter, densely clothed in persistent scales; scales clearly in 2 ranks, pale, 5–25 mm long. Culms solitary to tufted, subterete or flattened, (0.3–) 1–2 m tall, 3–7 mm in diameter, smooth, not leafy. Leaves equitant, in 2 ranks; cauline sheaths scale-like below, up to 20 cm long above, usually bladed above; blades unifacial, shorter than to slightly over-

Fig. 165. *Machaerina rubiginosa*: a, base of plant (4 cm); b, top of inflorescence (1 cm); c, spikelet (1 mm); d, nut (1 mm).

topping the inflorescence, biconvex with obtuse edges to terete, (2–) 4–7 mm wide, smooth, spongy inside, sharply pointed. Involucral bracts sheathed below bearing short blades up to 6 cm long. Inflorescence a compound panicle with short branches, usually interrupted below. Spikelets in almost globose clusters of 3–6; each spikelet 4–7 mm long, 2–2.5 mm in diameter, rusty brown, 1- or 2-fruited. Glumes 5, spiral, imbricate, the lower ones empty, the flowering ones boat-shaped, rusty brown, 4–6.5 mm long, 1.2–2.2 mm wide. Perianth absent. Stamens 3; anthers 1.5–2 mm long; pollen polyphorate (unique in the Cyperaceae). Style (2.5–) 3–7 mm long, 3-cleft; stigmas 3, about as long as the style; style base swollen forming a whitish, papillose cap on the top of the nut. Nuts oblong-ellipsoid to ellipsoid, obscurely 3-sided, 3–5 mm long (including cap), 1.5–2 mm in diameter, orange to reddish-brown, shiny except for the whitish cap.

Perennial. In swamps and shallow water, often building large stands in shallow water at the edge of lakes. A very polymorphic species. The plants from Sri Lanka are rather large and robust and are often recognized as *Machaerina rubiginosa* subsp. *crassa* (Thwaites) T. Koyama.

{Ten/Hyp}. Warm to tropical regions of S, SE and E Asia and Australasia: Cey, Ind: As, Mg.

Mapania [*Pandanophyllum*]: ± 45 species: pantropics: 2 rather similar wetland species in Sri Lanka and India.

Robust perennials. Leaves radical or nearly so. Involucral bracts much longer than the culms. Inflorescence a terminal head or reduced to a single spikelet. Each glume subtending an abbreviated cyme; the cymes consisting of a single, terminal (central), naked ovary surrounded by 4–6 glume-like scales (squamellae); squamellae mostly bearing a single stamen; lowest pair of squamellae opposite and lateral, folded with an acute keel, dissimilar to the remainder.

Mostly found as undergrowth in humid forests. It is mostly poorly collected and there are numerous taxonomic problems to be solved.

Excluded: the following have been recorded from India or Myanmar, but are not wetland plants: ***Mapania cuspidata*** (Miquel) Uittien [*M. humilus* F. Vill] An; ***M. cuspidata*** var. ***angustifolia*** (Uittien) Uittien: An; ***M. cuspidata*** var. ***petiolata*** (C. B. Clarke) Uittien: An; ***M. palustris*** (Hasskarl ex Steudel) F. Vill [*M. andamanica* C. B. Clarke, *M. silhetensis* C. B. Clarke].

1. Culms 3–5 cm long, clothed in scales for nearly the entire length; bracts at base of spike with linear blades, up to twice as long as the spike; glumes and other floral scales 7–8 mm long ***M. immersa***

1. Culms 10–30 cm long, naked (scales only at base); bracts at base of spike scale-like, much shorter than the spike; glumes and other floral scales ± 10 mm long ***M. zeylanica***

Mapania immersa (Thwaites) Bentham ex C. B. Clarke [*Pandanophyllum immersum* Thwaites]: Fig. 166

Like *M. zeylanica* but culms 3–5 cm long, clothed in scales for nearly the entire length; bracts at base of spike with linear blades, up to twice as long as the spike; glumes and other floral scales 7–8 mm long.

Perennial. In running water in lowland forest, very rare. It is perhaps identical to *M. enodis* (Miquel) C. B. Clarke from SE Asia.

{Hyp}. Endemic to Sri Lanka: ♦Cey.

Mapania zeylanica (Thwaites) Bentham ex C. B. Clarke [*Lepironia zeylanica* (Thwaites) Miquel]: Fig. 166

Rhizomes short, knotty, woody, erect or ascending, 3–8 cm long, 7–10 mm in diameter, dark brown. Culms lateral, 3–6 to a shoot, obscurely 3-angled, flattened, 6–30 cm tall, 2–3 mm in diameter, clothed only at the base with 5–9 scales. Leaves in basal rosettes; sheaths 2–6 cm long, leathery; blades linear, much longer than

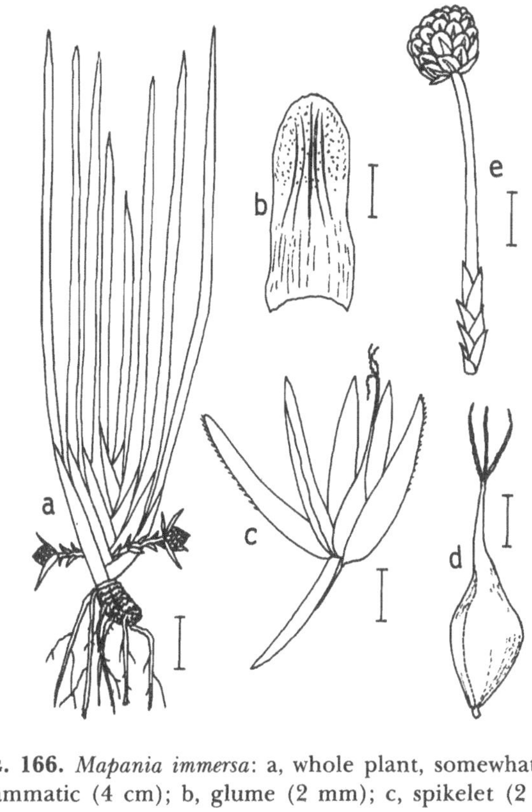

FIG. 166. *Mapania immersa*: a, whole plant, somewhat diagrammatic (4 cm); b, glume (2 mm); c, spikelet (2 cm); d, fruit (2 mm). *M. zeylanica*: e, culm with inflorescence (2 cm).

the culms, 60–110 cm long, 7–20 mm wide. Involucral bracts scale-like, ovate to ovate-oblong, 1–1.5 cm long. Inflorescence a single terminal, head-like spike. Spike globose to globose-ellipsoidal, 1.5–2.5 cm in diameter, pale green tinged with brown. Spikelets shorter than the glumes, each with 1-staminate male flowers below and a terminal female flower. Glumes and glume-like scales 6. Perianth absent. Anthers 4–4.5 mm long. Style 11–12 mm long, 3-cleft; stigmas 3, 2–2.5 mm long; style base slightly thickened, ± persistent on the nut. Nuts obovoid, terete, 4.5–6 mm long (excluding beak), 2.8–3.2 mm in diameter, brown to yellowish-brown, smooth, with slender beak.

Perennial. Often in or near water or on moist ground in low-altitude forests.

{Hyp/Ten/Hel}. Endemic to Sri Lanka and S India: ♦Cey, ♦Ind (I have not been able to confirm the record for India).

Mariscus [*Courtoisia*, '*Courtoisinia*' orth. mut., *Indocourtoisia*, excluded from *Cyperus*]: ± 200 species: pantropics and subtropics, most in S America.

Annual or perennial, tufted or rhizomatous. Culms usually 3-sided and erect. Leaves mostly radical. Involucral bracts usually long and leaf-like. Inflorescence terminal, simple or compound panicles, with branches bearing clusters of spikelets or sometimes contracted and head-like; bracts at base of rays (prophylls) keeled, tubular at base. Spikes cylindrical, ovate or globose. Spikelets flattened or subterete, 1- or more-flowered; rachilla with a joint above the prophyll and falling before the glumes (the whole spikelet falling in entirety). Glumes distichous, clearly in 2 opposite rows, often imbricating, margins usually hyaline; keel distinct. Flowers bisexual, all fertile except sometimes the very lowest ones sterile. Perianth bristles absent. Stamens usually 3; connectives often crested. Style 3-cleft, not swollen at the base; stigmas 3; nuts 3-sided.

Excluded: the following may sometimes grow on wet soil but they are not considered to be wetland plants: ***Mariscus cyperinus*** (Retzius) Vahl subsp. ***cyperinus*** [*Kyllinga cyperina* Retzius, *Cyperus cyperinus* (Retzius) Valckenier-Suringar]; ***M. cyperinus*** (Retzius) Vahl subsp. ***laxtus*** (C. B. Clarke) T. Koyama [*M. tenuifolius* Schrader ex Nees]; ***M. javanicus*** (Houttuyn) Merrill et Metcalf [*M. albescens* Gaudichaud-Beaupré, *Cyperus pennatus* Lamarck]; ***M. konkanensis*** (T. Cooke) Sedgwick [*M. sieberianus* Nees var. *subcompositus* C. B. Clarke]; ***M. pedunculatus*** (R. Brown) T. Koyama; *Cyperus steadii* Kükenthal].

1. Spikes globose or nearly so
  2. Robust perennial; spikelets cylindrical, linear, 3- to 14-flowered; glumes obtuse, not mucronate ***M. compactus***
  2. Slender annual; spikelets ovoid, 1- or 2-flowered; glumes acute, with a well developed mucro ***M. cyperoides***
1. Spikes not globose, distinctly longer than wide or sometimes if sessile wider than long
  3. Culms with swollen, bulb-like bases with chestnut brown sheaths; inflorescence simple
    4. Inflorescence a digitate cluster of cylindrical spikes; stolons present; spikelets with 1 nut ***M. clarkei***
    4. Inflorescence a single hemispherical or lobed head; stolons absent; spikelets with 2–6 nuts ***M. dubius***
  3. Culms not swollen and bulb-like (but sometimes bearing woody corms) at the base; inflorescence compound or sometimes reduced to a single spikelet
    5. Spikelet with 1 or rarely 2 fruits; glumes without awns; stoloniferous or rhizomatous perennials
      6. Spikes (10–) 20–40 mm long, densely bearing very many (25–150) spikelets, greenish-yellow often with a brownish tinge; spikelets whitish-green, spreading at maturity, mostly 2–3 mm long; plants rhizomatous and developing stolons; inflorescence usually compact ***M. paniceus***
      6. Spikes 8–15 mm long, densely bearing up to 40 spikelets, light glaucous green; spikelets green to straw-coloured, some patent to ascending at maturity, mostly 4–5 mm long; plants rhizomatous not developing stolons; inflorescence usually umbel-like ***M. sumatrensis***
    5. Spikelet with 6–25 fruits; glumes with recurved awns; slender annuals
      7. Inflorescence a simple panicle of 1 or 2 sessile spikes and 1–7 stalked spikes or sometimes reduced to a single hemispherical spike ***M. squarrosus***
      7. Inflorescence of 1 central group of spikelets and usually 1–4 stalked, digitate spikelet clusters, each with 3–15 spikelets (each spikelet with a short axis – not true spikes)
        8. Glumes 1.4–2 mm long including awn, keel green with purple stripes, the sides rusty to reddish-brown ***Cyperus cuspidatus***
        8. Glumes ± 2.5 mm long including awn, green with red streaks ***M. maderaspatanus***

Mariscus clarkei (T. Cooke) T. Koyama [*Cyperus clarkei* T. Cooke, *M. bulbosus* C. B. Clarke, *M. pictus* Nees]: Fig. 167

Rhizomes short, with stolons; stolons up to 10 cm long and 1 mm in diameter. Culms solitary, (6–) 10–30 cm tall, 1–2.5 mm thick, 3-sided, smooth throughout, the base thickened and bulb-like. Leaves 5–7 to a

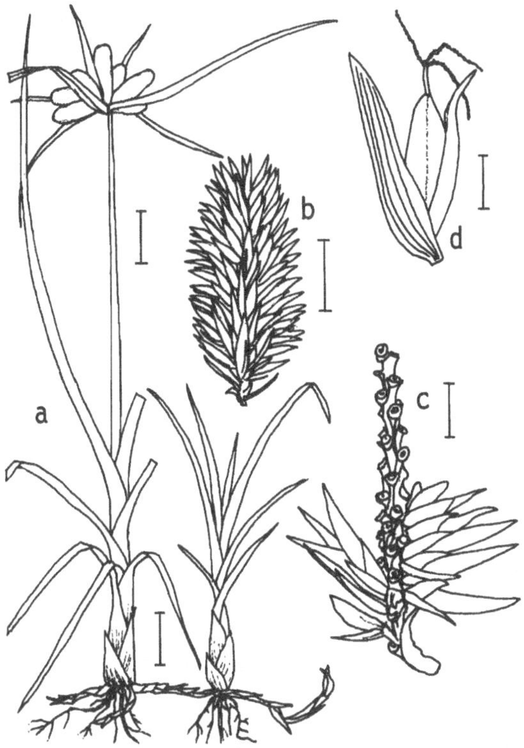

Fig. 167. *Mariscus clarkei*: a, whole plant (1 cm); b, spike at anthesis (4 mm); c, spike at maturity (1 mm); d, spikelet with nut (1.5 mm).

culm; sheaths cylindrical, dusky brown, wider than the blade; blades 7–20 cm long, 2–4 mm wide, flattish-plicate, channelled, glaucous green. Involucral bracts 5–9, spreading or reflexed, the lower ones much overtopping the inflorescence, the longest 10–20 cm long. Inflorescence simple, often congested in a digitate cluster; cluster 1.5–5 cm long, 1.5–3 mm in diameter. Spikes cylindrical, often subsessile, 8–15 mm long, 5–7 mm in diameter, glaucous green, the rachis hidden by densely disposed spikelets. Spikelets spreading, lanceolate-elliptical, 3-sided, 3–3.3 mm long, 0.7–0.8 mm wide, bearing 2 glumes and 1 flower. Lower glume ovate or narrowly ovate, ± 3 mm long, 1.3–1.5 mm wide, subacute to acute, conspicuously 11- or 13-nerved, the midrib green and slightly keeled; upper glume lanceolate, acute, ± 2.5 mm long, ± 1.2 mm wide. Stamens 2. Style 1.7–2 mm long, 3-cleft to halfway down. Nuts oblong-ellipsoid to oblong, 3-sided, 1.5–1.8 mm long, ± 0.8 mm wide, maturing dark brown.

Perennial. Seasonally flooded places and on wet sand in river beds. It has bulb-shaped stem bases like *M. dubius* but a digitate inflorescence.

{Ten/Hel}. Endemic in Sri Lanka and India: ♦ Cey, ♦Ind: Kt, Mh, Rj, Tn. 0–1500 m.

Mariscus compactus (Retzius) Boldingh [*Cyperus compactus* Retzius, *C. spinulosus* Roxburgh, *C. haenkianus* Kunth, *C. dilutus* Vahl, *M. microcephalus* Presl]: Fig. 168

Rhizomes short, corm-like. Culms solitary or a few together, (10–) 30–100 (–180) cm tall, 3–6 mm thick, obtusely 3-sided or terete, rather soft. Leaf sheaths cylindrical, purple brown; blades slightly longer or shorter than the culm, 5–9 (–12) mm wide, channelled, glaucous green. Involucral bracts 3–5 (–8), upright or spreading, the lower ones much overtopping the inflorescence, the longest 60–100 cm long. Inflorescence of compound panicles. Spikes nearly globose, 8–20 (–40) mm in diameter, bearing many (up to 80) stellately spreading spikelets, the rachis not more than 4 mm long. Spikelets jointed at base, narrowly lanceolate to linear-subulate, subterete, 5–15 mm long, 0.8–1.5 mm wide, 3- to 8- (-14)-flowered; rachilla jointed above the prophyll. Glumes linear-lanceolate to linear-oblong, obtuse or subobtuse, 3–4.5 mm long, 1–1.5 mm wide, pale brown and often tinged with reddish-brown, 5- to 7-nerved; keel greenish. Style 0.5–1 mm long; stigmas 1.2–1.5 mm long. Nuts linear-oblong, sometimes slightly curved, 3-sided, 1.5–2 mm long, 0.4–0.5 mm wide, minutely puncticulate.

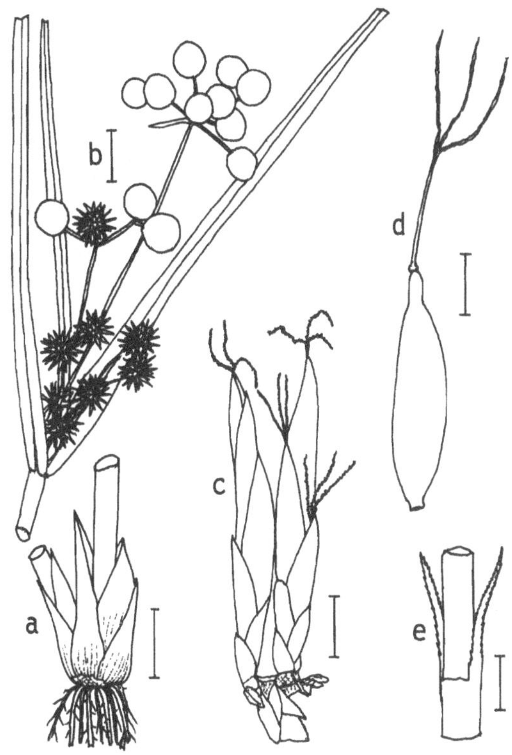

Fig. 168. *Mariscus compactus*: a, base of shoot (2 cm); b, part of inflorescence (1 cm); c, two spikelets (1 mm); d, nut (0.5 mm); e, base of ray (1 mm).

Perennial. In shallow water, in seasonally flooded places, swamps, wet rice fields, marshes, ditches, river banks, often gregarious and forming stands. Distinguished by the large, reddish-brown inflorescences with globose spikes.

{Ten/Hel}. Tropical Asia, introduced in Reunion and Mauritius: Ban, Bur, Cey, Ind, Nep, Pak: An, Ap, Ar, As, Bh, Go, Kl, Kt, Lk, Mg, Mn, Mp, Mz, Or, Pj, Rj, Sk, Tn, Wb. 100–1000 m.

Mariscus cyperoides (Roxburgh) A. Dietrich [(*Cyperus cyperoides* (Linnaeus) O. Kuntze is a synonym of *Mariscus sumatrensis* (Retzius) J. Raynal); *Kyllinga cyperoides* Roxburgh, *Courtoisia cyperoides* (Roxburgh) Nees, *Indocourtoisia cyperoides* (Roxburgh) Bennet et Raizada, *Cyperus pseudokyllingioides* Kükenthal]: Fig. 169

Rhizome absent; stolons absent. Culms tufted, 12–80 cm tall, 1–3 mm thick, 3-sided, smooth throughout. Leaves 1 or 2 to a culm; sheaths loose, thickened below, brownish; blades half as long to almost equalling the culm, 5–40 cm long, 2–6 mm wide, flat, scabrid on midrib and margins. Involucral bracts 3–7, upright to spreading, all or most overtopping the inflorescence, the longest 7–40 cm long. Inflorescence a compound panicle, with many, sessile and stalked, almost spherical spikes; rays 3–8, 1–10 cm long. Spikes globose, 4–12 mm in diameter, densely bearing 4–12 spikelets, light glaucous green. Spikelets stellately divergent, 1- or 2-flowered, ovate, (3–) 3.8–4.2 mm long, 1.8–2.2 mm wide, falling off entire when mature. Glumes elliptical, 4 mm long, yellowish-grey with a green and strongly winged midrib: midrib excurrent in a 0.5 mm long straight or slightly curved mucro. Nuts linear-lanceolate to linear-oblong, 3-sided, 2.7–3 mm long, 0.4–0.5 mm wide, maturing reddish-brown.

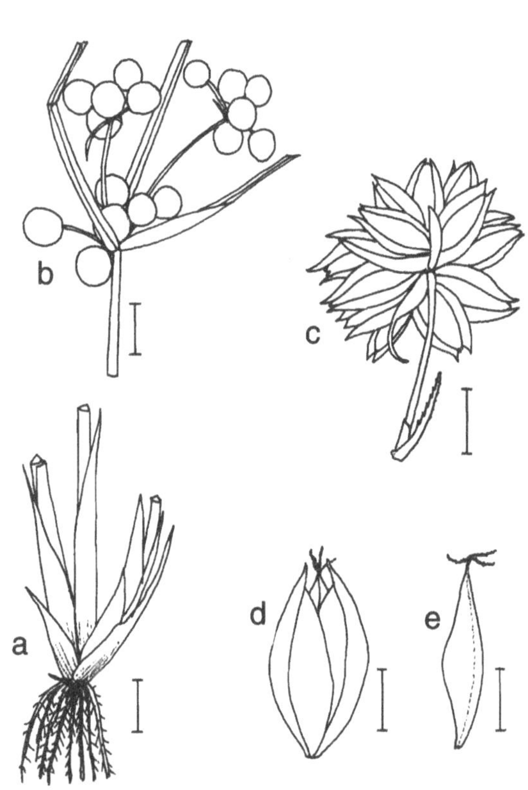

Fig. 169. *Mariscus cyperoides*: a, base of shoot (1 cm); b, small inflorescence (1 cm); c, spike (2 mm); d, spikelet (1 mm); e, nut (1 mm).

Annual. Seasonally wet places, swamps, temporary pools and rice fields; not found only in wetlands but also in open or shady grassland, a common weed of fields and gardens. Clearly distinguishable from other *Cyperus*-like plants by the combination of spherical spikes and winged glumes. Some authors place it in a separate genus *Indocourtoisia* [*Courtoisia* pro parte].

{⊙ Ten/Hel}. Tropical Old World, naturalized in the West Indies: Ban, Bhu, Bur, Ind: An, Ar, As, Bh, Go, Kl, Kt, Mg, Mh, Mp, Ng, Or, Rj, Sk, Tn, Up.

Mariscus dubius (Rottbøll) Kükenthal ex C. E. C. Fischer [*Cyperus dubius* Rottbøll, *C. kyllingaeoides* Vahl, *M. dregeanus* Kunth, *M. blatteri* C. McCann]: Fig. 170

Rhizomes short, without stolons. Culms tufted, (5–) 15–40 (–70) cm tall, 0.5–3 mm thick, 3-sided, smooth throughout, the base thickened and bulb-like. Leaves several to a culm; sheaths somewhat inflated towards the base; blades much shorter than to almost equalling the culm, 5–30 (–80) cm long, 1–5 (–6) mm wide, flat or with incurved margins, soft, scabrous along the upper midrib and margins. Involucral bracts 3–6 (–9), erect to reflexed, the lower ones much overtopping the inflorescence, the longest 8–35 cm long. Inflorescence a dense, hemispherical or ± lobed head; head ovoid-deltoid, 5–20 mm in diameter, greenish-white to dirty white, bearing 1–6 confluent spikes. Spikes broadly ovoid, densely bearing 6–24 spikelets, light greenish, the rachis subulate, up to 3 mm long. Spikelets spreading, ovate to lanceolate-oblong, terete to slightly flattened, (2–) 4–8 mm long, 1–2.3 mm wide, bearing 4–9 glumes and 3–6 flowers. Glumes broadly ovate to lanceolate-elliptical, (2–) 2.8–3.8 mm long, (1.5–) 2–3 mm wide, subacute to acute, conspicuously 11- to 19-nerved, pale brown, the midrib green and slightly keeled. Stamens 2 or 3. Style ± 1 mm long; stigmas 3, shorter than the style. Nuts ellipsoid to obovoid-ellipsoid, 3-sided, (1.2–) 1.6–2 mm long, 0.6–0.9 mm wide, maturing light brown with brown angles.

Perennial. In seasonally flooded areas and in pockets of soil in rocks. Not confined to wetlands and

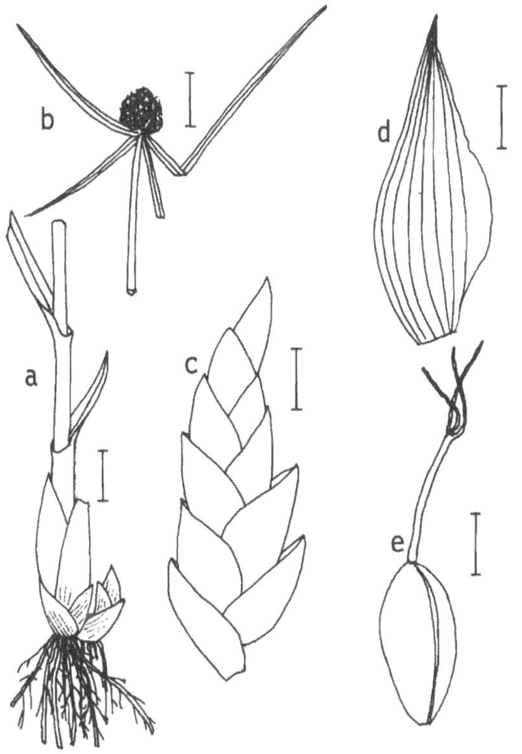

Fig. 170. *Mariscus dubius*: a, base of shoot (1 cm); b, inflorescence (1 cm); c, spikelet (1 mm); d, glume (0.5 mm); e, nut (1 mm).

sometimes found as a weed in fields and near the sea on sandy beaches. Distinguished by its greenish white congested head. It is very variable; rather large plants from Mahableshwar, Mh (culms up to 70 cm long, inflorescence up to 24 mm in diameter; involucral bracts up to 9, ± 6 mm wide) have been named *Mariscus blatteri*, I doubt that it is specifically distinct.

{Ten/Hel}. Old World tropics: Ban, Cey, Ind: Kl, Kt, Lk, Tn. 0–2000 m.

Mariscus maderaspatanus (Willdenow) Napper [*Cyperus maderaspatanus* Willdenow]: Fig. 171

Like *M. squarrosus* but inflorescence of 1 central group of spikelets and usually 1–4 stalked, digitate spikelet clusters, each with 3–7 spikelets (each spikelet with a short axis – not true spikes); glumes ± 2.5 mm long, green with red streaks, 3-nerved; stamens 1 or 2.

Annual. In seasonally wet habitats and shallow soil on rocks; also found in damp places in forests. In many older Floras it is considered to be a synonym of *M. squarrosus*.

{⊙ Ten/Hel}. Tropical E Africa to SE Asia: Ind: Kt.

Mariscus paniceus (Rottbøll) Vahl [including var. *roxburgianus* (Kükenthal) C. B. Clarke, *Kyllinga panicea* Rottbøll]: Fig. 172

Rhizomes short, 1–3 cm long, 2–3 mm thick, knotty, woody, developing stolons; stolons up to 10 cm long, 0.3–0.5 mm in diameter. Culms usually solitary, 8–45 (–75) cm tall, 1–2 mm thick, 3-sided to 3-angled above, smooth throughout. Leaves 3–7 to a culm; sheaths cylindrical, reddish to purplish-brown; blades half as long to almost equalling the culm, 1–3 mm wide, flattish-plicate. Involucral bracts 3–7, upright to spreading, all or most overtopping the inflorescence, the longest 10–30 cm long. Inflorescence simple, somewhat open or digitate contracted. Spikes sessile or shortly stalked, cylindrical to oblong-cylindrical, 8–15 mm long, 4–6 mm in diameter, densely bearing up to 40 spikelets, light glaucous green. Spikelets horizontally divergent, ovate to lanceolate-oblong, slightly recurved, 2.5–3.5 mm long, 0.8–1 mm wide, bearing 4 glumes and 1 flower. Lower 2 glumes sterile, ovate, ± 1 mm long, pale whitish, acute, dorsally 3- or 5-nerved; third glume fertile, ovate to lanceolate, keeled, acute, 2.5–3.5 mm long, 1.4–1.7 mm wide, the sides membranous, pale whitish, obscurely 2-nerved on each side, the keel greenish-grey and 3-nerved; fourth

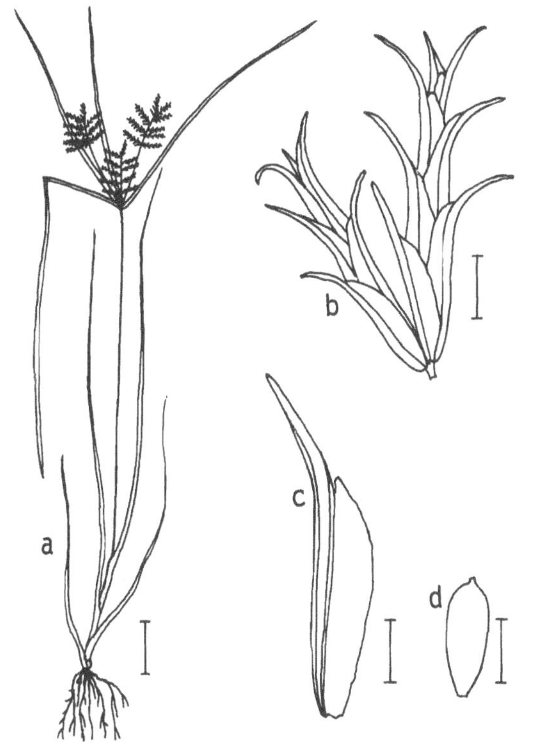

Fig. 171. *Mariscus maderaspatanus*: a, whole plant (1 cm); b, two spikelets (1 mm); c, glume (0.5 mm); d, nut (1 mm).

wide, flat, soft, gradually tapering to an acute apex. Involucral bracts 2–4, upright, the lower ones much overtopping the inflorescence, the longest 3–8 cm long. Inflorescence a simple panicle of 1 or 2 sessile spikes and 1–7 stalked spikes or sometimes reduced to a single hemispherical spike, at first green turning yellowish or reddish-brown at maturity. Spikes broadly ellipsoid to subglobose, 5–20 mm long, 5–12 mm in diameter, echinate, with 20–60 spikelets, light yellow green, the rachis 1–5 mm long. Spikelets densely spicate and appearing subradiate, linear, flattened, (3–) 5–13 mm long, (2–) 3–4 (–5) mm wide (including awns), 6- to 15-flowered; rachilla wingless. Glumes elliptical, erect and recurved towards the tip, 1–2 mm long, 0.5–0.9 mm wide; sides yellow to reddish-brown, with (2) 3 or 4 nerves on each side of the midrib; keel green, strongly excurrent, forming a 0.4–1.3 mm long, recurved awn. Stamens 1. Style nearly as long as the nut; stigmas 3, shorter than the style. Nuts narrowly oblong to almost linear, 3-sided, 0.6–1 mm long, 0.2–0.4 mm wide, dark greyish-brown.

Annual. Often found in seasonally flooded areas, usually flowering as the water recedes. It is also found

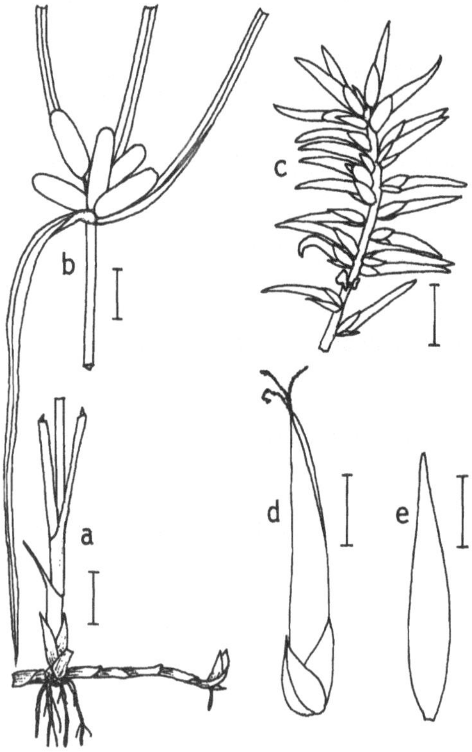

FIG. 172. *Mariscus paniceus*: a, base of shoot (1 cm); b, inflorescence (1 cm); c, spike (2 mm); d, spikelet (1 mm); e, nut (1 mm).

glume sterile, reduced. Stamens 2 or 3. Style very short; stigmas 3, longer than the style. Nuts ellipsoid-oblong, 3-sided, 1.8–2.2 mm long, 0.6–0.8 mm wide, maturing yellowish-brown.

Perennial. Seasonally flooded places, marshes and roadsides. A common plant not confined to wetlands. Differs from *M. cyperoides* by having stolons; culms shorter and more slender (3–30 cm long, ± 0.5 mm in diameter); leaf blades narrower (1–2 mm wide); spikes sessile or subsessile, shortly cylindrical; spikelets finally spreading horizontally; third glume only fertile.

{(Ten)/Hel}. India and Sri Lanka to Cochinchina: Ban, Bhu, Bur, Cey, Ind, Nep, Pak: Ar, As, Gj, Hp, Kl, Kt, Mh, Rj, Sk, Tn, Wb. 0–2000 m.

Mariscus squarrosus (Linnaeus) C. B. Clarke [*Cyperus squarrosus* Linnaeus, *C. aristatus* Rottbøll]: Fig. 173

Rhizome absent, root system minute. Culms solitary or tufted, (1–) 3–10 (–20) cm tall, 0.5–2 (–3) mm thick, 3-sided with almost wing-like angles, smooth. Leaves 3–5 to a culm; sheaths membranous, greenish stained with purple; blades shorter than or nearly equalling the culm, 2–10 cm long, 1–3 (–4) mm

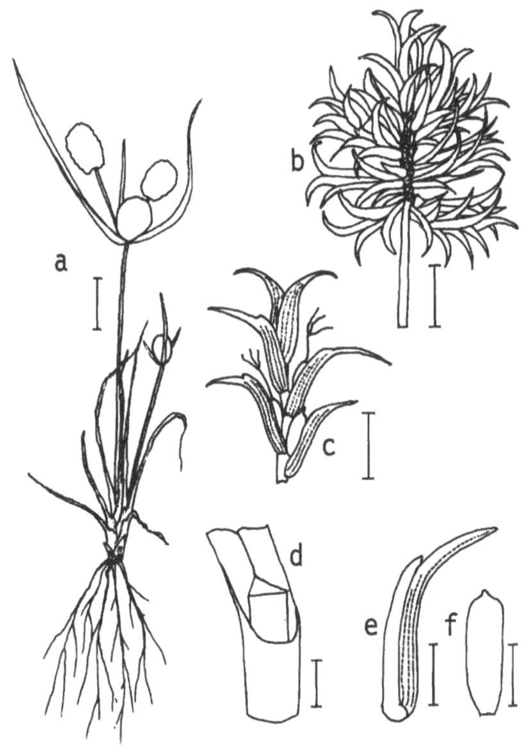

FIG. 173. *Mariscus squarrosus*: a, whole plant (1 cm); b, spike with some spikelets removed (2 mm); c, spikelet (1 mm); d, culm emerging from sheath (2 mm); e, glume (0.5 mm); f, nut (1 mm).

at the edges of tanks, pools, streams, ditches and in rice fields. It is not confined to wetlands and may be found in relatively dry places. It is easily recognized by its short crowded spikes with flattened spikelets bearing recurved awns. The rachillas are not always caducous and it is therefore intermediate between *Cyperus* and *Mariscus*; I place it in the latter to conform with most Indian Floras.

{⊙ Ten/Hel}. Pantropics to warm temperate: Bhu, Bur, Cey, Ind, Nep: An, Gj, Go, Hp, Kl, Kt, Mh, Pj, Rj, Sk, Tn, Up. ?–1800 m.

Mariscus sumatrensis (Retzius) J. Raynal [*Kyllinga sumatrensis* Retzius, *Cyperus cyperoides* (Linnaeus) O. Kuntze, *Scirpus cyperoides* Linnaeus, *Kyllinga umbellata* Rottbøll, *M. sieberianus* Nees ex C. B. Clarke]: Fig. 174

Rhizomes short and knotty, woody, clothed with brown fibres, not developing stolons. Culms solitary or tufted, (10–) 30–80 cm tall, 1–5 mm thick, 3-angled, smooth throughout. Leaves 3–7 to a culm; sheaths cylindrical, pale becoming reddish or reddish-brown; blades half as long to almost equalling the culm, up to 25 cm long, 3–6 (–9) mm wide, flat or slightly plicate. Involucral bracts 3–15, leaf-like, upright to spreading, all or most overtopping the inflorescence, the longest up to 30 cm long. Inflorescence simple or rarely sub-compound, open and umbel-like; rays (3–) 6–18, up to 12 cm long. Spikes cylindrical, not attenuate at the base, (1–) 2–4 cm long, 6–12 mm in diameter, densely bearing numerous (25–150) spikelets, greenish-yellow often with a brownish tinge. Spikelets spreading at right angles or the lower ones ± reflexed, some patent to ascending at maturity, linear to linear-lanceolate, subterete, (2.5–) 3–5 (–6) mm long, 0.6–1 mm wide, bearing 4 or 5 glumes and (1) 2 (3) flowers, rarely more than one developing a nut. Lower 2 glumes sterile, ovate, ± 1 mm long, pale whitish, acute, dorsally 3- or 5-nerved; 3rd or/and 4th glumes fertile, oblong to ovate-oblong, keeled, acute, 2–3.5 mm long, (1–) 1.4–1.7 mm wide, the sides membranous, pale green, obscurely several-nerved on each side, the keel 3-nerved; 4th or 5th glumes sterile, reduced. Style longer than the nut, 3-cleft. Nuts linear-oblong, 3-sided, 1.5–2.3 mm long, ± 0.5 mm wide, maturing straw-coloured.

Perennial. Seasonally flooded places, marshes and roadsides. A common plant not confined to wetlands. Easily recognized by its umbel-like heads with cylindrical spikes crowded with numerous small spikelets. It is very variable and numerous infraspecific variants have been described.

{(Ten)/Hel}. Tropical Old World, introduced in the West Indies: Ban, Bur, Cey, Ind, Nep: As, Hp, Kt, Mh, Sk. 0–1500 m.

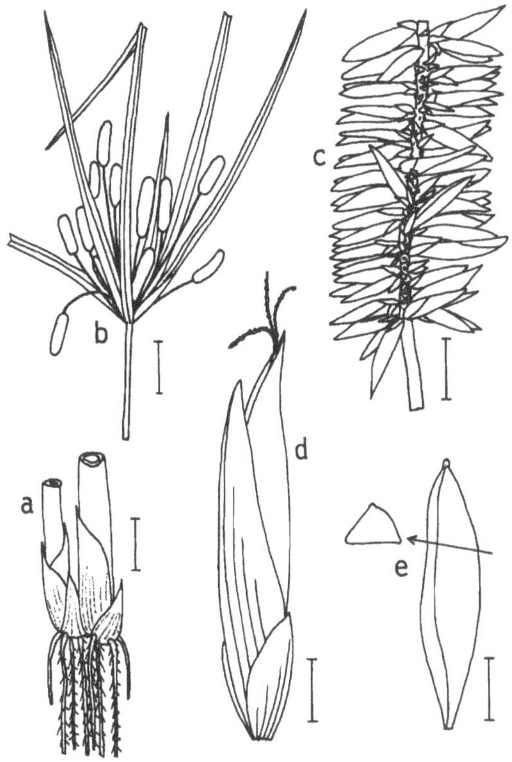

Fig. 174. *Mariscus sumatrensis*: a, base of shoot (1 cm); b, inflorescence (2 cm); c, basal portion of spike with some spikelets removed (2 mm); d, spikelet (0.5 mm); e, nut (0.5 mm) with transverse section (inset).

Pycreus [excluded from *Cyperus*] ± 70 species: pantropics, most in Africa.

Annual or perennial. Culms usually erect, tufted. Leaves mostly radical, well-developed and leaf-like; blades flat, channelled or grooved. Involucral bracts usually long and leaf-like, often overtopping the inflorescence. Inflorescence terminal, rarely pseudolateral, simple or compound panicles, with branches bearing clusters of spikelets, rarely head-like; bracts at base of rays (prophylls) keeled, tubular at base. Spikes elongate, with many spikelets. Spikelets strongly flattened laterally, many-flowered; rachilla straight or flexuous, persistent. Glumes usually all alike, distichous, clearly in 2 opposite rows, often imbricating, margins usually hyaline, with distinct keel, caducous (glumes and nuts are shed leaving the ± deeply pitted and usually winged rachillas). Flowers bisexual, all fertile except sometimes the very lowest ones sterile. Perianth bristles absent. Stamens 1 or 2; connectives usually crested. Style not swollen at the base, usually 2-cleft; stigmas (1) 2 (3). Nuts flattened (biconvex or plano-convex), with one angle facing the rachilla; dark brown or black, often apiculate.

Excluded: ***Pycreus flavescens*** (Linnaeus) Reichenbach [*Cyperus flavescens* Linnaeus]; a widespread species from temperate regions of the North Hemisphere, reaching the western Himalayas but does not extend into Peninsular India; records from Peninsular India are usually referrable to *P. stramineus* or sometimes to *P. flavidus.*

The following species may occasionally grow on wet soil but are they not considered to be wetland plants: ***P. atroglumosa*** (E. Govindarajalu) P. Singh et V. Singh: ♦Ind: Kt; ***P. curvibracreatus*** (E. Govindarajalu) P. Singh et V. Singh: ♦Ind: Tn; ***P. diaphanus*** (Schrader ex Roemer et Schultes) S. Hooper et T. Koyama [*Cyperus diaphanus* Schrader ex Roemer et Schultes, *C. latespicatus* Boeckeler, *C. pseudobromoides* Boeckeler, *P. pseudobromoides* (Boeckeler) C. B. Clarke, *C. latespicatus* var. *gracilescens* Kükenthal, *P. diaphanus* var. *laxus* (Kükenthal) Nainthani et Sas]; ***P. dwarkensis*** (Sahni et Naithani) Hooper: ♦?Ind: Gj; ***P. govindarajalui*** V. S. Raju [*P. decumbens* (E. Govindarajalu) P. Singh et V. Singh]: ♦Ind: Mh; ***P. latovaginatus*** (E. Govindarajalu) P. Singh et V. Singh: ♦Ind: Tn; ***P. lurida*** (E. Govindarajalu) P. Singh et V. Singh: ♦Ind: Tn; ***P. malabaricus*** C. B. Clarke: ♦Ind: Kl; ***P. plumbeonuceus*** (E. Govindarajalu) P. Singh et V. Singh: ♦Ind: Tn; ***P. plurinodosus*** (E. Govindarajalu) P. Singh et V. Singh: ♦Ind: Kt; ***P. rubiglumosus*** (E. Govindarajalu) P. Singh et V. Singh: ♦Ind: Tn; ***P. stricticulmis*** (E. Govindarajalu) P. Singh et V. Singh: ♦Ind: Tn; ***P. sulcinux*** C. B. Clarke.

1. Glumes at least 3 mm long; spikelets lanceolate or oblong, somewhat turgid, more than 3.5 mm wide ***P. unioloides***

1. Glumes less than 3 mm long; spikelets linear to linear-oblong, flat, less than 3.5 mm wide

2. Glumes with a furrow on each side of the keel; culms decumbent, few- to several-noded below the middle, the lower part decumbent or obliquely ascending, branching and rooting at lower nodes, usually clothed for one-quarter to half their length by leaf sheaths ***P. sanguinolentus***

2. Glumes without a furrow on each side; culms without nodes above the base, erect from the base, not branching, not rooting, leaves only at base

3. Glumes truncate to emarginate at the apex; the midvein excurrent beyond the glume apex into a recurved mucro; culms usually less than 12 cm tall; slender annual ***P. pumilus***

3. Glumes acute to obtuse at the apex, never emarginate; the midvein not excurrent beyond the glume apex or if so then only minutely so and the mucro not curved; culms usually more than 15 cm tall; annual or perennial

4. Glumes with a very distinct white margin about 0.3 mm wide, the midrib stopping before the white margin ***P. macrostachyos***

4. Glumes without a white margin or white and hyaline only towards the apex and less than 0.3 mm wide, the midrib ending beyond the hyaline margin

5. Nuts narrowly oblong, at least twice as long as wide ***P. polystachyos***

5. Nuts ellipsoid or obovoid to orbicular, less than one and a half times as long as wide

6. Nuts with 13–17, discontinuous, undulate, transverse wrinkles; epidermal cells longitudinally oblong; glumes without any hyaline margin ***P. stramineus***

6. Nuts smooth or minutely puncticulate; epidermal cells isodiametrical or hexagonal, usually appearing dotted; glumes with a narrow but distinct hyaline margin towards the tip

7. Leaf blades channelled, grooved and folded, ± needle-like, 1–2 (–3) mm wide; glumes oblong-ovate, 1–1.5 mm wide; rachilla obscured by the glumes at maturity ***P. flavidus***

7. Leaf blades flat or plicate, 4–8 (–10) mm wide; glumes ovate-orbicular, 1.8–2 mm wide; rachilla visible between the glumes at maturity ***P. puncticulatus***

Pycreus flavidus (Retzius) T. Koyama [*Cyperus flavidus* Retzius, *C. capillaris* J. König ex Roxburgh, *C. globosus* Allioni, *C. strictus* Roxburgh, *Pycreus globosus* (Allioni) Reichenbach, *C. nilagiricus* Hochstetter ex Steudel, *P. capillaris* (J. G. König ex Roxburgh) Nees ex C. B. Clarke, *P. bolei* S. M. Almeida, ?*P. lancelotii* S. M. Almeida]: Fig. 175

Rhizomes inconspicuous, roots yellow. Culms densely tufted, smooth, (5–) 8–80 cm tall, 1–1.5 mm thick, obtusely 3-angled. Leaves few, shorter than the culm; sheaths elongated, tinged with reddish-brown; blades channelled, grooved or folded, rigid, often needle-like, up to 15 cm long, 1–2 (–3) mm wide, gradually acuminate. Involucral bracts 2–4, the lower 2 overtopping the inflorescence, the longest up to 25 cm long. Inflorescence simple or partly compound, open or rarely contracted into a single cluster of spikelets; primary rays 1–6, up to 7 cm long. Spikes ovoid to broadly ovoid, up to 2.5 cm long, pale or fuscous, with (5–) 10–15 (–40) spikelets. Spikelets spreading, linear or linear-oblong, parallel-sided, strongly flattened, 5–10 (–30) mm long, 1–2 (–3) mm wide, with 20–40 (–60) flowers, reddish-brown to purplish-brown; rachilla straight, persistent, wingless, 4-angled. Glumes oblong-ovate, 1.5–2.5 mm long, 1–1.5 mm wide, membranous, reddish to purplish or

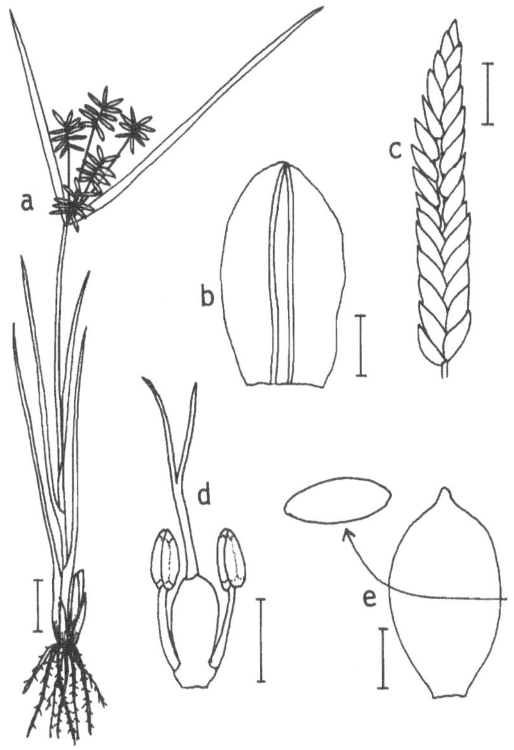

Fig. 175. *Pycreus flavidus*: a, whole plant (2 cm); b, glume (0.5 mm); c, spikelet (2 mm); d, flower (0.5 mm); e, nut (0.5 mm) with transverse section (inset).

black, the sides nerveless, hyaline margin whitish but very narrow and only towards the tip, the tip obtuse; keel distinct, green, 3-nerved. Stamens always 2. Style slightly shorter than the nut; stigmas 2, longer than the style. Nuts biconvex, laterally flattened, ellipsoid or obovoid, (0.4–) 0.8–1.2 mm long, stipitate, apiculate, maturing dark brown, minutely puncticulate, with isodiametric epidermal cells.

Annual or occasionally perennial in permanently wet habitats. In open, wet places, swamps, marshes, margins of pools and ditches, in rice fields and other irrigated areas. Often forming large clumps. Locally very common. It is very variable as reflected in the synonymy. Superficially it resembles *Cyperus pygmaeus* but the glumes are without excurrent midribs. It is sometimes confused with *P. sanguinolentus* but is distinguished by the erect, non-rooting, unswollen culm base, parallel-sided spikelets and the ellipsoid to obovoid nut. It has also been confused with *Pycreus flavescens* (Linnaeus) Reichenbach but this species is very different with transversely wrinkled nuts and in India it occurs only in the western Himalayas.

{⊙/(⊙) Hel}. Warm to tropical Old World: Ban, Bhu, Bur, Cey, Ind, Pak: An, Ap, As, Bh, Go, Hp, Jk, Kl, Kt, Mg, Mh, Mp, Ng, Or, Pj, Rj, Sk, Tn, Up, Wb. 0–2800 m.

Pycreus macrostachyos (Lamarck) J. Raynal [*Cyperus macrostachyos* Lamarck, *P. albomarginatus* Nees]: Fig. 176

Rhizome absent. Culms solitary or tufted, 15–90 cm tall, 0.5–5 mm thick, glabrous, 3-sided. Leaves few; sheaths loose, reddish-brown, covering the lower quarter to half of the culm, splitting as the culms grow, becoming pierced by emergent roots and new buds; blades flat or folded, 8–40 cm long, 1–8 mm wide. Involucral bracts 3–7, erect or spreading, the longest 2–35 (–50) cm long. Inflorescence a simple or compound umbel-like panicle of large spikelets; primary rays, 4–12, unequal, up to 15 cm long. Spikelets 3–30 loosely spreading along the rachis, narrow lanceolate, flattened, 7–24 mm long, 1.5–3 mm wide, with 12–24 flowers, gradually narrowed to an acute apex, golden to reddish-brown; rachilla flexuous, narrowly winged. Glumes obovate, 2–3 mm long, ± 1.3 mm wide, golden to red-brown; margin very distinct, white, hyaline

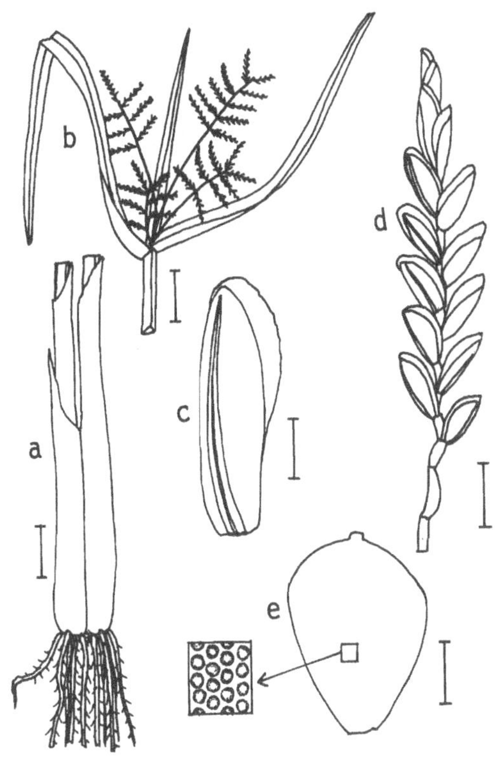

Fig. 176. *Pycreus macrostachys*: a, base of plant (1 cm); b, inflorescence, rather depauperate (2 cm); c, glume (0.5 mm); d, spikelet (2 mm); e, nut (0.5 mm) with surface pattern (inset).

± 0.3 mm wide; the tip rounded; keel distinct, green, 3-nerved, stopping before the white hyaline margin. Stamens 2 or 3. Style deeply 2- or 3-cleft. Nuts biconvex, laterally flattened, oblong to obovate, 1.4–2 mm long, 0.6–1 mm wide, stipitate, apiculate, maturing grey, surface with minute papillae in distinct longitudinal rows.

Annual. Mostly found in seasonally flooded areas, in pools, streams and buffalo wallows; common in rice fields. In open places it tends to be rather short and robust and in closed communities it is tall and slender.

{⊙ Ten/(Hel)}. Pantropics: Bur, Ind: Kl, Kt, Mh.

Pycreus polystachyos (Rottbøll) Palisot de Beauverd [‘*polystachus*’ orth. mut., including var. *laxiflorus* (Bentham) C. B. Clarke (ferrugineous glumes), var. *circinatus* (Ridley) Naithani et Shah, *Cyperus polystachyos* Rottbøll, *C. paniculatus* Rottbøll, *Pycreus odoratus* Urban (not the same as *Cyperus odoratus* Linnaeus)]: Fig. 177

Rhizomes short or absent. Culms tufted, stiffly erect, (10–) 20–60 (–80) cm tall, 1–2 (–3) mm thick, smooth, 3-sided becoming 3-angled just below the inflorescence. Leaves few, shorter than the culm; sheaths reddish-brown; blades flat or slightly plicate, 5–40 cm long, 1.5–3 (–5) mm wide. Involucral bracts 3–5, at least the lowest overtopping the inflorescence, the longest up to 20 cm long. Inflorescence loose or compact panicle or contracted in a subglobose or irregularly lobed head-like cluster of spikelets; primary rays 0–8, spreading, up to 7 cm long. Spikelets digitately arranged, linear to linear-lanceolate, flattened, (0.5–) 1–2.5 cm long, 1.5–2 mm wide, with 10–50 flowers, gradually narrowed to an acute apex, yellowish to red-brown; rachilla flexuous, narrowly winged. Glumes oblong-ovate to ovate-elliptical, 1.7–2.5 mm long, 0.8–1 mm wide, yellowish to red-brown, the sides nerveless, margin whitish hyaline, the tip acutish or obtuse with a minute mucro; keel distinct, green, 3-nerved. Stamens (1) 2. Style ± twice as long as the nut. Nuts biconvex, laterally flattened, narrowly oblong to oblong, 1–1.5 mm long, 0.3–0.5 mm wide, stipitate, subtruncate at top, minutely apiculate, minutely puncticulate.

Annual or short-lived perennial. Seasonally flooded areas, damp places, marshes, rice fields, river banks and at the margins of pools. Sometimes found near the sea in brackish conditions. A common and very variable species but distinguishable by: culms erect, leafy only at base; leaves flat; spikelets linear, gradually tapering to an acute tip; rachilla winged; glumes at most 2.5 mm long, acute, muticous, without depressions; nuts with flat or slightly convex sides, narrowly oblong, subtruncate at the top, minutely apiculate.

{⊙/(⊙) Ten/(Hel)}. Warmer parts of the world: Ban, Bur, Cey, Ind, Pak: <An, As, Kl, Lk, Mg, Mp, Rj, Sk, Tn, Up, Wb.

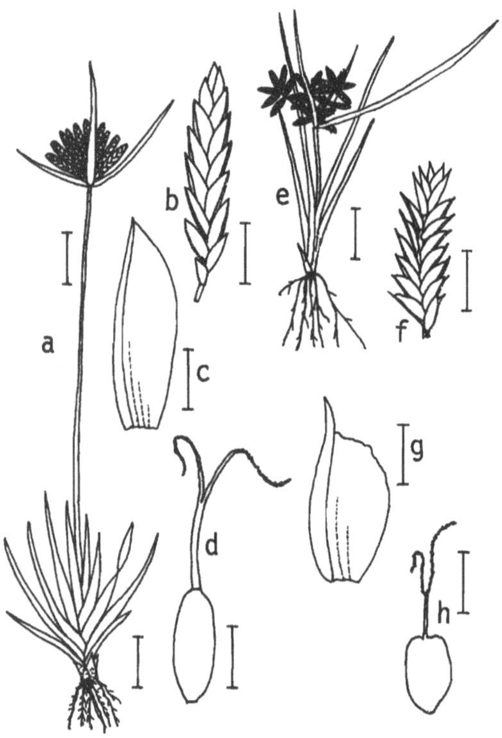

FIG. 177. *Pycreus polystachyos*: a, whole plant, rather small specimen (1 cm); b, spikelet (2 mm); c, glume (0.5 mm); d, nut (0.5 mm). *P. pumilus*: e, whole plant, rather small specimen (1 cm); f, spikelet (2 mm); g, glume (0.5 mm); h, nut (0.5 mm).

Pycreus pumilus (Linnaeus) Nees ex C. B. Clarke [*Cyperus pumilus* Linnaeus, *C. membranaceus* Vahl, *P. nitens* Retzius, *C. punctatus* Roxburgh, *C. pulvinatus* Nees et Meyen ex Nees]: Fig. 177

Roots very thin and short, yellowish. Culms tufted, slender, smooth, (1–) 6–12 (–20) cm long, 0.3–1 mm wide, 3-sided. Leaves few, basal, shorter or overtopping the inflorescence; sheaths pale stained with reddish-brown; blades flat or channelled, 2–8 cm long, 1–2 mm wide, gradually acuminate, scabrid above. Involucral bracts 3–5, erect to spreading, overtopping the inflorescence, the longest 2.5–15 (–20) cm long. Inflorescence simple or compound, of 1 sessile and 1–6 stalked, spike-like clusters or sometimes (in depauperate plants) reduced to 1 head-like cluster; primary rays 3–5 (–7). Spike-like clusters loose, ovoid to subglobose, 1–2 cm in diameter, with 3–15 (–30) spikelets. Spikelets divergent, oblong to linear, 4–10 (–13) mm long, 1–2.5 mm wide, with 8–20 (–40) flowers; rachilla straight, unwinged. Glumes ovate to obovate-elliptical, 0.6–1.8 mm long, membranous, sometimes remaining closed otherwise spreading and exposing the nuts, grey to reddish-brown, the sides nerveless and silvery, truncate to emarginate at tip; keel sharp, 3- or 5-nerved, green or reddish-brown, excurrent beyond the glume, the tip mucronate, the mucro erect or recurved and 0.3–1 mm long. Stamens 1 or 2. Stigmas 2, longer than the style. Nuts biconvex, laterally compressed, oblong-obovoid, 0.5–0.8 mm long, 0.3–0.5 mm wide, almost white when young becoming brown or grey with a metallic shine when mature, surface with minute tubercles, stipitate, apiculate.

Annual. In seasonally flooded places, wet hollows, marshes, shallow pools, moist meadows, along drainage ditches and in rice fields. It is common and rather variable; plants with looser spikelets showing the rachilla internode between the glumes and obovate-elliptical and thinly membranous glumes are sometimes recognized as var. *patens* (Vahl) Kükenthal or subsp. *membranaceus* (Vahl) T. Koyama.

{⊙ Ten/Hel}. Old World tropics: Ban, Bhu, Bur, Cey, Ind, Nep, Pak: An, As, Bh, Gj, Go, <Hp, Jk, Kl, Kt, Mh, Mp, Mz, Pj, Rj, Sk, Tn, Up, Wb. 0–1800 m.

Pycreus puncticulatus (Vahl) Nees [*Cyperus puncticulatus* Vahl, *C. baccha* (Nees) Kunth]: Fig. 178

Roots fibrous, brown. Culms mostly solitary, smooth, 30–60 cm tall, 2–4 mm thick, 3-sided, often somewhat enlarged at base. Leaves 3–5, slightly overtopping the culm; sheaths elongated, broader than the blade, pale or rusty brown; blades flat to plicate, 15–40 cm long, 4–8 (–10) mm wide, gradually acuminate. Involucral bracts 3–5, most overtopping the inflorescence, the longest 20–35 cm long. Inflorescence compound, lax; primary rays 4–11, up to 14 cm long. Spikes lax, cylindrical, 2–7 cm long, distantly bearing 7–20 spikelets. Spikelets spreading to weakly reflexed, linear-oblong to oblong-lanceolate, flattened, 8–20 mm long, 2–3 mm wide, with 12–34 flowers, pale red; rachilla weakly flexuous, hardly winged. Glumes ovate-orbicular, 2–2.3 mm long, 1.8–2 mm wide, membranous, purplish to reddish, the sides nerveless, margin whitish hyaline, the tip rounded and shallowly erose, mucronulate at the sinus; keel distinct, green, 3-nerved. Stamens 2. Style 1.5 mm long, 2-cleft to one-third from the tip. Nuts biconvex, strongly flattened laterally, obovate-orbicular, 1–1.2 mm long and wide, stipitate, apiculate, densely puncticulate, with isodiametric epidermal cells.

Annual. In wet and seasonally flooded places; common in rice fields. Within *Pycreus* it is very distinct with its large panicles with reddish spikelets; it superficially resembles *Cyperus procerus*, a stoloniferous perennial with 3-sided nuts.

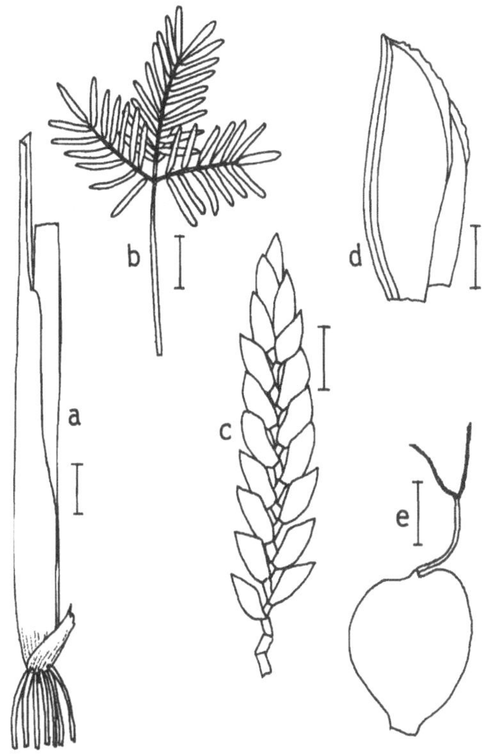

Fig. 178. *Pycreus puncticulatus*: a, base of culm (1 cm); b, inflorescence branch (1 cm); c, spikelet (2 mm); d, glume (0.5 mm); e, nut (0.5 mm).

{⊙ Ten/Hel}. S India and SE Asia: Cey, Ind: Kl, Kt, Mh, Rj? Tn.

Pycreus sanguinolentus (Vahl) Nees ex C. B. Clarke [*Cyperus sanguinolentus* Vahl, *C. cruentus* Retzius, *C. albidus* Lamarck, *C. eragrostis* Vahl, non Lamarck]: Fig. 179

Rhizomes inconspicuous, without stolons. Culms tufted, the lower third to half leafy, smooth, (5–) 10–60 cm tall, 0.5–2 mm thick, 3-sided, decumbent at base; culm base usually somewhat swollen, 1- to 4-noded, rooting and branching at the nodes, clothed with sheaths. Leaves on up to the lower half of the culm; sheaths elongated, yellowish and tinged with reddish-brown; blades flat or channelled, sometimes with a red midrib, 1–4 mm wide, gradually acuminate, scabrid on the upper part. Involucral bracts 2–5, erect to spreading, the lower ones overtopping the inflorescence, the longest up to 15 cm long. Inflorescence simple, either of a few stalked spikes or contracted into a dense head or reduced to single cluster of spikelets; primary rays 1–5, up to 5 cm long. Spikes ovoid to broadly ovoid, 0.7–1.5 (–2) cm long, pale or fuscous, with 3–18 spikelets. Spikelets divergent, ovate to oblong-lanceolate, flattened, 5–10 (–20) mm long, 1–3

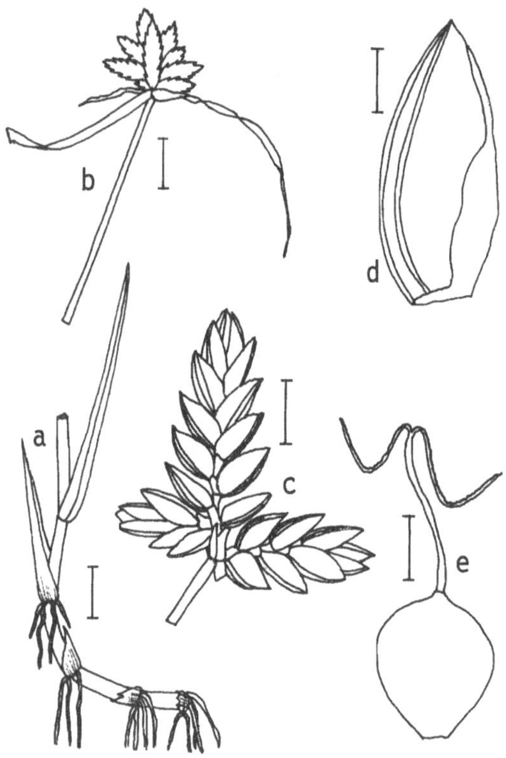

Fig. 179. *Pycreus sanguinolentus*: a, base of culm (1 cm); b, inflorescence (1 cm); c, three spikelets (2 mm); d, glume (0.5 mm); e, nut (0.5 mm).

(–3.5) mm wide, with 8–30 flowers, red-brown; rachilla straight, persistent, wingless, 4-sided. Glumes membranous, obliquely erect or patulous, ovate, 1.5–2.5 mm long, 1–2 mm wide, the sides nerveless, whitish or purplish to black, margin whitish hyaline; keel distinct, curved, green, 3- or 5-nerved, with a furrow or groove on each side. Stamens 2 or 3 (the number constant in the same specimen). Style up to 3 times longer than the nut; stigmas 2. Nuts biconvex, laterally flattened, broadly obovate to orbicular, 0.8–1.5 mm long, 0.6–1.3 mm wide, stipitate, apiculate, smooth, epidermal cells isodiametric.

Annual or perennial in permanent water. Wet open places, swamps, margins of pools, river banks and wet rice fields. Locally very common. It is rather variable and numerous infraspecific taxa have been published. Characteristic are the glumes with arcuate keels with groove-like depression on each side; culms swollen and decumbent at base, rooting at nodes, the lower third to half leafy.

{⊙/(⊙) Ten/Hel}. Old World tropics and subtropics: Ban, Bhu, Bur, Cey, Ind, Nep, Pak: An, Ap, As, Bh, Gj, Go, Hp, Jk, Kl, Kt, Mh, Mg, Mp, Or, Pj, Rj, Sk, Tn, Up, Wb. 0–3000 m.

Pycreus stramineus (Nees) C. B. Clarke [*Cyperus substramineus* Kükenthal]: Fig. 180

Roots fibrous, yellowish. Culms tufted, slender but hard, often weakly curved, smooth, 8–35 cm tall, 0.5–1 mm thick, obtusely 3-sided. Leaves 2 or 3, basal; sheaths elongated, yellowish and tinged with reddish-brown; blades shorter than to nearly equalling the culm, channelled, subrigid, 0.8–1 mm wide, gradually acuminate, scabrid on the upper part. Involucral bracts (1) 2 (3), erect or suberect, the lower up to 10 cm long, overtopping the inflorescence. Inflorescence simple, usually a single cluster of 5–16 spikelets or rarely paniculate with 1–3 short rays each bearing a cluster of spikelets; rays not more than 1 cm long. Spikelet clusters rather loose. Spikelets erect to spreading, lanceolate-oblong to linear-oblong, flattened, (0.8–) 15–35 mm long, 1.8–2.2 mm wide, with 14–70 flowers, yellowish; rachilla straight, persistent, wingless. Glumes membranous, patent, ovate, acute, mucronulate, ± 2 mm long, 1.2–1.5 mm wide, the sides nerveless, straw-coloured to golden yellow, margin hyaline; keel distinct, green, 3-nerved. Stamens 2. Style ±0.8 mm long; stigmas 2, as long as the style. Nuts biconvex, laterally flattened, broadly elliptical to obovate-orbicular, 0.8–1 mm long,

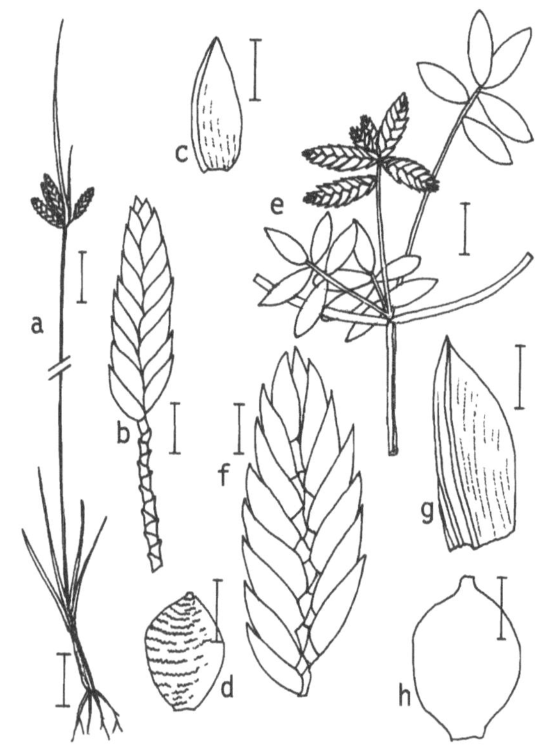

Fig. 180. *Pycreus stramineus*: a, whole plant with culm shortened (1 cm); b, spikelet (2 mm); c, glume (0.25 mm); d, nut (0.5 mm). *P. unioloides*: e, inflorescence (1 cm); f, spikelet (2 mm); g, glume (0.25 mm); h, nut (0.5 mm).

0.6–0.7 mm wide, stipitate, apiculate, smooth, maturing dark purplish-brown to black, finely transversely undulate with 13–17 discontinuous wrinkles, epidermal cells longitudinally oblong.

Annual. In seasonally flooded places, rice fields; sometimes in brackish water. It is not confined to wetlands and found in waste places and roadsides. The roots are aromatic. It closely resembles *P. flavescens* (a species from N temperate regions) which has spikelets shorter (6–10 mm long) and glumes fewer and obtuse.

{⊙ Ten/Hel}. Indomalaysia: Ban, Bhu, Bur, Cey, Ind: As, Kl, Kt, Mg, Mz, Tn, Up, Wb. 0–900 m.

Pycreus unioloides (R. Brown) Urban, [*Cyperus unioloides* R. Brown, *C. bromoides* Link, *C. angulatus* Nees]: Fig. 180

Rhizomes inconspicuous, without stolons. Culms crowded, slender but rigid, 30–100 cm tall, 0.7–3 mm thick, 3-angled, smooth or scabrous above. Leaves 1–4, only at base of culm; sheaths reddish to blackish-brown; blades flat or folded, 10–60 cm long, 2–6 mm wide, with cartilaginous margins, very gradually acuminate, scabrid on the upper part. Involucral bracts 2–4, erect to spreading, the lower ones overtopping the inflorescence, the longest 8–35 cm long. Inflorescence simple panicle with 1 sessile and 1–7 stalked heads of spikelet clusters, rarely reduced to single, sessile cluster of spikelets; primary rays 3–6, up to 10 cm long. Spikelets spreading, lanceolate to oblong, flattened but somewhat turgid, 8–16 mm long (elongating to 25 mm when fruiting), (3.5–) 4–5 mm wide, with 10–20 (–30) flowers, yellowish-brown to light reddish-brown; rachilla slightly flexuose, persistent, wingless. Glumes ovate to oblong, 3–4 mm long, 2.5–3 mm wide, the sides nerveless, shining yellowish-green to brown, margin not or only slightly hyaline and undulate when dry; keel acute, green, 3-nerved, slightly excurrent. Stamens 3. Style 1.5–3.5 mm long; stigmas 2, up to 3 mm long. Nuts biconvex, laterally flattened, shortly elliptical to obovate, 0.8–1.5 mm long, 0.7–1 mm wide, stipitate, apiculate, epidermal cells isodiametric.

Perennial. In seasonally flooded areas, swamps, marshes and borders of lakes and rivers. It is often found in standing water. It is not confined to wetlands. The distinguishing features are: culms erect, leafy only at base; spikelets lanceolate to oblong, 3.5–5 mm wide, somewhat turgid; glumes 3–4 mm long, muticous with at most a very short, indistinct mucro; stamens 3.

{(Hyp/Ten)/Hel}. Pantropics: Bhu, Bur, Ind, Nep: As, Kt, Mg, Tn, Wb. 0–2200 m.

Rhynchospora ± 200 species: almost cosmopolitan, mainly tropical America

Large or small, annuals or perennials. Culms usually leafy. Leaves numerous, usually stiff, densely crowded at base or well-spaced along the culm, usually with spine-like marginal teeth; ligule absent or poorly developed. Involucral bracts leaf-like, usually exceeding the inflorescence. Inflorescence of terminal or axillary spikelets in 1 or more heads or paniculate often clustered. Spikelets sessile or stalked. Lower glumes often distichous or spiral; upper glumes spirally arranged; the lowest 3 (or very rarely more) glumes usually sterile, the succeeding 1–40 glumes with bisexual flowers, the uppermost male or sterile. Perianth bristles usually 6, rarely more or absent. Stamens 1–3. Style entire or 2-cleft, swollen at the base. Nuts oblong or obovoid, beaked by the persistent style base.

Excluded: the following species may grow on wet soil or in damp places but are not considered to be wetland plants: ***Rhynchospora chinensis*** Nees et Meyen; ***R. rubra*** (Loureiro) Makino [*R. wallichiana* Kunth]; ***R. triflora*** Vahl [*R. ceylonica* (Nees) Kunth]; ***R. wightiana*** Steudel.

1. Blades of lower leaves at least 8 mm wide; style entire or slightly 2-lobed ***R. corymbosa***
1. Blades of lower leaves not more than 5 mm wide; style deeply 2-cleft to halfway down ***R. rugosa***

Rhynchospora corymbosa (Linnaeus) Britton [*R. aurea* Vahl]: Fig. 181

Rhizomes thick, creeping and rather soft. Culms solitary or tufted, 60–120 (–250) cm tall, (1–) 2.5–3.5 mm thick above (up to ± 10 mm thick below), 3-sided with shallow longitudinal grooves, the base covered with closely imbricated tough leaf bases, split by emerging buds. Leaves numerous and densely crowded; sheaths closed, greenish but with a transparent greyish area opposite the blade; blades very tough except at base where they are thick and soft, the lower ones 50–100 cm or more long and 10–20 mm wide, the margin and midrib with minute spine-like teeth. Involucral bracts of main corymbs leaf-like, 10–50 cm or more cm long, 5–20 mm wide; base auricled; margin and midrib with minute spine-like teeth. Inflorescence of 1 terminal and several lateral corymbs set on the upper half of the culm. Corymbs with a 1–6 cm long main axis with numerous erect-spreading branches, at the base of each bract a narrow leaf-like bract 2–4 cm long and a tubular and hairy prophyll 5–10 mm long. Branches up to ± 10 cm long, laterally flattened, the edges with spine-like teeth; the top of the branches with a number of short secondary branches carrying clusters of spikelets (all branches with filiform bracts and tubular prophylls at their bases). Spikelets narrowly lanceolate, subterete, 6–10 mm long, 1.5–2.5 mm wide, brown, grouped into stalked clusters; new generations of spikelets arising from prophylls of the old (which disarticulates as the nut ripens). Each spikelet with 1 (or rarely more) bisexual flower and usually 1 male flower above. Glumes reddish-brown, glabrous, with narrow midrib; lower sterile glumes 2–4 mm long, the 2 upper ones 6–9 mm long. Bisexual flowers with 6, subequal perianth bristles, 4–6 mm long, each set with minute spine-like teeth. Stamens 3; anthers 2–3 mm long, apiculate above. Style unbranched or slightly 2-lobed at the tip. Nuts oblanceolate, 2.5–4 mm long, 1–2 mm wide, brown, coarsely nodular above; style base 4–5 mm long (longer than the nut), brown, long-conical with 1 longitudinal groove on each of the 2 flattened sides.

Perennial. A common plant growing in and along streams, on the shores of lakes, pools and rivers, often found in shallow water in swamps and rice fields. It is tolerant of light shade. Sometimes used for making mats, baskets and screens.

{Hyp/Ten/Hel}. Pantropics: Cey, Ind: An, As, Kl, Kt, Mg, Ng, Tn, Tp. 0–1300 m.

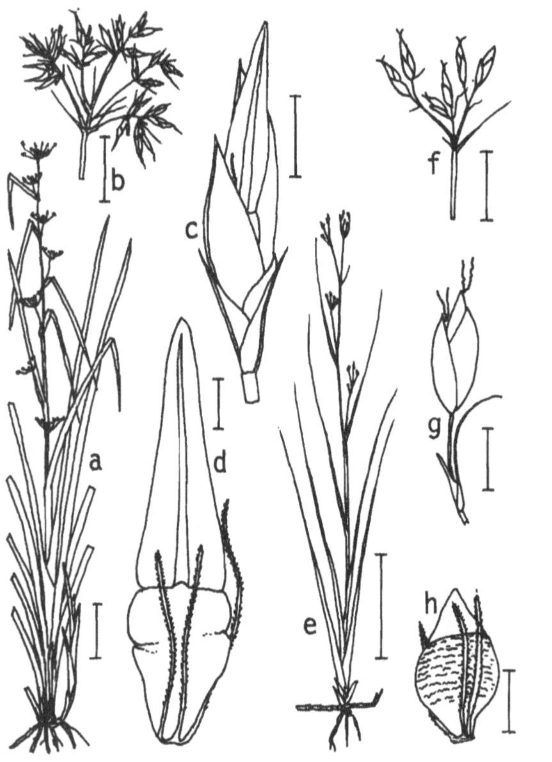

Fig. 181. *Rhynchospora corymbosa*: a, whole plant (10 cm); b, inflorescence branch (2 cm); c, spikelet (2 mm); d, fruit (1 mm). *R. rugosa*: e, whole plant (10 cm); f, inflorescence branch (5 mm); g, spikelet (2 mm); h, fruit (1 mm).

Rhynchospora rugosa (Vahl) Gale subsp. ***brownii*** (Roemer et Schultes) T. Koyama [*R. laxa* R. Brown, *R. glauca* Vahl var. *condensata* Kükenthal]: Fig. 181

Differing from *R. corymbosa* by being generally smaller; culms 30–70 cm tall; leaf blades 1.5–3 (–5) mm wide; inflorescence bearing 2, 3 or rarely 4 corymbs, the branches filiform; spikelets 3–4.5 mm long; style 2-cleft to halfway down; nuts broadly obovate, 1.5–2 mm long, 1–1.5 mm wide, light brown, weakly transversely wrinkled; style base 0.7–1 mm long, shorter than the nut.

Perennial. Usually in hills in seasonally flooded areas, marshes, boggy places and wet grassland. The subspecies *rugosa* is confined to subtropical and tropical America.

{Ten/Hel}. Old World tropics: Cey, Ind: Mg, Mz, Ng, Tn. 1000–3500 m.

Schoenoplectus [*Actinoscirpus, Isolepis* pro parte, *Scirpus* pro parte; excluding *Bolboschoenus*]: ± 50 species: cosmopolitan

Large perennials, up to 4 m tall or annual and smaller. Culms terete, 3-sided or 3-winged, solid but usually somewhat spongy, sometimes with many transverse septa and numerous approximately equal gas canals, often nearly or quite leafless at maturity. Leaves basal, reduced to scales or elongate and ribbon-like when totally submerged. Lowest involucral bract usually appearing as a continuation of the culm or occasionally leaf-like. Inflorescence pseudolateral or sometimes terminal, simple or compound, spreading or head-like. Spikelets many-flowered; rachilla persistent, not winged. Glumes spirally arranged, caducous, the lower ones empty. Flowers bisexual. Perianth segments 6, usually bristle-like, sometimes fewer or scale-like or rarely absent. Stamens 1–3. Stigmas 2 or 3; styles not swollen at the base, wholly caducous. Nuts 3-sided or biconvex.

Most are aquatic, rooting in the bottom with erect, emergent culms. Many species are important: used in weaving, some eaten, others used in water purification, while others are weeds in aquatic systems. There is considerable taxonomic disagreement within this genus; a critical revision on a worldwide basis is much needed. In doubtful cases I have tended to err on the side of the splitters rather than take the easier way and lump the difficult taxa together.

Excluded: the following species are wetland plants in the Himalayas which do not extend into Peninsular India: ***Schoenoplectus fuscorubens*** (T. Koyama) T. Koyama: ♦Nep; ***S. tabernaemontani*** (C. C. Gmelin) Palla [*Scirpus lacustris* subsp. *glauca* (Smith) Hartman, *S. lacustris* var. *tabernaemontani* (C. C. Gmelin) Doell]; ***S. triqueter*** (Linnaeus) Palla.

1. Inflorescences appearing terminal; involucral bracts 3 or more, leaf-like, spreading or erect

2. Glumes more than 5 mm long, emarginate or frayed at the apex, distinctly hairy on the back; inflorescence simple or 1 times compound, spreading; at least some terminal spikelets in sessile clusters **Bolboschoenus maritimus** p. 96.

2. Glumes not more than 3 mm long, entire at the apex, glabrous or minutely hairy towards the tip on the margin, midnerve and mucro; inflorescence 1, 2 or 3 times compound, spreading; terminal spikelets solitary and stalked ***S. grossus***

1. Inflorescences appearing lateral; lowest involucral bract culm-like, erect, its base continuing down the culm, the other bracts absent or scale-like

3. Culms and main involucral bracts sharply 3-angled ***S. mucronatus***

3. Culms and main inflorescence bracts terete or obscurely 3-sided or involucral bracts flattened

4. Inflorescence an open panicle with elongated rays; at least some spikelets or spikelet clusters stalked

5. Perianth scales plumosely fringed ***S. subulatus***

5. Perianth scales absent

6. Rhizomatous perennial; culms robust, usually more than 50 cm tall; spikelets pale to dark brown; glumes with glabrous margins ***S. corymbosus***

6. Tufted annual; culms rarely more than 30 cm tall; spikelets greenish to yellow or pink; glumes with microscopically hairy margins ***S. lateriflorus***

4. Inflorescence a dense lateral cluster of few to many, sessile spikelets

7. Culms and main inflorescence bracts with distinct transverse septa

8. Thickest culms usually more than 2 mm thick; nuts smooth; spikelets somewhat lax, variegated grey, greenish or reddish-brown ***S. articulatus***

8. Thickest culms not more than 1.7 mm thick; nuts transversely wavy on the sides; spikelets very compact, golden-brown to bronze-coloured ***S. senegalensis***

7. Culms and main inflorescence bracts without transverse septa or if culms septate then bracts not septate

9. Perianth bristles 4–6, with backwardly directed, hair-like teeth along the margins; nuts 1.8–2 mm long; glumes 2.5–4 mm long

10. Bristles shorter or some slightly longer than the nut, the longest 2–2.5 mm long; nuts unequally biconvex ***S. juncoides***

10. Bristles all distinctly longer than the nut, at least 3 mm long; nuts plano-convex (the ventral side flat) ***S. wallichii***

9. Perianth bristles absent; nuts not more than 1.3 mm long; glumes usually less than 3 mm long

11. Nutlets flattened; glumes with microscopically hairy margins; main involucral bract shorter than the culm ***S. lateriflorus***

11. Nutlets equally 3-sided; glumes with entire margins; main involucral bract equal to or much longer than the culm

12. Glumes 2–2.5 mm long, keel, scarcely prominent and not forming a mucro; nuts with angles the same colour as the sides ***S. roylei***

12. Glumes 2.5–3.2 mm long, often ending in a short and indistinct mucro; nuts with angles paler than the sides ***S. senegalensis***

Schoenoplectus articulatus (Linnaeus) Palla [*Scirpus supinus* auct. Ind., non Linnaeus; *Scirpus incurvatus* Roxburgh, *Scirpus subarticulatus* Roxburgh, *Isolepis prolongata* Nees; *Schoenoplectus supinus* auct. Ind., non (Linnaeus) Palla]: Fig. 182

Roots shallow. Culms in tufts of 3–10, terete but with shallow ridges, (4–) 10–40 (–80) cm tall including the

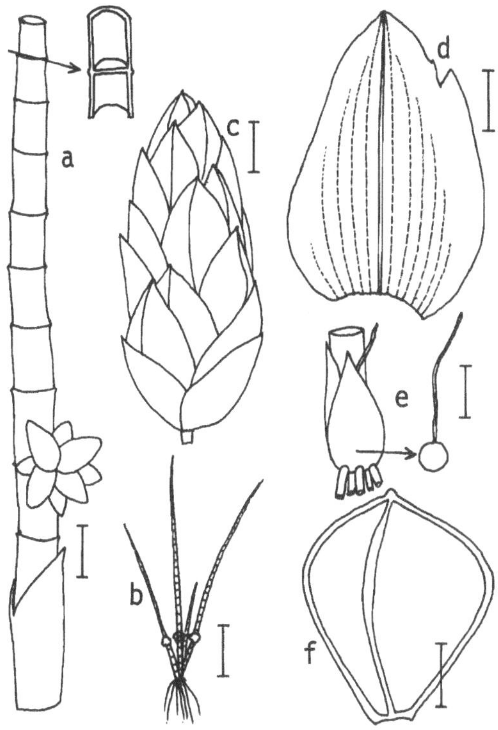

FIG. 182. *Schoenoplectus articulatus*: a, basal portion of culm with inflorescence (1 cm) with longitudinal section (inset); b, habit (10 cm); c, spikelet (5 mm); d, glume (1 mm); e, basal sheath with amphicarpous flower (inset) (1 cm); f, nut (0.5 mm).

bract, (1–) 3–10 mm in diameter, at base hardened, above hollow or pith-filled, with transverse septa at intervals of 5–20 mm, smooth, deep green, slightly shiny. Leaves reduced to 1 or 2 (3), bladeless sheaths, the lower scale-like and brown, the upper cylindrical, 3–20 cm long, much wider than the culm, pale green becoming yellow later, often reddish below, ending in a broad lobe. Main inflorescence bract appearing as a continuation of the culm, much longer than the culm proper, 6–40 (–60) cm long, somewhat flattened above with very conspicuous transverse septa. Inflorescence pseudolateral, of 1–30 (–60) sessile, apparently lateral spikelets in a dense, head-like cluster (1–4 cm in diameter). Spikelets sessile, ovoid or oblong-ovoid, terete, 6–18 mm long, 4–6 mm in diameter (or up to 10 mm with glumes spreading), variegated grey, greenish and reddish-brown. Glumes relatively stiff, deltate to broadly ovate, concave, 3.5–6.6 mm long and wide, acute, apiculate, many-nerved, pale green to golden or reddish-brown, with a green and protruding midrib, longitudinal ridges distinct; keel scarcely prominent. Perianth segments absent. Style 1.5–2 mm long; stigmas 3, shorter than the style. Nuts 3-angled, with slightly concave sides, obovoid, 1.5–2 mm long, 1.5–1.7 wide, smooth, whitish when young becoming dark brown to black when mature.

Solitary female flowers sometimes develop at the base of the culms in the axils of sheaths. These flowers have very long unbranched styles (up to 5 cm long) which develop large globular fruits (4–5 mm long, 3–4 mm in diameter). The combination of heterocarpy (amphicarpy) and geocarpy is usually associated with desert plants.

Annual or perhaps sometimes perennial. In seasonally flooded places, flowering as the water recedes, also in swamps and rice fields; sometimes in brackish water. Locally abundant. It is a very polymorphic species and some botanists consider it to be no more than a synonym of the Eurasian species *S. supinus* (Linnaeus) Palla [*Scirpus supinus* Linnaeus]. It also is rather similar to *S. senegalensis* but the nuts are larger (1.5–2 (–5)) mm long and smoother; the glumes larger (3.5–6 mm long) and with more acuminate tips; the spikelets are not golden-brown or bronze but rather dull grey to green or reddish-brown. It has been reported that the tubers are used in medicine in Bengal — it is annual and does not have tubers! Perhaps it is the geocarpous fruits which are used.

{☉/(☉) Hyp/Ten}. Mediterranean through Africa to Asia and Australia: Ban, Ind: As, Bh, Dl, Gj, Go, Kl, Kt, Mh, Or, Rj, Tn, Up, Wb. 0–1000 m.

Schoenoplectus corymbosus (Roth ex Roemer et Schultes) J. Raynal [*Isolepis corymbosus* Roth ex Roemer et Schultes, *Scirpus corymbosus* auct. non Linnaeus, *Scirpus inclinatus* (Raffenau-Delile) Ascherson et Schweinfurt, *Scirpus brachyceras* Hochstetter ex A. Richard]: Fig. 183

Rhizomes horizontal, short, thick and woody. Culms in a row or occasionally tufted, obscurely 3-sided to almost terete but with numerous longitudinal ridges, 50–350 cm tall, 6–8mm wide below the middle, 2–4 mm wide below the inflorescence, pith-filled. Leaves reduced scales and sheaths, the lowest scale-like and blackish and polished, the upper sheath-like often splitting to give a filigree across the split, sometimes ending in a stiff, erect lobe. Main involucral bract 1–3 cm long, stiff and leaf-like, boat-shaped and continuing in the direction of the culm. Inflorescence a simple or compound panicle with clusters of spikelets on very unequal branches; main branches up to 12 cm long, but frequently only 2–3 cm long; secondary branches up to 4 cm long. Spikelets sessile, ovoid, pale to dark brown, 3–8 mm long, 0.5–2.5 mm wide; occasionally some spikelets replaced by viviparous shoots. Glumes ovate, 2–4 mm long, ± 2 mm wide, pale to dark reddish-brown; margins without hairs; midrib raised at least above, usually densely set with

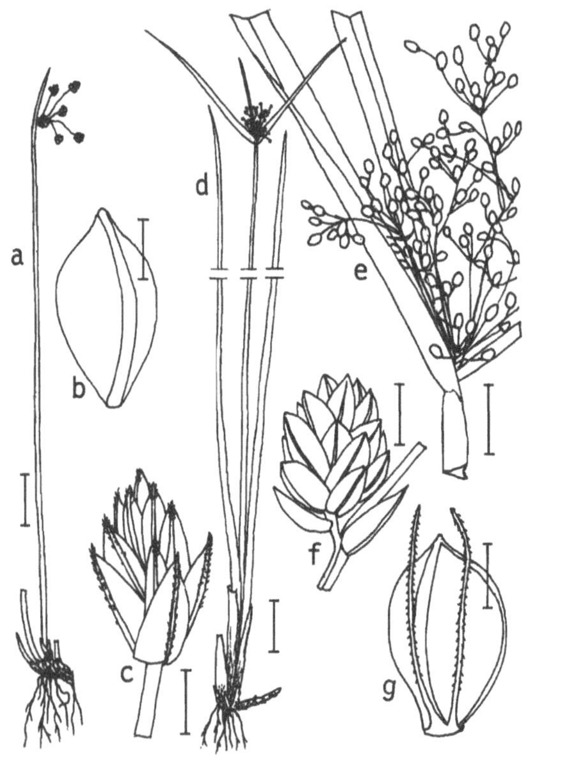

Fig. 183. *Schoenoplectus corymbosus*: a, habit (10 cm); b, spikelet (2 mm); c, nut (0.5 mm). *S. grossus*: d, habit, slightly shortened (10 cm); e, portion of inflorescence (2 cm); f, spikelet (2 mm); g, nut (0.5 mm).

spine-like hairs at least in the excurrent mucro-like apex (the lowest glumes prominently keeled and spiny throughout entire length). Perianth segments absent. Styles (2) 3 (4)-cleft, sometimes irregularly split. Nuts 3-sided, broadly elliptical, 1.2–1.5 mm long, 0.9–1.3 mm wide, smooth, when young yellowish-white, becoming dark brown to black at maturity.

Perennial. In open water at the edges of lakes and tanks, along streams and rivers, also in swamps and bogs. Often forming characteristic dark green clumps.

{Hyp}. Old World tropics: Ind: Kt, Rj, Up, Wb.

Schoenoplectus grossus (Linnaeus fil.) Palla [*Scirpus grossus* Linnaeus fil., *Scirpus maximus* Roxburgh, *Actinoscirpus grossus* (L. fil.) Goetghebeur et D. A. Simpson]: Fig. 183

Stolons slender, up to ± 5 mm in diameter, ending in small tubers. Culms solitary (often growing in straight lines) or a few in tufts, corm-like at base, 80–200 cm tall, 2–2.5 cm in diameter towards the base, 0.7–1 cm in diameter below the inflorescence, sharply 3-angled, the faces concave, smooth. Leaves few to a culm, linear, 50–180 cm long, 1.5–3cm wide, mostly equalling the culm. Involucral bracts 3 or 4, leaf-like, at least 2 exceeding the inflorescence, the longest 15–80 cm long. Inflorescence a large terminal corymb with primary, secondary and tertiary rays. Spikelets solitary, the terminal ones sessile, the rest pedicellate, ellipsoid, (4–) 6–10 mm long, 3.5–4 mm in diameter, many-flowered, light brown to rusty-coloured. Glumes broad-elliptical to oval-elliptical, boat-shaped, 2–3 mm long, 2–2.5 mm wide, rusty brown, ciliate along the margins, obtuse at apex, nerveless; keel green. Perianth bristles (5) 6, light brown, slightly exceeding the nut, subequal, the longest 2 mm or more long, retrorsely scabrous. Stamens 3; anthers 1.5 mm long, connective deltoid. Style ± 3 mm long, the upper half 3-cleft. Nuts somewhat flattened, broadly obovate, 1–1.5 mm long, up to 1 mm wide, yellowish-brown.

Perennial. Marshy places both inland and near the coast, often standing in shallow water. Common and gregarious, often dominant over large areas. Found as a weed in rice fields. It used for making mats, bags and baskets; the culms are dried then after removal of one of the ribs, they are flattened and bleached in the sun. Differs from other *Schoenoplectus* species by having a truly terminal inflorescence subtended by involucral bracts and the rachilla becomes characteristically pitted in later stages of fruiting. It is nowdays sometimes called *Actinoscirpus grossus*.

{Hyp/Ten}. Tropical SE Asia: Ban, Cey, Ind: An, As, Bh, Go, Kl, Mg, Mn, Ng, Rj, Tn, Up, Wb. 0–1600 m.

Schoenoplectus juncoides (Roxburgh) Palla [*Scirpus juncoides* Roxburgh, *Scirpus junciformis* Nees, *Scirpus erectus* auct. Ind., non Poiret]: Fig. 184

Culms tufted, subterete or obtusely several-angled and ribbed, 15–70 (–120) cm tall, 1–4 mm thick, light green, dull. Leaves reduced to 2 or 3 sheaths, the lower ones scale-like and brown, the upper one cylindrical, 5–15 cm long, membranous, pale green, with a small, mucronate, rudimentary blade. Main inflorescence bract 5–15 cm long, shorter than the culm, 1-furrowed on the inner side, rather suddenly subacute at the callous tip. Inflorescences pseudolateral, of (1–) 2–10 (–12), sessile, apparently lateral spikelets in a dense head-like, cluster. Spikelets ovoid to oblong-ovoid, terete, 5–18 mm long at anthesis, lengthening to 30mm in fruit, 2–6 mm in diameter, densely many-flowered, brownish to straw-coloured. Glumes broadly ovate, 2.5–4mm long, 1.8–3.5 mm wide, concave; sides faintly many-nerved, purplish; margins minutely hairy towards the tip; midrib not prominent, excurrent or not and keeled, green, 1- or 3-nerved, minutely mucronate. Perianth bristles 4–6, with backwardly directed, hair-like teeth along the margins, shorter or some slightly longer than the nut, the longest rarely more than 2.5 mm long. Style ± 2 mm long; stigmas

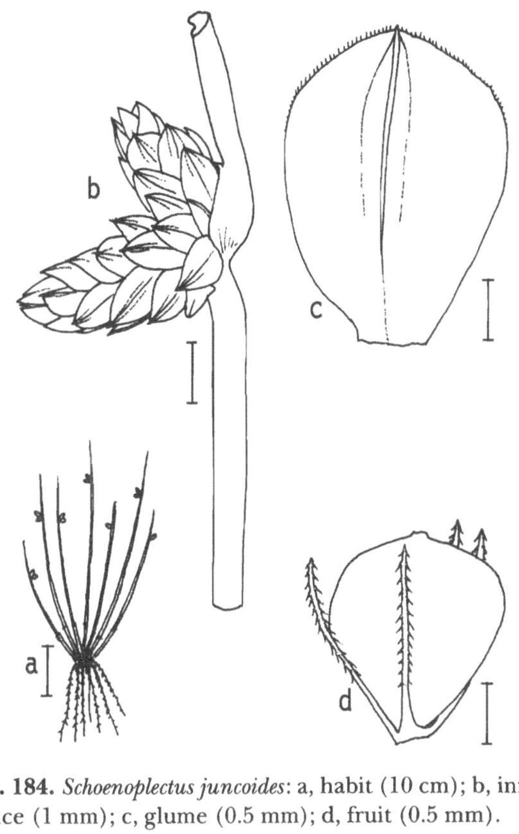

FIG. 184. *Schoenoplectus juncoides*: a, habit (10 cm); b, inflorescence (1 mm); c, glume (0.5 mm); d, fruit (0.5 mm).

2 or sometimes a third short branch present. Nuts broadly obovoid, unequally biconvex, 1.8–2 mm long, 1.5–1.8 mm wide, with faint pits or transverse wrinkles, shining dark brown to black.

Annual. Wet places, swamps, river banks, often in shallow water, particularly common in inundated rice fields. A rather polymorphic species that grades into *S. wallichii*.

{⊙ Hyp/Ten/(Hel)}. Madagascar, India to China and Australia, perhaps naturalized in Fiji and Hawaii: Cey, Ind, Nep: An, Ar, As, Bh, Hp, Kl, Kt, Mg, Mp, Pj, Rj, Sk, Tn, Wb. 0–2200 m.

Schoenoplectus lateriflorus (J. F. Gmelin) K. Lye [*Scirpus lateriflorus* J. F. Gmelin, *Scirpus lateralis* Retzius; *Isolepis oryzetorum* Steudel; *Isolepis uninodis* Raffenau-Delile, *Scirpus supinus* pro parte, auct. Ind., non Linnaeus, *Schoenoplectus supinus* subsp. *lateriflorus* (J. F. Gmelin) T. Koyama]: Fig. 185

Culms tufted, slender, weak, angular or subterete, 5–30 cm tall, 0.5–1.7 mm thick. Leaves reduced to 2 or 3, sheaths or sometimes with well-developed blades; sheaths 5–15 cm long, membranous, pale green, yellowish or pinkish-grey, slightly widened above, mucronate or the upper one sometimes developing a blade; blades flattened triangular, up to 10 cm long and 0.2 cm wide with a membranous ligule. Involucral bracts 1–3, the lower erect, similar to and continuous with the culm, always distinctly shorter than the culm proper, 3–15 cm long, the second and third when present very much shorter and spreading. Inflorescences pseudolateral, capitate or corymbose, with 1–4 primary rays and 3–10 (–20) spikelets in clusters. Spikelets congested or in clusters of 2–7, elongate-ovate to oblong, subterete, 5–10 (–15) mm long, ± 2.5 mm in diameter, greenish to straw-coloured or reddish-brown. Glumes elliptical to ovate, 2–2.5 mm long, 1.7–2 wide, concave, yellow or green to reddish-brown, with a reddish blotch at the tip; sides nerveless; margins hyaline, microscopically hairy towards the tip; midrib prominent and keeled throughout, 1-nerved, mucro distinct. Perianth segments absent. Style 0.5–0.8 mm long; stigmas 3, ± as long as the nut. Nuts obovoid, flattened, very unequally 3-sided, 1–1.3 mm long, 0.9–1 mm wide, obscurely or strongly transversely wrinkled, black, apiculate.

Solitary female flowers sometimes develop at the base of the culms in the axils of sheaths. These flowers have very long 3-cleft styles.

Annual. Common in seasonally flooded places, in ditches, rice fields and other wet places. It is highly polymorphic as is reflected in the synonymy and it should, perhaps, be united with the Eurasian species *Schoenoplectus supinus* (Linnaeus) Palla [*Scirpus supinus Linnaeus*] which usually has larger nuts (1.7–2.3 mm long).

{⊙ Ten/(Hel)}. Asian tropics and Australia: Cey, Ind, Nep: As, Gj, Go, Kl, Kt, Mp, Pj, Rj, Tn, Up, Wb. 0–1200 m.

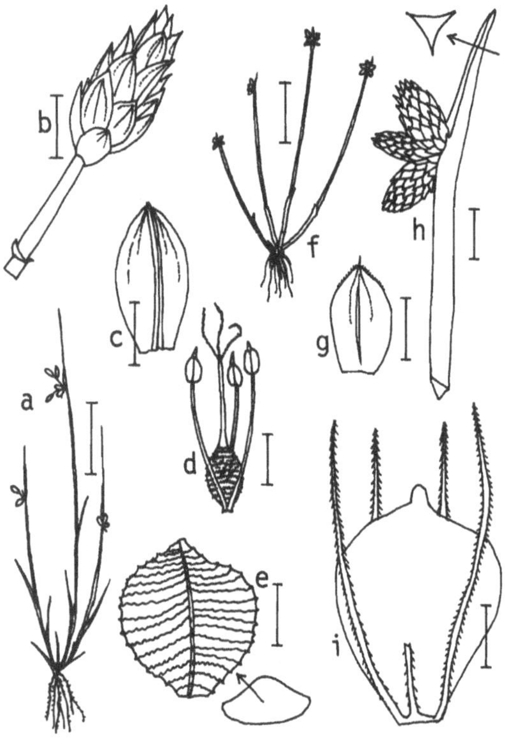

FIG. 185. *Schoenoplectus lateriflorus*: a, habit (10 cm); b, spikelet (2 mm); c, glume (1 mm); d, flower (0.5 mm); e, nut (0.5 mm) with transverse section (inset). *S. mucronatus*: f, habit, rather small plant (10 cm); g, glume (2 mm); h, inflorescence (1 cm) with transverse section of bract (inset); i, fruit (0.5 mm).

Schoenoplectus mucronatus (Linnaeus) Palla ex Kerner [*Scirpus mucronatus* Linnaeus, *Scirpus triangulatus* Roxburgh, *Scirpus muticus* D. Don, *Scirpus javanus* Nees]: Fig. 185

Rhizomes horizontal, short or sometimes elongated, but without stolons. Culms tufted or in a row along the rhizome, sharply 3-angled with ± concave sides, (25–) 40–100 (–120)cm tall, (2–) 3–9 mm wide below the middle, deep green. Leaves reduced to 1–3, yellowish to brown, bladeless sheaths, the lowest scale-like and brown, the upper cylindrical, 3-angled and lighter brown with a triangular apex or mucro. Main inflorescence bract at first appearing as a continuation of the culm but finally often patent to reflexed, very much shorter than the culm, 1.5–2.5 (–10) cm long, sharply

3-angled, the tip rather blunt. Inflorescence pseudolateral, congested, hemispherical, with (1–) 4–10 (–40) spikelets. Spikelets sessile, ovoid to oblong-ovoid, terete, densely many-flowered, rusty to brown, ± 5 mm long at anthesis, but stretching to ± 30 mm long and 6 mm wide when fruiting. Glumes firm, tightly appressed, ovate to broadly ovate, acutish, shortly mucronate, 2.8–4 mm long, 2–3 mm wide; sides many-nerved, pale or brown with reddish-brown stripes; margins minutely hairy towards the tip; midrib prominent, green, briefly awn-like and often incurved at the tip. Perianth bristles (5) 6, unequal, slightly to distinctly longer than the nut, retrorsely scabrid. Stigmas 2 or 3. Nuts 3-sided or biconvex, strongly dorsoventrally flattened, broadly obovoid, 1.7–2.3 mm long, 1.8–2 mm wide, with minute transverse wrinkles or smooth, shining black.

Perennial. Mostly found in shallow water in pools and ditches, on river banks, also in marshes and swamps where it is often dominant or co-dominant. It prefers nutrient-rich muddy soils. Frequently found with *Leersia hexandra* in inundated rice fields. It is a very widespread and polymorphic species.

{Hyp/Ten?}. Warmer parts of the Old World, naturalized in N America: Bhu, Cey, Ind: Ar, As, Bh, Hp, Kl, Kt, Mg, Mh, Mn, Mz, Pj, Rj, Sk, Up, Wb. 600–2700 m.

Schoenoplectus roylei (Nees) Ovczinnikov et Czukavina [*Isolepis roylei* Nees, *Scirpus quinquefarius* F. Hamilton ex Boeckeler]: Fig. 186

Roots shallow. Culms tufted but sometimes only 4–6 together, terete but with longitudinal ridges, 5–25 cm long (including the bract), 0.5–0.9 mm in diameter, transverse septa present but not prominent. Leaves reduced to 1–3, bladeless sheaths, the lower scale-like, pale to reddish-brown, the upper cylindrical, without any lobe since the midrib ends below the apex. Main inflorescence bract appearing as a continuation of the culm, usually much longer than the culm proper or rarely somewhat shorter, 6–17 cm long, 0.7–1 mm wide, distinctly flattened, with longitudinal ridges. Inflorescence pseudolateral, of 2–10 sessile, apparently lateral spikelets in a dense, head-like cluster. Spikelets sessile, ovoid, 4–6 mm long, 2–3 mm in diameter, variegated pale green or reddish-brown to yellow or golden (see note below). Glumes ovate, 2–2.5 mm long, midrib broad and 1-nerved, lateral nerves prominent; keel scarcely prominent; apex rounded or apparently acute when the margins incurved. Perianth segments absent. Stigmas 3. Nuts 3-sided, obovoid, 0.8–1.2 mm long, 0.6–0.8 mm wide, transversely wavy even on the angles, pale brown when young becoming dark brown to black, with angles same colour as the sides.

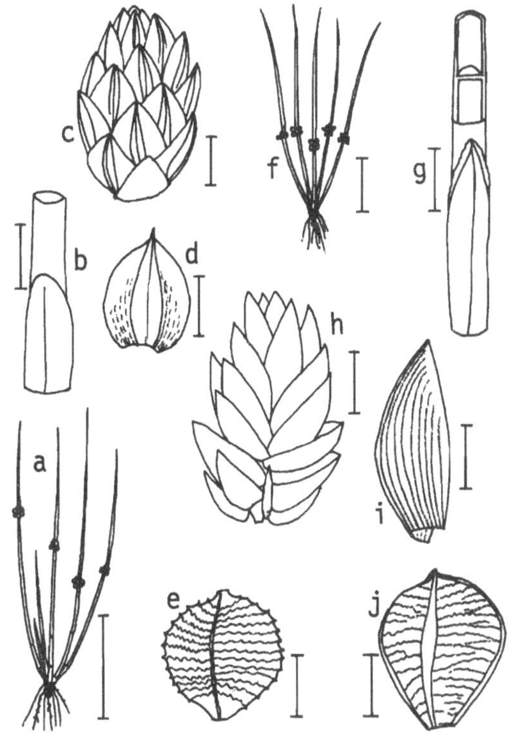

FIG. 186. *Schoenoplectus roylei*: a, habit (10 cm); b, tip of leaf sheath (2 mm); c, spikelet (2 mm); d, glume (1 mm); e, nut (0.5 mm). *S. senegalensis*: f, habit (10 cm); g, tip of sheath (2 mm); h, spikelet (2 mm); i, glume (1 mm); j, nut (0.5 mm).

Annual. Mostly in seasonally flooded places, flowering as the water recedes. Common in marshlands, ponds and rice fields. Sometimes cleistogamous flowers develop within the basal sheaths; the cleistogamously developed nuts are usually somewhat larger (up to 2 mm long) than chasmogamous ones. In Rajasthan the most common colour sequence of the spikelet is like the neck of a peacock: golden-yellow to pale brown to green; the local name 'morlo' means peacock-like. Like S. senegalensis but glumes smaller (2–2.5); angles of the nuts transversely wavy and the same colour as the sides.

{⊙ Hyp/Ten}. India and Africa: Ind: As, Dl, Hp, Pj, Rj, Up.

Schoenoplectus senegalensis (Hochstetter ex Steudel) Palla ex J. Raynal [*Isolepis senegalensis* Hochstetter ex Steudel, *Scirpus jacobii* C. Fischer]: Fig. 186

Roots shallow. Culms numerous, tufted, terete to angular, with longitudinal ridges, 6–60 cm long (including the bract), 0.3–1.6 mm in diameter, hollow, with transverse septa present. Leaves reduced to 1–3, bladeless sheaths, pale to reddish-brown, the upper cylindrical, ending in a narrow triangular lobe with an acute apex. Main inflorescence bract appearing as a continuation of the culm or leaf-like, as long as or much longer than the culm proper, 5–30 cm long, 5–25 mm wide, distinctly flattened, with occasional transverse septa. Inflorescence pseudolateral, of 1–25 sessile, apparently lateral spikelets in a dense, head-like cluster. Spikelets sessile, ovoid to subglobose, 3–9 mm long, 2–4 mm in diameter, obtuse, golden-brown to bronze-coloured. Glumes deltate to broadly ovate, 2.5–3.2 mm long, often ending in a short and indistinct mucro; midrib 3-nerved, lateral nerves prominent; keel scarcely prominent. Perianth segments absent. Stigmas 3. Nuts sharply 3-angled, obovoid, 1.2–1.5 mm long, 0.8–1.2 mm wide, transversely wavy on the sides but not on the angles, pale brown when young becoming dark brown to black, with angles paler than the sides.

Annual. In seasonally flooded places or standing in shallow permanent water. Sometimes cleistogamous flowers develop within the basal sheaths; the cleistogamously developed nuts are usually somewhat larger (up to 2.5 mm long) than chasmogamous ones. It is easily recognized by its dense clusters of golden-brown or bronze-coloured spikelets and the very concave triangular glumes. It is rather like *S. roylei* but is distinguished by the larger glumes (2.5–3.2 mm long), as well as the more distinct transverse septa in the culm. It is sometimes considered to be no more than a synonym of *S. articulatus*.

{⊙ Hel/Ten}. Africa and India: Ind: Ap, Gj, Kt, Rj, Tn.

Schoenoplectus subulatus (Vahl) K. Lye [*Scirpus subulatus* Vahl; *Scirpus (Schoenoplectus) litoralis* auct. Ind., non Schrader, *'littoralis'* orth. mut., *Scirpus thermalis* Trabut]: Fig. 187

Rhizomes with stolons; stolons up to 11 mm thick, usually terminating in tubers. Culms loosely tufted, terete but sometimes obtusely 3-sided just below the inflorescence, glaucous, 40–200 (–450) cm tall, 6–12 mm in diameter below, 2–3 mm thick below the inflorescence. Leaves basal, reduced to scales or bladeless or sometimes the upper sheath bladed; sheaths greyish-brown to rust-coloured, more or less splitting into fibres; blade when present shorter than the culm, V-shaped below, 3-angular and winged at the tip. Main inflorescence bract appearing as a continuation of the culm, 2–5 (–10) cm long, shorter or just overtopping the inflorescence, 3-sided but somewhat flattened, channelled along the inner side, the tip obtuse. Inflorescence pseudolateral, spreading, rarely simple, usually compound, with 3 or more, unequal, ascending rays, terminating in umbels or panicles of spikelets; rays 1–3 (–6) cm long. Spikelets usually stalked and solitary, ovoid to oblong-ovoid, terete, densely many-flowered, pale to medium brown, 6–15mm long, 2–4.5 mm in diameter. Glumes scarious, appressed, very broadly oblong, strongly concave, obtuse or slightly notched, 3–4 mm long, ± 2 mm wide; sides nerveless, pale with darker brown lines or patches; margin paler, minutely hairy or erose at the tip; midrib prominent and excurrent into a 0.3–0.5 mm long, whitish, hairy or scabrid mucro. Perianth of (3) 4 or 5, scales, each ligulate-spathulate, plumosely fringed, with moliniform hairs, rust-coloured, as long as or slightly longer than the nut. Stamens (2) 3; filaments flattened and persisting on the nut; anthers 1.5–2 mm long; connective with a white, fan-like, fimbriate appendage. Style longer than the nut; stigmas 2. Nuts unequally biconvex, strongly dorsiventrally flattened, elliptical to obovate, ± 2 mm long (excluding the 0.5 mm long apex), 1.5–1.7 mm wide, smooth, chestnut brown to black.

Perennial. Found in wet places, often standing in water, tolerating depths of up to 2 m. It is particularly common near the coast and is found in saline pools.

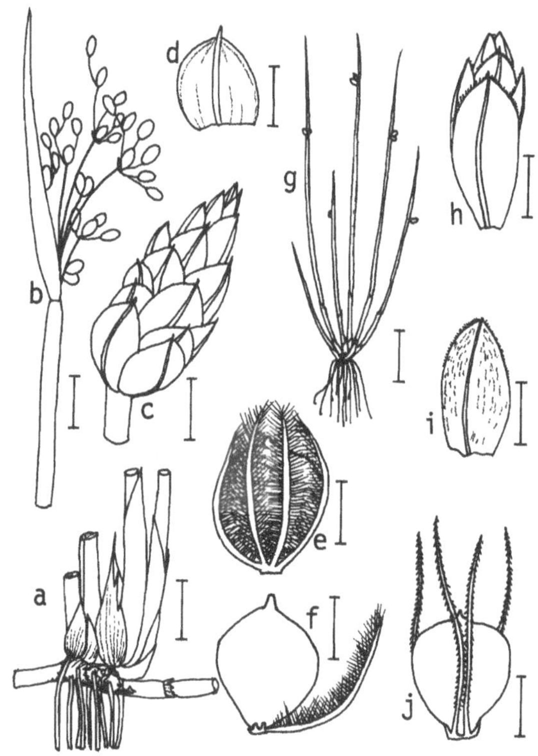

Fig. 187. *Schoenoplectus subulatus*: a, base of plant (2 cm); b, inflorescence (2 cm); c, spikelet (2 mm); d, glume (2 mm); e, fruit (1 mm); f, nut with all but one perianth segment removed (1 mm). *S. wallichii*: g, habit (10 cm); h, spikelet (2 mm); i, glume (2 mm); j, fruit (1 mm).

Used locally for making mats. It superficially resembles *S. corymbosus* but is distinguished by its plumose perianth segments. In most Indian Floras it is considered conspecific to *S. litoralis* which is a temperate species from Europe and W Asia with 3-angled culms and bracts and broader perianth segments.

{Hyp/Ten}. India, SE Asia and Micronesia: Cey, Ind: An, Dl, Hp, Kl, Kt, Pj, Rj, Tn.

Schoenoplectus wallichii (Nees) J. Sojak [*Scirpus wallichii* Nees]: Fig. 187

Like *S. juncoides* but smaller and more slender (culms up to 40 cm tall, 0.7–1 mm wide); inflorescence with 1 or 2 (3) spikelets; spikelets 3–4 mm wide; perianth bristles 4–5, all distinctly longer than the nut, not less than ± 3 mm long; anthers 0.5–0.8 mm long; style always 2-cleft; plano-convex (the ventral side flat), ± 1.8 mm long, ± 1.5 mm wide.

Annual. In wet open places and rice fields. It seems to intergrade with *S. juncoides* and may not be worth recognition at the rank of species.

{⊙ Hyp/Ten}. NE India, SE and E Asia extending into Korea: Ind: As? Wb?

Scirpus in the strict sense found in the Himalayas; most of the species cited in older Floras are to be found in *Bolboschoenus, Eleocharis, Eleogiton, Lipocarpha, Schoenoplectus* or *Websteria* in this work. Users should refer to the key on page 94.

Scleria: 200 species: cosmopolitan

Verma, D. M. and Chandra Veena. The genus *Scleria* Berg. in India. Bulletin of the Botanical Survey of India 32(1–4): 57–79 (dated 1990, published 1992).

Excluded: the following may grow in damp places on wet soil but they are not considered to be wetland plants: ***Scleria biflora*** Roxburgh [*S. tessellata* Willdenow, *S. steudeliana* Miquel]; ***S. levis*** Retzius [*S. hebecarpa* Nees]; ***S. neesii*** Kunth [*S. stricta* Moon ex Pfeiffer, *S. capitata* (Nees) Pfeiffer]; ***S. parvula*** Steudel [*S. uliginosa* Hochstetter ex Boeckeler, ***S. fenestrata*** Franchet et Savigny]; ***S. pergracilis*** (Nees) Kunth; ***S. rugosa*** R. Brown [*S. lateriflora* Boeckeler, S. *flaccida* sensu C. B. Clarke non Steudel]; ***S. sumatrensis*** Retzius; ***S. swamyii*** E. Govindarajalu.

Scleria poaeformis Retzius [*S. oryzoides* Presl]: Fig. 188

Rhizomes creeping, 4–10 mm thick. Culms arising solitary along the rhizome, 3-angled, with concave sides, leafy, (80–) 130–200 cm tall, 3–10 (–13) mm thick. Leaf sheaths reddish towards the mouth; blades linear, usually flat above, up to 120 cm long, 8–25 (–50) mm wide, tough, glabrous but scabrous on the margins and the three prominent nerves. Inflorescence a single, terminal panicle, with compound branches bearing numerous spikelets, without leafy bracts; panicle 10–20 cm long, 5–12 cm broad, elliptical in outline. Spikelets solitary, with 1–4 spirally arranged glumes, bisexual or unisexual; bisexual spikelets with 1 terminal female flower and 1–several male flowers; female spikelets with 1 female flower surmounted by 1 or 2 empty glumes; male spikelets numerous, 4–5 mm long, with several male flowers. Glumes straw-coloured or brown, glabrous or hairy. Perianth bristles absent. Stamens 3. Stigmas 3. Nuts broadly ovoid to subglobose, 2.5–3.5 mm long, 2–3 mm broad, smooth, glabrous, glistening white, with 3 depressions at the base and with a persistent stalk or cup-shaped structure ± 2 mm across at the base bearing 3 short and triangular lobes.

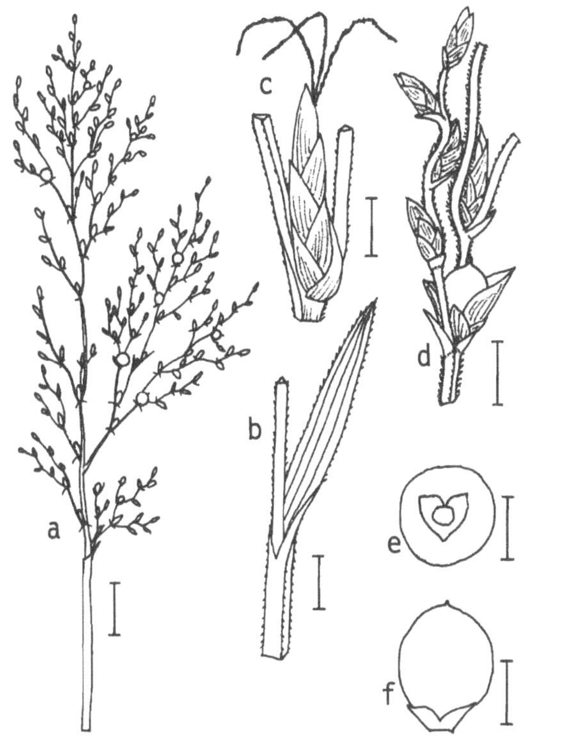

FIG. 188. *Scleria poaeformis*: a, portion of inflorescence (1 cm); b, uppermost leaf blade (1 cm); c, female spikelet at anthesis (1.5 mm); d, one female spikelet in fruit with five male spikelets (3 mm); e, nut viewed from below (2 mm); f, nut viewed from the side (2 mm).

Perennial. In shallow water at the edge lakes and pools, in streams and in swamps. It often forming large colonies. Recorded as a weed in rice fields. Recognized by the large terminal panicle without any lateral panicles.

{Hyp?/Ten?/Hel}. Tropical E Africa, Asia and Australia: Cey, Ind: As.

Websteria [*Baeothryon, Eleocharis* pro parte, *Schoenus* pro parte, *Scirpus* pro parte]: 1 species: widespread in the tropics

Websteria confervoides (Poiret) S. Hooper [*Eleocharis confervoides* (Poiret) T. Koyama, *Scirpus confervoides* Poiret, *Scirpus submersus* C. H. Wright ex Sauvalle; *Rhynchospora ruppioides* Bentham]: Fig. 189

Stems up to 100 cm long, with branches arising in whorl-like clusters, usually subtended by 2 bracts or scale-like leaf sheaths (prophylls); internodes between successive whorls progressively shorter except for the last whorls; the terminal branches longer than the lower internodes, 2–6 cm long, less than 0.1 mm thick, simple, hair-like and often very numerous. Leaves reduced to scales or tubular sheaths, which quickly decay (the hair-like terminal branches have often been considered to be leaves or are called phyllodes). Inflorescence of 1–4, long, peduncled (2–5 cm long) spikelets borne in umbels at the ends of branches but among the hair-like branches, each subtended by 2 scale-like sheaths. Spikelets oblong-lanceolate, 5–12 mm long, 1.2–2mm wide, 1-flowered. Glumes 2, enclosing the spikelet, the lowest sterile, the next bisexual, male or sterile. Perianth bristles 6–14, 2–5.5 mm long, filiform but flattened at base, whitish, bearing backwardly directed barbs, persisting in fruit. Stamens 3; anthers ± 2.2 mm long. Style 5–6 mm long; stigmas 2, ± 2 mm long; style base persistent, 2–3 mm long, flattened caducous. Nuts ovoid-trigonous to almost round, 2–2.6 mm long (excluding the persistent style base), 1.3–1.5 mm wide, grey to brownish, minutely reticulate.

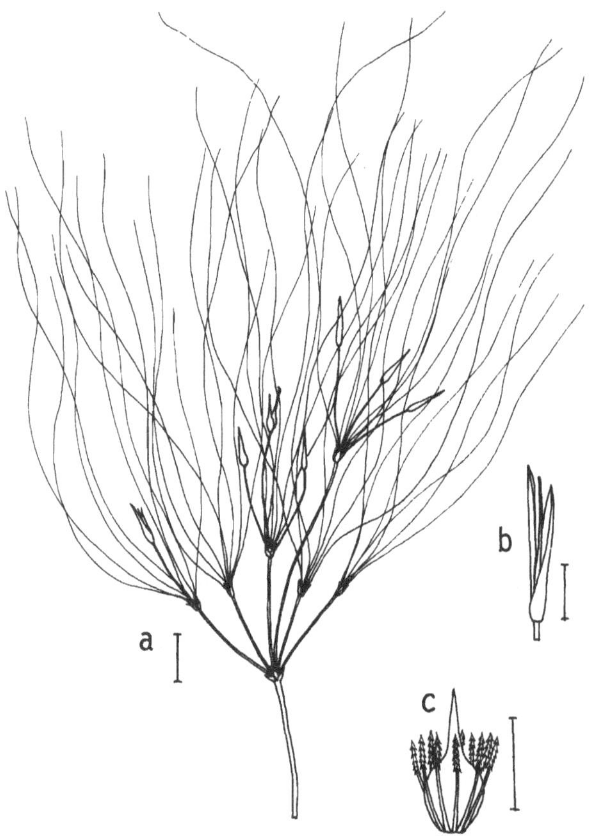

FIG. 189. *Websteria confervoides*: a, habit (5 mm); b, spikelet (1 mm); c, fruit (1 mm).

Perennial or perhaps exceptionally annual. A submerged, free-swimming or floating, slender, somewhat alga-like plant with emergent spikelets. Found in standing or slowly flowing, oligotrophic water. It seems to be rather rare throughout its range. The disseminules have persistent and barbed perianth segments and are thus probably dispersed by animals. It is sometimes placed in *Eleocharis* and sometimes in *Scirpus* but it has also been placed in *Baeothryon, Rhynchospora* and *Schoenus*; recent work indicates that it is better placed in a genus of its own. Eiten, L. T. The morphology of some critical Brazilian species of Cyperaceae. Annals of the Missouri Botanical Garden 63: 181–199 (1976).

{(⊙)? Vit/Hyp}. Tropical America, Africa, Malay Peninsular, Sri Lanka and perhaps S India: Cey, Ind.

DROSERACEAE: 4 genera

Carnivorous herbs. Leaves alternate or in whorls; blades rolled in bud. Flowers bisexual, radially symmetrical, usually 5-merous. Sepals united below. Petals free. Stamens as many as or up to 4 times as many as the petals. Ovary superior, 1-locular; styles 2–5. Fruit a capsule; seeds small and numerous.

1. Rootless planktonic herb; leaves in whorls, each terminating in 4–6 bristles and a hinged orbicular lobe ***Aldrovanda***

1. Rooted herbs; leaves in basal rosette or alternate along the stem, blades flattened bearing glandular tentacles ***Drosera***

Aldrovanda: 1 species

Aldrovanda vesiculosa Linnaeus: Fig. 190

Planktonic, rootless herb. Stems free-swimming, up to 60 cm long but usually less, 0.6–0.7 mm in diameter, irregularly branched. Leaves 6–17 mm long, in crowded whorls of 5–9; each leaf with a cuneate basal part and terminating in 4–7 bristles and an orbicular lobe, hinged along the midrib, which can close rapidly and trap small animals. Flowers axillary, solitary, shortly stalked. Sepals 5, ovate, 3–4 mm long. Petals 5, obovate, 4–5 mm long, greenish-white. Stamens 5, ± as long as the sepals. Styles 5, each with branched and fringed stigmas. Seeds ellipsoid, ±1.5 mm long, black and shiny.

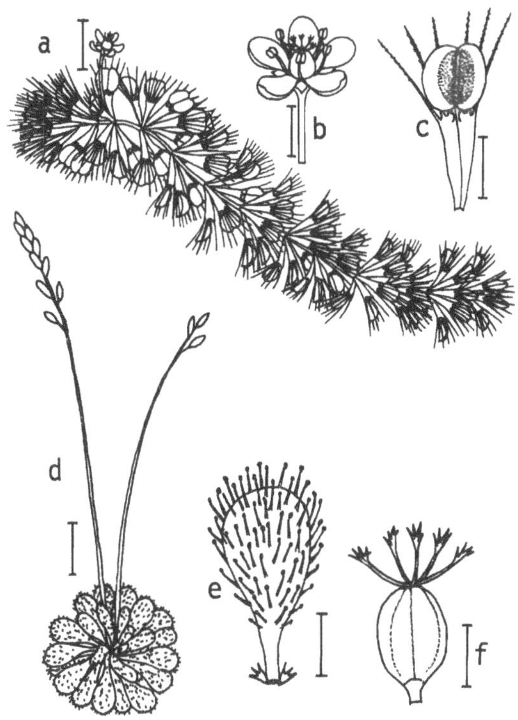

Fig. 190. *Aldrovanda vesiculosa*: a, flowering shoot (1 cm); b, flower (4 mm); c, leaf with trap (5 mm). *Drosera burmannii*: d, flowering plant (5 mm); e, young leaf with stipules (1 mm); f, fruit (2 mm).

Perennial. Totally submerged in water, usually found among reeds. Gregarious and forming large colonies in West Bengal where it is sometimes considered to be a pest. Flowers emergent and entomophilous or autogamous or submerged and cleistogamous. The disseminules are seeds or turions (specialized apical buds). It is a carnivorous plant with active traps. Degreef, J. D. *Aldrovanda vesiculosa*. Association Française d'Amateurs de Plantes Carnivores (Rouen) 1–40 (undated, published shortly after 1986).

{Pla}. Africa, Asia and Australia, probably introduced in Europe where it apparently does not flower: Ban, Ind: Mn, Wb (it has been reported to be extinct in Salt Lake near Calcutta).

Drosera: ± 85 species

Terrestrial herbs. Leaf blades flattened, bearing sensitive, sticky, gland-tipped tentacles. Flowers in spirally wound (helicoid), raceme-like inflorescences. Capsule loculicidal, opening by 3–5 valves.

Perennial or perhaps sometimes annual. Pollinated by insects or autogamous. The disseminules are small seeds.

1. Cauline leaves absent, all leaves in basal rosettes; styles 5 ***D. burmannii***

1. Leaves borne along the stem, not all in basal rosettes; styles 3

2. Leaf blades linear, longer but hardly broader than the petioles; subterranean tubers absent; styles bipartite to base ***D. indica***

2. Leaf blades peltate, much broader than the petioles; subterranean tubers present; styles fimbriate (much branched) ***D. peltata***

Drosera burmannii Vahl: Fig. 190

Leaves in basal rosettes, cauline leaves absent; stipules 3-partite, scarious; blades obovate-orbicular, 5–8 mm long, 3.5–6 mm wide, reddish to green. Inflorescences up to 10 cm long, (2-) 10- to 20-flowered. Sepals oblong. Petals oblong-oblanceolate, ± 4 mm long, pink to whitish. Ovary of 5 carpels; styles 5. Capsule 5-valved, 1–2 mm long.

Perennial. In permanently wet places or regularly flooded areas. Usually on sandy soil in open marshes amidst grass and around tanks. Flowers open only in the morning.

{Hel}. W Africa to NE Australia: Cey, Ind, Nep: Bh, Go, Kl, Kt, Mp, Mz, Tn, Wb.

Drosera indica Linnaeus: Fig. 191

Stem erect, up to 5 cm tall. Leaf blades linear, longer but hardly broader than the petioles; the lower leaves spreading or deflexed, up to 2.5 cm long; the upper leaves erect, 1–3 cm long. Inflorescences up to 1.5 cm long, 3- to 20-flowered. Sepals ovate, margin entire. Petals oblong-oblanceolate, ± 2 mm long, pink to whitish. Ovary of 3 carpels; styles 3, each 2-cleft to the base. Capsule 3-valved, ± 3 mm long.

Perennial or perhaps sometimes annual. In permanently wet or regularly flooded places. Found in swamps, shallow pools and ditches, often in or near rice fields. Usually on clayey soil. Sometimes grows with *Drosera burmannii*.

{(⊙)? Hel}. Tropical Africa to Australia: Bur, Cey, Ind: Bh, Go, Kl, Kt, Mh, Mp, Wb.

Drosera peltata J. E. Smith ex Willdenow [*D. peltata* var. *lunata* (F. Hamilton ex A. P. de Candolle) C. B. Clarke]: Fig. 191

Stem erect or straggling, up to 25 cm long, with a reddish tuber. Basal leaves in a rosette; stem leaves alternate or subopposite; leaf blades peltate, lunate, 2.5–4 mm long, 1.5–2.5 mm wide; tentacles 1.5–2.5 mm long. Inflorescences up to 5 cm long, ± 10-flowered. Sepals ovate, margin toothed toward the tip. Petals spathulate-obovate, ± 2 mm long, white.

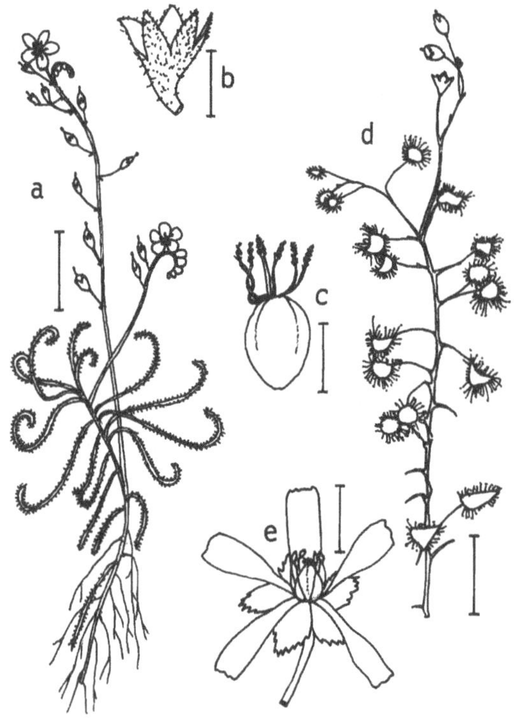

Fig. 191. *Drosera indica*: a, flowering plant (2 cm); b, sepals (4 mm); c, young fruit (2 mm). *D. peltata*: d, flowering plant (1 cm); e, flower (4 mm).

Ovary of 3 carpels; styles 3, each fimbriate (several time branched). Capsule 3-valved, ± 3 mm long.

Perennial. In moist places, particularly around exposed rocks on open hill sides, on wet grassy patches and along the banks of waterways. It is confined to hilly regions above 1000 m. Used for dying silk brown and used locally in medicine.

{Hel}. S and SE Asia and Australia: Cey, Ind, Nep: Kt, Tn, Up. 1000–3600 m.

ELATINACEAE: 2 genera

Annual and perennial, sometimes woody. Leaves opposite, rarely whorled, simple; margin entire or toothed; stipules paired. Flowers bisexual, radially symmetrical, axillary, solitary or in cymes, often inconspicuous. Sepals 2–5, free or united at base. Petals as many as sepals, free. Stamens as many or twice as many as sepals. Ovary superior, 2- to 5-locular; styles 2–5; placentation axile; fruit a septicidal capsule, with an axile placenta; seeds numerous, usually curved, mostly with a net-like surface pattern.

Tucker, G. C. The genera of Elatinaceae in the southeastern United States. Journal of the Arnold Arboretum 67(4): 471–483 (1986) — literature survey.

1. Sepals free, acute, with a distinct nerve; plants usually robust and conspicuous; flowers rarely solitary, 2 or more, usually in dense axillary clusters; leaf margins toothed ***Bergia***
1. Sepals united at the base, obtuse, without a distinct nerve; individual plants usually small and inconspicuous; flowers solitary in leaf axils; leaf margins entire ***Elatine***

Bergia: 24 species: tropical and warm regions of the world; centres of diversity in Australia and S Africa.

Usually conspicuous and robust, herbaceous or woody. Stems when in water often swollen. Leaves decussate, never whorled; margin frequently toothed; stipules persistent. Flowers often in dense, axillary clusters. Sepals 3 or 5, acute, free, with a distinct nerve, often glandular. Petals 3 or 5, obovate, membranous. Stamens (15–) 10 or less. Ovary 3- or 5-locular; capsule ovoid to globose.

Annual or perennial. Helophytic or seasonally submerged. Pollinated by insects or autogamous, cleistogamy is common. The disseminules are small seeds, presumably dispersed in mud by birds. Leach, G. J. Taxonomic revision of *Bergia* (Elatinaceae) in Australia. Journal of the Adelaide Botanic Gardens 11(2): 75–100 (1989).

Excluded: shrubs or subshrubs with 10 stamens sometimes found along rivers but not really wetland plants: ***Bergia aestivosa*** Wright et Arnott: ♦Ind, ♦Pak (glabrous; leaves keeled and narrow; flowers 1–4 in axils of leaves); ***Bergia polyantha*** Sonder; ***B. suffruticosa*** (Raffeneau-Delile) Fenzel [*B. odorata* Edgeworth]: tropical Old World (hispid; leaves flat, not keeled and elliptical; flowers rarely solitary usually 4 or more in axillary clusters).

1. Stems glabrous, usually inflated; capsule valves greenish or pale brown; stamens 10–15 ***B. capensis***
1. Stems with hairs, wire-like, not inflated; capsule valves reddish; stamens 3–5
  2. Seeds smooth or faintly striate, 0.2–0.25 mm long; indumentum of simple and glandular hairs; sepals, petals and stamens 5 ***B. ammannioides***
  2. Seeds distinctly reticulate, (0.25–) 0.3–0.4 mm long; indumentum of simple, eglandular hairs; sepals, petals and stamens 3, 4 or rarely 5 ***B. trimera***

Bergia ammannioides Roxburgh ex Roth: Fig. 192

Stems prostrate or erect, simple or branched below, up to 50 cm long but usually less, never inflated, reddish, with glandular and simple hairs. Stipules erect, elongate-triangular, 1.5–2 mm long, ± 0.5 mm wide at the base, membranous, entire or with toothed margins. Leaves sessile; blades narrow-elliptical to elliptical, 8–25 mm long, 3–8 mm wide, rather leathery,

FIG. 192. *Bergia ammannioides*: a, habit (2 cm); b, flower (1 mm); c, base of leaf (0.5 mm); d, seed (0.1 mm). *B. trimera*: e, seed (0.1 mm); f, base of leaf (0.5 mm); g, flower (1 mm).

with glandular hairs when young, margin serrate with glandular hairs, nerves prominent, apex acute. Flowers in dense axillary cymes, pedicels up to 4 mm long. Sepals 5, narrow-ovate, 1.2–2 mm long, ± 0.5 mm wide, acute to acuminate, pubescent, green, margin fimbriate. Petals 5, narrow-ovate, hardly exceeding the sepals, erect at anthesis, whitish-pink. Stamens 5; ± 1 mm long, filaments swollen at the base. Styles 5, 0.25–0.3 mm long. Capsules globose, 1.5–2 mm in diameter, bright red, opening by 5 valves. Seeds ellipsoid, 0.2–0.25 mm long, ± 0.15 mm in diameter, smooth or faintly striate, never reticulate.

Annual. Found at the edge of tanks, ditches, creeks and in rice fields. Sometimes in saline areas. It flowers as the water recedes.

{⊙ Hel}. Warmer and tropical regions of Africa, Asia and Australia: Ban, Cey? Ind, Nep, Pak: Ap, Bh, Dl, Go, Hr, Jk, Kl, Kt, Mh, Mp, Or, Pj, Rj, Tn, Tp, Up, Wb.

Bergia capensis Linnaeus [*B. verticillata* Willdenow]: Fig. 193

Stems erect or creeping and rooting at nodes, usually branched, up to 50 cm long but usually less, red or pink, glabrous, inflated but constricted at nodes when submerged or floating. Stipules erect, elongate-triangular, 1–3 mm long, membranous, with toothed margins. Leaves sessile; blades lanceolate to elongate-lanceolate, 1.5–5 cm long, 0.75–2.5 cm wide, serrate, green with red teeth. Flowers in dense axillary cymes, pedicels up to ± 4 mm long. Sepals 5, pale green with red flecks and tips, 1–2 mm long. Petals 1.3–2 (–5) mm long, spreading, translucent white or pink. Stamens 15 or less; filaments unequal, 0.5–1 mm long. Styles 0.25–0.3 mm long. Capsules globose, ± 2.5 mm in diameter, green or pale brown, opening by 5 valves. Seeds oblong, 0.4–0.6 mm long, slightly curved, with reticulate seed coat.

Annual or perhaps sometimes perennial. Usually found in seasonally inundated areas, characteristically found around tanks, in rice fields and along irrigation ditches. Flowers at any time of year.

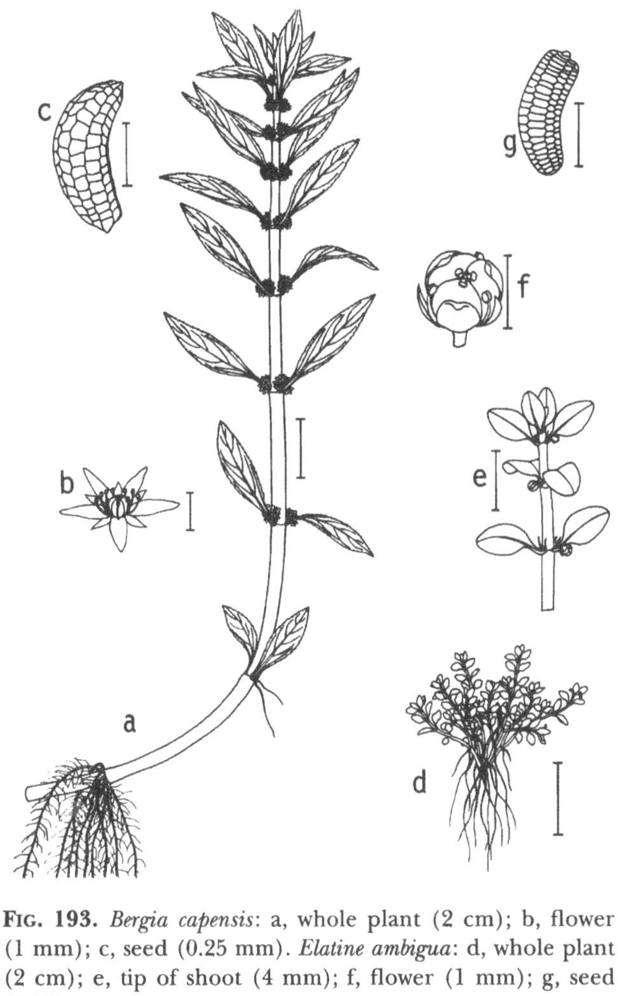

FIG. 193. *Bergia capensis*: a, whole plant (2 cm); b, flower (1 mm); c, seed (0.25 mm). *Elatine ambigua*: d, whole plant (2 cm); e, tip of shoot (4 mm); f, flower (1 mm); g, seed (0.25).

{(⊙)? Ten}. Tropical and subtropical Africa and Asia, introduced in Europe and S America: Ban, Bur, Cey, Ind: Ap, Bh, Dl, Go, Kl, Kt, Mh, Mp, Or, Tn, Tp, Up, Wb.

Bergia trimera Fischer et C. A. A. Meyer [*B. ammannioides* var. *trimera* (Fischer et C. A. A. Meyer) Bentham]: Fig. 192

Shoots and sepals without red-tipped glandular hairs. Stems prostrate when young becoming erect later, usually branched below, up to 15 cm long, never inflated, often somewhat woody, reddish, with simple eglandular hairs. Stipules erect, elongate-triangular, 1.2–2.6 mm long, 0.5–0.9 mm wide at the base, membranous, entire or with toothed margins. Leaves sessile or nearly so; leaf blades narrow-obovate to narrow-elliptical, 3.5–30 mm long, 1.5–8 mm wide, rather leathery, glabrous, margin serrate, nerves obscure, apex acute. Flowers 4–20, in dense axillary cymes, pedicels 0–2.5 mm long. Sepals 3, 4 or rarely 5, narrow-elliptical, 0.7–1.5 mm long, ± 0.3 mm wide, acute, glabrous or rarely pubescent, green or pink, margin entire or fimbriate. Petals 3, 4 or rarely 5, elliptical, ± equalling the sepals, erect at anthesis, whitish-pink. Stamens 3 or 4 (5); 0.5–1 mm long, filaments swollen at the base. Stigmas 3 or 4 (5), sessile or nearly so. Capsules globose, 1–1.2 mm in diameter, bright red, opening by 3–5 valves. Seeds ovoid to reniform, 0.3–0.4 mm long, ± 0.2 mm in diameter, reticulate.

Perennial or annual. Marshes, borders of tanks, seepage areas and in rice fields. Often confused with *B. ammannioides* but differs in lacking glandular hairs; with nerves on leaf blade obscure; usually 3-merous flowers; stigmas sessile or nearly so, 3 or rarely 4; seeds ovoid to reniform, 0.3–0.4 mm long, ± 0.2 mm in diameter, reticulate. It also tends to be perennial. It is probably more widely distributed in India than shown here, the Rajasthan record is based on my own collections.

{(⊙) Hel}. India, Sri Lanka and Australia: Cey, Ind: Rj.

Elatine: 12 species, all aquatic: cosmopolitan.

Usually small and inconspicuous. Leaves opposite (or in *E. alsinastrum* whorled), glabrous; margin not toothed. Flowers not in dense clusters, usually solitary. Sepals 2, 3 or 4, obtuse, united below, without a distinct vein, not glandular. Ovary 2- to 4-locular.

Annual or short-lived perennials. Seasonally or totally submerged. Autogamous or perhaps sometimes entomophilous. Disseminules seeds, probably dispersed in mud or by water. Commonly found in fish ponds and rice fields but usually considered beneficial as it consolidates mud.

Excluded: ***Elatine gracilis*** Mason; a N American species, it has been reported at 1387 m in Himachal Pradesh, it is not known to have spread south.

1. Flowers distinctly stalked; the capsules turning to one side away from the leaf axils at maturity ***E. ambigua***

1. Flowers sessile or nearly so; the capsules remaining in the leaf axils ***E. triandra***

Elatine ambigua Wight: Fig. 193

Stems creeping and rooting at nodes, densely branched, turf-forming; internodes (1.5–) 3–4 (–11) mm long, ±0.5 mm in diameter. Stipules minute, membranous, elongate-triangular, irregularly toothed or fimbriate. Leaves in opposite pairs, blades lanceolate to elongate-lanceolate, (3–) 4–6 (–8) mm long, (0.5–) 1–1.5 (–2) mm wide, margins entire, apex rounded. Flowers solitary, usually opening, distinctly pedicellate; pedicels 1–2 mm long, with 2 membranous bracteoles at the base; bracteoles minute, irregularly toothed or fimbriate. Sepals 3, ovate, ± 0.5 mm long. Petals 3, ovate to elongate ovate, ± 1 mm long, reddish-white, spreading at anthesis but lying on the fruit later. Stamens 3, shorter than the sepals. Styles 3, curved. Capsule globose, ±1.5 mm in diameter, when ripe lying to the side of and not in the leaf axil, opening by 3 valves. Seeds ± 0.5 mm long, ±0.1 mm in diameter, slightly curved or almost straight, with a net-like pattern on the surface, the individual reticulations mostly hexagonal.

Annual. Found in seasonally inundated areas, along rivers, around tanks and in irrigation ditches. It prefers eutrophic conditions and is mostly found on fine mineral soils; it is frequent in rice fields. Underwater it is flaccid with some branches unrooted, on land it highly branched and forms a tight turf very close to the ground. The flowers are chasmogamous or with post-anthesis cleistogamy.

{⊙ Ten}. Warm and tropical Asia, introduced in Europe and North America: Ban, Ind: Ap, Jk, Kl, Kt, Mg, Mp, Rj, Tn.

Elatine triandra Schkuhr [? *E. americana* (Pursh) Arnott]

Like *E. ambigua* but usually somewhat larger: stems (2–) 6–18 mm long; leaf blades (3–) 5–12 (–15) mm long, 1–2 mm wide; flowers sessile or nearly so, usually cleistogamous; sepals (2–) 3 (–4), usually 2 large and 1 small, 0.4–0.8 mm long; petals hardly exceeding the sepals; seeds rarely straight, with mostly rectangular reticulations.

Ecology like that of *E. ambigua*; both species are sometimes found growing intermingled; *E. triandra* is in these circumstances usually larger in all its parts and is the more aquatic of the two. The flowers are usually cleistogamous. It is often misidentified and some of these records are perhaps wrong. It is unlikely to occur in Peninsular India.

{⊙ Ten}. Warmer areas of the Northern Hemisphere: Ban, Ind, Nep: Jk. 1400–2400 m.

ERIOCAULACEAE: 9 genera

Eriocaulon: ± 400 species: almost cosmopolitan

Stems simple, corm-like or elongate. Leaves spiral, without petioles, in rosettes or cauline, in submerged aquatic species usually dense and filiform. Scapes simple, often twisted, enclosed at the base by a tubular sheath, usually exceeding the leaves. Inflorescence a dense head, subtended by an involucre of bracts. Involucral bracts scarious to leathery, spreading or reflexed at maturity. Flowers unisexual, male and female mixed or in separate bands in the head, each subtended by a floral bract. Floral bracts frequently white hairy towards the tip. Perianth 2- or 3-merous; outer segments (sepals) free or united into a spathe-like tube split down one side; inner segments (petals) free, each bearing a black or rarely red nectar gland towards the apex (may be absent in female flowers). Stamens 4 or 6, one whorl on the petals the other on the sinuses between the petals; anthers 4-locular, black or white. Ovary superior 2- or 3-locular; stigmas 2 or 3, simple, free, elongate; ovules solitary and pendulous. Fruit a membranous capsule, loculicidally dehiscent. Seeds ellipsoid to globose, yellow to brown, sometimes appearing somewhat translucent, smooth or distinctively patterned.

Perennial or annual. Most species are found in wet places in marshes, swamps and seasonally inundated areas, while others have totally submerged leaves. Pollination by insects or autogamous. The disseminules are minute seeds, dispersal unknown but probably transported in mud on the feet of birds. Often found in rice fields and irrigation ditches but rarely considered to be seriously detrimental.

The genus *Eriocaulon* has many local endemic species in the subcontinent of India; it is also rather confused from a taxonomic point of view. Most Indian species grow in wet places in hills rather than in wetlands. I have, therefore, concentrated my effort on the aquatic species and the lowland wetland species. The species growing in India as a political area have been revised by Ansari, R. and Balakrishnan, N. P. The family Eriocaulaceae in India. Bishen Singh Mahendra Pal Singh, Dehra Dun 1–195 (1994); this work has been used as the basis of several illustrations. For the Indian subcontinent there is the work of Fyson, P. F. The Indian species of *Eriocaulon*. Journal of Indian Botany, 2: 133–150, 192–207, 259–266, 307–320 (1921); 3: 12–18, 91–115 (1922) — this work is out of date and rather unreliable. For Sri Lanka — Phillips, Sylvia M. Notes on some *Eriocaulon* species from Ceylon. Kew Bulletin 49(2): 287–303 (1994). I would particularly like to thank Dr R. Ansari (Kozhikode), Dr Sylvia M. Phillips (Kew) and Prof T. Stützel (Bochum) for their help.

Excluded species of ***Eriocaulon***:

Endemic to the Himalayas: ***E. bhutanicum*** Noltie: ♦Bhu; ***E. katmanduense*** Satake: ♦Nep; ***E. pseudoquinquangulare*** Ruhland: ♦Ind: Up; ***E. pumilio*** J. D. Hooker: ♦Ind: Up; ***E. staintonii*** Satake: ♦Nep; ***E. trisectoides*** Satake: ♦Nep; ***E. viride*** Körnicke.

Endemic to the hills of NE India and the Himalayas: ***E. alpestre*** J. D. Hooker et Thomson ex Körnicke: ♦Ind, ♦Nep: As, Mg, Sk, Up; ***E. edwardii*** Fyson [*E. edwardii* var. *clarkei* Haines]: Ban? ♦Ind: Bh, Sk, Up, Wb.

Endemic to the hills of NE India and or Myanmar: ***E. barba-caprae*** Fyson: ♦Ind: Mg; ***E. cherrapunjianum*** R. Ansari et N. P. Balakrishnan: ♦Ind: Mg; ***E. colletii*** J. D. Hooker: ♦Bur, ♦Ind: As, Mg; ***E. cristatum*** Martius ex Körnicke [*E. cristatum* var. *mackii* J. D. Hooker]: ♦Ind: As, Mg; ***E. gracile*** Martius [*E. gracile* var. *kurzii* Fyson, *E. infirmum* Steudel var. *kurzii* (Fyson) Moldenke]: ♦Bur; ***E. gregatum*** Körnicke: ♦Ind: Mg; ***E. miserum*** Körnicke: ♦Ind: Mg; ***E. mitophyllum*** J. D. Hooker [*E. miserum* var. *mitophyllum* (J. D. Hooker) Fyson]: ♦Ban? ♦Ind: As, Mg.

Endemic to the Himalayas, the hills of NE and S India: ***E. nepalense*** Prescott ex Bongard [*E. luzulifolium* Martius ['*luzulaefolium*' orth. mut., ?*E. viride* Körnicke]: ♦Ind, ♦Nep: As, Hp, Mg, Sk, Tn, Up.

Endemic to NE India, Bangladesh and SW India: ***E. achiton*** Körnicke: ♦Ban, ♦Ind: Go, Kt, Mh, Mg (records from Lk, Mp based on incorrect identifications).

Endemic to the highlands of Sri Lanka: ***E. atratum*** Körnicke: ♦Cey; ***E. catopsioides*** S. M. Phillips [*E. brownianum* sensu Thwaites, non Martius]: ♦Cey; ***E. ceylanicum*** Körnicke ['*zeylanicum*' orth. mut., *E. subcaulescens* J. D. Hooker, *E. ceylanicum* var. *subcaulescens* (J. D. Hooker) Fyson]: ♦Cey; ***E. philippo-coburgi*** Szyszyowicz ex Wawra [*E. atratum* Körnicke var. *major* Thwaites ex Ruhland, *E. caulescens* J. D. Hooker et Thomson, non Poiret, *E. robustum* var. *caulescens* Fyson]: ♦Cey; ***E. subglaucum*** Ruhland: ♦Cey; ***E. trimeni*** J. D. Hooker: ♦Cey .

Endemic to the hills of S India and the hills of Sri Lanka: ***E. collinum*** J. D. Hooker: ♦Cey, ♦Ind; ***E. longicuspe*** J. D. Hooker ['*longicuspis*' orth. mut., *E. polycephalum* J. D. Hooker, *E. longicuspe* var. *polycephalum* J. D. Hooker) Fyson]: ♦Cey, ♦Ind: Kl, Kt, Mp, Tn; ***E. odoratum*** Dalzell [*E. christopheri* Fyson, *E. oliveri* Fyson, *E. collinum* var. *nanum* Moldenke, *E. infirmum* sensu Gandhi (Fl. Hassan), non Steudel]: ♦Cey, ♦Ind: Ap, Kt, Kl, Go, Mh, Tn; ***E. thwaitesii*** Körnicke [*E. neesianum* auct. Ind., non Körnicke, *E. mariae* Fyson, *E. gamblei* C. E. C. Fischer]: ♦Cey, ♦Ind: Kl, Tn.

Endemic to the hills of Peninsular India: ***E. conicum*** (Fyson) C. E. C. Fischer [*E. dianae* var. *conicum* Fyson]: ♦Ind: Kl, Kt, Tn; ***E. elenorae*** Fyson: ♦Ind: Kl, Kt, Mh, Tn; ***E. ensiforme*** C. E. C. Fischer: ♦Ind: Kl, Tn; ***E. margaretae*** Fyson: ♦Ind: Kt, Mh; ***E. meeboldii*** R. Ansari et N. P. Balakrishnan [*E. hamiltonianum* var. *minimum* Fyson]: ♦Ind: Kt; ***E. leucomelas*** Steudel [*E. melaleucum* Martius non Bongard]: ♦Ind: Kl, Tn. ***E. minutum*** J. D. Hooker [*E. humile* Moldenke]: ♦Ind: Mh, Rj; ***E. nairii*** Chandrabose et Chandrasekan: ♦Ind: Kl, Tn; ***E. palghatense*** R. Ansari et N. P. Balakrishnan: ♦Ind: Kl; ***E. panagudianum*** R. Ansari et N. P. Balakrishnan: ♦Ind: Tn; ***E. pectinatum*** Ruhland [*E. geoffreyi* Fyson]: ♦Ind: Tn; ***E. rajendrababui*** R. Ansari et N. P. Balakrishnan: ♦Ind: Mp; ***E. ramnadense*** R. Ansari et N. P. Balakrishnan: ♦Ind: Tn; ***E. ritchieanum*** Ruhland [*E. horsley-kondae* Fyson, '*horsley-kundae*' orth. mut.]: ♦Ind: Ap, Kt, Mh, Tn; ***E. robustum*** Steudel [*E. bracteosum* Steudel]: ♦Ind: Tn; ***E. rhodae*** Fyson: ♦Ind: Kt; ***E. sharmae*** R. Ansari et N. P. Balakrishnan: ♦Ind: Mh; ***E. talbotii*** R. Ansari et N. P. Balakrishnan [*E. horsley-kondae* var. *megalocephalum* Fyson]: ♦Ind: Kt, Kl, Mh, Tn; ***E. tuberiferum*** A. R. Kulkarni et M. H. Desai: ♦Ind: Mh; ***E. vasudevanii*** R. Ansari et N. P. Balakrishnan: ♦Ind: Kl.

Widespread on hills of the subcontinent: ***E. brownianum*** Martius ex Wallich [*E. nilagirense* Steudel, *E. brownianum* var. *latifolium* Moldenke, *E. nilagirense* var. *parvifolium* Moldenke]: Ban, Bur, Cey, Ind: As, Kl, Mg, Mn, Tn.

Distribution and identity not discovered in time for publication: ***E. glaucum*** Griffith; ***E. rouxianum*** Steudel: ♦Ind? Mh?

Ansari and Balakrishnan (1994) make much use of seed coat patterns in their identification key. To see the pattern the seed must be observed under a microscope. I have, therefore, tried to use characters that need no more than a dissecting microscope and a good lens. The measurements of leaf width cited in the key are taken about halfway up the leaf.

1. Stems elongate; leaves filiform or ribbon-like; submerged aquatics
- **2.** Leaves clothing the elongate stems, not widened below, not more than 0.5 mm wide, 1-nerved, fenestrate; anthers black (widespread) ***E. setaceum***
- **2.** Leaves clustered at the tip of a slender rooting rhizome, distinctly widened below, more than 0.5 mm wide, 3- to 7-nerved, opaque; anthers white (S India and Sri Lanka)
  - **3.** Leaves ribbon-like, gradually widening towards the base ***E. breviscapum***
  - **3.** Leaves filiform, abruptly widening towards the base
    - **4.** Scapes clustered; sheaths closely nerved, the slit 4–8 mm long, the tip truncate and toothed; heads creamy white; female sepals linear pallid; seeds 0.3–0.5 mm long, smooth and glossy ***E. dalzellii***
    - **4.** Scapes solitary; sheaths thin with spaced nerves and obvious transverse lamellae, the slit 10–15 mm long, the tip acute; heads grey or blackish; female sepals oblong-ovate, boat-shaped, grey; seeds 0.48–0.6 mm long, faintly reticulate and wrinkled
      - **5.** Male sepals obdeltate, ± truncate and not lobed at the tip; female petals without black glands ***E. barbeyanum***
      - **5.** Male sepals distinctly 3-lobed; female petals with black glands ***E. fluviatile***

1. Stems abbreviated with leaves in a rosette or if elongate then leaves linear and 1–6 mm wide; leaves more than 0.5 mm wide; temporarily submerged or marsh plants
- **6.** Leaves oblong (at least 4 mm wide) up to the rounded and cuspidate tip ***E. cuspidatum***
- **6.** Leaves linear or triangular, gradually narrowing towards the tip
  - **7.** Anthers white or pale yellow; leaves needle-like in a neat and dense cushion-like cluster, 1–2 (–5) cm long, not more than 4 mm wide at the base (a small tufted annual, probably the commonest and most widespread species in India) ***E. cinereum***
  - **7.** Anthers black; leaves usually flattened, usually forming spiky rosettes, mostly more than 2 cm long, usually more than 4 mm wide at the base
    - **8.** Leaves, sheaths and scapes flushed with red, purple or pink ***E. quinquangulare***
    - **8.** Leaves, sheaths and scapes green, whitish or brown, not flushed with red, purple or pink
      - **9.** Floral bracts terminating in long, erect or recurved, awn-like cusps; cusps as long as or exceeding the bracts, much exceeding the flowers; heads appearing spiny ***E. echinulatum***
      - **9.** Floral bracts without cusps or if with cusps then not awn-like; cusps when present much shorter than the bracts and the flowers; heads not spiny
        - **10.** Robust perennials with short, tough rootstocks; scapes 10–80 cm tall
          - **11.** Leaves densely woolly (Sri Lanka) ***E. fergusonii***
          - **11.** Leaves glabrous or with some hairs
            - **12.** Female petals leathery, distinctly clawed; involucral bracts snow white (Myanmar) ***E. wightianum***
            - **12.** Female petals membranous, not clawed; involucral bracts grey, yellowish-grey or greenish
              - **13.** Involucral bracts hairy, membranous, hairy or glabrous, the outer ones obovate;

heads 8–25 mm in diameter, appearing rough through prominent floral bracts **E. robusto-brownianum**

13. Involucral bracts leathery in the centre, usually with membranous margins, glabrous, the outer ones orbicular or nearly so; heads 6–10 mm in diameter, smooth without prominent floral bracts

14. Floral bracts leathery, tightly overlapping; heads hard; sepals broadly winged; female petals ± alike, linear or thread-like

15. Female petals thread-like, with a brush of hairs near the tip; the petal hairs septate, 0.8–1 mm long (widespread) **E. sexangulare**

15. Female petals narrowly linear, with a few stout hairs at the middle and papillose hairs at the tip; the petal hairs not septate, 0.3–0.6 mm long (Myanmar, Andaman Islands and Sri Lanka) **E. willdenovianum**

14. Floral bracts membranous, not overlapping; heads soft; sepals not winged; female petals unequal, oblanceolate, the median narrower, hairy

16. Floral bracts cuneate-spathulate, with terminal cusp; male sepals deeply lobed, the lobes truncate (Sri Lanka) **E. psammophilum**

16. Floral bracts oblong or oblanceolate, acuminate; male sepals shallowly lobed, the lobes rounded (SW India) **E. richardianum**

10. Slender annuals without rootstocks; scapes usually considerably less than 30 cm long

17. Floral bracts totally leathery, tightly overlapping; the heads hard; flowers conspicuously flattened; sepals broadly winged

18. Female petals narrowly linear, with a few stout hairs at the middle and papillose hairs at the tip; the petal hairs not septate, 0.3–0.6 mm long **E. willdenovianum**

18. Female petals thread-like, with a brush of hairs near the tip; the petal hairs septate, 0.8–1 mm long

19. Sheaths much longer than the leaves; small annual; sepals of male and female flowers 2; sepals in male flower free; scapes 4-ribbed **E. eurypeplon**

19. Sheaths as long as or shorter than the leaves; large perennial; sepals of male and female flowers 3; sepals in male flower united except at the tip; scapes 5- to 8-ribbed **E. sexangulare**

17. Floral bracts at least partly membranous, loosely overlapping; the heads usually soft; flowers not or just slightly flattened; sepals with or without wings

20. Involucral bracts longer than the heads, spreading and showing beyond the margin of the head when viewed from above (head flower-like)

21. Floral bracts longer than the involucral bracts, stellately spreading **E. stellulatum**

21. Floral bracts shorter than involucral bracts, not stellately spreading

22. Male sepals free, 2; female petals 2; involucral bracts 2.5 mm or more wide **E. fysonii**

22. Male sepals united below and spathe-like, 2- or 3-lobed; female petals 2 or 3; involucral bracts less than 2.5 mm wide

23. Seeds with longitudinal, white, ribbon-like bands; male sepals 2-lobed; female sepals 2 **E. truncatum**

23. Seeds without longitudinal bands; male sepals 3-lobed; female sepals 3

24. Male and female sepals glabrous (Madhya Pradesh and West Bengal) **E. martianum**

24. Male and female sepals bearing hairs at or towards the tips

25. Involucral bracts 4–6 mm long, obtuse or subacute; floral bracts acute or acuminate, black **E. heterolepis**

25. Involucral bracts less than 4 mm long, acute, acuminate or terminating in a cusp; floral bracts truncate, yellowish-grey

26. Involucral bracts obovate-cuneate, ± 2 mm wide, with a cusp at the tip; receptacle hairy; male sepal lobes rounded at the tips; seeds ± 0.6 mm long (Kerala) **E. sivarajanii**

26. Involucral bracts linear-oblong to oblanceolate, 0.25–1 mm wide, acute to acuminate at the tip; receptacle glabrous; male sepal lobes truncate at the tips; seeds ± 0.4 mm long (widespread) **E. xeranthemum**

20. Involucral bracts as long as or shorter than the heads, not showing beyond the the margin of the head when viewed from above

27. Male sepals free, not united below, not spathe-like; female petals 2

28. Female sepals boat-like, broadly keeled; scape sheaths longer than the leaves **E. eurypeplon**

28. Female sepals flat or folded but not keeled; scape sheaths shorter than the leaves

29. Involucral bracts, floral bracts, sepals and petals glabrous or nearly so; female petals without black glands **E. duthiei**

29. Involucral bracts, floral bracts, sepals and petals densely hairy at least towards the tips; female petals each with a black gland ***E. sedgwickii***

27. Male sepals united below and spathe-like; female petals 3 or 2

30. Female petals leathery, with a claw at the base (Andaman Islands) ***E. helferi***

30. Female petals membranous, without claws

31. Involucral bracts hairy ***E. lanceolatum***

31. Involucral bracts glabrous

32. Floral bracts with apical cusps

33. Pedicels 1–3 ***E. oryzetorum***

33. Pedicels more than 3

34. Leaves more than 2.5 mm wide; receptacle with short hairs; petals with papillose white hairs; seeds appearing horizontally flaky (widespread) ***E. quinquangulare***

34. Leaves not more than 2 mm wide; receptacle with densely woolly hairs; petals with woolly hairs; seeds appearing smooth (Sri Lanka) ***E. walkeri***

32. Floral bracts acute or obtuse, without cusps

35. Sepals of male and female flowers 2; involucral bracts as wide as the head, spreading upwards at maturity

36. Scapes filiform, ± 0.3 mm in diameter; petals of the female flowers linear, glabrous except for 1 or 2 hairs at the apex ***E. minimum***

36. Scapes stout, 0.4–0.8 mm in diameter; petals of the female flowers spathulate, hairy ***E. truncatum***

35. Sepals of male and female flowers 3; involucral bracts narrower than the head, spreading or reflexed at maturity

37. Female sepals ± equal in shape and size ***E. trilobum***

37. Female sepals unequal, two boat-like and keeled, 1–1.5 mm long, the third linear, 0.5–1 mm long, flat

38. Female petals hairy; heads ± 5 mm in diameter; male sepal lobes deep with acute or obtuse tips; seeds oblong to ovoid ± 0.38 mm long (C and W India) ***E. parviflorum***

38. Female petals glabrous; heads ± 3 mm in diameter; male sepal lobes shallow with obtuse tips; seeds ovoid-globose ± 0.45 mm long (Maharashtra) ***E. santapaui***

Eriocaulon barbeyanum Ruhland: Fig. 200

Like *E. fluviatile* but male sepals obdeltate, ± truncate and not lobed at the tip; female petals without black glands. Perennial or perhaps also annual. In flowing water. {⊙? Ros}. Endemic to SW India: ♦Ind: Kt.

Eriocaulon breviscapum Körnicke ['*breviscapon*' orth. mut.]: Fig. 194

Stems simple, elongate, slender, up to 1 cm or perhaps more long, the leaves clustered towards the tip. Leaves linear, up to ± 30 cm long, 1–2 mm wide, from an expanded and woolly base (up to 7 mm wide at the base), glabrous. Scapes solitary or 2, up to ± 30 cm long, scarcely twisted; sheaths up to ± 10 cm, much shorter than the leaves, loose, the tip obtuse. Head ± hemispherical, 7–9 mm in diameter, grey; receptacle convex, hemispherical, glabrous. Involucral bracts oblong-ovate to suborbicular, ± 1.75 mm long glabrous, yellowish-grey. Floral bracts oblong-oblanceolate, ±3 mm long, membranous, black, with white woolly hairs, acuminate. Flowers 3-merous. Male flowers: sepals obovate, ± 1.75 mm long, united below, 3-lobed above, black, with white woolly hairs; petals 3, oblong, one ± 1 mm long, protruding from the sepals, the other two much smaller, all glandular and with white woolly hairs; anthers white. Female flowers:

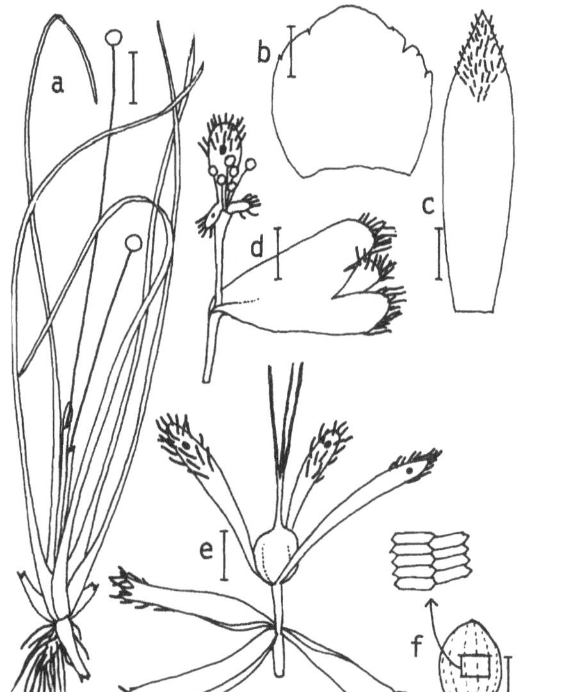

Fig. 194. *Eriocaulon breviscapum*: a, flowering plant (2 cm); b, involucral bract (5 mm); c, floral bract (5 mm); d, male flower (5 mm); e, female flower (5 mm); f, seed (0.2 mm) with seed coat pattern (inset).

sepals 3, subequal, boat-like, 1.5–2 mm long, yellowish-grey, with white woolly hairs; petals spathulate, ±2 mm long, obtuse or laciniate, with a subapical gland and with white woolly hairs at the tips, one slightly larger than the other two. Seeds subglobose, ± 0.48 mm long; seed coat cells transversely elongate, aligned in vertical rows, appearing smooth and faintly reticulate; appendages absent.

Perennial or perhaps also annual. It is an aquatic species growing submerged in flowing water in streams and rivers, often attached to rocks. Frequent and sometimes abundant.

{⊙? Ros}. Endemic to S India: ♦Ind: Kt, Mp, Or. November–February. Mostly about 1000 m.

Eriocaulon cinereum R. Brown [*E. sieboldianum* Siebold et Zuccarini ex Steudel, '*sieboldtianum*' orth. mut., *E. redactum* Ruhland, '*reductum*' orth. mut.]: Fig. 195

Plants tufted, glabrous. Leaves numerous, forming a neat, dense cluster, capillary or subulate, 1–2 (–5) cm long, 0.2–0.4 mm in diameter, up to 4 mm wide at base but usually less, green. Scapes few or many, ± 4–8 (–15) cm long, 0.3–0.4 mm in diameter, 5-ribbed, scarcely to tightly twisted; sheaths shorter than the leaves, slightly inflated, shortly obliquely split, acute. Heads globose to conical or ovoid, 2–4 mm in diameter, whitish, greyish or straw-coloured; receptacle thinly pilose to almost glabrous. Involucral bracts shorter than the head, obovate-oblong, 1.2–1.5 mm long, ascending or spreading, glabrous, pale yellowish-grey. Floral bracts linear-elliptical to lanceolate-oblong, 1.5–2.25 mm long, membranous, loosely erect (not overlapping), glabrous, pale or dark, but usually with a dark central band and scarious margins. Flowers 3-merous, the female around the periphery. Male flowers: the 3 sepals united into a spathe with a 3-pointed tip, 0.75–1 mm long, dark grey, glabrous or sparsely hairy at the tip; petals included within the sepals, the lobes minute, unequal; anthers white or pale yellow. Female flowers: much reduced; sepals 3, 2 or absent, filiform, 0.6–1 mm long, simple or branched, caducous; petals absent or small and linear; stipes strikingly linear, eglandular. Seeds ovoid-ellipsoid, 0.25–0.3 mm long; seed coat cells transversely elongate, aligned in vertical rows, appearing almost translucent, glossy, very faintly reticulate; appendages absent.

A small annual. Mostly found in shallow, temporary water or in seepage areas. It germinates under water and flowers and fruits as the water recedes. It seems to tolerate considerable disturbance and can almost be called ruderal. It is particularly common in rice fields but can hardly be called noxious. Plants with floral bracts oblanceolate, sepals of male flowers with some papillose hairs at the tips and sepals of the female

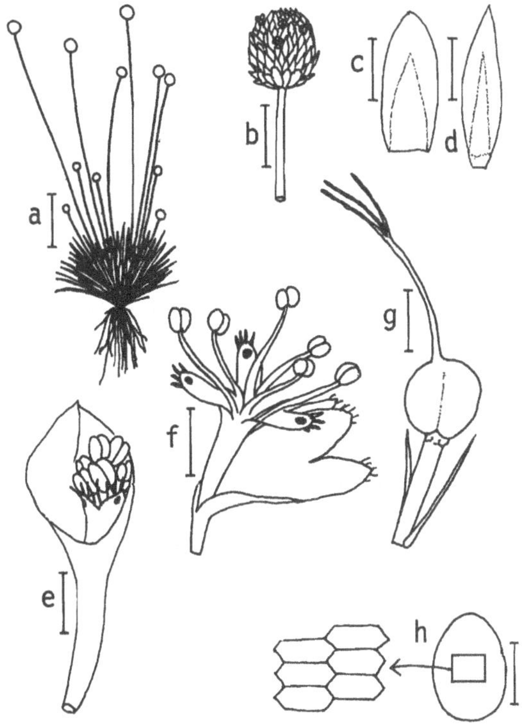

FIG. 195. *Eriocaulon cinereum*: a, flowering plant (1 cm); b, head (2 mm); c, involucral bract (0.5 mm); d, floral bract (0.5 mm); e, young male flower (0.25 mm); f, old male flower (0.25 mm); g, female flower (0.25 mm); h, seed (0.5 mm) with seed coat pattern (inset).

flowers reduced to branched hairs are considered by Ansari and Balakrishnan (1994) to represent a distinct species called *E. redactum.*

{⊙ Ten(Ros)}. Tropical and warm Asia and Australia, naturalized in Europe and N America: Ban, Cey, Ind, Nep: As, Go, Hp, Kl, Kt, Mh, Mp, Or, Rj, Sk, Tn, Wb. 0–2300 m.

Eriocaulon cuspidatum Dalzell: Fig. 196

Tufted. Leaves oblong to linear-oblong (longest when submerged), abruptly rounded at the tip with a terminal cusp, (5-) 7- to 19-nerved, 2.5–8 (–27) cm long, 4–6 (–14) mm wide, glabrous. Scapes few to many, 15–30 cm long, 7-ribbed; sheaths up to 10 cm long, glabrous, mouth oblique, acute, entire or cut. Heads globose to hemispherical, 4–6.25 mm in diameter, yellowish-grey to white; receptacles convex, glabrous. Involucral bracts broadly obovate-cuneate, ± 2.25 mm long, ± 1.5 mm wide, obtuse, glabrous. Floral bracts like involucral bracts but slightly narrower, membranous, with some hairs towards the tip. Male flowers: sepals 2, free, obovate-lanceolate, ± 1 mm long, hyaline, slightly winged and puberulous

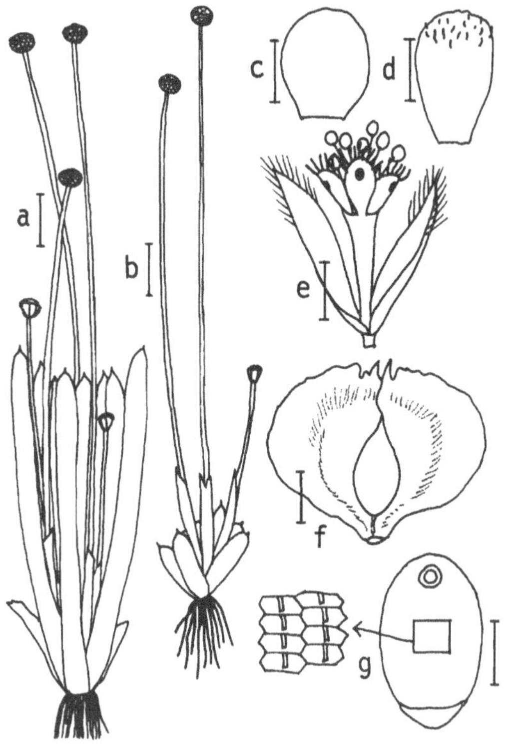

Fig. 196. *Eriocaulon cuspidatum*: a, flowering plant from deep water (1 cm); b, flowering plant from shallow water (1 cm); c, involucral bract (1 mm); d, floral bract (1 mm); e, male flower (0.5 mm); f, capsule with persistent sepals (0.5 mm); g, seed (0.25 mm) with seed coat pattern (inset).

near the tip; petals 3, unequal; anthers black. Female flowers: sepals 2, oblanceolate, ± 1 mm long, folded and boat-shaped, glabrous, keeled, acute; petals 3, subequal, narrowly linear-lanceolate, ± 1.25 mm long, tip with woolly hairs and with a conspicuous black gland. Seeds oblong-obovoid, 0.6–0.75 mm long, 0.35–0.4 mm in diameter; seed coat cells transversely elongate, aligned in vertical rows; appendages 1 or 2 from the middle of the transverse radial walls and occasionally 1 from a vertical wall, appearing as vertical lines on the surface of the seeds, bristle-like, truncate at the tip.

Annual. Usually partly submerged. Often abundant and locally dominant in pools and rice fields.

{⊙ Hyp/Ros/(Hel)}. Endemic to lowlands of S India: ♦Ind: Go, Kl, Kt, Mh, Mp? July–December. 0–250 m.

Eriocaulon dalzellii Körnicke [*E. rivulare* Dalzell, non G. Don ex Bentham, *E. dalzellii* var. *glabratum* Moldenke]: Fig. 197

Stems elongate, slender, up to 35 cm or more long, densely clothed in a tangle of fine roots together with leaf remnants, the numerous leaves clustered towards the tip and surrounding the terminal bunch of up to 25 emergent scapes. Leaves linear from a broadened translucent base, 8–25 cm long, 0.6–1 mm wide (2 mm at the base), 5- or 7-nerved. Scapes 12–30 cm long, 0.8–1 mm in diameter, 8- to 10-ribbed, twisted; sheaths shorter than the leaves, loose, parallel-sided, the slit 4–8 mm long, the tip truncate, often toothed. Heads globose to shortly conical, 5–6 mm in diameter, creamy white; receptacle columnar, glabrous. Involucral bracts ± as wide as the head, obovate-oblong, ±2.25 mm long, rounded, glabrous, straw-coloured sometimes flushed with grey, turned down at maturity, in 3 or 4 whorls, the inner subtending male flowers. Floral bracts narrowly oblanceolate, ± 2.25 mm, membranous, with white woolly hairs, subacute, the black-spotted petals visible among them. Flowers 3-merous. Male flowers 1.7–2.2 mm long; sepals free in flowers subtended by involucral bracts, otherwise pallid, united and spathe-like, ± 1.5 mm long, 3-lobed, the lobes obtuse, with woolly hairs around the upper margin; petals oblong, glandular and with white woolly hairs, two ± 0.5 mm long, the third ± 0.8 mm long; anthers creamy white,

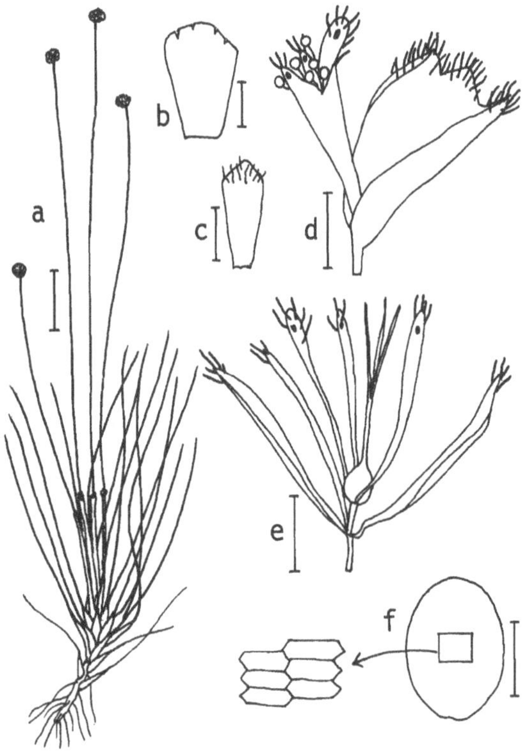

Fig. 197. *Eriocaulon dalzellii*: a, flowering plant (2 cm); b, involucral bract (1 mm); c, floral bract (1 mm); d, male flower (0.5 mm); e, female flower (0.5 mm); f, seed (0.25 mm) with seed coat pattern (inset).

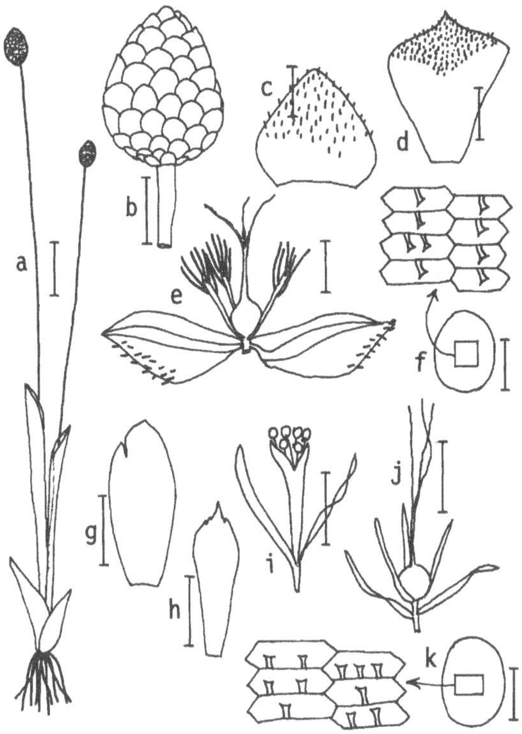

Fig. 198. *Eriocaulon duthiei*: g, involucral bract (1 mm); h, floral bract (1 mm); i, male flower (0.5 mm); j, female flower (0.5 mm); k, seed (0.2 mm) with seed coat pattern (inset). *E. eurypeplon*: a, flowering plant (2 cm); b, head (5 mm); c, involucral bract (0.5 mm); d, floral bract (0.5 mm); e, female flower (0.5 mm); f, seed (0.35 mm) with seed coat pattern (inset).

± 0.2 mm long. Female flowers ± 2.5 mm long; sepals linear, falcate, pallid, with some white woolly hairs, two slightly keeled 1.5–1.75 mm long, the third slightly smaller and flat; petals narrowly linear-spathulate with a subapical gland and some white woolly hairs at the tips, two 1.5–1.7 mm long, the third slightly larger. Seeds oblong-ovoid, 0.3–0.5 mm long; seed coat cells transversely elongate, aligned in vertical rows, appearing smooth and glossy; appendages absent.

Perennial or perhaps sometimes annual. An aquatic species growing between and on submerged rocks in streams, apparently never seen in still water.

{⊙? Ros}. Endemic to coastal W India: ♦Ind: Go, Kt, Mh. September–December.

Eriocaulon duthiei J. D. Hooker: Fig. 198

Like *E. eurypeplon* but leaves rarely more than 3 cm long; receptacles glabrous; involucral bracts oblong-lanceolate, ±2.25 mm long; floral bracts oblong-lanceolate; male petals without black glands; female sepals linear-elliptical, ± 0.75 mm long, not keeled; female petals glabrous; seeds oblong-ovoid, ± 0.42 mm long; seed cell appendages appearing to be in transverse rings.

Annual. Rare in wet grasslands.

{⊙ Hel}. Endemic to W India: ♦Ind: Mh, Mp.

Eriocaulon echinulatum Martius: Fig. 199

Tufted. Leaves linear-subulate, 1–3 (–6) cm long, ± 1 mm wide, glabrous, 5- to 7-nerved, acute or acuminate. Scapes few or many, fine, 3–15 cm long, glabrous, 4-ribbed, twisted; sheaths obliquely spathe-like at the mouth, obtuse at the tip. Heads globose but flattened at base, appearing spiny, 4–6 mm in diameter, pale yellowish-grey; receptacle pilose. Involucral bracts broadly ovate, ± 3 mm long, terminating in a cusp, pale straw colour, shiny, glabrous, becoming reflexed in fruit. Floral bracts broadly ovate, longer than the flowers, terminating in long, erect or recurved, awn-like cusps; cusps as long as or exceeding the bracts, much exceeding the flowers. Male flowers: sepals 2, united on one side and spathe-like, obovate-cuneate, white, glabrous; petals absent. Female flowers: sepals 3, free, lanceolate, glabrous, dorsally winged, yellowish; petals absent.

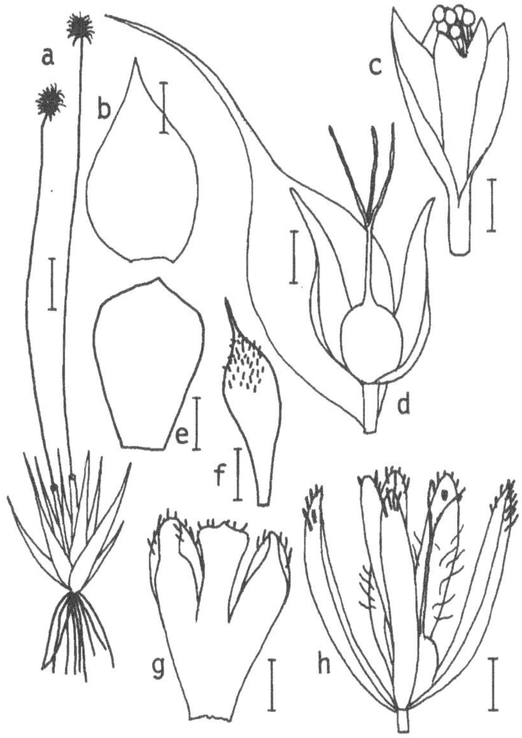

Fig. 199. *Eriocaulon echinulatum*: a, flowering plant (1 cm); b, involucral bract (0.75 mm); c, male flower (0.75 mm); d, floral bract with female flower (0.75 mm). *E. psammophilum*: e, involucral bract (1 mm); f, floral bract (1 mm); g, male sepals (0.5 mm), h, female flower (0.5 mm).

Annual. Inundated places, marshland and in rice fields. It is easily recognizable by its long-cusped involucral and floral bracts which give the head a spiny appearance.

{⊙ Hel}. S and SE Asia: Bur, Ind: Mg, Or.

Eriocaulon eurypeplon Körnicke [*E. vanheurckii* Müller Argoviensis, *E. bombayanum* Ruhland, *E. thomasii* Fyson, *E. indicum* Moldenke, *E. vanheurckii* forma *minimum* Moldenke, *E. keralense* R. Ansari, V. J. Nair et N. C. Nair]: Fig. 198

Tufted. Leaves linear to oblong-lanceolate, 3–15 cm long, 4–10 mm wide (to 15 mm wide at base), many nerved, glabrous, obtusely acuminate to acute at the tip. Scapes several, slender, (4–) 10–25 cm long, 2–5 mm in diameter, glabrous, acutely 3-, 4- or 5-ribbed, slightly twisted; sheaths loose, 2–16 cm long, longer than the leaves, widened in the upper half, half as long as the scapes, glabrous, mouth about three-quarters of its length, prolonged and long-acuminate above. Heads cylindrical-globose, 3–5 (–8) mm in diameter, grey to white; receptacle ovoid to ellipsoid, ± hairy. Involucral bracts closely overlapping, almost semicircular, 1.25–1.9 mm long, blackish to dark brown or yellowish-grey, the backs with a powdery white pubescence. Floral bracts like involucral bracts but narrower and sharper, cuneate-obovate or rhomboid, 1.5–1.7 mm wide, leathery, closely overlapping, tips acute or acuminate and arched inwards. Male flowers: sepals 2 (3) free, narrow boat-shaped, 1.2–1.5 mm long, ± 0.4 mm wide, black to olive, curved, powdery pubescent upwards; petals 3, minute, papillose, with or without a black gland; anthers black. Female flowers: sepals 2 (3), free, falcate-ovate or semi-rhomboid, boat-shaped, ± 1.5 mm long, powdery pubescent upwards; petals 3, linear-spathulate, ± 0.75 mm long, hairy, with or without a black apical gland at the tip. Seeds oblong-ellipsoid, ± 0.7 mm long; seed coat cells transversely elongate, aligned in vertical rows; appendages 1 or 2 from the middle of the transverse radial walls, appearing as vertical lines on the surface of the seeds, bristle-like, swollen at the tip.

Annual. In marshes, wet rice fields, at the edges of drying pools and in crevices between wet rocks. It is very variable in size.

{⊙ Hel}. Endemic to W India: ♦Ind: Go, Kl, Kt, Mh. July–December. 0–1200 m.

Eriocaulon fergusonii (Moldenke) S. M. Phillips [*E. willdenovianum* var. *fergusonii* Moldenke]: refer to Fig. 210 and 214

Like *E. sexangulare* (Fig. 219) but the female petals lack a brush of hairs; like *E. willdenovianum* (Fig 214) but it has 3-merous flowers and is more robust; it differs from both in having dense woolly hairs on the leaves.

Perennial. Marshlands.

{Hel}. Endemic to the western lowlands of Sri Lanka: ♦Cey.

Eriocaulon fluviatile Trimen: Fig. 200

Stems elongate, slender, up to 10 cm long but usually shorter, clothed in fine roots and a mesh of fibres from decaying leaves, the leaves clustered towards the tip and surrounding the single terminal emergent scape. Leaves filiform from an expanded and woolly base, often ± crescentic, drying black, up to 55 cm long, 0.5–1.5 mm wide (± 4 mm at the base), 3- or 5-nerved. Scapes solitary, 14–50 cm long, 0.8–1.8 mm in diameter, 6- to 9-ribbed, scarcely twisted; sheaths up to ± 8 cm, much shorter than the leaves, loose, with spaced nerves and transverse lamellae, the slit 10–15 mm long, the tip acute. Head depressed-globose, 5–9 mm in diameter, black to greyish-brown; receptacle hemispherical, glabrous. Involucral bracts ± as wide as the head, in 3 or

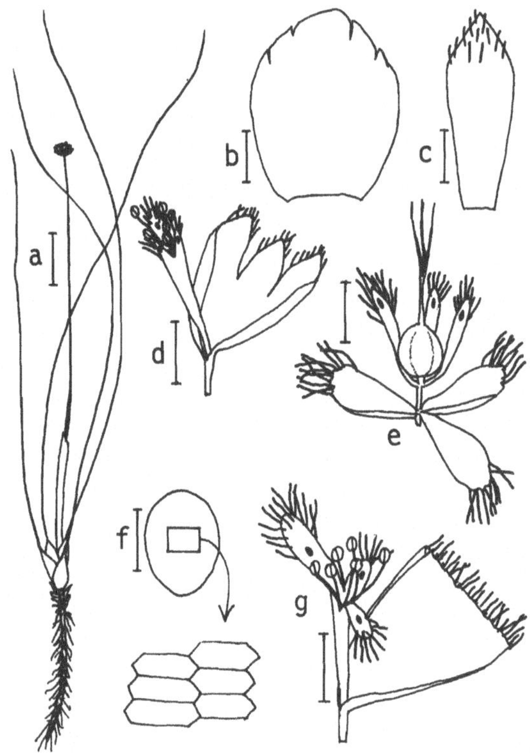

Fig. 200. *Eriocaulon barbyanum*: g, male flower (0.5 mm). *E. fluviatile*: a, flowering plant (2 cm); b, involucral bract (0.5 mm); c, floral bract (0.5 mm); d, male flower (0.5 mm); e, female flower (0.5 mm); f, seed (0.25 mm) with seed coat pattern (inset).

4 whorls, the outer ones orbicular, ± 2 mm long glabrous, black or grey to yellowish-grey, the inner ones oblong, thinly white papillose, subtending flowers, turned down at maturity, the inner subtending male flowers. Floral bracts narrowly cuneate, ± 2 mm long, greyish brown to blackish, with white woolly hairs, shortly acuminate, the black-spotted petals visible among them. Flowers 3-merous. Male flowers: sepals obovate, 1.5–2 mm long, united below, 3-lobed above, grey or black, with white woolly hairs, two keeled, the third flat; petals 3, oblong, glandular and with white woolly hairs, two ± 0.6 mm long, the third ± 1 mm long, protruding from the sepals; anthers white. Female flowers: sepals boat-like, grey or blackish, with white woolly hairs, two 1–1.5 mm long, the third similar but narrower; petals and ovary on a stalk up to 0.8 mm long; petals narrowly oblanceolate, with a subapical gland and with white woolly hairs at the tips, one slightly larger than the other two. Seeds oblong-ovoid, 0.48–0.6 mm long; seed coat cells transversely elongate, aligned in vertical rows, appearing faintly reticulate, somewhat wrinkled to almost smooth; appendages absent.

Perennial or perhaps also annual. It grows submerged in swiftly flowing streams, rooting on rocks and in mud, apparently never seen in still water. The heads only emerging above the water surface. Often growing in colonies.

{⊙? Ros}. Endemic to Sri Lanka and India south of 22 N: ♦Cey, ♦Ind: Kl, Kt, Mp. December–April. Up to 1000 m.

Eriocaulon fysonii R. Ansari et N. P. Balakrishnan [*E. cuspidatum* var. *bracteatum* Fyson]: Fig. 201

Like *E. cuspidatum* but involucral bracts up to 5 mm long, distinctly exceeding the head, 2.5–3.5 mm wide; male sepals keeled at the back; female sepals hairy; appendages of the seeds swollen at the tips.

Annual. In wet and marshy places.

{⊙ Hyp/Ros/(Hel}. Endemic to W India: ♦Ind: Kl, Kt, Mh.

Eriocaulon helferi J. D. Hooker [*E. wightianum* auct. Ind, non Martius, *E. wightianum* var. *helferi* (J. D. Hooker) Fyson, *E. wightianum* Martius forma *viviparum* Moldenke]: Fig. 201

Tufted. Stems short, without a rhizome. Leaves sword-shaped, up to 6 cm long, 8–20 cm long, ± 1 cm wide at base, many nerved, glabrous, attenuate towards the apex, the tip obtuse or subobtuse. Scapes few, up to 18 cm long, sparsely hairy, indistinctly ± 7-ribbed, usually twisted; sheaths up to 5 cm long, ribbed, sparsely hairy, mouth oblique. Heads globose, ± 8 mm, densely white villous; receptacle hairy. Involucral bracts cuneate-obovate, the outer obtuse, the inner subacute, scarious, spreading or ultimately reflexed, hairy at the back in the upper part, ± 2 mm long, ± 1.5 mm wide. Floral bracts rhomboid-cuneate or obovate-cuneate, ±3 mm long, membranous, acuminate, densely clothed in white hairs in the upper part. Flowers 3-merous. Male flowers: sepals 3, obovate-oblong, cuneate at the base, united and spathe-like and split down one side, ± 1.25 mm long, hyaline, the tips hairy; petal tube ± 1.2 mm long, the lobes all densely hairy at the apex, without a black gland, one slightly larger than the other 2, ± 0.8 mm long; anthers black. Female flowers: sepals 3, free, slightly unequal, linear-obovate, bearded towards the tip; petals linear-spathulate, 2–2.5 mm long, with long white hairs and lacking black apical glands. Seeds ovoid-globose, ± 0.55 mm long; seed coat cells transversely elongate, aligned in vertical rows; appendages solitary from the centre of each transverse radial wall, elongated, curved and joined to the adjacent ones of the same vertical row forming longitudinal parallel ribs.

Annual. Ecology unknown. Usually misidentified as *E. wightianum.*

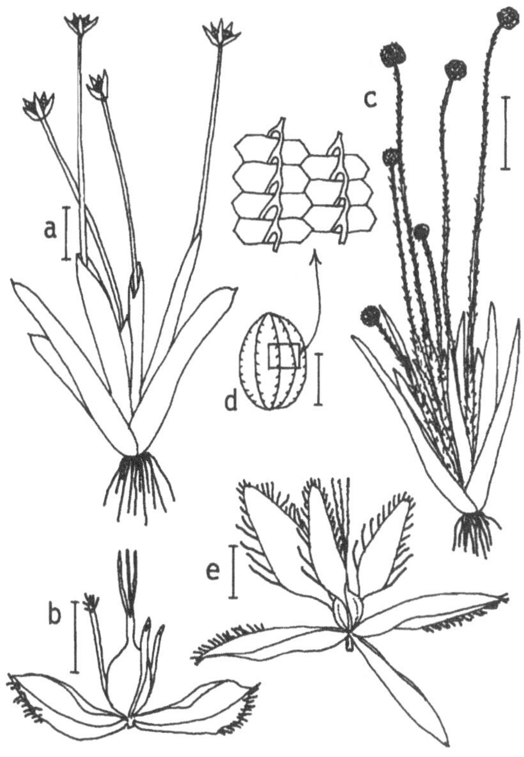

FIG. 201. *Eriocaulon fysonii*: a, flowering plant (1 cm); b, female flower (1 mm). *E. helferi*: c, flowering plant (2 cm); d, seed (0.25 mm) with seed coat pattern (inset); e, female flower (0.5 mm).

{⊙ Hel?}. Endemic to Myanmar and the Andaman Islands: ♦Bur, ♦Ind: An.

Eriocaulon heterolepis Steudel [*E. dianae* Fyson, *E. dianae* var. *longibracteatum* Fyson]: Fig. 202

Tufted. Leaves linear or linear-lanceolate, (0.5–) 5–15 cm long, 4–8 mm wide, flat, glabrous, with 5–13 nerves, attenuate towards an obtuse to acute tip. Scapes many (up to 40 or more), 6–22 cm long, 6–10 mm in diameter, ± 5-ribbed, glabrous; sheaths lax, up to 6 cm long, mostly half as long as the leaves, acute. Heads globose to hemispherical, 5–10 mm in diameter, from white to dark grey; receptacle ovoid or cylindrical, with short hairs. Involucral bracts oblong-lanceolate, 4–6 mm long, ± 1 mm wide, spreading, much longer than the floral bracts and flowers, straw-coloured, glabrous and glistening, often showing beyond the margin of the head when viewed from above, often becoming reflexed in fruit, subacute or obtuse. Floral bracts oblanceolate-cuneate, 1.25–2 mm long, membranous, acute to acuminate, black, the tips papillose hairy. Flowers 3-merous. Male flowers: sepals united below, spathe-like and split down one side, 1–1.5 mm long, black, 3-lobed, the tips papillose hairy; petals very small, with white papillae and a black apical gland; anthers black. Female flowers: sepals free, 1.2–2 mm long dark, 2 narrowly 'boat-shaped' and keeled, with white hairs along the keel and at the tip, the third flat, linear-spathulate, ± 1 mm long, black, with white hairs at the tip; petals linear-oblanceolate to spathulate, 1–2 mm long, hairy at the tip, with or without black glands. Seeds oblong, 0.4–0.6 mm long; seed coat cells transversely elongate, aligned in vertical rows; appendages 5 or 6 from the transverse radial walls, appearing as transverse rings on the surface of the seeds, bristle-like, flattened at the tip.

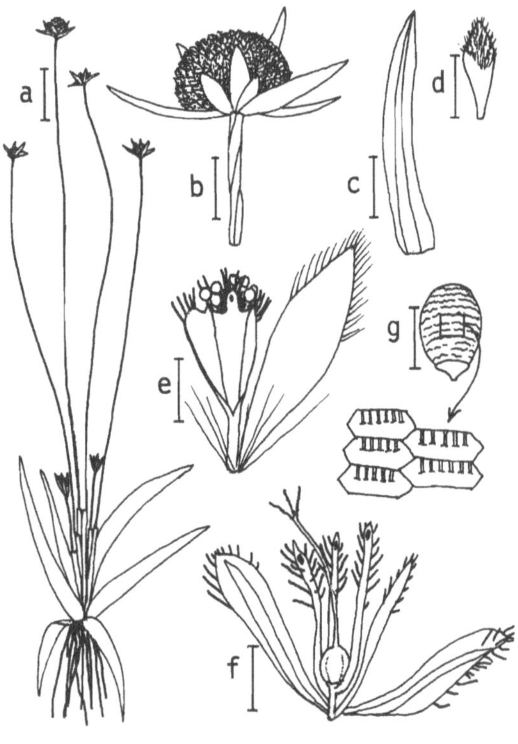

FIG. 202. *Eriocaulon heterolepis*: a, flowering plant (1 cm); b, head (2 mm); c, involucral bract (1 mm); d, floral bract (1 mm); e, male flower with floral bract and some receptacular hairs (0.5 mm); f, female flower (0.5 mm); g, seed (0.25 mm) with seed coat pattern (inset).

Annual. In marshes and seasonally inundated places; locally abundant. Common in rice fields; flowering after the harvest. Very variable in length of the involucral bracts, shape and colour of the heads and form of the female sepals. It is perhaps conspecific with the African species *E. transvaalicum* N. E. Brown [*E. hanningtonii* N. E. Brown].

{⊙ Hel}. Probably endemic to W India: ♦Ind: Go, Kl, Kt, Mh, Rj. September–March. 0–200 m.

Eriocaulon lanceolatum Miquel ex Steudel [*E. lanceolatum* var. *pilosum* Moldenke, *E. kasaragodense* R. Ansari, V. J. Nair et N. C. Nair]: Fig. 203

Tufted. Leaves oblong-lanceolate, 2.5–6 (–10) cm long, 5–15 mm wide, many nerved, glabrous or slightly hairy, attenuate towards the apex. Scapes few, slender, 5–20 (–30) cm long, ± 5 mm in diameter, glabrous or sparsely hairy, indistinctly 6- or 7-ribbed, scarcely twisted; sheaths up to ± 9 cm long, loose, glabrous or hairy, mouth oblique, acute. Heads globose or hemispherical, (3–) 5–6 mm in diameter, white; receptacle convex, hairy. Involucral bracts broadly obovate or quadrate, 2–3 mm long, yellowish-grey, membranous, not reflexed, rounded at the tip, hairy upwards and ciliate at the tip. Floral bracts cuneate-oblong, ± 2 mm long, membranous, shorter than the female sepals, obtuse to subacute, hairy on the back. Flowers 3-merous. Male flowers: sepals spathe-like and split down one side, 1.25–3 mm long, dark, 3-lobed, glabrous or sparsely hairy towards the tip; petals minute, glabrous, with a black gland; anthers black. Female flowers: sepals 3, free, deeply boat-shaped, 1.75–3.3 mm long, yellowish-grey, 1 sometimes winged on the back, 1 sometimes flat, glabrous or sparsely hairy towards the tip; petals spathulate, hairy, with a black apical gland at the tip. Seeds oblong-ellipsoid or ovoid, 0.6–0.8 mm long; seed coat cells transversely elongated, aligned in vertical rows; appendages solitary from the centre of transverse radial walls, bristle-like, forked at the tip.

Annual. In marshes, rice fields, on wet rocks and around drying pools. The more hairy plants have

Heads globose to cup-shaped, 2.5–4 mm in diameter, glistening pale yellowish-grey; receptacle cylindrical, hairy. Involucral bracts as wide as the head, petal-like, oblong to cuneate, ± 1.5 mm long, with broadly rounded to truncate and toothed tips. Outer floral bracts like involucral bracts, ± 2 mm long; inner floral bracts obovate, grey, deeply concave. Flowers 1.2–1.3 mm long, glabrous. Male flowers with stipe; sepals 2, oblong, ±1 mm long, grey or black, truncate, lightly keeled, united on one side at the base; petals 3, very small with white papillae or absent; anthers black. Female flowers with a ±0.25 mm long stipe; sepals 2 (3), linear-subulate, black, ±1.25 mm long, obtuse; petals 3, narrowly oblanceolate, two ± 0.75 mm long, the third slightly longer. Seeds ovoid-ellipsoid, 0.3–0.4 mm long; seed coat cells transversely elongate, aligned in vertical rows; appendages in the form of transverse and vertical radial walls, appearing coarsely reticulate.

Annual. A poorly known and perhaps here wrongly interpreted species. It is reported to grow on wet rock faces and marshy places. Like *E. truncatum* but with a more slender habit, fewer and thinner scapes, smaller

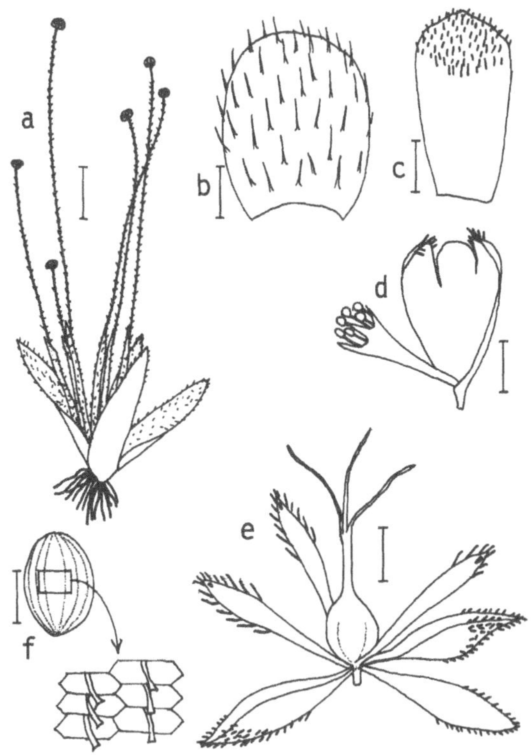

Fig. 203. *Eriocaulon lanceolatum*: a, flowering plant (2 cm); b, involucral bract (0.5 mm); c, floral bract (0.5 mm); d, male flower (0.5 mm); e, female flower (0.5 mm); f, seed (0.35 mm) with seed coat pattern (inset).

been recognized as *E. lanceolatum* var. *pilosum* or *E. kasaragodense.*

{⊙ Hel}. Endemic to SW India: ♦Ind: Gj? Kl, Kt, Mh. July–February. 0–200 m.

Eriocaulon martianum Wallich ex Körnicke [*E. quinquangulare* var. *martianum* (Wallich ex Körnicke) Fyson]: Fig. 206

Like *E. quinquangulare* but involucral bracts elliptical, ±3.5 mm long, acute, longer than the floral bracts; male and female sepals glabrous.

Annual. Rare. In marshy fields.

{⊙ Hel}. Endemic to India: ♦Ind: Mp, Wb.

Eriocaulon minimum Lamarck [*E. hamiltonianum* Martius, *E. oryzetorum* sensu J. D. Hooker non Martius, *E. hamiltonianum* var. *minor* Fyson]: Fig. 204

Delicate, tufted. Leaves linear, 1–6 cm long, 0.5–2 mm wide, spongy with obvious cells and thin margins, subacute. Scapes capillary, 1–4, 3–5 (–10) cm long, ± 0.3 mm in diameter, 5-ribbed, slightly twisted; sheaths up to 3.5 cm shorter than the leaves, papery, inflated at the mouth, obliquely split, obtuse or acute.

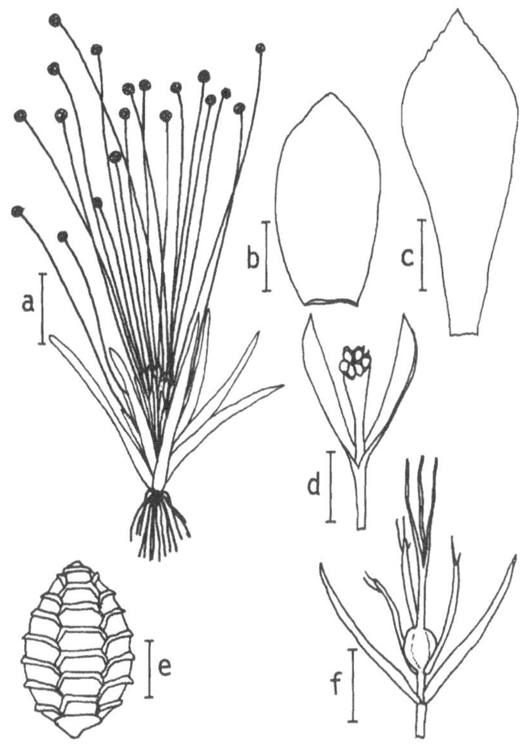

Fig. 204. *Eriocaulon minimum*: a, flowering plant (2 cm); b, involucral bract (0.5 mm); c, floral bract (0.5 mm); d, male flower (0.5 mm); e, seed (0.25 mm); f, female flower (0.5 mm).

flowers with female petals linear and slightly smaller shining seeds with a reticulate surface pattern, the reticulations wider than long.

{⊙ Hel}. S and SW Asia: Ban, Bur, Cey, Ind: As, Gj, Kt, Mg? Mp, Wb.

Eriocaulon oryzetorum Martius: Fig. 205

Tufted. Leaves linear to linear-lanceolate, 4–12 cm long, 3–8 mm wide, flat, glabrous, attenuate towards the apex, obtuse at the tip, with 8–10 nerves. Scapes 1–3, 18–40 cm long, 5- or 6-ribbed, gently twisted; sheaths loose, obtuse, 7–10 cm long, mouth oblique, tip obtuse. Heads subglobose, 5–7 mm in diameter, straw-coloured; receptacle hairy. Involucral bracts obovate, ± 3 mm long, the outer obtuse, the inner becoming subacute, yellowish-grey, shining, glabrous. Floral bracts cuneate-obovate, cuspidate to acuminate, ± 3.25 mm long, membranous, exceeding the flowers, yellowish-grey, the backs hairy. Flowers 3-merous. Male flowers: sepals 3, obovate, united below, spathe-like, split down one side, ± 2 mm long, 3-lobed, the lobes acute or acuminate, greyish-yellow, hairy at apex; petals 3, subequal, the lateral ones minute, the remaining one linear, up to 0.5 mm long, without glands, sparsely hairy; anthers black. Female flowers: sepals 3, free, oblanceolate, ± 2 mm long, the lateral ones boat-like, keeled, middle one flat and not keeled, acute, hairy at the apex; petals 3, subequal, linear spathulate, ± 2 mm long, without glands, white, sparsely hairy. Seeds oblong-ovoid or globose, ± 0.53 mm long, purple; seed coat cells transversely elongate, aligned in vertical rows, appearing to have transverse rings on the surface; appendages 3–6, arising from transverse walls and 0 or 1 from the vertical wall, bristle-like, truncate at the tip.

Annual. Found in wet places.

{⊙ Hel}. Malaya, Myanmar, NE and N India: Bur, Ind, Nep: As, Bh, Sk, Up. ?–± 600 m.

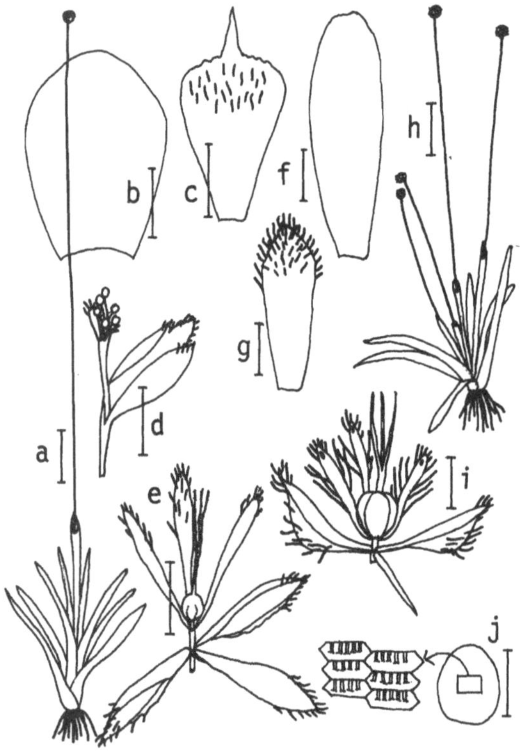

Fig. 205. *Eriocaulon oryzetorum*: a, flowering plant (3 cm); b, involucral bract (1 mm); c, floral bract (1 mm); d, male flower (1 mm); e, female flower (1 mm). *E. parviflorum*: f, involucral bract (0.5 mm); g, floral bract (0.5 mm); h, flowering plant (2 cm); i, female flower (0.5 mm); j, seed (0.2 mm) with seed coat pattern (inset).

Eriocaulon parviflorum (Fyson) R. Ansari et N. P. Balakrishnan [*E. dianae* var. *parviflorum* Fyson, *E. dianae* var. *triloboides* Fyson]: Fig. 205

Tufted. Leaves linear, up to 8 cm long, up to 6 mm wide, flat, glabrous, attenuate towards an obtuse to acute tip. Scapes few or many, up to 20 cm long, glabrous; sheaths lax, up to 7 cm long. Heads globose to hemispherical, ± 5 mm in diameter, grey; receptacle globose to ovoid, with short hairs. Involucral bracts oblong-lanceolate, ± 2.25 mm long, straw-coloured, glabrous and glistening, not showing beyond the margin of the head when viewed from above, spreading or reflexed in fruit, obtuse. Floral bracts oblanceolate, ±1.5 mm long, membranous, subacute to acuminate, black, the tips papillose hairy. Flowers 3-merous. Male flowers: sepals united below, spathe-like and split down one side, ±1 mm long, black, 3-lobed, the lobes acute or obtuse, the tips papillose hairy; petals very small, with white papillae and a black apical gland; anthers black. Female flowers: sepals free, 1–1.5 mm long dark, 2 narrowly 'boat-shaped' and keeled, with white hairs along the keel and at the tip, the third flat, linear-spathulate, 0.5–0.75 mm long, black, with white hairs at the tip; petals linear-oblanceolate to spathulate, 1–2 mm long, hairy at the tip, with black glands. Seeds oblong to ovoid, ± 0.38 mm long; seed coat cells transversely elongate, aligned in vertical rows; appendages 5–7 from the transverse radial walls, appearing as transverse rings on the surface of the seeds, bristle-like, flattened at the tip.

Annual. In wet and muddy places, rice fields and in ditches; also in crevices between wet rocks. Ansari feels it is perhaps conspecific with *E. pseudoquinquangulare* Ruhland.

{⊙ Hel}. Endemic to C and W India: ♦Ind: Kl, Kt, Mh, Mp. September–February. 0–600 m.

Eriocaulon psammophilum S. M. Phillips [*E. wrightianum* sensu Trimen, non Martius]: Fig. 199

Like *E. sexangulare* but scape sheaths very short (not more than ± 5 cm long); floral bracts cuneate-spathulate, with cuspidate tips; male petals deeply lobed, the lobes truncate; female sepals ± alike, oblanceolate, flat; female petals oblanceolate, the median fractionally longer, densely hairy; seeds subglobose, ± 0.55 mm long, yellow, glossy, longitudinally striate.

Perennial. Sandy stream margins in lowland rain forest.

{Hel}. Endemic to lowland Sri Lanka: ♦Cey. February.

Eriocaulon quinquangulare Linnaeus [*E. argenteum* Martius, ?*E. roseum* Fyson, *E. quinquangulare* forma *viviparum* Moldenke]: Fig. 206

Tufted, vegetative parts often but not always flushed with red, purple or pink (at least when dry). Leaves elongate-triangular to linear, 2.5–5 (–11) cm long, (1–) 1.5–5 mm wide, attenuate towards the apex, thin, with 5–11 well-spaced nerves, dying back at maturity to expose the central bunch of scape sheaths. Scapes many (up to 30), 5–40 cm long, 0.4–0.7 mm in diameter, 5- or 6-ribbed, gently twisted; sheaths loose and papery, obtuse, up to 8 cm long, shorter than the leaves, often purplish. Heads globose or nearly so, 3–7 mm in diameter, pale grey to white; receptacle with short hairs. Involucral bracts linear-oblong to obovate, very obtuse, up to 1.5 mm long, shorter than the floral bracts and flowers, light brown or yellowish, glabrous and glistening, becoming reflexed and often concealed under the head in fruit. Floral bracts rhomboid to obovate, cuspidate-acuminate, 1.5–2 mm long, membranous, glabrous or dorsally hairy. Flowers 3-merous, pale, 1.3–1.6 mm long, on a short glabrous stipe. Male flowers: sepals variable, free or united and spathe-like and split down one side, ± 1.25 mm long, white, hairy at apex; petals very small, with white papillae and black glands; anthers black. Female flowers: sepals free, linear-oblanceolate, ± 1.5 mm long, white, each hairy at the apex; petals narrowly oblanceolate-oblong, dimorphic with 1 large (up to 3 mm long) and 2 small, ciliate lobes. Seeds oblong to ovoid, 0.3–0.4 mm long; seed coat cells transversely elongated, aligned in vertical rows; appendages 2–4 arising from the transverse radial walls, making the seed look horizontally flaky.

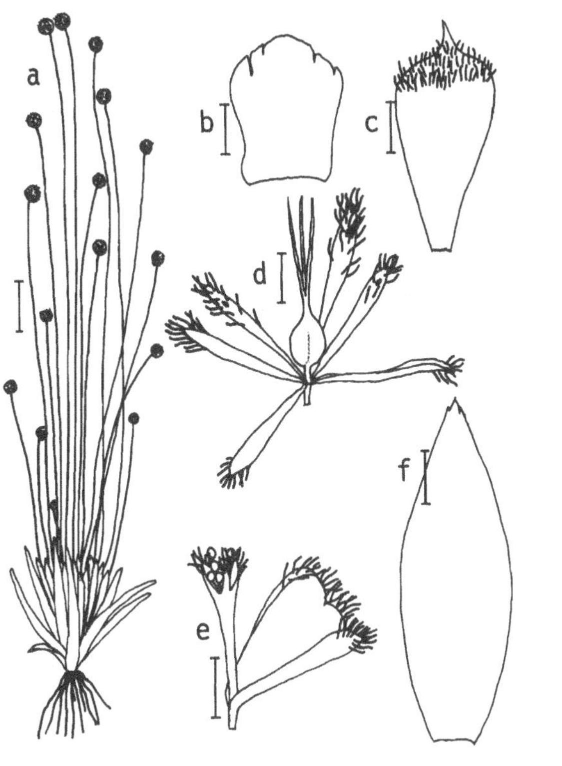

FIG. 206. *Eriocaulon martianum*: f, involucral bract (0.5 mm). *E. quinquangulare*: a, flowering plant (2 cm); b, involucral bract (0.5 mm); c, floral bract (0.5 mm); d, female flower (0.5 mm); e, male flower (0.5 mm).

Annual. Common and often very abundant. Found in seasonally inundated places, marshy grassland, wet rocks and wet sandy flats. It often grows in rice fields but is not considered to be noxious. The red colour when it develops is very distinctive for this species but not all plants go red.

{⊙ Ten/Hyp(Ros)}. From Sri Lanka through India to Bangladesh and Myanmar: ♦Ban, ♦Bur, ♦Cey, ♦Ind, ♦Nep no recent record: Ap, As, Bh, Hp, Kl, Kt, Mh, Mp, Or, Rj, Tn, Up, Wb. 0–1500 m.

Eriocaulon richardianum (Fyson) R. Ansari et N. P. Balakrishnan [*E. dianae* var. *richardianum* Fyson]: Fig. 207

Like *E. heterolepis* but larger and rhizomatous; rhizome ±1.5 cm long, ± 8 mm thick; leaves up to 25 cm long; scapes up to 35 cm long; sheaths up to 8.5 cm long; involucral bracts ± 3 mm long, not showing beyond the margin of the head when viewed from above; floral bracts ± 2.25 mm long; seeds oblong-ellipsoid, ± 0.48 mm long, pale yellow.

Probably perennial. In wet fields.

{Hel}. Endemic to SW India: ♦Ind: Kl, Kt. October–January. 0–400 m.

Eriocaulon robusto-brownianum Ruhland [*E. mysorensis* Fyson]: Fig. 207

Tufted. Leaves lanceolate or linear, 20–30 (–50) cm long, 1.2–2 cm wide (to 2.5 cm wide at base), glabrous, attenuate towards the apex, the tip subacute or obtuse. Scapes several to many, stout, 30–45 (–80) cm long,

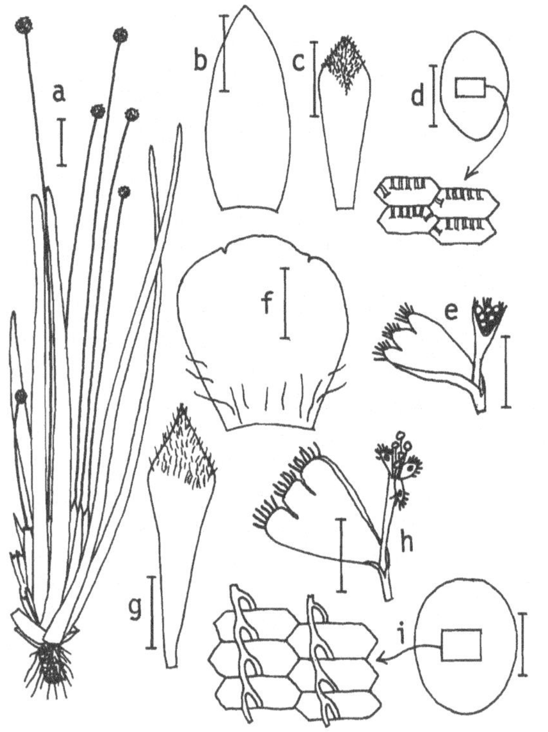

Fig. 207. *Eriocaulon richardianum*: a, flowering plant (2 cm); b, involucral bract (1 mm); c, floral bract (1 mm); d, seed (0.25 mm) with seed coat pattern (inset); e, male flower (1 mm). *E. robusto-brownianum*: f, involucral bract (1 mm); g, floral bract (1 mm); h, male flower (1 mm); i, seed (0.25 mm) with seed coat pattern (inset).

1 mm or more in diameter, glabrous or sparsely hairy, 6- to 9-ribbed, twisted; sheaths up to ± 15 cm long, loose, glabrous or sometimes hairy, mouth very obtuse, acuminate. Heads globose or nearly so, 0.8–1 (–2.5) cm in diameter, snowy white, appearing rough through prominent floral bracts; receptacle flat to narrowly columnar, with short hairs. Involucral bracts obovate to obovate-oblong, the outer ones 2.5–3 (–5) mm long, rounded at the tip, the inner ones with acute tips, yellowish-grey or pale brown, glabrous or with brown silky hairs at the base, up to 2.75 mm long, mostly shorter than the floral bracts and flowers. Floral bracts cuneate-obovate, acuminate, pale or dark greyish-green, ±3.75 mm long, membranous, hyaline, the tip densely covered with white papillose hairs. Flowers 3-merous. Male flowers: sepals spathe-like and split down one side, ±1.75 mm long, dark, lobes 3 and ± equal, white papillose hairy; petals triangular-ligulate, with white papillose hairs and a black gland; anthers black. Female flowers: sepals 3, free, ± 2.25 mm long, oblanceolate-oblong, concave or nearly flat, often keeled, each with dense white hairs at the tip; petals oblanceolate-oblong, shortly clawed, hairy, with or without a black apical gland. Seeds oblong-ellipsoid, ±0.6 mm long, ± 0.45 mm wide; seed coat cells transversely elongated, aligned in vertical rows; appendages solitary from the centre of each transverse radial wall, elongated, curved and joined to the adjacent ones of the same vertical row forming longitudinal parallel ribs.

Annual or perhaps sometimes perennial. In marshes and along the edges of ponds.

{(⊙)? Hel}. Endemic to India: ♦Ind: Go? Kl, Kt, Mh? Tn. November–March. 0–1300 m.

Eriocaulon santapaui Moldenke: Fig. 208

Like *E. parviflorum* but smaller; heads ± 3 mm in diameter; involucral bracts ± 2 mm long; male sepal lobes shallow with obtuse lobes; female petals 0.75–1 mm long, glabrous, without black glands; seeds ovoid-globose, ± 0.45 mm long.

Annual. A rare plant found along streams at low altitudes.

{⊙ Hel}. Endemic to W India: ♦Ind: Mh.

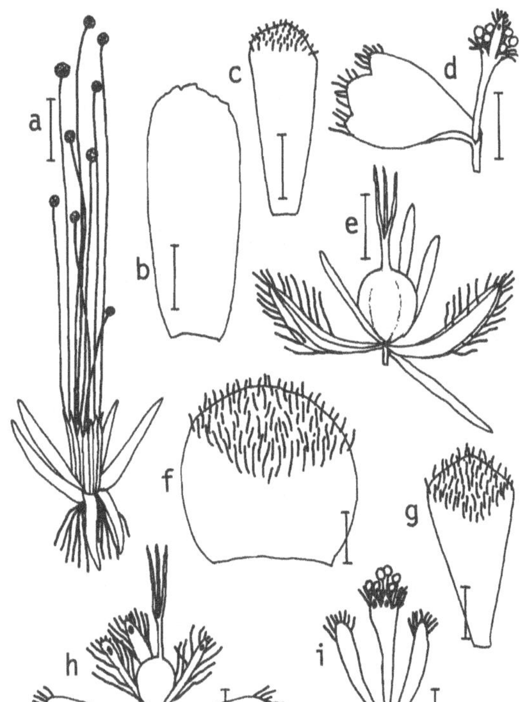

Fig. 208. *Eriocaulon santapaui*: a, flowering plant (2 cm); b, involucral bract (0.5 mm); c, floral bract (0.5 mm); d, male flower (0.5 mm); e, female flower (0.5 mm). *E. sedgwickii*: f, involucral bract (0.5 mm); g, floral bract (0.5 mm); h, female flower (0.5 mm); i, male flower (0.5 mm).

Eriocaulon sedgwickii Fyson: Fig. 208

Like *E. heterolepis* and *E. quinquangulare* but petals of female flowers broadly lanceolate, with large glands; sepals 2; floral bracts with very dense covering of white hairs making the heads snow white; involucral bracts not larger than floral bracts, blackish.

Annual. Wet places and rocks on hill sides.

{⊙ Hel}. Endemic to SW India: ♦Ind: Gj? Kt, Mh. 450–? m.

Eriocaulon setaceum Linnaeus [*E. intermedium* Körnicke, *E. capillus-naiadis* J. D. Hooker]: Fig. 209

Stems elongate, 10–35 (–160) cm long, spongy, densely clothed in numerous leaves, terminating in an umbel of many scapes, the heads floating on or emerging above the water surface. Leaves filiform, 4–10 cm long, 0.1–0.4 mm wide, broadening towards the base, glabrous, semi-translucent, 1-nerved. Scapes 7–20 cm long, ± 0.4 mm in diameter, 5- to 7-ribbed, twisted; sheaths up to 3.5 cm long, loose, glabrous, parallel-sided, obliquely split, acute. Heads hemispherical to conical, whitish, pale green, brown, grey or black, 2.5–4.5 mm in diameter; receptacle conical to columnar, pilose. Involucral bracts shorter than the head width, obovate-oblong, ± 1.25 mm long, rounded, glabrous, with scarious and sometimes cut margins, turned slightly downwards at maturity. Floral bracts oblong-cuneate, ±1.25 mm long, black, white-papillose on the centre back or glabrous, acute. Flowers 3-merous, 0.9–1.4 mm long, enclosed within the bracts. Male flowers shortly stalked; sepals united and spathe-like to almost free, ± 1 mm long, 3-lobed, black or yellowish-grey glabrous or with some white papillae near the rounded tips; petals minute, enclosed within the sepals, without black glands; anthers black. Female flowers shortly stalked; sepals subequal, variable, boat-like, swollen, ± 0.75 mm long, with a thickened and sometimes winged keel, glabrous or white papillose; petals narrowly oblanceolate to almost linear, ± 0.75 mm long, white papillose at the tip, glands poorly developed or absent. Seeds oblong-ellipsoid, 0.4–0.6 mm long; seed coat cells transversely elongated, aligned in vertical rows; appendages 3–8 arising from the transverse radial walls, appearing as transverse rings on the seed surface, bristle-like, truncate at the tip, with a mucilaginous coat above the appendages.

Annual or perhaps sometimes perennial. Vegetative parts totally submerged. Found in a wide variety of aquatic habitats.

It is rather variable in many characters. *Eriocaulon capillus-naiadis* J. D. Hooker is sometimes considered to be distinct (it differs in having floral bracts, male sepals and female sepals clothed in hairs; in the Indian subcontinent it is confined to the NE and the Himalayas). Plants with glabrous floral bracts and black heads are sometimes called *E. intermedium* but numerous intermediates exist. The differences between the tropical American species *E. melanocephalum* Kunth and the tropical African species *E. bifistulosum* Van Heurck et Müller Argoviensis [*E. fluitans* Baker, *E. limosum* Engler et Ruhland, *E. schweinfurthii* Engler et Ruhland] are by no means clear. If these prove to be conspecific then *E. melanocephalum* is the earliest name.

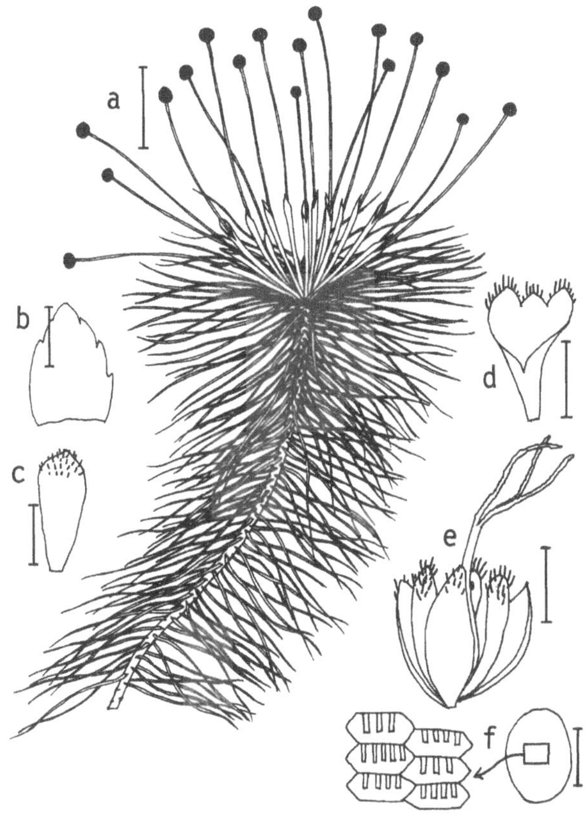

FIG. 209. *Eriocaulon setaceum*: a, flowering plant (2 cm); b, involucral bract (0.5 mm); c, floral bract (0.5 mm); d, male sepals (0.5 mm); e, female flower (0.5 mm); f, seed (0.18 mm) with seed coat pattern (inset).

{(⊙) Vit}. From Sri Lanka through India to Myanmar and Vietnam. Ban, Bur, Cey, Ind: An, Ap, Ar, As, Go, Kl, Kt, Mg, Rj, Wb. 0–1800 m.

Eriocaulon sexangulare Linnaeus [*E. wallichianum* Martius, *E. myriophyllum* Wallich, *E. quinquangulare* forma *viviparum* Moldenke]: Fig. 210

Rhizome short and tough or absent. Leaves broadly linear to linear-lanceolate, 10–45 cm long, 5–11 mm wide (up to 25 mm at base), papery, subacute. Scapes 30 or more per rosette, 20–60 cm long, 0.7–1 mm in

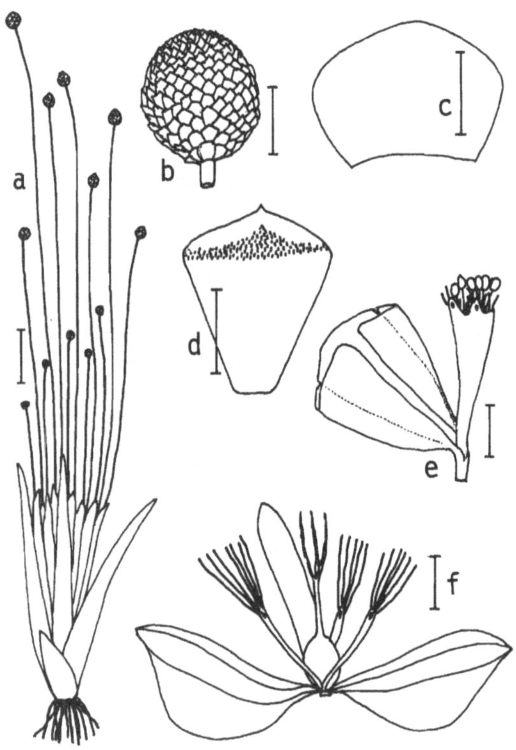

Fig. 210. *Eriocaulon sexangulare*: a, flowering plant (2 cm); b, head (3.5 mm); c, involucral bract (1 mm); d, floral bract (1 mm); e, male flower (0.5 mm); f, female flower (0.5 mm).

diameter, 5- to 8-ribbed, strongly twisted; sheaths up to ± 25 cm long, the mouth oblique, acute or acuminate. Heads hemispherical to conical, 6–10 mm in diameter, grey to white, sometimes pseudoviviparous; receptacle hairy. Involucral bracts as wide as the head, obovate to orbicular, ± 2 mm long, yellow to yellowish-grey or reddish-brown, glabrous, shiny, leathery with scarious margins, spreading outwards in fruit. Floral bracts broadly cuneate, cuspidate, ± 2.5 mm long, leathery, tightly overlapping, inflexed and enclosing the flowers, grey with a ± triangular patch of powdery white hairs. Flowers 3-merous, strongly laterally compressed, 2–2.5 mm long, pale. Male flowers: sepals tubular, united except at the tip, wedge-shaped, ± 2 mm long, indistinctly lobed, enclosing the petals, broadly winged upwards, a few white papillae on the upper wing margin; petals 0.3 mm long, glandular and white-papillose; anthers black. Female flowers: sepals 3, free; two sepals boat-shaped, ± 2 mm long, swollen, firm, broadly winged on the keel, narrowed to a basal claw, a few white papillae near the tip, the third sepal tongue-like, thin and flat; petals thread-like, ± 1 mm long, with a brush of septate hairs 0.8–1 mm long towards the glandular (but not papillose) tip. Seeds ovoid to globose, 0.55–0.7 mm long; seed coat cells transversely elongated, aligned in vertical rows; appendages solitary, arising from the transverse radial walls, appearing as vertical lines on the seed surface, bristle-like, swollen at the tip.

Perennial or perhaps sometimes annual. Swampy places, marshes, along streams, rice fields, often on sandy soil. Common but rarely forming large colonies. Among the lowland species of *Eriocaulon* in Peninsular India the combination of filiform female petals with an apical brush and the perennial habit are distinguishing features; the heads exceeding 6 mm and the triangular patches of powdery white hairs on the tightly overlapping floral bracts are useful field characters.

{(⊙) Hel}. Lowlands of Sri Lanka, western Peninsular India, Myanmar and SE Asia: Bur, Cey, Ind, Nep: An, Ap, Go, Kl, Kt, Mh. September to March. 0–175 m.

Eriocaulon sivarajanii R. Ansari et N. P. Balakrishnan: Fig. 215

Like *E. xeranthemum* but involucral bracts obovate-cuneate with a terminal cusp, ± 3 mm long, ± 2 mm wide; receptacle hairy; male sepal lobes rounded at the tips; seeds ± 0.6 mm long; seed coat appendages longer than the seed coat cells.

Annual. In moist places and wet flushes on the coastal plains of Kerala. Rare but locally abundant.

{⊙ Ten(Ros)/Hel}. Endemic to SE India: ♦Ind: Kl.

Eriocaulon stellulatum Körnicke: Fig. 211

Tufted. Leaves oblong-lanceolate, up to 15 cm long, ±1 cm wide, many-nerved, glabrous, attenuate towards the apex. Scapes few, slender, up to 20 cm long, sparsely hairy, indistinctly ribbed, scarcely twisted; sheaths up to ± 6 cm long, loose, glabrous or sparsely hairy, mouth oblique, acute. Heads globose, ± 1 cm in diameter, appearing spiny, yellowish-grey; receptacle ovoid-cylindrical, hairy. Involucral bracts oblong-obovate, ± 3 mm long, yellowish-grey, reflexed, acuminate. Floral bracts lanceolate, ± 5 mm long, membranous, spreading, yellowish-grey, glabrous outside, glabrous or papillose inside. Flowers 3-merous. Male flowers: sepals spathe-like and split down one side, ± 1.5 mm long, dark, 3-lobed, with acute or acuminate lobes, sparsely hairy towards the tip; petals minute, glabrous, with a black gland; anthers black. Female flowers: sepals 3, free, oblong-lanceolate, folded and keeled, ± 1.75 mm long, yellowish-grey, sparsely papillose along the keels; petals linear-spathulate, ± 1.25 mm long, hairy, with a black apical gland at the tip. Seeds oblong-ovoid to globose, ± 0.63 mm long; seed coat cells transversely elongated, aligned in vertical rows; appendages 1–3 from the

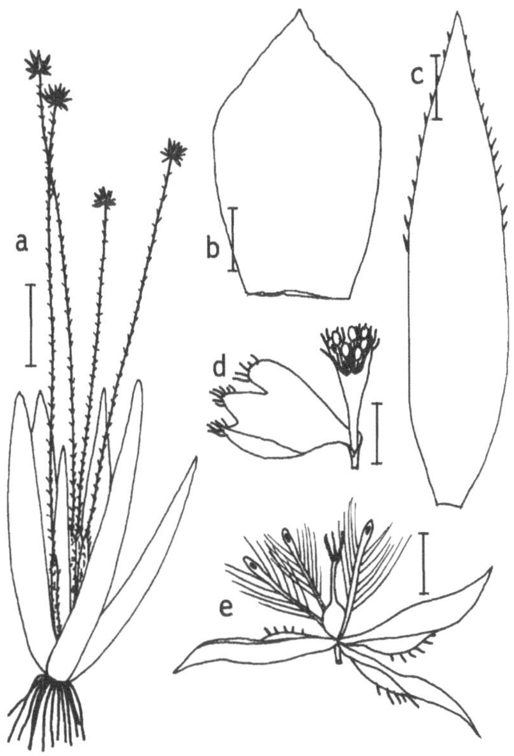

Fig. 211. *Eriocaulon stellulatum*: a, flowering plant (2 cm); b, involucral bract (0.5 mm); c, floral bract (0.5 mm); d, male flower (0.5 mm); e, female flower (0.5 mm).

transverse and 0–1 from the vertical radial walls, appearing to be in transverse rings on the surface of the seeds, bristle-like, truncate at the tip.

Annual. On the banks of streams and along the shores of lakes, often found in wet rocky places. Characteristic is that the involucral bracts are acuminate and longer than the flowers; the head resembles a 'flower'.

{⊙ Hel}. Endemic to SW India: ♦Ind: Kt, Kl, Mh. September–April. 0–1000 m.

Eriocaulon trilobum F. Hamilton ex Körnicke [*E. sollyanum* Royle, nom. illeg. (Royle, J. F. Illustrations of the botany of the Himalaya mountains, 409, t.?7 (1839) has an illustration but no description)]: Fig. 212

Leaves caespitose, linear to elongate-triangular, attenuate towards the apex, 4–11 cm long, 2.5–6 (–8) mm wide, glabrous, 9- to 11-nerved, the tip acute or subobtuse. Scapes few to many, 7–20 cm long, glabrous, 5- or 6-ribbed, twisted; sheaths loose, 2–6 cm long, with oblique mouth. Heads globose to hemispherical, 3–6 mm in diameter, black to dark grey; receptacle with short hairs. Involucral bracts linear-oblong or cuneate, 1.5–2 (–3.3) long, 0.5–1.5 mm wide, yellowish-grey, the tip obtuse, glabrous and shining, at first radiating beyond the head becoming reflexed in fruit. Floral bracts cuneate-obovate, 1.25–2 mm long, membranous, blackish, obtuse or subacute but never acuminate, hairy at the apex. Flowers 3-merous. Male flowers: sepals united below and spathe-like, 1–1.5 (–3) mm long, the lobes 3, obtuse; petals minute; anthers black. Female flowers: sepals 3, ± equal, oblanceolate, boat-shaped, 1–1.5 (–3) mm long, each hairy at the apex; petals 3 free, narrowly spathulate, 1–1.25 mm long, apically hairy lobes, each with a black apical gland. Seeds oblong-ellipsoid, ±0.4 mm long; seed coat cells transversely elongated, aligned in vertical rows; appendages 3–6 from the transverse radial walls, appearing to be in transverse rings on the surface of the seeds, bristle-like, minutely flattened at the tip.

Annual. Wet places, particularly in rice fields, sprouting after the harvest. Close to *E. quinquangulare* but not drying red, heads dark grey, smaller (3–5 mm in diameter) and involucral bracts becoming reflexed at maturity. It differs from *E. heterolepis* in having female sepals equal; it differs from *E. odoratum* in having male petals equal and receptacle ± glabrous.

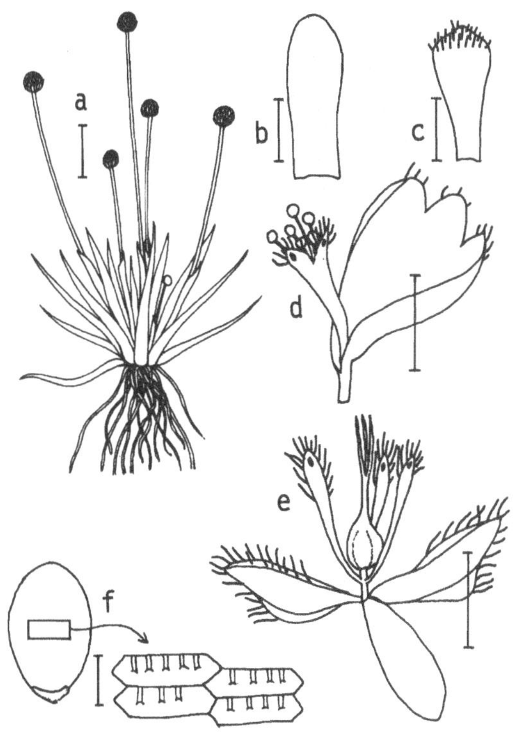

Fig. 212. *Eriocaulon trilobum*: a, flowering plant (1 cm); b, involucral bract (0.5 mm); c, floral bract (0.5 mm); d, male flower (0.5 mm); e, female flower (0.5 mm); f, seed (0.17 mm) with seed coat pattern (inset).

{⊙ Ten(Ros)}. Probably endemic to the Indian subcontinent: Ban? Ind, Pak: Hp? Kt, Kl, Mg, Mh, Mp, Or, Rj, Up, Wb. October–February. 0–800 m.

Eriocaulon truncatum F. Hamilton ex Martius [*E. truncatum* var. *disepalum* Fyson, *E. truncatum* var. *malaccense* J. D. Hooker]: Fig. 213

Tufted. Leaves linear to lanceolate, 1–6 cm long, 1.5–3.8 mm wide, glabrous, rather broad at the base, 10- to 18-nerved, subacute and apiculate, dying back at maturity to expose the central bunch of scape sheaths. Scapes few or many (4–30), stout, 5–15 (–25) cm long, 0.4–0.8 mm in diameter, 5- to 8-ribbed, twisted; sheaths loose, up to 4.5 cm long, shorter than the leaves, obliquely spathe-like at the mouth, often splitting and becoming 2-lobed. Heads cup-shaped to flattened-globose, 3–5.5 mm in diameter, glistening white or yellowish-grey; receptacle thinly pilose to almost glabrous. Involucral bracts as wide as the head, at or before anthesis sometimes visible when viewed from above, broadly obovate or oblong, 2–3 mm long, truncate or with an obtusely triangular tip, pale yellowish-grey, shiny, glabrous and glistening, spreading upwards in fruit. Floral bracts closely overlapping and enclosing the flowers, oblong-ovate to spathulate, 1.75–2.5 mm long, membranous, glabrous or papillose at the tip, scarious and shiny. Flowers 1.5–1.6 mm long, male and female intermingled. Male flowers: stipe 0.25–0.7 mm long; sepals 2, oblong, ± 1.25 mm long, grey, glabrous, united on one side at the base; petals 3, very small, with white papillae; anthers black. Female flowers: sepals 2 (3), free, linear-spathulate, ± 1.3 mm long; petals 3, narrowly oblanceolate, 1.25–1.5 mm long, thinly pilose to almost glabrous. Seeds oblong to ovoid, 0.35–0.5 (–0.7) mm long; seed coat cells rectangular, vertically elongated, aligned in vertical rows; appendages in the form of ribbon-like bands from the vertical walls and solitary incomplete bands from the transverse radial walls, the seed surface appearing coarsely reticulate with longitudinal white, ribbon-like bands.

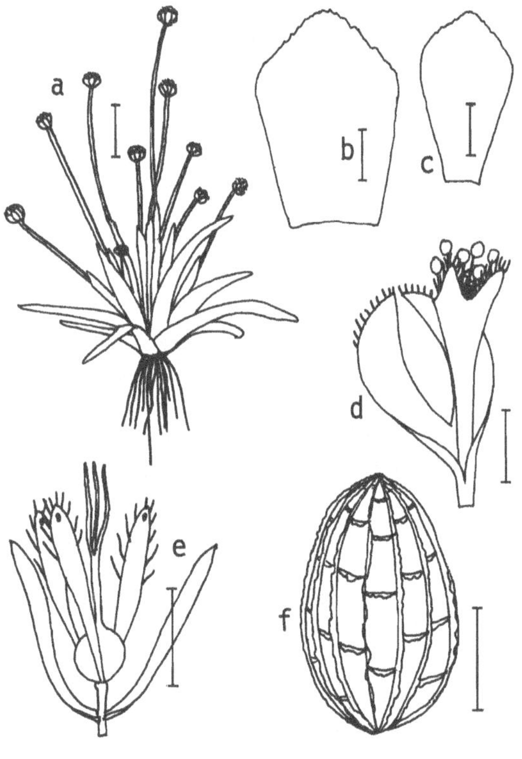

Fig. 213. *Eriocaulon truncatum*: a, flowering plant (1 cm); b, involucral bract (0.5 mm); c, floral bract (0.5 mm); d, male flower (0.5 mm); e, female flower (0.5 mm); f, seed (0.2 mm).

Annual. Inundated places, marshland, wet rocks, common in rice fields. It is easily recognizable by its pallid, shining heads on relatively stout scapes; bracts scarious and completely enclosing the flowers; seeds with longitudinal white, ribbon-like bands, with the cells longer than wide.

{⊙ Ten/(Ros)/Hel}. S and SE Asia: Ban, Bhu, Bur, Cey, Ind: An, Ap, As, Go, Kl, Kt, Mg, Mh, Mn, Mp, Sk, Tn, Wb. July–January. 0–1250 m.

Eriocaulon walkeri J. D. Hooker [*E. quinquangulare* var. *walkeri* (J. D. Hooker) Fyson]: refer to Fig. 206

Like *E. quinquangulare* but lacks the red colour; leaves thicker and narrower (1–2 mm wide); receptacle densely woolly-hairy; petals with woolly hairs; seeds smooth.

Annual. Locally abundant, on damp sand and in rice fields.

{⊙ Hel}. Endemic to the lowlands of Sri Lanka: ♦Cey.

Eriocaulon wightianum Martius: refer to Fig. 201

Like *E. helferi* but larger and rhizomatous; the rhizome often as thick as a thumb; leaves 8–20 cm long, 5–18 mm wide, slightly hairy; Scapes few or many, 16–38 cm long; heads ± 1.75 cm in diameter, appearing white; involucral bracts 3–4 mm long; floral bracts ± 4 mm long, snow white, the outer ones nearly glabrous; female petals elliptical-spathulate; seeds ellipsoid, ± 0.5 mm long, smooth, minutely papillose.

{Hel}. Endemic to Myanmar: ♦Bur.

Eriocaulon willdenovianum Moldenke [*E. longifolium* Nees ex Kunth non Rafinesque, *E. sexangulare* var. *longifolium* (Kunth) J. D. Hooker, nom illeg.]: Fig. 214

Rhizome short and tough or absent. Leaves broadly linear to linear-lanceolate, 7–30 cm long, 3.5–7 mm wide (up to ±15? mm at base), papery, subacute. Scapes 8–15 per rosette, 15–25 cm or more long,

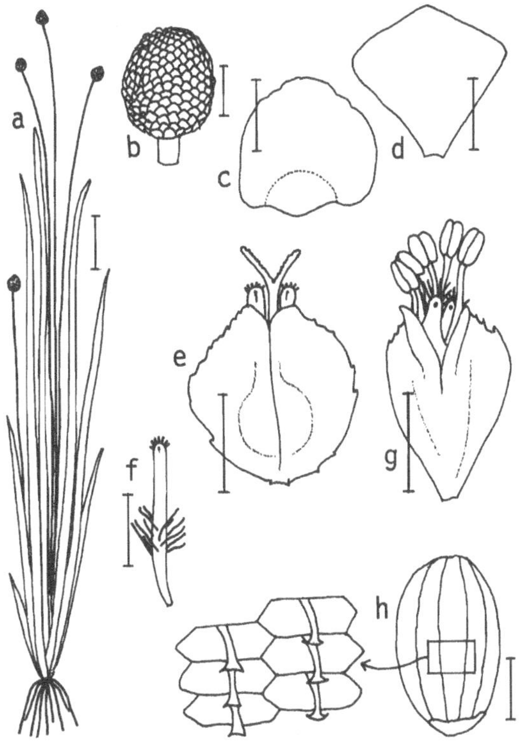

FIG. 214. *Eriocaulon willdenovianum*: a, flowering plant (2 cm); b, head (1.5 mm); c, involucral bract (1 mm); d, floral bract (1 mm); e, female flower (1 mm); f, female petal (1 mm); g, male flower (0.5 mm); h, seed (0.25 mm) with seed coat pattern (inset).

0.6–1 mm in diameter, 5- or 6-ribbed, strongly twisted; sheaths up to ± 20 cm long, the mouth oblique, acute or acuminate. Heads hemispherical to conical, 4–6 mm in diameter, grey to white, sometimes pseudoviviparous; receptacle glabrous. Involucral bracts as wide as the head, obovate to orbicular, ± 2 mm long, yellow to yellowish-grey or reddish-brown, glabrous, shiny, leathery with scarious margins, spreading outwards in fruit. Floral bracts broadly cuneate, cuspidate, ± 2.5 mm long, leathery, tightly overlapping, inflexed and enclosing the flowers, white or grey with a ± triangular patch of powdery white hairs. Flowers 2-merous, strongly laterally compressed, ±2 mm long, pale, mostly female. Male flowers: sepals tubular, united except at the tip, wedge-shaped, ± 1.75 mm long, indistinctly lobed, enclosing the petals, 2, broadly winged upwards, a few white papillae on the upper wing margin; petals 2, ± 0.35 mm long, glandular and white papillose; anthers 4, black or brown. Female flowers suborbicular: sepals 2, free; boat-shaped, ± 1.75 mm long, broadly winged on the keel, glabrous, greyish-yellow; petals 2, linear, ± 1.75 mm long, hyaline, with a bunch of hairs about the middle and small papilla-like hairs at the tip. Seeds ellipsoid, 0.6–0.7 mm long, 0.48–0.5 mm wide; seed coat cells transversely elongated, aligned in vertical rows; appendages solitary, arising from the transverse radial walls, appearing as vertical lines on the seed surface, bristle-like, swollen at the tip.

Perennial or annual. In seasonally inundated places, swamps and slow-moving water. It is sometimes confused with *E. sexangulare*, the best diagnostic features are the 2-merous flowers and female petals lacking the terminal tuft of hairs; leaves narrower (3.5–7 mm wide); scapes more slender (0.6–1 mm in diameter); heads smaller (4–6 mm wide); female sepals rounded, without a clawed base and with a pronounced recurved tip; female petals equalling the sepals, somewhat broader (linear as opposed to filiform) and lacking the terminal brush of long hairs; seeds longer and narrower (± ellipsoid, 0.6–0.7 mm long, 0.48–0.5 mm wide).

{(⊙) Hyp/Hel}. Madagascar, Mauritius, Sri Lanka, Andaman and Nicobar Islands, Burma, SE Asia and Indonesia (lacking from Peninsular India): Bur, Cey, Ind: An.

Eriocaulon xeranthemum Martius [*E. pygmaeum* Dalzell non Solander]: Fig. 215

Tufted. Leaves linear or linear-lanceolate, rarely more than 4 cm long; ± 3 mm wide, flat, glabrous, attenuate towards an acute or acuminate tip. Scapes many or few, rarely more than 6 cm long, 4–7 mm in diameter, 5- to 7-ribbed, glabrous; sheaths lax, up to 3 cm long, acute or subacuminate. Heads hemispherical to flattened, ± 6 mm in diameter, yellowish-grey; receptacle convex, glabrous. Involucral bracts linear-oblong to oblanceolate, 2–3.5 mm long, 0.25–0.75 mm wide, erect or spreading, much longer than the floral bracts and flowers, straw-coloured to almost white, glabrous, often showing beyond the margin of the head when viewed from above, acuminate. Floral bracts cuneate, ± 1.25 mm long, membranous, truncate, hyaline, the tips papillose hairy. Flowers 3-merous. Male flowers: sepals united below, spathe-like and split down one side, cuneate, ± 1 mm long, black, 3-lobed, the lobes truncate, the tips papillose hairy; petals very small, with white papillae and a black apical gland; anthers black. Female flowers: sepals free, ± 1 mm long dark, 2 lateral ones oblanceolate, yellowish-grey obtuse, with white hairs at the tip, the third linear, glabrous or with a few white hairs at the tip; petals spathulate, 0.75–1 mm long, hyaline, hairy at the tip, with black glands. Seeds oblong-ellipsoid, ± 0.42 mm long; seed coat cells transversely elongate, aligned in vertical rows; appendages 2–4 from the transverse radial walls, shorter than the seed coat cells, appearing as transverse rings on the surface of the seeds, bristle-like, truncate at the tip.

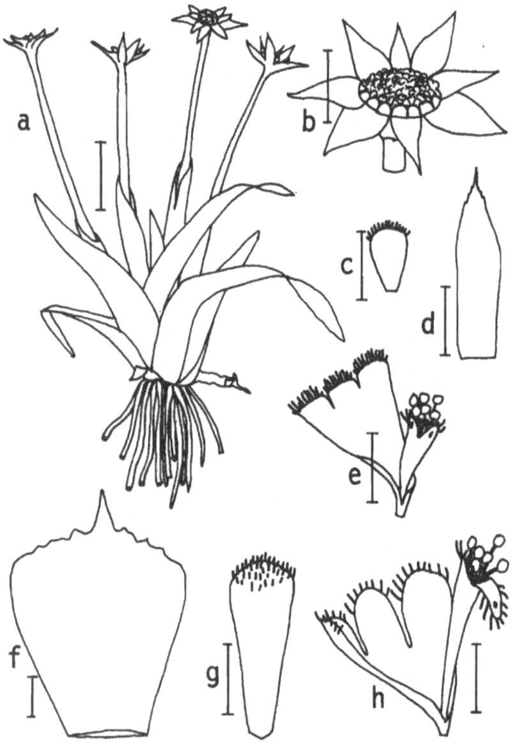

Fig. 215. *Eriocaulon xeranthemum*: a, flowering plant (1 cm); b, head (1 mm); c, floral bract (1 mm); d, involucral bract (0.5 mm); e, male flower (1 mm). *E. sivarajanii*: f, involucral bract (1 mm); g, floral bract (1 mm): h, male flower (1 mm).

Annual. In marshes, pools on laterite, rice fields and other seasonally inundated places; also found in the crevices between wet rocks; locally abundant. It is easily recognizable by its shining pale yellow to white flower-like heads. The heads are superficially like some races of *E. heterolepis* but it has floral bracts rounded and not cuspidate and with rounded tips to the floral bracts and male sepal lobes.

{⊙ Ten(Ros)/Hel}. Tropical Africa, C Himalayas and India: Ban, Bur, Ind, Nep: An, As, Bh, Go, Kl, Kt, Mg, Mh, Mn, Mp, Tn, Wb. July–December. 0–700 m.

EUPHORBIACEAE [BISCHOFIACEAE]

excluded see p. 12

FABACEAE [LEGUMINOSAE]: 657 genera

Herbs, shrubs or trees. Leaves stipulate, alternate, usually compound. Flowers usually bisexual, solitary or in racemes. Sepals 5 free or united. Petals 5, usually free and unequal. Stamens 10 or indefinite; filaments free or united. Ovary superior, solitary, with 1 or more ovules. Fruit a dry pod, either splitting or separating into 1-seeded joints.

A very large and widespread family with relatively few aquatic and wetland species. Hutchinson, J. The genera of flowering plants (Oxford University Press) 1: 221–516 (1964); the following work, despite its title, is no more than a list: Sanjappa, M. Legumes of India. Bishen Singh Mahendra Pal Singh, Dehra Dun i–iv, 1–338 (1991).

Excluded: the following are weedy plants sometimes found in moist situations but they are not considered to be wetland species: ***Alysicarpus hamosus*** Edgeworth; ***A. vaginalis*** (Linnaeus) A. P. de Candolle; ***Cassia tora*** Linnaeus; ***Desmodium ferrugineum*** Wallich ex Thwaites [*D. rufescens* Wright et Arnott]; ***D. heterophyllum*** (Willdenow) A. P. de Candolle; ***D. laxiflorum*** A. P. de Candolle; ***D. triflorum*** (Linnaeus) A. P. de Candolle; ***Indigofera cordifolia*** Heyne ex Roth; ***Smithia bigemina*** Dalzell; ***S. conferta*** J. E. Smith; ***S. sensitiva*** W. Aiton; ***S. setulosa*** Dalzell; ***Trigonella occulta*** Raffenau-Delile.

1. Leaves 2-pinnate; flowers radially symmetrical, lower flowers of inflorescence sterile ***Neptunia***
1. Leaves 1-pinnate; flowers bilaterally symmetrical, all fertile
  2. Leaves with 2 pairs of leaflets; bracts ± concealing the pods ***Geissaspis***
  2. Leaves with many more than 2 pairs of leaflets; bracts not concealing the pods
    3. Pods breaking into regularly-shaped, 1-seeded portions ***Aeschynomene***
    3. Fruits longitudinally dehiscent, not breaking into portions ***Sesbania***

Aeschynomene: 150 species: warm regions

Shrubs or herbs. Stems erect, creeping or floating, internodes often swollen when submerged or floating. Leaves alternate, odd-pinnate with numerous leaflets; stipules usually persistent. Inflorescence racemose, axillary or terminal, simple or branched. Flowers bisexual, bilaterally symmetrical. Sepals tubular, either 5-lobed or 2-lipped. Petals yellow, often with purple streaks; adaxial petal orbicular; lateral petals obliquely obovate or oblong; keel either obovate and slightly curved or often narrow and much incurved. Stamens 10, united below, usually in 2 bundles of 5 or rarely 9 united and 1 free. Ovary shortly stalked, 2 to many ovules; style filiform, incurved. Pod transversely septate between seeds; joints 1-seeded, separating, indehiscent.

Free-floating, partly floating, emergent or seasonally submerged. The stems are often pithy and swollen. Flowers entomophilous. The disseminules are flattened, one-seeded parts of a pod, dispersed by water and perhaps also by other means. Sometimes cultivated for the pith-like aerenchyma, decoration, green manure and forage. Reynolds, S. T. Aeschynomeneae (Benth.) Hutch. (Leguminosae) in Australia. Austrobaileya 3(2): 177–202 (1990).

1. Leaflets with 2 or more nerves, somewhat bent, with asymmetrical oblique tips; pod borne on a ± 2 mm long stalk; dorsal part of the pod with a cut margin ***A. americana***

1. Leaflets with 1 nerve, straight with symmetrically obtuse tips; borne on a 4–20 mm long stalk; dorsal part of the pod with a smooth margin

2. Flowers 12–18 mm long; sepals hairy, ± 9 mm long; pod falling apart in (1–) 3–5 (–7) joints, each 10–15 mm long usually; adaxial (upper) suture of the pod indented as much as the abaxial, rough on the faces above the seeds; stems usually thick and spongy ***A. aspera***

2. Flowers usually less than 10 mm long; sepals glabrous, 4–6 mm long; pod falling apart in (3–) 6–10 (–13) joints, each 3–6 mm long; adaxial (upper) suture of the pod straight, smooth or papillose on the faces above the seeds; stem slender or thick and fistular but rarely spongy and then only at the base ***A. indica***

Aeschynomene americana Linnaeus: Fig. 216

Like *A. indica* but leaflets somewhat curved, asymmetrically oblique at the tip, with 2 or more nerves; pod borne on a ± 2 mm long stalk; dorsal part of the pod with a cut margin; joints (3–) 5–7, semi-orbicular.

Perennial. Sometimes in wet or moist places; it can tolerate temporary flooding but is rarely found in wetlands.

{Hel}. Pantropics, native in S America: <Cey? <Ind: <An, <Bh, <Kl, <Tn, <Wb.

Aeschynomene aspera Linnaeus: Fig. 216

Stems erect and up to 2 m tall or trailing and floating, in water developing spongy tissue up to 5 cm in diameter. Stipules narrowly ovate, 8–19 mm long, 1.5–3 mm wide, attenuate, the margin ciliate or toothed, produced below their point of insertion to a spur; lower part ± 5 mm long, ± 2 mm wide, usually notched or toothed. Leaves shortly petiolate; rachis 6–22 cm long, ending in a minute stipe; leaflets in (10–) 30–50 pairs, linear-oblong, 5–10 (–16) mm long, those near the base and apex shorter than those in the centre, 1.5–3 mm wide, with 1 median, usually black vein, the base asymmetrical, obtuse, the tip subacute but minutely mucronate. Flowers 1–4, 12–18 mm long. Peduncles (0.6–) 1–3 (–5) cm long, hairy; bracts cordate, 3–5 mm long, 1.5–3 wide, with or without hairs, the hairs yellowish; pedicels 1–5 mm long, hairy; bracteoles ovate, 4–5 mm long, ± 2 mm wide, hairy. Sepals hairy, 2-lobed almost to the base, each 6–8 (–10) mm long. Petals partly hairy, yellow; adaxial petal obovate, 10–15 mm long, 14–17 mm wide, with a 1–2 mm long claw, the upper margin sometimes toothed; lateral petals 8–10 (–16) mm long, ± 3 mm wide; keel 14–20 mm long, ± 6 mm broad. Stamens in 2 bundles of 5, 15–20 mm long. Pods borne on 10–20 mm long stalks, strap-shaped, 4–6 cm long, 7–8 mm wide, flattened, the adaxial (upper) suture of the pod indented as much as the abaxial, rough on the faces above the seeds; falling apart in (1–) 3–5 (–7) almost quadrangular, 1-seeded joints, each 10–15 mm long, 7.5–9 mm wide. Seeds kidney-shaped, ± 7 mm long, 4–5 mm wide, ± 1.5 mm thick, black, glossy.

Perennial. Growing in and along tanks and lakes. The pith-like secondary xylem is employed in the making of such varied items as artificial flowers, toys, sunhats, life-belts and carbon for gunpowder. It is also cultivated for decoration, green manure and forage.

{Hel/Ten/Hyp}. Warm and tropical Asia, naturalized in Australia: Ban, Bhu, Bur, Cey, Ind, Nep, Pak: As, Bh, Dl, Jk, Kl, Kt, Mh, Mp, Or, Sk, Tn, Up, Wb. 0–1300 m.

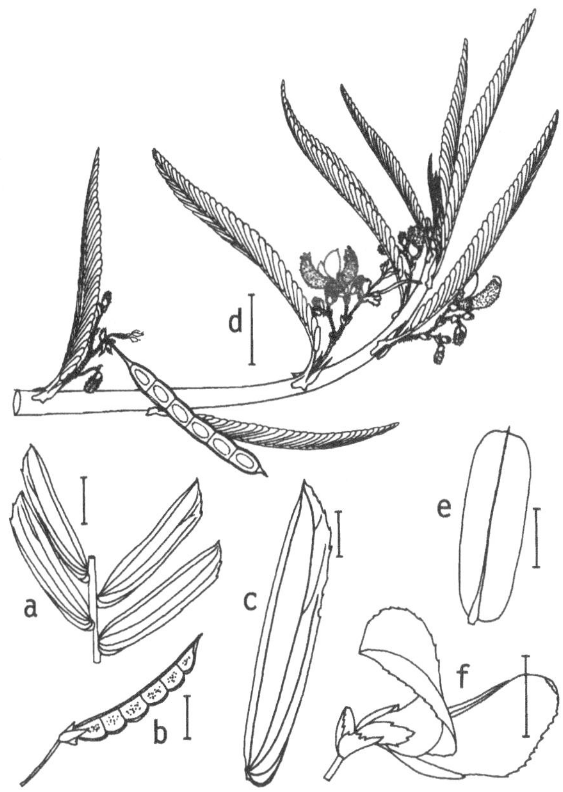

FIG. 216. *Aeschynomene americana*: a, four leaflets (2 mm); b, pod (1 cm); c, one leaflet (1 mm). *A. aspera*: d, flowering shoot (4 cm). *A. indica*: e, one leaflet (2 mm); f, flower (4 mm).

Aeschynomene indica Linnaeus [*Smithia aspera* Roxburgh, *Hedysarum neli-tali* Roxburgh]: Fig. 216

Stems erect, often somewhat woody at the base, branched, 30–250 cm tall; main stem cylindrical and warty at the base; the warts bearing caducous glandular hairs, rarely spongy and then only at the base. Stipules narrowly ovate or elliptical, 5–18 mm long, 1–3.5 mm wide, acuminate, produced below their point of insertion; lower part usually notched or toothed. Leaves petiolate; petioles 0.5–2 cm long; rachis 4–14 cm long, ending in a minute stipe; leaflets glabrous, in (4–) 8–10 (–45) pairs, linear-oblong, 2–10 (–13) mm long, 1–2.5 (–4) mm wide, with 1 median vein, the base asymmetrically obtuse, the tip obtuse but minutely mucronate. Flowers 1–4 (–6) (peduncle often bears a leaf with another raceme in the axil), 7–9 mm long. Peduncles 3–8 cm long, bearing minute tubercles; bracts ovate, 1.5–4, subentire or serrate, acute; pedicels 1.5–6 mm long; bracteoles lanceolate-ovate to oblong, 2–4 mm long, subentire or serrate. Sepals glabrous, 2-lobed almost to the base, each 4–6 mm long. Petals glabrous, yellow with orange flush, often with purple streaks; adaxial petal obovate, 6.5–10 mm long, 4–7 mm wide, with a 1–2 mm long claw, the upper margin sometimes ciliate; lateral petals and keel subequal, 7–9 mm long, 2–3 mm wide. Stamens in 2 bundles of 5, 6–9 mm long. Pods borne on 4–11 mm long stalks, strap-shaped, (1.2–) 2.5–5 cm long, 3–6 mm wide, flattened, the adaxial (upper) suture of the pod straight, smooth or papillose on the faces above the seeds, the abaxial suture crenate, falling apart in (3–) 6–10 (–13) almost quadrangular, 1-seeded joints, each 3–6 mm long. Seeds oblong kidney shaped, 3–4 mm long, 2–3 mm wide, dark brown to black, glossy.

Perennial or annual. In wet places, along ditches, around pools, in irrigated fields and waste places. It makes a good green manure and is used as fodder but it can become a toublesome weed in irrigated fields.

{(⊙) Hel/Ten/Hyp}. Pantropics, probably native to southeastern N America: <Ban, <Bhu, Bur, <Cey, <Ind, <Nep, <Pak: <An, <As, <Bh, <Dl, <Gj, <Co, <Jk, <Kl, <Kt, Lk, <Mh, <Mp, <Pj, <Rj, <Sk, <Tn, <Tp, <Up, <Wb. 0–1600 m.

Geissaspis: 2 or 3 species: tropical Asia

Roots swollen and spongy when submerged. Stems diffuse, creeping or floating, internodes often pithy and swollen when submerged or floating. Leaves abruptly even-pinnate, with 2 pairs of leaflets; leaflets obovate, without stipels; stipules membranous, peltate, produced below the insertion. Inflorescences long-peduncled, axillary, simple. Flowers bisexual, bilaterally symmetrical; bracts large, obliquely reniform, imbricate, persistent, more or less concealing the flowers and fruit; bracteoles absent. Sepal tube deeply 2-lipped, the adaxial (upper) lip entire, the abaxial lip obscurely 3-toothed. Petals yellow, orange or red, longer than the sepals; adaxial petal suborbicular, shortly clawed; lateral petals oblong or obovate or oblong; keel incurved, obtuse, a little shorter than the adaxial petal. Stamens 10, united, the upper third free; anthers all alike. Ovary shortly stalked; ovules 2; style incurved; stigma minute. Pod of 1 or 2, 1-seeded joints, the joints swollen with a flat border, indehiscent.

Annual or perennial. Flowers entomophilous or perhaps autogamous. The disseminules are flattened, 1-seeded parts of a pod, probably dispersed by water, surface contact with animals or in mud.

1. Bract margins bearing many long, stiff, brown hairs; pods shorter than the bracts ***G. cristata***
1. Bract margins mucronate-dentate, without hairs; pods longer than the bracts ***G. tenella***

Geissaspis cristata Wight et Arnott: Fig. 217

Stipules lanceolate, long-spurred, with long hairs, persistent. Leaflets mostly 8–15 mm long. Bracts orbicular, the bases long-sagittate, the margins bearing many long, stiff, brown hairs. Flowers ± 5 mm long,

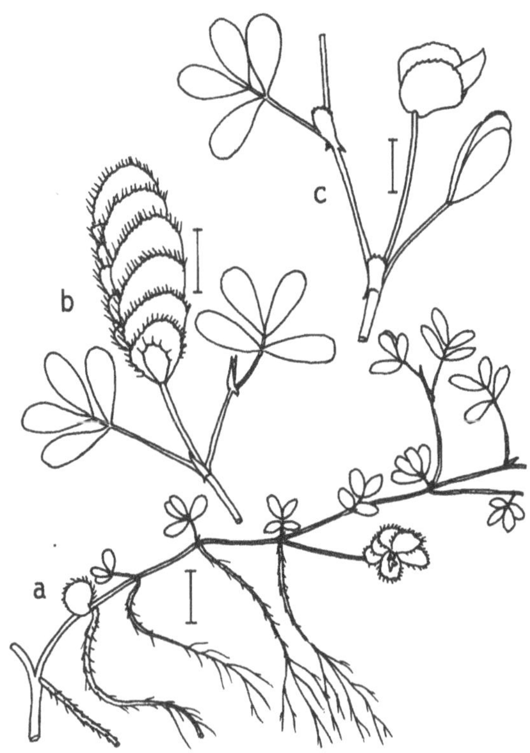

FIG. 217. *Geissaspis cristata*: a, flowering plant (1 cm); b, fruiting shoot (5 mm). *G. tenella*: c, flowering shoot (5 mm).

very much shorter than the bracts at anthesis. Petals yellow to orange, red or purple. Pods shorter than the bracts.

Annual or sometimes perennial in permanently wet places. Usually found in and around temporary ponds and rice fields. It often grows in shallow water but is sometimes found in deep water where it develops finger-thick pithy stems. Common and often locally abundant.

{(⊙) Hel/Eph}. Indochina, Malay Peninsular, Burma and India: Bur, Cey, Ind, Nep: An, Gj, Go, Kl, Kt, Mh, Tn, Wb. 0–1000 m.

Geissaspis tenella Bentham [*G. tenella* var. *malabarica* Sivarajan et Babu]: Fig. 217

Stipules oblanceolate, shortly spurred, with spiny hairs, deciduous. Leaflets mostly 5–8 mm long. Bracts ovate, the bases shortly sagittate, the margins mucronate-dentate. Flowers 5–10 mm long, as long or longer than the bracts at anthesis. Petals yellow or maroon. Pods longer than the bracts.

Annual or sometimes perennial in permanently wet places. Found in wet places, often in and around temporary ponds and rice fields. Common and often locally abundant. Like *G. cristata* but smaller and more delicate and tends to prefer dryer places.

{(⊙) Hel/Eph}. Endemic to SW India: ♦Ind: Kl, Kt. October–January. 0–1000 m.

Neptunia: 11 species: warm and tropical regions, most in Australia

Stems prostrate, ascending or floating, terete or occasionally angled when young, simple or sparsely branched; internodes swollen and pithy when in water. Leaves alternate, 2-pinnate and usually sensitive; stipules present but not evident on floating stems. Inflorescence a pedunculate head. Flowers radially symmetrical, dimorphic, the upper bisexual, the lower male either fertile or sterile with large staminodes. Sepals campanulate, small, 5-lobed, green or rarely yellow. Petals 5, small, green or rarely yellow. Stamens 10 or 5; staminodes 10, petaloid, yellow. Fruit a flattened marginally dehiscent pod, usually with several compressed seeds.

Floating or emergent, sometimes locally dominant. Pollinated by insects. The disseminules are flattened seeds, dispersal unknown, perhaps by surface contact to animals. Windler, D. R. A revision of the genus *Neptunia*. Australian Journal of Botany 14: 379–420 (1966).

Excluded: ***Neptunia acinaciformis*** (Spannagel) Miquel: Bur; ***N. triquetra*** (Vahl) Bentham: ♦Ind; sometimes found in moist places but not wetland plants.

1. Stems rarely branched, usually floating; pinnae 2–6, rachis glandless; leaflets 8–20 pairs; pods up to 2.5 cm long; seeds 4–10 ***N. oleracea***

1. Stems freely branched, usually erect or ascending, terrestrial or emergent; pinnae 4–10, rachis with a sessile gland below the lowest pair of pinnae; leaflets 9–40 pairs; pods up to 6 cm long; seeds 5–20 ***N. plena***

Neptunia oleracea Loureiro [*Mimosa natans* auct. Ind., non Linnaeus fil., *N. prostrata* (Lamarck) Baillon]: Fig. 218

Stems horizontal, hardly branched, with thickened nodes, developing white, spongy tissue when in water, rooting at nodes. Petioles glandless; stipules obliquely ovate-cordate, thin, persistent, 5–9 mm long. Pinnae 2–6, 5–8 cm long. Leaflets sessile, 8–14 (–20) pairs, narrowly oblong 3–18 mm long, 1.5–5 mm wide, the lower ones smaller. Peduncles erect, very accrescent, 6–25 cm long. Flower head oblong, up to 2.5 cm long and 0.8 cm wide. Bracts obovate, trapezoid. ± 2 mm long, ± 1 mm wide. Sepals ± 2 mm long. Petals

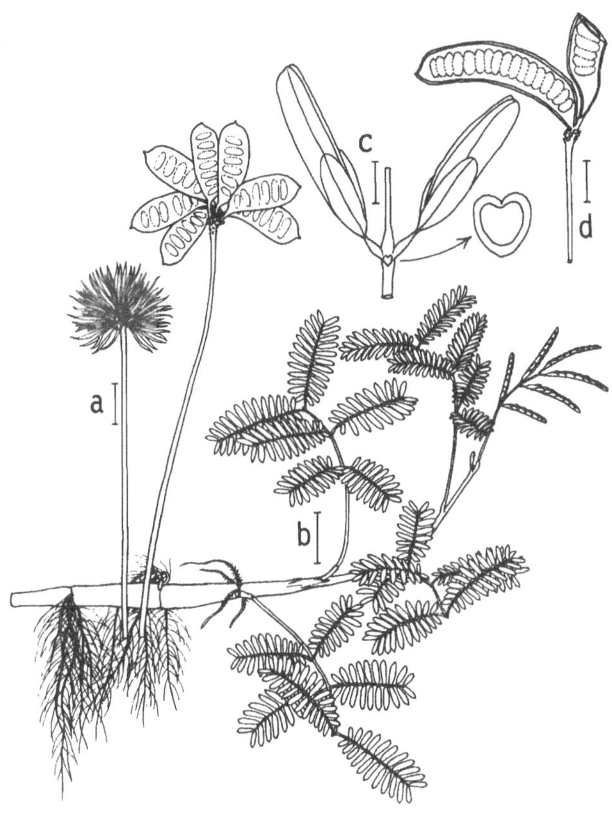

FIG. 218. *Neptunia oleracea*: a, flowering head and fruiting head (1 cm); b, floating shoot (2 cm). *N. plena*: c, base of rachis at first pair of pinnae (2 mm) with gland (inset); d, pods (1 cm).

± 12 mm long. Staminodes of the sterile flowers strap-shaped, ± 10 mm long. Pod flat, up to 2.5 cm long; seeds 4–10, ovate to elliptical, ± 5 mm long.

Annual or perennial. Freely floating on the surface of water, often forming floating mats with other plants. When the water recedes it may persist a while as a land plant; the rooted land plants have non-swollen stems and smaller leaves and flowers. Young shoots are eaten as a pot herb. It is used locally in medicine to cure earache and syphilis.

{(⊙) Ple/Hyp}. Pantropics: Ban, Bur, Cey, Ind: As, Bh, Gj, Go, Kl, Mh, Mp, Pj, Rj, Tn, Tp, Up, Wb.

Neptunia plena Bentham [*Mimosa adenanthera* Roxburgh]: Fig. 218

Like *N. oleracea* but stems much branched, erect or ascending, very rarely prostrate, up to 2 m tall; pinnae 4–10, rachis with a sessile gland below the lowest pair of pinnae; leaflets 9–40 pairs, 4–18 mm long, 1.5–3 mm wide; pods up to 6 cm long; seeds 5–20.

Perennial. It is not confined to wetlands but tolerates considerable flooding and even develops spongy tissue on the stems when in water.

{(Ple)/Hel}. Coasts of tropical and warm America, naturalized in Asia: <Ban, <Cey, <Ind: <Go, <Kt, <Rj, <Wb.

Sesbania: 50–55 species: tropical to warm regions

Shrubs or herbs. Stems erect or floating, the internodes often swollen when in water. Leaves alternate, 1-pinnate; stipules caducous. Inflorescence of lax axillary racemes. Flowers bisexual, bilaterally symmetrical. Sepals broadly tubular, 5-lobed or 5-toothed. Petals yellow, purplish, spotted or white; adaxial petal orbicular to ovate, spreading or reflexed; lateral petals oblong, falcate; keel incurved with long claws. Stamens 10, 9 united, the adaxial (upper) 1 free. Fruit a straight or slightly curved pod, transversely septate between the seeds but not breaking into joints, marginally dehiscent; seeds numerous, transversely oblong or subquadrate.

Annual or perennial. Some grow in wet or regularly inundated places. They often develop swollen and pithy stems. Pollinated by insects. The disseminules are seeds, dispersal unknown. Some used as green manure, others locally in medicine or for fibres and sometimes reported to be weeds.

This account is unsatisfactory. It is difficult to find adequately collected material — most specimens have either flowers or fruits, also the leaves tend to loose their leaflets. I have more or less accepted the treatment of Prain, D. Journal of the Asiatic Society of Bengal 56, part 2, no. 2: 366–370 (1897).

Excluded: ***Sesbania cannabina*** (Retzius) Persoon; ***S. concolor*** Gillett; ***S. sesban*** (Linnaeus) Merrill [*S. aegyptiaca* Poiret]; sometimes found in moist or damp situations but not in wetlands.

1. Adaxial petal more than 2 cm long, with 2 appendages at the base which run as keels into the claw; stems and leaf rachides smooth, without prickles; pods twisted, pendulous ***S. javanica***

1. Adaxial petal not more than 1 cm long, without appendages or appendages not running as keels into the claw; stems and leaf rachides usually bearing prickles; pods straight or curved but not twisted, erect or ascending

- **2.** Leaflets and branches bearing silky hairs; prickles minute and confined to the base of the leaf rachis and angles of the stem below the leaves ***S. sericea***
- **2.** Leaflets and branches without hairs; prickles scattered over stems, branches and rachides or occasionally absent
  - **3.** Petals ± 10 mm long; pod 20–30 cm long; seeds ± 40 ***S. bispinosa***
  - **3.** Petals ± 6 mm long; pod 5–10 cm long; seeds 12–20.
    - **4.** Leaflets in 15–20 pairs; flowers solitary or occasionally in pairs ***S. procumbens***
    - **4.** Leaflets in 30–50 pairs; flowers 6–8 in racemes ***S. uliginosa***

Sesbania bispinosa (Jacquin) W. F. Wight [*S. aculeata* (Willdenow) Poiret]: Fig. 219

Stems erect, branched, up to 3 m tall, bearing short prickles. Stipules linear-lanceolate, 6–10 mm long. Leaves up to 35 cm long; stipules lanceolate, deciduous; rachis and petiole with prickles; leaflets in 25–55 pairs, oblong, 1–2 cm long, up to ± 3 mm wide, glabrous. Racemes erect, 3- to 12-flowered, up to 13 cm long. Sepals glabrous except for margin, tube ± 3 mm long, teeth triangular ± 1 mm long. Petals brownish-yellow with violet flecks; adaxial petal ± 10 mm long, usually slightly longer than wide, cuneate at the base, the appendages wedge-shaped, the claw ± 3 mm long; keel ± 7 mm long, the basal tooth narrow, very acute. Filament sheaths 9–13 mm long, free parts 2–4 mm long. Pods subterete, ± erect, up to 30 cm long, ± 2 mm in diameter, not twisted, slightly constricted between the seeds, the margins not grooved. Seeds ± 40, brown, ± 3 mm long.

Annual or biennial. Not confined to wetlands but often found in swamps, marshy wastelands, waterlogged places, banks of ponds and rivers, rice fields and seasonally inundated areas. It is also tolerates saline and alkaline conditions.

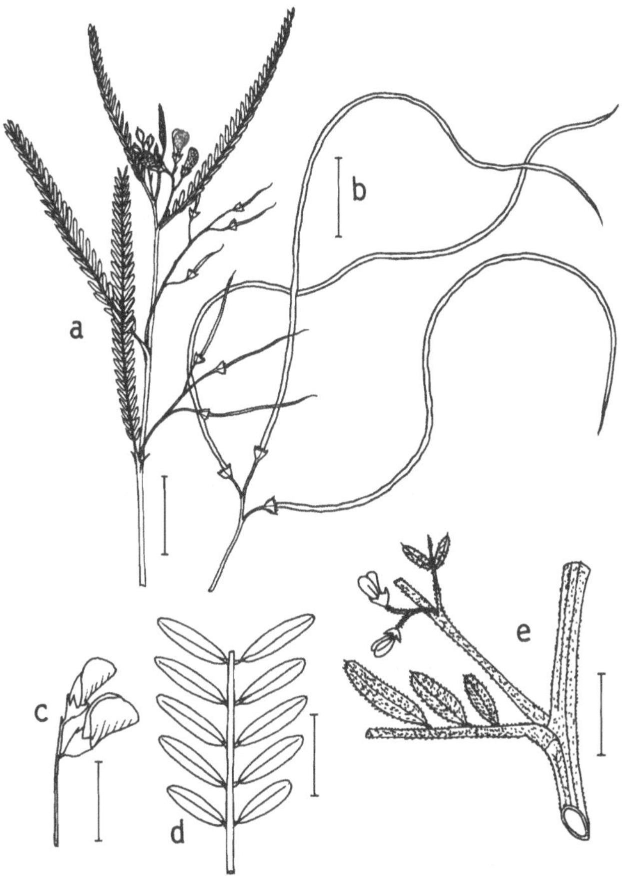

FIG. 219. *Sesbania bispinosa*: a, flowering shoot (2 cm); b, three pods (2 cm). *S. javanica*: c, two flowers (2 cm); d, base of rachis with five pairs of pinnae (2 cm). *S. sericea*: e, part of a flowering shoot (2 cm).

{⊙ or biennial, Hel}. Tropical Africa and Asia: Cey, Ind, Nep, Pak: Dl, Gj, Go, Kl, Kt, Lk, Mh, Mp, Pj, Rj, Tn, Tp, Up, Wb. 0–700 m.

Sesbania javanica Miquel [*Aeschynomene paludosa* Roxburgh, *S. aculeata* var. *paludosa* (Roxburgh) J. G. Baker, *S. roxburghii* Merrill]: Fig. 219

Stems up to 3 m tall, swollen at the base (up to 5 cm in diameter) with white pith-like aerenchyma. Stems and leaf rachides smooth, without prickles. Stipules ± 8 mm long. Leaves 15–30 cm long, with 10–32 pairs of leaflets; leaflets linear to oblong, 0.4–2.5 (–4) cm long, 2–7 mm wide, tapering and apiculate at the tip. Flowers in racemes of 5–12; pedicels 5–13 mm long. Sepal tube 6–8 mm long, the teeth 1.5–2 mm long. Adaxial petal 2–3 cm long, 2.4–3 cm wide, yellow inside with dark brown or purple dots and streaks outside, bearing 2 appendages at the base which run as keels into the claw; Pods usually pendulous, 25–30 cm long, somewhat twisted, slightly constricted between the seeds.

Annual. It is not confined to wetlands but is found in stagnant pools, marshy places, swamps and rice fields. Mostly along the coast. Sometimes cultivated as a green manure and for shade in plantations. It bears a remarkable superficial similarity to *Aeschynomene aspera*.

{⊙ Hel}. S and SE Asia: Ban, Bur, Ind: Kl, <Kt, Wb.

Sesbania procumbens Wright et Arnott: refer to Fig. 219

Like *S. bispinosa* but stems very much smaller and more branched, usually prostrate; stipules up to 3.5 mm long; leaves 5–7 cm long; leaflets 15–20 pairs, each up to 7 mm long; flowers solitary or in pairs; petals 6–8 mm long, ± 3 times longer than the sepals; pods 5–10 cm long, 1.5–2 mm wide, clearly constricted between the seeds; seeds 12–20.

Annual. A weed of wet places along the coast, particularly common in rice fields.

{⊙ Hel}. Endemic to Peninsular India: ♦Ind: Ap, Gj, Kt, Mh, Mp, Rj, Tn.

Sesbania sericea Link: Fig. 219

Like *S. bispinosa* but leaflets bearing silky hairs; prickles either absent or minute and confined to the base of the leaf rachis and angles of the stem below the leaves; adaxial petal wider than long, cordate at the base; the basal tooth of the keel broad and blunt; pods without constrictions between the seeds.

Annual. Wet places.

{⊙ Hel}. Tropical Africa, naturalized in America and Sri Lanka: <Cey 1860.

Sesbania uliginosa (Roxburgh) Sweet: refer to Fig. 219

Like *S. procumbens* but leaflets in 30–50 pairs; racemes with 6–8 flowers.

{⊙ Hel}: ♦Ban? ♦Ind: Bh, Or, Wb.

GENTIANACEAE: 74 genera

Herbs. Stems 4-angular or 4-winged. Leaves decussate or rarely whorled; blades simple. Flowers 4- or 5-merous, usually in cymes. Sepals united below, persisting in fruit. Petals tubular, twisted to the right in bud. Stamens as many and alternating with the petal lobes. Ovary superior, 1- or 2-locular; ovules numerous. Fruit a 2-valved capsule. Seeds numerous, minute.

Excluded: the following are annuals found in moist and often shady places but are not considered to be wetland plants: ***Canscora decurrens*** Dalzell; ***C. decussata*** (Roxburgh) Schultes et Schultes fil.; ***C. diffusa*** (Vahl) R. Brown ex Roemer et Schultes; ***Enicostema axillare*** (Lamarck) A. Raynal [*Exacum hyssopifolium* Willdenow, *Enicostema littorale* Blume]; ***Exacum pedunculatum***

Linnaeus; ***E. pumilum*** Griesbach; ***E. sessile*** Linnaeus; ***E. tetragonum*** Roxburgh; ***Swertia corymbosa*** (Grisebach) Wight ex C. B. Clarke; ***S. minor*** (Griesebach) T. Cooke.

Hoppea: endemic to Indian subcontinent, 2 species

Stems erect, sparsely to profusely branched. Leaves sessile, 1- or 3-nerved from the base. Flowers in terminal or axillary, lax cymes; bracts leaf-like. Sepals tubular below, 4-angular; sepal lobes 4, unequal. Petals tubular below, ± equally 4-lobed, yellow or greenish-yellow. Stamens inserted on the petal tube, 1–4, 1 fertile, 0–3 staminodes; anther with longitudinal dehiscence. Ovary 1-locular; ovules numerous, on 2 parietal placentas; stigma sessile or nearly so, 2-lobed.

1. Petals not exceeding the sepal lobes; capsule globose; branches spreading *H. dichotoma*
1. Petals exceeding the sepal lobes; capsule ellipsoid; branches in upright bunches *H. fastigiata*

Hoppea dichotoma Willdenow: Fig. 220

Stems 4-angled or slightly 4-winged, up to 6 (–12) cm; branches spreading. Leaves elliptical-ovate, 6–8 mm long. 2.5–4 mm wide, base cuneate. Cymes lax; peduncle up to 10 mm long; pedicel up to 3 mm long. Sepal lobes elliptical-lanceolate, ± 3 mm long, shortly acuminate, marginal nerves thick. Petal lobes oblong-ovate, ± 2 mm long, not exceeding the sepal lobes, ridged within. Capsule globose, ± 4 mm in diameter.

Annual. In moist and exposed places, along river banks, at the edges of pools and tanks and in rice fields.

{⊙ Hel}. Endemic to India: ♦Ind: Ap, Bh, Kt, Mh, Mp, Rj, Tn, Wb. 0–900 m.

Hoppea fastigiata (Grisebach) C. B. Clarke: Fig. 220

Stems distinctly 4-winged, up to 8 cm; branches clustered and growing upright. Leaves broadly ovate, 4–7 mm long. 3–5 mm wide, base rounded to truncate. Cymes clustered; peduncle up to 5 mm long; pedicels less than 3 mm long (flowers sessile or nearly so). Sepal lobes lanceolate, ± 4 mm long, distinctly acuminate, marginal nerves thin. Petal lobes oblong, ± 2.5 mm long, exceeding the sepal lobes, smooth within. Capsule ellipsoid, ± 3.5 mm in diameter.

Annual. In seasonally inundated places, such as natural depressions, river flood plains and the edges of pools, tanks and rice fields; it usually grows in more or less closed communities often associated with various species of *Eriocaulon* and *Utricularia*. Not confined to wetlands

{⊙ Hel}. Endemic to the Indian subcontinent: ♦Cey, ♦Ind: Ap, Kl, Kt, Mp? Rj, Tn.

GRAMINEAE see POACEAE

HALORAGACEAE: 8 genera

Herbs or sometimes small shrubs. Leaves alternate, opposite or whorled, entire or pinnately or palmately divided, often heterophyllous; stipule-likeappendages rarely present. Inflorescence paniculate, spicate, axillary clusters of cymes or flowers solitary. Flowers bisexual or unisexual, radially symmetrical, usually inconspicuous. Sepals 2–4, in female flowers sometimes absent. Petals 2–4, boat-shaped, hooded, keeled, deciduous or absent. Stamens as many as or twice as many as the sepals. Ovary inferior, of 1–4 united carpels; each carpel with 1 or rarely 2 pendulous ovules; fruit a 1- to 4-seeded nut or breaking into 1-seeded mericarps.

Annual or perennial herbs or sometimes small shrubs. Monoecious or dioecious, pollinated by wind. Mostly found in wet places, some aquatic.

Excluded: ***Haloragis philipinensis*** Merrill [*H. isomera* Parker]: found in high land in Myanmar and Assam; ***Laurembergia coccinea*** (Blume) Kanitz [*Serpicula brevipes* Wight et Arnott, *S. hirsuta* Wight et Arnott, *S. indica* Thwaites, *S. zeylanica* Arnott ex C. B. Clarke]: found in

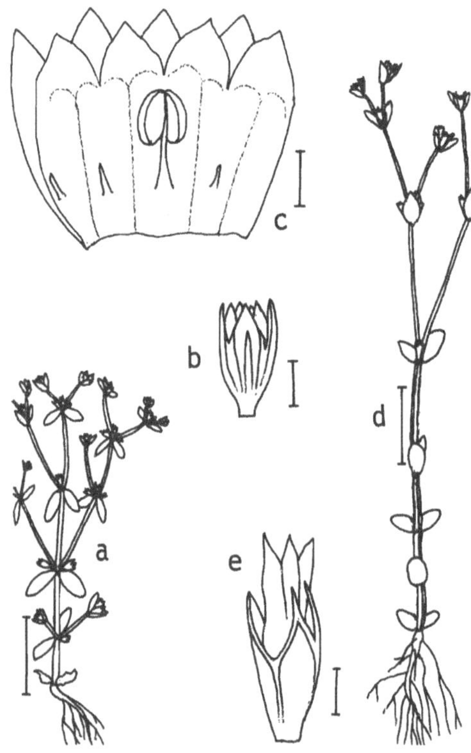

Fig. 220. *Hoppea dichotoma*: a, flowering plant (1 cm); b, flower viewed from outside (1 mm); c, flower dissected (0.5 mm). *H. fastigiata*: d, whole plant (1 cm); e, flower viewed from outside (1 mm).

the mountains of S India and Sri Lanka; although both are found on moist soil neither are considered to be wetland plants.

Myriophyllum: ± 60 species: almost cosmopolitan, most in Australia, absent from most of Africa, Asia Minor and northeastern S America.

Stems floating or ascending. Leaves usually whorled, sometimes opposite or alternate, usually heterophyllous with submerged leaves distinct from emergent ones, pinnately divided, entire or rarely reduced to scales, sometimes flanked by stipule-like outgrowths. Inflorescence mostly terminal, spike-like, rarely branched or flowers solitary in leaf axils. Flowers bisexual or unisexual, sessile or nearly so. Sepals 4, very small or absent. Petals 2, 4 or absent (usually reduced or absent in female flowers). Stamens (1-) 4 or 8; anthers linear-oblong; connectives without apical appendages. Ovary of (2) 4 carpels. Fruit dry, breaking up in to 1-seeded mericarps.

Annuals or rhizomatous perennials. Submerged, emergent or some seasonally terrestrial, often very abundant and locally dominant. Monoecious or dioecious, pollinated by wind. The disseminules are mericarps, dispersed in water, in mud and perhaps also by animals; vegetative fragments also play a role in the dispersal. Some species are serious weeds, particularly *M. spicatum* in N America and *M. aquaticum* in Asia, others eaten as pot herbs. Meijden, R. van der and Caspers, N. Flora Malesiana ser. 1, 7(1): 248–259 (1972).

Excluded: found in the Himalayas but not extending southwards in India: ***Myriophyllum spicatum*** Linnaeus; ***M. verticillatum*** Linnaeus.

1. Plants dioecious; aerial leaves glaucous, light bluish-green, densely covered with minute, transparent, hemispherical glands, all pinnately divided with linear segments ***M. aquaticum***
1. Plants monoecious; aerial leaves not glaucous, light to dark green, not obviously glandular, at least the upper ones simple
  2. Bracteoles pinnately or digitately dissected or lobed
    3. Stamens 8; anthers linear, 1.5–1.8 mm long; petals 1.5–2 mm long ***M. indicum***
    3. Stamens 4; anthers oblong, 0.6–0.8 mm long; petals ± 1 (–1.5) mm long ***M. tetrandrum***
  2. Bracteoles simple, not dissected or lobed
    4. Flowers (1–) 3–6 together in the leaf axil; the middle flower bisexual and distinctly stalked; mericarps dorsally rounded ***M. oliganthum***
    4. Flowers solitary in the leaf axil; all flowers sessile; mericarps dorsally acute or ridged ***M. tuberculum***

Myriophyllum aquaticum (Velloso) Verdcourt [*M. brasiliense* Cambessèdes, *M. proserpinacioides* Gilles ex W. J. Hooker]: Fig. 221

Differs from all Asian species by: plants dioecious (females only found); leaf bases somewhat swollen; aerial leaves glaucous, light bluish-green, densely covered with minute, transparent, hemispherical glands, never entire, all pinnately divided with linear segments; bracteoles subulate with 1 or 2 lateral lobes.

Perennial. In ditches, usually most of the plant emergent.

{Vit/Hyp}. Subtropical SE America, introduced in Asia, Africa and N America: <Ind: <Jk (around Dal Lake), might be expected elsewhere in uplands.

Myriophyllum indicum Willdenow [*M. intermedium* A. P. de Candolle]: Fig. 221

Stems submerged, floating or ascending, upper part usually emergent. Leaves in whorls of (4) 5 or 6; sub-

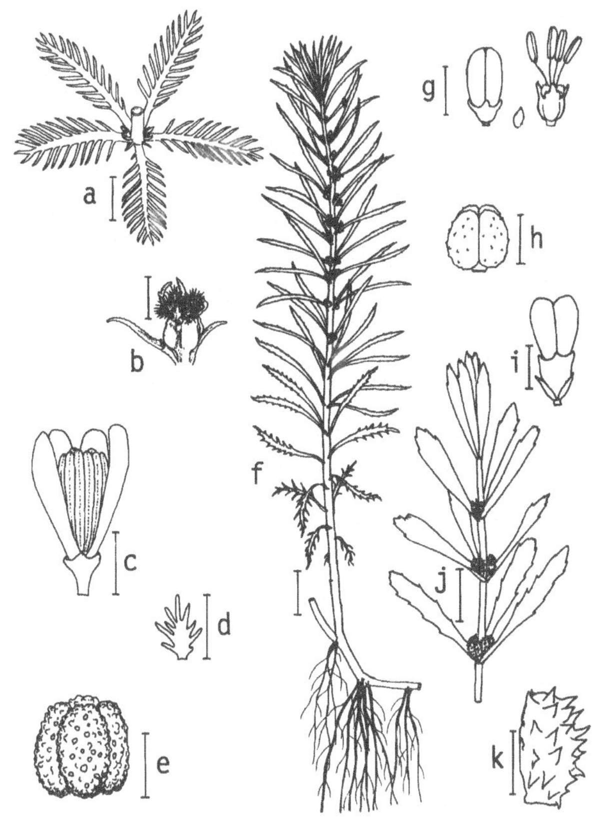

Fig. 221. *Myriophyllum aquaticum*: a, whorl of aerial leaves (5 mm); b, female flower (1 mm). *M. indicum*: c, male flower (1 mm); d, bracteole (1 mm); e, fruit (1 mm). *M. oliganthum*: f, flowering plant (1 cm); g, male flower: left, before anthesis; centre, bracteole; right, at anthesis (1 mm); h, fruit (1 mm). *M. indicum*: i, male flower (1 mm); j, flowering shoot (5 mm); k, one mericarp (2 mm).

merged leaves, 1.5–4 cm long, with 12–24 pairs of filiform segments, the segments 1–2 mm long, brown-tipped; aerial leaves pinnate below with stiff segments becoming lanceolate above with a few teeth or entire. Bracts leaf-like, but smaller, rarely more than 1.25 cm long, usually reflexed in fruit; bracteoles pinnately or digitately dissected or lobed, 0.6–1 mm long. Flowers sessile, solitary, usually bisexual below and male above. Sepals deltate, 0.1–0.3 mm long, entire or finely serrate. Petals 1–2 mm long, deciduous. Stamens 8; anthers linear, 1.5–1.8 mm long. Fruit ± 2 mm long and wide; mericarps ovate, rounded on the back, with ± flat sides, finely tuberculate or usually smooth.

Perennial or perhaps sometimes annual. In ponds, tanks and lowland marshes. Mostly near the coast. The name *M. indicum* has been used rather indiscriminately in Indian Floras

{Vit/Hyp}. Endemic to SE India and Sri Lanka: ♦Cey, ♦Ind: Ap, Go, Mh, Or, Tn.

Myriophyllum oliganthum (Wight et Arnott) F. von Mueller [*M. intermedium* auct. Ind., non A. P. de Candolle]: Fig. 221

Stems submerged, floating or ascending, upper part usually emergent, somewhat succulent and red. Leaves usually heterophyllous: submerged leaves sometimes absent, in whorls, pinnately divided, with filiform lobes, 1.5–3 cm long; aerial leaves alternate or opposite, usually linear, 1.5–3 cm long, entire to serrate or pinnately divided but then much smaller than the submerged ones. Bracts leaf-like; bracteoles linear, ± 1 mm long, margin slightly toothed. Flowers 3–6 in axillary clusters or rarely solitary, sessile or central flower shortly stalked, unisexual or in the centre of the cluster bisexual. Sepals 4, 0.5 mm long. Petals 4, pink or yellowish. Stamens 4; anthers linear-oblong, up to 2 mm long; connectives without apical appendages. Fruit ± 1 mm long and wide; mericarps rounded on the back, finely tuberculate.

Perennial or perhaps sometimes annual. Usually found in permanent water in tanks and lakes, often gregarious.

{Vit/Hyp/Ten?}. Endemic to India: ♦Ind, Pak?: Kl, Kt, Mp, Tn. 800–1600 m.

Myriophyllum tetrandrum Roxburgh: refer to Fig. 221

Like *M. indicum* but stamens 4; anthers oblong, 0.6–0.8; petals 1–1.5 mm long.

Perennial. In shallow open water in pools, ditches and canals. In most Indian floristic works it is considered to be a synonym of *M. indicum.*

{Vit/Hyp/Ten}. Northern India and SE Asia: Ban, Ind: As, Bh, Mg, Or, Up, Wb.

Myriophyllum tuberculatum Roxburgh [*M. spathulatum* Blatter et Hallberg]: Fig. 221

Stems submerged, floating or ascending, upper part usually emergent. Leaves usually heterophyllous; submerged leaves in whorls of 4 or 5, 2.5–4 cm long, 1–1.5 cm wide, pinnately divided, with 8–25 pairs of filiform lobes, the lobes 1–2 cm long; emergent leaves in lower part like the submerged ones but smaller, the upper ones ultimately alternate, with less and shorter lobes, the uppermost ones entire, spathulate to linear, 5–20 mm long. Bracts leaf-like; bracteoles ± rhomboid, serrate, ± 1.2 mm long, ± 0.8 mm wide, acute. Flowers solitary, the lowest sometimes female, followed by bisexual ones, with male ones above. Sepals ± orbicular, 0.1–0.25 mm long and wide, finely serrate or entire. Petals 4, 0.5–1.5 mm long, white. Stamens 4; filaments ± 1 mm long; anthers elliptical-oblong, 0.5–1 mm long. Fruits quadrangular, 2–3.5 mm long and wide, with sharp longitudinal ribs, both ribs and furrows with pointed tubercules.

Perennial or perhaps sometimes annual. In water in tanks, lakes and marshes.

{Vit/Hyp/Ten}. S and SE Asia and N Australia: Ban, Bur, Ind: As, Kl, Kt, Mh, Mp, Or, Rj, Wb.

HANGUANACEAE: 1 genus

Hanguana: 1 species

Hanguana malayana (Jacquin) Merrill: Fig. 223

Dioecious robust herb. Stems erect up to 2 m tall, ascending or floating, often with long, creeping or floating stolons. Leaves stiff, erect, mostly in basal rosettes, the lower ones petiolate, the upper sessile, variable, 20–120 cm long and 1.5–15 cm wide, usually lanceolate, longitudinally nerved, with many finecross-nerves; leaf bases sheathing and somewhat petiole-like. Inflorescence a panicle, 10–120 cm long, 1–3 times branched; branches subtended by large bracts. Flowers unisexual, sessile. Perianth 2-seriate, 3-merous, persistent, green or yellowish, united below; outer segments 2–2.5 mm long; inner segments 2.5–3 mm long, often red-dotted. Male flower with stamens 6, inserted on the perianth. Female flower with staminodes 6 (the 3 outer ones smaller than the inner), dorsally compressed. Ovary 3-locular, with 1 axile ovule in each locule; stigma sessile, 3-lobed. Fruit an oblong, 2 cm long, red, fleshy drupe.

Perennial. It shows a very wide ecological amplitude: from free-floating or emergent to terrestrial. When growing in water it rarely flowers. The pollination mechanism is unknown. The disseminules are red and fleshy drupes, presumably dispersed by animals. The economic importance is unknown. Airy Shaw, H. K. Hanguanaceae. Kew Bulletin 18: 260–261 (1965).

{Ple/Hel}. Sri Lanka, SE Asia and Pacific: Cey.

HIPPURIDACEAE excluded see p. 11

HYDATELLACEAE: 2 genera

Monoecious or dioecious. Leaves all basal, capillary, without a ligule, base not sheathing. Inflorescence a terminal head with few opposite bracts and few to many flowers. Flowers in heads, minute, unisexual, subtended by 1–6 hyaline bracts or numerous hairs. Sepals and petals absent. Stamen 1; filament capillary. Ovary 1-locular; 1-ovulate, crowned by stigmatic hairs. Fruit indehiscent or a capsule opening by 2 or 3 valves.

Annuals or creeping perennials.

Trithuria [*Juncella*]: 3 species in temperate Australasia, 1 in India

Trithuria konkanensis S. R. Yadav et M. K. Janarthanam: Fig. 222

Tufted, green to red, moss-like, up to 1.3 cm high. Stems reduced. Leaves erect or spreading, tufted, linear, up to 1.2 cm long, up to 0.8 mm wide, entire, 1-nerved; uniseriate hairs present among the leaves. Male flower solitary in the centre of each head, consisting of a single red stamen; filaments up to 6 mm long; anthers basifixed, up to 1 mm long. Female flowers 15–20 in each head, each a solitary carpel. Ovary 1-locular, ovoid, up to 0.1 mm long; stigmatic hairs 2–5 (–6), up to 0.5 mm long, red. Seeds translucent with dark tip.

Fig. 222. *Trithuria konkanensis*: a, flowering plant (2 mm); b, head with bracts (1 mm); c, bracts (1 mm); d, female flower with four hairs (0.5 mm).

Annual. Seasonally submerged on sandy and gravelly areas. Presumably pollinated by wind. The disseminules are seeds, dispersal unknown. Discovered in 1993, known only from Maharashtra, Sindhudurg District.

{⊙ Ten?}. Endemic to W India: ♦Ind: Mh.

HYDROCHARITACEAE [excl. NAJADACEAE]: 17 genera of which 3 are marine: almost cosmopolitan

Monoecious, dioecious, gynodioecious or trioecious herbs. Stems elongate or contracted, simple or complexly branched. Leaves very variable, with or without petioles and stipules, with (1–) 2–10 or more small scales (squamulae intravaginales) in the leaf axils. Inflorescence complex or flowers solitary, subtended by 1 or 2, free or united bracts (collectively called a spathe). Flowers bisexual or unisexual, the males sometimes abscising from the parent. Sepals 3 or less, rarely absent. Petals 3 or less, minute to large and showy or absent. Stamens 1 to numerous, often staminode-like. Ovary inferior, 3–20 or more carpels; placentas parietal or without dissepiments; styles simple or forked. Fruit a berry or capsule.

Annual or perennial. All are submerged aquatics, marine or freshwater. Pollination and dispersal is dealt with under each genus.

Egeria densa Planchon: {Vit}. From subtropical S America has become established in most warm regions of the world but is apparently not found in the Indian subcontinent but may be expected on higher ground.

Excluded (keyed out but not described and illustrated):
The following are marine aquatics: ***Enhalus*** (1 species); ***Halophila*** (± 5 species); ***Thalassia*** (1 species).

The following is found in the Himalayas but does not extend to Peninsula India: ***Hydrocharis dubia*** (Blume) Backer [*H. morsus-ranae* auct. Ind., non Linnaeus].

1. Leaves clearly differentiated into petiole and blade at maturity

2. Stipules present (median, ± 3 cm long); branched roots usually present; leaf blades thick, not translucent (blade reniform to almost orbicular with a pad of honeycomb-like aerenchymous tissue below and 4 or 6 primary veins originating from the point of petiolar attachment, Himalayas) ***Hydrocharis***

2. Stipules absent; branched roots absent; leaf blades thin, translucent

3. Leaves rarely exceeding 3 cm long; stems elongate regularly branched; perianth uniseriate or absent; the blade lanceolate to ovate and flat; marine ***Halophila***
3. Leaves much more than 3 cm long; stems contracted and corm-like; perianth clearly biseriate with sepals and petals; the blade usually cordate with a quilt-like pattern; freshwater ***Ottelia***
1. Leaves not differentiated into petiole and blade
4. Plants rhizomatous (horizontal axes wholly buried in the substrate, relatively thick, fleshy and with short internodes); marine
5. Rhizomes dimorphic, with extended, horizontal, scale-bearing long shoots and erect, foliage-bearing short shoots; flowers wholly submerged; leaf veins in a single series ***Thalassia***
5. Rhizomes monomorphic, bearing foliage leaves at the apex; flowers rising to the water surface for anthesis; leaf veins in 2 series, the adaxial series inverted ***Enhalus***
4. Plants not rhizomatous (horizontal axes if present not wholly buried in the substrate, thin or if thick and fleshy then short, rarely more than 5 cm long); freshwater
6. Leaves confined to basal rosettes, stems truncated, rhizomatous or corm-like, with or without elongated runners or stolons
7. Leaves gradually attenuate to fine point at the apex; stems monomorphic corm-like, without stolons; flowers bisexual ***Blyxa***
7. Leaves mostly parallel-sided coming abruptly to a rounded or pointed tip; stems dimorphic corm-like and with elongated runners or stolons ***Vallisneria***
6. Leaves arranged along elongated stems; stems not truncated rhizomatous or corm-like but sometimes bearing apical or axillary tubers (turions)
8. Leaves in whorls of 3 or more, arranged in regular and repeating patterns along elongated stems; squamulae intravaginales fringed with finger-like, orange-brown hairs; stigmas linear or club-shaped ***Hydrilla***
8. Leaves in two rows even if sometimes spirally wound, arranged somewhat irregularly along the stem (sometimes solitary or sometimes in bunches); squamulae intravaginales smooth, without hairs; stigmas flattened ***Nechamandra***

Blyxa: 9 species: tropical Old World, naturalized in N America and Europe

Monoecious or dioecious. Roots unbranched. Stems either elongate or contracted. Leaves spirally arranged, submerged, flat, ribbon-like, not differentiated into stipule, petiole and blade, midrib usually prominent, margin usually dentate. Inflorescence with 1–22 or more flowers; spathe tubular, of 2 united bracts. Petals band-shaped and scarcely wider than the sepals, capillary or rudimentary. Stamens 3, 6 or 9, all fertile, in female flowers staminodes minute. Ovary of 3 carpels, distally elongated into a hypanthium in bisexual and female flowers; styles 3, simple, elongate, terete. Fruit an elongated, irregularly dehiscent capsule. Seeds smooth or spiny.

Annual or perhaps sometimes perennial. The vegetative parts are submerged; the flowers are emergent or submerged. Cook, C. D. K. and Lüönd, Ruth. A revision of the genus *Blyxa*. Aquatic Botany 15: 1–52 (1983).

1. Stems elongate, flaccid; leaves cauline; spathes sessile or nearly so ***B. japonica***
1. Stems contracted to a simple or forked rootstock or corm; leaves radical; spathes distinctly pedunculate; peduncles up to ± 50 cm long
2. Flowers bisexual; stamens 3
3. Seeds ellipsoidal, 1.25–1.8 mm long, smooth or with up to 12, somewhat irregular, longitudinal non-spiny ridges or ribs, without apical and basal tails ***B. aubertii*** var. ***aubertii***
3. Seeds with up to 12, irregular but distinctly spiny ridges and with apical and/or basal tails ***B. aubertii*** var. ***echinosperma***
2. Flowers unisexual; stamens 9
4. Leaf tips gradually attenuate to a fine and flat point with microscopic, regularly spaced, unicellular spines; the lowest 3 anthers ± 3 mm long, sessile or nearly so; pollen grains 33.6 ± 4.4 μm in diameter; seeds 1.3–2 mm long, with up to 12, somewhat irregular, longitudinal ridges, each bearing blunt peg-like spines or almost smooth ***B. octandra***
4. Leaf tips distinctly obtuse and hooded and without spines; the lowest 3 anthers ± 2 mm long with ± 1.5 mm long filaments; pollen grains 56.7 ± 3.3 μm in diameter; seeds 3–3.5 mm long, with 2 dorsal and 2 or rarely 3 ventral ridges ***B. quadricostata***

Blyxa aubertii L. C. Richard var. ***aubertii***: Fig. 223

Stems corm- or rootstock-like. Leaves radical, 2.5–60 cm or more long, 0.2–1.2 cm wide, linear but gradually attenuate to a fine point at the apex, midrib distinct and prominent with up to 10 secondary and up to 28 tertiary parallel veins, faint cross-veins often present; margin (at least towards the apex) with microscopic, regularly spaced, unicellular spines. Spathes 1- or rarely 2-flowered; distinctly pedunculate; peduncles up to ± 50 cm long (flowers carried above the leaves). Flowers bisexual, usually emergent but some-

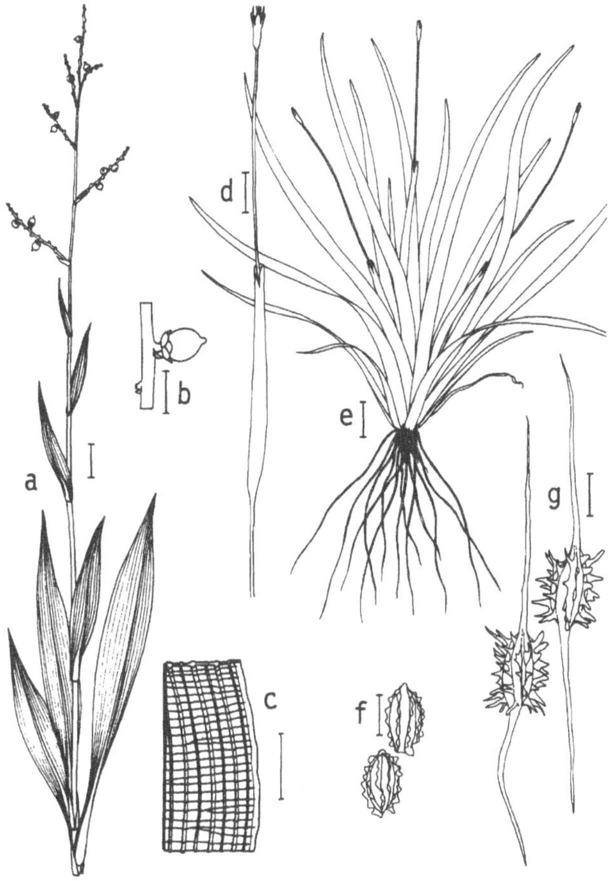

Fig. 223. *Hanguana malayanum*: a, flowering shoot (3 cm); b, fruit (1 cm); c, detail of leaf nervature (1 cm). *Blyxa aubertii* var. *aubertii*: d, inflorescence (1 cm); e, flowering plant (2 cm); f, two seeds (1 mm). *B. aubertii* var. *echinosperma*: g, two seeds (1 mm).

times submerged. Petals linear, band-like, often remaining folded and twisted within the calyx. Stamens 3, antesepalous. Stigmas laterally papillose. Capsules (2–) 3–5 (–8) cm long. Seeds ellipsoidal, 1.25–1.8 mm long, smooth or with up to 12, somewhat irregular, longitudinal ridges or ribs, long spines and apical and/or basal tails absent.

Usually annual but may persist longer in permanent water. The leaves are submerged. It usually grows in temporary water and is often found in rice fields and irrigation ditches, but is not considered to be a serious weed. It flowers and fruits throughout the year as long as water is present. The emergent flowers are strongly autogamous with pre-anthesis cleistogamy, the submerged flowers are totally cleistogamous. Seed-set is usually good. The disseminules are seeds or complete capsules, dispersion unknown but it is probably transported in mud on the feet of birds or with rice seed.

{(⊙) Ros}. Tropical and subtropical Asia, Australia, Madagascar and E Africa, introduced in N America. (lacking in NW India): Ban, Bhu, Bur, Cey, Ind, Nep: An, Ap, Bh, Jk, Kl, Kt, Mp, Or, Rj, Tn, Wb. 0–1200 m.

Blyxa aubertii L. C. Richard var. ***echinosperma*** (C. B. Clarke) C. D. K. Cook et R. Lüönd: Fig. 223

Like var. *aubertii* but the seeds have up to 12, irregular but distinctly spiny ridges and with apical and/or basal tails.

Ecologically indistinguishable from var. *aubertii.*

{(⊙) Ros}. Tropical and subtropical Asia and Australia: Ban, Cey, Ind, Nep?: Ap, As, Go, Kl, Kt, Mg, Mp.

Blyxa japonica (Miquel) Maximovicz ex Ascherson et Gürke var. ***japonica***: Fig. 224

Stems flaccid, slender, (1–) 10–20 (–60) cm long, usually branched below and simple above. Leaves cauline, lanceolate, 15–50 (–70) mm long, 0.75–3 (–4) mm wide, semi-amplexicaul at base, gradually

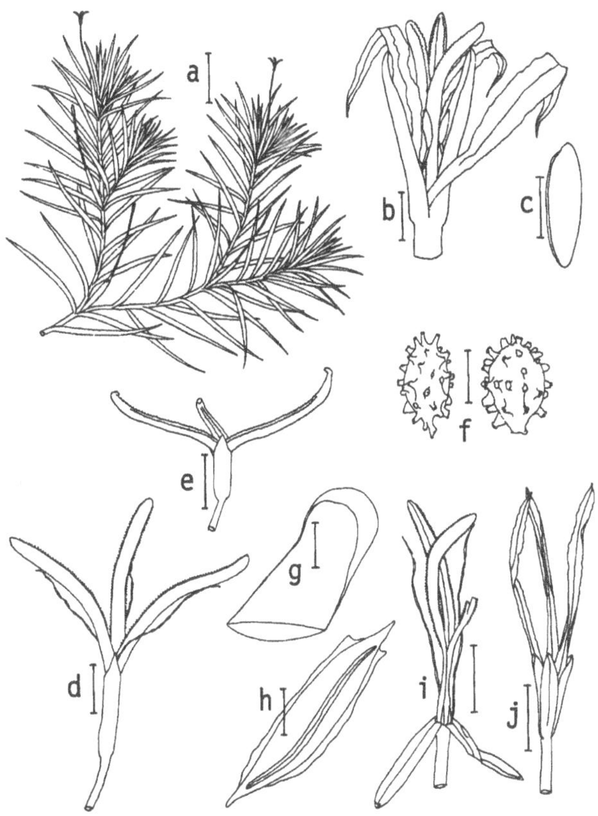

Fig. 224. *Blyxa japonica*: a, flowering shoot (2 cm); b, flower with one petal moved to the side (2 mm); c, seed (1 mm). *B. octandra*: d, female flower, note thread-like petals (5 mm); e, male flower (5 mm); f, two seeds (1 mm). *B. quadricostata*: g, leaf tip (0.5 mm); h, seed (0.5 mm); i, female flower with sepals bent back (4 mm); j, male flower at anthesis (4 mm).

attenuate to a blunt point at the tip, midrib distinct and prominent with up to 12 secondary parallel veins; margins with microscopic, regularly spaced, unicellular spines. Spathes 1- or rarely 2-flowered, sessile or nearly so. Flowers bisexual, usually emergent but sometimes submerged. Petals lanceolate, 3–8 (–10) mm long, 0.2–0.8 (–1) mm wide, attenuate to a fine point. Stamens 3, antesepalous. Stigmas linear, 3–4.5 mm long, white. Capsules 1–2 (–2.5) cm long. Seeds fusiform, (1–) 1.5–2 (–2.5) mm long, smooth.

Annual. Found in temporary pools and streams, frequently in rice fields but hardly considered to be a serious weed. The flowers show pre-anthesis cleistogamy.

{⊙ Vit}. S and E Asia, introduced in rice fields of Italy: Ban, Bur, Ind, Nep: Mg. 700–1300 m.

Blyxa octandra (Roxburgh) Planchon ex Thwaites: Fig. 224

Stems corm- or rootstock-like. Leaves radical, 6–40 cm or more long, 0.2–0.8 cm wide, linear, somewhat narrowed below, gradually attenuate to a fine and flat point at the tip, midrib distinct and prominent with up to 20 secondary parallel veins, faint cross-veins often present; margin (at least towards the apex) with microscopic, regularly spaced, unicellular spines. Spathes distinctly pedunculate; peduncles up to ± 50 cm long (flowers carried above the leaves); the male with 22 or more flowers in each spathe, 1 or 2 opening each day; the female with 1 or rarely 2 flowers. Flowers unisexual. Male petals linear, up to ± 15 mm long, white, the adaxial surface with 3 longitudinal furrows and row of fine pollen carrying papillae. Stamens 9, the lowest 3 anthers ± 34 mm long, almost sessile; pollen grains 33.6 ± 4.4 μm in diameter. Female petals filiform, about as long as the stigmas, withering as the stigmas spread. Stigmas superficially resembling the petals of male flowers. Capsules 2.3–5.4 cm long. Seeds oblong-ellipsoidal, 1.3–2 mm long, with up to 12, somewhat irregular, longitudinal ridges, each bearing blunt peg-like spines or almost smooth.

Annual. In pools that dry out in the dry season, sometimes also in rice fields. Male plants often outnumber female ones. It is pollinated by insects, the pollen is presented on modified petals. Cook, C. D. K., Lüönd, Ruth and Bhadran Nair. The floral biology of *Blyxa octandra.* Aquatic Botany 10: 61–68 (1981).

{⊙ Ros}. India and Sri Lanka, New Guinea and Australia: Ban, Cey, Ind: An? Ap, Kl, Mh, Mp, Or, Tn, Wb.

Blyxa quadricostata Den Hartog: Fig. 224

Like *B. octandra* but the leaf tips distinctly obtuse and hooded and without spines; the lowest 3 anthers ± 2 mm long with ± 1.5 mm long filaments; pollen grains 56.7 ± 3.3 μm in diameter; seeds 3–3.5 mm long, with 2 dorsal and 2 or rarely 3 ventral ridges.

Annual. In pools and slowly flowing streams.

{⊙ Ros}. Endemic to Myanmar and Thailand: Bur.

Hydrilla: 1 species

Hydrilla verticillata (Linnaeus fil.) Royle: Fig. 225

Gynodioecious. Roots unbranched. Stems elongate, branched regularly but at distant intervals, horizontal and stoloniferous below, erect and spreading above; bulbil-like hibernacula (turions) develop either underground terminally on stolons or terminally or axillary on erect stems or their branches. Leaves cauline, scale-like and opposite or foliate in whorls of 3–12, sessile, linear to lanceolate or rarely widely ovate, regularly spaced along the stem but contracted towards the apex, margin serrate or toothed, apex acute and terminating in a single spine cell; squamulae intravaginales paired (or more at branches), ± 5 mm long, narrowly triangular to lanceolate, fringed with finger-like, orange-brown hairs. Inflorescences in the axils of

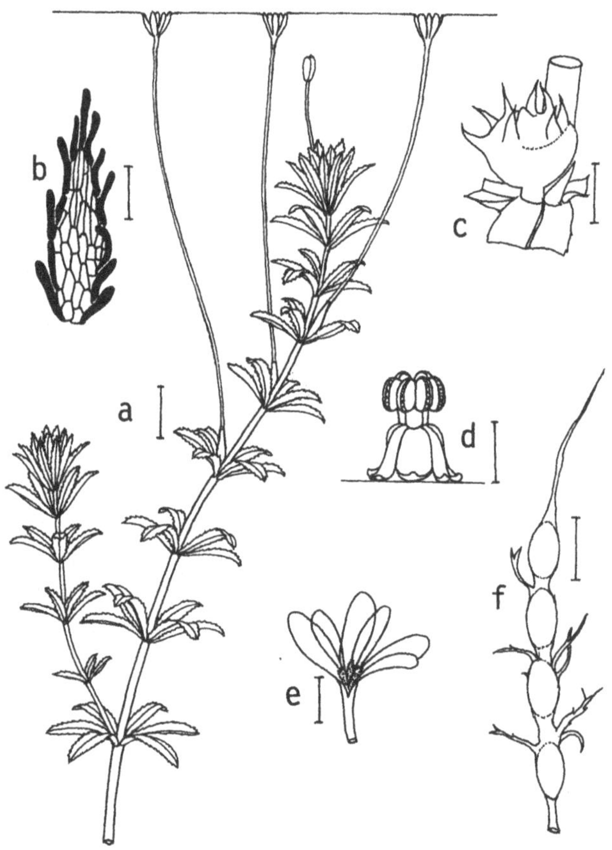

Fig. 225. *Hydrilla verticillata*: a, female flowering shoot (1 cm); b, squammula intravaginalis (0.1 mm); c, male spathe (1 mm); d, male flower after anthesis (1 mm); e, female flower (1 mm); f, fruit (0.25 mm).

normal foliage leaves; spathe sessile or subsessile, of 2 united bracts. Male flowers solitary, small, abscising as buds and opening explosively on the water surface; sepals 3, ovate and reflexed; petals 3, linear and reflexed; stamens 3, staminodes absent. Female flowers solitary, subsessile but with a long thread-like hypanthium carrying the flower to the surface; perianth of 3 oblong-ovate sepals and 3 elongate-ovate petals, spreading and floating; staminodes 3, minute. Carpels 3; styles 3, simple, very small (up to 0.75 mm long). Fruit a cylindrical capsule, indehiscent, armed or smooth. Seeds rarely more than 5, fusiform and smooth.

Mostly perennial but sometimes annual. Totally submerged. Growing in still or slowly flowing water. Often very abundant and dominant over large areas. Pollination anemophilous with pollen explosively liberated from free-floating male flowers and caught by floating female flowers. Monoecious or female. Disseminules are indehiscent capsules or seeds or specialized buds and tubers (turions), presumably dispersed by water, the capsules are sometimes spiny and probably dispersed by animals. It is considered to be a very serious weed in N America. In India it can reach pest proportions and is particularly troublesome in tanks used for breeding fish. It is readily eaten by grass carp but mostly chemical control is used. Cook, C. D. K. and Lüönd, Ruth. A revision of the genus *Hydrilla*. Aquatic Botany 13: 485–504 (1982); Pieterse, A. H. *Hydrilla verticillata* – a review. Abstracts on Tropical Agriculture (Koninklijk Instituut voor de Tropen, Amsterdam), 7(6): 9–34 (dated 1981, published 1982).

{(⊙) Vit}. Old World, naturalized in America: Ban, Cey, Ind, Nep, Pak: Ap, Ar, As, Bh, Dl, Go, Hp, Hr, Jk, Kl, Kt, Mg, Mh, Mn, Mp, Or, Pj, Rj, Tn, Up, Wb. 0–1600 m.

Nechamandra [*Lagarosiphon* auct. Ind.]: 1 species

Nechamandra alternifolia (Roxburgh ex Wight) Thwaites, [*Lagarosiphon alternifolia* (Roxburgh ex Wight) Druce]: Fig. 226

Dioecious. Roots unbranched. Stems elongate up to 1 m or more long, irregularly branched, internodes long or contracted at apparently irregular intervals. Leaves cauline, sometimes clustered, distichous, flat, band-shaped, 1.5–6 cm long, veins parallel without a prominent midrib, sheathing and somewhat amplexicaul at the base, the margin and apex with unicellular spines. Inflorescences axillary, often in clusters; spathe of 2 united bracts. Male flowers very small, very numerous, abscising as buds and opening on the water surface; sepals 3, becoming reflexed at anthesis; petal 1 rudimentary or absent; staminode 1 or absent; stamens 2, borne parallel to the water surface. Female flowers solitary, sessile, with long thread-like hypanthia, floating on the surface at maturity; sepals 3; petals rudimentary or absent; staminodes rudimentary or absent. Ovary of 3 carpels; styles 3, with flattened irregularly obdeltoid lobes. Fruit breaking irregularly. Seeds numerous, ovoid, ± 1 mm long, with reticulate seed coat.

Annual or sometimes persisting longer in permanent water. Leaves submerged, mostly found in temporary water; it appears to be replaced by *Vallisneria* in permanent water although they do sometimes grow together. Pollination is epihydrophilous with mobile male flowers caught by floating but attached female flowers. The disseminules are seeds, their dispersal mechanism is unknown, but they are probably transported in mud. Economic importance unknown. *Nechamandra* is patristically related to *Vallisneria* and not to the African genus *Lagarosiphon*. Cook, C. D. K.

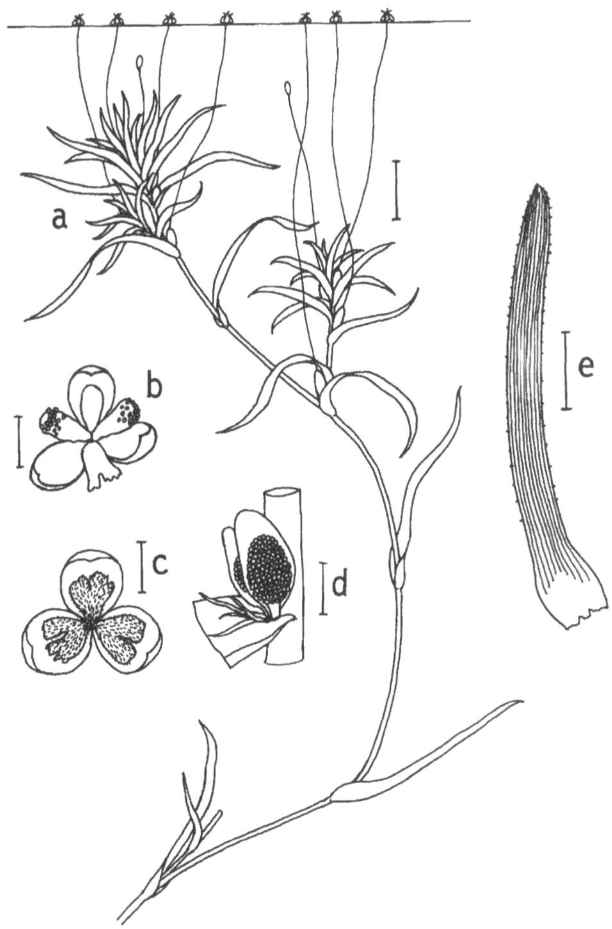

Fig. 226. *Nechamandra alternifolia*: a, female flowering shoot (1 cm); b, male flower (1 mm); c, female flower (1 mm); d, two male spathes before dehiscence (2.5 mm); e, leaf (5 mm).

and Lüönd, Ruth. A revision of the genus *Nechamandra*. Aquatic Botany 13: 505–513 (1982).

{(⊙) Vit}. India, China and Vietnam, naturalized in Sudan. Ban, Cey, Ind, Nep: Ap, As, Bh, Kt, Mh, Mn, Mp, Or, Pj, Rj, Tn, Up, Wb. 0–900 m.

Ottelia: 21 species: Tropical and subtropical Africa, Asia, Australasia and S America, naturalized in Europe and N America

Monoecious, dioecious or trioecious. Roots unbranched. Stems usually contracted but sometimes creeping. Leaves distichous or spirodistichous, without stipules, usually radical; juvenile leaves submerged and linear; adult leaves usually differentiated into petiole and blade, with or without spines; blade submerged or floating, from linear to orbicular or cordate. Flowers numerous or solitary; spathe of 2 united bracts, often armed or winged. Petals white, yellow or tinged with lilac or pink, large and showy or reduced in cleistogamous flowers. Stamens 3–15, in female flowers all staminodial but in male and bisexual flowers some usually staminodial. Carpels 3–20 or more; styles bifid. Fruit a fleshy capsule, usually topped by persistent sepals. Seeds very numerous released in a pulpy mass.

Cook, C. D. K., Symoens J.-J. and Urmi-König, Katherina. A revision of the genus *Ottelia*, part 1. Generic considerations. Aquatic Botany 18: 263–274 (1984); Cook, C. D. K. and Urmi-König, Katherina. part 2. The species of Eurasia, Australia and America. Aquatic Botany 20: 131–177 (1984).

1. Spathes membranous, with 3 or more wings; mature leaf blades submerged, membranous, almost translucent (widespread) ***O. alismoides***
1. Spathes leathery, without wings, smooth, warty or spiny; mature leaf blades floating, thick and leathery (Myanmar and SE Asia) ***O. cordata***

Ottelia alismoides (Linnaeus) Persoon: Fig. 227

Leaves totally submerged; petiole up to 50 cm or more long; leaf blades usually cordate with a distinct midrib and 2–10 longitudinal veins connected by numerous cross-veins joining at an angle of 60–70, giving the lamina the appearance of a quilt with rhomboidal patches. Spathes membranous with 3 or more wings. Flowers polygamous but in India mostly bisexual. Stamens 3–12, without staminodes. Ovary of 3–9 (–10) carpels; styles bifid to the base. Fruit ovoid to cylindrical, (15–) 20–40 (–50) mm long. Seeds up to 2000 or more, fusiform, 0.9–1.1 (–1.8) mm long, ±0.3 mm wide, dark purple to black when mature, densely covered in whitish, unicellular hairs.

Annual, may persist longer in permanent flowing or very deep water. Leaves submerged, found in permanent, still or slowly flowing water or in temporary pools. Flowers usually emergent, sweet-scented perhaps sometimes entomophilous but usually autogamous (pre-anthesis cleistogamy or true cleistogamy). The disseminules are seeds embedded in pulp (rather like bananas), probably dispersed by fish and also perhaps by other animals and by water. The leaves and inflorescences are eaten in SE Asia. It is often found in rice fields and irrigation ditches but seldom considered to be a serious pest. *Ottelia alismoides* is exceptionally variable in form and in chromosome numbers ($2n$=22 to $2n$=132); no satisfactory intraspecific classification exists.

{(⊙)? Ros}. Tropical Asia and Australia, introduced in Europe and N America: Ban, Bur, Cey, Ind, Nep: As, Bh, Go, Kl, Kt, Mh, Mn, Mp, Rj, Tn, Up, Wb.

Ottelia cordata (Wallich) Dandy: Fig. 227

Dioecious. Juvenile leaves linear, up to 70 cm or more long, up to 5 cm wide; mature leaves petiolate; petioles up to 100 cm long; mature leaf blades floating, cordate,

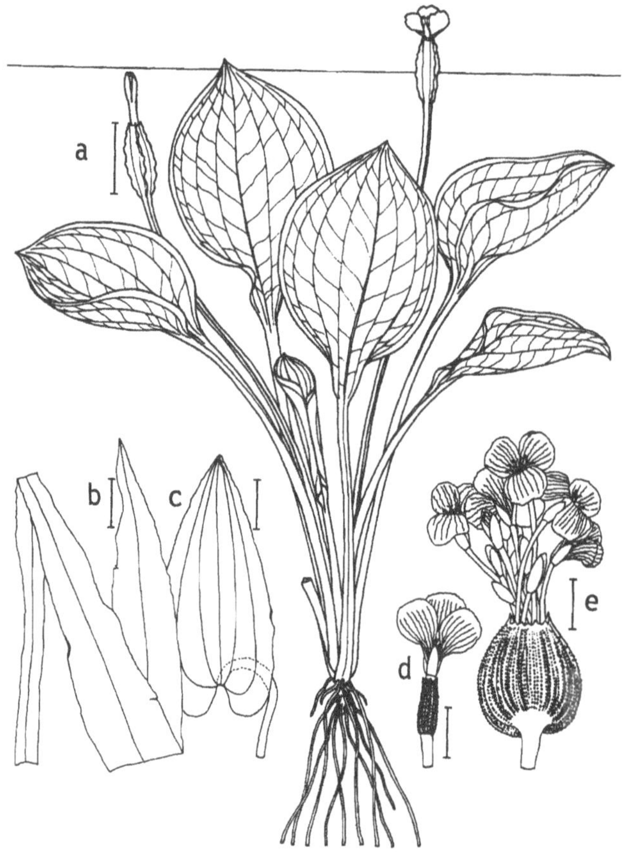

Fig. 227. *Ottelia alismoides*: a, flowering plant (3 cm). *O. cordata*: b, submerged leaf (3 cm); c, floating leaf (3 cm); d, female inflorescence (3 cm); e, male inflorescence (3 cm).

thick and leathery, never appearing quilt-like. Female spathes very hard and thick, usually covered with warts and/or spines; male spathes softer and thinner. Flowers unisexual. Male flowers numerous in each spathe, with 3–4 staminodes and (9–) 12 (–15) stamens. Ovary of 9–20 carpels; styles half to one-third bifid. Fruit obovoid to ellipsoid, developing within the spathe.

Perennial and perhaps also sometimes annual. Usually found in permanent water in lakes, pools and slowly flowing rivers. Leaves are sold in markets in Myanmar as a pot herb. Female *O. cordata* has been crossed with *O. alismoides*. The hybrid offspring were bisexual and self-fertile.

{⊙? Ros/Eph}. Myanmar and SE Asia: Bur.

Vallisneria: ± 4 species: tropical to warm regions of the world

Dioecious, annual or perennial. Roots unbranched. Stems either contracted or elongated, mostly developing stolons or runners, the branching system complex. Foliage leaves distichous or spirodistichous, narrowly strap-shaped, up to 2 m or more long, the margin usually with unicellular spines. Inflorescences axillary; spathe of 2 united bracts. Male flowers very small, very numerous, abscising as buds and opening on the water surface; sepals unequal (2 large, 1 smaller), becoming reflexed at anthesis; petal rudimentary or absent; staminode 1 or absent; stamens 2 or 1 (probably derived from 3 united stamens). Female flowers solitary, long pedicelled, floating on the surface at maturity; sepals 3 or 2, relatively large; petals 3 or 2, minute; staminodes 3 or 2, alternating with the petals. Ovary of 3 or 2 carpels; styles 3 or 2, each with 2 flattened papillose lobes. Fruit breaking irregularly. Seeds numerous, ellipsoidal, up to 2 mm long.

Lowden, R. M. An approach to the taxonomy of *Vallisneria* L. Aquatic Botany 13(3): 269–289 (1982).

Vallisneria spiralis Linnaeus [*V. natans* (Loureiro) Hara, *V. spiraloides* Roxburgh, *V. minor* Royle, *V. denseserrulata* (Makino) Makino]: Fig. 228

Stamens of the staminate flowers 2, free and obliquely extended forming a 'V', without any hairs at the base. Female flowers with an incision deepest between the matching stigmatic lobes (opposite the petal rudiment). Stigmas fringed along the margin.

Perennial or occasionally annual. Leaves totally submerged. It grows in still and flowing water (up to ± 1 m/sec) and is often very common and sometimes locally dominant. The pollination is epihydrophilous, with mobile male flowers on the water surface caught by floating but attached female flowers. The disseminules are seeds and vegetative fragments. The leaves are eaten in Japan and used as animal fodder in China. It is often a troublesome plant in irrigation and drainage ditches and channels, the leaves may reach 2 m long. It is eaten by grass carp but they rarely effectively control it, therefore, chemicals are usually used. It is also a very popular aquarium plant.

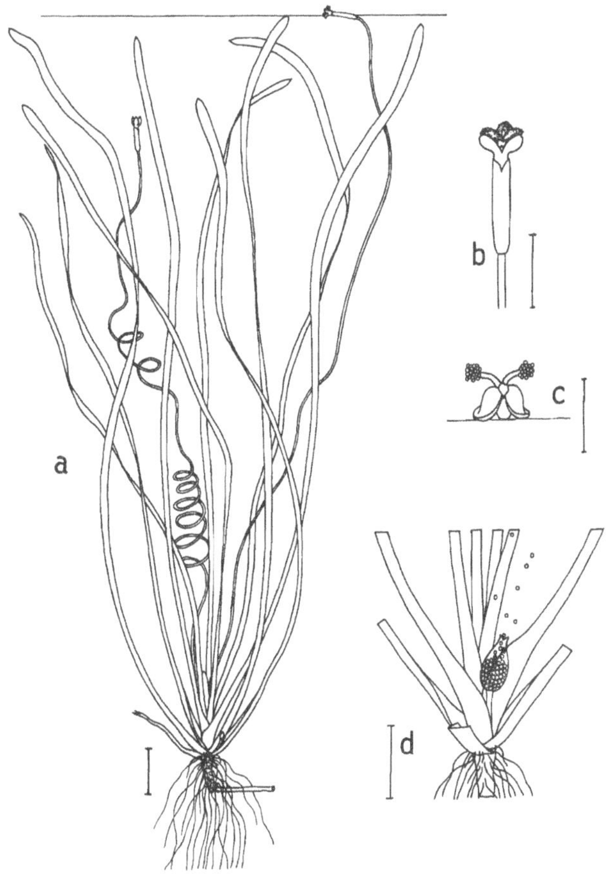

FIG. 228. *Vallisneria spiralis*: a, female flowering plant (2 cm); b, female flower (1 cm); c, free-floating male flower (1 mm); d, male spathe releasing flower buds (1 cm).

{(⊙) Ros}. Widespread in the Old World, also found in Australia and Oceania: Ban, Bur, Cey, Ind, Nep, Pak: Ap, As, Bh, Dl, Go, Hp, Kl, Kt, Mn, Mp, Pj, Rj, Up, Wb.

HYDROPHYLLACEAE: 22 genera

Hydrolea: 11 species: pantropics, most in America

Spiny or unarmed herbs. Stems terete, often swollen and spongy. Leaves alternate, simple, entire, usually narrow; stipules absent. Flowers bisexual, radially symmetrical, solitary or in terminal or axillary cymes. Sepals 5, almost free to the base, persistent. Petals 5, united at the base, blue. Stamens 5, inserted at the base of the petals and alternating with their lobes. Ovary superior, 2- or rarely more-locular, with parietal placentas; styles 2 or rarely more; stigmas capitate; ovules numerous. Fruit a capsule splitting by valves or opening irregularly.

Davenport, L. J. A monograph of *Hydrolea.* Rhodora 90(862): 169–208 (1988).

1. Stems bearing axillary spines at some nodes, the entire stem with patent, gland-tipped and very sticky hairs; petals 10–12 mm long; placentas of each locule lobed, divided into 2 wings (Sri Lanka) ***H. spinosa***
1. Stems not bearing spines, glabrous or with glandular hairs towards the tip; petals 4.5–7 mm long; placentas of each locule entire, not divided into 2 wings (widespread) ***H. zeylanica***

Hydrolea spinosa Linnaeus: Fig. 229

Like *H. zeylanica* but shrub-like, up to 2 m tall; stems and branches bearing 1 or 2 axillary spines at some nodes, the entire stem with patent, gland-tipped and very sticky hairs; sepal lobes 6–14 mm long; petals 10–12 mm long; placentas of each locule lobed, divided into 2 wings; seeds 0.4–0.7 mm long, 0.2–0.3 mm broad.

Perennial. Usually in wet places.

{Hel}. Native to America, naturalized in Sri Lanka: <Cey.

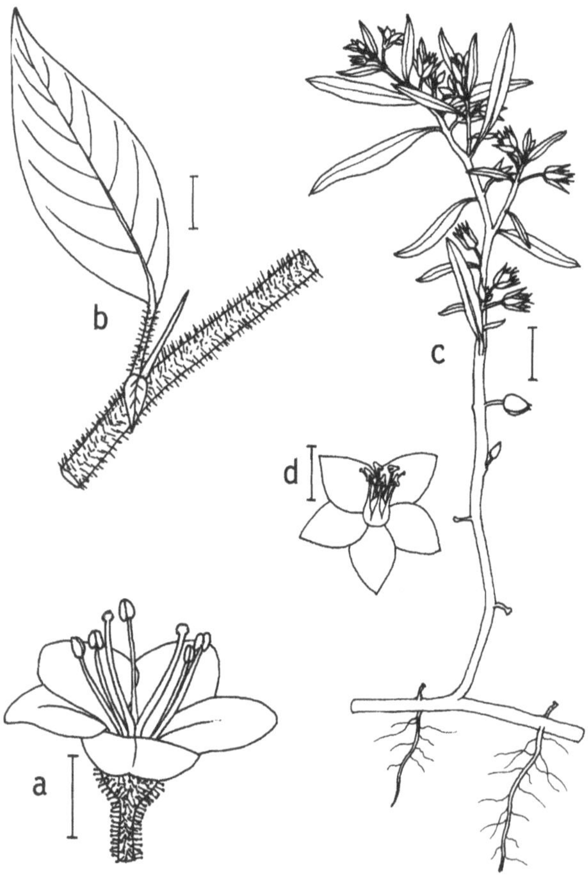

Fig. 229. *Hydrolea spinosa*: a, flower (4 mm); b, node with a spine (4 mm). *H. zeylanica*: c, flowering plant (2 cm); d, flower (5 mm).

Hydrolea zeylanica (Linnaeus) Vahl [Steris aquatica N. L. Burman]: Fig. 229

Stems prostrate or erect, up to 60 cm tall, green to brown often tinged with purple, without spines, glabrous or hairy, sometimes glandular towards the tip. Leaf blades ovate to lanceolate, 2–10 cm long, 3–25 mm wide, glabrous or hairy. Flowers in terminal panicles. Sepal segments lanceolate, 4.5–8 mm long, 1–2 mm wide. Petal lobes 4.5–8 mm long, 1.5–3.5 mm wide, bright blue. Capsules globose to cylindrical, 4–5 mm long, 2.5–3.5 mm broad; placenta of each locule entire, not winged. Seeds ovoid to oblong, 0.3–0.4 mm long, less than 0.2 mm broad, with longitudinal and transverse ridges.

Perennial or sometimes annual. Seasonally submerged, emergent or helophytic, often gregarious and forming a distinct zone above receding water. It tends to flower as the water recedes. Pollinated by insects or autogamous. The disseminules are seeds, their dispersal is unknown. Young shoots are eaten as a vegetable, they are reported to have antiseptic properties and are used in medicine. Sometimes reported to be weeds in rice fields and irrigation ditches but not considered to be very troublesome.

{(⊙) Hel(Hyp)}. Old World tropics: Ban, Bur, Cey, Ind, Nep, Pak: An, As, Dl, Gj, Go, Kl, Kt, Mh, Mp, Mz, Pj, Rj, Up, Wb. 0–1200 m.

IRIDACEAE excluded see p. 11

JUNCACEAE: 8 genera

Juncus: 255 species: cosmopolitan but mostly in temperate zones

Stems erect, floating or emergent. Leaves alternate, terete, capillary or flat or reduced to minute scales or absent, occasionally sheathing at the base; sheath base open. Inflorescence a cluster of terminal or apparently lateral cymes, sometimes reduced to a 1- or 2-flowered head, subtended by scale-like bracts. Flowers bisexual. Perianth segments 6, in 2 whorls, scale-like, thin and scarious, greenish or brownish. Stamens 3 or 6, shorter than the perianth. Ovary superior, of 3 united carpels, with parietal placentas; stigmas 3, linear, twisted. Fruit a 3-valved loculicidal capsule. Seeds numerous and small.

Perennial or annual. Most species occur in wet places and some are submerged. Pollinated by wind or autogamous. The disseminules are seeds or specialized bulbils, dispersed in water and perhaps also by animals.

Excluded: about 40 species occur at high altitudes in the Himalayas; ***Juncus maritimus*** Lamarck: from brackish conditions in NE India and Pakistan.

1. Leaves flat, not hollow, without transverse septae; flowers solitary or in clusters of up to 5 ***J. bufonius***
1. Leaves terete or flattened, hollow, with transverse septae; flowers 4–10 in dense heads ***J. prismatocarpus***

Juncus bufonius Linnaeus: Fig. 230

Tufted. Stems terete, (2–) 10–35 (–60) cm tall. Leaves basal and on stem, grass-like, flat, 1–12 cm long, 0.5–1 (–1.5)mm wide, not hollow, not transversely septate; sheaths scarious up to 3.5 cm long. Flowers solitary or in clusters of 2–5; floral bracts 3 (1 bract, 2 bracteoles), ovate to lanceolate, shorter than the perianth. Perianth segments 3–8 mm long, 1–1.5 mm wide, the outer ones longer than the inner. Stamens 6 or less in cleistogamous flowers, much shorter than the perianth. Capsules ellipsoid to ovoid, 3–5mm long, 1–2 mm wide. Seeds obovoid, 0.3–0.5 mm long, reddish-brown.

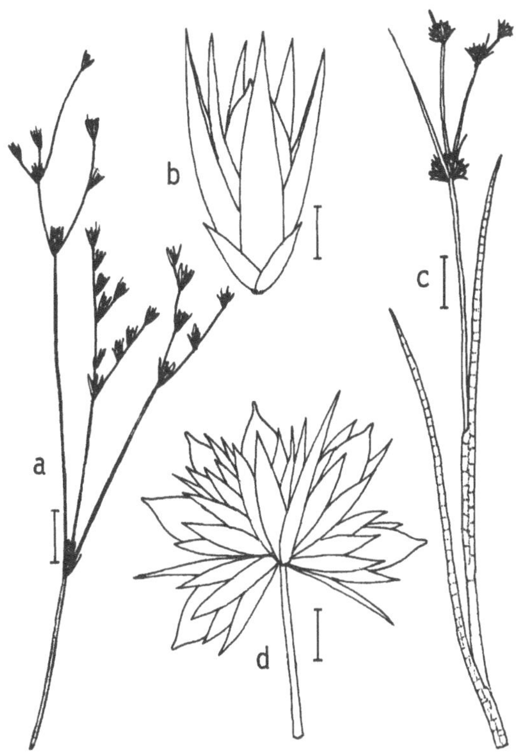

FIG. 230. *Juncus bufonius*: a, flowering shoot (1 cm); b, fruiting head (1.6 mm). *J. prismatocarpus*: c, flowering shoot (1 cm); d, fruiting head (1.6 mm).

Annual or very rarely perennial. A pioneer plant of wet mud and sand, flowering as the water recedes, not confined to wetlands. Often locally abundant and gregarious in marshy ground, on the margins of tanks. In young seedlings the seeds are elevated considerably above the ground so the plant apparently looks like a moss, having a capsule at the apex. It is a highly variable species. Mostly found on higher land.

{(⊙) Hel}. Temperate and warm regions of Eurasia and America, probably introduced in Sri Lanka: Bhu, <Cey, Ind, Nep, Pak: Dl, Hp, Rj, Sk. 200–2400 m.

Juncus prismatocarpus R. Brown, non sensu J. D. Hooker: Fig. 230

Tufted. Stems terete or flattened, (8–) 20–60 (–100) cm tall. Leaves basal and on stem, linear to lanceolate or filiform, terete or flattened, (3.5–) 10–20 cm long, up to 5 mm wide, the lower ones longer than the upper, hollow and transversely septate within; sheaths scarious up to 3.5 cm long. Heads hemispherical to subglobose, ± 1 cm in diameter, with 4–10 flowers; floral bract solitary (bracteoles absent), ovate to lanceolate, shorter than the perianth. Perianth segments 3.5–5 mm long, all ± alike, greenish or brown. Inflorescence sometimes pseudoviviparous. Stamens 3, much shorter than the perianth. Capsules up to 5 mm long, 3-sided, with prominent angles. Seeds ellipsoid to obovoid, 0.4–0.8 mm long,

Perennial or sometimes annual. In marshes, along streams, in muddy places and waterlogged fields. Mostly on hills.

{(⊙) Hel}. Subtropical and tropical E and SE Asia, Australia and New Zealand: Ban, Bur, Cey, Ind, Pak: Kl, Kt, Mn, Mp, Tn. 100–2500 m.

JUNCAGINACEAE excluded see p. 11.

LAMIACEAE [LABIATAE]: 180 genera

Herbs, sometimes woody at base, often aromatic. Stems usually square in transverse section. Leaves opposite or whorled. Inflorescence whorl-like, the whorls often crowded at the stem apex and forming a spike. Sepals 5 or rarely 4, united below, often 2-lipped, with 3 adaxial lobes and 2 abaxial ones. Petals 5 or 4, united below in to a tube, the 2 adaxial lobes forming a lip, the abaxial 3 forming a 3-lobed lower lip or lobes ± equal. Stamens usually 4, rarely reduced to 2. Ovary of 2 carpels, each with 2 ovules, finally appearing 4-lobed; style bifid above. Fruit of 4 nutlets.

Mukerjee, S. R. Records of the Botanical Survey of India 14(1): i–vii, 1–228 (1940).

Excluded: wetland plants not extending south of the Himalayas: ***Lycopus europaeus*** Linnaeus; ***L. exaltatus***

Linnaeus fil.; ***Mentha aquatica*** Linnaeus; ***M. pulegium*** Linnaeus; ***Scutellaria galericulata*** Linnaeus; ***Teucrium scordium*** Linnaeus.

Plants sometimes found in moist or damp situations but not wetland species: ***Ajuga bracteosa*** Wallich ex Bentham; ***A. macrosperma*** Wallich ex Bentham; ***Mesona palustris*** Blume [*Geniosporum parviflorum* Wallich ex Bentham]; ***Nepeta hindostana*** (Roth) Haines [*N. calminthoides* Bentham, *N. ruderalis* F. Hamilton ex Bentham]; ***Salvia plebeia*** R. Brown.

Pogostemon [*Dysophylla, Eusteralis*]: ± 71 species: Asia and Africa

Herbs. Leaves opposite or whorled. Flowers small, in dense, ± continuous, terminal spikes. Sepal tube ovoid or bell-like, the throat naked; sepal lobes 5, equal. Petal lobes 4, ± equal. Stamens 4, exerted, straight or slightly curved; filaments bearded; anthers confluent, 1-locular.

Annual or perennial. Pollinated by insects. The disseminules are nutlets, the dispersal is unknown. The aquatic and wetland species in this group are highly variable and in great need of critical revision; previous authors have, I believe, recognized too many species. Press, J. R. Taxonomic studies in the Labiatae tribe Pogostemoneae. Bulletin of the British Museum (Natural History) Botany series 10(1): 1–89 (1982).

Excluded: found in the Himalayas, not extending south: ***Pogostemon helferi*** (J. D. Hooker) Press [?*Dysophylla kachinensis* S. K. Mukerjee]: ♦Bur, ♦Ind.

Species found on periodically flooded stream sides (rheophytes): ***P. quadrifolius*** (Bentham) Kuntze: ♦Ban, ♦Bur, ♦Ind; ***P. salicifolius*** (Dalzell ex J. D. Hooker) Kuntze: ♦Ind.

Species sometimes found in moist or damp places but not wetland species: ***P. auricularius*** (Linnaeus) Hasskarl; ***P. cruciatus*** (Bentham) Press [*Dysophylla tetraphylla* Wight]: ♦Ban, ♦Ind, ♦Nep; ***P. linearis*** (Bentham) Kuntze: ♦Ban? ♦Ind; ***P. paludosus*** Bentham: ♦Ban, ♦Ind; ***P. peguanus*** (Prain) Press: ♦Bur; ***P. stocksii*** (J. D. Hooker) Press: ♦Ind; ***P. wightii*** Bentham: ♦Ind.

1. Leaves mostly less than 1 cm long; whorls mostly with many more than 5 leaves; floral bracts at the base of the inflorescence ± as long as the leaves, persisting after the fruit falls ***P. erectum***

1. Leaves mostly more than 2 cm long; whorls with rarely more than 5 leaves (at least upper nodes with 4 or less); floral bracts much shorter than the leaves, not persisting after the fruit falls

  2. Sepals in fruit stellately spreading and disclosing the nutlets (open sepal tubes look like holes in the solid spike) ***P. stellatus***

  2. Sepals in fruit erect or incurved, not disclosing the nutlets

    3. Sepal tube obconic, deeply 5-grooved, glabrous, chestnut brown, shining below, with 5 elevated ridges; sepals in fruit ± 2.4 mm long ***P. griffithii***

    3. Sepal tube broadly campanulate, not grooved, usually hairy, greenish, not shining, without ridges; sepals in fruit 1.5–2 mm long ***P. pumilus***

Pogostemon erectum (Dalzell) Kuntze [*Mentha stellata* F. Hamilton ex Roxburgh non Loureiro, *Dysophylla tomentosa* Dalzell non *Pogostemon tomentosa* Hasskarl, *Dysophylla gracilis* Dalzell non *Pogostemon gracilis* Hasskarl, *Dysophylla stellata* var. *gracilis* (Dalzell) Cooke, *Pogostemon deccanensis* (Panigrahi) Press]: Fig. 231

Stems creeping or decumbent and much branched below, erect part up to 6–20 (–35) cm tall, flexible, with short internodes, glabrous or with white hairs. Leaves sessile or nearly so, in whorls of (5–) 8–20, narrowly linear, 4–10 mm long, ± 2.5 mm wide (when submerged filiform, ± 1 mm wide), glabrous or slightly hairy, often revolute, margins entire, tips obtuse or acute. Floral bracts linear, as long as the upper leaves, persisting even after the fruit falls. Spikes 2–4 (–8) cm long; 3–4 (–5) mm in diameter in fruit. Sepals very small, hemispheric or broadly campanulate, glandular

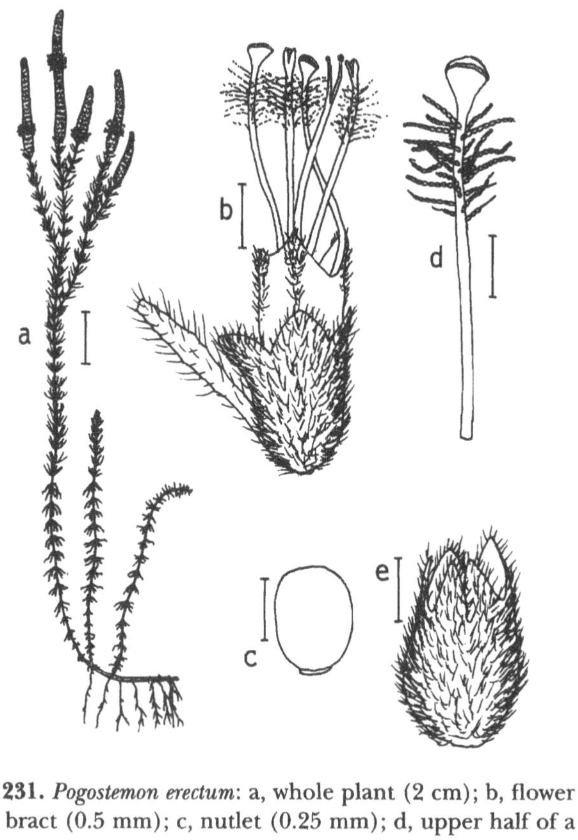

FIG. 231. *Pogostemon erectum*: a, whole plant (2 cm); b, flower with bract (0.5 mm); c, nutlet (0.25 mm); d, upper half of a stamen (0.25 mm); e, sepals in fruit (0.5 mm).

hairy; tube ± 1 mm long; lobes subequal, deltate, ±0.25 mm long, erect or incurved in fruit. Petals up to 2 mm long, lilac to purple; tube ± 1.5 mm long; lobes ± equal, ovate-oblong, hairy outside. Stamens conspicuous; filaments, lilac to purple, with long moliniform hairs towards the tip; anthers 1-locular. Nutlets discoid, ± 0.4 mm in diameter, pale yellowish-brown, smooth and shining.

Perennial in permanently wet habitats or annual. Found along the margins of ditches and ponds, usually seasonally submerged, often forming large clumps which are very conspicuous when in flower. Frequently found in brackish water near to the coast. This species is very variable; also there is considerable nomenclatural confusion. It is incorrectly called *Dysophylla stellata* in most Indian Floras.

{(⊙)? Hyp/Ten}. Endemic to S India: ♦Ind: Kl, Kt, Mh.

Pogostemon griffithii (J. D. Hooker) Press: Fig. 232

Like *P. pumilus* but leaves longer (up to ± 7 cm long), rarely more than 4 in a whorl, margins entire; floral bracts linear, exceeding the flowers; sepal tube obconic, glabrous, chestnut brown, shining below, with 5 grooves and 5 elevated ridges; sepal lobes incurved, villous with white hairs; sepals in fruit ± 2.4 mm long.

Annual. Wet places.

{⊙ Hyp/Hel}. Endemic to India: ♦Ind: Ap, Or.

Pogostemon pumilus (Graham) Press [*Mentha verticillata* auct. non Linnaeus, *Dysophylla crassicaulis* Bentham]: Fig. 232

Stems ascending or erect, simple or with some simple branches, 10–50 cm tall, often somewhat fleshy and inflated below, glabrous or with glandular hairs. Leaves sessile, opposite or in whorls of (3–) 4–6 (–10) (the number deceasing towards the stem apex), linear to linear-lanceolate, 2–5 cm long, ± 0.5 cm wide, densely gland-dotted below, ± cordate at the base, margins entire or distantly serrate. Floral bracts ovate-lanceolate, 1.5–2 mm long, as long as the sepals, purplish, deciduous. Spikes 1.7–6 (–10) cm long, ± 5 (–8) mm in diameter. Sepals broadly campanulate, 1.5–2 mm long, greenish, eglandular or with glandular hairs; lobes 4, subequal, deltate. Petals up to 3–4 mm long, lilac to purple; lobes 4, short, subequal, glabrous. Stamens conspicuous; filaments, pink to lilac, with long purplish hairs in the middle. Style as long as the stamens, 2-lobed at the tip. Nutlets ovoid, ± 0.8 mm long, ±0.7 mm in diameter, pale brown.

Annual. Wet places, rice fields, ditches and along the edges of lakes and pools.

{⊙ Hel]}. Endemic to the Indian subcontinent: ♦Ban, ♦Bhu, ♦Bur, ♦Cey, ♦Ind, ♦Nep: As, Jk, Sk.

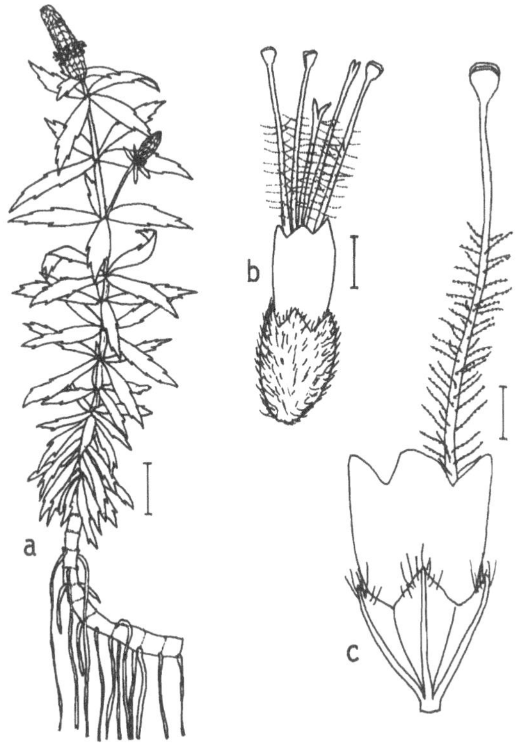

FIG. 232. *Pogostemon griffithii*: c, sepals and petals with one stamen (1 mm). *P. pumilus*: a, whole plant (8 mm); b, flower (1 mm).

Pogostemon stellatus (Loureiro) Kuntze [*Dysophylla verticillata* (Roxburgh) Bentham, *D. benthamiana* Hance]: Fig. 233

Stems glabrous or hairy, usually creeping below and erect above, simple or branched, 10–60 (–100) cm tall, often swollen below. Leaves sessile, spreading, opposite or 3–8 (–10) in a whorl, linear to lanceolate, 1.5–7 (–16) cm long (submerged leaves long and many in a whorl, aerial ones short and fewer in a whorl), 2–13 mm wide, narrowed at the base, spreading or deflexed, glabrous or nearly so, margins subentire to sharply serrate towards the tips, tips very acute. Floral bracts linear-lanceolate, 1.5–2.5 mm long, much smaller than the leaves. Spikes 2–6 (–12) cm long, 4–5 mm in diameter., densely hairy. Sepals broadly campanulate, glandular and densely hairy; tube ± 1 mm long; lobes 5, subequal, triangular-ovate, ± 0.5 mm long, stellately spreading and disclosing the nutlets (open sepal tubes look like holes in the solid spike. Petals up to 2 mm long, pale pink to lilac, hairy outside; lobes 4, ± equal, ovate, ± 0.5 mm long, glabrous or hairy. Stamens 4, exerted, conspicuous; filaments ± 3 mm long, lilac, tufted with long moliniform hairs from the middle upwards. Nutlets slightly flattened globose, 0.6–0.7 mm in diameter, smooth, pale brown.

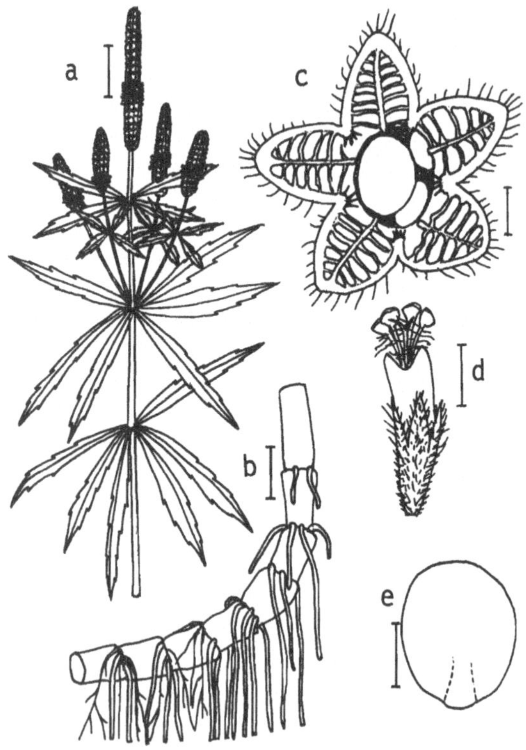

FIG. 233. *Pogostemon stellatus*: a, whole plant (6 mm); b, base of the stem (1 cm); c, fruiting sepals viewed from above (0.25 mm); d, flower (0.5 mm); e, nutlet viewed from the side (0.25 mm).

Annual, perhaps sometimes perennial when growing in water. Mostly found in seasonally inundated places flowering as the water recedes, sometimes in permanent water. It is offered for sale to be cultivated in aquaria for decoration.

{⊙ Ten(Hyp/Hel)}. Tropical to warm regions of Asia and Australia: Ban, Bur, Cey, Ind, Nep: As, Bh, Go, Kl, Kt, Mp, Sk, Up, Wb.

LEGUMINOSAE see FABACEAE

LEMNACEAE: 4 genera: almost cosmopolitan

Plant body of small thalloid fronds, not or very rarely attached to a solid substrate, mostly free-swimming or free-floating, partly floating or submerged, the fronds solitary or connected in groups. Roots simple, 1–several or absent. Flowers unisexual, naked or enclosed in a sheath (spathe). Perianth absent. Male flower consisting of 1 stamen with a 1- or 2-locular anther. Female flowers consisting of 1, sessile, 1-locular ovary. Fruit a nut or 1- to 4-seeded capsule.

The frond has variously been interpreted as a modified stem, a modified leaf or part stem and part leaf (the latter is the most plausible; I suspect it may be a modified hypocotyl). Pollinated by insects or, exceptionally, epihydrophilous; the Indian species are probably highly autogamous. The disseminules are seeds or complete fronds, dispersal by water and perhaps also by surface adhesion to animals (mostly water fowl). Landolt, E. The family of Lemnaceae — a monographic study. Veröffentlichungen des Geobotanischen Institutes der Eidg. Techn. Hochschule, Stiftung Rübel, in Zürich 71: 1–566 (1986).

1. Fronds bearing roots; budding pouches 2; inflorescences of 1 female and 2 male flowers, enclosed in a membranous sheath; anthers 2-locular, transversely dehiscent

  2. Roots solitary on each frond segment; fronds usually 1–5 mm long, without dorsal and ventral scales, the under surface green, reddish or streaked with reddish-brown; nerves 1–5 (–7), often indistinct ***Lemna***

  2. Roots several from each frond segment; fronds usually 4–10 mm long, with 1 dorsal and 1 ventral scale, the under surface usually reddish; nerves (3–) 5–15 (–21), usually distinct ***Spirodela***

1. Fronds without roots; budding pouch 1; inflorescences of 1 female and 1 male flower, not enclosed in a membranous sheath; anther 1-locular, apically dehiscent

  3. Fronds globose, solitary or sometimes mother and daughter connected; budding pouch funnel-shaped with a circular opening ***Wolffia***

  3. Fronds flat, ovate to linear, sometimes connected in groups, often asymmetrically curved; budding pouch deltate ***Wolffiella***

Lemna: 13 species: almost cosmopolitan

Fronds solitary or usually remaining in groups of 2–10 or more, lanceolate to ovate or suborbicular in outline, flat or inflated, without discreet pigment cells in the epidermis; veins 1–5 (–7); dorsal and ventral scales absent. Roots solitary on each frond. Inflorescences and daughter fronds borne in 2 marginal budding pouches, one on each side at the base of the frond. Inflorescence of 1 female and 2 male flowers enclosed in a membranous sheath. Anthers 2-locular. Carpel with 1–7 ovules. Seeds ovoid, up to 2 mm long, with 8–70 longitudinal ribs, with transverse striations or rarely smooth.

Free-floating or *L. tenera* and *L. trisulca* free-swimming. Usually highly gregarious, large populations may reach pest proportions, sometimes eaten by humans in SE Asia, a valuable food for wildfowl, also used for fodder and as a source of methane and alcohol.

The following species, although appearing in the account, do not extend into Peninsular India: *L. minor* and *L. turionifera* are found in the Himalayas and *L. tenera* is found in Myanmar.

1. Fronds toothed towards the apex and narrowed at the base to a persistent green stalk, forming branched chains of 3–50 fronds, submerged except when flowering **L. trisulca**

1. Fronds with entire margins and rounded at the base with a minute and short-lived white stalk, never forming branched chains, rarely more than 3 fronds together, floating on the surface or submerged

2. Fronds 2.5–4 times as long as wide, tapering to a long point at the apex, submerged except when flowering or fruiting **L. tenera**

2. Fronds 1–2 times as long as wide, ± rounded at the apex, floating on the surface of the water

3. Root sheath winged at the base; root tip sharply pointed (fresh material needed); roots usually less than 3 cm long; fronds without a reddish tinge or reddish flecks **L. aequinoctialis**

3. Root sheath not winged; root tip mostly rounded (fresh material needed); roots often more than 3 cm long; fronds often with a reddish tinge or with reddish spots

4. Fronds with distinct papillae along the median line of the upper surface; the papilla at the tip smaller than the proximal papillae; seeds with 30–60 longitudinal ribs; under unfavourable conditions the plants form small, olive to brown, rootless turions that sink **L. turionifera**

4. Fronds without distinct papillae along the median line of the upper surface or if present then the papilla at the tip distinctly larger than the proximal papillae; seeds with 10–16 longitudinal ribs; turions never developed

5. Lower surface of the frond red; fronds usually gibbous, more than 1 mm thick, largest lacunae more than 0.3 mm in diameter (seen from below); ovary with 1–7 ovules **L. gibba**

5. Lower surface of the frond not coloured or very slightly reddish; fronds usually flattened rarely gibbous, not more than 1 mm thick, largest lacunae spaces less than 0.3 mm in diameter; ovary with 1 ovule **L. minor**

Lemna aequinoctialis Welwitsch [*L. gibba* auct. Ind., non Linnaeus, *L. minor* auct. Ind., non Linnaeus, *L. perpusilla* auct. Ind., non Torrey, *L. paucicostata* auct. Ind., non Hegelmaier, *L. minima* Blatter et Hallberg, *L. blatteri* McCann, *L. eleonorae* McCann]: Fig. 234

Fronds floating, flattened, ovate to lanceolate, 1–6.5 mm long, 0.8–4.5 mm wide, 1–2 (–3) times as long as wide; the upper surface green, bearing 2 papillae, the one above the node smaller than the one at the tip; veins 3. Roots up to 3.5 cm long; root sheath winged, the wing 1–2.5 times as long as wide; root cap sharply pointed (fresh material needed). Seeds 1 per fruit, 0.45–0.8 mm long, 0.3–0.7 mm in diameter, brownish, with 8–26 longitudinal ribs, released from the fruit when ripe.

Perennial, perhaps sometimes annual. Usually found in eutrophic conditions, roadside pools, village ponds, buffalo wallows, tanks (particularly those used for cultivating *Trapa*), frequently found in rice fields and in sluggish canals and ditches. Often growing together with *Spirodela* or *Wolffia*.

{⊙? Ple}. Subtropical regions of the world: Ban, Bur, Cey, Ind, Nep, Pak: An, Ap, As, Bh, Dl, Gj, Hp, Jk, Kl, Kt, Mh, Pj, Rj, Up, Wb.

Lemna gibba Linnaeus: Fig. 234

Fronds floating, gibbous, up to 4 mm thick, 1–8 mm long, 0.8–6 mm wide, 1–1.5 times as long as wide; the upper surface without distinct papillae or sometimes papillae visible near the tip of small fronds, often with red flecks; veins (3) 4 or 5 (7), the lateral veins originating at the node; lower surface gibbous, usually red, largest lacunae more than 0.3 mm across. Roots usually exceeding 3 cm long; root sheath not winged; root cap rounded (fresh

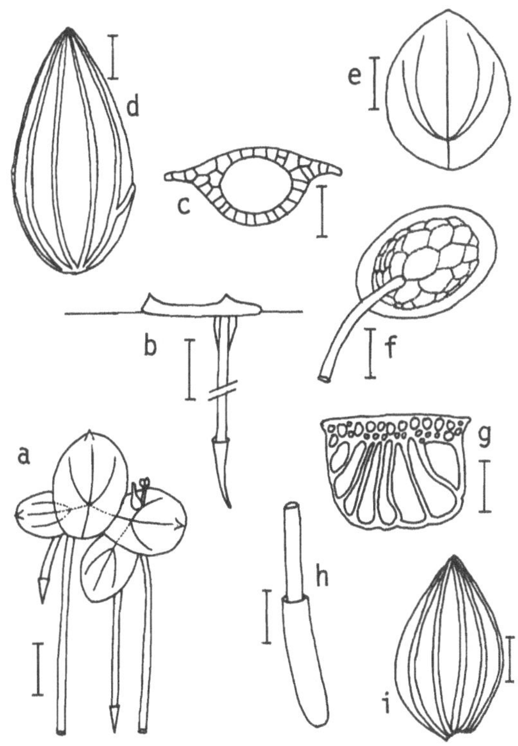

FIG. 234. *Lemna aequinoctialis*: a, fronds viewed from above (2 mm); b, frond viewed from the side (2 mm); c, root sheath in transverse section (0.1 mm); d, seed (0.1 mm). *L. gibba*: e, frond viewed from above (2 mm); f, frond viewed from below (2 mm); g, frond in transverse section showing lacunae (2 mm); h, root tip (0.5 mm); i, seed (0.2 mm).

material needed). Seeds 1–5 per fruit, 0.7–0.9 mm long, 0.4–0.6 mm thick, whitish, with 8–16 longitudinal ribs, released from the fruit when ripe.

Perennial or exceptionally annual. Found in a very wide range of aquatic habitats. Usually in mesotrophic to eutrophic water but not as tolerant as *L. aequinoctialis* to extremely eutrophic or organically loaded water.

{⊙? Ple}. Almost throughout the world in warm and Mediterranean climatic regions, naturalized in Japan: Ind, Pak: Gj, Jk, Pj.

Lemna minor Linnaeus: Fig. 235

Fronds floating, flattened, not more than 1 mm thick, 1–8 (–10) mm long, 0.6–5 mm wide, 1.3–2 times as long as wide; the upper surface with indistinct papillae along the midvein, the papillae near the tip and near the node larger than the ones in between, green or tinged with red but without red flecks; veins 4 or 5, the lateral veins originating away from the node; lower surface flat or rounded but not gibbous, very slightly reddish, largest lacunae less than 0.3 mm across. Roots usually exceeding 3 cm long; root sheath not winged; root cap rounded (fresh material needed). Seeds 1 per fruit, 0.7–1 mm long, 0.4–0.6 mm in diameter, whitish, with 10–16 longitudinal ribs, retained in the fruit when ripe.

Perennial. Found in a very wide range of aquatic habitats. Usually in mesotrophic to eutrophic water.

{Ple}. Cooler regions of the N Hemisphere, naturalized in Australasia: Ind, Nep, Pak: Hp, Jk, Sk, Up.

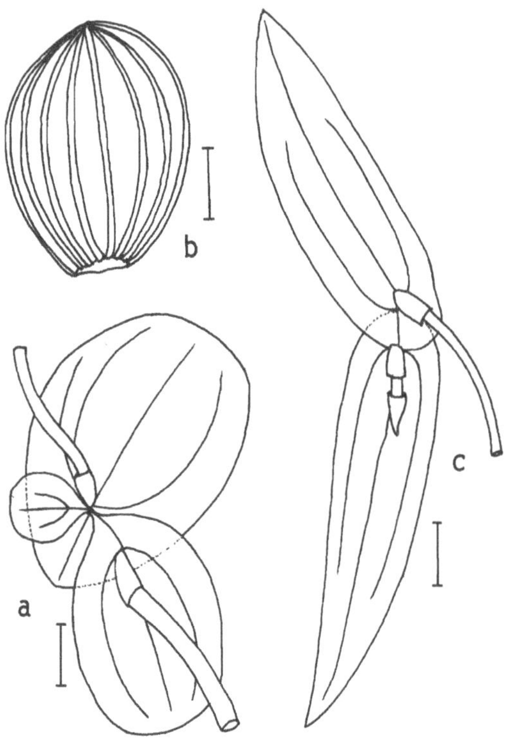

FIG. 235. *Lemna minor*: a, fronds viewed from below (1 mm); b, seed (0.2 mm). *L. tenera*: c, fronds viewed from below (1 mm).

Lemna tenera S. Kurz: Fig. 235

Fronds submerged except when flowering or fruiting, flattened, elongate, tapering to a long pointed tip, rounded at the base with a small white stipe that soon decays, 3.5–9 mm long, 1.2–3 mm wide, 2.5–4 times as long as wide; the upper surface without papillae, green; veins 3, indistinct; lower green. Roots up to 2.5 mm long; root sheath not winged; root cap rounded. Fruits unknown.

{Pla/Ple}. Humid tropical SE Asia: Bur (Pegu).

Lemna trisulca Linnaeus [*L. cruciata* Roxburgh]: Fig. 236

Fronds submerged except when flowering or fruiting, remaining together and often forming branched chains, flattened, narrowly ovate, with the margin toothed towards the tip, narrowed at the base to a persistent green stalk, 3–15 mm long (excluding the stalk), 1–5 mm wide, 2–3.5 times as long as wide, the stalk 2–20 mm long; the upper surface without papillae, green or reddish; veins (1) 3. Roots up to 2.5 mm long; root sheath not winged; root cap pointed (fresh material needed). Seeds 1 per fruit, 0.6–1.1 mm long, 0.5–0.8 mm in diameter, with 12–18 longitudinal ribs, retained in the fruit when ripe.

Perennial. Found in still water in rather sheltered situations, often between emergent reeds. It is ± confined to mesotrophic water. Sometimes may develop turions.

{Pla/Ple}. Cooler regions of the world: Ban, Ind, Pak: As? Dl? Jk, Mn, Rj, Up, Wb.

Lemna turionifera E. Landolt: Fig. 236

Fronds floating, gibbous, up to 2 mm thick, 0.8–3.5 mm long, 0.8–3.5 mm wide, 1–1.5 times as long as wide; the upper surface with distinct papillae along the veins, the papilla near the tip not larger than the others, green and sometimes with red flecks especially near the tip; veins 3; lower surface gibbous, reddish, largest lacunae usually less than 0.3 mm across. Roots usually exceeding 3 cm long; root sheath not winged; root cap usually rounded but sometimes pointed or rarely rootless when developing olive to brown turions. Seeds 1 per fruit, 0.5–0.8 mm long, ± 0.5 mm in diameter, whitish, with 30–60, indistinct, longitudinal ribs, retained in the fruit when ripe.

Perennial. Turions sink and remain at the bottom during the cold period.

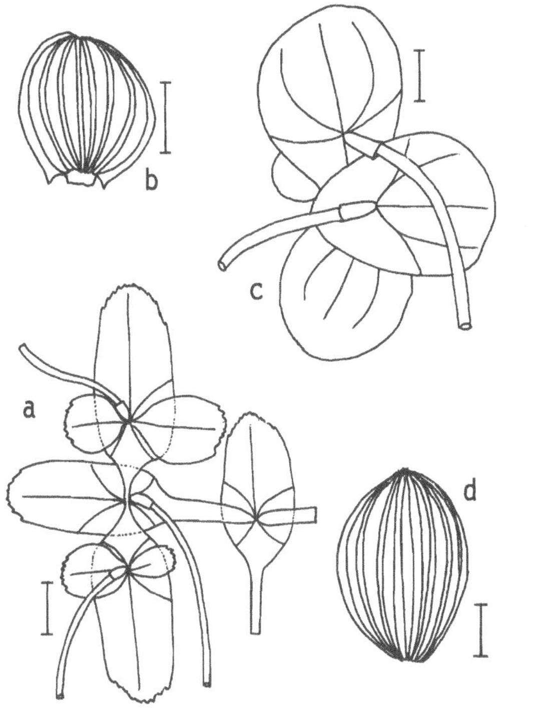

Fig. 236. *Lemna trisulca*: a, fronds viewed from below (1 mm); b, seed (0.4 mm). *L. turionifera*: c, fronds viewed from below (1 mm); d, seed (0.1 mm).

{Ple}. Temperate N America and continental Asia: Ind: Hp, Jk.

Spirodela: 3 species: almost cosmopolitan

Fronds floating on the surface of the water, several cohere together (often rosette-like), lanceolate to ovate or suborbicular, flat or inflated, underside usually red, with brown pigment cells in the epidermis; veins (3–) 5–16 (–21); dorsal and ventral scales present. Roots 2–21 on each frond. Inflorescences and daughter fronds borne in 2 ventral budding pouches. Inflorescence of 1 female and 2 male flowers enclosed in a membranous sheath. Anthers 2-locular. Carpel with 1–5 ovules. Seeds ovoid, longitudinally ribbed, with transverse striations or smooth with a spongy outer layer.

Free-floating herbs, often forming large colonies and frequently growing with *Lemna* and *Wolffia*.

1. Fronds 1.5–2 times as long as wide, the upper surface with papillae on the median line; veins 3–7; roots 2–7 ***S. punctata***

1. Fronds 1–1.5 times as long as wide, the upper surface without papillae; veins 7–16 (–21); roots 7–21

2. The lobe which covers the point of attachment of the roots rarely perforated by more than 1 root; upper leaf surface usually with a red fleck above the node; turions may develop ***S. polyrhiza***

2. The lobe which covers the point of attachment of the roots perforated by 2–5 roots; upper leaf surface without any red fleck; turions never develop ***S. intermedia***

Spirodela intermedia W. Koch: Fig. 237

Like *S. polyrhiza* but 2–5 roots perforate the lobe which covers the point of attachment of the roots; no red spot on the upper surface of the frond; turions never develop.

{Ple}. Tropical and subtropical S America, introduced in India: <Ind: <Dl (known only in botanic garden in Delhi).

Spirodela polyrhiza (Linnaeus) Schleiden [*'polyrrhiza'* orth. mut., *Lemna orbiculata* Roxburgh, *S. maxima* (Blatter et Hallberg) McCann]: Fig. 237

Fronds ovate to orbicular, 1–1.5 times as long as wide, 1.5–10 mm long, 1.5–8 mm wide, usually rounded at the tip but sometimes pointed; upper leaf

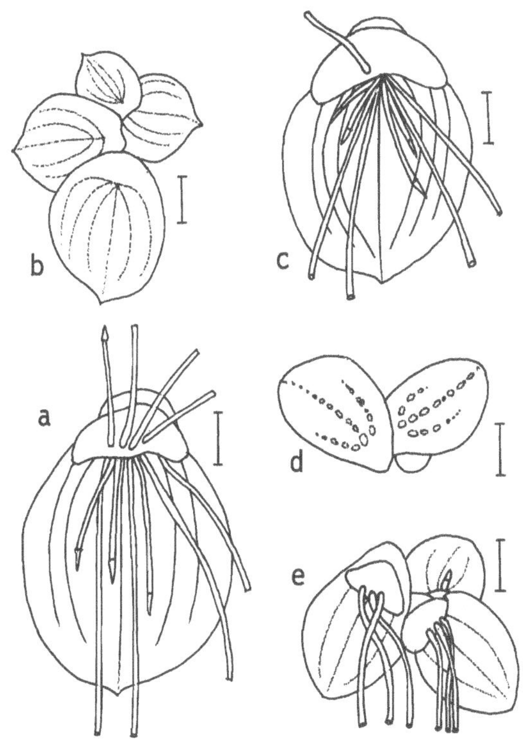

Fig. 237. *Spirodela intermedia*: a, frond viewed from below (1.5 mm). *S. polyrhiza*: b, fronds viewed from above (2.5 mm); c, frond viewed from below (2 mm). *S. punctata*: d, fronds viewed from above (2 mm); e, fronds viewed from below (2 mm).

surface usually with a red fleck above the node; veins 7–16 (–21). Roots 7–21, rarely more than 1 perforates the lobe which covers the point of attachment of the roots. Seeds 0.7–1 mm long, ± 0.7 mm in diameter, with 12–20 longitudinal ribs. Turions may develop.

Perennial. Usually found in eutrophic conditions, mostly in lakes and large tanks (particularly those used for cultivating *Trapa*) or among reeds, often growing together with *Lemna* or *Wolffia*. Plants very rarely flower and fruit. In unfavourable conditions the fronds sometimes develop turions (small orbicular to reniform, brown to olive green fronds without roots which sink), they may remain viable when buried in mud during the dry period.

{Ple}. Almost cosmopolitan: Ban, Bur, Cey, Ind, Nep, Pak: As, Bh, Dl, Hp, Jk, Kl, Mg, Mh, Mp, Pj, Rj, Sk, Tn, Up, Wb.

Spirodela punctata (G. F. W. Meyer) Thompson [*S. oligorhiza* (S. Kurz) Hegelmaier], Fig. 237

Fronds lanceolate to ovate, 1.5–2 times as long as wide, 1.5–8 mm long, 1–5 mm wide, usually pointed at the tip; veins 3–7. Roots 2–7 (–12), all of which perforate the lobe which covers the point of attachment of the roots. Seeds 0.8–1 mm long, 0.5–0.6 mm in diameter, with 10–15 longitudinal ribs.

Perennial. A poorly known species in India. Flowers and fruits occasionally, fronds do not develop turions.

{Ple}. Probably originally a species from the Southern Hemisphere and E Asia, now almost cosmopolitan in warm regions: Ind: Dl, Mh, Mn, Wb.

Wolffia: 10 species: almost cosmopolitan, absent from colder regions

Fronds floating upon or swimming just below the water surface, solitary or 2 remaining attached, inflated, globose to ovoid or sometimes conical or nutshell-shaped, with or without pigment cells in the epidermis; dorsal and ventral scales absent. Roots absent. Daughter fronds borne in 1 funnel-shaped basal pouch with a circular opening. Inflorescence of 1 female and 1 male flower, without a membranous sheath. Anthers 1-locular. Carpel with 1 ovule. Seeds globose, smooth with a spongy outer layer.

Very small, rootless, free-floating or free-swimming herbs. Of no known economic importance but frequently cultivated as a curiosity because it is the smallest flowering plant. Landolt, E., Taxonomy and ecology of the section *Wolffia* of the genus *Wolffia* (Lemnaceae). Berichte des Geobotanischen Institutes der Eidg. Techn. Hochschule, Stiftung Rübel, Zürich 60: 137–151 (1994).

The name *Wolffia arrhiza* has been used rather indiscriminantly in India; in the correct sense it is a species found in the Himalayas and does not extend to Peninsular India. It is possible that *W. angusta* does not occur in India; it has been confused with *W. neglecta*.

1. Fronds subglobose, with an up to 3 mm long tapered projection on the lower surface, the margin toothed **W. microscopica**

1. Fronds ellipsoidal or boat-shaped, without a tapered projection on the lower surface, the margin entire

2. Fronds with their greatest width at the surface of the water, at least 1.5 times as deep as wide; stigma with pigments cells

3. Fronds 2–3 times as deep as wide, whitish-green on the upper surface with bright green margins **W. angusta**

3. Fronds 1.5–2 times as deep as wide, intensely green on the upper surface without green margins **W. neglecta**

2. Fronds with their greatest width just below the surface of the water, 1–1.5 times as deep as wide; stigmas not pigmented

4. Fronds shiny on the upper surface, bright green (not transparent), with (10–) 30 (–100) stomata **W. arrhiza**

4. Fronds not shiny on the upper surface, pale and transparent green, with 1–10 (–15) stomata **W. globosa**

Wolffia angusta E. Landolt [*W. arrhiza* auct. Ind., non (Linnaeus) Horkel, nec Wimmer]: Fig. 238

Fronds floating at or below the surface of the water, boat-shaped, with the greatest width at the surface of the water, 0.5–0.8 mm long, 0.2–0.4 mm wide, 1.7–2.5 times as long as wide, 2–3 times as deep as wide, with 8–10 (–25) stomata, whitish-green at the surface with more intensely green margins. Seeds 0.3–0.4 mm long, ± 0.25 mm in diameter.

Probably perennial but perhaps also annual. Found in tanks and village pools in rather eutrophic water.

{⊙? Ple/Pla}. Tropical and subtropical Australia, Singapore, Malaysia and perhaps West Bengal: Ind?: Wb?

Wolffia arrhiza (Linnaeus) Horkel: Fig. 238

Fronds floating at or just below the surface of the water, spherical to ellipsoid, with greatest width just below the surface of the water, 0.5–1.5 mm long, 0.4–1.2 mm wide, 1–1.3 times as long as wide, 1.25–1.5 times as deep as wide, with 10–100 stomata, bright green at the surface (not transparent). Seeds 0.4–0.5 mm long, ± 0.4 mm in diameter.

{Ple/Pla}. Europe, Africa, W Asia and NW Himalayas, perhaps introduced in Brazil: Ind: Jk (Srinagar).

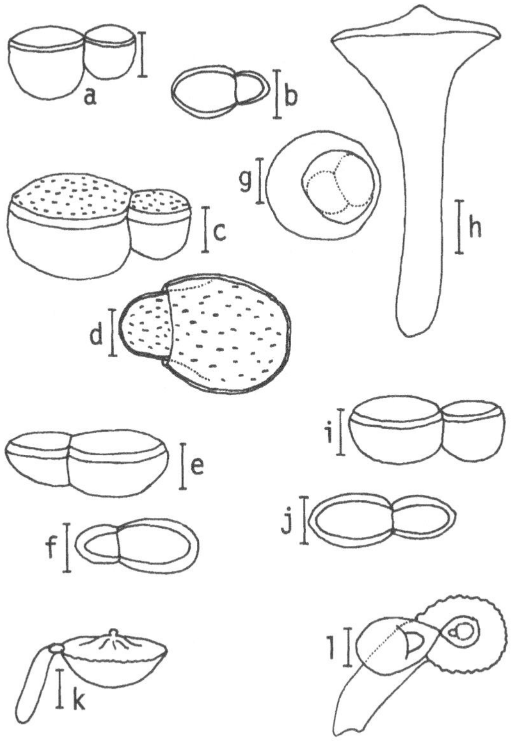

FIG. 238. *Wolffia angusta*: a, frond viewed from the side (0.4 mm); b, frond viewed from above (0.4 mm). *W. arrhiza*: c, frond viewed from the side (0.4 mm); d, frond viewed from above (0.4 mm). *W. globosa*: e, frond viewed from the side (0.4 mm); f, frond viewed from above (0.4 mm). *W. microscopica*: g, frond viewed from above (0.5 mm); h, frond viewed from the side (0.3 mm). *W. neglecta*: i, frond viewed from the side (0.4 mm); j, frond viewed from above (0.4 mm). *Wolffiella hyalina*: k, frond viewed from the side (0.5 mm); l, frond viewed from above (0.5 mm).

Wolffia globosa (Roxburgh) den Hartog et van der Plas [*W. arrhiza* auct. Ind., non (Linnaeus) Horkel, nec Wimmer, *W. delilii* var. *schleideni* S. Kurz]: Fig. 238

Probably perennial. Fronds floating at or just below the surface of the water, ellipsoidal, with greatest width slightly below the surface of the water, 0.4–0.8 mm long, 0.3–0.5 mm wide, 1.3–2 times as long as wide, 1–1.5 times as deep as wide, with 1–10 (–15) stomata, rather pale and transparently green at the surface. Seeds unknown.

{Ple/Pla}. Tropical and warm regions of E Asia and Africa, naturalized in N America: Ban, Bur, Cey, Ind, Nep, Pak: Ap, As, Dl, Gj, Kl, Mp, Pj, Rj, Tn, Up, Wb.

Wolffia microscopica (Griffith) S. Kurz: Fig. 238

Fronds floating on the surface of the water, 0.4–1 mm long, 0.3–0.8 mm wide, 1–1.5 times as long as wide, pale green, the upper surface suborbicular with 5–10 teeth along the margin, with 5–30 stomata, the lower surface tapering downwards into a 0.4–3 mm long and conical to cylindrical (sometimes almost root-like) projection, not developing turions. Seeds ± 0.3 mm long, ± 0.2 mm in diameter.

Probably annual perhaps also sometimes perennial. It flowers and fruits profusely, presumably passing the dry season as seed.

{(⊙)? Ple/Pla}. Endemic to Bangladesh, India and Pakistan, particularly in warm regions with a dry season; N Indian plain: ♦Ban, ♦Ind, ♦Pak: Ap, Dl, Gj, Hr, Tn, Up.

Wolffia neglecta E. Landolt: Fig. 238

Fronds floating at or below the surface of the water, boat-shaped, with the greatest width at the surface of the water, nearly no translucent edge visible from above, 0.6–0.9 mm long, 0.4–0.6 mm wide, 1.5–1.7 times as long as wide, 1.5–2 times as deep as wide, with 8–20 stomata. Stigma with pigment cells. Seeds unknown.

A poorly known species. Probably perennial but perhaps also annual. Found in tanks and village pools in rather eutrophic water.

{⊙? Ple/Pla}. Endemic to the Indian subcontinent: ♦Cey, ♦Ind, ♦Pak: Rj, Wb?

Wolffiella: 9 species

Wolffiella hyalina (Raffeneau-Delile) Monod: Fig. 238

Fronds floating upon or swimming just below the water surface or sometimes just the base in contact with air, solitary or 2 remaining attached, thin, ovate to trapezoid, 1–3 mm long, 0.8–2 mm wide, with or without 1-celled marginal teeth, without pigment cells in the epidermis; nerves absent; dorsal and ventral scales absent. Roots absent. Daughter fronds borne in a deltate and flat pouch. Inflorescence 1 on each frond, without a membranous sheath. Anthers 1-locular. Carpel with 1 ovule. Seeds globose or ovoid, 0.35–0.4 mm long, 0.25–0.3 mm in diameter, smooth with a spongy outer layer.

Perennial or perhaps sometimes annual. Not very gregarious, usually found mixed with other floating plants.

{⊙? Ple/Pla}. Drier regions of Africa, introduced in India: <Ind: <Ap 1969 (Hydrabad).

LENTIBULARIACEAE: 4 genera, 1 with wetland species in India

Utricularia: 214 species: cosmopolitan

Rootless, free-swimming, free-floating or fixed by rhizoids (root-like appendages) or emergent, mostly stoloniferous, sometimes developing tubers or turions (modified buds). Leaves either capillary and usually much branched or entire then linear to spathulate, reniform or peltate. Traps bladder-like, ovate or globose, usually shortly stalked, develop at stolon nodes, leaves or sometimes at the base of the inflorescence stalk. Inflorescence a bracteate, usually solitary, simple or rarely branched raceme. Flowers bisexual, bilaterally symmetrical, usually showy and brightly coloured but sometimes insignificant and cleistogamous. Sepals 2. Petals 5, united at the base forming 2 lips; abaxial (lower) lip spurred or saccate. Stamens 2, attached to the base of the petals; anthers dorsifixed. Ovary superior. Fruit a capsule, 1 to many seeds; seeds very variable in size and shape, sometimes with glochidia (small hooks).

Perennials or annuals. All bear bladder-like animal catching traps. Submerged, planktonic, free-floating with specialized floats, anchored to the substrate by specialized organs or emergent (tenagophytes or helophytes) with laminate leaves. Pollination entomophilous or autogamous, sometimes cleistogamous. The disseminules are seeds or specialized tubers or buds (turions), some are dispersed by water, others by animals and perhaps also by wind. Many are cultivated for decoration or curiosity. Taylor, P. The genus *Utricularia*. Kew Bulletin, Additional Series 14: 1–724 (1989). This work has been used as the basis for the illustrations.

Excluded: the following species are found in the Himalayas or in Myanmar and do not extend into Peninsular India: ***Utricularia brachiata*** Oliver; ***U. christopheri*** P. Taylor: ♦Ind, ♦Nep; ***U. corynephora*** P. Taylor; ***U. forestii*** P. Taylor; ***U. minor*** Linnaeus [*U. rogersiana* Lace, *U. nepalensis* Kitamura]; ***U. furcellata*** Oliver: ♦Ind; ***U. involens*** Ridley; ***U. kumaonensis*** Oliver; ***U. multicaulis*** Oliver; ***U. punctata*** Wallich; ***U. recta*** P. Taylor; ***U. salwinensis*** Handel-Mazzetti.

1. Leaves divided into narrowly linear or capillary segments; submerged aquatics, usually planktonic; traps borne on leaves floating below the surface; bracteoles absent

2. Leaves sparsely forked with rarely more than 6 ultimate segments; adaxial (upper) lip of the petals as large or larger than the abaxial (lower) one and usually 3-lobed ***U. gibba***

2. Leaves many times forked with 10 or many more ultimate segments; adaxial (upper) lip of the petals smaller than the abaxial (lower) one and entire or 2-lobed

3. Bristles on the margins of the ultimate leaf segments and the ultimate rhizoid segments each arising from the apex of a short acute tooth or multicellular swelling; outer surface of the petals glabrous; inflorescence stalk bearing scales below the flowers and without a whorl of inflated organs at or near the base ***U. australis***

3. Bristles on the margins of the ultimate leaf segments and ultimate rhizoid segments absent (terminal only) or arising directly from the flat margin; outer surface of the petals pubescent with simple or glandular hairs; inflorescence stalk without scales but with a whorl of inflated organs at or near the base

4. Petals white with violet veins; sepals forming a flat disk in fruit, completely enclosing the capsule; style in fruit more than 1.5 mm long ***U. inflexa***

4. Petals yellow; sepals spreading and reflexed in fruit, not enclosing the capsule; style in fruit rarely more than 1 mm long

5. Inflated organs on the inflorescence stalk stalked, arising from near the base, fusiform bearing leaf-like or rhizoid-like capillary segments at intervals along their entire length ***U. aurea***

5. Inflated organs on the inflorescence stalk sessile, arising from the distal half, short, ellipsoidal bearing leaf-like capillary segments only from the distal half ***U. stellaris***

1. Leaves entire, linear to obovate or orbicular; emergent aquatics or helophytes, firmly fixed in the soil; traps borne in or near the substrate; bracteoles present or absent

6. Bracts subtending flowers, attached at some point above the base or produced below the point of attachment to the base (basisolute)

7. Leaves peltate; bracts and bracteoles very shortly basisolute; trap mouth with a short abaxial projection fringed (as in the abaxial lip of the trap) with radiating rows of gland-tipped trichomes ***U. pubescens***

7. Leaves not peltate; bracts at least distinctly basisolute; trap mouth without a short adaxial projection and without radiating rows of gland-tipped trichomes

8. Bracteoles absent; leaves narrowly linear, not differentiated into petiole and blade ***U. subulata***

8. Bracteoles present, resembling bracts; leaves usually petiolate with leaf blade obovate to transversely elliptic

9. Sepals slightly unequal, almost alike; abaxial (lower) petal lip entire; traps with a funnel-shaped rim, the adaxial side produced into a beak; seeds without glochidia ***U. caerulea***

9. Sepals very dissimilar, adaxial sepal much larger than the abaxial (lower); abaxial petal lip 5-lobed; traps without a rim, without an adaxial beak; seeds with glochidia

10. Adaxial petal lip longer than the adaxial sepal; plant developing ellipsoid to fusiform tubers along the stolons like a string of beads ***U. moliniformis***

10. Adaxial petal lip shorter than the adaxial sepal; plant without tubers *U. striatula*

6. Bracts subtending flowers attached by the base (basifixed)

11. Trap mouth without 2 adaxial appendages; bracteoles as long as the bracts

12. Sepals densely covered with long to short multicellular hairs *U. hirta*

12. Sepals glabrous

13. Raceme axis bearing bracts which subtend dormant or aborted buds, in addition to those which subtend flowers; trap mouth with a short obtuse adaxial projection, naked or fringed with simple or gland-tipped trichomes *U. minutissima*

13. Raceme axis bearing only bracts which subtend flowers; traps with a single adaxial appendage and with abaxial or lateral appendages or both *U. arenaria*

11. Trap mouth with 2 adaxial, subulate or narrowly deltoid, simple or rarely branched appendages; bracteoles usually smaller than the bract sect. Oligocista

14. Petals yellow

15. Adaxial sepal broadly ovate, obtuse at apex; inflorescence stalk not twining; pedicels deflexing in fruit; capsule wall of uniform thickness *U. bifida*

15. Adaxial sepal ovate, acute to acuminate at apex; inflorescence stalk twining; pedicels remaining erect in fruit; capsule wall thickened on either side of the line of dehiscence *U. scandens*

14. Petals violet, mauve, blue, pink or white

16. Inflorescence stalk bearing very numerous, relatively large and conspicuous scales, scale-like below and bract-like towards the inflorescence *U. wightiana*

16. Inflorescence stalk without scales or if present then few and scale-like and not bract-like

17. Capsule wall thickened either side of the line of dehiscence

18. Petals very small, 3–4 mm long; traps with branched (antler-like) adaxial appendages (sepals very accrescent and pedicels sharply deflexed in fruit) *U. foveolata*

18. Petals larger, more than 6 mm long; traps with simple, unbranched adaxial appendages

19. Leaves 1-nerved

20. Inflorescence stalk erect, not twining; inflorescence mostly 5–20 cm long; petals 7–10 mm long *U. polygaloides*

20. Inflorescence stalk twining; inflorescence mostly 20–100 cm long; petals 15–20 mm long *U. reticulata*

19. Leaves at least 3-nerved

21. Adaxial petal lip shorter than the adaxial sepal; abaxial (lower) petal lip with 2, obtuse, tooth-like projections on the distal margin of the palate; spur strongly curved, very slender (± 1 mm wide at the middle), not much widened at the base, about as long as and ± parallel with the abaxial lip *U. arcuata*

21. Adaxial petal lip longer than the adaxial sepal; abaxial (lower) petal lip without tooth-like projections on the palate; spur straight or slightly curved, distinctly widened at the base, longer than and widely diverging from the abaxial lip *U. graminifolia*

17. Capsule wall of uniform thickness

22. Seeds globose, seed coat cells isodiametric

23. Inflorescence stalk twining; petals 12–15 mm long; distal margin of the fruiting sepals entire (under the microscope) *U. smithiana*

23. Inflorescence stalk erect, not twining; petals 5–8 mm long; distal margin of the fruiting sepals minutely toothed (× 20 magnification needed) *U. uliginosa*

22. Seeds ellipsoid or ovoid, seed coat cells elongate

24. Petals blue and white; swelling at the base of the abaxial (lower) petal lip 2-lobed

25. Petals 12–15 mm long; adaxial petal lip circular, wider than the adaxial sepal; periclinal wall of the seed coat cells smooth *U. albocaerulea*

25. Petals 8–10 mm long; adaxial petal lip oblong, narrower than the adaxial sepal; periclinal wall of the seed coat cells rough (minutely warty) *U. lazulina*

24. Petals violet or mauve, never truly blue; swelling at the base of the abaxial (lower) petal lip rounded, entire or obscurely 3-lobed

26. Petals 10–15 mm long; adaxial petal lip broadly obovate, wider than the adaxial sepal *U. cecilii*

26. Petals 5–8 mm long; adaxial petal lip oblong, narrower than the adaxial sepal *U. praeterita*

Utricularia albocaerulea Dalzell: Fig. 239

Like *U. reticulata* but leaves few, arising from inflorescence stalk base and from the stolons, up to 0.5 cm long, 3-nerved, the lamina obovate, ± 1 mm wide, apex rounded. Inflorescence stalk erect, not twining, 8–15 cm long. Bracts ± 1.5 mm long. Bracteoles much shorter and very much narrower than the bract. Flowers 2–5. Petals 12–15 mm long, pale blue with a white adaxial lip; adaxial petal lip circular, wider than the adaxial sepal; abaxial (lower) petal lip with a 2-lobed swelling at the base; palate with a raised hairy rim; spur subulate, slightly curved, ± 6 mm long, very much shorter than and held at ± 90 to the abaxial (lower) lip. Capsule broadly ellipsoid, dorsiventrally flattened, ± 3 mm long, completely concealed by the sepals.

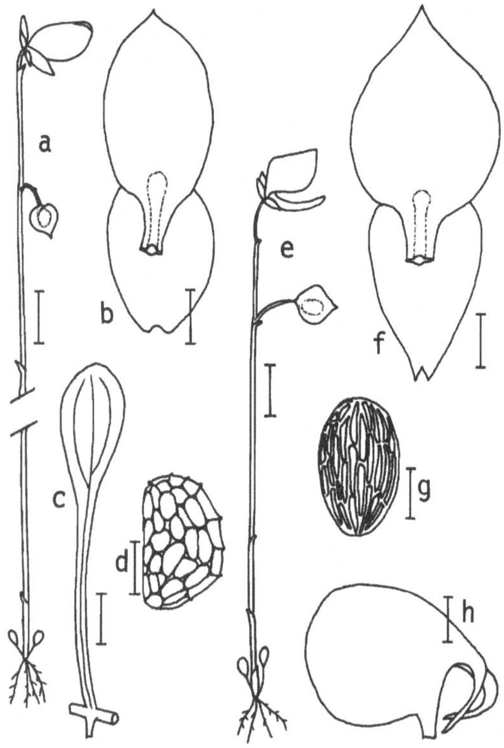

FIG. 239. *Utricularia albocaerulea*: a, flowering plant with inflorescence stalk shortened (1 cm); b, sepals at anthesis (1 mm); c, leaf (1 mm); d, seed (0.1 mm). *U. arcuata*: e, flowering plant (1 cm); f, sepals at anthesis (1 mm); g, seed (0.1 mm); h, trap (0.5 mm).

Seeds obliquely ovoid, 0.3–0.35 mm long, seed coat cells elongate, with the anticlinal boundaries much raised, ± straight, with distinct, triangular intercellular spaces, the periclinal walls concave, smooth.

Probably annual. On wet rocks and damp soil. Very similar to *U. lazulina* but adaxial petal lip wider and seeds different.

{⊙? Hel/Ten}. Endemic to India: ♦Ind: Mh. Low to medium altitudes.

Utricularia arcuata Wight [*U. ogmosperma* Blatter et McCann]: Fig. 239

Like *U. graminifolia* but rhizoids few, branching, basally thickened. Stolons few, branched, capillary. Leaves few, very small, up to 5 mm long, 3-nerved, petiolate, the lamina obovate, 0.5–1.5 mm wide, apex rounded to subacute. Inflorescence erect, 6–20 cm long. Bracts deltoid, ± 2 mm long. Flowers 2–6; pedicels ascending at anthesis, ± spreading or decurved in fruit, up to 17 mm long. Sepals subequal, accrescent, 4–6 mm long. Petals 12–15 mm long, blue-violet, with a white patch at the base of the abaxial (lower) lip; adaxial lip not or not much longer than the adaxial sepal; palate with a raised hairy rim and with a prominent tooth-like projection at either side of its distal margin; spur subulate, strongly curved, very slender (± 1 mm wide at the middle), about as long and ± parallel with the abaxial (lower) lip, apex acute. Capsule ± 2.5 mm long, completely concealed by the accrescent sepals. Seeds narrowly ellipsoid or ovoid, ± 0.3–0.35 mm long; seed coat cells very elongate and relatively narrow, the anticlinal boundaries ± straight, raised, smooth, the periclinal walls tabular, smooth or rough, with a few transverse wrinkles.

Perennial and perhaps sometimes annual. In seasonally flooded areas and wet grassland.

{⊙? Ten}. Endemic to India: ♦Ind: Go, Kt, Kl? Mh, Or? Low to medium altitudes.

Utricularia arenaria A. P. de Candolle: Fig. 240

Rhizoids arising from the base of the inflorescence stalk, numerous, simple, capillary. Stolons few, sparsely branched, capillary. Leaves arising from the base of the inflorescence stalk and from stolons, few, 0.2–1.5 cm long, 1-nerved, petiolate, the lamina obovate to nar-

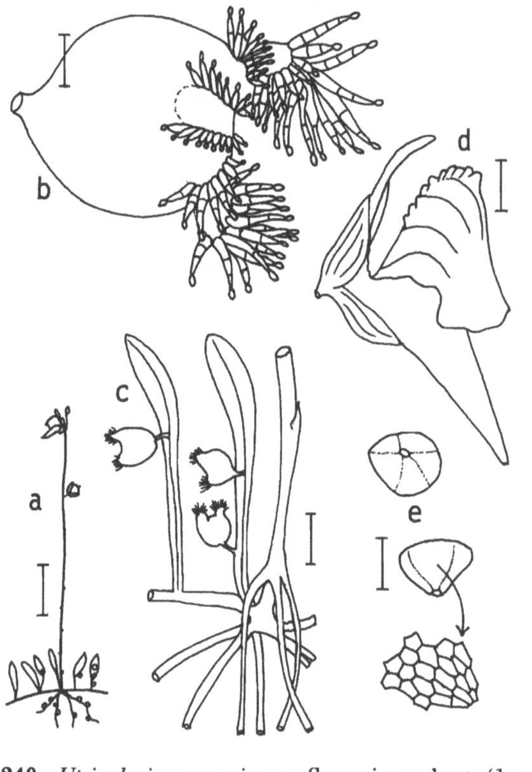

FIG. 240. *Utricularia arenaria*: a, flowering plant (1 cm); b, trap (0.1 mm); c, base of plant with stolons, rhizoids, leaves, traps and base of inflorescence stalk (0.13 mm); d, flower (0.13 mm); e, seed viewed from above and from the side (0.25 mm) with surface pattern (inset).

rowly obovate, 0.3–2 mm wide. Traps ovoid, 0.6–1 mm long, the mouth terminal, with a dorsoventrally flattened, rounded, adaxial appendage and a similar but much shorter abaxial appendage, both with radiating comb-like rows of stalked glands. Inflorescence stalk erect, solitary, 2–20 cm long, glabrous above, usually papillose below. Bracts attached by the base, ovate, ± 1 mm long, apex acute, some subtending dormant or aborted buds. Bracteoles narrowly oblong, ± as long the bract. Flowers 1–5 (–8); pedicels erect, filiform, terete, 0.5–1 mm long. Sepals subequal, 1–2 mm long, glabrous; adaxial sepal broadly ovate, acute apex; abaxial (lower) sepal slightly narrower, apex obtuse or rounded. Petals 3–7 mm long, violet, lilac or white; adaxial lip yellowish, constricted above the middle, with the inferior part deltoid and the superior part narrower, oblong, apex truncate or emarginate; abaxial (lower) lip with a yellow blotch at the base, ± circular, with 2, longitudinal, slightly raised, smooth or transversely wrinkled crests, apex rounded, entire or crenate; palate margin papillose; spur narrowly conical-subulate, straight or curved, ± 2 times as long as the abaxial (lower) lip. Filaments curved, ± 1 mm long; anther thecae almost distinct. Capsule globose, 1.5–2.5 mm long, dehiscing by adaxial and abaxial, longitudinal, marginally thickened slits. Seeds obovoid-conical, slightly angular, ± 0.25 mm long.

In seasonally flooded areas, mostly in shallow sandy or peaty soil but also in marshes and swampy grassland. {⊙ Hel/Ten}. Africa, with 1 record in India: Ind: Mp (Pachmarhi).

Utricularia aurea Loureiro [*U. flexuosa* Vahl, *U. fasciculata* Roxburgh, *U. confervifolia* Jackson ex D. Don]: Fig. 241

Rhizoids usually present at the base of the inflorescence stalk, either short and linear or elongated and inflated, up to 10 cm long, bearing leaf-like branches. Stolons filiform, terete, branched, papillose, up to 1 m long, 0.5–2 mm thick, sometimes bearing capillary air shoots. Leaves very numerous, 1–8 cm long, divided from the base into 3–5 primary filiform or somewhat inflated segments, the secondary segments pinnate, each pinna forked from the base into numerous further segments, the ultimate segments capillary, usually with some bristles; stipule-like auricles usually present at the base of the primary segments. Traps dimorphic, mostly lateral on the secondary or tertiary leaf segments, obliquely ovoid, 1–4 mm long, the mouth lateral, naked or with 2 sparsely branched bristle-like appendages; others sparsely clustered at the base of primary segments, subglobose, the mouth basal, usually without oral appendages. Inflorescence stalk erect, emergent, 1–20 (–30) cm long, terete, glabrous. Scales absent. Bracts basifixed, broadly ovate to circular, apex rounded, 1–2 mm long, obscurely nerved. Bracteoles absent. Flowers 5–10; the axis initially short, becoming elongated in fruit; pedicels dorsoventrally compressed, 0.4–2 cm long, erect at anthesis, usually sharply deflexed and thickened, especially distally, in fruit. Sepals subequal (abaxial sepal often slightly broader), ovate, apex rounded to subacute, 2–3 mm long at anthesis, becoming ± fleshy and much larger, up to 9 mm long and spreading or reflexed in fruit. Petals pale yellow with reddish-brown veins, 1–1.5 cm long, glabrous to densely pubescent; adaxial lip broadly ovate, apex rounded; abaxial (lower) lip limb transversely elliptical, the base with a prominent, 2-lobed swelling, apex rounded or emarginate; palate pubescent; spur cylindrical from a narrowly conical base, often somewhat constricted near the middle, apex obtuse or subacute, about as long and parallel with the abaxial (lower) lip. Filaments curved, 1–1.5 mm long; anther thecae ± confluent. Capsule globose, up to 5 mm in diameter, the wall relatively thick and fleshy, circumscissile, the persistent style greatly enlarged, ± as long or longer than the

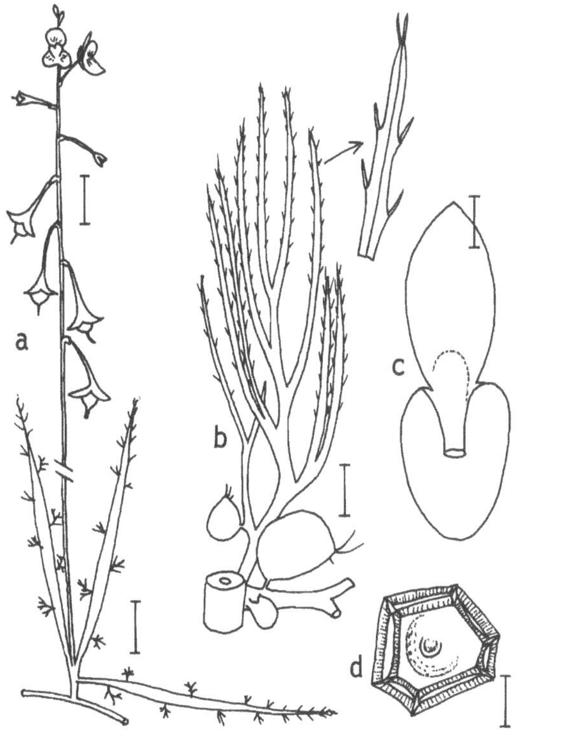

FIG. 241. *Utricularia aurea*: a, plant with floats and inflorescence stalk shortened (1 cm); b, secondary leaf segment with large and small traps, with enlarged tip of a leaf segment (inset) (1 mm); c, sepals at anthesis (1 mm); d, seed viewed from above (0.4 mm).

capsule. Seeds disk-shaped, ± 5 angled, 1.5–2 mm in diameter, narrowly winged on all angles.

Perennial perhaps sometimes annual. Planktonic or anchored by rhizoids or stolons. In tanks, pools, backwaters, rice fields, ditches and canals. Usually found growing entangled in other aquatic plants. Flowering throughout the year but reaching a peak after the rains. It is considered to be a pest in pools used for fish culture.

{⊙? Pla}. S and E Asia and Australia: Ban, Bur, Cey, Ind, Nep, Pak: Ap, As, Bh, Dl, Go, Jk, Kl, Kt, Mg, Mh, Mp, Or, Pj, Rj, Tn, Wb.

Utricularia australis R. Brown [*U. vulgaris* auct. Ind., non Linnaeus]: Fig. 242

Rhizoids usually present, few, filiform, bearing numerous short, forked branches with narrowly ovoid, papillose ultimate segments bearing apical bristles. Stolons filiform, branched, terete, glabrous, up to 50 cm long, 0.3–1 mm thick. Leaves very numerous, 1.5–4 cm long, divided from the base into 2 ± equal, primary segments, each ± pinnately divided, the secondary segments forked into further segments, the ultimate segments capillary, somewhat flattened and bearing minute bristles, the lateral bristles each arising from the apex of a short acute tooth. Traps dimorphic, arising laterally from secondary to penultimate leaf segments, also at base of the primary segments; the lateral traps ovoid, 0.5–2.5 mm long, the mouth lateral, with 2 adaxial, simple or sparsely branched bristle-like appendages and usually with the basal traps ovoid, the mouth basal, naked or with 2 very short bristle-like appendages. Inflorescence stalk weakly erect, emergent, 10–30 (–100) cm long, filiform, terete, glabrous, 1–2 mm thick, at first straight later becoming flexuous. Scales 1 (–3), always present on the adaxial half of the inflorescence stalk, similar to the bracts. Bracts basifixed, ± circular, the base auriculate, apex rounded or obscurely 3-toothed, 3–5 mm long, many-nerved. Bracteoles absent. Flowers 4–10; the axis initially short, becoming elongated in fruit; pedicels filiform, terete, 1.5–3 cm long, erect at anthesis, decurved in fruit. Sepals subequal, ovate, 3–4 mm long; adaxial sepal with rounded apex; abaxial (lower) sepal emarginate. Petals yellow with the basal swollen part of the abaxial (lower) lip much darker and with reddish-brown lines and spots, 1–2.2 cm long; adaxial lip very broadly ovate, apex retuse; abaxial lip limb reniform or transversely elliptical, the base with a very prominent swelling, the distal part ± flat, up to 1.8 cm wide, apex rounded or retuse; palate glabrous; spur broadly conical, apex obtuse, slightly curved, considerably shorter than the abaxial (lower) lip. Filaments curved, ± 2 mm long; anther thecae distinct. Capsule extremely rare (never seen in India), globose, up to 4 mm in diameter, circumscissile. Seeds prismatic, 4–6 angled, ± 0.5 mm long, narrowly winged on all angles.

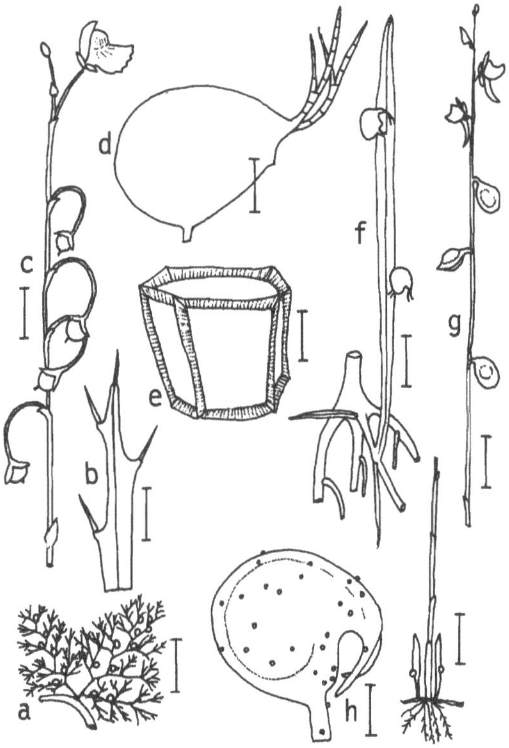

FIG. 242. *Utricularia australis*: a, leaf (1 cm); b, tip of ultimate leaf segment (0.2 mm); c, inflorescence (1 cm); d, trap (0.4 mm); e, seed (0.3 mm). *U. bifida*: f, base of plant with stolons, rhizoids, leaf, traps and inflorescence stalk base (1 mm); g, whole plant with inflorescence stalk shortened (1 cm); h, trap (0.2 mm).

Perennial. Planktonic. In lakes, pools, river backwaters and ditches. It grows in both still and slowly flowing water. Apparently remaining sterile in India.

{Pla}. Widely distributed in the Old World (Europe, Africa, Asia, Australasia) in temperate and tropical regions: Bur, Cey, Ind, Nep, Pak: Hp, Jk, Kt, Mg, Mp, Rj, Tn, Up.

Utricularia bifida Linnaeus [*U. humilis* Vahl, *U. ramosa* Vahl pro parte, *U. antirrhinoides* Wallich, *U. alata* L. Benjamin, *U. wallichiana* L. Benjamin, non Wight]: Fig. 242

Rhizoids numerous, branching, capillary. Stolons few, branched, capillary. Leaves arising from stolon nodes, up to 2 cm long, 1-nerved, petiolate, the lamina narrowly linear, 0.5–1 mm wide, apex rounded or subacute. Traps ovoid, 0.6–1 mm long, the mouth basal, with 2 simple, subulate, reflexed, adaxial appendages and with a rounded swelling on the abaxial side of the mouth or on the adjacent stalk. Inflorescence stalk erect, 3–20 cm long, glabrous. Scales few, like bracts. Bracts attached by the base, ovate, 1–3 mm long, 1- to 5-nerved, apex obtuse to acute. Bracteoles subulate, much shorter than the bract. Flowers 1–10; pedicels spreading at anthesis, decurved in fruit, capillary, broadly winged, 2–5 mm long. Sepals slightly unequal, 2.5–7 mm long, broadly ovate, glabrous; adaxial sepal with obtuse apex; abaxial (lower) sepal slightly smaller, apex rounded or very shortly 2-lobed. Petals 6–10 mm long, yellow; adaxial lip slightly constricted below the middle, the superior part oblong or oblong-ovate, apex rounded, the inferior part broadly ovate-deltoid; abaxial (lower) lip ± circular, the base with a prominent rounded swelling, apical margin rounded; palate margin hairy; spur subulate, curved, ± as long as and diverging from the abaxial (lower) lip. Filaments straight, ±1 mm long; anther thecae distinct. Capsule broadly ellipsoid, dorsoventrally flattened, 2.5–3 mm long, dehiscing by adaxial and abaxial, longitudinal slits. Seeds obliquely obovoid, 0.4–0.5 mm long.

Small annual. In seasonally flooded places, wet soil between rocks and in rice fields.

{⊙ Hel/Ten}. India to Japan and Australia: Ban, Bur, Cey, Ind, Nep: Ap, As, Bh, Kl, Kt, Mg, Mh, Mp, Or, Tn, Wb.

Utricularia caerulea Linnaeus ['*coerula*' orth. mut., *U. nivea* Vahl, *U. ramosa* Vahl pro parte, *U. capillaris* D. Don, *U. racemosa* Wallich pro parte, *U. filicaulis* Wallich, *U. rosea* Edgeworth, *U. paucifolia* L. Benjamin, *U. albiflora* Griffith, *U. roseopurpurea* Stapf ex Gamble, *U. sampathii* Subramanyam et Yoganarasimhan]: Fig. 243

Rhizoids arising from the base of the inflorescence stalk, numerous, branched, capillary. Stolons branched, capillary. Leaves arising from base of inflorescence stalk and from stolons, often in rosettes, up to 1 cm long, 1-nerved, petiolate, the lamina narrowly obovate, up to 2 mm wide. Traps ovoid, dimorphic, the larger ones 1–1.5 mm long, the smaller ones 0.25–0.5 mm long, the mouth lateral, surrounded by an oblique funnel-shaped rim, the margin and inner surface of which is covered with radiating rows of long-stalked glands; the adaxial side of the rim of the large traps produced into a long, narrowly deltoid beak, reaching 2 times as long as the body of the trap, the beak bearing long-stalked glands. Inflorescence stalk erect, (2–) 5–30 (–50) cm long, glabrous above, usually papillose below. Bracts basisolute, attached just above the middle, ± 1 mm long, superior part ovate with apex acute, the inferior part narrower, apex very acute. Bracteoles ± basisolute, attached very near the base, superior part like superior part of bract. Flowers 1–20 or more, distant to congested; pedicels erect at anthesis, spreading to sharply deflexed in fruit, filiform, terete, 0.5–1 mm long. Sepals unequal, 1.5–3 mm long, glabrous; adaxial sepal ovate-oblong, apex rounded; abaxial (lower) sepal broader, transversely elliptical, apex retuse. Petals 2–8 (–12) mm long, white or yellowish to pink or pale to dark violet marked with yellow at the base of the abaxial (lower) lip; adaxial lip constricted near the middle, with the superior part oblong or ovate-oblong, apex retuse; inferior part broadly deltoid; abaxial (lower) lip broadly ovate, with a basal, usually 4-ridged swelling, apex rounded, entire or emarginate; palate pubescent with a distal, transverse crest, the crest terminating at each end in an acute conical process; spur narrowly conical-subulate, often constricted towards the apex, usually considerably, rarely shorter than the abaxial (lower) lip. Filaments straight, 1–2 mm long; anther thecae almost distinct. Capsule globose or ellipsoid, 1.5–2 mm long, dehiscing by a short, abaxial, proximal, marginally thickened slit. Seeds obovoid, 0.2–0.25 mm long.

Annual and perhaps also perennial in permanently wet habitats. Seasonally flooded places, between rocks,

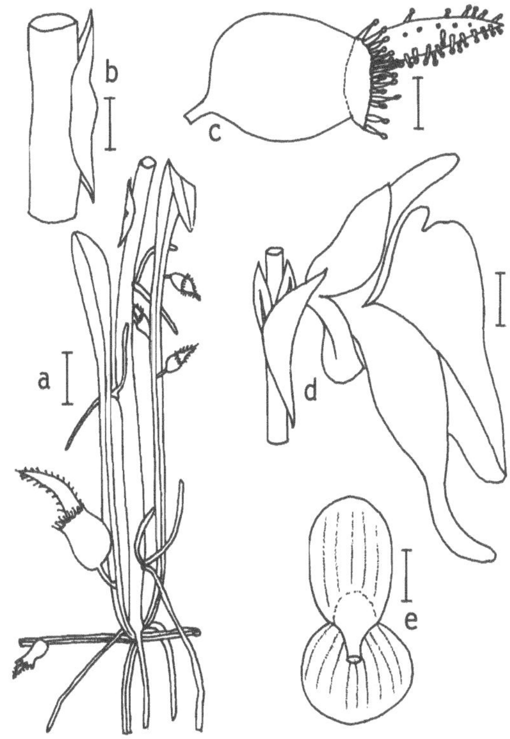

FIG. 243. *Utricularia caerulea*: a, base of plant with stolons, rhizoids, leaves, traps and base of inflorescence stalk (1 mm); b, scale (0.8 mm); c, trap (0.4 mm); d, flower with bract and bracteoles (1 mm); e, sepals (1 mm).

in wet grassland and along stream sides. Although very variable it is distinguished from all other species by the combination of distinctly basisolute (attached just above the middle) bracts, the presence of bracteoles and unequal sepals.

{(☉) Hel/Ten}. Madagascar, W India to Japan and Australia: Ban, Bur, Cey, Ind, Nep: Ap, As, Bh, Gj, Go, Kl, Kt, Mg, Mh, Mp, Or, Tn, Wb.

Utricularia cecilii P. Taylor: Fig. 244

Like *U. reticulata* but leaves up to 1 cm long, 3-nerved, the lamina narrowly obovate, ± 1 mm wide, apex rounded. Inflorescence stalk erect, 5–17 cm long. Bracts ± 1.5 mm long. Bracteoles shorter than the bract. Flowers 1–5; pedicels erect at anthesis, spreading in fruit, 2–5 mm long. Sepals slightly unequal, 5–10 mm long; adaxial sepal broadly ovate, 3–5 mm long, apex acute; abaxial (lower) sepal slightly shorter. Petals 10–15 mm long; palate with a hairy rim; spur ± as long and held at an obtuse angle to the abaxial lip. Capsule ± 3 mm long, uniformally membranous, dehiscing by a single abaxial, longitudinal slit. Seeds obovoid, ± 0.25 mm long, seed coat cells slightly elongate with the anticlinal boundaries much raised, straight, with conspicuous intercellular spaces, the periclinal walls concave, smooth.

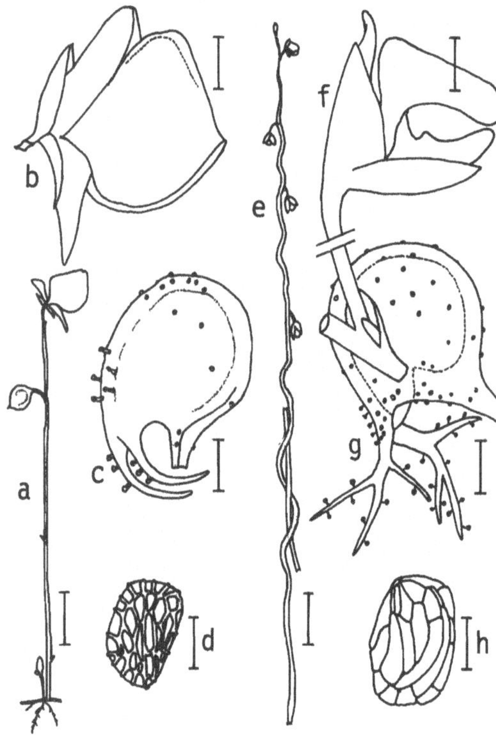

Fig. 244. *Utricularia cecilii*: a, flowering plant (1 cm); b, flower (2 mm); c, trap (0.2 mm); d, seed (0.1 mm). *U. foveolata*: e, flowering plant (1 cm); f, flower with peduncle shortened (0.8 mm); g, trap (0.3 mm); h, seed (0.1 mm).

Seasonally flooded places on laterite. It is distinguished from *U. uliginosa* by larger petals, entire sepal margins and different seeds.

{☉ Hel/Ten}. ♦Ind: Kt (Mangalore).

Utricularia foveolata Edgeworth: Fig. 244

Rhizoids few, branching, capillary. Stolons few, branched, capillary. Leaves few, up to 3 cm long, 1-nerved, petiolate, the lamina linear, 0.4–1 mm wide, apex acute or rounded. Traps few, on stolon nodes and leaves, shortly stalked or subsessile, globose, 0.8–1.2 mm long, the mouth basal, with 2 sparsely branched, antler-like, adaxial appendages bearing stalked glands. Inflorescence twining, up to 20 cm long, glabrous. Scales few, like bracts. Bracts attached by the base, ovate or ovate-oblong, ± 1.2 mm long, apex shortly acuminate to obtuse or truncate. Bracteoles linear-subulate, ± half as long as the bract. Flowers 2–5; pedicels erect or spreading at anthesis, sharply deflexed in fruit, filiform, flattened and narrowly winged, 2–4 mm long. Sepals subequal, 2–4 mm long, ovate, glabrous; adaxial sepal with obtuse or subacute apex; abaxial (lower) sepal slightly longer and narrower, apex obtuse or subacute. Petals 3–4 mm long, mauve; adaxial lip oblong, apex truncate, slightly longer than the sepals; abaxial (lower) lip ± circular, the basal swelling obscurely 2-lobed, apex entire or obscurely 3-toothed; palate with a scarcely raised, minutely hairy rim; spur narrowly conical, longer than and held at ±90 to the abaxial lip, apex obtuse. Filaments straight, ±1 mm long; anther thecae subdistinct. Capsule broadly ovoid, slightly dorsoventrally flattened, ± mm long, dehiscing by a single, abaxial, longitudinal, marginally thickened slit. Seeds obliquely ovoid or ellipsoid, ± 0.3 mm long; seed coat cells elongate, the anticlinal boundaries straight or sinuate, raised, smooth, the periclinal walls at first tabular, almost smooth or obscurely striate, the outer wall is shed to expose a concave, very conspicuously and deeply longitudinally striate surface beneath.

Probably annual, perhaps sometimes perennial. In shallow water, on wet soil, in seasonally flooded habitats, in rice fields and in wet grassland. A distinctive species recognized by the following character combination: twining inflorescences; petals very small (3–4 mm long) and mauve; traps with branched (antler-like) adaxial appendages; pedicels sharply deflexed in fruit.

{(☉)? Ten/Hel}. Tropical Old World: Ban, Ind: Bh, Kt, Mp, Or, Tn.

Utricularia gibba Linnaeus [*U. exoleta* R. Brown, *U. biflora* Roxburgh non Lamarck, *U. roxburghii* Sprengel, *U. khasiana* Joseph et Mani], Fig. 245

Rhizoids absent or filiform, sometimes stolon-like. Stolons usually numerous, much branched and often mat-forming, filiform, terete, glabrous, up to 20 cm long or longer. Leaves numerous, 0.5–1.5 cm long, ± uniform, the primary segments (1) 2, sparsely forked into up to 4 (rarely 8) ultimate segments or sometimes 1 branch of each fork replaced by a trap, the ultimate segments capillary, usually with some bristles; stipule-like auricles absent. Traps lateral on the primary to penultimate segments, ovoid, 1–1.25 mm long, the mouth lateral with 2 adaxial and bristle-like branched appendages. Inflorescence stalk erect, emergent (or sometimes submerged with cleistogamous flowers), 1–20 (–30) cm long, solitary or sometimes 2 or more together or developing in succession, filiform. Scales 1 or sometimes 2, at or near the middle of the stalk, similar to the bracts or frequently absent. Bracts basifixed, semi-amplexicaul, ± semicircular, apex rounded, truncate or obscurely toothed, ± 1 mm long, minutely glandular, obscurely nerved. Bracteoles absent. Flowers (1–) 2–6 (–12); pedicels erect or spreading, filiform, terete, 0.2–3 cm long. Sepals subequal, broadly ovate to circular, 1–3 mm long, apex rounded. Petals yellow, often with reddish-brown veins, 4–25 mm long; adaxial lip broadly ovate to almost circular, somewhat 3-lobed; abaxial (lower) lip smaller, circular to transversely elliptical, the base with a prominent, rounded, 2-lobed swelling, apex rounded; palate densely pubescent with simple 1-celled hairs; spur narrowly cylindrical from a conical base, apex obtuse or ± obscurely bifid, shorter than the abaxial lip. Filaments curved, 1–1.5 mm long; anther thecae confluent. Capsule globose, 2–3 mm in diameter. Seeds thickly lenticular, 0.8–1 mm in diameter, with a relatively broad wing surrounding the seed.

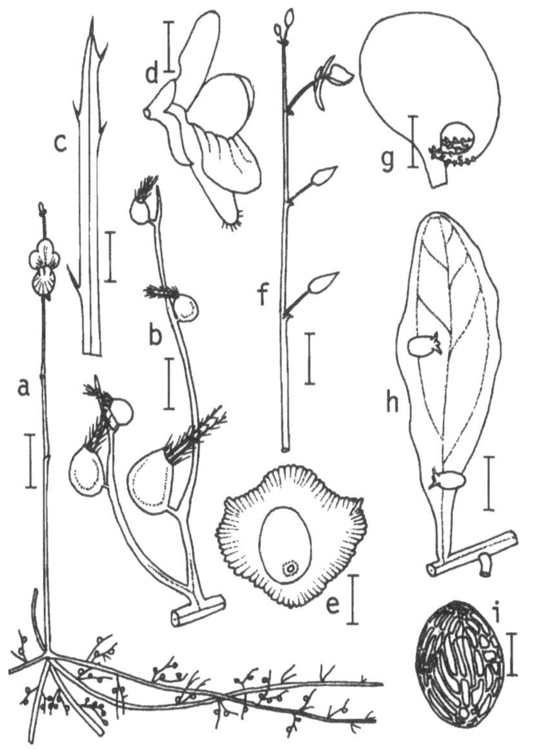

FIG. 245. *Utricularia gibba*: a, plant (1 cm); b, leaves with traps (1.3 mm); c, ultimate leaf segments (0.2 mm); d, flower (1.3 mm); e, seed (0.3 mm). *U. graminifolia*: f, inflorescence (1 cm); g, trap (0.4 mm); h, leaf with traps and part of stolon and rhizoid (1 mm); i, seed (0.1 mm).

Annual or perennial. Affixed or planktonic, in humid regions sometimes found growing terrestrially. Found in a very wide spectrum of aquatic habitats. Often very common but usually found growing intertwined between other plants. Frequent in rice fields but rarely considered harmful.

{(⊙) Pla/Ten}. Almost cosmopolitan in warm to tropical regions: Ban, Bur, Cey, Ind, Nep: An, Ap, Kl, Kt, Mg, Mh, Mp, Rj, Up, Wb.

Utricularia graminifolia Vahl [*U. parviflora* F. Hamilton, *U. purpurescens* J. Graham, *U. acuta* L. Benjamin, *U. conferta* Wight, *U. uliginoides* ('*uligenoides*' orth. mut.) Wight, *U. pedicellata* Wight, *U. suberecta* Lace, *U. equiseticaulis* Blatter et McCann, *U. caerulea* var. *graminifolia* (Vahl) Bhattacharyya]: Fig. 245

Rhizoids numerous, branching, capillary. Stolons very numerous, branched, capillary, mat-forming. Leaves usually very numerous, up to 4 cm or more long, 3- or more-nerved, petiolate, the lamina linear to narrowly obovate, up to 6 mm wide, apex rounded to subacute. Traps globose, 0.5–2 mm long, the mouth basal, with 2 simple, subulate, adaxial appendages bearing sessile and stalked glands. Inflorescence erect or sometimes twining, 4–30 cm long, glabrous. Scales few to many, like bracts. Bracts attached by the base, ovate-deltoid, ± 1 mm long, apex acute. Bracteoles subulate, much narrower and shorter than the bract. Flowers 2–6 (–10); pedicels ascending at anthesis and in fruit, filiform, narrowly winged, (3–) 5–10 mm long. Sepals slightly unequal, 2.5–6 mm long, narrowly ovate, glabrous; adaxial sepal with a very acute apex; abaxial (lower) sepal slightly smaller, apex minutely 2-toothed. Petals 7–15 mm long, various shades of mauve and violet; adaxial lip very slightly constricted below the middle, the superior part ± circular, apex rounded, the inferior part quadrate or broadly deltoid, longer than the adaxial sepal; abaxial (lower) lip ± circular, the base with a conspicuous swelling, apex rounded; palate with a raised hairy margin, without tooth-like projections; spur subulate, straight or slightly curved, distinctly widened at the base, longer than and widely diverging from the abaxial lip, apex acute. Filaments straight, ± 1.5 mm long; anther thecae distinct. Capsule ellipsoid, slightly dorsoventrally flattened, ± 2.5 mm long, dehiscing by a single, abaxial, longitudinal, marginally thickened slit. Seeds ovoid, ± 0.3–0.6 mm long; seed coat cells ± elongate, 2–6 times as long as wide, the anticlinal boundaries ± straight, raised, smooth, the periclinal walls concave, smooth.

Perennial or perhaps sometimes annual. In shallow water, in marshes, swamps, by springs, along streams and on wet soil among rocks.

{⊙? Hel/Ten}. S and SE Asia: Bur, Cey, Ind: Go, Kl, Kt, Mh, Mp, Or, Tn. 0–1500 m.

Utricularia hirta Klein ex Link [*U. tayloriana* Joseph et Mani]: Fig. 246

Rhizoids arising from the base of the inflorescence stalk, numerous, simple, capillary. Stolons few, sparsely branched, capillary. Leaves arising from stolons, few, 0.8–1.5 cm long, 1-nerved, petiolate, the lamina narrowly obovate, 0.3–0.5 mm wide. Traps long-stalked, ovoid, 0.2–0.25 mm long, the mouth lateral with 1 adaxial, subulate, multicelled appendage and an abaxial and sometimes lateral pair of rows of ± united gland-tipped trichomes. Inflorescence stalk erect, (3–) 5–15 (–30) cm long, densely covered with multicellular hairs 0.1–1.5 mm long. Bracts attached by the base, narrowly ovate-deltoid, apex acute, 0.5–1 mm long, densely hairy, each bract subtending a flower with another subtending a dormant or aborted bud a variable distance above it. Bracteoles subulate as long or longer than the bract. Flowers (1) 2–6; pedicels erect, filiform, terete, as long or longer than the bract. Sepals subequal, 1–2 mm long, densely hairy; adaxial sepal ovate, obtuse apex; abaxial (lower) sepal slightly larger, apex truncate or emarginate. Petals 0.25–1 cm long, violet, mauve or white; adaxial lip narrowly oblong, apex rounded or emarginate; abaxial lip larger, ± circular, the base with a rounded, ± 4-lobed swelling, apex rounded; palate shortly pubescent within; spur subulate from a conical base, ± straight, much longer than the abaxial (lower) lip, apex obtuse or subacute. Filaments curved, 0.5–0.7 mm long; anther thecae almost distinct. Capsule obliquely ovoid, about as long as the sepals, dehiscing by a single, longitudinal, abaxial, marginally thickened slit. Seeds ovoid, 0.17–0.2 mm long.

Probably perennial. Damp or wet places, marshes, but also in woodland. Like *U. minutissima* but has hairs extending to the sepals.

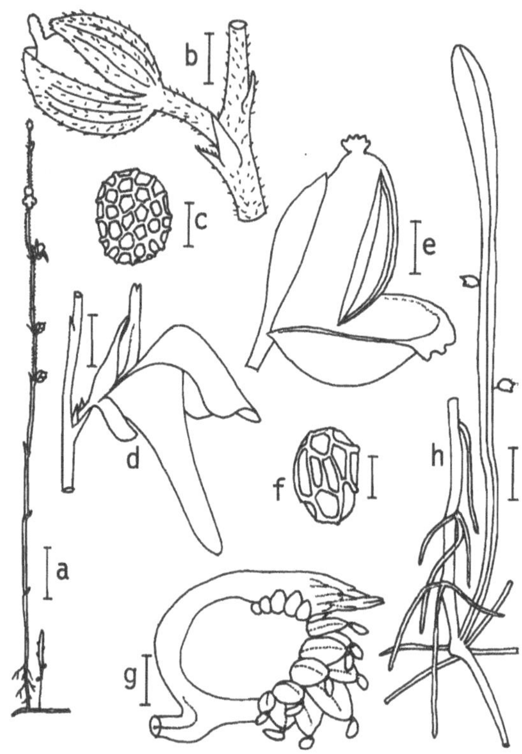

Fig. 246. *Utricularia hirta*: a, flowering plant (1 cm); b, capsule with sepals, bract, bracteole and sterile bract (0.8 mm); c, seed (0.1 mm). *U. minutissima*: d, flower (0.8 mm); e, dehisced capsule with sepals (0.4 mm); f, seed (0.1 mm); g, trap (0.4 mm); h, base of plant with stolons, rhizoids, leaves, traps and base of inflorescence stalk (1 mm).

{Hel/Ten}. India and SE Asia: Ban, Cey, Ind: Bh, Kt, Mg, Mp, Tn. 0–100 m.

Utricularia inflexa Forsskål, Fig. 247

Very like *U. stellaris* but petals white with violet nerves. Floats on inflorescence stalk long and cylindrical. Leaf auricles, if present, marginally toothed. Sepals somewhat fleshy, up to 1 cm long in fruit, completely enclosing and concealing the capsule. Seeds narrower than long. Style long (more than 1.5 mm, up to 3 mm).

Probably perennial perhaps sometimes annual. Planktonic with emergent flowers. In shallow to deep, still and slowly flowing water in lakes, pools, backwaters, ditches, swamps and rice fields.

{⊙? Pla}. Almost throughout Africa, central India and Assam: Ind: As, Dl, Up.

Utricularia lazulina P. Taylor: Fig. 248

Like *U. albocaerulea* but petals clear blue, smaller (8–10 mm long); adaxial petal lip oblong, narrower than the adaxial sepal. Seeds obovoid, 0.2–0.3 mm long, seed coat cells elongate, with the anticlinal boundaries much raised and thickened like a string of beads, almost straight to slightly sinuate, without intercellular spaces, the periclinal walls convex, at first appearing rough but shedding the outer skin to expose a warty surface, the warts numerous, subspherical.

Small annual. Seasonally flooded places, wet soil over laterite and in wet grassland.

{⊙ Hel/Ten}. Endemic to India: ♦Ind: Kt. Low altitudes

Utricularia minutissima Vahl: Fig. 246

Like *U. hirta* but leaves up to 3 cm long, 1-nerved, petiolate, the lamina linear to very narrowly obovate, 0.4–0.8 mm wide. Traps long-stalked, ovoid, 0.2–0.4 mm long. Inflorescence stalk 2–15 (–22) cm long, glabrous to distinctly hairy. Flowers 1–5 (–10), frequently cleistogamous. Sepals 2–3.3 mm long, glabrous; abaxial (lower) sepal elliptical, with truncate or rounded. Petals 0.6–1 cm long, the apical margin distinctly 3-lobed; palate shortly pubescent within; spur straight or curved, slightly longer than or up to 2 times as long as the abaxial (lower) lip, apex acute. Filaments ± 0.8 mm long. Seeds globose or broadly ellipsoid, ± 0.25 mm long.

Small to very small annual or perhaps sometimes perennial. Wet sandy soil or mud in open, grassy places. Like *U. hirta* but sepals glabrous.

{(⊙)? Hel/Ten}. India to Japan and Australia: Bur, Cey, Ind: Kl, Kt, Tn.

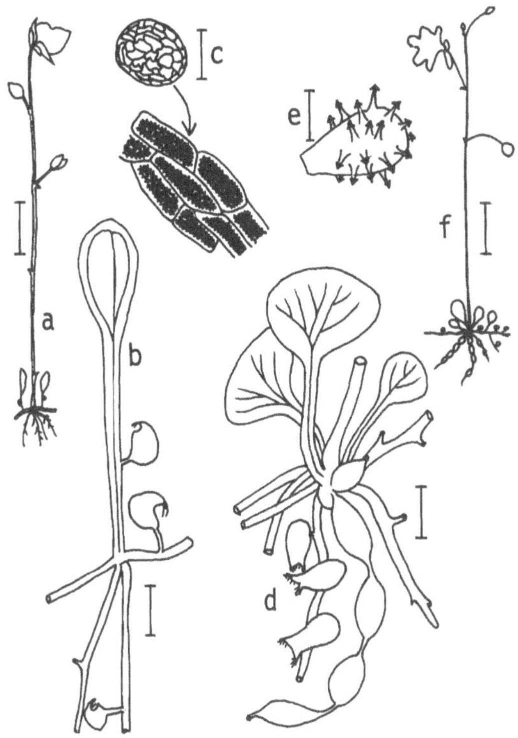

Fig. 248. *Utricularia lazulina*: a, stolon with leaf, traps and rhizoids (1 mm); b, flowering plant (1 cm); c, seed (0.2 mm) with some seed coat cells (inset). *U. moliniformis*: d, base of plant with tubers, stolons, leaves, traps and base of inflorescence stalk (1 mm); e, seed (0.1 mm); f, flowering plant (1 cm).

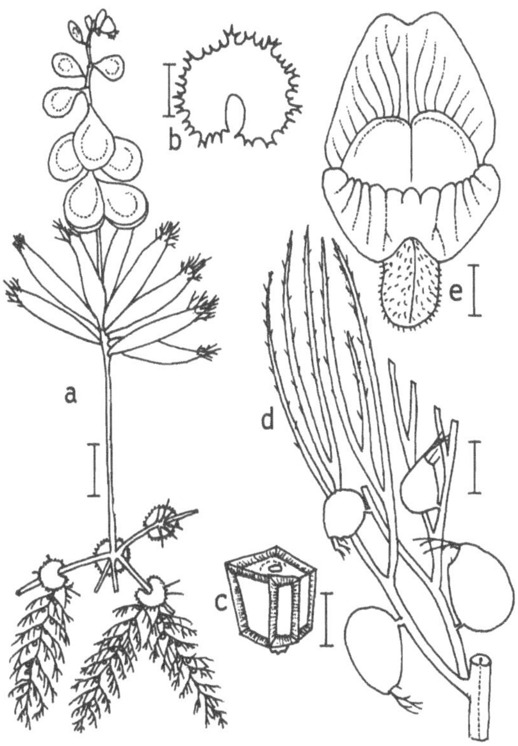

Fig. 247. *Utricularia inflexa*: a, flowering plant, with floats (1 cm); b, leaf auricle (1.3 mm); c, seed viewed from the side (0.3 mm); d, secondary leaf segment with traps (1 mm); c, sepals at anthesis (1 mm); e, flower (1.3 mm).

Utricularia moliniformis P. Taylor: Fig. 248

Like *U. striatula* but plant developing ellipsoid to fusiform tubers along the stolons, arranged like strings of beads; adaxial petal lip very broadly obovate, longer than the adaxial sepal, apex irregularly 6-lobed; spur club-shaped, much shorter than the abaxial (lower) lip.

Perennial. Wet rocks. It has frequently been misidentified as *U. striatula.*

{Hel-lithophyte}. Endemic to Sri Lanka: ♦Cey. 750–2300 m.

Utricularia polygaloides Edgeworth [*U. stricticaulis* (J. G. König ex Oliver) Stapf ex Gamble: Fig. 249

Like *U. reticulata* but rhizoids capillary. Leaves up to 2 cm long, the lamina 0.2–0.3 mm wide, apex acute. Inflorescence erect, rarely branched, not twining, 5–20 (–30) cm long. Bracts 1–1.5 mm long. Pedicels erect at anthesis, spreading in fruit, broadly winged, 3–6 mm long. Petals 7–10 mm long, deep violet; palate with a distinct rectangular, raised, hairy rim; spur only a little longer than the abaxial (lower) sepal and at a very obtuse angle to the abaxial lip. Capsule 3–4 mm long. Seeds obliquely ovoid, ± 0.3 mm long; seed coat cells elongate with the anticlinal boundaries raised, longitudinally striate, slightly sinuate, the periclinal walls tabular, longitudinally striate.

Probably annual. Open wet habitats, rice fields and coastal pools. It can be distinguished from *U. uliginosa* by the cuneate bases and smooth margins of the sepals and the different kind of trap; from *U. graminifolia* by its 1-nerved leaves, shorter pedicels, the more cuneate base of the sepals.

FIG. 249. *Utricularia polygaloides*: a, flowering plant (1 cm); b, stolon with leaf and traps (1 mm); c, flower (2 mm); d, seed (0.1 mm). *U. reticulata*: e, flowering plant with inflorescence stalks shortened (1 cm); f, flower (2 mm); g, trap (0.4 mm); h, seed (0.1 mm) with some seed coat cells (inset).

{⊙? Ten/?Hyp}. Endemic to India and Sri Lanka; ♦Cey, ♦Ind: Ap, Bh, Kt, Mh, Mp, Or, Tn, Wb. 0–1000 m.

Utricularia praeterita P. Taylor: Fig. 250

Like *U. uliginosa* but leaves 0.2–1.5 cm long, 3-nerved, petiolate, the lamina obovate, 1–2 mm wide, apex rounded. Inflorescence erect, 2–15 cm long. Bracts ± 1 mm long, 1-nerved, apex subacute. Flowers 1–6; pedicels erect at anthesis, spreading in fruit, 1–5 mm long. Sepal margins minutely papillose or toothed (× 20 magnification needed). Petals 5–8 mm long; adaxial petal lip oblong, narrower than the adaxial sepal. Capsule dehiscing by a single, longitudinal, abaxial slit. Seeds ovoid, ± 0.35 mm long, seed coat cells elongate, with anticlinal boundaries much raised and obscurely thickened like a string of beads, sinuate, smooth, with distinct triangular intercellular spaces, the periclinal walls concave, smooth.

Small annual. In seasonally flooded places, shallow wet soil over laterite and stream banks.

{⊙ Hel/Ten}. Endemic to S India: ♦Ind: Kl, Kt, Mh, Mp, Tn.

Utricularia pubescens J. E. Smith: Fig. 250

Like *U. arenaria* but vegetative part very distinctive: Leaves with filiform petioles 2–10 mm long, the lamina peltate, ± circular, somewhat fleshy, 1–8 mm in diameter, the margin entire, the nerves numerous, forked but not or scarcely anastomosing, the distal surface mucilaginous-glandular. Bracts basisolute, attached just above the base, ± 1 mm long, hairy; the superior part ovate, apex acute; the inferior part quadrate or semicircular. Sepals hairy. Petals white or pale lilac with a yellow blotch and darker violet marks on the abaxial (lower) lip or entirely violet. Capsule dehiscing by 1, abaxial, longitudinal, marginally thickened slit.

Probably annual. Damp peaty soil in boggy grassland, by stream sides, found in both open and shady situations.

{⊙? Hel/Ten}. C and S America, India and tropical Africa: Ind: Mg, Mp, Up. 0–1900 m.

Utricularia reticulata J. E. Smith: Fig. 249

Rhizoids numerous, branching, tapering. Stolons few, branched, capillary. Leaves arising 1 from each stolon node, up to 3 cm long, 1-nerved, petiolate, the lamina narrowly linear, up to 0.5 mm wide, apex rounded. Traps globose, 0.5–1.2 mm long, the mouth basal, with 2 simple, subulate, glandular, adaxial appendages. Inflorescence stalk twining or rarely erect, sometimes branched, 10–100 cm long, glabrous. Scales numerous, like bracts. Bracts attached by the base, broadly ovate-deltoid, 2–3 mm long, 3-nerved, apex

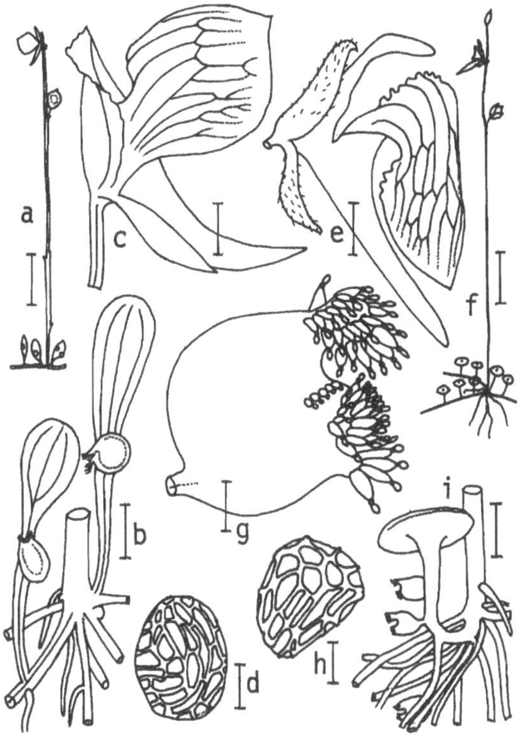

Fig. 250. *Utricularia praeterita*: a, flowering plant (1 cm); b, base of plant with stolons, rhizoids, leaves, traps and base of inflorescence stalk (1 mm); c, flower (1 mm); d, seed (0.1 mm). *U. pubescens*: e, flower (1.3 mm); f, flowering plant (1 cm); g, trap (0.17 mm); h, seed (0.1 mm); i, base of plant with stolons, rhizoids, leaves, traps and base of inflorescence stalk (1 mm).

acute. Bracteoles subulate, scarcely shorter than the bract. Flowers 2–10 or more; pedicels erect at anthesis, sharply deflexed in fruit, filiform, narrowly winged, 4–15 mm long. Sepals equal, 5–10 mm long, narrowly elliptical, glabrous, the base cuneate and decurrent on the pedicel; adaxial sepal with an acuminate apex; abaxial (lower) sepal with a minutely 2-toothed apex. Petals 15–20 mm long, dark violet with a paler or white, violet-nerved patch at the base of the abaxial lip; adaxial lip constricted below the middle but narrowest at the base, the superior part ± circular, apex rounded, the inferior part shortly truncate-cuneate; abaxial (lower) lip ± circular, the base with a very prominent, laterally flattened, obscurely 3-ridged swelling, apex rounded; palate hairy; spur subulate, slightly curved, ± as long and widely diverging from the abaxial lip, apex very acute. Filaments slightly curved, ± 1.5 mm long; anther thecae distinct. Capsule ovoid, dorsoventrally flattened, 4–6 mm long, dehiscing by adaxial and abaxial, longitudinal, marginally thickened slits. Seeds obliquely ovoid, ± 0.35 mm long; seed coat cells very elongate, the anticlinal boundaries ± straight, raised, smooth or longitudinally striate, the periclinal walls tabular, longitudinally striate.

Probably annual. In shallow water, marshy grassland and rice fields. Often very abundant and decorative with its large violet-veined flowers. It can be distinguished from *U. uliginosa* by the cuneate bases and smooth margins of the sepals and the different kind of trap; from *U. graminifolia* by its 1-nerved leaves, shorter pedicels, the more cuneate base of the sepals.

{⊙? Ten/Hyp}. Endemic to India and Sri Lanka: ♦Cey, ♦Ind: Ap, Bh? Go, Kl, Kt, Mh, Mp, Or? 0–750 m.

Utricularia scandens L. Benjamin [*U. volubilis* Wight, *U. wallichii* Wight, *U. macrolepis* Wight, *U. wallichiana* Wight, *U. bifida* var. *wallichiana* (Wight) Thwaites]: Fig. 251

Rhizoids numerous, branching, capillary. Stolons often numerous, sparsely branched, capillary, mat-forming. Leaves few to many, up to 2 cm, 1-nerved, petiolate, the lamina narrowly linear, 0.5–1 mm wide, apex rounded to subacute. Traps globose, 0.5–1 mm

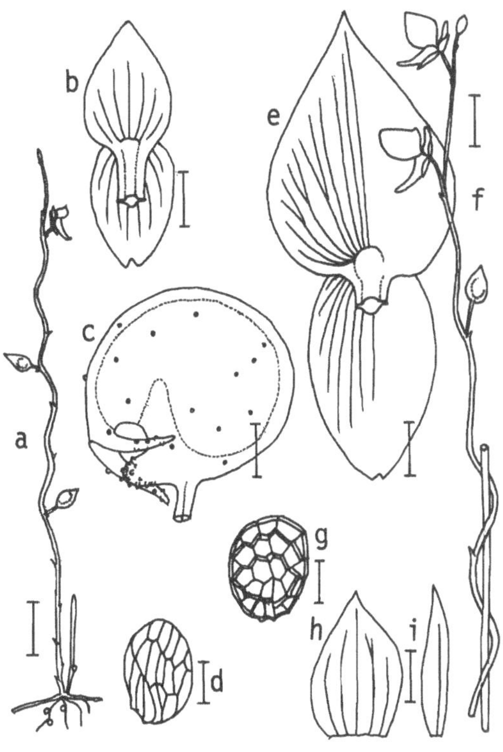

Fig. 251. *Utricularia scandens*: a, flowering plant (1 cm); b, sepals at anthesis (1 mm); c, trap (0.17 mm); d, seed (0.1 mm). *U. smithiana*: e, sepals at anthesis (1 mm); f, inflorescence (1 cm); g, seed (0.2 mm); h, bract (1 mm); i, bracteole (0.1 mm).

long, the mouth basal, with 2 simple, subulate, adaxial appendages bearing sessile glands and a single, truncate or apically 2-lobed, glandular abaxial appendage. Inflorescence erect or twining, 3–35 cm long, glabrous. Scales few to many, like bracts. Bracts attached by the base, ovate-deltoid, 1–1.5 mm long, apex acute or acuminate. Bracteoles linear or subulate, as long or somewhat shorter than the bract. Flowers 1–8; bracts subtending aborted or dormant buds alternating with those which subtend flowers; pedicels erect at anthesis and in fruit, filiform, broadly winged, up to 15 mm long, 1–5 times as long as the flowering sepals. Sepals subequal, 2–4 mm long at anthesis, up to 6 mm long in fruit, ovate, glabrous; adaxial sepal with an acute or acuminate apex; abaxial (lower) sepal slightly smaller, apex minutely 2-toothed. Petals 5–15 mm long, yellow; adaxial lip constricted below the middle, the superior part oblong to transversely elliptical, apex rounded or emarginate, the inferior part quadrate; abaxial (lower) lip ± circular, the base distinctly swollen, with 2 or 4 rounded, longitudinal ridges, apex rounded or obscurely 2- or 3-toothed; palate papillose; spur subulate, straight or curved, longer than and widely diverging from the abaxial (lower) lip, apex acute. Filaments straight, ± 1 mm long; anther thecae distinct. Capsule ovoid, dorsoventrally flattened, 2–2.5 mm long, dehiscing a single, abaxial, longitudinal, marginally thickened slit. Seeds obliquely ovoid, ± 0.2 mm long; seed coat cells elongate, the anticlinal boundaries straight or sinuate, raised, microscopically tuberculate, the periclinal walls concave, microscopically tuberculate.

Probably annual. Bogs and wet grassland, mostly on sandy or clayey soil.

{⊙? Hel/Ten}. Tropical Old World: Ban, Bur, Cey, Ind, Nep: Ap, Bh, Kl, Kt, Mh, Mp, Or, Tn, Up, Wb. 0–2300 m.

Utricularia smithiana Wight [*U. caerulea* var. *smithiana* (Wight) C. B. Clarke: Fig. 251

Like *U. uliginosa* but leaves up to 10 cm long, 3-nerved, petiolate, the lamina linear, up to 1.5 mm wide, apex rounded. Inflorescence twining or sometimes erect, (10–) 20–50 cm long. Flowers 2–8; pedicels ascending at anthesis and in fruit, broadly winged, 0.5–1.5 mm long. Sepals subequal, 4–5 mm long (in fruit), ovate, margins entire; adaxial sepal with an acute apex. Petals 15–20 mm long, various shades of mauve or violet; adaxial lip slightly constricted near the base, longer than the adaxial sepal. Capsule broadly ellipsoid, 3–4 mm long, dehiscing by a single abaxial, longitudinal slit.

Probably perennial. Shallow water or in very wet ground.

{Ten/Hel?}. Endemic to S India: ♦Ind: Kl, Kt, Tn.

Utricularia stellaris Linnaeus fil. [*U. macrocarpa* Wallich]: Fig. 252

Rhizoids absent. Stolons filiform, terete, sparsely branched, glabrous, up to 1 m long, 0.5–1.5 mm thick, sometimes bearing capillary air shoots, with apical, minute, elliptical membranous scales. Leaves very numerous, semicircular in outline, 1–6 cm long, divided from the base into 3–6, semi-verticillate, primary, filiform segments, the secondary segments pinnate, alternate, each pinna forked from the base into numerous further segments, the ultimate segments capillary, somewhat flattened, usually with some bristles; stipule-like auricles usually present at the base of the primary segments. Traps numerous, lateral on the secondary segments, obliquely ovoid, 1–3 mm long, the mouth lateral, naked or with 2 simple or sparsely branched bristle-like appendages. Inflorescence stalk erect, emergent, 3–30 cm long, glabrous, 0.5–1.5 mm thick, with a whorl of 3–8 spongy floats considerably above the middle; the floats narrowly ovoid, 0.5–2 cm long, sessile, with a few short, capillary, leaf segments at or near the apex only. Scales absent. Bracts basifixed, broadly ovate to circular, apex obtuse or rounded, membranous, ± 2 mm long,

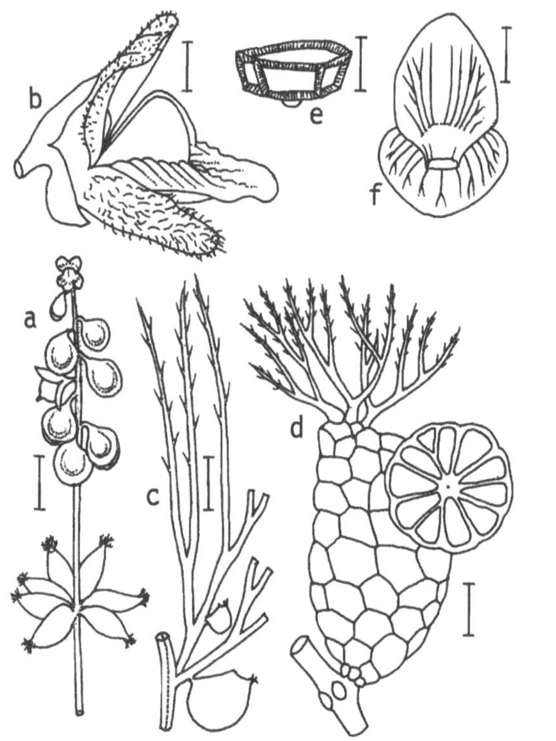

FIG. 252. *Utricularia stellaris*: a, inflorescence (1 cm); b, flower (1.3 mm); c, secondary leaf segment with traps (1.3 mm); d, float with transverse section (inset) (1.3 mm); e, seed (0.3 mm); f, sepals at anthesis.

obscurely many-nerved. Bracteoles absent. Flowers 2–12; the axis short, not greatly elongating in fruit; pedicels filiform, dorsoventrally compressed and winged, 1–5 mm long, erect at anthesis, usually sharply deflexed and thickened, especially distally, in fruit. Sepals subequal, broadly ovate to circular, ± 3 mm long at anthesis, reflexed and accrescent in fruit but scarcely as long as the capsule with stigma; adaxial sepal with a rounded apex; abaxial (lower) sepal with a retuse apex. Petals yellow, 0.7–1 cm long, ± densely covered with fine, multicellular, gland-tipped hairs; adaxial lip broadly ovate with a retuse or emarginate apex; abaxial (lower) lip limb transversely elliptical, the base with a prominent, 2-lobed swelling, apex rounded or emarginate; palate pubescent; spur cylindrical from a broad, conical base, apex retuse or emarginate, shorter than the abaxial lip. Filaments slightly curved; anther thecae ± confluent. Capsule globose, up to 3–4 mm in diameter, the wall somewhat fleshy, circumscissile, the persistent style up to 1.5 mm long. Seeds disk-shaped, 4–6 angled, 0.8–1.2 mm in diameter, usually 3 or more times as wide as long, narrowly winged on all angles.

Probably perennial perhaps sometimes annual. Planktonic with emergent flowers. In shallow to deep, still or slowly flowing water in lakes, pools, backwaters, ditches, swamps and rice fields. It can be a serious weed in newly planted rice fields.

{⊙? Pla}. Almost throughout Africa, tropical Asia from India to Vietnam but absent from Malaysia and N Australia: Ban, Bur, Cey, Ind, Nep: Ar, As, Bh, Dl, Go, Hr, Kl, Kt, Mh, Mp, Pj, Rj, Tn, Up, Wb.

Utricularia striatula J. E. Smith [*U. pusilla* J. Graham non Vahl, *U. orbiculata* Wallich, *U. glochidiata* Wight]: Fig. 253

Rhizoids few, simple, capillary. Stolons simple, capillary. Leaves forming a rosette at the base of the inflorescence stalk, also on stolons, up to 8 mm long, with numerous forked nerves, subsessile or petiolate, the lamina obovate to transversely elliptical, 1–5 mm wide. Traps borne on rhizoids and stolons but never on leaves, ovoid, 0.6–1 mm long, the mouth lateral, with an adaxial deeply 2-lobed appendage, the distal margin of which fringed with long-stalked glands. Inflorescence stalk erect, 1–20 cm long, glabrous. Bracts basisolute, attached near or very near the inferior extremity and slightly constricted at that point, 1–1.5 mm long, narrowly ovate or oblong, the superior apex subacute or rounded, the inferior apex truncate, irregularly toothed. Bracteoles like the bracts but slightly smaller. Flowers 1–10; pedicels spreading at anthesis, pendulous and withered in fruit, capillary, slightly flattened, 2–6 mm long. Sepals unequal, membranous, glabrous; adaxial sepal circular to broadly obcordate, 1.5–2.5 mm long, apex rounded to emarginate; abaxial (lower) sepal very much smaller, oblong-elliptical, apex rounded to emarginate. Petals 4–10 mm long (or much smaller when cleistogamous), white tinged with violet with a yellow patch at the base of the abaxial (lower) lip; adaxial lip ovate, apex shortly 2-lobed, very much shorter than the sepals; abaxial lip transversely elliptical to almost circular, apex ± regularly 5-lobed; palate fringed with hairs; spur subulate, about as long as the abaxial lip. Filaments straight, ± 0.6 mm long; anther thecae distinct. Capsule globose, obliquely dorsoventrally flattened, keeled on the abaxial surface, ± 2.5 mm long, dehiscing by a single abaxial slit. Seeds narrowly to broadly obovoid, ± 0.25 mm long, the surface covered with glochidia (hooked processes).

Annual or perennial. Permanently wet places, under waterfalls, dripping rocks and tree trunks. It is often abundant and forms small mats; when in flower it is very attractive.

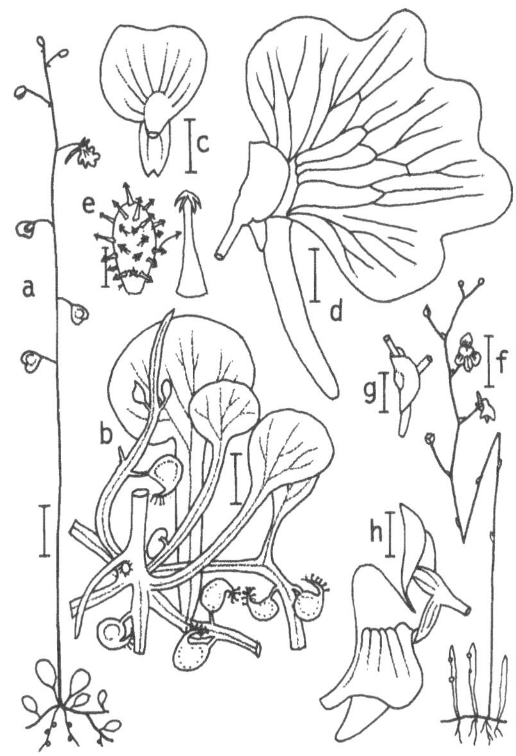

FIG. 253. *Utricularia striatula*: a, flowering plant (1 cm); b, base of plant with stolons, rhizoids, leaves, traps and base of inflorescence stalk (1.3 mm); c, sepals at anthesis (0.8 mm); d, flower (1 mm); e, seed (0.1 mm) with an enlarged glochidium (inset). *U. subulata*: f, flowering plant (1 cm); g, bract (1 mm); h, flower (1 mm).

{(⊙) Ten/Hel-lithopyte, epiphyte}. Africa to New Guinea: Bhu, Bur, Ind: An, Ap, Go, Kl, Kt, Mg, Mh, Mp, Sk, Tn, Up, Wb. 0–3300 m.

Utricularia subulata Linnaeus: Fig. 253

Rhizoids ± 6, tapering from thickened base, usually less than 1 cm long, branched. Stolons numerous, capillary, branched. Leaves often numerous, narrowly linear, up to 2 cm long and 1 mm wide, 1-nerved, gradually narrowed to the base. Traps ovoid, 0.2–0.7 mm long, the mouth lateral with 2, adaxial subulate, sparsely branched appendages. Inflorescence stalk erect, (2–) 10–50 cm long, glabrous above, usually papillose to shortly hairy below. Bracts basisolute, peltate, circular to transversally elliptical, with both extremities rounded, 1–2 mm long. Bracteoles always absent. Flowers 1–25; pedicels ascending, capillary, terete, 2–10 mm long. Sepals subequal, 1–1.5 mm long, glabrous, broadly ovate to circular, apex rounded or subtruncate. Petals 5–10 mm long (or much shorter in cleistogamous flowers), yellow (or in cleistogamous flowers white to reddish); adaxial lip broadly ovate, apex rounded; abaxial (lower) lip rhombic to broadly cuneate in outline, the base with a very prominent rounded and 2-lobed swelling, apex deeply 3-lobed; palate pubescent; spur subulate or cylindrical, slightly shorter than the abaxial lip. Filaments almost straight, ± 1 mm long; anther thecae confluent. Capsule globose, 1–1.5 mm in diameter, the wall thin, dehiscing by 1 abaxial, ovate pore. Seeds obovoid, ± 0.23 mm long, ± 0.16 mm wide.

Annual. Damp situations, seasonally flooded places, ditches, stream and pool sides. It is the most widespread of all *Utricularia* species but is known only from one collection in India; it is often very inconspicuous and might well be more common.

{⊙ Hel/Ten/Hyp}. Almost pantropical: Ind: Mg.

Utricularia uliginosa Vahl [*U. affinis* Wight, *U. brachypoda* Wight, *U. liliacina* Griffith, *U. decipiens* Dalzell, *U. elachista* Goebel, *U. nayarii* Janarthanam et Henry]: Fig. 254

Rhizoids numerous, branching, capillary. Stolons branched, capillary. Leaves arising 1 from each stolon node, up to 5 cm long, 3- to many-nerved, petiolate, the lamina linear, 1.5–6 mm wide, apex rounded to acute. Traps globose, 1–1.5 mm long, the mouth basal, with 2 simple, subulate, adaxial appendages bearing stalked glands, the stalk sometimes distally swollen and bearing stalked glands. Inflorescence erect, 3–30 cm long, glabrous. Scales few, like bracts. Bracts attached by the base, broadly ovate-deltoid, ±2 mm long, 5-nerved, apex acute. Bracteoles narrowly linear or subulate, slightly shorter than the bract. Flowers 2–10; pedicels suberect at anthesis and in fruit, filiform, narrowly winged, 1.5–6 mm long, not or not much longer than the sepals. Sepals subequal, 3–5 mm long (in fruit), very broadly ovate to almost circular, distal margins minutely toothed (× 20 magnification needed); adaxial sepal with an obtuse to subacute apex; abaxial (lower) sepal slightly smaller, apex obtuse and minutely 2-toothed. Petals 5–8 mm long, usually lilac or violet, with the swollen basal part of the abaxial (lower) lip usually paler, sometimes streaked with darker purple and often with a blue margin or sometimes petals wholly white; adaxial lip slightly constricted at about the middle, not or scarcely longer than the the adaxial sepal, the superior part broadly oblong, apex truncate or emarginate, the inferior part broadly ovate; abaxial lip ± circular, the base much swollen, apex rounded or emarginate; palate with a hairy margin; spur subulate, slightly curved, ± as long as and widely diverging from the abaxial lip, apex acute. Filaments straight, ± 1 mm long; anther thecae distinct. Capsule ovoid, dorsoventrally flattened, 2.5–3.5 mm long, the wall uniformly membranous, dehiscing by adaxial and abaxial, longitudinal slits. Seeds globose, ± 0.3 mm in diameter, seed coat cells isodiametric, with anticlinal boundaries much raised,

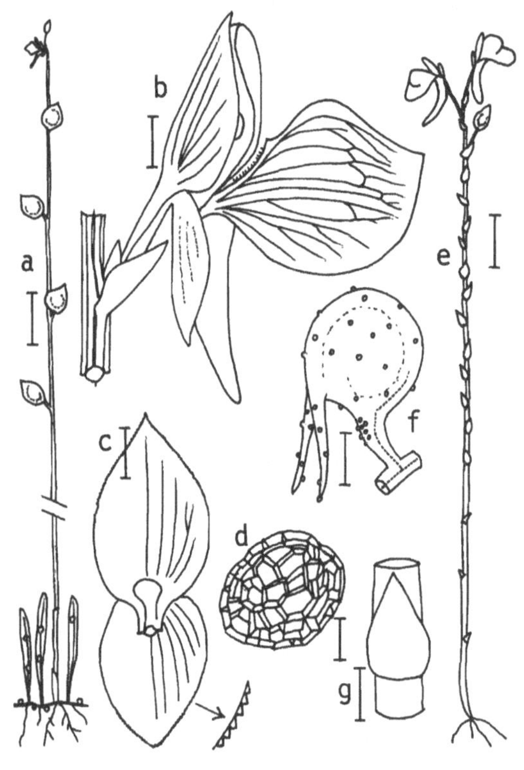

FIG. 254. *Utricularia uliginosa*: a, flowering plant with inflorescence stalk shortened (1 cm); b, flower (1 mm); c, sepals at anthesis (1 mm) with enlarged marginal teeth (inset); d, seed (0.1 mm). *U. wrightiana*: e, flowering plant (1 cm); f, trap (0.4 mm); g, scale (1 mm).

straight, smooth, with distinct triangular intercellular spaces, the periclinal walls concave, smooth.

Small annual or perhaps sometimes perennial. Seasonally flooded ground and along banks and beds of streams. It is often confused with *U. graminifolia* which has ovoid seeds with elongate seed coat cells, a much longer pedicel (up to 10 mm long), capsule with thickened edges to the dehiscence slits and smooth sepal margins. It is very difficult to distinguish it from *U. praeterita* without seeds. *Utricularia uliginosa* may be distinguished from *U. lazulina* and *U. albocaerulea* as they have bright true blue petals with a much more ample abaxial (lower) lip with a distinctive 2-lobed swelling at its base; *U. smithiana* has a twining inflorescence stalk and larger petals (15–20 mm long); *U. cecilii* has a more ample abaxial (lower) petal lip and very different seeds and leaves; *U. wrightiana* has numerous and conspicuous scales on the inflorescence stalk and larger petals (10–15 mm long).

{(⊙) Ten/Hel?}. India to Australia: Bur, Cey, Ind: Ap, Go, Kl, Kt, Mp, Or, Tn. Low altitudes.

Utricularia wightiana P. Taylor [*U. squamosa* auct. Ind., non L. Benjamin]: Fig. 254

Like *U. uliginosa* but trap appendages with sessile glands. Inflorescence stalk up to 40 cm tall, bearing very numerous and conspicuous scales, scale-like towards the base and bract-like above. Sepals 3–5 mm long, ovate, adaxial sepal acute or acuminate. Petals larger (10–15 cm long).

Probably perennial. A poorly known species; probably in marshes and wet grassland.

{Hel?/Ten?}. Endemic to the Western Ghats: ♦Ind: Kl, Kt, Tn. 100–2200 m.

LECYTHIDACEAE see BARRINGTONIACEAE excluded see p. 12.

LIMNOCHARITACEAE: 3 genera

Annual or perennial, with latex in specialized tubes. Juvenile leaves linear; adult leaves differentiated into petiole and blade, the blades lanceolate to ovate or cordate with parallel nerves. Inflorescence umbel-like, axillary clusters or flowers solitary. Flowers bisexual, actinomorphic. Sepals 3, green, persistent, with latex tubes. Petals 3, white or yellow, delicate, not persistent. Stamens 6–9 or numerous. Carpels superior, 3 to numerous, free, in 1 or rarely 2 whorls. Fruit a head of follicles or thin-walled nutlets. Seeds numerous, scattered over the inner surface of the carpel wall or solitary; embryos curved or folded.

Juveniles submerged, adults emergent. Pollinated by insects or autogamous. The disseminules are seeds or thin-walled nutlets.

1. Leaf blades lanceolate to oblanceolate, the apex tipped by a hard, blunt mucro; petals white; stamens about 9, staminodes absent ***Butomopsis***
1. Leaf blades ovate to suborbiculate, the apex rounded or emarginate; petals yellow; stamens more than 9, surrounded by staminodes ***Limnocharis***

Butomopsis [*Tenagocharis*]: 1 species

Butomopsis latifolia (D. Don) Knuth [*Tenagocharis latifolia* (D. Don) Buchenau, *Butomopsis lanceolata* (Roxburgh) Kunth]: Fig. 255

Leaves radical, erect; petioles up to 20 cm or more long; leaf blades lanceolate to oblanceolate, 3–12 (–15) cm long, 1.5–3.5 (–5) cm wide, cuneate at base, acute or obtuse at apex, tipped by a hard blunt mucro. Inflorescence borne above the leaves, of 1 or 2 whorls. Petals white, deliquescent. Stamens ± 9, all fertile. Carpels 4–9, united at the base. Fruit 9–12 mm long, exceeding the sepals. Seeds horseshoe-shaped, less than 0.5 mm long, minutely warted.

FIG. 255. *Butomopsis latifolia*: a, flowering plant (2 cm); b, flower (4 mm). *Limnocharis flava*: c, flowering plant (5 cm).

Annual. In shallow water, in marshes and in seasonally inundated places. It is often found in rice fields but is not considered to be a serious pest.

{⊙ Hyp}. Old World tropics and W Australia: Ban, Ind, Nep: As, Bh, Mn, Mp, Or, Rj, Sk, Up, Wb.

Limnocharis: 2 species

Limnocharis flava (Linnaeus) Buchenau [*Damsonium flava* (Linnaeus) Miller]: Fig. 255

Leaves radical; petioles thick, triangular in transverse section; blades ovate to suborbicular, inrolled when young, rounded or emarginate at apex, cuneate at base. Inflorescence a terminal, 2- to 15-flowered umbel, shorter than the leaves. Petals pale yellow with darker base, ovate to rhomboidal. Stamens numerous, surrounded by a whorl of staminodes. Carpels numerous, laterally flattened into globose heads. Fruit of flat and thin-walled nutlets. Seeds horseshoe-shaped.

Perennial or annual. In shallow water, drying mud and in floating mats of vegetation. The disseminules are flattened nutlets, dispersed by adhesion to animals and by water. It is eaten and used as fodder in Asia. Large populations in S India have become a pest. It is apparently increasing its range.

{(⊙) Hyp/(Ple)}. Tropical America: <Ban, <Bur 1955, <Cey, <Ind: <Kl 1961.

LOBELIACEAE see CAMPANULACEAE

LYTHRACEAE: 30 genera

Annual, perennial or shrubby. Stems often inflated when submerged or floating. Leaves whorled, opposite or sometimes alternate, simple, entire, without stipules. Inflorescence simple or complex, usually based on repeated axillary cymes. Flowers bisexual, radially or bilaterally symmetrical, sometimes heteromorphic, sometimes cleistogamous. Sepals 4–6 (–7), united below into a tube, the lobes valvate, often with folds or appendages between the lobes. Petals 4–6 (–7) or absent, caducous, often crumpled in bud. Stamens usually as many or twice as many as the sepal lobes, rarely more or less. Ovary superior, free from the sepal tube, (1) 2- to 6-locular; septae separating the locules often incomplete; placentation axile or free-central. Fruit a capsule.

Scrivastava, R. C. and Singh, A. N. The family Lythraceae in Madhya Pradesh. Journal of Economic and Taxonomic Botany 14(2): 409–420 (1990). I would particularly like to thank Mr P. Thomas Mathew, NBPGR, Thrissur, Kerala for his help and for allowing me to use his unpublished work.

Excluded: ***Lythrum salicaria*** Linnaeus and ***L. tribracteum*** Salzmann ex Sprengel; — found in NW Himalayas but not extending south into India.

1. The placenta of the capsule running up into the style; septae of the ovary complete; capsule at first opening with a lateral slit or lid, later opening irregularly below; petals usually brick red or absent; (flowers in sessile or stalked axillary cymes or sometimes some solitary or in terminal heads) ***Nesaea***

1. The placenta of the capsule not running up into the style; septae of the ovary incomplete; capsule opening by longitudinal slits or irregularly but not opening with a lateral slit or lid; petals usually pink to purple or absent

- **2.** Flowers in axillary cymes; capsules without fine horizontal lines, dehiscing irregularly ***Ammannia***
- **2.** Flowers solitary in leaf axils or in simple terminal spikes; capsules regularly dehiscent, opening by valves, each with fine horizontal lines ***Rotala***

Ammannia: 25 species: tropical to warm regions of the world

Stems 4-angled, usually swollen if growing in standing water. Leaves decussate, sessile. Flowers radially symmetrical, in sessile or stalked, (1-) 3- to 15-flowered, axillary cymes. Sepal tube campanulate to urn-shaped, often becoming globose in fruit; sepal lobes 4 or rarely 5, with or without folds or appendages between the lobes. Petals 4 (5) or absent. Ovary (1-) 4- (5)-locular, without a disk at the base; style very short to longer than the ovary; stigma capitate. Capsule dehiscing irregularly Seeds numerous, less than 1 mm long.

Annual or short-lived perennials. Seasonally submerged, emergent or floating. Pollinated by insects or autogamous. The disseminules are sticky seeds, dispersed by animals or in mud. Graham, Shirley A. A revision of *Ammannia* in the western Hemisphere. Journal of the Arnold Arboretum 66: 395–420 (1985).

1. Styles (0.5–) 1–5 mm long

- **2.** Sepal tube 4-winged, at least 4 mm wide; capsule shorter than and completely enclosed in the sepal tube ***A. octandra***
- **2.** Sepal tube not winged, 3.5 mm or less wide; capsule as long as or longer than the sepal tube, at least partly exposed
  - **3.** Leaves papillose; axillary cymes sessile or nearly so; sepal tube leathery; capsule ellipsoid ***A. desertora***
  - **3.** Leaves glabrous; axillary cymes distinctly pedunculate; sepal tube membranous; capsule globose
    - **4.** Petals 2 mm or more long; capsules 2–3.5 mm in diameter; styles (1.5–) 2–3 mm long; sepal lobes erect in fruit ***A. auriculata***
    - **4.** Petals not more than 1.5 mm long; capsules 1–1.5 mm in diameter; styles 0.5–1 (–1.5) mm long; sepal lobes reflexed in fruit ***A. multiflora***

1. Styles less than 0.5 mm long
  5. Cymes and flowers distinctly stalked at anthesis; cymes lax *A. senegalensis*
  5. Cymes and flowers sessile or nearly so at anthesis; cymes head-like
    6. Leaves and sepals glabrous; sepal appendages absent *A. baccifera*
    6. Leaves and sepals hairy; sepal appendages distinct, usually longer than the sepal lobes *A. verticillata*

Ammannia auriculata Willdenow [*A. senegalensis* auct. ind., non Lamarck]: Fig. 256

Stems 10–30 (–57) cm tall, erect or decumbent, usually 4-angled, simple or with ascending branches. Leaf blades linear-lanceolate to lanceolate, (0.6–) 1.5–6 (–9) cm long, (1.7–) 3–8 (–14) mm wide, mostly longer than the stem internodes, base cordate with auricles, tips acute. Cymes pedunculate, (1–) 3- to 15-flowered; peduncles (1.5–) 2–4 (–6) mm long; pedicels of the central flowers 3–17 mm long. Sepal tube ± campanulate, becoming ± globose in fruit, (1–) 2–3 mm long, 1.5–3.5 mm broad, with 8–10 green longitudinal ribs at anthesis; sepal lobes reduced to 3-angled teeth; sepal appendages smaller than the lobes or absent. Petals purple, violet or white, 2–5 mm long, deciduous. Stamens 4 (8). Styles (1.5–) 2–3 mm long. Capsules globose, 2–3.5 mm in diameter, exceeding the sepals. Seeds discoid.

Annual or perhaps sometimes perennial. In wet places, marshes, river banks and rice fields. Used as a counter irritant for rheumatic pains.

{(⊙)? Hel/Ten}. Warm and tropical regions of the world: Ind, Nep, Pak: Bh, Hp, Hr, Jk, Kl? Kt? Pj, Rj, Up, Wb. ?–1400 m.

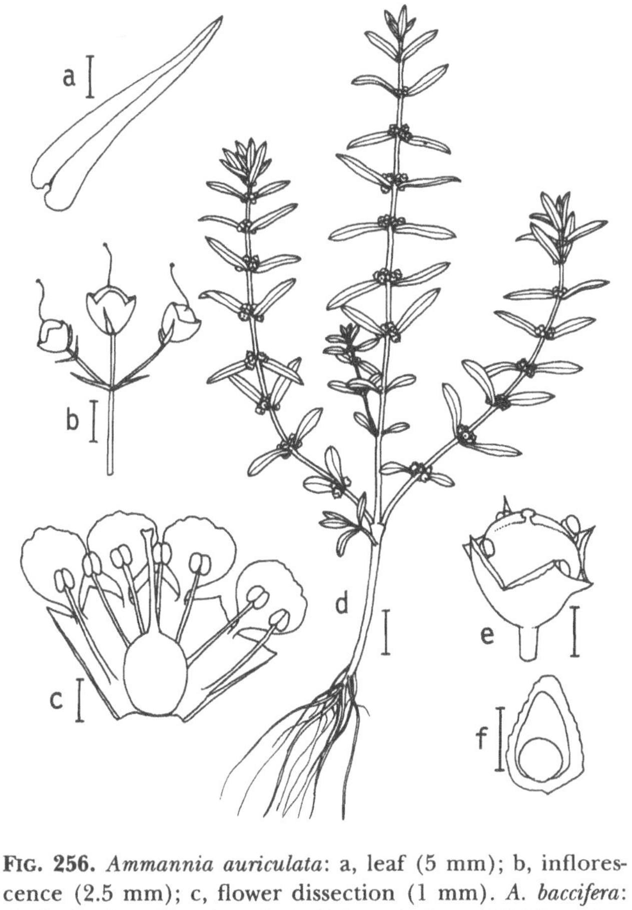

FIG. 256. *Ammannia auriculata*: a, leaf (5 mm); b, inflorescence (2.5 mm); c, flower dissection (1 mm). *A. baccifera*: d, whole plant (1 cm); e, fruit (0.5 mm); f, seed (0.25 mm).

Ammannia baccifera Linnaeus [*A. salicifolia* auct. Ind, non Monti, *A. baccifera* subsp. *aegyptiaca* (Willdenow) Koehne]: Fig. 256

Stems (7–) 25–50 (–100) cm tall, glabrous, decumbent or erect, usually much branched, the branches ascending and shorter than the main axis. Leaf blades linear-lanceolate to oblanceolate, (4–) 10–40 (–70) mm long, (0.5–) 1–8 (–16) mm wide, glabrous, mostly longer than the stem internodes, becoming smaller towards the apex of the stem, base gradually narrowed, rarely truncate or cordate, apex acute to subobtuse. Cymes compact, globose, sessile or nearly so or flowers solitary. Flowers sessile or nearly so; pedicels at centre rarely more than 1 mm long. Sepal tube broadly bell-shaped or obpyramidal, (0.5–) 1–2.5 mm long, 1–2.5 mm in diameter, with 8 longitudinal ribs; sepals 4 (5), lobes deltate, ± two-thirds as long as the tube, purple-tipped; sepal appendages absent. Petals minute or absent. Stamens 4, included; filaments red; anthers yellow. Styles 0.1–0.3 (–0.5) mm long. Fruits bright red to purple. Capsules globose, 1–1.5 (–2) mm in diameter, partly exerted from the sepal tube. Seeds ± 20, ± globose, concave on the inner side, rounded on the back, 0.3–0.4 mm long, bright red.

Annual. In wet or regularly inundated places such as marshes, river banks, rice fields, etc., but rarely found in permanent water. Often abundant and gregarious. The ripe fruits form small purple clusters making the plants visible from a distance. Used locally in medicine as a counter-irritant and considered to be poisonous but used as to reduce the sexual libido of animals. The larger more branched plants with subcordate leaves are often recognized as subspecies *aegyptiaca*.

{⊙ Hel/Ten}. Tropical and warm parts of Asia and Africa and perhaps also Australia, introduced in the New World (Caribbean) and Europe: Bhu, Bur, Ind, Nep, Pak?: <An, Ap, As, Dl, Gj, Go, Kl, Kt, Mh, Mp, Or, Pj, Rj, Tn, Tp, Up, Wb. 0–1800 m.

Ammannia desertora Blatter et Hallberg: Fig. 257

Stems 12–70 cm tall, erect, profusely branched, usually 4-angled, simple or with ascending branches, scabrid papillose. Leaf blades linear-lanceolate to lanceolate, 12–70 mm long, 1–8 mm wide, papillose, base cordate with auricles, midribs prominent, margin reflexed, tips acute. Cymes compact, sessile or nearly so, in almost every leaf axil from base to tip, (1–) 3- to 7-flowered; pedicels ± 1 mm long. Sepal tube ± campanulate, leathery, 8-nerved, becoming semi-globose in fruit, 1.5–2 mm long at anthesis, becoming ± 2.5 mm long in fruit, 1.5–3.5 mm broad, with 8–10 green longitudinal ribs at anthesis; sepal lobes (3) 4, tooth-like, triangular, ± 0.5 mm long, apiculate; sepal appendages smaller or absent. Petals 4, purple, ± 1 mm long, deciduous. Stamens 8, inserted on the lower third of the sepal tube; anthers borne level with the top of the sepal lobes. Styles 0.75–1.5 mm long. Capsules ellipsoid, 2.5–3 mm long, 2–2.5 mm broad, exceeding the sepals, reddish-brown. Seeds angular-subglobose, yellowish-brown, shining.

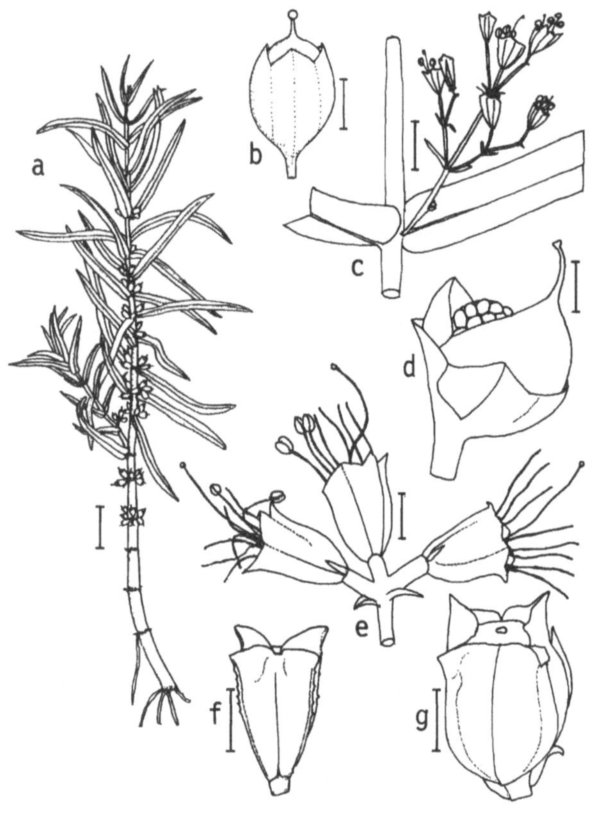

Fig. 257. *Ammannia desertora*: a, whole plant (1 cm); b, fruit (1 mm). *A. multiflora*: c, part of inflorescence (2 mm); d, fruit (0.5 mm). *A. octandra*: e, part of inflorescence (2 mm); f, sepals (2 mm); g, fruit (2 mm).

Annual. Found in seasonally inundated areas. Flowering and fruiting in October and November. It has been included in the Indian Red Data Book 2: 148 (1988).

{⊙ Hel}. Endemic to NE India and S Pakistan: ♦Ind, ♦Pak: Gj, Rj. September–October.

Ammannia multiflora Roxburgh [*A. nagpurensis* T. Mathew et M. P. Nayar]: Fig. 257

Stems 3–30 (–65) cm tall, simple or sparsely branched, not or only slightly swollen when growing in water. Leaf blades linear-lanceolate to oblanceolate or elliptical, (3–) 10–50 mm long, 0.75–3 (–10) mm wide, base subcordate to auriculate, apex obtuse. Cymes stalked and lax, with 3–5 (–15) flowers; peduncles up to 10 (–15) mm long; pedicels up to 2 mm long but usually some flowers sessile or nearly so. Sepal tube broadly campanulate, 1–2 mm in diameter, 1–1.5 mm long, 8-ribbed at anthesis; sepals lobes 4, pinkish, deltate to rotund, up to 0.5 mm long, usually reflexed in fruit, often mucronate; sepal appendages minute and horn-like or absent. Petals 4, 0.5–1.5 mm long, pink to magenta. Stamens 4. Styles 0.5–1 (–1.5) mm long. Capsules globose, 1–2 mm in diameter, partly exerted or remaining enclosed in a greenish-pink sepal tube. Seeds ± 70, ovoid to obconical, 0.3–0.45 mm long, 0.2–0.3 mm broad, brown.

Annual. Wet places, river banks, rice fields, often standing or partly floating in water. Common but not very gregarious. It is very variable; sometimes plants are unbranched and not more than 3 cm tall or bush-like and up to 50 cm tall. In NW India plants with long pedicels have been recognized as distinct and called *A. nagpurensis*.

{⊙ Hel/Ten}. Tropical and warm Old World: Ban, Ind, Nep, Pak: An, Ap, As, Gj, Kl, Kt, Mh, Mp, Pj, Rj, Tn, Up, Wb. 0–1800 m.

Ammannia octandra Linnaeus fil.: Fig. 257

Stems (15–) 25–50 (–100) cm tall, often strongly branched. Leaf blades linear-lanceolate, 2–8 cm long, 2.5–10 mm wide, base broad subcordate to auriculate, apex acute. Flowers solitary or in 2- to 4-flowered cymes; cymes sessile or nearly so, pedicels 1.5–5 mm long. Sepal tube broadly campanulate, 3.5–6 mm long, 4-angled to 4-winged; sepals lobes 4, very short, green; sepal appendages folded, horn-like in bud or absent. Petals 4, orbicular, 2.5–7 mm long, rose to flame coloured. Stamens 8 or rarely 4; filaments ± 5 mm long; anthers borne well above the sepals. Styles 4–5 mm long. Capsules globose to elongate-globose, 4.5–5 mm long and broad, ± equal to and included in the sepal tube. Seeds ± globose, concave inside, rounded on the back, ± 0.5 mm long, brownish-yellow.

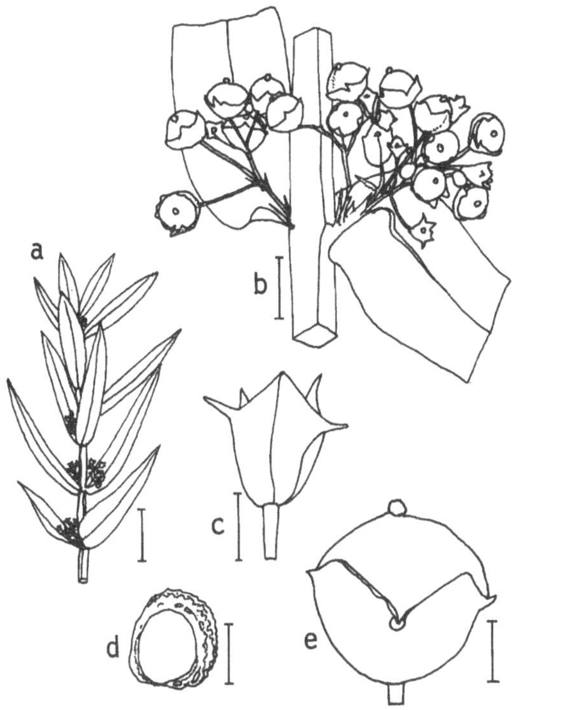

Fig. 258. *Ammannia senegalensis*: a, flowering shoot (1 cm); b, inflorescence (2 mm); c, sepals (0.5 mm); d, seed (0.25 mm); e, fruit (0.5 mm).

Annual. In low-lying moist areas, edges of tanks, marshes and rice fields.

{⊙ Hel}. S and SE Asia: Bur, Cey, Ind: Ap, Kt, Tn.

Ammannia senegalensis Lamarck: Fig. 258

Stems 10–30 cm tall, erect or suberect, simple or sparsely branched, square in transverse section. Leaf blades sublinear to oblanceolate, 5–30 mm long and 1–6 mm wide, base auriculate-cordate to subcordate or rarely gradually narrowed, tip acute or obtuse. Cymes axillary, compact, globose, subsessile; peduncles 1–4 mm long. Flowers 3–7 (–15) in axillary cymes; pedicels (0.5–) 1–1.5 mm mm long. Sepal tube broadly campanulate, becoming globose in fruit, 1–1.25 mm long, with 8 longitudinal ribs; sepal lobes 4, deltate, ± 1 mm long; sepal appendages horn-like in bud or absent. Petals 4, obovate, up to ± 1 mm long. Stamens 4, inserted at the middle of the sepal tube. Styles minute. Fruits bright red to purple. Capsules globose, ± 1.7–2.5 mm in diameter, partly exerted from the sepal tube. Seeds ± globose, flattened on 1 side, ± 0.4 mm long, rather rough on the curved part.

Annual. In wet or regularly inundated places in waste places.

{⊙ Hel}. Warm and tropical Africa, naturalized in Europe and India, apparently confined to the Pune and Satara Districts, discovered in 1895 persisting until at least 1948: <Ind: <Mh 1895.

Ammannia verticillata (Ardino) Lamarck [*A. salicifolia* Monti]: Fig. 259

Stems 10–30 (–50) cm tall, sparsely branched, square in transverse section, with wings on the angles, not swollen when growing in water. Leaves linear to lanceolate or elliptical, (8–) 15–35 mm long, (2.5–) 4–17 mm wide, scabrid, base of upper leaves usually rounded, apex acute. Cymes lax, sessile or nearly so; pedicels 0.5–3 mm long. Sepal tube 4-angled, 2–2.5 mm long, scabrid; sepal lobes 4, tooth-like; sepal appendages usually longer than the sepal lobes or obscure. Petals 4, minute or absent. Stamens 4. Fruits remaining enclosed in the sepal tube. Styles 0–0.25 mm long. Capsules globose, 0.5–1 mm in diameter, shorter than the sepal tube. Seeds ± globose with 1 side flattened, ± 3.5 mm long, rather rough on the curved part.

Annual. Wet and marshy places, river banks, sometimes standing in temporary water and rice fields. Flowering and fruiting from July to October. It does not occur in India; published reports are misidentified

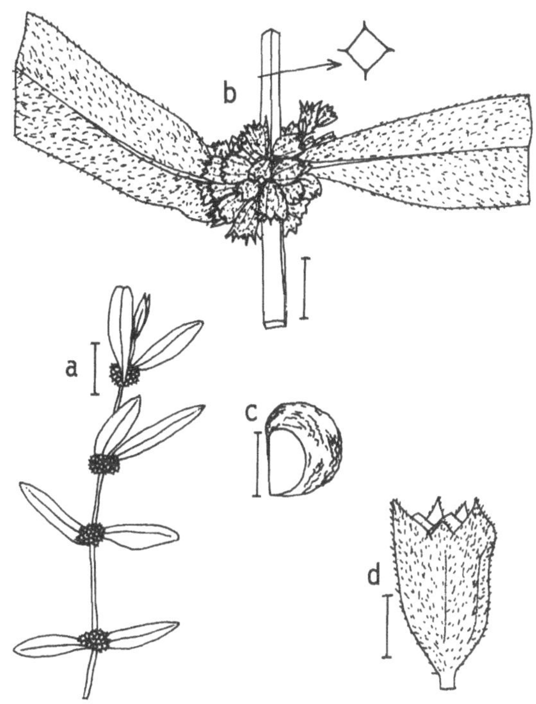

Fig. 259. *Ammannia verticillata*: a, flowering shoot with hairs omitted (1 cm); b, flowering node (2 mm) with transverse section of the stem (inset); c, seed (0.25 mm); d, sepals (1 mm).

Ammannia baccifera (which has glabrous leaves and sepals and lacks sepal appendages) or misidentified *Nesaea prostrata* (which has large bracteoles).

{⊙ Hel}. Central and SE Asia absent in India but present in Sind (Pakistan), naturalized in S Europe: Pak.

Nesaea: 50 species: tropical to warm regions of the world, most in Africa

Stems 4-angled or sometimes inflated and almost leafless. Leaves opposite, rarely whorled or alternate, sessile, linear to ovate. Flowers in sessile or stalked axillary or terminal, many-flowered, bracteate heads or rarely solitary. Sepal tube campanulate to urn-shaped or semi-globose; sepal lobes 4–8, with or without appendages. Petals 0–8, corrugated in bud. Stamens 4–23. Ovary 2- to 5-locular, septum complete and therefore continuous with the style; style very short to longer than the ovary; stigma capitate. Capsules opening with a lid, later irregularly below. Seeds numerous.

Annual or perennial. Pollinated by insects or autogamous. The disseminules are sticky seeds, probably dispersed by animals and in mud. Some are cultivated for decoration.

1. Leaves cordate and subamplexicaul at the base: petals absent ***N. brevipes***
1. Leaves attenuate at the base: petals pink, suborbicular or very rarely absent ***N. prostrata***

Nesaea brevipes Koehne [*Ammannia cordata* Wight et Arnott]: Fig. 260

Stems erect 9–40 cm tall or procumbent and rooting below, simple or sparingly branched. Leaf blades elliptical-oblanceolate to oblong, (1–) 2–3.5 cm long, (0.25–) 0.8–1 cm wide, cordate to subamplexicaul at the base, obtuse or subacute at the apex. Cymes 1- to 3-flowered. Flowers subsessile; bracteoles ± as long as the sepals. Sepal tube broadly bell-shaped, ± 2 mm long, glabrous, longitudinally ribbed; sepal lobes 4 (5), up to 1.5 mm long, pinkish; sepal appendages ciliate. Stamens 4 (5), ± 1.5 mm long. Petals minute or absent. Styles up to 0.5 mm long. Capsules globose, 1.5–2 mm long, red. Seeds 36–40, ovoid, ± 0.2 mm long.

Annual. In swamps, seasonally flooded places, along river banks and on wet soil.

{⊙ Hel}. Perhaps endemic to the Indian subcontinent: ♦?Ban, ♦?Cey, ♦?Ind: Ap, Gj, Go, Kl, Kt, Lk, Mp, Or, Pj, Rj, Tn, Up, Wb. 0–1000 m.

Nesaea prostrata (F. Hamilton ex Dillwyn) Suresh [*N. lanceolata* (Heyne ex C. B. Clarke) Koehne]: Fig. 260

Stems erect or decumbent, 6–25 cm tall. Leaf blades linear-lanceolate to oblanceolate or oblong, 0.7–3.5 cm long, 2–11 mm wide, glabrous or hairy. Cymes sessile or nearly so, 1- to 5-flowered. Flowers sessile or nearly so; bracteoles lanceolate to almost oblong, as long or longer than the sepals. Sepal tube bell-shaped, 4- (5)-angular, 2.5–3 mm long, glabrous or hairy; sepal lobes 4 (5), as long or shorter than the tube; sepal appendages shorter or longer than the lobes, usually with hairs at the tips. Petals 4 (5), obovate to suborbicular, ± half as long as or shorter than the sepal lobes, pink or rarely absent. Stamens 4 (5). Stigma sessile or nearly so. Capsules globose or nearly so, ± 2 mm in diameter. Seeds 35–40, globose, flattened or concave on one side, 0.4–0.5 mm long, brown to reddish-brown, somewhat rough.

Annual or perennial. Not confined to wetlands but sometimes found in wet low-lying areas, banks of ponds, along streams and in fallow rice fields. It is superficially like *Ammannia verticillata* but is easily distinguished by the bracteoles which are almost as long as the sepals.

{(⊙) Hel}. India, Sri Lanka and Australia: Cey, Ind: Ap, Kl, Kt, Lk, Rj, Tn, Up, Wb.

Rotala [*Ameletia*]: 46 species: almost cosmopolitan

Stems erect, ascending, creeping or floating. Leaves opposite, whorled or rarely alternate, sessile or shortly petiolate, capillary to linear or suborbicular, cuneate

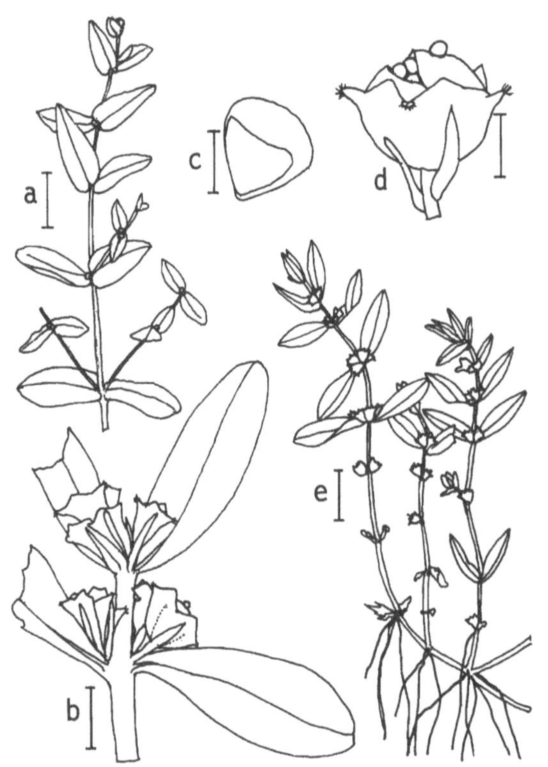

FIG. 260. *Nesaea brevipes*: a, flowering shoot (2 cm). *N. prostrata*: b, two flowering nodes (2 mm); c, seed (0.25 mm); d, fruit (0.5 mm); e, three flowering shoots (1 cm).

to cordate at the base. Flowers radially symmetrical, often inconspicuous, sometimes cleistogamous, usually solitary in leaf axils, sometimes terminally congested into spikes. Sepal tube campanulate to urn-shaped or globose in fruit; sepal lobes (3) 4 (–6), with or without appendages. Petals 0–6, minute or large and showy. Stamens (1–) 4 (–6). Ovary 2- to 4-locular, without a disk at the base; style very short or long (some distylic); stigma capitate. Capsule dehiscing by 2–4 valves; each valve with fine horizontal lines. Seeds numerous or few, 0.3–0.6 (–1.5) mm long.

Annual or perennial. Pollinated by insects or autogamous; some species distylic. The disseminules are seeds which are sticky when wet; dispersed by adhesion to animals (mostly birds) or in mud. Some are weeds in rice fields and irrigation ditches, others cultivated for decoration. Cook, C. D. K. A revision of the genus *Rotala*. Boissiera 29: 1–156 (1979); Joseph, K. T. and Sivarajan, V. V., *Rotala* Linn. (Lythraceae) in Peninsular India. Proceedings of the Indian Academy of Sciences (Plant Sciences) 99(3): 179–197 (1989).

1. Lower leaves alternate (upper leaves occasionally irregularly arranged in clusters); upper bracts scale-like, resembling bracteoles; stigmas 2-lobed **R. floribunda**

1. Lower and upper leaves decussate or whorled; bracts not scale-like, not resembling bracteoles or bracteoles absent; stigmas capitate, not 2-lobed

2. Leaves on lower part of the stem in whorls of 3 or more

3. Petals absent

4. Bracteoles leaf-like, partly or completely enclosing the flower at anthesis, at least twice as long as the sepals; sepal tube ± 1 mm long **R. occultiflora**

4. Bracteoles scarious, linear, not enclosing the flower at anthesis, rarely exceeding the sepals; sepal tube 0.5–0.7 mm long **R. mexicana**

3. Petals present

5. Sepal tube at anthesis urn-shaped (constricted just below or at the mouth), ± 2 mm long; sepals, petals and stamens 3 or very rarely 4; leaves in whorls of 4–8 (–10) **R. verticillaris**

5. Sepal tube at anthesis campanulate (widest at the mouth), 0.7–1.5 mm long; sepals, petals and stamens 4; leaves in whorls of 7–20 or more

6. Petals rather insignificant, white, up to 0.8 mm long, shorter than the sepal lobes; anthers inserted level with the point of attachment of the petals; nectar scales not lobed **R. cookii**

6. Petals showy, pink or white, ± 2.5 mm long, much longer than the sepal lobes; anthers inserted below the middle of the sepal tube; nectar scales 2-lobed **R. wallichii**

2. Leaves on lower part of stem in decussate pairs (rarely in whorls of 3 at the tips)

7. Stipule-like outgrowths present on the nodes between the leaves or bracts: sepals, petals and stamens 6 **R. hexandra**

7. Stipule-like outgrowths absent; sepals and petals 5, 4 or 3; stamens 1–5

8. Petals pinnately divided into linear segments (petals 2–3 mm long) **R. fimbriata**

8. Petals entire or lobed, not pinnately divided or absent

9. Bracteoles absent; dwarf, mat-forming plants with creeping stems **R. simpliciuscula**

9. Bracteoles 2; dwarf or not, stems erect above, not mat-forming

10. Appendages present between the sepal lobes

11. Seeds bright red; sepal lobes 3, 4 or rarely 5, bright red **R. malampuzhensis**

11. Seeds yellow or brown; sepal lobes 4 or 5, green to pink or brown

12. Sepal lobes 4

13. Leaves and bracts broadly ovate-cordate; sepal lobes acutely triangular, much longer than wide; sepal appendages ± 1 mm long, apiculate **R. illecebroides**

13. Leaves and bracts linear to orbicular, never cordate; sepal lobes shallowly triangular, much wider than long; sepal appendages not more than 0.25 mm long, obtuse **R. ritchiei**

12. Sepal lobes 5

14. Bracteoles longer than the total sepals (tube, lobes and appendages), with a mid rib; style at least 0.5 mm long, petals obovate, flowering branches compact, sepal lobes sharply triangular **R. densiflora**

14. Bracteoles equal to or shorter than the sepal tube, without a midrib; style not more than 0.25 mm long, petals elliptical, flowering branches lax, sepal lobes shallowly triangular **R. rosea**

10. Appendages between the sepal lobes absent

15. Sepal tube constricted above (urn-shaped); capsule valves induplicate (flowers in remote inflorescences, sepal lobes 4, stamens 2, capsule valves 4) **R. serpyllifolia**

15. Sepal tube not constricted above (campanulate); capsule valves not induplicate

16. Leaves and sepal lobes with a cartilaginous margin; capsules opening by 2 valves

17. Flowers distinctly stalked; anthers borne above the petals; petals showy, exceeding the sepal lobes (NE India and Myanmar) **R. subrotunda**

17. Flowers sessile or subsessile; anthers included within the sepals; petals insignificant, shorter than the sepal lobes (widespread) ***R. indica***

16. Leaves and sepal lobes without a cartilaginous margin; capsules opening by 3 or 4 valves

18. Capsules opening by 4 valves; floral bracts distinctly different from the foliage leaves; inflorescences remote, distinctly separated from the vegetative part

19. Anthers and stigmas included within the sepals; bracts longer than wide, ± remote in fruit; bracteoles ± as long as the sepal tube; stigmas discoid ***R. rotundifolia***

19. Anthers and stigmas exerted beyond the sepals; bracts wider than long, imbricate in fruit; bracteoles half or less as long as the sepal tube; stigmas minutely capitate ***R. macrandra***

18. Capsules opening by 3 valves; floral bracts leaf-like, at least at base of inflorescence; inflorescence continuous with the vegetative part

20. Bracteoles leaf-like, partly or completely enclosing the flower at anthesis, at least twice as long as the flower ***R. occultiflora***

20. Bracteoles scarious, linear, not enclosing the flower at anthesis, not more than twice as long as the flower

21. Sepal lobes 5

22. Petals absent; sepals not exceeding 0.75 mm long ***R. mexicana***

22. Petals present; sepals at least 1 mm long

23. Capsule not exceeding the sepals; bracteoles longer than the sepals ***R. densiflora***

23. Capsule exceeding the sepals; bracteoles ± as long as the sepal tube

24. Petals suborbicular and clawed, ± 0.5 mm long; stigma 0.25–0.5 mm long; seeds ± 0.5 mm long, brown (Kerala, Kannur District) ***R. malabarica***

24. Petals ovate to rhomboid, cuneate, rudimentary or up to 0.25 mm long; stigma absent or up to 0.25 mm long; seeds 0.3–0.35 mm long, yellowish (widespread) ***R. rosea***

21. Sepal lobes 4 or 3

25. Sepal lobes 4

26. Petals absent; bracteoles at least half as long as the sepal tube; sepal tube not exceeding 0.75 mm long ***R. mexicana***

26. Petals present, as long or longer than the sepal lobes; bracteoles not more than one-quarter as long as the sepal tube; sepal tube 1–1.5 mm long

27. Capsule included within the sepal tube; style 1–1.5 mm long (NE India) ***R. cordata***

27. Capsule exerted beyond the sepal tube; style rudimentary or up to 0.1 mm long (Nepal) ***R. rubra***

25. Sepal lobes 3

28. Capsule included within the sepal tube; petals absent ***R. mexicana***

28. Capsule exerted beyond the sepal tube; petals present ***R. malampuzhensis***

Rotala cookii K. T. Joseph et V. V. Sivarajan [*R. vasudevanii* K. T. Joseph et V. V. Sivarajan]: Fig. 261

Stems creeping and branched below, erect or floating and usually simple above, up to 50 cm or more long. Leaves in whorls of 7–12; submerged leaves capillary, up to 15 mm long, at tip microscopically bifid; aerial leaves linear, 5–6 mm long, in whorls of 7–12 (± the same number as the submerged leaves), nar-

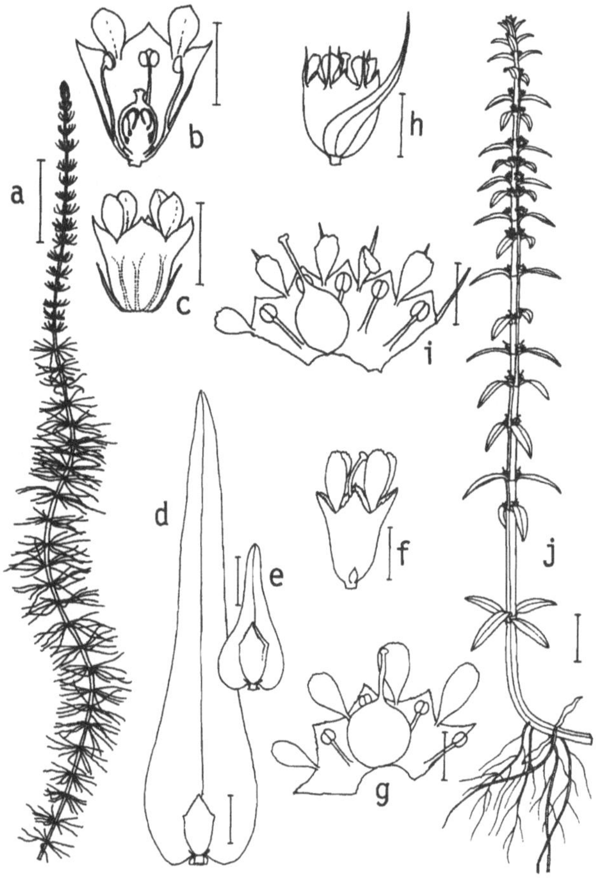

FIG. 261. *Rotala cookii*: a, flowering shoot (2 cm); b, flower in longitudinal section (1.5 mm); c, flower with two bracteoles (1 mm). *R. cordata*: d, bract and flower from the main axis (1 mm); e, bract and flower from a lateral branch (1 mm); f, flower with bracteole (1 mm); g, flower dissection (1 mm). *R. densiflora*: h, flower (1 mm); i, flower dissection (1 mm); j, flowering shoot (1 cm).

rowed at the base, minutely bifid at the tip. Floral bracts like foliage leaves; bracteoles capillary, up to 0.5 mm long. Flowers monomorphic, cleistogamous or chasmogamous, sessile. Sepal tube campanulate, membranous, translucent, 0.7–1 mm long, splitting and disintegrating as the fruit ripens; sepal lobes 4, deltate, 0.5 mm long, pink; sepal appendages absent. Petals 4, obovate, pale pink, up to 0.8 mm long. Stamens 4, inserted on the lower half of the sepal tube; anthers borne level with the base of the petals. Nectar scales entire. Style no more than 0.25 mm long. Capsule globose, ± 1.2 mm in diameter, opening by 2 valves. Seeds 4–8, semi-ellipsoid, ± 0.5 mm long, brown.

Annual. In coastal wetlands and flooded paddy fields. Flowering during and shortly after the rainy season. It is locally abundant, found growing with other aquatic annuals such as *Limnopoa meeboldii, Wiesneria triandra, Nymphoides indica* and *Hydrilla verticillata.* The description above is based on plants and differs slightly from the type description. *Rotala vasudevanii* is based on a single collection from Kerala, Alwaye. It seems to be no more than a somewhat small-flowered *R. cookii,* it has monomorphic leaves but this is likely to be environmentally induced; it is not a convincingly distinct species.

{⊙ Vit(Ten)}. Endemic to SW India: ♦Ind: Kl Jul–Nov (Ernakulam and Malappuram Districts).

Rotala cordata Koehne [*R. diversifolia* Koehne]: Fig. 261

Superficially like *R. densiflora* but stems with white wings running directly onto the margins of the leaves; bracteoles minute and very much shorter than the sepals; flowers constantly 4-merous; sepals without appendages; sepal lobes usually minutely serrate; styles 1–1.5 mm long; stigmas exerted.

Annual or perhaps sometimes perennial. In wet places.

{(⊙)? Ten/Hyp?}. From NE India and E Nepal through tropical SE Asia to Irian Barat: Ban, Bur, Ind, Nep: Bh, Or, Sk, Wb.

Rotala densiflora (Roth ex Roemer et Schultes) Koehne: Fig. 261

Stems erect, sometimes creeping below, up to 40 cm or more long, 4-angled, 4-winged, usually branched or sometimes simple. Leaves decussate below, sometimes in whorls of 4 above, very variable in shape and size (even on a single plant), linear-lanceolate to ovate or oblong, 2–35 mm long, base obtuse to cordate, tips acute or acuminate. Floral bracts almost dimorphic: on main stem and lower branches like foliage leaves, on branches above bract-like and scarcely exceeding the flowers; bracteoles longer than the sepals, with distinct midribs. Flowers monomorphic, sessile. Sepal tube campanulate, ± 1 mm long; sepal lobes 5 or rarely 4, triangular to shallowly triangular, ± 0.5 mm long; sepal appendages linear or capillary, as long or up to 2 times longer than the sepal lobes or very rarely rudimentary. Petals 5 or rarely 4, bright pink or rarely white, 0.5–1 mm long, at least as long as the sepal lobes. Stamens 5 or rarely less, inserted on the lower half of the sepal tube; anthers borne level with the sepal lobes. Style 0.5–1 mm long. Capsule as long as the sepal tube, opening by 3 valves. Seeds ± 0.5 mm long.

Annual. In shallow water at the edges of tanks and ditches, in floating mats of vegetation, in marshes and in rice fields. It is a rather attractive plant with showy pink petals.

{⊙ Ten/Hyp}. From the Pamirs through northern Pakistan and northern India to Assam, absent from the N Indian Plain but widespread in Peninsular India, extending to Sri Lanka and Northern Australia, naturalized in N Italy: Ban, Bur? Cey, Ind, Nep? Pak: Ap, As, Bh, Go, Hp, Jk, Kl, Kt, Mh, Mn, Mp, Or, Rj, Tn, Up, Wb.

Rotala fimbriata Wight: Fig. 262

Stems erect, sometimes creeping below, up to ± 50 cm long, weakly 4-angled. Leaves decussate, lanceolate to oblong, 1–3 cm long, tips obtuse; lower leaves cuneate to cordate at base; upper leaves cordate to auriculate at base. Floral bracts like upper leaves, usually somewhat auriculate at base; bracteoles acute triangular to subulate, rarely more than 1 mm long. Flowers monomorphic, sessile. Sepal tube campanulate, 2–2.5 mm long; sepal lobes 5 or rarely 6, deltate, ± 0.5 mm long; sepal appendages absent. Petals 5 or rarely 6, bright pink, pinnately divided into linear segments, 2–3 mm long. Stamens 5 or rarely 6, inserted at the base of the sepal tube; anthers exerted, borne level with the stigma. Style ± 2 mm long. Capsule elongate-ellipsoid, ± 3 mm long, opening by 3 valves: Seeds semi-ellipsoidal, ± 1.5 mm long.

Annual. In shallow water in ponds, temporary pools or in marshes. An attractive plant with showy flowers.

{⊙ Hyp/Ten}. Endemic to S India: ♦Ind: Ap, Go, Kt, Tn.

Rotala floribunda (Wight) Koehne: Fig. 262

Stems shortly creeping at base, erect above, up to 40 cm long, slender, branched towards the tips. Leaves alternate, opposite or in clusters (on the same plant); submerged leaves linear, up to 50 mm long and 2 mm wide, flaccid; aerial leaves linear to lanceolate, rarely more than 5 mm long and 0.5 mm wide, rigid and somewhat ericoid in appearance, at base cuneate to

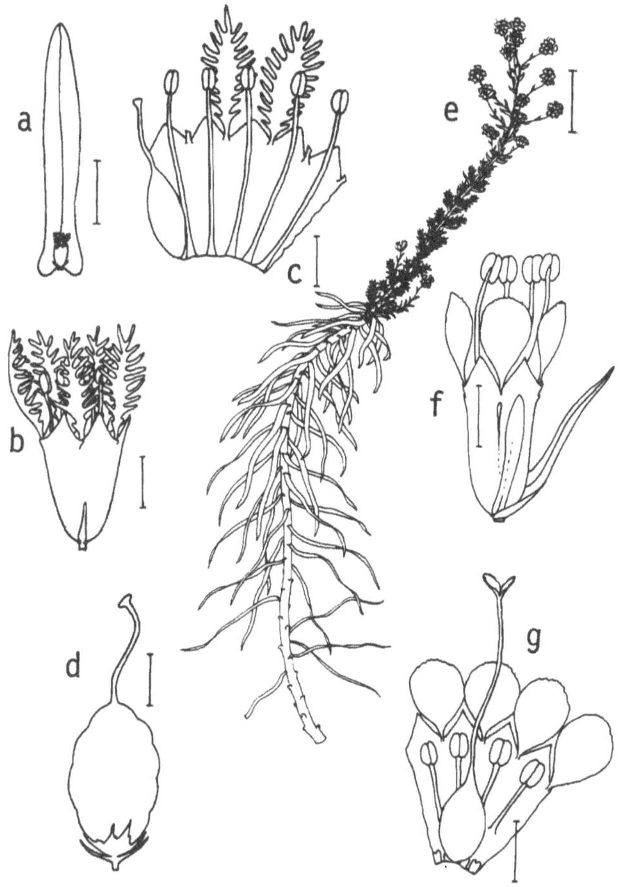

Fig. 262. *Rotala fimbriata*: a, bract with flower (5 cm); b, flower with bracteole (1 mm); c, flower dissection (1 mm); d, fruit (1 mm). *R. floribunda*: e, flowering shoot (2 cm); f, thrum flower with bract and bracteole (1 mm); g, pin flower dissection (1 mm).

auriculate, at tips acute. Floral bracts at base of inflorescence leaf-like, at tips bracteole-like; bracteoles linear to lanceolate, relatively thick, up to 2 mm long. Flowers in dense, terminal, ± remote, globose racemes, dimorphic, heterostylic, subsessile or with an up to 1.5 mm long pedicel. Sepal tube elongate-campanulate, ± 2 mm long; sepal lobes 4, deltate, ± 0.5 mm long; sepal appendages absent. Petals 4, showy, bright rose, obovate, 1–1.5 mm long. Stamens 4, inserted on the lower third of the sepal tube; filaments either ± 1 mm or ± 4 mm long; anthers either included or exerted and borne above the petals. Style either ± 0.5 mm or ± 3.3 mm long; stigmas deeply 2-lobed. Capsule ovoid to ellipsoidal, opening by 2 valves; seeds semi-ellipsoidal, ± 0.4 mm long.

Perennial or perhaps also annual. It grows in dense patches on rocks and in small rock pools. It is very attractive, flowering in December and January; it is called Mahabaleshwar heather. It is remarkable for its combination of aquatic and xerophytic characters. It shows relationships to the Asiatic species *R. rotundifolia* and to the E African species *R. repens*.

{⊙? Vit/Hyp/Ten?}. Endemic to W India: ♦Ind: Mh (Mahabaleshwar). ± 1500 m.

Rotala hexandra Wallich ex Koehne: Fig. 263

Stems erect, sometimes creeping below, up to 40 cm or more long, sharply 4-angled to 4-winged. Leaves decussate, sessile, lanceolate to oblong, 1–3 cm long, base obtuse to cordate or auriculate, tips obtuse, with stipule-like appendages on the nodes near the leaf bases, up to 2 mm long, similar but smaller structures on the nodes between the leaf pairs. Floral bracts leaf-like; bracteoles scarious, linear, not more than 1 mm long; a second pair of small bracteoles sometimes found in the axils of normal bracteoles. Flowers probably monomorphic, shortly pedicellate, often cleistogamous. Sepal tube campanulate, ± 2 mm long; sepal lobes 6, shallowly triangular, ± 0.5 mm long; sepal appendages absent. Petals 6, obovate to oblanceolate, ± 1 mm long, usually with erose

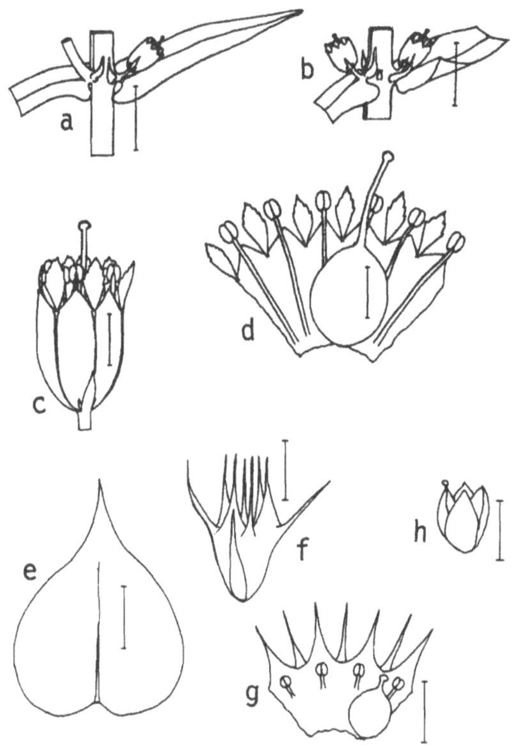

Fig. 263. *Rotala hexandra*: a, node with one flower (5 mm); b, node with two flowers showing stipels (5 mm); c, flower with bracteole and two bractlets (1 mm); d, flower dissection (1 mm). *R. illecebroides*: e, bract (1 mm); f, flower with bracteole (1 mm); g, flower dissection (1 mm); h, dehiscing capsule (1 mm).

margins. Stamens 6, inserted near the base of the sepal tube; anthers exerted. Style ± 2 mm long. Capsule subglobose, ± 2.5 mm long, apparently breaking up at maturity. Seeds semi-pyriform, 0.6–1 mm long.

Annual or perhaps sometimes perennial. In shallow water and in marshes.

{⊙/(⊙)? Ten}. SE Asia, from Myanmar to Java, China and the Philippines: Bur. The report for Kerala in Gopal (1990) must be a mistake.

Rotala illecebroides (Arnott ex C. B. Clarke) Koehne [*R. fysonii* Blatter et Hallberg]: Fig. 263

Stems erect, sometimes creeping below, up to 10 cm long, 4-angled to 4-winged. Leaves decussate, sessile, regularly spaced along the stem and very constant in shape and size, often red above, broadly ovate-cordate, ± 4 mm long and 3 mm wide, base semi-amplexicaul, at tips acute. Floral bracts leaf-like; bracteoles whitish, linear, equal to or longer than the sepal tube. Flowers monomorphic, sessile. Sepal tube, campanulate, ± 1 mm long, whitish; sepal lobes 4, acutely triangular, ± 1 mm long; sepal appendages 4, linear, ± as long as sepal lobes, spreading in bud. Petals minute or absent. Stamens 4 or rarely less, inserted on the upper half of the sepal tube; anthers borne level with the top of the sepal tube. Style very short. Capsule globose, ± 1 mm in diameter, opening by 4 or rarely 3 valves. Seeds semi-ellipsoidal, ± 0.3 mm long.

Annual. Terrestrial or amphibious, in marshes, on wet rocks, beside streams, in springs and wet flushes. Rajagopal, T. and Ramayya, N. *Rotala illecebroides* — a little known flowering plant from India. Current Science 37: 386–388 (1968).

{⊙ Ten/Hel}. Endemic to Peninsular India: ♦Ind: An? Ap, Kt, Mp, Tn.

Rotala indica (Willdenow) Koehne [*Ammannia nana* Roxburgh, *Ammannia repens* Rottler ex Martius, *Ameletia polystachya* Wallich ex Wight et Arnott, *Ameletia uliginosa* Miquel, *R. andamanensis* S. P. Mathew et P. Lakshminarasimhan]: Fig. 264

Stems erect or decumbent, usually 1, 2 or more times branched, occasionally simple; erect flowering shoots 4–40 (–86) cm long, weakly 4-angled or terete, when old somewhat woody. Leaves decussate, obovate to spathulate-oblong, suborbicular or narrowly oblong, 4–20 mm long, tips acute or obtuse, base cuneate to obtuse or somewhat cordate; leaf margins cartilaginous (distinct on dried leaves). Floral bracts polymorphic: leaf-like on main axes, scale-like on some lateral branches; bracteoles linear to lanceolate, 1.25–2.25 mm long, usually about as long as the sepal tube. Flowers either solitary in leaf axils or in spike-like lateral branches. Sepal tube campanulate, angled

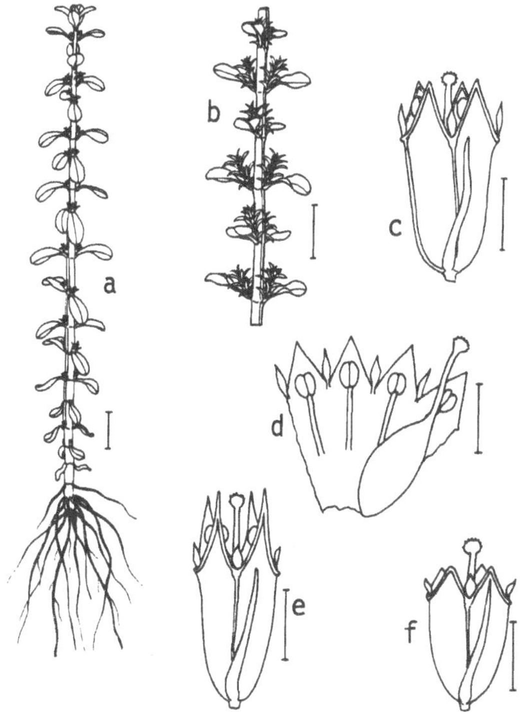

Fig. 264. *Rotala indica*: a, unbranched flowering plant (1 cm); b, branched inflorescence (1 cm); c, flower with bracteole (1 mm); d, flower dissection (1 mm); e, flower with long sepal lobes (1 mm); f, flower with short sepal lobes (1 mm).

on the nerves, 1.5–2.25 mm long; sepal lobes 4, deltate to acutely triangular, 0.5–1.5 mm long; sepal appendages absent. Petals 4, persistent, linear to narrowly ovate, less than half as long as the sepal lobes or minute and rudimentary. Stamens 4, reaching the mouth of the sepal tube or somewhat shorter. Style 0.5–1 mm long. Capsule ellipsoidal, ± 2 mm long, opening by 2 valves. Seeds almost hemispherical, 0.3–0.4 mm long.

Annual, perhaps sometimes persisting in permanently wet places. Common, found in a wide variety of wet habitats. It is a very successful weed of rice fields but is not really harmful. I have not seen material of *R. anadamanensis* but from the original description and illustrations — the 'sepal appendages' are probably rudimentary petals — I can find no character to distinguish it from *R. indica*.

{⊙/(⊙)? Ten/(Hyp)}. Central, southern and eastern Asia, extending northwards to Korea and southwards to Sri Lanka, naturalized in tropical Africa, Europe and N America: Ban, Bhu, Bur, Cey, Ind, Nep, Pak: An, Ap, As, Bh, Go, Hr, Jk, Kl, Kt, Mg, Mh, Mn, Mp, Mz, Or, Pj, Rj, Sk, Tn, Up, Wb.

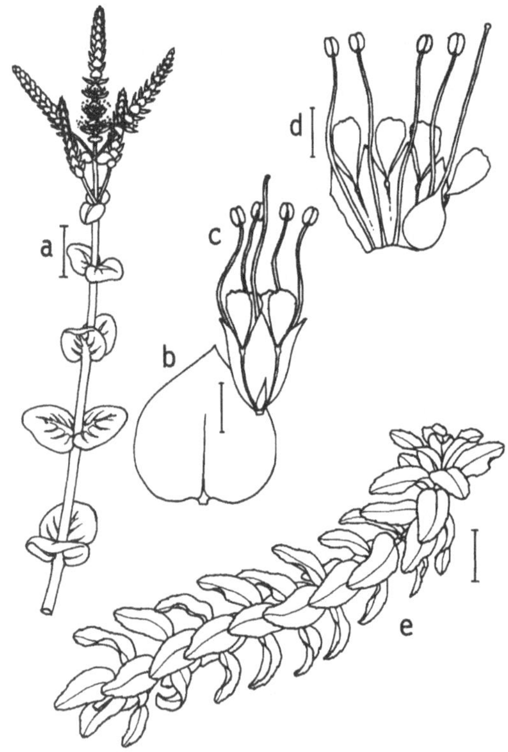

Fig. 265. *Rotala macrandra*: a, flowering shoot (1 cm); b, bract (1 mm); c, flower with bracteole (1 mm); d, flower dissection (1 mm); e, submerged shoot (1 cm).

Rotala macrandra Koehne: Fig. 265

Like *R. rotundifolia* but inflorescence remaining compact in fruit (stem not visible between the flowers); bracteoles ±0.5 mm long, ± half as long as the sepal tube; filaments ±4 mm long; anthers exceeding the sepal lobes and petals; style ± 3.5 mm long; stigma minutely capitate and scarcely wider than the style.

Perennial or sometimes annual. It is essentially a species of coastal wetlands. It is a common plant found in streams, lagoons, temporary ponds and flooded paddy fields. When submerged the leaves are very thin and pale green to pink or red. It flowers as the water recedes. It is cultivated as a decorative aquarium plant but is usually sold under the name *Rotala rotundifolia*. Benl, R. *Rotala macrandra* Koehne. Aquarien- und Terrarien-Zeitschrift 25(6): 198-201 (1972).

{(⊙) Vit/Hyp}. Endemic to S India: ♦Ind: Kl, Kt, Tn. September–January. Near sea level.

Rotala malabarica A. K. Pradeep, K. T. Joseph et V. V. Sivarajan: Fig. 266

Like *R. malampuzhensis* but stems 4-angled to 4-winged, with longer internodes; leaves narrower and prominently bimucronate at the tip; flowers 5-merous; petals orbicular to suborbicular, ± 0.5 mm long, clawed at the base; stigma 0.25–0.5 mm long; seeds ± 0.5 mm long, brown. Differs from *R. rosea* and *R. densiflora* by lacking sepal appendages, but having prominent nectar scales and bimucronate leaves.

Short-lived annual. Found in seasonally wet places. It flowers shortly after the onset of the monsoon.

{⊙ Ten}. Endemic to SW India: ♦Ind: Kl (Kannur District). July–November.

Rotala malampuzhensis R. Vasudevan Nair ex C. D. K. Cook: Fig. 266

Like *R. rosea* but flowers 3-, 4- or rarely 5-merous; sepals bright red; seeds bright red, ± 0.6 mm long.

It often grows intermingled with *R. rosea*, in the field it may be distinguished by its tufted habit, smaller and narrower leaves and the intense red of its sepals and capsules.

Short-lived annual. It is common and often gregarious in flooded rice fields, temporary pools, beside streams and in wet pockets in rocks; often forming a dense turf. It flowers within two weeks of the onset of the monsoon.

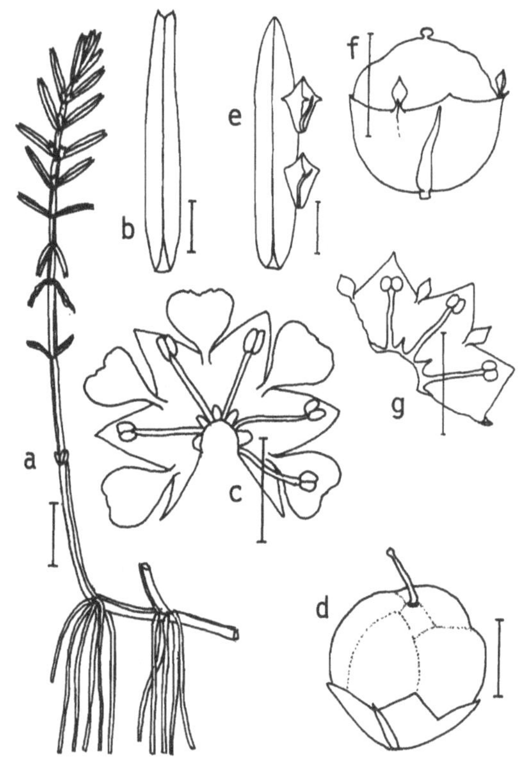

Fig. 266. *Rotala malabarica*: a, flowering plant (1 cm); b, bract (1 mm); c, flower dissection (1 mm); d, fruit with bracteole (0.5 mm). *R. malampuzhensis*; e, bract with two flower buds (1 mm); f, fruit with bracteole (1 mm); g, flower dissection (1 mm).

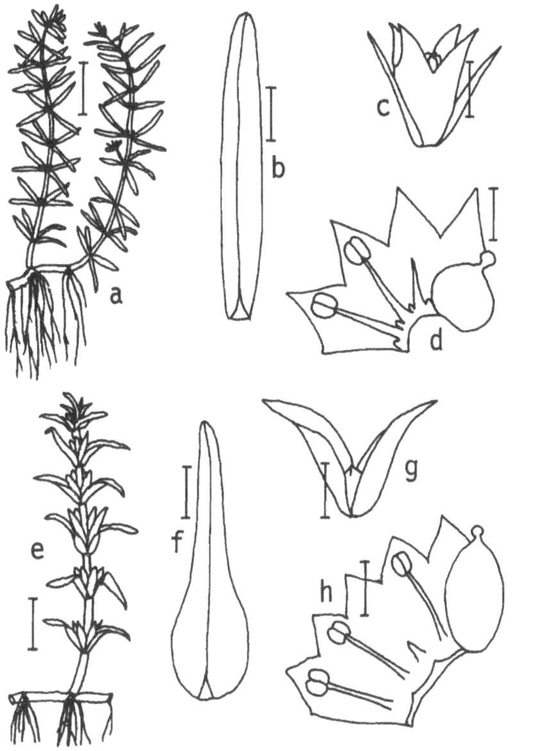

Fig. 267. *Rotala mexicana*: a, flowering plant (5 mm); b, bract (1 mm); c, flower with two bracteoles (0.5 mm); d, flower dissection (0.5 mm). *R. occultiflora*: e, flowering plant (3 mm); f, bract (0.8 mm); g, flower with two bracteoles (0.8 mm); h, flower dissection (0.5 mm).

{⊙ Ten}. Endemic to SW India, between 9° N and 12° N: ♦Ind: Kl. July–Sepember.

Rotala mexicana Chamisso et Schlechtendal [*R. pygmaea* (S. Kurz) Rajagopal et Ramayya, *Hoshiarpuria minutiflora* P. K. Hajra, P. Daniel et D. Philcox — originally described in the Scrophulariaceae]: Fig. 267

Stems branched or simple, creeping, ascending, erect or floating, often tuft- or turf-forming. Leaves at base of the stem in whorls of 3–8, sometimes decussate above; submerged leaves linear, up to 15 mm long but usually less, rarely more than 0.5 mm wide; aerial leaves linear to lanceolate or ovate, up to 10 mm long but usually less. Floral bracts either leaf-like or ovate to obspatulate and somewhat inflated below; bracteoles linear, somewhat variable but usually ± as long as the sepal tube. Flowers solitary, sessile, often cleistogamous. Sepal tube semi-globose, 0.5–0.75 mm long, red or pink; sepal lobes (3) 4 (5), widely deltate to triangular, 0.25–0.5 mm long; sepal appendages absent. Petals absent. Stamens 1–4, usually included within the sepal tube but sometimes exerted. Style up to 0.3 mm long. Capsule globose, opening by 3 or rarely 2 valves. Seeds semi-ovoid, ± 0.3 mm long.

Annual. Usually found in shallow water or on wet mud following rain. It is sometimes very short-lived and can complete its generative history in less than a month. It passes the dry period as seed.

{⊙ Vit/Ten}. Tropical and warm parts of the world but absent from the Congo Basin, NE Africa, Arabia and the Pacific Islands: Ban, Bur, Ind: Ap, As, Bh, Kl, Kt, Mp, Or, Rj, Sk, Up.

Rotala occultiflora Koehne: Fig. 267

Differs from all other *Rotala* species in having large, leaf-like bracteoles that partly enclose the flower.

It has the habit of *R. mexicana* but may be distinguished by sepal tube slightly constricted at the mouth, transparent; sepal lobes consistently 5, not spreading at anthesis; capsules ellipsoidal, somewhat longer than the sepals.

Inconspicuous annual. Found in marshes and temporary wet places, such as the margins of pools and streams. Mostly found along the coast, sometimes in brackish water.

{⊙ Vit/Ten}. SW India and N Australia: Ind: Go, Kl, Kt, Mh.

Rotala ritchiei (C. B. Clarke) Koehne: Fig. 268

Stems creeping or floating, usually simple above, up to 30 cm or more long. Leaves decussate, sessile, very variable in form; submerged leaves linear to orbicular, up to 3 cm long but usually less, almost translucent, green to red; aerial leaves obovate to orbicular, up to 1 cm long. Floral bracts like foliage leaves and exceeding the flowers; bracteoles subulate, 0.2–0.4 mm long. Flowers monomorphic, distinctly pedicellate; pedicels in fruit 1–1.5 mm long. Sepal tube subcylindrical, ± 2 mm long, pink when young; sepal lobes 4, very shallowly triangular (almost flat in fruit); sepal appendages 4, acute, somewhat exceeding the sepal lobes. Petals 4, less than 1 mm long, usually minute and caducous or rudimentary. Stamens 4, inserted in the lower half of the sepal tube; anthers included within the sepal tube. Style very short. Capsule ellipsoidal, up to 3 mm long, opening by 4 valves. Seeds semi-ellipsoidal, 0.3–0.8 mm long, straw-coloured.

Annual or perhaps sometimes perennial. Most of the plant is submerged, only the flowering tips are emergent. Flowering from August to November. It is considered to be vulnerable and is included in the 'Red Book' for India.

{(⊙) Hyp/Ten?}. Endemic to SW India: ♦Ind: Kl, Kt, Mh, Tn.

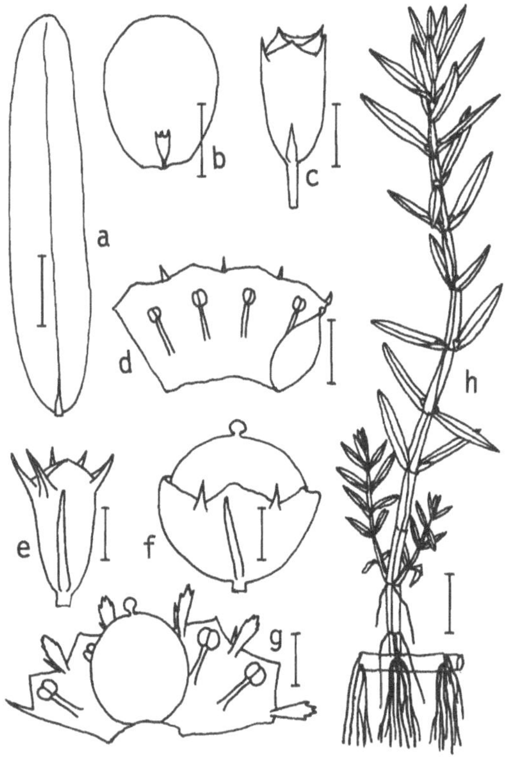

Fig. 268. *Rotala ritchiei*: a, submerged leaf (5 mm); b, bract with flower (5 mm); c, flower with bracteole (1 mm); d, flower dissection (1 mm). *R. rosea*: e, flower bud with bracteole (1 mm); f, fruit (0.8 mm); g, flower dissection (0.8 mm); h, flowering plant (1 cm).

Rotala rosea (Poiret) C. D. K. Cook [*R. pentandra* (Roxburgh) Blatter et Hallberg, *R. leptopetala* (Blume) Koehne]: Fig. 268

Stems erect or sometimes creeping below, rarely more than 30 cm long, simple or branched from below. Leaves decussate, linear-lanceolate to ovate, tips obtuse or acute, base cuneate or obtuse. Floral bracts like foliage leaves and exceeding the flowers; bracteoles shorter than the sepal tube, without a midrib. Flowers monomorphic, sessile. Sepal tube campanulate; sepal lobes 5 or rarely 4, greenish or pinkish, shallowly triangular, ± 0.25 mm long; sepal appendages subulate, longer than to shorter than the sepal lobes or sometimes rudimentary. Petals 5 or rarely 4 or rudimentary, pink, distinctly longer than wide, ± 0.25 mm long, as long or shorter than the sepal lobes. Stamens 5 or less, included within the sepal tube. Style absent or up to 0.25 mm long. Capsule about 2 times as long as the sepal tube, opening by 3 valves. Seeds 0.3–0.35 mm long, yellow to straw-coloured.

Annual. Usually found growing in shallow water or on wet mud. Common and often locally dominant in rice fields and around tanks. The size and form of the flowers and the habit are very variable. The yellow to straw colour of the seeds is characteristic.

{⊙ Ten/(Hyp)}. S and SE Asia, extending in Japan to 40° N: Ban, Bur, Cey, Ind: Ap, As, Bh, Go, Kl, Kt, Mh, Or, Pj, Rj, Tn, Up, Wb.

Rotala rotundifolia (F. Hamilton ex Roxburgh) Koehne: Fig. 269

Stems branched and somewhat woody below, creeping or floating, up to 40 cm or more long. Leaves decussate or rarely in whorls of 3, sessile or shortly petiolate, rather variable; submerged leaves linear to orbicular, up to 2 cm long, green to grey, red or purple; aerial leaves obovate to orbicular, up to 2 cm long and 2.5 cm wide, green or occasionally red on the abaxial surface; leaf nerves distinct, pinnate, green or occasionally red. Floral bracts distinct from foliage leaves, ovate, 1–4 mm long; bracteoles linear-lanceolate, ± equal to the sepal tube in length. Flowers monomorphic, subsessile; inflorescence a distinct, many-flowered, pedunculate raceme; racemes dense at early anthesis becoming lax in fruit (stem visible between nodes). Sepal tube

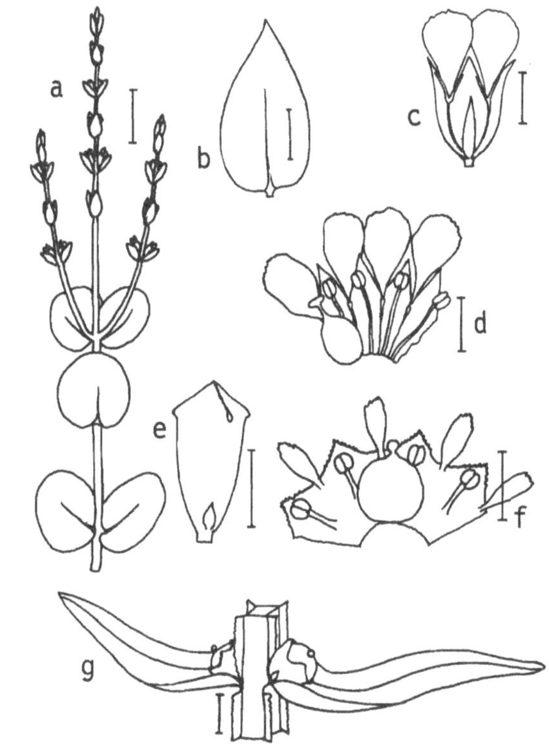

Fig. 269. *Rotala rotundifolia*: a, flowering shoot (1 cm); b, bract (1 mm); c, flower with bracteole (1 mm); d, flower dissection (1 mm). *R. rubra*: e, flower bud with bracteole (1 mm); f, flower dissection (1 mm); g, fruiting node (1 mm).

campanulate, ± 1 mm long, with 4 rib-like nerves; sepal lobes 4, triangular, up to 1 mm long; sepal appendages absent. Petals 4, obovate, bright rose, 2–3 times as long as the sepal lobes. Stamens 4; filaments inserted near the base of the sepal tube, ± 1.5 mm long; anthers included within the sepal tube. Style ± 1 mm long. Capsule ± 1.5 mm in diameter, opening by 4 valves. Seeds semi-ellipsoidal, ± 0.5 mm long.

Perennial or perhaps sometimes annual. It is found growing submerged, partly floating, emergent or terrestrial in wet places. It is mostly found in mountainous regions, reaching 2650 m in China.

{(⊙)? Vit/Hyp/(Ten/Ple)}. S and SE Asia, from India to Japan: Ban, Bhu, Bur, Ind, Nep, Pak: Ap, As, Bh, Kl, Kt, Mh, Mn, Mp, Or, Pj, Sk, Up, Wb.

Rotala rubra (F. Hamilton ex D. Don) Hara [*R. alata* Koehne]: Fig. 269

Like *R. cordata* but leaves cuneate to obtuse at the base (never cordate); petals are smaller (not exceeding 1 mm long); styles very short (± 0.1 mm long); capsules exceed the sepal tube.

Annual or perhaps sometimes perennial. Wet places.

{(⊙)? Ten/Hyp}. Endemic to Nepal: ♦Nep (Tunmingar).

Rotala serpyllifolia (Roth) Bremekamp [*R. tenuis* (Roth) C. B. Clarke]: Fig. 270

Stems creeping below and ascending above or with erect axillary shoots; erect flowering shoots up to 25 cm long, slender, terete, red and shining. Leaves decussate, when submerged linear and deep reddish-brown, when aerial linear to broadly ovate or even orbicular at base, 2–15 mm long, green, tips acute or obtuse, base cuneate. Floral bracts green or tinged with red, smaller than foliage leaves, rarely exceeding 4 mm long, lanceolate to oblong; bracteoles caducous, linear-lanceolate, ± 1.5 mm long. Flowers in dense, terminal, pedunculate, spike-like racemes; racemes up to 35 mm long. Sepal tube urn-like, constricted at the top, 2.5–3 mm long; sepal lobes 4, widely deltate, ± 0.5 mm long; sepal appendages absent. Petals 4, obovate, somewhat longer than the sepal lobes. Stamens 4, included within the sepal tube. Style ± 1 mm long. Capsule shorter than the sepal tube, opening by 2 valves. Seeds ± 0.7 mm long.

Annual or perennial in permanent water. It grows submerged in rivers, streams and lakes but is also found in standing water in seasonally inundated areas. It flowers as the water recedes. The shoots are often tightly packed and turf-forming.

{(⊙) Ten}. C and NW India and Pakistan: Ban, Ind, Pak: Bh, Gj, Kt, Mh, Mp, Rj, Up, Wb.

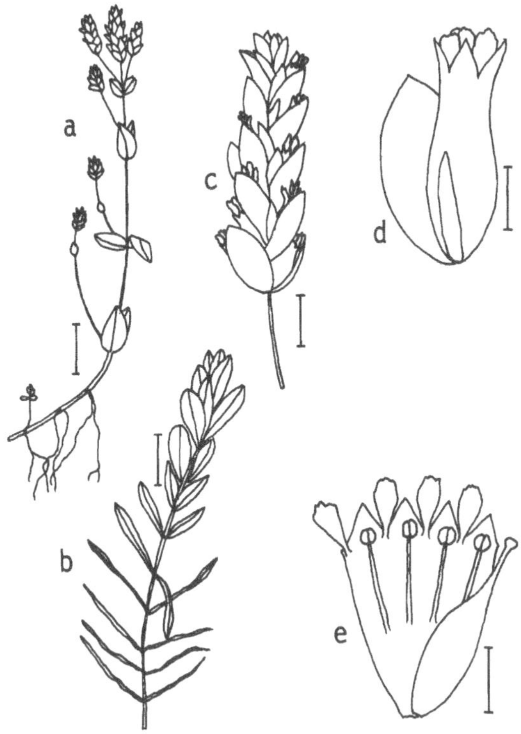

FIG. 270. *Rotala serpyllifolia:* a, flowering shoot (1 cm); b, submerged shoot (1 cm); c, inflorescence (2 mm); d, flower with bract and bracteole (1 mm); e, flower dissection (1 mm).

Rotala simpliciuscula (S. Kurz) Koehne: Fig. 271

Stems branched and creeping below, ascending and usually simple above, 1–4 cm long, 4-angled. Leaves somewhat petiolate; lower leaves narrowly oblong to oblong, 2.5–5 mm long, 0.5–2 mm wide; upper leaves somewhat smaller, oblong to ovate-oblong, tips obtuse or retuse. Floral bracts leaf-like, usually shortly staked; bracteoles absent. Flowers monomorphic, distinctly pedicellate; pedicel 0.5–1 mm long. Sepal tube campanulate at anthesis becoming hemispherical in fruit, 0.25–0.25 mm long, bright red; sepal lobes 3 or rarely 4, deltate at anthesis becoming shallowly triangular in fruit, 0.25–0.35 mm long; sepal appendages 3 or rarely 4, erect in bud. Petals absent. Stamens 1 or occasionally 2, inserted on the lower half of the sepal tube; anthers level with the sepal lobes. Style very short. Capsule globose, ± 1.4 mm in diameter, ± twice as long as the sepals, opening by 3 valves. Seeds semi-pyriform to hemispherical, ± 0.5 mm long.

Annual. It is a dwarf, mat-forming, amphibious plant that superficially resembles *Elatine ambigua*. Apparently not collected in this century.

{⊙ Ten}. Endemic to Bangladesh: ♦Ban (Chittagong).

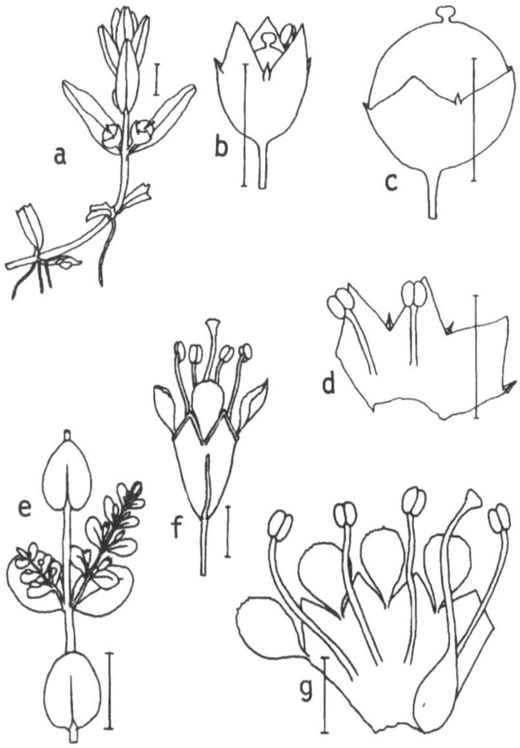

FIG. 271. *Rotala simpiciuscula*: a, fruiting shoot (1 mm); b, flower (1 mm); c, fruit (1 mm); d, flower dissection (1 mm). *R. subrotunda*: e, part of inflorescence (1 cm); f, flower with bracteole (1 mm); g, flower dissection (1 mm).

Rotala subrotunda (Wallich ex S. Kurz) Koehne: Fig. 271

Like *R. indica* but shoots usually taller and more branched; flowers distinctly pedicellate and borne only on lateral branches (never on the main axis); petals relatively large and showy (up to 1 mm long); anthers and stigmas exerted and borne above the petals.

{(⊙) Ten/Hyp?}. Endemic to NE India and Upper Myanmar, Sagaing: ♦Bur, ♦Ind: Mn. December–April.

Rotala verticillaris Linnaeus: Fig. 272

Stems creeping and somewhat woody below, floating, ascending or erect above, up to 20 cm long but usually less. Leaves in whorls of 4–10, sometimes decussate at the base; submerged leaves capillary, 10 mm or more long, rarely more than 1 mm wide; aerial leaves linear to oblong or oblanceolate, up to 2.5 mm wide; leaf tips truncate or minutely bifid. Floral bracts leaf-like, usually in whorls of 4 or rarely 8 or 2; bracteoles capillary, up to 0.8 mm long. Flowers monomorphic, sessile. Sepal tube membranous, constricted at the mouth (urn-shaped), 2–3 mm long; sepal lobes 3 or rarely 4, deltate, ± 0.5 mm long; sepal appendages absent but small folds sometimes present. Petals 3 or rarely 4, ovate to cordate, pink, ± as long as the sepal lobes, usually persisting in fruit. Stamens 3 or rarely 4, inserted at about the middle of the sepal tube; anthers scarcely reaching the sinuses of the sepal lobes. Style very short. Capsule globose-ellipsoidal, opening by 3 valves; seeds ovoid, 0.25–0.4 mm long.

Annual or perhaps also sometimes perennial. Found in standing water in pools and along the marshy margins of ponds. It flowers as the water recedes.

{(⊙) Vit/Hyp/Ten}. Endemic to SE India and Sri Lanka: ♦Cey, ♦Ind: Ap, Mp, Tn.

Rotala wallichii (J. D. Hooker) Koehne: Fig. 272

Stems creeping and branched below, erect or floating and usually simple above, up to 30 cm or more long. Leaves in whorls of up to 20 or more; submerged leaves linear or capillary, thin, translucent, tinged with red, up to 20 mm long, at tip bifid; aerial leaves linear to oblong, rarely more than 5 mm long, in whorls of 3–12 (less than the submerged leaves) or rarely decussate, at base scarcely narrowed, at tip obtuse or bifid. Floral bracts leaf-like, ovate to oblong, 1.5–3 mm long, at tip acute or bifid; bracteoles capillary, up to 0.5 mm

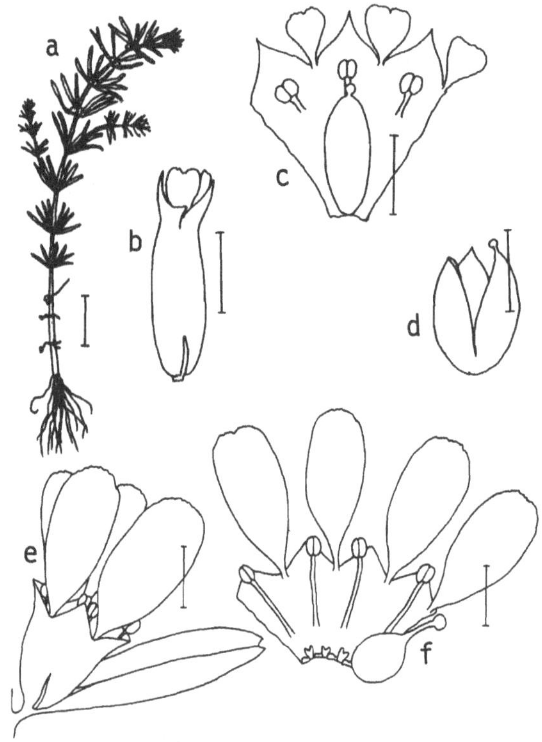

FIG. 272. *Rotala verticillaris*: a, flowering plant (1 cm); b, flower with bracteole (1 mm); c, flower dissection (1 mm); d, dehiscing capsule (1 mm). *R. wallichii*: e, flower with bract and bracteole (1 mm); f, flower dissection (1 mm).

long. Flowers monomorphic, chasmogamous, shortly pedicellate; pedicel adnate to the bract, the combined organ up to 1 mm long. Sepal tube campanulate, membranous, translucent, ± 1.5 mm long; sepal lobes 4, deltate, 0.5 mm long; sepal appendages absent. Petals 4, obovate, showy, pink fading to white, ± 2.5 mm long. Stamens 4, inserted on the lower half of the sepal tube; anthers borne level with the top of the sepal lobes. Nectar scales 2-lobed. Style up to 0.5 mm long. Capsule globose, ± 1 mm in diameter, opening by 2 valves. Seeds semi-ovoid, ± 0.7 mm long (perhaps larger — ripe material rare)

Perennial. Usually found in relatively deep water but can be cultivated terrestrially. Effectively self-incompatible and rarely found with seeds. Cultivated as a decorative aquarium plant.

{Vit}. NE India to Malaya and SE China: Ban, Bur: Wb. The record from the Aravalli Hills (Rj) is surely an error.

MARANTACEAE: excluded see p. 12

MENYANTHACEAE: 5 genera

Monoecious or dioecious, perennial with tufted rootstock or rhizome or annual. Leaves alternate; petioles sheathing at the base; blades linear to orbicular or trifoliate. Inflorescence of simple or branched racemes or cymes or in dense heads or clusters. Flowers bisexual, radially symmetrical, often heterostylous. Sepals (4) 5–7 (8), united at the base. Petals (4) 5–7 (8), united at the base, often fimbriate inside. Stamens 5, united to the petals. Ovary superior, 1-locular with 2 parietal placentas. Fruit a capsule or sometimes berry-like.

Excluded: found in wetlands in the Himalayas but does not extend further south in India: ***Menyanthes trifoliata*** Linnaeus.

Nymphoides: 20 species: cosmopolitan

Dioecious, gynodioecious or monoecious, annual or perennial, some rhizomatous, some with banana-shaped root tubers. Flowering shoots sometimes with a single node and a single leaf, thus resembling petioles. Leaf blades submerged or floating, broadly ovate to orbicular or reniform, entire or crenate, green above, often dark below. Flowers solitary or in clusters, supported by floating leaves, often appearing to arise from the petiole or from the base of the leaf blade, unisexual or bisexual, usually distylic; adventitious root tubers sometimes present at the base of the inflorescence. Sepals triangular. Petals white or yellow, mostly fimbriate inside, often with crested midnerve and margins. Capsule sometimes regularly dehiscent but often opening irregularly or sometimes indehiscent.

Bottom rooted with leaves floating on the surface, submerged leaves may or may not be present. Pollinated by insects or wind, some species autogamous, others distylic, gynodioecious or dioecious. Disseminules seeds, dispersed by adhesion to animals or by water or indehiscent capsules with unknown dispersal. The tubers of some species eaten locally. Seeds are used locally in medicine. Some are cultivated for ornament. Sivarajan, V. V. and Joseph, K. T. The genus *Nymphoides* Séguier in India. Aquatic Botany 45(2–3): 145–170 (1991).

Excluded: an aquatic plant found in the western Himalayas which does not extend further south in India: ***Nymphoides peltata*** (S. G. Gmelin) O. Kuntze [*Limnanthemum nymphoides* (Linnaeus) Link].

1. Flowering stems with several nodes (alternating leaves bearing flowers at intervals or the node at end of the petiole-like stem bearing 2 or more leaves); petals yellow
- **2.** Flower clusters subtended by 1 leaf only (bracts scale-like); leaves clearly dotted with black glands beneath; seeds densely hairy and almost spherical (tropical S India) ***N. aurantiaca***
- **2.** Flower clusters subtended by 1 leaf and 1 or 2 leaf-like bracts; leaves dotted with minute (not visible with naked eye) glands beneath; seeds disk-like with marginal hairs (Himalayas) ***N. peltata***

1. Flowering stems 1-noded (flowers apparently arising from the end of the petiole); petals white
- **3.** Leaves monomorphic, all floating and associated with inflorescences
  - **4.** Petal lobes not fimbriate and without hairs (except at the throat), each lobe with a smooth but undulate longitudinal crest along the centre on the inner surface and with smooth, crest-like, undulate margins ***N. hydrophylla***
  - **4.** Petal lobes either fimbriate or with hairs, crests (when present) fimbriate or toothed along the margins
    - **5.** Flowers bisexual; petal lobes usually 1 cm or more long, uniformly and densely fimbriate with multicellular, cotton-like outgrowths on the inner surface and along the margins, without any longitudinal folds or crests; seeds 1–2 mm in diameter ***N. indica***
    - **5.** Flowers unisexual; petal lobes rarely more than 0.6 cm long, with fimbriae confined to midribs, longitudinal crests and margins; seeds 3–6 mm in diameter ***N. macrosperma***
- **3.** Leaves dimorphic; basal leaves in rosettes, not associated with inflorescences, usually submerged; fertile leaves not in rosettes, all floating and each with an inflorescence
  - **6.** Flowers unisexual; stigmas with 2 whorls of long, radiating hairs; anthers blue to purple ***N. krishnakesara***

6. Flowers bisexual; stigmas without hairs; anthers yellow

7. Petal lobes 3–5 mm long, the marginal fimbriae tooth-like, less than 0.5 mm long; anthers exerted beyond the petal tube ***N. parvifolia***

7. Petal lobes ± 8 mm long, the marginal fimbriae hair-like, mostly more than 1 mm long; anthers included within the petal tube ***N. sivarajanii***

Nymphoides aurantiaca (Dalzell) O. Kuntze [*Limnanthemum biflorum* Thwaites]: Fig. 273

Rhizome purplish-red, elongate bearing leaves and flowers at regular intervals. Petiole-like shoots with several nodes, up to 45 cm long, simple, flexible. Petioles 1.5–6.5 cm long, purplish-red; leaves on flowering stems alternate; blades of floating leaves oval-orbicular, 2–5 (–8) cm in diameter, dotted with blackish-purple glands beneath, deeply cordate at the base, with a narrow sinus, purplish-brown on the abaxial (lower) surface. Flowers bisexual, weakly distylic, in lax clusters of (1) 2–4 flowers. Pedicels up to 5 cm long. Bracts sheathing, oblong-lanceolate, up to 6 mm long and 2 mm wide, dotted with reddish glands. Sepal lobes (4) 5, ovate-lanceolate, ± 5 mm long, purplish-red, glandular, margins membranous. Petal tubes 5–7 mm long, orange; lobes (4) 5, obovate-oblong, up to 4–7 mm long, orange, fimbriate at base and each with a small tuft of multicellular, yellow, plumose hairs, margins fimbriately toothed, apex notched. Anther filaments 1–2.5 mm long. Disk glands minute or absent. Capsule depressed-globose to obovoid, 3–6 mm long, 3.5–4.5 mm in diameter, usually 12–15 seeds. Seeds flattened-globose, ± 2 mm in diameter, densely covered with minute spines.

Annual or perennial. Shallow ponds, particularly near the coast. It comes into flower with the southwest monsoon.

{(⊙) Eph}. Endemic to S India and Sri Lanka: ♦Cey, ♦Ind: Kl, Kt, Mh. Records from Bengal are probably wrong.

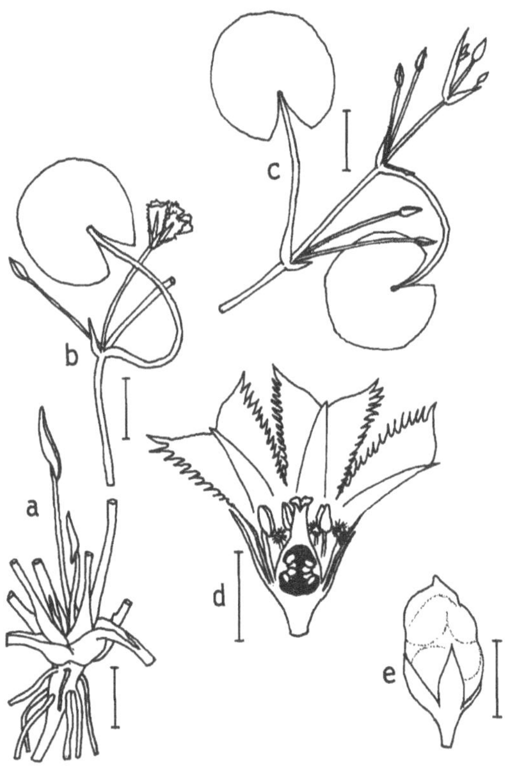

FIG. 273. *Nymphoides aurantiaca*: a, rootstock (2 cm); b, flowering node (2 cm); c, shoot tip (2 cm); d, flower in longitudinal section (5 mm); e, fruit (2 mm).

Nymphoides hydrophylla (Loureiro) O. Kuntze [*N. cristata* (Roxburgh) O. Kuntze]: Fig. 274

Rhizome short, erect. Petiole-like shoots up to 85 cm long, slender. Petioles up to 2.5 cm long; blades of floating leaves oval-orbicular to orbicular, 4–13 cm long and 3.5–12 cm wide, green sometimes with purplish blotches, without glands, rather thin, deeply cordate to hastate at the base, coarsely crenate; nerves obscure above, prominent on abaxial (lower) surface. Flowers bisexual or sometimes female, weakly distylic, in umbel-like clusters of 10–20 or more flowers, borne below a floating leaf. Pedicels 2.8–6 cm long. Bracts ovate, ± 2.5 mm long. Sepal lobes 5 (6), linear-oblong, 3–5 mm long, tips subacute. Petal tubes up to 2 mm long, with a ring of glandular hairs above, the throat yellow or sometimes white within; lobes 5 (6), oval-oblong, 7–10 mm long, pure white, with a longitudinal crest along the centre on the inner surface and with crest-like undulate margins, without hairs but with some hairy glands near the base, obtuse. Stamens inserted halfway up the petal tube, anthers subsessile, ± 1.5 mm long. Disk glands 5 or 6, each suborbicular, with papillose hairs at the tips. Capsule broadly ovoid, 3–6 mm long, 3.5–4 mm in diameter, dehiscing irregularly. Seeds 4–6 (–10), globose, ± 2 mm in diameter, yellowish-white, with small but prominent, slightly barbed tubercles.

Mostly annual, sometimes perennial. Deeply rooted in mud in lakes, tanks and temporary pools, sometimes in slowly flowing water. The roots may give rise to new shoots. It flowers and fruits as long as it has sufficient water. Gregarious and common, sometimes growing with *Nymphoides indica*. It is monoecious and gynodioecious but bisexual flowers are self-compatible, see Vasudevan Nair, R. Heterostyly and breeding system of *Nymphoides cristatum* (Roxb.) O. Kuntze [sic]. Journal of the Bombay Natural History Society 72: 677–682 (1976). The leaves are used in medicine for ulcers and bites, it is also claimed to be an helminthic.

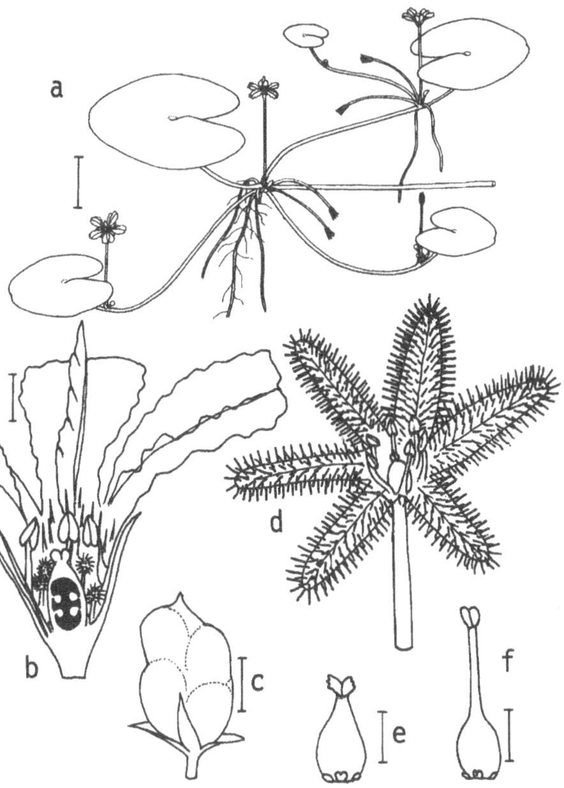

Fig. 274. *Nymphoides hydrophila*: a, shoot with floating leaves and inflorescences (2 cm); b, flower in longitudinal section (1 mm); c, fruit (2.7 mm). *N. indica*: d, thrum flower (6 mm); e, ovary with thrum style (2 mm); f, ovary with pin style (2 mm).

{(⊙) Eph}. Tropical Asia: Ban, Cey, Ind, Nep, Pak: An, Ap, As, Bh, Dl, Go, Hp, Kl, Kt, Mh, Mn, Mp, Or, Pj, Rj, Tn, Up, Wb.

Nymphoides indica (Linnaeus) O. Kuntze [*Limnanthemum wrightianum* Grisebach, *L. kleinianum* Grisebach]: Fig. 274

Rhizome oblique, up to 5 cm thick, with numerous scars and scale leaves, sending out long stolons which develop new plantlets. Petiole-like shoots up to 150 cm long, simple, flexible. Petioles up to 16 cm long but when flowering rarely more than 2 cm long; blades of floating leaves oval-orbicular to orbicular, up to ± 27 cm long and 30 cm wide, glossy green above, pale and gland-dotted below, rather thick and leathery, deeply cordate at the base, coarsely crenate; nerves obscure above, prominent on abaxial (lower) surface. Flowers bisexual, distylic, in umbel-like clusters of (3–) 15–50 flowers, borne below a floating leaf. Pedicels up to 12 cm long. Bracts sheathing, oblong-lanceolate, up to 8 mm long and 5 mm wide, rather thick. Sepal lobes (4–) 5–7 (–8), ovate-lanceolate, up to 7 mm long, margins hyaline, tips obtuse, reflexed with age. Petal tubes up to 2.5 mm long, the throat yellow or sometimes white; lobes (4–) 5–7 (–8), linear-lanceolate to oblong-lanceolate, up to ± 17 mm long, pure white or white with a yellow or orange base (sometimes entirely yellow in other regions of the world), inner surface and margins densely fimbriate, with multicellular, cotton-like outgrowths. Anther filaments broadened at the base, up to 2.5 mm long, yellow; anthers ± 3 mm long, yellow. Disk glands 4–8, each suborbicular, ciliolate at the tips with papillose hairs. Styles either ± 1 mm or ± 3 mm long. Capsule ellipsoid to obovoid, 6–8 mm long, 4.5–6 mm in diameter, dehiscing irregularly. Seeds globose, 1.2–2.1 mm in diameter, yellowish-white, smooth and shining or with a few small tubercles.

Annual or perennial. Generally common and abundant in fresh and brackish water. In lakes, tanks and pools but rarely in flowing water. Flowering and fruiting as long as enough water is available.

{(⊙) Eph}. Pantropics: Ban, Cey, Ind, Nep: An, Ap, As, Bh, Gj, Go, Jk, Kl, Kt, Mh, Mn, Mp, Or, Pj, Rj, Tn, Up, Wb.

Nymphoides krishnakesara K. T. Joseph et V. V. Sivarajan: Fig. 275

Vegetative parts like *N. parvifolia* with sterile radical leaves that may remain submerged but differ in having winged or sheathing petioles. Florally like *N. macrosperma* with flowers unisexual; petals folded backwards and arched upwards with a short dorsal crest; filaments 1–1.5 mm long; anthers ± 2 mm long, spirally wound, blue to purple; stigmatic lobes 15–20, linear, ± half as long as the carpel, papillose; seeds 5–10, discoid, ± 3 mm in diameter, rough.

Annual and perhaps also sometimes perennial. Found during the monsoon season in shallow, temporary pools on laterite. It has been reported that male and females usually grow about 1:1. It flowers from August to November. In the original description Joseph and Sivarajan misinterpreted the shape of the petals and the morphology of the stigmas. The flowers are pollinated by wind — a unique feature in this family.

{(⊙)? Eph}. Endemic to northern Kerala: ♦Ind: Kl (Kannur District).

Nymphoides macrosperma R. Vasudevan (Nair): Fig. 276

Rhizome up to 2 cm thick, developing elongate stolons. Leaf blades up to ± 20 cm in diameter, green and gland-dotted below, rather thick and leathery. Fertile shoots dimorphic; primary shoots petiole-like, up to 2.5 m long, each bearing a single leaf; secondary

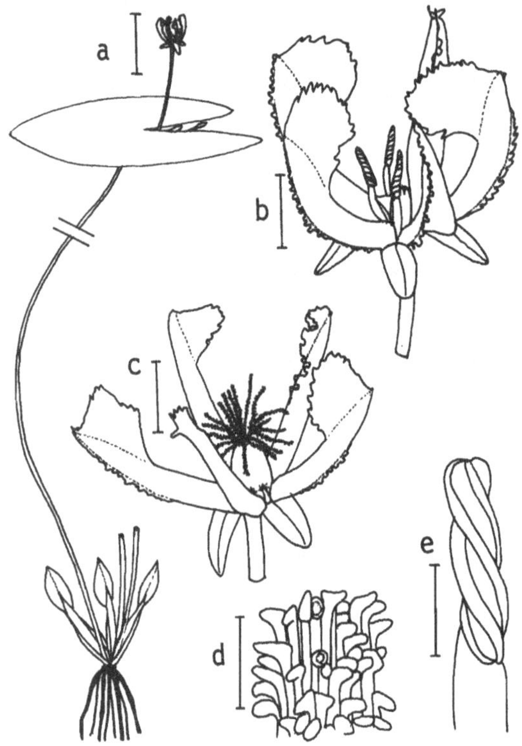

FIG. 275. *Nymphoides krishnakesara*: a, flowering plant with axillary shoot shortened (2.5 cm); b, male flower (4 mm); c, female flower (4 mm); d, stigmatic papillae (90 μm); e, anther (720 μm).

shoots arising from the primary ones, each terminating in a leaf and cluster of flowers. Flowers unisexual, clustered at proximal and distal end of the stem, up to ± 50 flowers in a cluster. Pedicels 2–6 (–10) cm long. Sepal lobes oblong, 1.5–2 mm long, 3–4 mm broad, acute. Petal lobes oblong, ± 6 mm long, ± 2 mm broad, pure white, basal part of upper surface and the margins fimbriate, with glands at base. Stamens inserted at the petal tube throat; filaments up to 2 mm long; anthers ± 1 mm long, pale purple, rarely pale yellow before anthesis. Pistillode in male flowers prominent, ± 3 mm long, resembling a gynoecium. Staminodes in female flowers reduced to short, white filaments. Stigmas with 1 series of long, radiating hairs. Capsule irregularly subglobose, reddish, ± 6 mm long, ± 8 mm in diameter, usually 6-seeded, dehiscing irregularly. Seeds 2–6, obovate-lenticular, ± 3.5 mm long, ± 4.5 mm in diameter, slightly rough.

Annual or perennial. Usually found in deep water (2 m or more deep) in lagoons and slowly flowing streams at sea level. It is also sometimes in temporary water in ditches and rice fields. Dwarf plants are sometimes found flowering and fruiting on wet mud after the water has receded. It is dioecious but male plants are sometimes more abundant than females.

{(⊙) Eph}. Endemic to coastal Kerala: ♦Ind: Kl. September–December.

Nymphoides parvifolia (Grisebach) O. Kuntze [*Limnanthemum moonii* Thwaites]: Fig. 277

Rhizomes small, erect, slender. Stems bearing radical leaves and petiole-like stems, not rooting at nodes. Radical leaves with petioles 2–4 cm long, usually winged at the base; blades spathulate-orbicular to deltate, 7–15 (–30) mm long, 6–11 (–15) mm wide, narrowed at the base. Floating leaves on petiole-like stems with petioles 2–5 mm long; blades broadly oval to cordate-orbicular or obovate, 1–4.2 cm long, 1–3 cm wide, with a deep triangular sinus, green. Flowers bisexual, homostylic, in clusters of 4–8. Pedicels very slender, 0.4–4 cm long. Sepal lobes 4 (5), oblong-lanceolate, 1.75–3 mm long, faintly 3-nerved. Petal

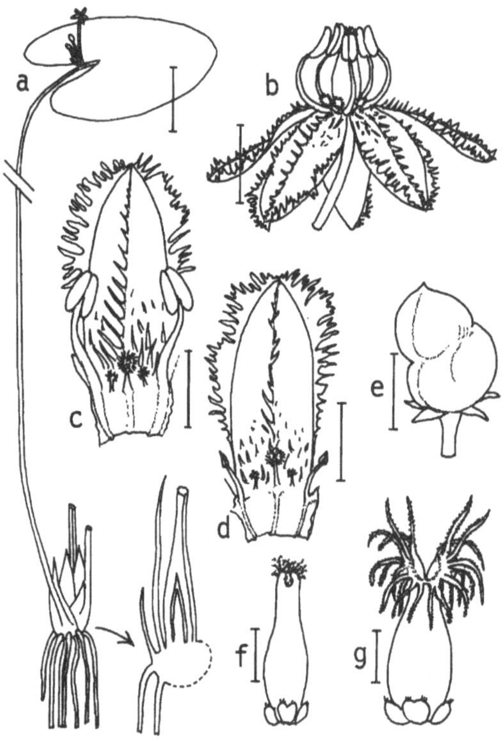

FIG. 276. *Nymphoides macrosperma*: a, flowering plant (2 cm) with axillary shoot shortened and a longitudinal section of the base of an axillary branch (inset); b, male flower (3 mm); c, petal of male flower with two stamens and three glands (2 mm); d, female petal with two staminodes and three glands (2 mm); e, fruit (4 mm); f, pistillode (1 mm); g, carpel (1 mm).

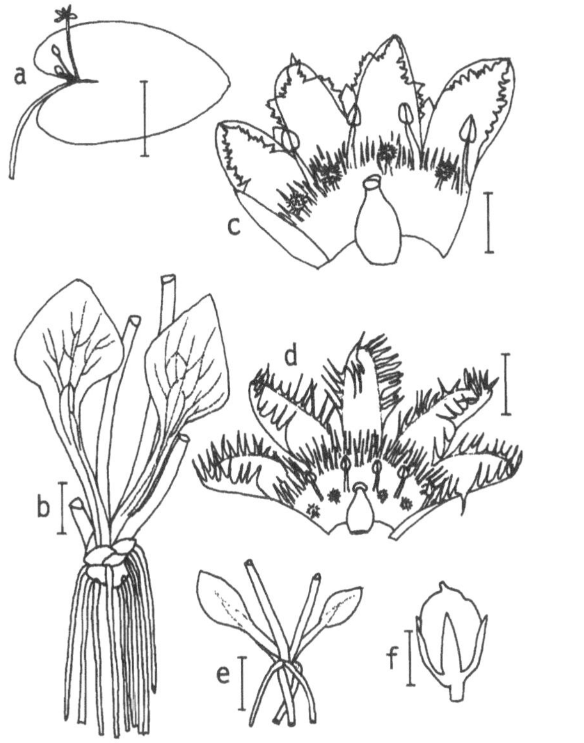

FIG. 277. *Nymphoides parvifolia*: a, floating leaf with inflorescence (2 cm); b, base of stem with submerged leaves (5 mm); c, flower dissection (2 mm). *N. sivarajanii*: d, flower dissection (5 mm); e, base of young plant (5 mm); f, fruit (2 mm).

tube 2–3 mm long, the throat usually yellow and with a ring of hyaline hairs; lobes 4 (5), 3–8 mm long, pure white, inner surface and margins densely fimbriate (the fimbriae not exceeding 0.5 mm long), notched at the tip. Disk glands 4 or 5, minute, each tufted and minutely stalked. Stamens inserted towards the top of the throat; filaments 1–1.5 mm long, anthers up to ± 2 mm long, exerted beyond the petal tube. Capsule obovoid to broadly conical, 2–3.5 mm long. Seeds 8–15 or more, globose, 0.6–1 mm in diameter, white, tuberculate.

Annual or perhaps also perennial. In shallow ponds, rice fields and wet places near to the coast. Rather rare. It sometimes grows terrestrially on drying mud; these terrestrial plants are often very small and resemble *Cochliaria*.

{(⊙)? Eph}. S and SE Asia and N Australia: Ban, Cey, Ind: Gj, Kl, Kt, Mh, Mn.

Nymphoides sivarajanii K. T. Joseph: Fig. 277

Like *N. parvifolia* but more profusely branched, flowers usually 5-merous, in clusters of 5–15; sepals ± 4 mm long; petal lobes ± 8 mm long with hair-like marginal fimbriae mostly more than 1 mm long; anthers not exerted from the petal tube; seeds 15–20.

Annual. Locally abundant in flooded lowlands during the rainy season.

{⊙ Eph}. Endemic to S India: ♦Ind; Kl (Malappuram District, not found in recent years). August–October.

NAJADACEAE: 1 genus

Najas: 39 species: cosmopolitan

Dioecious or monoecious, annual or sometimes perennial, entirely submerged, bottom-rooted but parts often becoming detached and free-swimming; roots simple. Stems usually much branched. Leaves alternating but often appearing to be opposite, pseudowhorled or in axillary clusters; base sheathing, open, occasionally auriculate; leaf blades simple, linear, the margins usually toothed. Flowers inconspicuous, unisexual, axillary, solitary or in small clusters. Male flower enclosed within 2 membranous sheaths, the outer sheath (spathe) terminating in 2 or 4 teeth or spathe absent, the inner sheath (perianth or involucre) terminating in 2 thickened lips; stamen 1; anther subsessile, 1- or 4-sporangiate. Female flowers naked or surrounded by a membranous sheath (spathe). Ovary of 1 carpel tapering to a short style; stigmas 2–4, linear, often unequal; ovules solitary; fruit a capsule with a thin, separable, membranous pericarp. Seed coats usually pitted.

Totally submerged, sometimes very abundant and locally dominant, some in brackish water. Pollination hypohydrophilous (under water). The disseminules are seeds which sink, they are dispersed by water or in mud and in the digestive tract of animals (fish and waterfowl). Some species are serious weeds in water reservoirs and irrigation channels, others are used for green manure, fish food and packing material. Snails carrying bilharzia parasites have been reported to feed on *Najas*. Triest, L. A revision of the genus *Najas* in the Old World. Academie Royale des Sciences d' Outre-Mer, Classe des Sciences Naturelles et Médicales (Bruxelles). Mèmoires in 8°, Nouvelle Série, 22(1): 1–178 (1988); it was used as the basis for most illustrations.

The names *Najas foveolata*, *N. indica* and *N. minor* have been used very indiscriminately in Indian floristic works.

Excluded: ***Najas foveolata*** A. Braun; it is from eastern tropical Asia and N Australia and does not even occur in India.

1. Stems and the abaxial side of the midribs armed with spines; leaf blades fleshy and often brittle; seeds at least 6 mm long ***N. marina***

1. Stems and the abaxial side of the midribs without spines; leaf blades flat and usually flexible; seeds less than 4 mm long

2. Leaf sheaths sloping, rounded to broadly rounded, never auriculate

3. Male and female flowers borne on the same branches (male flowers towards the apex but also together with female flowers); female flower enclosed within a spathe; seed coat pits hexagonal (South Andaman Island) ***N. kingii***

3. Male and female flowers borne on different branches (or even plants); female flower naked or enclosed within a spathe; seed coat pits squarish to hexagonal (widespread) ***N. tenuis***

2. Leaf sheaths truncate or lacerate to auriculate

4. Seed coat pits ± 0.12 mm long, irregularly arranged in longitudinal rows of 12 (Sri Lanka) ***N. grossareolata***

4. Seed coat pits less than 0.1 mm long, arranged in regular longitudinal rows of more than 20

5. Seed coat pits rectangular, wider than long (±0.03 mm long, ± 0.1 mm wide), arranged ladder-like, more than 60 in each longitudinal row (N India)

6. Seeds 1.74–2.74 mm long; anther 1-sporangiate, 0.56–1.3 mm long ***N. minor***

6. Seeds 2.6–3.5 mm long; anther 4-sporangiate, 1.3–1.8 mm long ***N. oguraensis***

5. Seed coat pits squarish to hexagonal, ± as wide as long (0.03–0.08 mm long and wide), not arranged like ladders, less than 60 in each longitudinal row (widespread)

7. Male flower enclosed within a spathe (female flower naked)

8. Anther 1-sporangiate, 0.5–0.6 mm long; leaf blades without septa, marginal teeth cells not borne on multicellular swellings; leaf sheaths slightly auricled; spathe of male flower 0.7–1 mm long (Madhya Pradesh) ***N. kurziana***

8. Anther 4-sporangiate, (0.8–) 1.1–1.4 mm long; leaf blades with or without septa, marginal teeth cells borne on multicellular swellings; leaf sheaths with large auricles; spathe of male flower (1.4–) 1.6–2.3 mm long (widespread) ***N. indica***

7. Male flower naked (female flower naked)

9. Anther 4-sporangiate; leaf margin teeth minute and hardly visible with the unaided eye ***N. graminea***

9. Anther 1-sporangiate; leaf margin teeth visible with the unaided eye ***N. malesiana***

Najas graminea Raffeneau-Delile: Fig. 278

Monoecious, slender or robust, often appearing feather-like above because of closely packed leaves. Stems without spines, 0.4–1.5 mm in diameter. Leaf sheaths (1.4–) 2–4 (–5) long (including auricle and spine cells), 0.9–2 (–2.7) mm wide, deeply auriculate; auricles (0.4–) 0.8–1.2 (–2.6) mm long (including the spines), (0.16–) 0.2–0.3 (–0.5) mm wide, serrulate with 3–14 spines on each side and 1–7 on the inner edge, apex acute. Leaf blades, flat, (9–) 14–25 (–33) mm long, (0.24–) 0.5–0.9 (–1.1) mm wide (including the teeth), (0.2–) 0.4–0.8 (–0.9) mm wide (excluding teeth); margins minutely serrulate, with unicellular teeth, invisible to the unaided eye; the teeth (18–) 34–60 (–74) on each margin, 0.02–0.07 (–0.12) mm long, not borne on multicellular swellings; midrib without teeth; septa absent; fibres mostly present along the midrib and along the margins. Male flowers naked; anther 4-sporangiate, 0.7–1.3 (–2.7) mm long. Female flowers naked, 1.6–3.7 mm long. Seeds elliptical-oblong, (1.26–) 1.5–2.4 (–4.2) mm long, (0.42–) 0.5–0.7 (–0.9) mm in diameter; seed coat pits squarish

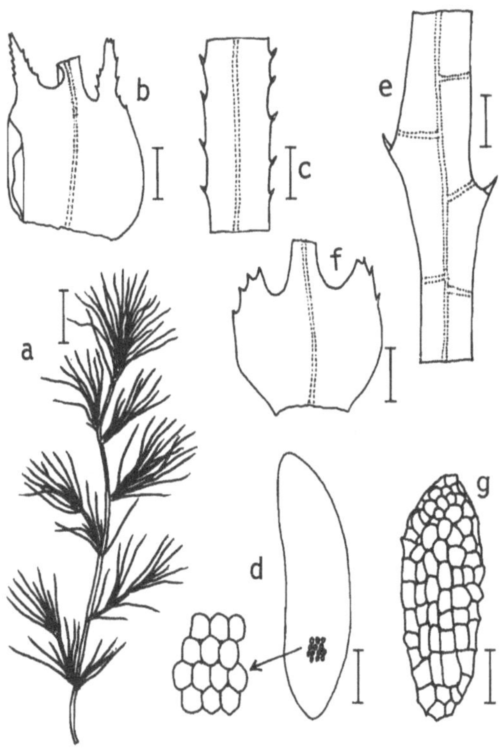

FIG. 278. *Najas graminea*: a, habit (1 cm); b, leaf sheath (1 mm); c, part of leaf blade (0.3 mm); d, seed (0.6 mm) with surface pattern (inset). *N. grossareolata*: e, part of leaf blade (0.3 mm); f, leaf sheath (1 mm); g, seed (0.3 mm).

to hexagonal or rectangular, (0.05–) 0.06–0.08 mm long, in regular longitudinal rows of (23–) 25–35 (–60).

Perennial and probably also annual. Often locally dominant in standing fresh or brackish water. It can be distinguished from the other species by the usually fine, 'spineless' leaves which fan out like feathers at the tips of the stems.

{(⊙)? Vit}. Warm and tropical Europe, Africa, Asia and Australia, naturalized in N America: Ban, Bur, Cey, Ind, Nep: Ap, Kl, Kt, Mp, Rj, Tn, Wb.

Najas grossareolata L. Triest: Fig. 278

Monoecious, slender. Stems without spines, ± 0.5 mm in diameter. Leaf sheaths truncate to lacerate, ± 2.4 mm long (including the spine cells), ± 2.25 mm wide, serrulate or laciniate with 4–6 spines on each side; auricles 0.3–0.4 mm long (including the spine cells), 0.25–0.3 mm wide, serrulate or laciniate with 4–6 spines on each side but none on the inner side, apex acuminate. Leaf blades flat, flexible, 13–17 mm long, 0.5–0.8 mm wide (including the teeth), 0.35–0.4 mm wide (excluding the teeth); margins serrulate, with 9–15, conspicuous spiny teeth; the teeth 0.08–0.17 mm long, borne on rather small multicellular swellings; midrib without teeth; septa clearly visible up to the margins; fibres absent. Male flowers naked, anther 1-sporangiate, ± 0.8 mm long. Female flowers naked. Seeds elliptical-oblong, ±1.45 mm long, ± 0.6 mm in diameter, clearly pitted; seed coat pits ± 0.12 mm long, irregularly shaped, irregularly arranged in longitudinal rows of 12, cell walls raised.

Annual or perennial.

{(⊙) Vit}. Endemic to Sri Lanka: ♦Cey (Ellawewa).

Najas indica (Willdenow) Chamisso [*N. heteromorpha* Griffith, *N. falciculata* A. Braun, *N. minor* var. *spinosa* Rendle, *N. lacerata* Rendle, *N. foveolata* var. *minor* Rendle]: Fig. 279

Probably dioecious but perhaps monoecious, slender. Stems without spines, ± 1–1.3 mm in diameter. Leaf sheaths auriculate, truncate or lacerate, 1.6–3.7 mm long (including the spine cells), 1.2–3.2 mm wide, serrulate or laciniate with 4–10 spines on each side; auricle 0.22–1 mm long (including the spine cells), 0.2–0.64 (–0.8) mm wide, serrulate or laciniate with 4–10 spines on each side but none on the inner edge, apex rather obtuse. Leaf blades flat, flexible, 10–26 mm long, 0.5–1.2 mm wide (including the teeth), 0.2–0.5 (–0.8) mm wide (excluding the teeth); margins serrulate, with 8–20, conspicuous spiny teeth, the teeth 0.14–0.4 mm long, borne on broad multicellular swellings, clearly visible to the naked eye; midrib mostly without teeth; septa clearly visible up to the margins; fibres absent. Male flowers borne in spathes, (1.3–) 2–3 mm long; anther 4-sporangiate, (0.8–) 1.4–1.6 mm long. Female flowers naked. Seeds elliptical-oblong, 1.5–2.6 mm long, 0.6–0.8 mm in diameter; seed coat pits squarish to hexagonal, overlapping but in regular longitudinal rows of 25–35.

Perennial and perhaps also annual. A poorly understood species in the past and frequently misapplied and has even been considered synonymous with *N. minor*. It may be distinguished from *N. foveolata* and *N. minor* by leaves with large teeth and large septa; leaf sheath with auricles; anther 4-sporangiate; seed coat pits overlapping (looking like a brick wall, the other two species have the pits arranged in rows like ladders).

{(⊙)? Vit}. India, Sri Lanka, Myanmar and perhaps Vietnam: Ban, Bur, Cey, Ind: As? Gj, Kt, Mh, Or? Tn, Up, Wb.

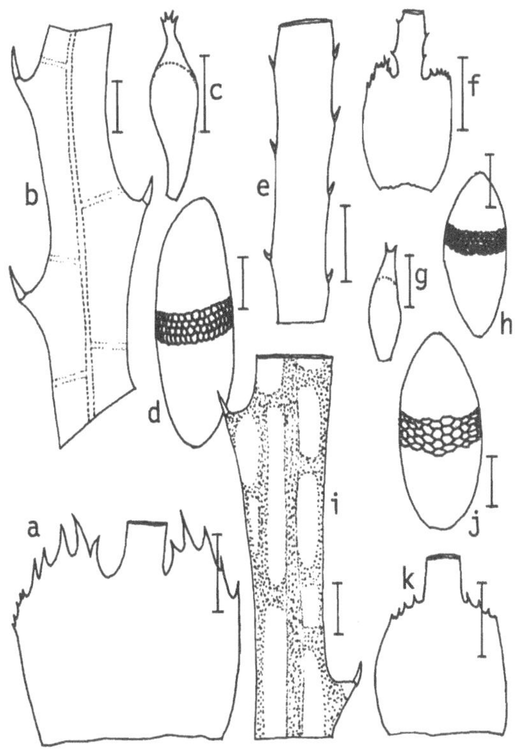

FIG. 279. *Najas indica*: a, leaf sheath (1 mm); b, part of leaf blade (0.3 mm); c, male flower in spathe (1 mm); d, seed (0.3 mm) with surface pattern (inset). *N. kurziana*: e, part of leaf blade (1 mm); f, leaf sheath (1 mm); g, male flower in spathe (0.3 mm); h, seed (0.3 mm) with surface pattern (inset). *N. kingii*: i, part of leaf blade (0.3 mm); j, leaf sheath (1 mm); k, seed (0.3 mm) with surface pattern (inset).

Najas kingii Rendle: Fig. 279

Like *N. tenuis* but male and female flowers borne on the same branches (male flowers towards the apex but also together with female flowers); female flower in a spathe; seeds 1.6–1.9 mm long, 0.5–0.8 mm in diameter; seed coat pits hexagonal, regularly arranged, (20–) 24–30 in each longitudinal row.

Perennial or perhaps sometimes annual. Recorded in rice fields in Vietnam. Cultivated as a ornamental in aquaria.

{(⊙)? Vit}. SE Asia: Ind: An (South Andaman).

Najas kurziana Rendle: Fig. 279

Like *N. indica* but leaf blades lack septa and fibres or some fibres on or near both sides of the midrib and eventually near each margin; male spathe 0.7–1 (–1.8) mm long; anther 1-sporangiate, 0.5–0.65 (–1) mm long; seeds 1–1.3 mm long, 0.4–0.5 mm in diameter, with 16–20 rows of squarish to hexagonal seed coat pits.

Perennial or perhaps sometimes annual.

{(⊙)? Vit}. Endemic to India: ♦Ind; Mp.

Najas malesiana de Wilde [*N. graminea* var *minor* Rendle, *N. graminea* var. *angustifolia* Rendle]: Fig. 280

Monoecious, slender. Stems without spines, 0.5–1 mm in diameter. Leaf sheaths (1.37–) 1.5–2.5 (–3.4) long (including spine cells), 0.95–1.5 (–1.9) mm wide, auriculate, serrulate or lacerate with 3–10 spines on each side; auricles 0.34–0.65 (–1.13) mm long (including the spines), 0.23–0.48 (–0.56) mm wide, serrulate or lacerate with 3–10 spines on each side and 0–3 on the inner edge, apex acuminate. Leaf blades flat, 10–19 mm long, (0.29–) 0.4–0.5 (–0.56) mm wide (including the teeth), (0.15–) 0.25–0.35 (–0.44) mm wide (excluding teeth); margins serrulate, with (12–) 20–40 (–50), somewhat inconspicuous teeth; the teeth 0.05–1 mm long, not borne on multicellular swellings; midrib without teeth; septa absent or rarely present then visible only up to half way to the margin; fibres absent or rarely present along the midrib and along the margins. Male flowers naked; anther 1-sporangiate, 0.5–0.8 (–1.1) mm long. Female flowers naked. Seeds elliptical-oblong, (0.8–) 1.2–1.3 (–1.4) mm long, 0.34–0.48 mm in diameter; seed coat pits squarish to hexagonal or rectangular, 0.03–0.04 (0.05) mm long, in regular longitudinal rows of (20–) 30–35 (–40).

Annual. Often common and locally dominant. It is usually wrongly identified in floristic works and is much more widely distibuted than the cited list suggests.

{⊙ Vit}. Tropical Asia and Australia: Ban, Bur, Ind: Kl, Kt, Pj.

Najas marina Linnaeus: Fig. 280

Dioecious, robust, rather brittle and very spiny. Stems with spines. Leaf bases rounded, not auriculate, 1–5.5 mm long, 1–5.5 mm wide, entire or serrulate with 0 or 1 (–3) spines on each side. Leaf blades fleshy, usually undulate, usually hard and brittle, (4–) 30–50 (–65) mm long, (1–) 2.5–4.5 mm wide (including the teeth), 0.4–2.5 mm wide (excluding spines); margins serrulate, with 8–10 (–14), conspicuous spiny teeth, the teeth up to 2 mm long, the spine cells visible to the naked eye, borne on large multicellular swellings; the abaxial side of the midrib with (0–) 2–8 (–40) teeth; septa absent; fibres absent. Male flowers borne in spathes, ±2.5 mm long. Female flowers naked, 2–5 mm long. Seeds ovate, slightly asymmetrical, 1.9–7.5 mm long, 0.8–3.3 mm in diameter; seed coat pits very irregular in shape and number, not arranged in regular longitudinal rows.

Mostly annual. In India there are surprisingly few records; it is found in still and flowing fresh water.

{(⊙) Vit}. Almost throughout the Old World. A very variable species. Three subspecies are known from India.

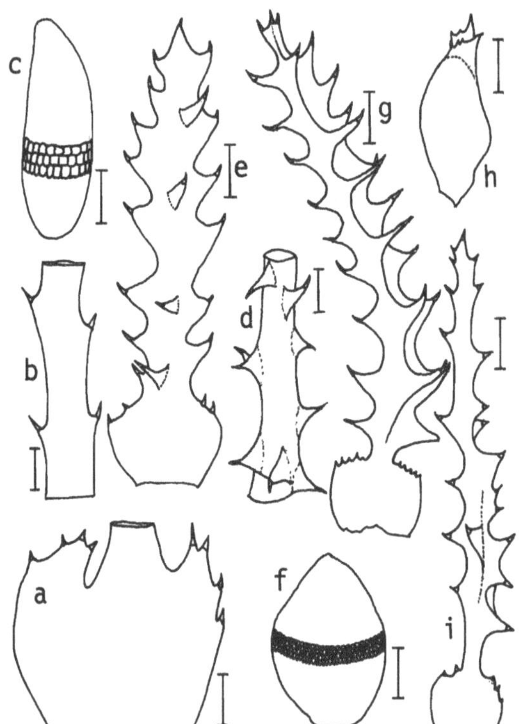

Fig. 280. *Najas malesiana*: a, leaf sheath (0.3 mm); b, part of leaf blade (0.3 mm); c, seed (0.3 mm) with surface pattern (inset). *N. marina* subsp. *marina*: d, part of stem (2 mm); e, leaf of var. *kashmirensis* (1.7 mm); f, seed (1 mm) with surface pattern (inset). *N. marina* subsp. *armata*: g, leaf (2 mm); h, male flower in spathe (1 mm). *N. marina* subsp. *intermedia*: i, leaf (2 mm).

Najas marina subsp. ***marina*** var. ***kashmirensis*** L. Triest: Fig. 280

Stems not very spiny, ± 1–5 (–12) spines in 2 cm. Leaf blades 30–50 (–65) mm long, (2–) 3–4 (–5) mm wide (including teeth), (0.7–) 1.2–1.7 (–2.4) mm wide (excluding the teeth); margins each with 8–10 (–14) teeth; seeds (6.1–) 6.6–7.3 (–7.5) mm long, 2.2–2.6 (–3) mm in diameter.

Annual. In lakes and rivers. Sometimes locally abundant but mostly rather rare.

{⊙ Vit}. Endemic to India: ♦Ind: Jk, Rj.

Najas marina subsp. ***intermedia*** (Gorski) Casper: Fig. 280

Stems not very spiny, (0–) 2–10 (–15) spines in 2 cm. Leaf blades (4–) 8–20 (–50) mm long, 1.1–3.3 (–4) mm wide (including teeth), 0.4–1.5 (–2) mm wide (excluding teeth); margins each with (2–) 4–8 (–12) teeth; seeds (2.8–) 3–4 (–5.1) mm long, (0.9–) 1.2–2 (–2.8) mm in diameter.

{⊙ Vit}. Temperate to tropical Eurasia: Ind, Pak: Gj, Mp.

Najas marina subsp. ***armata*** (Lindberg fil.) Horn af Rantzien: Fig. 280

Stems very spiny, (10–) 15–25 spines in 2 cm. Leaf blades 6–20 (–24) mm long, (1.1–) 1.5–5.3 mm wide (including teeth), 0.5–2.3 mm wide (excluding teeth); margins each with 4–10 teeth; seeds 2.5–3.5 mm long, 0.8–2.1 mm in diameter.

Annual or perennial. It sometimes develops turions for surviving the dry period.

{(⊙) Vit}. S Europe, Africa, Asia Minor, Sri Lanka and Australia: Cey.

Najas minor Allioni: Fig. 281

Monoecious, slender or robust. Stems without spines, less than 0.5 mm in diameter, plumose or bushy towards the apex because of closely packed curved leaves. Leaf sheaths truncate to lacerate, 1–2.26 mm long (including auricle and teeth), 1–2.4 mm wide, serrulate or laciniate; auricles 0.13–0.37 mm long (including teeth), 0.3–0.4 mm wide, serrulate or laciniate with (3–) 8–13 spines on each side but none on the inner side, apex rather obtuse. Leaf blades flat, flexible, (5–) 7–22 mm long, (0.3–) 0.6–1 (–1.8) mm wide (including teeth), (0.2–) 0.4–0.5 (–0.7) wide (excluding teeth); margins serrulate, with (6–) 10–18, conspicuous, spiny teeth; the teeth (0.07–) 0.2–0.3 (–0.5) mm long, borne on small or broad, multicellular swellings; midrib occasionally with teeth; septa clearly visible up to the margins; fibres absent. Male flowers enclosed within a spathe; anther 1-sporangiate, 0.56–1.3 mm long. Female flowers naked. Seeds elliptical-oblong, slightly curved at apex, ± 1.74–2.74 mm long, 0.45–0.65 (–0.73) in diameter, clearly pitted; seed coat pits rectangular, broader than long, ± 0.03 mm long, ± 0.1 mm wide, arranged regularly and ladder-like in longitudinal rows of (60–) 80–100.

Annual. Characteristically found in habitats which show great fluctuations in water level or even dry out annually such as tanks, reservoirs and rice fields.

{⊙ Vit}. Temperate and warm areas of Europe, N Africa, Asia and N America: Ind, Pak: As? Bh, Pj.

Najas oguraensis Miki: Fig. 281

Like *N. minor* but leaf sheath only slightly auricled; anther 4-sporangiate, 1.3–1.8 mm long; seeds 2.6–3.5 mm long, 0.6–0.8 mm in diameter; seed coat pits 80–100 in each longitudinal row.

Annual.

{⊙ Vit}. N India, Nepal, China and Japan: Ind, Nep, Pak?: Jk.

Najas tenuis A. Braun ex Magnus: Fig. 281

Monoecious or perhaps also dioecious, slender. Stems without spines. Leaf bases sloping, laciniate, rounded or broadly rounded, never auriculate, 1.8–3

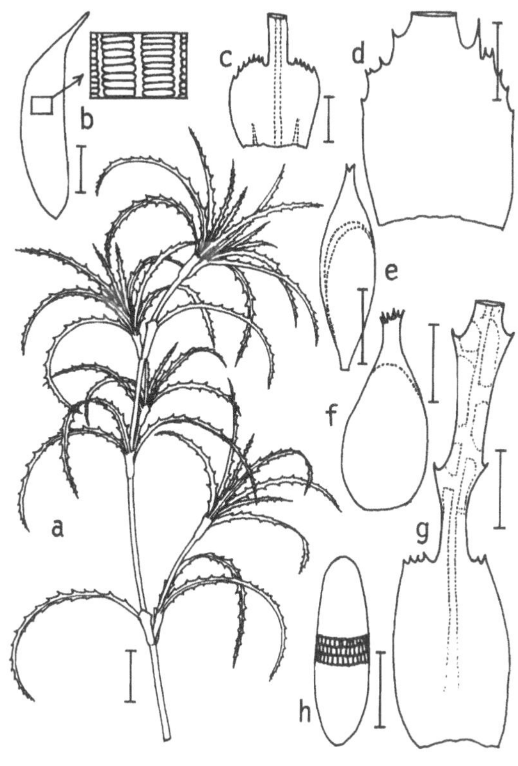

FIG. 281. *Najas minor*: a, habit (2 cm); b, seed (0.5 mm) with surface pattern (inset); c, leaf sheath (1 mm). *N. oguraensis*: d, leaf sheath (1 mm); e, male flower in spathe (1 mm). *N. tenuis*: f, male flower in spathe (1 mm); g, leaf sheath and base of blade (1 mm); h, seed (1 mm) with surface pattern (inset).

(–3.5) mm long, 1.3–2.4 (–3.1) mm wide, serrulate or laciniate with 4–12 (–18) spines on each side. Leaf blades flat, 15–25 mm (9.3–30 mm) long, 0.5–1 (–1.3) mm wide (including teeth); margins serrulate, with 5–15 (–30), conspicuous, spiny teeth; the teeth 0.14–0.4 mm long, the spine cells visible to the naked eye, borne on multicellular swellings; midrib without teeth; septa clearly visible up to the margins; fibres absent. Male flowers borne in spathes. Female flowers naked or borne in spathes, 1.6–3.7 mm long. Seeds elliptical-oblong, 1.2–2.4 mm long, 0.45–0.8 mm in diameter; seed coat pits squarish to hexagonal, in regular longitudinal rows of 25–35.

Perennial and annual. In lakes, tanks, pools, ditches and rice fields. Common and widespread, often mixed with *N. graminea*, sometimes locally dominant.

{(⊙) Vit}. India, Sri Lanka, Myanmar and SW China: Bur, Cey, Ind: Ap, Bh, Kl, Kt, Mp, Pj, Rj, Tn.

NELUMBONACEAE: 1 genus

Nelumbo: 2 species and numerous varieties (*Nelumbo lutea*, N America; *N. nucifera*, warm Asia and Australia, naturalized elsewhere)

Ni, Xiue-min *et al. Lotus.* Wuhan Institute of Botany, Acadamia Sinica, Science Press: i–vii, 1–163, pl. 1–105 (1987) — in Chinese

Nelumbo nucifera Gaertner [*Nelumbium speciosum* Willdenow]: Fig. 282

Large, perennial, with milky latex. Stems dimorphic; slender horizontal, vegetative stolons or rhizomes or thickened storage rhizomes. Leaves alternate; petioles up to 3 m or more long, bearing numerous spines; leaf blades when immature floating, when mature raised above the water surface, peltate, reniform to orbicular (eccentrically peltate when very young), 10–100 cm or more in diameter, glaucous, nerves prominent but without spines. Flowers 8–20 (–40) cm in diameter; pedicels exceeding the petioles. Perianth segments 14–30, the outer 2–5 somewhat persistent, greenish-white and somewhat sepal-like, the inner ones caducous, petal-like, white to pink or red. Stamens up to 200 or more; filaments up to 1 cm long; anthers elongate, up to 1.2 cm long, connectives each with a conspicuous and fleshy terminal appendage, up to 4.5 mm long. Carpels (9–) 12–30 (–40), borne singly in cavities in the pithy obconical receptacle; fruit a head of very hard-coated nutlets; nutlets ovoid, ± 2 cm long, ± 1 cm wide.

Widely known as lotos or lotus. The juveniles are submerged, the adults have floating or emergent leaves. It flowers almost throughout the year. The flowers have thermo-genetic staminal connectives and are strongly scented; pollination is by insects, it is particularly attractive to bees; *Nelumbo* honey is much in demand. The disseminules are large and hard nutlets, the dispersal mechanism is not documented but the nutlets are probably transported by water, in mud and by animals. The nutlets may remain viable for several hundred years in mud — Priestly, D. A. and Posthumus, M. A. Extreme longevity of lotus seeds from Pulantien. Nature 299: 148–149 (1980). It is widely cultivated as a crop for its edible rhizomes, edible nuts and for honey, it is also grown for ornament, the leaves are used for serving food. It is sacred for Hindus ('podma' devoted to Brahma); the buds are collected daily early in the morning for worship in temples. It is used in medicine: the seeds for skin diseases and diarrhoea, the rhizomes for piles and ringworm.

It is very widely cultivated in India and it is difficult to determine where it is truly native.

{Eph/Hyp}. Warm Asia and Australia, naturalized elsewhere: Ban, Bhu, Ind, Nep, Pak: Bh, Dl, Go, Hr, Jk, Kl, Kt, Mh, Mp, Pj, <Rj, Tn, Up, Wb.

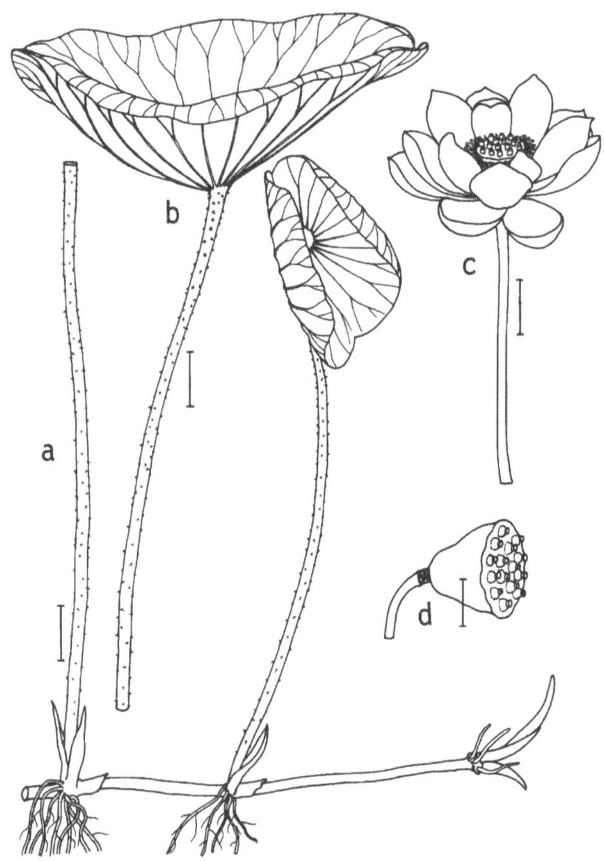

FIG. 282. *Nelumbo nucifera*: a, stolon with two leaves (3 cm); b, leaf blade (3 cm); c, flower (3 cm); d, fruiting head (2 cm).

NYMPHAEACEAE [including BARCLAYACEAE, excluding CABOMBACEAE]: 6 genera

Stems submerged, contracted and corm-like or elongate, rhizomatous or stoloniferous. Leaves alternate, usually radical; petioles submerged, long and flexible; leaf blades submerged or floating, rarely emergent, cordate or peltate, linear to orbicular or sagittate. Flowers solitary, bisexual, radially symmetrical. Sepals 4–6, free or fused to the ovary. Petals rarely absent, usually numerous, often gradually passing into stamens or staminodes. Stamens numerous. Carpels 3–30 or more, united or partially united; stigmas on a disk with radiating rays. Fruit an irregularly dehiscent berry. Seeds numerous, sometimes arillate.

Perennial or annual. All species are submerged or epihydates, occasionally some leaves may become emergent when plants are overcrowded. Mitra, R. L. Barclayaceae and Nymphaeaceae. Fascicles of Flora of India. 20: 1–2, 11–25 (1990).

Excluded: occasionally cultivated but apparently not becoming naturalized: ***Nuphar***, ***Victoria***.

1. Plants bearing spines; petioles of mature leaves inserted at the centre of the blade
   2. Leaf blades bearing spines on both surfaces; leaf margins not turned upwards; staminodes absent; carpellary appendages absent ***Euryale***
   2. Leaf blades bearing spines on the lower (abaxial) surface only; leaf margins turned upwards; staminodes replacing the inner stamens; carpellary appendages present ***Victoria***
1. Plants without spines; petioles of the mature leaves inserted at the broader and notched end of the blade
   3. Stamens fused to the petals; staminodes absent; leaves mostly submerged ***Barclaya***
   3. Stamens free from the petals; staminodes present; leaves mostly floating, sometimes emergent or submerged
      4. Sepals 5 or 6; ovary superior; petals not spreading, scale or stamen-like; seeds without arils ***Nuphar***
      4. Sepals 4; ovary semi-inferior; petals spreading, petaloid, showy; seeds with bell-shaped arils ***Nymphaea***

Barclaya [*Hydrostemma*]: 2 species: tropical SE Asia
Barclaya longifolia Wallich: Fig. 283

Stems with corms and slender rhizomes. Petioles 6-25 cm long; leaf blades usually totally submerged but occasionally emergent for short periods, linear-oblong, 12–28 (–40) cm long, 2–6 cm wide, membranous, the base cordate to almost sagittate, the margins undulate, the tip obtuse. Peduncles 10–35 (–80) cm long. Sepals hypogynous, 5, linear-elliptical, 1.5–2.5 cm long, greenish-purple outside, purple inside, with midrib extended into a 3–5 mm long tail. Petals 8 to numerous, perigynous, united below into a tube, shorter than the sepals, greenish outside becoming pink, red or purple inside. Stamens numerous, fused to the petals; staminodes absent. Carpels up to 10, united, inferior; stigmatic disk conical. Fruit an irregularly dehiscent, 50- to 500-seeded berry, 1-1.5 cm across. Seeds globose, ± 1 mm in diameter, covered in fine hooked spines, released in white pulp.

Perennial. In streams or rarely in pools, usually in forested regions or rubber plantations. Juveniles and adults are usually submerged, but sometimes seasonally emergent. It is pollinated by insects and is also sometimes autogamous; the flowers sometimes do not even open. The disseminules are seeds with hooked spines released in pulp, dispersed by animals (wild pigs in Malaysia). It is cultivated for decoration. Küfner, J. Erfahrung zur Haltung und Vermehrung von *Barclaya longifolia* Wallich. Aqua-Planta 94(3): 119–126 (1994).

{Ros}. SE Asia: Bur, Ind: An.

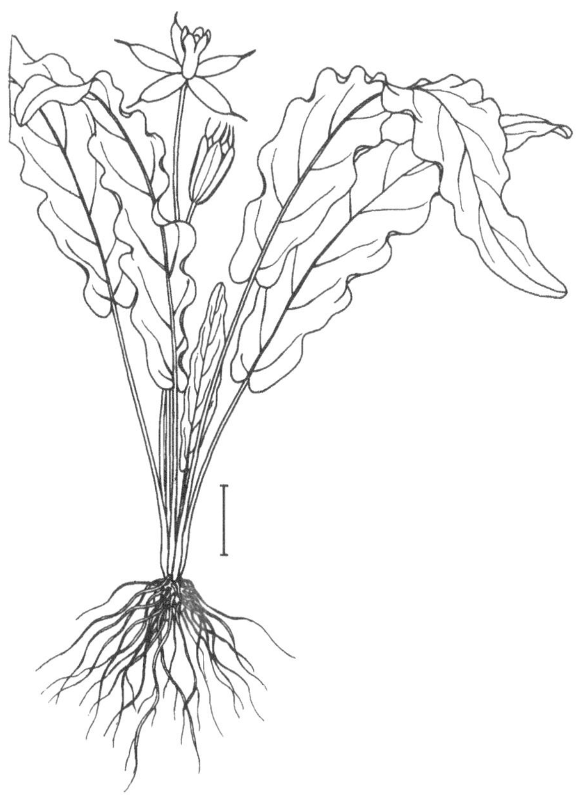

FIG. 283. *Barclaya longifolia*: whole plant with flowers and submerged leaves (5 cm).

Euryale: 1 species

Euryale ferox Salisbury: Fig. 284

Stem a short, thick rhizome. Leaves in rosettes; petioles densely prickly; mature leaf blades floating, peltate, orbicular, up to 1 m or more in diameter, both surfaces with spines, margin not upturned. Flowers often submerged and cleistogamous, 5–6 cm in diameter. Sepals 4, with prickles, epigynous. Petals 30–35, epigynous, shorter than the sepals, blue to violet or red. Stamens 78–92; staminodes absent. Carpels 8–16, united, inferior; stigmatic disk obconical. Fruit crowned with persistent sepals and covered with stout prickles. Seeds with pulpy aril.

Annual or short-lived perennial. Found in water up to 5 m deep in lakes, pools or in slowly flowing water. Juveniles are submerged; the adults have floating leaves. Although apparently specialized for pollination by insects (with scent and nectar) it is highly autogamous, often with pre-anthesis cleistogamy. The disseminules are large seeds with pulpy arils. The dispersal is probably by animals (fish and birds) and also by water. It is extensively cultivated in northern Bihar. In West Bengal and Kashmir it is becoming very rare and may now be extinct. The seeds, fruits and rhizomes are eaten locally by humans; it is also used in medicine and cultivated for decoration.

{(⊙) Eph}. N India to China, Manchuria and Japan: Ban, Ind: As, Bh, Jk, Mg, Mn, Or, Rj, Tp, Up, Wb.

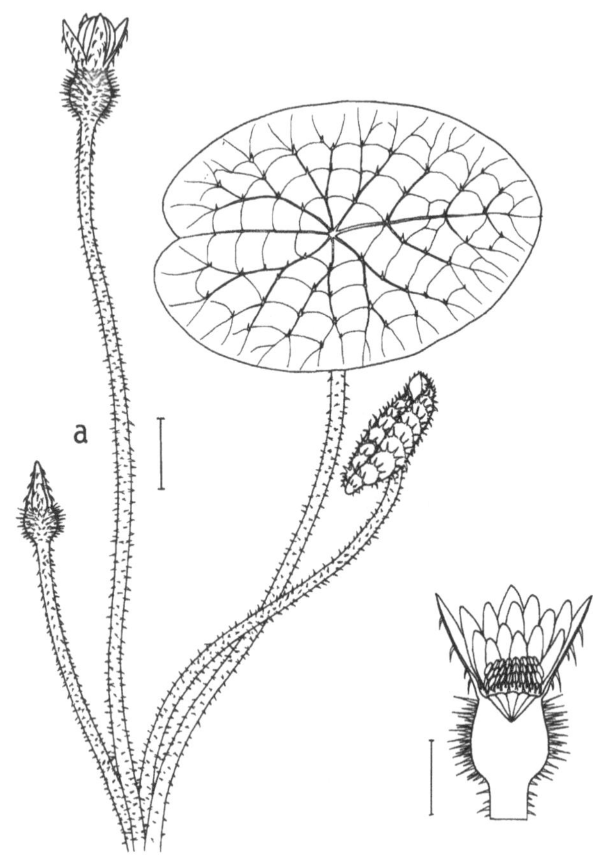

Fig. 284. *Euryale ferox*: a, flower bud, flower at anthesis, leaf blade and leaf buds (2 cm); b, flower in longitudinal section (1 cm).

Nymphaea: ± 40 species plus numerous hybrids and varieties: cosmopolitan

Plants without spines. Stem a slender or stout rhizome, a corm or a tuber. Petioles smooth; mature leaf blades with a deep sinus, mostly floating, ovate to orbicular, leathery, if submerged usually sagittate, filmy and delicate. Flowers borne above the water surface, strongly protogynous. Sepals 4, green or streaked with red. Petals numerous, showy, sometimes passing into stamens. Stamens numerous, with or without appendaged connectives. Carpels 3–35, united or partially united; fruit a berry ripening below water. Seeds with bell-shaped arils.

Perennial or annual. Found in still or flowing water up to ± 2 m deep, in a wide variety of habitats. The juveniles are submerged; the adults are submerged or floating. Pollination is by insects (night- or day-flowering, each with different and bizarre pollination strategies). The disseminules are seeds with pulpy arils, dispersed by animals and by water, some species also spread by specialized buds and bulbils. The rhizomes, flower buds and seeds of some species eaten for food, medicine or to cause hallucinations. The rhizomes are used as a mordant in dyeing and tanning. Many species and hybrids are cultivated for ornament.

The nomenclature of *Nymphaea* in India has been very confused and some names are often used in the wrong sense.

Excluded: the following have been recorded from the Himalayas but do not extend south into India: ***Nymphaea alba*** Linnaeus, ***N. candida*** C. Presl, ***N. odorata*** Aiton, ***N. tetragona*** Georgi, ***N. tuberosa*** Paine.

1. Leaf blades entire to bluntly dentate, glabrous; anthers with terminal appendages; sepals obscurely veined, usually with purple streaks, persistent in fruit; day-flowering ***N. nouchali***

1. Leaf blades sharply dentate, pubescent below (hairs sometimes restricted to veins); anthers without appendages; sepals with 5–9 white veins, decaying as the fruit develops; night-flowering ***N. pubescens***

Nymphaea nouchali N. L. Burman [*N. stellata* Willdenow]: Fig. 285

Rhizomes horizontal or erect, up to ± 3 cm in diameter. Leaf blades mostly floating, entire to bluntly toothed, waxy, ± elliptical-orbicular to round, 5–25 (–45) cm long and wide, the radial slit to the centre or to within 6 mm of the centre, glabrous on both surfaces, the adaxial (upper) surface green (sometimes blotched with purple), the abaxial surface reddish-purple, the margin entire or with broad obtuse teeth. Flowers barely emergent. Sepals lanceolate to ovate-lanceolate, 2.5–5 (–9) cm long, green often with purple streaks, obscurely veined, persistent in fruit. Petals 6–16, white, pinkish, blue or bluish-purple, the outer ones lanceolate to oblong-lanceolate, 2.5–7.5 cm long. Stamens (15–) 20–40 (–65), 1.5–3 cm long, not separated from the petals by a gap; anthers not flattened, yellow with white or blue apical connective appendages up to 4 mm long (usually less than one-quarter as long as the anther). Stigmatic disk surrounded by short (1–3.5 mm long), incurved, carpel appendages. Fruits 1.5–4 cm in diameter. Seeds ellipsoid-globose, 0.5–1 (–1.5) mm long.

Perennial or annual. Common and often locally dominant in permanent and temporary water. It grows in shallow pools or water up to ± 2 m deep. The flowers are fragrant; they open during the day and close at night; after pollination they tend to remain closed day and night. Leaves, petioles and peduncles are eaten locally; most parts of the plant are used in medicine to cure various complaints.

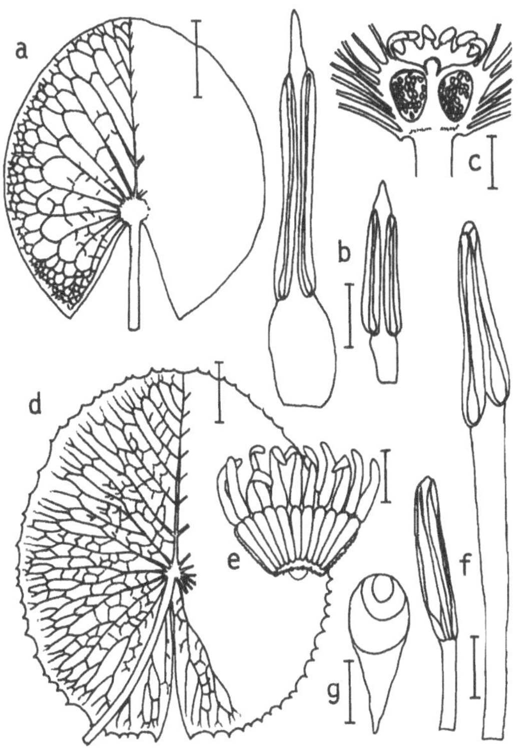

Fig. 285. *Nymphaea nouchali*: a, leaf blade (6 cm); b, largest and smallest stamens (2 mm); c, flower in diagramatic longitudinal section (1 cm). *N. pubescens*: d, leaf blade (3.5 cm); e, stigmas with appendages (1 cm); f, largest and smallest stamens (2 mm); g, seed with aril (1 mm).

Plants resembling *N. nouchali* with the abaxial (lower) leaf surface green with purple spots and growing in the hills may belong to the African species *N. caerulea* Savigny; plants proliferating from the sinus of the leaf blade may be *N.* × *daubeniana* O. Thomas a hybrid involving *N. micrantha* Guillarmod et Perrier de la Bâthie and perhaps *N. caerulea*.

{(⊙) Eph}. S and SE Asia: Ban, Bur, Cey, Ind, Pak: An, Ap, As, Dl, Go, Jk, Kl, Kt, Mh, Mp, Or, Pj, Rj, Tn, Up, Wb.

Nymphaea pubescens Willdenow [*N. rubra* Roxburgh ex Andrews, *N. nouchali* auct. Ind., non N. L. Burman, *N. lotus* auct. Ind., non Linnaeus]: Fig. 285

Leaf blades mostly floating, green above, brownish or purple below, ovate to round, 15–36 (–59) cm long and wide, the abaxial surface dark green and glabrous, the adaxial (lower) surface greenish, brownish or purple, densely covered with short hairs or the hairs sometimes restricted to veins, the margins sharply dentate, although the teeth sometimes absent on the apical part of old leaves; teeth 3–7 mm long, 7–15 mm apart. Flowers borne well above the water. Sepals ovate-lanceolate to obovate-oblong, 2.5–8.5 cm long, uniformly green with 5–9 white veins, decaying as the fruit develops. Petals 10–25, white to pink or deep red, the outer ones oblanceolate, 2–7 cm long. Stamens 25–70 (–90), yellow or red, separated from the petals by a gap; anthers flattened, 1–4 cm long, without appendages. Stigmatic disk surrounded by large (5-10 mm long), yellow, finger-like appendages which close over the stigmas after the first night of flowering. Fruits globose, 2.5–4 cm in diameter. Seeds ellipsoid to globose, ± 1.4 mm long, ± 1 mm wide, almost black with a white aril.

Perennial or annual. It is common and often locally dominant. It mostly grows in shallow permanent and temporary water. It is a very fragrant night-flowering species; after pollination the flowers usually remain open during the day but the stigmatic appendages lie over the stigmas. The rhizomes with roots are eaten as vegetables; the leaves are often used as fodder. Most parts of the plant are used medicinally.

I rather suspect that the Indian species should be merged with *N. lotus* Linnaeus. A cultivated race of *N. pubescens* which does not develop seeds and has red immature leaves, crimson petals and red anthers is sometimes found naturalized in permanent water and in decorative pools; is often called *N. rubra*.

{(⊙) Eph}. S and SE Asia: Ban, Bur, Cey, Ind, Pak: An, Ap, As, Bh, Dl, Gj, Go, Jk, Kl, Kt, Mh, Mp, Or, Pj, Rj, Tn, Tp, Up, Wb.

ONAGRACEAE: 24 genera, 2 with aquatic species

Herbs. Leaves sometimes with stipules, alternate, simple; leaf blades entire, pinnately veined. Flowers bisexual. Sepals 4 or 5, valvate. Petals 4 or 5, free. Stamens as many or twice as many as sepals. Ovary inferior, 4- or 5-locular; style simple; stigma capitate. Fruit a capsule, opening by valves or disintegrating irregularly. Seeds numerous.

Excluded: several species of ***Epilobium*** have been reported from wetland habitats in the Himalayas.

Ludwigia [*Fissendocarpa*, *Jussiaea*]: 75 species: almost cosmopolitan

Herbs, sometimes woody at base. Stems erect, ascending or creeping and often rooting at the nodes, when submerged or floating often swollen and spongy or bearing silver-white, inflated pneumatophores. Leaves cauline or in floating rosettes, opposite or alternate, mostly entire; stipules absent or reduced. Flowers bisexual, radially symmetrical, borne singly or clustered in the axils of leaves or in a terminal inflorescence. Sepals (3–) 4–5 (–7), adnate to the ovary, persisting in fruit. Petals as many as sepals or absent, yellow or white, entire or notched at tips, contorted in bud. Stamens as many or twice the number of sepals; anthers versatile or basifixed. Ovary inferior, as many locules as sepals; stigma slightly lobed, hemispherical or capitate; placentation axile; fruit a capsule, dehiscing by longitudinal slits or flaps, by terminal pores or irregularly; seeds numerous, sometimes dimorphic with some smooth and others corky.

Annual or perennial. Found in a large variety of aquatic habitats, submerged, floating, emergent or seasonally submerged. Pollinated by insects or autogamous. The disseminules are seeds (sometimes dimorphic), some dispersed by water, others by unknown means. Some species reported to be weeds in irrigation systems, others cultivated for decoration. Raven, P. H. The Old World species of *Ludwigia* (including *Jussiaea*), with a synopsis of the genus. Reinwardtia 6: 327–427 (1963); Sreemadawan, C. P. Notes on Indian species of *Ludwigia*. Bulletin of the Botanical Survey of India 8: 79–80 (1966).

Excluded species: ***Ludwigia repens*** Forster [*L. natans* Elliot]; from N America was reported from Bangladesh in 1966 but has probably not persisted or it was incorrectly determined.

1. Petals creamy white with yellow at the base, 5; stems usually floating and bearing bunches of spongy, gas-filled, spindle-shaped pneumatophores ***L. adscendens***

1. Petals yellow, 4 or rarely 5; stems usually erect or if floating then without bunches of spindle-shaped pneumatophores

2. Capsules enlarged and inflated above, narrow below; seeds dimorphic in several rows and free above, in 1 row and embedded in cubes of endocarp below; pollen grains shed singly ***L. linifolia***

2. Capsules ± uniform from top to bottom; seeds not dimorphic either several or in 1 row or either free or embedded in endocarp; pollen grains shed in tetrads

3. Stamens twice as numerous as the sepals; shrubs or subshrubs

4. Capsules terete, 8-ribbed; bracteoles borne above the base of the sepal tube, persisting in fruit; petals rather inconspicuous, usually not exceeding the sepals ***L. octovalvis***

4. Capsules 4-angled with 4 nearly flat sides, 4-ribbed; bracteoles at the base of the sepal tube, not persisting in fruit; petals showy, usually twice as long as the sepals ***L. peruviana***

3. Stamens as numerous as the sepals; herbs

5. Capsules rarely more than 1.5 cm long, ± terete, impressions of the seeds not visible through the capsule walls; seeds in several rows in each valve, 0.3–0.5 mm long; young stems and leaves usually green ***L. perennis***

5. Capsules usually more than 1.5 cm long, ± 4-angled, impressions of the seeds visible through the capsule walls; seeds in 1 row in each valve, 1.3–2.2 mm long; young stems and leaves usually tinged with red ***L. prostrata***

Ludwigia adscendens (Linnaeus) H. Hara [*Jussiaea repens* Linnaeus]: Fig. 286

Stems prostrate or ascending, usually floating and bearing bunches of silver-white, spongy, gas-filled, spindle-shaped pneumatophores, the flowering portions ascending, up to 50 cm or more long. Leaves with petioles 0.5–2 cm long; leaf blades broadly oblong-elliptical, 0.4–7 cm long, 0.7–4 cm wide, glossy green with pale nerves, narrowly cuneate at the base, acute or rounded at the tips. Flowers solitary in leaf axils; pedicels 2.5–5.5 4 cm long, hairy above the middle; bracteoles deltoid, ± 1.2 long, 1.3–1.5 mm wide, borne at base of sepal tube. Sepals 5, deltoid-acuminate, 5–13 mm long, 2–3.2 mm wide, hairy or glabrous. Petals 5, obovate, 9–18 mm long, 6–12 mm wide, creamy-white, yellow near the base. Stamens 10; pollen grains shed singly. Capsule glabrous or hairy, 1.2–3.5 cm long, 3–4 mm in diameter, terete, light brown, with 10 conspicuous darker brown ribs, thick-walled, irregularly dehiscent. Seeds in 1 row in each valve, visible through the capsule wall, ± 1.5 mm apart,

1.1–1.3 mm long, firmly embedded in cubes of woody endocarp 1.2–1.5 mm high, 1–1.2 mm thick.

Perennial or perhaps sometimes annual. Found in wet swampy places, usually floating on the water. It is often gregarious forming mats on the surface, the mats often become colonized by other species. It flowers and fruits throughout the year.

{(⊙)? Ple/Hyp}. Subtropical Himalayas, India, east to China, Malaysia and Australia: Ban, Bur, Cey, Ind, Nep: Ap, As, Bh, Dl, Go, Kl, Kt, Mh, Mp, Or, Pj, Rj, Tn, Up, Wb. 0–1500 m.

Ludwigia linifolia (Vahl) Rolla Rao [*L. hyssopifolia* (G. Don) Exell, *Jussiaea fissendocarpa* Haines]: Fig. 286

Stems erect, 5–300 cm tall, often becoming woody at the base, elongate, thread-like pneumatophores arising from buried and submerged roots. Leaves with petioles 2.5–18 mm long; leaf blades lanceolate to oblong-elliptical, 1–10 cm long, 0.2–3.5 cm wide, narrowly cuneate at the base, acuminate at the tips. Flowers solitary in leaf axils, subsessile or up to 4 mm long; bracteoles deltoid, ± 1 mm long, at the base of the sepal tube. Sepals 4 (5), lanceolate, 2–4 mm long. Petals 4 (5), elliptical, 2–5 mm long, 1–2 mm wide, yellow fading to orange-yellow. Stamens twice as many as the sepals; pollen grains shed singly. Capsule finely hairy, 1.5–3 cm long, 1–1.2 mm in diameter, subterete, enlarged in the upper sixth to third, thin-walled, irregularly dehiscent, 8-ribbed. Seeds in the lower part of the capsule in 1 row in each valve, visible through the lower part of the capsule wall, 0.7–0.85 mm long, each firmly embedded in cubes of woody endocarp; seeds in the upper inflated part of the capsule in several rows in each valve, free, ovoid, 0.35–0.5 mm long, paler brown than the lower seeds.

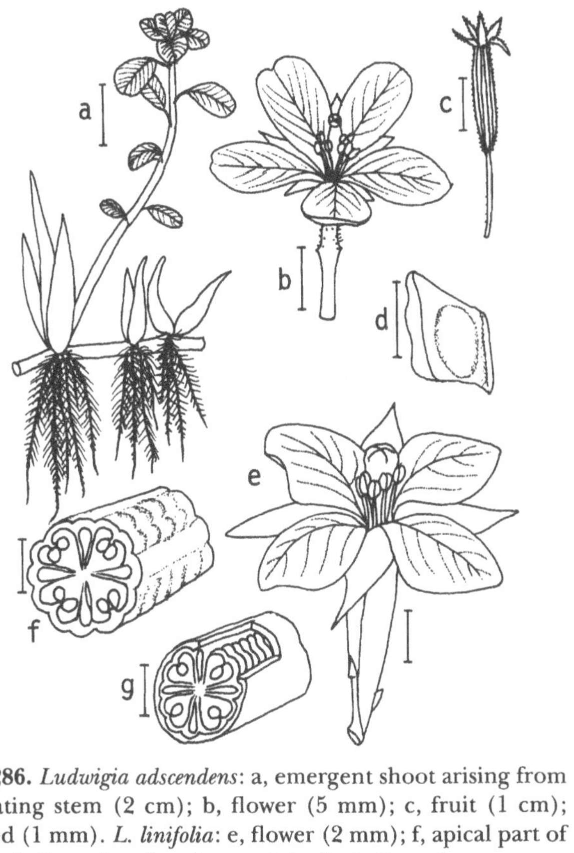

FIG. 286. *Ludwigia adscendens*: a, emergent shoot arising from a floating stem (2 cm); b, flower (5 mm); c, fruit (1 cm); d, seed (1 mm). *L. linifolia*: e, flower (2 mm); f, apical part of the fruit (1 mm); g, basal part of the fruit (1 mm).

Annual. Found in wet places, particularly in temporary pools. It often occupies shallow water where the main stem is often surrounded by fine, silver-white, thread-like, floating pneumatophores which may also arise from buried roots. Sometimes the plants are very small when growing on drying mud. Differs from other Indian species in having dimorphic seeds.

{⊙ Ten/Hyp}. Pantropics, probably native in America: Ban, Bhu, Bur, Cey, Ind, Nep: <An, As, Bh, Gj, Go, Kl, Mp, Mz, Or, Rj, Tn. 0–500 m.

Ludwigia octovalvis (Jacquin) Raven [*Jussiaea suffruticosa* Linnaeus, *L. octovalvis* subsp. *sessiliflora* (Michaux) Raven]: Fig. 287

Variable in pubescence either sparsely appressed hairy or densely patently hairy. Stems robust, usually much branched, up to 4 m tall, often becoming woody at the base. Leaves with petioles up to ± 1 cm long but usually less; leaf blades linear to subovate, 0.7–14.5 cm long, 0.1–4 cm wide, narrowly or broadly cuneate at the base, attenuate at the tips. Flowers solitary in leaf axils; pedicels up to 1 cm long but usually less; bracteoles reduced or subulate and up to 1 cm long, borne above the base of the sepal tube, persisting on the capsule. Sepals 4, ovate or lanceolate, 3–15 mm long, 1–7.5 mm wide. Petals 4, broadly obovate or cuneate, emarginate, 2–17 mm long, 2–17 mm wide, yellow. Stamens 8; pollen grains shed in tetrads. Capsule 1.7–4.5 (–6) cm long, 2–8 mm in diameter, terete or nearly so, thin-walled, pale brown with 8 darker ribs, regularly dehiscent. Seeds in several rows in each valve, not visible through the capsule wall, free, rounded, 0.6–0.75 mm long, 0.5–0.7 mm in diameter including the prominent and inflated raphe.

A pantropical weed. Found in wet and swampy places rivers, ditches, canals or tanks, mostly near cultivated land. It is a very variable species with four subspecies. Two of these are found in India: *Ludwigia octovalvis* subsp. *octovalvis* is characterized by being subglabrous or having appressed hairs and leaves that are lanceolate or linear; *Ludwigia octovalvis* subsp. *sessiliflora* (Michaux) Raven is characterized by having numerous, long, erect hairs on emergent parts and

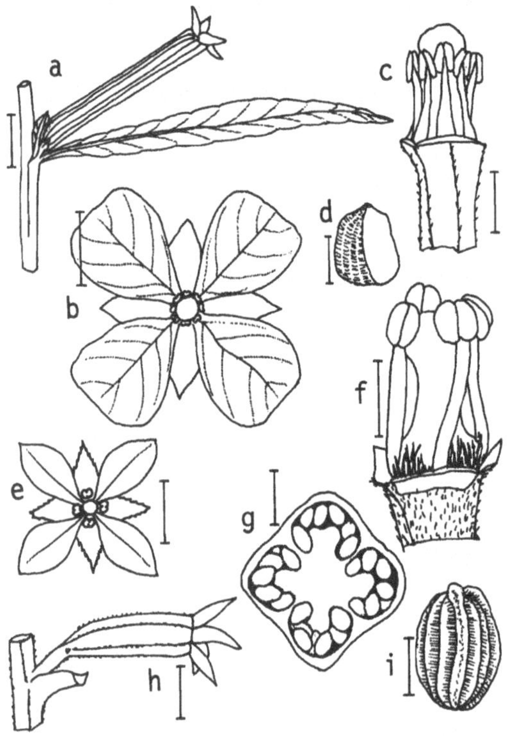

FIG. 287. *Ludwigia octovalvis*: a, fruit (1 cm); b, flower (5 mm); c, stamens and ovary (2 mm); d, seed (0.3 mm). *L. perennis*: e, flower (3 mm); f, stamens and ovary (1 mm); g, transverse section of the fruit (1 mm); h, fruit (5 mm); i, seed (0.2 mm).

leaves rather wide lanceolate to elliptical or almost ovate, up to 7 cm long and up to 1.5 cm wide. In India there are numerous intermediates so I have not tried to treat them separately.

{(⊙) Hel}. Tropics of the World. Ban, Bhu, Bur, Cey, Ind, Nep: An, As, Bh, Go, Kl, Kt, Mh, Mp, Mz, Or, Pj, Rj, Sk, Tn, Up, Wb. 0–1500 m.

Ludwigia perennis Linnaeus [*L. parviflora* Roxburgh]: Fig. 287

Stems up to 1 m tall, often becoming woody at the base. Leaves with petioles up to 2–15 mm long, winged; leaf blades narrowly elliptical to lanceolate, 1–11 cm long, 0.3–2.7 cm wide, narrowly cuneate at the base, subacute at the tips. Flowers solitary in leaf axils; pedicels absent or up to 6 mm long; bracteoles subulate, up to 1 mm long. Sepals 4 (5), deltoid, 1.5–3.5 mm long, 0.7–1.8 mm wide. Petals 4 (5), elliptical, 1–3 mm long, 0.7–2 mm wide, yellow. Stamens as many as the sepals; pollen grains shed in tetrads. Capsule 3–16 (–19) mm long, terete or nearly so, thin-walled, pale brown, regularly or irregularly dehiscent, often ± nodding. Seeds in several rows in each valve, purple, not visible through the capsule wall, free, ellipsoidal-rounded, 0.3–0.5 mm long, 0.2–2.2 mm in diameter, raphe very narrow and inconspicuous.

Annual. Found in wet places, sandy river beds, along streams, common in rice fields. Sometimes it rather resembles *L. linifolia* but may be distinguished by having as many stamens as sepal lobes and the seeds at the top of the capsule without corky disks.

{⊙ Hel}. Africa, Afghanistan, India, east to S Japan, Malaysia, Australia and New Caledonia: Ban, Bhu, Cey, Ind, Nep, Pak: <An, Ap, Ar, As, Bh, Gj, Go, Kl, Kt, Mh, Mp, Or, Pj, Rj, Sk, Tn, Up, Wb. 0–1500 m.

Ludwigia peruviana (Linnaeus) H. Hara [*Jussiaea speciosa* Ridley]: Fig. 288

Shrub-like; thread-like pneumatophores arise from submerged roots. Stems usually much branched, up to 4 m tall, the branches densely hairy. Leaves with petioles 3–15 mm long; leaf blades lanceolate to broadly lanceolate, (2–) 4–15 (–45) cm long, 0.3–4 (–10) cm wide, narrowly cuneate at the base, acute to acuminate at the tips. Flowers solitary in leaf axils; pedicels 2–4.5 (–6.5) cm long; bracteoles reduced or subulate and up

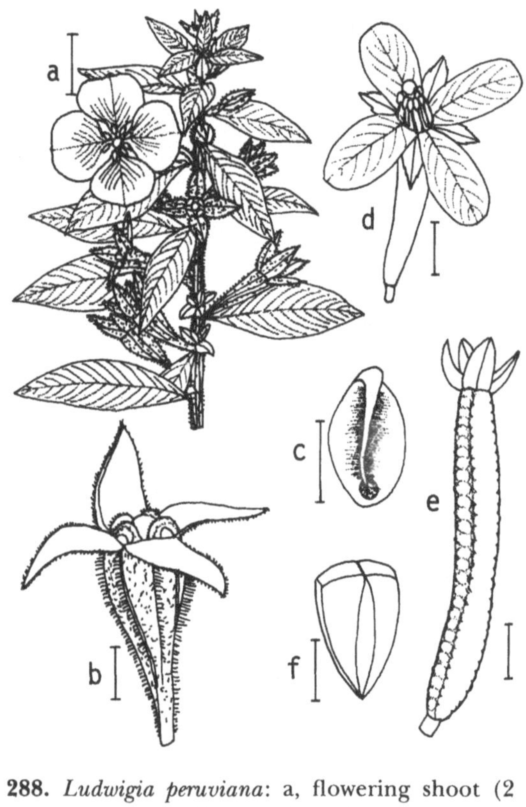

FIG. 288. *Ludwigia peruviana*: a, flowering shoot (2 cm); b, fruit (1 cm); c, seed (0.5 mm). *L. prostrata*: d, flower (1 mm); e, fruit (3 mm); f, seed (0.5 mm).

to 10 (–20) mm long, borne at the base of the sepal tube, usually deciduous and not persisting in fruit. Sepals 4 (5), lanceolate, 1–2 cm long, 4–9 mm wide, irregularly serrulate, villous. Petals 4 (5), suborbicular, shallowly emarginate, 1–4 cm long, bright yellow, clearly veined. Stamens twice as many as the sepals; pollen grains shed in tetrads. Capsule villous, 1–4 cm long, 0.5–1.3 cm thick, 4- (5)-angled, thin-walled, light yellowish-brown with 4 prominent dark brown ribs, regularly or irregularly dehiscent. Seeds in several rows in each valve, not visible through the capsule wall, free, rounded, 0.6–0.9 mm long.

Perennial. It is found in seasonally inundated areas or in wet places particularly along rivers, canals and ditches; sometimes growing on floating mats of vegetation.

{Hyp/Hel/Ple}. Native of the New World, spreading in Asia, known from the Nilgiri Hills in southern India and Sri Lanka: <Cey, <Ind: <Tn 1886.

Ludwigia prostrata Roxburgh: Fig. 288

Stems 10–60 cm tall, often reddish tinged. Leaves with petioles up to 4–25 mm long; leaf blades narrowly elliptical to elliptical, 1–13 cm long, 0.3–2.7 cm wide, narrowly cuneate at the base, acute at the tips. Flowers solitary in leaf axils; pedicels absent or up to 6 mm long. Sepals 4 (5), deltoid, 1.3–2.5 mm long, 0.4–0.9 mm wide. Petals 4 (5), narrowly spatulate, 1.3–2.2 mm long, 0.4–0.9 mm wide, yellow. Stamens (4) 8; pollen grains shed in tetrads. Capsule 12–22 mm long, 0.8–1.5 mm in diameter, somewhat 4-angled, thin-walled, pale brown, irregularly dehiscent. Seeds in 1 row in each valve, showing as indentations on the walls of the capsules, free, ovoid, apiculate at one end, 0.5–0.6 mm long, ± 0.3 mm in diameter, raphe narrow and linear.

Annual. In wet places, along ditches and streams and in rice fields. Resembles *L. perennis* but seeds in 1 row in each loculus, showing as indentations on the walls of the capsules

{⊙ Hel?}. Asia and Malaysia, from northern India and southern China to Sri Lanka, Indonesia and the Philippines: Ban, Bhu, Bur, Cey, Ind, Nep: An, As, Kt, Mg, Mp, Mz, Sk, Tn, Up, Wb. 0–800 m.

ORCHIDACEAE: 795 genera

Excluded: ***Peristylus lawii*** Wight; found in moist situations but not a wetland plant.

Zeuxine 77: Warm and tropical Old World

Zeuxine strateumatica (Linnaeus) Schlechter [*Z. sulcata* (Roxburgh) Lindley]: Fig. 289

Stems creeping and sometimes rhizomatous below, erect and slender above, 5–25 cm tall, leafy throughout. Leaves alternate, linear-lanceolate, 1.5–6 cm long, acute to subacuminate. Inflorescence a dense spike, up

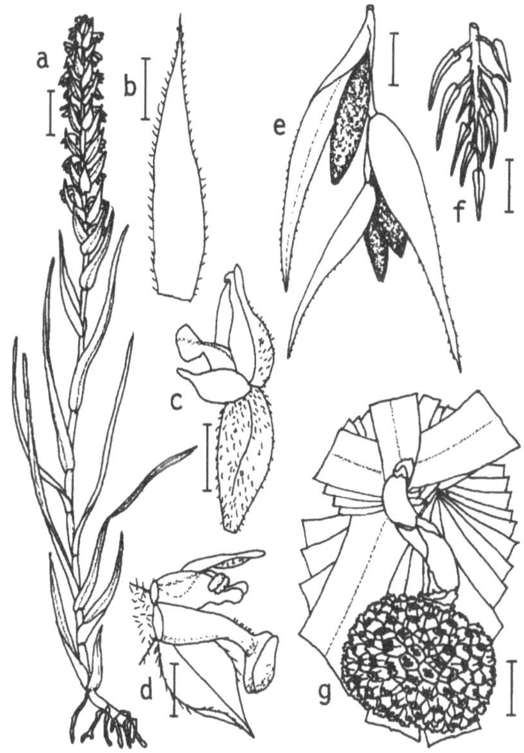

FIG. 289. *Zeuxine strateumatica*: a, flowering plant (1 cm); b, bract (2 mm); c, flower (2 mm); d, flower with sepals and petals removed (2 mm). *Pandanus fasicularis*: e, male inflorescence (4 cm); f, part of male flower with anthers arising from a common stalk (4 mm); g, fruiting head (4 cm).

to 5 cm long; bracts elongate-ovate to lanceolate, with ciliate margins. Flowers bilaterally symmetrical. Sepals 3, unequal, the lateral ones ovate-lanceolate, the dorsal broadly ovate, united with the petals to form a hood over the column. Petals 3, falcate-lanceolate to oblong, pale pink or white; petal lip saccate at the base, clawed, papillate. Column (of united androecium and gynoecium) very short. Pollen in sticky masses (pollinia). Capsules ellipsoid, ± 0.7 mm long.

Perennial or perhaps sometimes annual. Minute herbs in marshes, grassy places subjected to inundation and at the edge of ponds. It is not confined to wetlands. It is very variable and has many synonyms and is probably the commonest orchid in India.

{(⊙) Hel}. Tropical and subtropical Asia: Ban, Cey, Ind, Nep, Pak: As, Bh, Dl, Kt, Mp, Pj, Rj, Up.

PANDANACEAE: 3 genera

Pandanus ± 600 species: Old World tropics and subtropics

Much branched small trees or shrubs; branches spreading; trunk and lower branches with proproots.

Leaves spiral-tristichous, linear, sword-like to narrowly elliptical, margins prickly at least towards apex. Inflorescence subtended by leaf-like bracts, unisexual, of 1 or more heads. Flowers without bracts, bracteoles and perianth. Female flowers 1-locular; stigmas sessile or nearly so. Male flowers in dense spikes, not clearly recognizable as individual units. Fruit a globose to oblong mass of free or united woody or fleshy drupes. Seed enclosed in a hard, bone-like pyrene.

This genus is taxonomically complex. Many species are coastal others are in mountain forests but they hardly qualify as wetland plants. It is very difficult to know what species to include. The following species is widespread in the hotter and moister parts of India and is usually found in wet places; it is also often planted.

Pandanus fasicularis Lamarck [*P. odoratissimus* Linnaeus fil. nom. illeg., *P. tectorius* auct. Ind., non Solander ex S. Parkinson]: Fig. 289

Leaves usually less than 2 m long and 8 cm wide. Prickles on leaf margins white (a cultivated variety '*laevis*' is without spines). Inflorescence bracts creamy-white and fragrant. The female inflorescence a solitary, terminal head. Stigmas 'U'- or 'V'-shaped. Male inflorescence of ± 9–13, up to 12 cm long spikes.

Common in coastal regions in salt water marshes and mangroves but often found in fresh water marshes and swamps, along rivers and streams and in seasonally inundated places where it sometimes forms dense thickets. It is often cultivated: the leaves are used for weaving, a perfume is extracted from the bracts, male flowers are used as decoration and it also makes a useful fence. It is perhaps no more than a variety of the Pacific species *P. tectorius*.

{Hel}. S and SE Asia: Cey, Ind: Ap, <Dl, Gj, Go, Kl, Kt, <Rj, Tn.

PHILYDRACEAE: 4 genera

Philydrum: 1 species

Philydrum laniginosum Banks et Solander ex Gaertner: Fig. 290

Stem short, corm-like. Leaves unifacial, equitant, in 2 ranks, crowded at the stem base, linear, 40–80 cm long, glabrous, soft. Scapes 30–100 (–150) cm tall, woolly-hairy towards the apex. Flowers bisexual, bilaterally symmetrical, sessile, solitary in the axils of spathe-like bracts; bracts 2–7 cm long, reflexed during anthesis. Perianth petal-like, yellow, 2-seriate; outer segments 2, 1.2–2 cm long, ± 1 cm wide, acute at the tips; inner segments 2, united at the base, 4–8 mm long, ± 2 mm wide. Stamen 1, inserted at the base of the outer, adaxial perianth segment, 7–9 mm long; filament flattened; anther 2-locular, spirally twisted; pollen released in tetrads. Ovary superior, 1-locular; style 1, simple, elongate. Fruit a capsule, 3-valved, triangular-oblong, 9–10 mm long, 4–5 mm wide. Seeds numerous, ± 1 mm long, arranged on parietal placentas.

Annual or perennial. Occurs in ponds and marshes but also found in swamps and wet savannas. Also sometimes found in rice fields but not reported to be a serious weed. Pollinated by insects. The disseminules are seeds, dispersal unknown.

{(⊙) Hel}. SE and E Asia and NE Australia: Bur, Ind: An.

POACEAE [GRAMINEAE]: 651 genera

Stems creeping, ascending or erect, when perennial often differentiated into sterile shoots (rhizomes or stolons) and flowering shoots (culms); culms cylindrical, closed at nodes and usually hollow at internodes (in some aquatics filled with spongy aerenchyma). Leaves solitary at nodes, alternate and 2-rowed, usually consisting of sheath, ligule and blade; sheaths encircling the culm, open or closed, sometimes inflated;

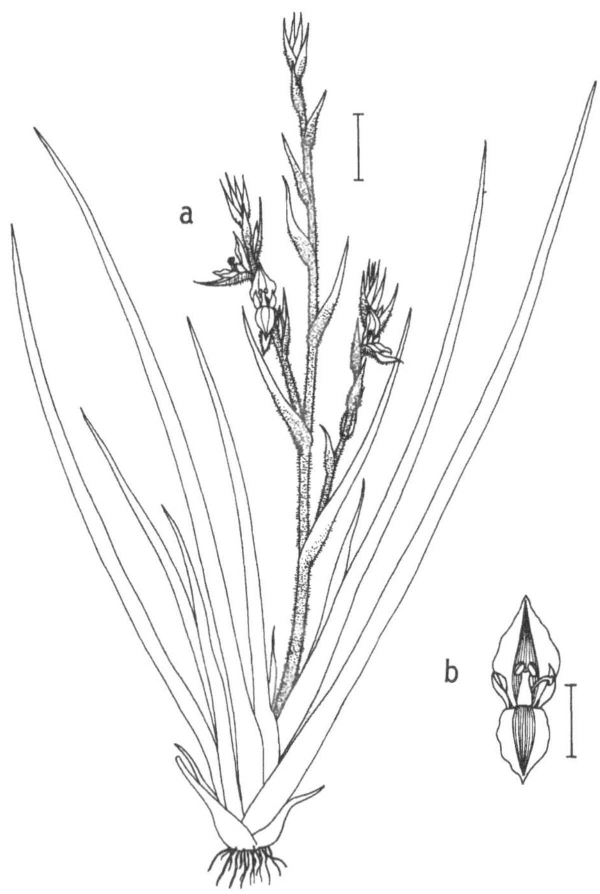

Fig. 290. *Philydrum laniginosum*: a, flowering plant (6 cm); b, flower (1 cm).

ligules placed at junction of sheath and blade, membranous, a fringe of hairs or rarely absent; leaf blades usually long and narrow (grass-like), rarely broad, rarely with a stalked base (pseudopetiole). Inflorescence very variable, made up of spikelets. The axis of the spikelet (rachilla) usually bears at the base 2 empty bracts (glumes); the axillary buds of the following bracts (lemmas) usually develop a branchlet bearing 1 leaflet (palea) and the subsessile flower. The unit consisting of lemma, palea and flower is called a floret. Flowers mostly bisexual, sometimes unisexual or sterile. Perianth of 2 or 3 minute, membranous or fleshy scales (lodicules) or absent. Stamens hypogynous, usually 3, rarely 1–14; filaments long and delicate; anthers opening by longitudinal slits, usually versatile. Ovary 1-locular; ovule solitary, anatropous; styles usually 2, rarely 1 or 3; stigmas usually plumose. Fruit mostly a caryopsis with a thin pericarp fused to the seed.

Annual or perennial. Pollinated by wind or autogamous. The disseminules are caryopses; a fruit usually enclosed in 2 bracts (lemma and palea). Because of the importance of the grasses there is an enormous amount of literature; the following is the best general survey: Watson, L. and Dallwitz, M. J. The grass genera of the World. C. A. B. International, Wallingford. 1–1038 (1992); for India: Bor, N. L., The grasses of Burma, Ceylon, India and Pakistan (excluding Bambuseae). Pergammon Press, Oxford, i–xviii, 1–767 (1960); Roy, G. P. Grasses of Madhya Pradesh. Botanical Survey of India, Howrah. 1–180 (1984); Sreekumar, P. V. and Nair, V. J. Flora of Kerala — Grasses. Botanical Survey of India, Calcutta. 1–470 (1991).

Excluded: the following wetland species are found in the Himalayas but do not extend south: ***Agrostis stolonifera*** Linnaeus; ***Arthraxon breviaristatus*** Hackel; ***Catabrosa aquatica*** (Linnaeus) Palisot de Beauvois; ***Glyceria plicata*** (Fries) Fries; ***G. tonglensis*** C. B. Clarke; ***Neyraudia arundinacea*** (Linnaeus) Henrard; ***Phalaris aquatica*** Linnaeus [*P. nodosa* Linnaeus, *P. tuberosa* Linnaeus]; ***P. arundinacea*** Linnaeus; ***Poa palustris*** Linnaeus; ***Spodiopogon cotulifer*** (Thunberg) Hackel.

The following may grow in damp or moist places but are not considered to be wetland plants: ***Arthraxon castratus*** (Griffith) Narayanaswami ex Bor; ***A. jubatus*** Hackel; ***A. lanceolatus*** (Roxburgh) Hochstetter; ***A. nudus*** (Steudel) Hochstetter; ***A. quartinianus*** (A. Richard) Nash; ***Bothriochloa pertusa*** (Linnaeus) A. Camus; ***Chrysopogon aciculatus*** (Retzius) Trinius; ***Crypsis schoenoides*** (Linnaeus) Lamarck; ***Cynodon arcuatus*** J. S. Presl ex C. B. Presl [*C. intermedius* Rangachari et Tadulingam]; ***C. dactylon*** (Linnaeus) Persoon; ***Dichanthium annulatum*** (Forsskål) Stapf; ***Digitaria ciliaris*** (Retzius) Koeler [*Panicum adscendens* Humbolt, Bonpland et Kunth., *D. marginata* Link]; ***D. setigera*** Roth ex Roemer et Schultes; ***D. stricta*** Roth ex Roemer et Schultes; ***Garnotia exaristata*** Gould; ***G. tenella*** (Arnott ex Miquel) Janowsky; ***Imperata cylindrica*** (Linnaeus) Raüschel; ***Iseilema anthephoroides*** Hackel; ***I. laxum*** Hackel; ***I. prostratum*** (Linnaeus) N. J. Andersson; ***Manisuris myuros*** Linnaeus: ♦Ind, ♦Pak; ***Pennisetum hordeoides*** (Lamarck) Steudel; ***P. purpureum*** Schumacher; ***Polypogon viridis*** (Gouan) Breistroffer [*P. littoralis* var. *mutica* J. D. Hooker, *Agrostis semiverticillata* (Forsskål) Christ, *A. verticillata* Villars]; ***P. monspeliensis*** (Linnaeus) Desfontaines; ***Rottboellia exaltata*** Linnaeus fil.; ***Setaria intermedia*** Roemer et Schultes [*S. tomentosa* (Roxburgh) Kunth]; ***S. paniculifera*** (Steudel) Fournier ex Hemsley; ***S. pumila*** (Poiret) Roemer et Schultes [*S. pallide-fusca* (Schumacher) Stapf et C. E. Hubbard; ***Sorghum halepense*** (Linneaus) Persoon; ***Sporobolus diander*** (Retzius) Palisot de Beauvois; ***Thysanolaena maxima*** (Roxburgh) O. Kuntze.

Confined to saline conditions: ***Myriostachya wightiana*** J. D. Hooker; ***Porteresia coarctata*** (Roxburgh) Teleoka [*Oryza triticoides* Griffith]; ***Sporobolus tremulus*** (Willdenow) Kunth; ***S. virginicus*** (Linnaeus) Kunth.

1. Leaf sheaths strongly inflated and spreading from the culm, usually functioning as floats (leaf blades 3–5 cm long and 0.7–2 cm wide, ovate to ovate-oblong, without a prominent midrib; spikelets awned; usually floating) ***Hygroryza***

1. Leaf sheaths not strongly inflated, but if inflated then enveloping the culm

2. Leaves membranous and translucent; mesophyll of only 2 layers of cells; spikelets of two, 7-nerved glumes enclosing two, 7- or 9-nerved lemmas, the uppermost containing a bisexual flower (W Ghats) ***Hubbardia***

2. Leaves not membranous, not translucent; mesophyll with many more than 2 layers of cells; spikelets not as above

3. Inflorescence partly enclosed in a nut-like, globose to fusiform utricle (a modified sheath) which has a pore at the apex only and sometimes bears a rudimentary blade; utricle hardened at maturity and enclosing the caryopsis; exerted part of the inflorescence bearing male spikelets only ***Coix***

3. Inflorescence not enclosed in a nut-like utricle

4. Inflorescence of clustered or solitary spike-like racemes which appear to arise in leaf axils along the culm (spike-like racemes terminating lateral branches, but branches often short and in clusters, so that several spikes may arise from each node, some enclosed by leaves at their bases; spikelets in opposite pairs) ***Hemarthria***

4. Inflorescence not of lateral spike-like racemes arising in leaf axils along the culm

5. Inflorescence a large (up to 1 m or more long), silky-haired panicle

6. Spikelets in pairs, 1 sessile, each with 2 flowers; lower floret sterile ***Saccharum***

6. Spikelets solitary, all pedicelled, each with (2–) 3–7 flowers; lowest floret bisexual, male or sterile

7. Lemmas with long silky hairs; base of leaf blades clasping the culm; rachilla glabrous or shortly hairy; lowest floret bisexual; rhizomes knotty ***Arundo***

7. Lemmas glabrous; base of leaf blades clasping the culm; rachilla with long, silky hairs; lowest floret male or sterile; rhizomes stout but not knotty ***Phragmites***

5. Inflorescence not a large, silky-hairy panicle

8. Tip of raceme axis prolonged into a bristle-like extension

9. Spikelets not arranged in neat rows along the rachis; rachis flattened ***Pseudoraphis***

9. Spikelets arranged in 1 or 2, neat rows along the rachis; rachis flattened

10. Lower glume well developed, although sometimes small ***Paspalidium***

10. Lower glume absent or reduced to a minute scale ***Paspalum***

8. Tip of the raceme axis not prolonged

11. Inflorescence 1–2.3 cm long, a single, simple, more or less spike-like raceme (culms delicate, usually floating, up to 180 cm long; spikelets in pairs; SW. India) ***Limnopoa***

11. Inflorescence usually more than 3 cm long, not a single, spike-like raceme

12. Inflorescence consisting of dense, globose or cylindrical clusters of spikelets on a central axis, each cluster being subtended by 2 or more bracts or sterile spikelets; in the lower and middle parts of the inflorescence the clusters being remote and distinct, above merging into an uninterrupted spike-like panicle ***Elytrophorus***

12. Inflorescence not consisting of globose or cylindrical clusters of spikelets

13. Inflorescence a terminal, dense, cylindrical, spike-like panicle

14. Leaf blades amplexicaul at the base; floating culms containing a white pith; lemma and palea of upper floret gaping at the apex ***Hymenachne***

14. Leaf blades not amplexicaul; floating culms hollow without pith; lemma and palea of the upper floret closed at the apex ***Sacciolepis***

13. Inflorescence not a terminal, dense, spike-like panicle

15. Spikelets in pairs (usually one sessile and the other pedicelled or both unequally pedicelled)

16. Lower glume awned; glume and awn flat; awn up to 2 mm wide and 30 mm long, with scabrous margins; inflorescence consisting of digitate or solitary, erect, glabrous, spike-like racemes (giant herb) ***Vossia***

16. Lower glume absent or if present then not awned

17. Inflorescence a large, loose panicle, composed of numerous verticillate, spike-like racemes or first order branches, which in turn bear spike-like racemes

18. Pedicelled spikelets 2-flowered, the lower floret sterile, the upper floret male; sessile spikelets slightly flattened laterally; lower glume keeled, 7-nerved, with numerous upwardly curved spines ***Vetiveria***

18. Pedicelled spikelets sterile, with 2 glumes which sometimes enclose a reduced lemma; sessile spikelets flattened abaxially; lower glume 2-keeled, with 8–11 nerves, scabrid on the keels ***Hemisorghum***

17. Inflorescence of 1 or more, simple, 1-sided, spike-like racemes arranged along the main axis or (1–) 2–7 spike-like racemes arranged in pairs or subdigitally on branches

19. Lower glume present; rachis disarticulating at the nodes; lemma of the upper floret membranous, usually awned ***Ischaemum***

19. Lower glume usually absent; rachis continuous and tough; lemma of the upper floret leathery, awnless ***Paspalum***

15. Spikelets not in pairs

20. Spikelets in 1-sided spikes or spike-like racemes; spikes or racemes usually digitate or distant along the axis

21. Back of the lemma of upper floret turned away from the rachis and away from the lower glume ***Brachiaria***

21. Back of the lemma of upper floret turned towards the rachis or towards the lower glume if glume present

22. Lower glume absent ***Paspalum***

22. Lower glume present, although sometimes small

23. Glumes acuminate or awned, rarely only acute; lemma of upper floret not mucronate ***Echinochloa***

23. Glumes without awns, if acuminate then with the lemma of upper floret mucronate

24. Lemma of upper floret acute, not mucronate; spikelets plano-convex ***Paspalidium***

24. Lemma of upper floret mucronate or shortly awned; spikelets terete or somewhat flattened ***Urochloa***

20. Spikelets in ± open panicles or panicles contracted and spike-like
25. Spikelets with 3–9 flowers
26. Lemmas entire at the tip; nerves on the lemmas glabrous ***Eragrostis***
26. Lemmas notched or toothed at the tip; nerves on the lemmas hairy
27. Spikelets sub-terete; lemmas rounded on the back ***Diplachne***
27. Spikelets laterally flattened; lemmas distinctly keeled on the back ***Leptochloa***
25. Spikelets with 1 or 2 flowers or spikelets sterile
28. Ligules absent ***Echinochloa***
28. Ligules membranous or a fringe of hairs
29. Glumes absent (spikelets strongly flattened abaxially) ***Paspalum***
29. Glumes present, sometimes unequal or the lower absent
30. Glumes and lemmas tipped by a thickened, laterally flattened crest, which appears as if nipped by pincers ***Acroceras***
30. Glumes and lemmas not tipped by a thickened, laterally flattened crest
31. Lower glume absent (spikelets plano-convex, awnless) ***Panicum***
31. Both glumes present
32. All spikelets unisexual; male and female spikelets on different branches of the same panicle; upper branches ascending, bearing female spikelets; male spikelets pendulous; style base not persistent ***Zizania***
32. All or most spikelets bisexual (not differing in size, shape, colour and position)
33. Glumes and lemmas tipped by a thickened, laterally flattened crest (appearing as if nipped by pincers) ***Acroceras***
33. Glumes and lemmas not tipped by a thickened, laterally flattened crest
34. Ligules reduced to a fringe of hairs
35. Glumes separately and precociously deciduous; spikelets 1-flowered ***Sphaerocaryum***
35. Glumes persistent or falling together; spikelets 2-flowered
36. Glumes unequal, upper one as long as the spikelet, lower one much shorter; spikelets slightly flattened abaxially ***Panicum***
36. Glumes equal to subequal, broad and convex; spikelets small, terete to globose
37. Glumes much shorter than the spikelet; lower floret bisexual, the upper one female ***Coelachne***
37. Glumes about as long as the spikelet or only slightly shorter; the lower or upper floret sometimes male ***Isachne***
34. Ligules membranous, with or without a fringe of hairs
38. Spikelets borne on a ring-like swelling (lower glume and top of the rachilla swollen to a ring-like callus) ***Eriochloa***
38. Spikelets not borne on ring-like swellings
39. Lower glume much shorter than the upper one; upper glume equalling the sterile lemma; upper glume and sterile lemma as long as the spikelet and enclosing the fertile floret (spikelets slightly flattened abaxially and awnless) ***Panicum***
39. Both glumes either absent or reduced to a 2-lobed rim or scale-like and then neither one as long as the spikelet; if sterile lemmas present then usually very small
40. Sterile lemmas absent; glumes completely absent or reduced to an entire or more rarely, obscurely 2-lobed, somewhat thickened rim ***Leersia***
40. Sterile lemmas present but usually very small; glumes present but very small and scale-like or reduced to an entire or 2-lobed rim
41. Leaf margins smooth or scabrid ***Oryza***
41. Leaf margins tuberculate-spiny ***Porteresia***

Acroceras: 15 species, tropical to warm regions of the world, most Madagascan endemics

Excluded: ***Acroceras munroanum*** (Balansa) Henrard; not found in wetlands.

Acroceras zizanioides (Kunth) Dandy [*Panicum latifolium* auct. Ind., non Linnaeus]: Fig. 291

Culms 60–120 cm tall, often stoloniferous and rooting at internodes. Leaves borne along the culm; ligules reduced to a rim fringed with hairs; leaf blades linear to lanceolate, 7–15 cm long, 8–30 mm wide, usually cordate at base. Inflorescence a panicle of 1–many, 1-sided racemes, 10–25 cm long; racemes inserted on a central axis, usually spreading only slightly. Spikelets 2-flowered, 5–6 mm long, disarticulating below the glumes, subterete, glabrous, awnless. Lower floret male or sterile, the upper bisexual. Glumes 2, unequal, each with a thickened, laterally flattened crest at the apex. Lemmas similar to the glumes, with laterally flattened apical crests (appearing as if nipped with pincers). Paleas with a more or less recurved callous tip. Caryopses oblong, abaxially flattened, adaxially convex, tightly enclosed in a hardened lemma and palea.

Perennial. Not confined to wetlands but mostly found in wet places and often growing seasonally submerged, otherwise floating or emergent.

{Hel/Ple}. Pantropics: Ban, Bur, Ind: As.

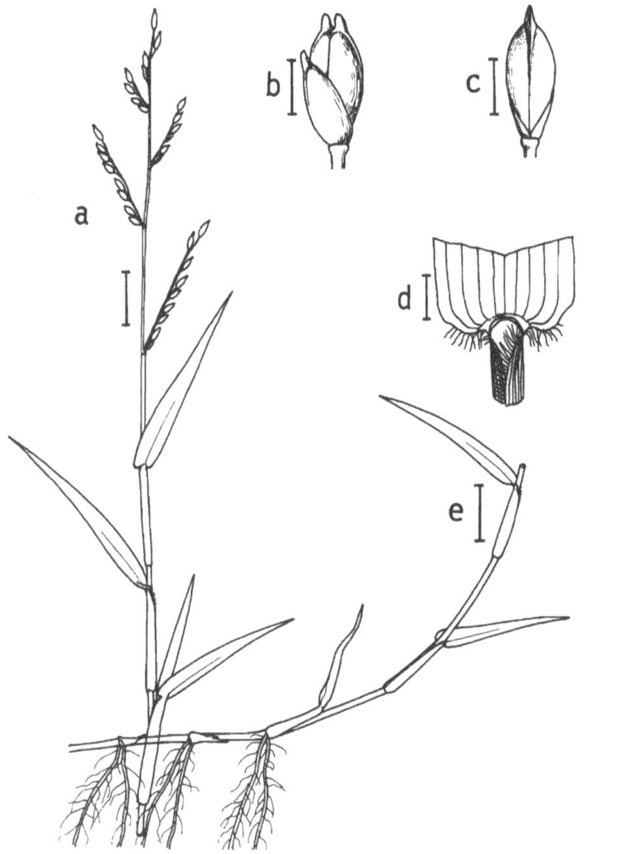

Fig. 291. *Acroceras zizanioides*: a, flowering culm (2 cm); b, spikelet viewed from the front (2 mm); c, spikelet viewed from the back (2 mm); d, ligule (1 mm); e, base of culm (2 cm).

Arundo: 3 species
Arundo donax Linnaeus: Fig. 292

Rhizomes creeping, knotty. Culms mostly erect, (0.4–) 2–4 (–6) m tall, often with stilt roots below. Leaves cauline; ligules short, membranous with minutely hairy margin, the hairs up to 0.3 mm long; leaf blades (10–) 50–75 (–100) cm long, (2.5–) 4–5 (–8) cm wide, glabrous, rounded to cordate and clasping the culm at the base, tapering above to a long fine point. Inflorescence a dense, erect panicle, 30–60 cm long, feathery, whitish to brown. Spikelets pedicelled, solitary, 2- to 5- (–7)-flowered, 8–18 mm long, laterally flattened, disarticulating above the glumes and between the florets; rachilla glabrous or shortly hairy. Florets bisexual. Glumes 2, almost equal, about as long as the spikelet, 3- or 5-nerved, acute. Lemmas with long silky hairs, awned; awn ± 3 mm long.

Perennial. Often found standing in water but mostly in areas subject to flooding or along ditches and canals. Caryopses dispersed by wind. It is used for hedging and for making lattices, mats, screens, reeds of musical instruments and as a source of industrial cellulose; sometimes a weed as it can block drainage ditches. It may be distinguished from *Phragmites* by the clasping leaf bases and hairy lemmas.

{Hel}. Tropical Asia and the Mediterranean Region, introduced in the New World, it is widely planted in India and is unlikely to be native throughout the subcontinent: <Cey, Ind: <An, <Dl, Go, Hp, Jk, Kl, Mn, Mp, Rj, Tn, Up, Wb.

Brachiaria [excluded from *Urochloa*]: ± 100 species: tropical to warm cosmopolitan

Culms erect, creeping or floating. Ligule a fringed membrane or a fringe of hairs. Inflorescence a panicle up to 25 cm long, of 1-sided, spike-like racemes along a main axis which itself terminates in a raceme. Spikelets 2-flowered, awnless or mucronate, terete or abaxially flattened, in the lower parts of the racemes in unequally pedicelled pairs or in clusters, in the upper

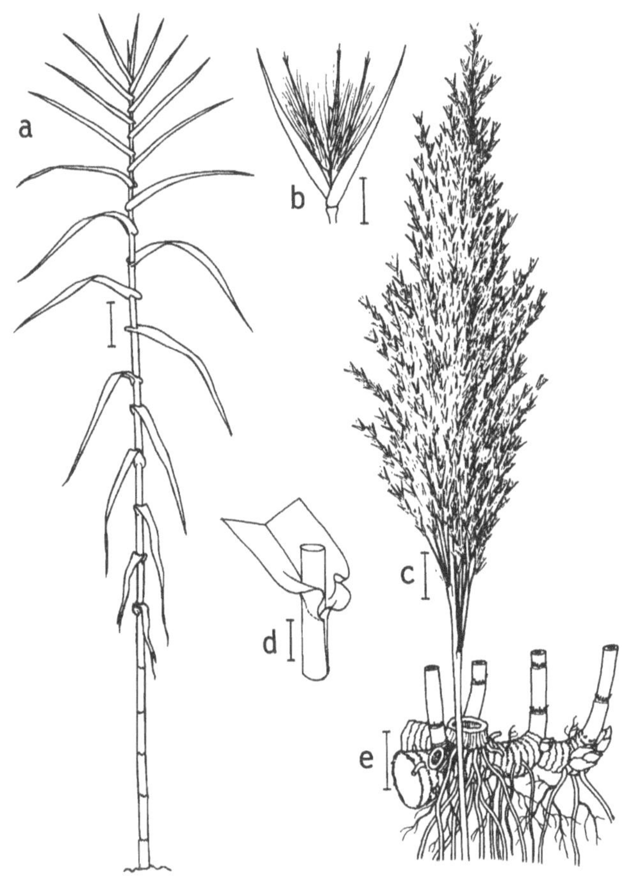

Fig. 292. *Arundo donax*: a, erect shoot (10 cm); b, spikelet (3 mm); c, inflorescence (4 cm); d, ligule (2 cm); e, rhizome (2 cm).

parts solitary, disarticulating below the glumes; lower floret male or sterile; upper floret bisexual. Glumes 2, unequal; lower glume next to the axis (adaxial), as long as the spikelet or very much shorter; upper glume as long as the spikelet. Lemma of lower floret as long as the spikelet, shorter or absent; lemma of upper floret turned away from the lower glume, leathery becoming hardened at maturity. Caryopses tightly enclosed by the hardened lemma and palea, abaxially flattened, broadly oblong to elliptical in outline.

Annual or perennial. Some species are important fodder plants but they can also become serious weeds in supply and drainage channels. The delimitation between *Brachiaria* and *Urochloa* is problematic; some botanists consider *Brachiaria* to be monotypic, with *B. eruciformis* as the only species. I have taken a more traditional position (*Brachiaria* has the lemma of the upper floret turned away from the lower glume and the rachis).

Excluded: the following may tolerate temporary flooding but are not considered to be wetland species: ***Brachiaria decumbens*** Stapf; ***B. deflexa*** (Schumacher) C. E. Hubbard; ***B. distachya*** (Linnaeus) Stapf; ***B. munae*** Basappa; ***B. nilgirica*** Bor; ***B. paspaloides*** (Presl) C. E. Hubbard; ***B. remota*** (Retzius) Haines; ***B. seminuda*** (Hochstetter) Stapf; ***B. setigera*** (Retzius) C. E. Hubbard [*Panicum setigerum* Retzius]; ***B. villosa*** (Lamarck) A. Camus.

1. Stoloniferous, very robust perennial; culms erect, 5–10 mm in diameter — ***B. mutica***
1. Stolons not developing, annual or not very robust perennial; culms decumbent or creeping below, not more than 3 mm in diameter
  2. Spikelets glabrous, rarely more than 2 mm long; leaves rarely more than 5 times as long as wide — ***B. reptans***
  2. Spikelets hairy, 2 mm or more long; leaves more than 6 times as long as wide
    3. Lower glume minute, less than one-third as long as the spikelet; spikelets 2–2.5 mm long; inflorescence ± linear, the branches erect or only slightly spreading — ***B. eruciformis***
    3. Lower glume one-third to half as long as the spikelet; spikelets 2.5–3 mm long; inflorescence panicle-like, the branches spreading — ***B. ramosa***

Brachiaria eruciformis (J. E. Smith) Grisebach [*Panicum isachne* Roth ex Roemer et Schultes]: Fig 293

Culms 10–75 cm tall, rarely more than 3 mm in diameter, ascending from a creeping base, rooting at lower nodes. Ligule 0.5–1 mm long, hairy; leaf blades 2–8 cm long, 3–6 mm wide, glabrous or somewhat hairy. Inflorescence ± linear, 4–6 (–10) cm long, of 2–15 ± erect spike-like racemes, each 1.5–3 cm long; rachis 3-angled, hairy. Spikelets solitary, borne in 2 rows, 1.8–2.5 mm long, up to 1.2 mm wide, sometimes tipped with purple. Lower glume minute (0.3–1 mm long), 1-nerved or without nerves; upper glume as long as the spikelet, 5-nerved. Lemma of lower floret ± like upper glume; lemma of upper floret slightly shorter than the spikelet, smooth, glossy and hardened. Palea hard like lemma of upper floret.

Annual. In rice fields, along ditches and canals and in marshy places. Not confined to wetlands and also found as a weed in cultivated fields.

{⊙ Hel}. Mediterranean eastwards to India: Cey, Ind, Pak: Ap, Bh, Gj, Go, Hp, Kt, Mh, Mp, Rj, Tn, Up, Wb. 0–2000 m.

Brachiaria mutica (Forsskål) Stapf [*Panicum amphibium* Steudel, *P. purpurescens* Raddi]: Fig 293

Stolons up to 5 m long. Culms 50–200 cm tall, 5–10 mm in diameter, erect, rooting at lower nodes. Ligule 0.7–1.9 mm long, membranous and fringed; leaf blades 10–30 cm long, (5–) 10–25 mm wide, glabrous or scarcely hairy. Inflorescence spreading, 8–30 cm long,

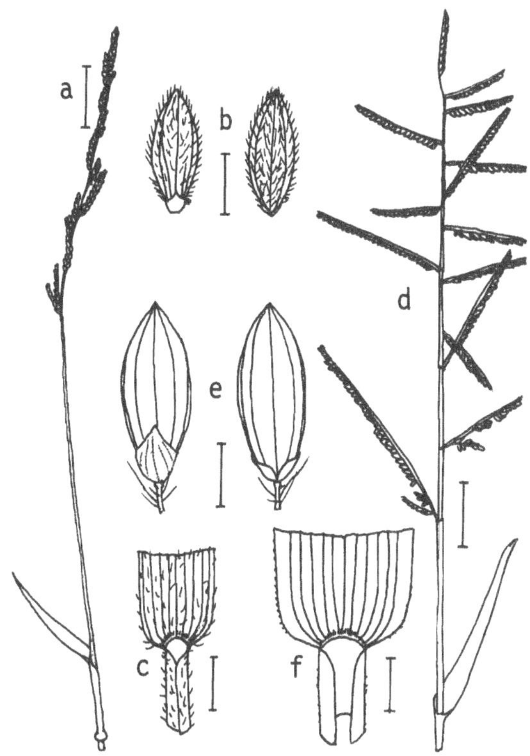

FIG. 293. *Brachiaria eruciformis*: a, inflorescence (2 cm); b, spikelet viewed from the front and from the back (1 mm); c, ligule (5 mm). *B. mutica*: d, inflorescence (2 cm); e, spikelet viewed from the front and from the back (1 mm); f, ligule (5 mm).

with 6–20 racemes, each 2–20 cm long; flattened and winged, ± 1 mm wide, hairy at the base. Spikelets solitary or paired, (2.5–) 3–3.5 (–5) mm long, 1.5–2 mm wide, arranged untidly in 3 or 4 rows, glabrous or hairy. Lower glume up to one-third as long as the spikelet, 1-, 3- or 5-nerved; upper glume as long as the spikelet, 5- or 7-nerved, hairy. Lemma of lower floret ± like upper glume; lemma of upper floret hardened, light yellowish-brown, ± 2.4 mm long. Palea hyaline.

Perennial. It is very vigorous and widely planted as a pasture grass. Usually known as paragrass in India. It is frequently found in aquatic biotopes and along the banks of streams and backwaters, sometimes becoming dominant in marshes, also sometimes forming floating mats. Under some conditions it may out compete *Typha* and is used to control it.

{Hel/(Ple)}. Tropical Africa and America, introduced in Asia and Australia: <Ind: <An, <As, <Kl, <Rj, <Tn, <Wb.

Brachiaria ramosa (Linnaeus) Stapf [*Panicum supervacuum* C. B. Clarke]: Fig. 294

Culms (15–) 40–60 (–100) cm tall, rarely more than 3 mm in diameter, ascending from a creeping base, rooting at lower nodes. Ligule membranous, 0.4–1 mm long, fringed with hairs; leaf blades 2–15 cm long, 6–8 (–15) mm wide, rarely more than 5 times as long as wide, glabrous or somewhat hairy. Inflorescence spreading, (2–) 5–16 cm long, of 3–16 racemes, each 1.5–5 cm long; rachis 3-angled, sparsely hairy. Spikelets paired, 2.5–3 (–4) mm long, up to 1.5 mm wide, hairy. Lower glume one-third to half as long as the spikelet (1–2 mm long), 5- or 7-nerved; upper glume as long as the spikelet, hairy, 7-nerved, the nerves green. Lemma of lower floret ± like upper glume; lemma of upper floret hardened and rough, 2–2.5 mm long. Palea delicate and hyaline.

Annual or perennial. Often common and abundant, almost always in wet places such as rice fields, along ditches and canals and in marshes. It is rather variable with glabrous and hairy races

{(⊙) Hel}. Tropics of the Old World: Cey, Ind: Ap, Gj, Hp, Kl, Kt, Mh, Mp, Rj, Tn. 0–2000 m.

Brachiaria reptans (Linnaeus) C. A. Gardner et C. E. Hubbard [*Panicum umbrosum* Retzius, *P. procumbens* Nees, *P. prostratum* Lamarck]: Fig. 294

Culms (10–) 15–35 (–50) cm tall, rarely more than 3 mm in diameter, ascending from a creeping base, rooting at lower nodes. Ligule membranous, 0.5–1 mm long, fringed with hairs; leaf blades ovate-lanceolate, 1.5–10 cm long, 4–20 mm wide, mostly hairy, at the base mostly clasping the culm, margin often wavy. Inflorescence spreading, 2–7 cm long, of 3–10 (–15) racemes, each 1.5–5 cm long; rachis flattened, sparsely hairy. Spikelets paired, 1.8–2.3 mm long, ± 0.8 mm wide, glabrous. Lower glume minute (0.2–0.5 mm long), faintly 1- or 3-nerved; upper glume as long as the spikelet, glabrous, 5- to 9-nerved, the nerves green. Lemma of lower floret ± like upper glume but not more than 5-nerved; lemma of upper floret boat-shaped, hardened, golden brown. ± 1.6 mm long. Palea like lemma of upper floret.

Annual or sometimes perennial when in water. A common weed of cultivated land, particularly in rice fields, on the banks of channels and other wet areas. It soon dies when the water recedes. The orientation of the spikelets is difficult to determine; this species would be perhaps better transferred to *Urochloa*.

{(⊙) Hel}. Tropical Old and New Worlds but usually considered naturalized in India: <Cey, <Ind, <Pak: <Dl, <Gj, <Go, <Hp, <Kl, <Mh, <Mp, <Rj, <Tn, <Wb.

Coelachne: 10 barely separable species: Old World tropics and subtropics

Low-growing, tufted, annual or perennial. Leaves cauline; ligules a fringe of hairs; leaf blades linear to ovate-lanceolate. Inflorescence a loose, open or con-

FIG. 294. *Brachiaria ramosa*: a, inflorescence (2 cm); b, spikelet viewed from the front and from the back (1 mm). *B. reptans*: c, flowering shoot (2 cm); d, spikelet viewed from the front and from the back (1 mm).

tracted panicle, up to ± 10 cm long. Spikelets 2-flowered, terete, awnless, disarticulating above the glumes; rachilla elongated between the florets; lower floret bisexual; upper floret female. Glumes 2, subequal, persistent, one-third to two-thirds the length of the spikelet. Lemma of lower floret as long as the spikelet, nerveless or obscurely 3- to 5-nerved, somewhat hardened at maturity; lemma of upper floret shorter than the lower, remaining membranous. Stamens of the lower floret 2 or 3. Caryopses ovoid, loosely enclosed by the lemma and palea.

Prakash, V. and Jain, S. K. Isachneae. Fascicles of Flora of India 14: 3–7 (1984).

1. Spikelets 1–1.3 (–1.5) mm long; stamens usually 2 ***C. minuta***
1. Spikelets more than 1.5 mm long; stamens 3
  2. Florets glabrous at the base; spikelets 1.5–2.5 (–3) mm long; anthers 0.6–1.1 mm long ***C. simpliciuscula***
  2. Florets densely bearded at the base; spikelets 2.5–4 mm long; anthers 0.6–1.1 mm long ***C. perpusilla***

Coelachne minuta Bor [*C. ghatica* Naik]: Fig. 295

Like *C. simpliciuscula* but spikelets 1–1.3 (–5) mm long; lemma of lower floret 1–1.2 mm long; lemma of upper floret and palea 0.8–1 mm long; stamens usually 2.

Annual of open rocky moist grassland, grows with *Eriocaulon, Utricularia* and *Murdannia*.

{⊙ Hel?}. Endemic to W India: ♦Ind: Mh (Mahabaleshwar).

Coelachne perpusilla (Arnott ex Steudel) Thwaites: refer to Fig. 295

Like *C. simpiciuscula* but larger and more robust; inflorescence lax and spreading; spikelets 2.5–4 mm long, widely spaced and seldom overlapping on the panicle branches, 1–3 (4) per branch; florets densely bearded at the base; lemma of lower floret 2–3 mm long. Anthers 0.6–1.1 mm long.

Densely tufted perennial in marshy places in highlands. A very poorly known species: Prakash and Jain (1984) recognize two varieties: var. *nilagirica* Prakash et Jain and var. *perpusilla*.

{Hel?}. ♦ India and Sri Lanka; ♦Cey, ♦Ind: Tn, Kl. July–November. 1200–2300 m.

Coelachne simpliciuscula (Wight et Arnott) Munro ex Bentham [*C. pulchella* auct. Ind., non R. Brown]: Fig. 295

Culms tufted, prostrate or ascending, 5–45 cm long. Leaf blades linear-lanceolate, 0.5–5 cm long, 1.5–5 mm wide. Inflorescence branches up to 1 (–1.5) cm long, densely crowded with overlapping spikelets. Spikelets 1.5–2.5 (–3) mm long; (3–) 5–9 per branch. Glumes glabrous; lower glume 0.5–1 mm long, 3-nerved; upper glume 1–1.5 mm long, with (4) 5 (6) nerves. Lemma of lower floret 1.5–2.5 mm long, without nerves, glabrous or sparsely hairy. Palea of lower floret like lemma but narrower. Stamens 3; anthers 0.3–0.4 mm long. Lemma and palea of upper floret 1–2 mm long, without nerves, 2-keeled, hairy along the keels and on the abaxial surface. Anthers 0.6–1.1 mm long.

Annual. Flooded areas, usually gregarious and forming dense mats. Frequently found in rice fields but not considered to be a serious pest. Usually found on higher ground. Flowering as long as it is wet.

{⊙ Hel}. SE Asia, China and Madagascar: Cey, Ind, Nep: Ap, Ar, As, Bh, Kl, Kt, Mg, Mh, Mp, Ng, Or, Tn, Wb. 550–2000 m.

Coix: ± 5 species, in need of revision: tropical Asia, through introductions now pantropical

Culms erect or straggling, up to ± 4 m tall, internodes solid. Leaves cauline; ligules short, with or without a fringe; leaf blades 8–100 cm long and 1.5–7 cm wide, linear to ovate-lanceolate; midrib prominent becoming

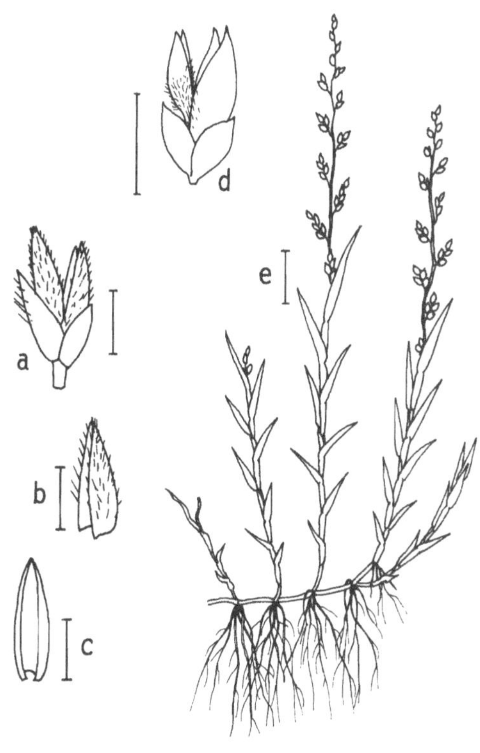

FIG. 295. *Coelachne minuta*: a, spikelet (0.5 mm); b, lower lemma (0.5 mm); c, lower palea (0.5 mm). *C. simpliciuscula*: d, spikelet (1 mm); e, flowering plant (1 cm).

wide and white towards the base. Inflorescence axillary, compound, comprising 2 racemes separated by a prophyll, the one sessile and female, the other pedunculate and male, subtended by a globose or elongated bony utricle (sometimes called involucre or spatheole) derived from a modified leaf sheath. Female raceme enclosed within the utricle, comprising 1 sessile spikelet and 2 pedicels; sessile spikelet female, with membranous glumes and lower floret reduced to a lemma; pedicels free, stout, with or without vestigial spikelets. Male raceme projecting from the mouth of the utricle, the spikelets in pairs or triplets and pedicels free; lower glume papery, laterally winged; both usually fertile.

Annual or perennial. Sometimes floating, usually emergent. Flowers unisexual, outbreeding or agamospermous. The disseminules are single female spikelets surrounded by a utricle, dispersal unknown. The taxonomy is somewhat confused and in need of revision. Used as food, fodder, medicine and decoration, see: Jain, S. K. and Deb Kumar Banerjee, Preliminary observations on the ethnobotany of the genus *Coix*. Economic Botany 28: 38–42 (1974).

Excluded: ***Coix puellarum*** Balansa; not reported from wetlands.

1. Annual; lower glume of the male spikelets narrowly or hardly winged; spikelets 2–4 mm broad; leaves linear-lanceolate, 10–40 cm long, 1–3.5 cm wide ***C. lachryma-jobi***

1. Perennial; lower glume of the male spikelets broadly winged; spikelets 5–6 mm broad; leaves linear-acuminate, 10–90 cm long, 1.5–6 cm wide

2. Culms creeping or floating, up to 30 m or more long; upper surface of the leaf blades with conspicuous glands which bear bristle-like hairs; utricle abruptly constricted at the neck into a beak ***C. aquatica***

2. Culms erect, up to 3 m or more tall; upper surface of the leaf blades glabrous; utricle not abruptly constricted at the neck, without a beak ***C. gigantea***

Coix aquatica Roxburgh: Fig. 296

Like *C. gigantea* but culms creeping or floating, up to 30 m or more long, when in water swollen with floating roots; upper surface of the leaf blades with conspicuous glands which bear bristle-like hairs; utricle abruptly constricted at the neck into a beak or at the middle, remains of the leaf blade often apparent.

Perennial. Found in water, often forming large floating mats. It may be no more than an aquatic state of *C. gigantea*. The distribution given below is based on literature and is probably inaccurate.

{Ple/Hyp/Eph}. SE Asia: Ind, Nep: <An, As, Hp, Kt, Mp, Rj, Or, Wb.

Coix gigantea J. G. König ex Roxburgh [*C. lingulata* Hackel]: Fig. 296

Like *C. lachryma-jobi* but perennial; culms erect, 1–3 m tall; leaves linear, 10–90 cm long, 1.5–6 cm wide, upper surface glabrous, without glands; utricle not abruptly constricted at the neck into a beak, (8–) 10–12 (–18) mm long, dull white; lower glume of the male spikelets broadly winged; spikelets 5–6 mm broad.

Perennial. Not confined to wet places but sometimes a serious weed in rice fields.

[Hel], SE Asia: Bur, Cey, Ind: Kl, Mp, Wb.

Coix lachryma-jobi Linnaeus: Fig. 296

Culms tufted, (10–) 25–100 (–150) cm tall. Leaf blades lanceolate to linear-lanceolate, 10–40 cm long, 1–3.5 cm wide, upper surface usually glabrous or with small tubercule-based hairs. Utricle surrounding the female spikelet cylindrical to ovoid or globose, (5–) 8–12 (–15) mm long, white to yellowish-white or bluish-grey, terete or partly flattened, smooth or with longitudinal ridges. Male spikelets (5–) 8–12 (–15) mm long,

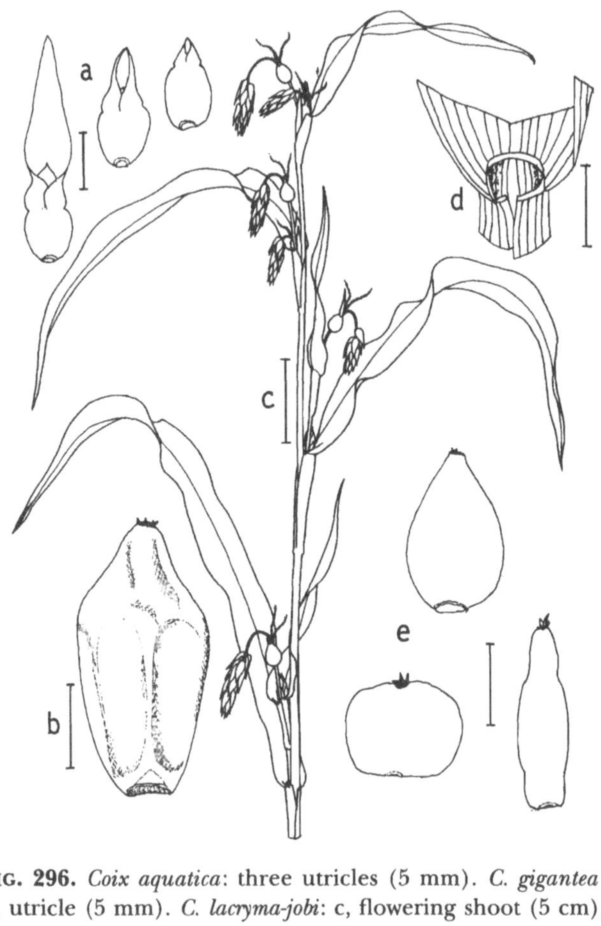

FIG. 296. *Coix aquatica*: three utricles (5 mm). *C. gigantea*: b, utricle (5 mm). *C. lacryma-jobi*: c, flowering shoot (5 cm); d, ligule (5 mm); e, three contrasting utricles (5 cm).

2–3 (–4) mm broad; lower glume papery, 5–15 mm long, 2–4 mm wide, 13- or more-nerved, with narrowly winged or hardly margins; upper glume thin, oblong lanceolate, 5–12 mm long, 2–3 mm wide, 7- or 9-nerved; lemmas oblong-lanceolate, 5–10 mm long, faintly 5- or 7-nerved; paleas elliptical to oblong-lanceolate, 5–10 mm long, 1–2 mm wide, 2-nerved, 2-keeled. Caryopses ovoid to globose, ± 5 mm long.

Annual. Found in a wide variety of wetlands, common in and around marshes, along water courses, often standing in shallow water in pools, backwaters and rice fields. It is very variable in habit and in colour of the utricles, several varieties are cultivated. The utricles are used for decoration, medicine and as food as flour or for beer. It is good fodder. The leaves sometimes used for thatching.

{⊙ Hel}. Warm and tropical Asia, cultivated or naturalized in Africa and America: Ban, Cey, Ind: As, Bh, Gj, Go, Hp, Kl, Mg, Mh, Mp, Ng, Or, Sk, Up.

Diplachne [excluding *Leptochloa*]: 14 species: tropical and subtropical regions

Stoloniferous or tufted perennials. Culms usually erect, 30–270 cm tall or floating. Leaves mostly basal; ligule membranous, sometimes surmounted by a ciliate fringe; leaf blades linear-acuminate, flat or rolled. Inflorescence an open panicle, composed of numerous, loosely spike-like racemes (rachis visible over the whole length on one side). Spikelets, 3- to 9- (to 20)-flowered, linear-oblong, 4–8 mm long, subterete, disarticulating above the glumes and between the florets. Florets laterally flattened. Glumes unequal to subequal, 1-nerved, lanceolate to ovate, keeled, with an acute or mucronate apex, usually shorter than the lowest lemma. Lemmas obtuse or acute, sometimes 2-toothed and mucronate or short-awned between the teeth, 3-nerved, the nerves sometimes pubescent, ± rounded on the back. Stamens 2 or 3. Caryopses ellipsoidal or obovoid.

Diplachne is very close to and sometimes united with *Leptochloa*; the best distinguishing features of *Diplachne* are the non-flattened rather loosely arranged spikelets and lemmas with rounded backs.

1. Lemmas with an up to 12 mm long awn; spikelets sessile; annual ***D. fascicularis***
1. Lemmas with a mucro or awn not more than 1.5 mm long; spikelets stalked; perennial ***D. fusca***

Diplachne fascicularis (Lamarck) Palisot de Beauvois: Fig. 297

Like *Diplachne fusca* but annual, spikelets sessile; lemmas with an up to 12 mm long awn.

{⊙ Hel}. Warm to tropical America, naturalized in Asia minor and parts of the Indian subcontinent.

Diplachne fusca (Linnaeus) Palisot de Beauvois ex Roemer et Schultes [*Leptochloa malabarica* (Linnaeus) Veldkamp]: Fig. 297

Culms tufted, erect, (30–) 60–200 cm high, rooting and branching on the lower nodes. Ligules membranous, tongue-shaped 2–8 mm long; leaf blades 8–50 cm long, 2–6 mm wide, glaucous. Panicles 10–40 cm long, loose, greyish. Spikelets stalked or subsessile, (4–) 8–9 (–15) mm long, 1–2 mm in diameter, 4- to 10- (to 20)-flowered. Lower glume 1.5–2 mm long; upper glume 2.5–3.5 mm long, mucronate. Lemmas 3–6 mm long, 3-nerved, 2-lobed, with central mucro or awn up to 1.5 mm long. Paleas 2–3 mm long.

Perennial. Not confined to wetlands but frequently found in fresh and brackish water, salt marshes, irrigation ditches, rice fields and along roadsides. The stems are spongy when in water.

{Hel/Ple}. Tropical Old World and Australia: Ban, Bur, Cey, Ind: Kl, Kt, Mp, Tn, Up, Wb.

Echinochloa [*Panicum* pro parte]: 30–40 often poorly defined species: cosmopolitan, probably introduced into America.

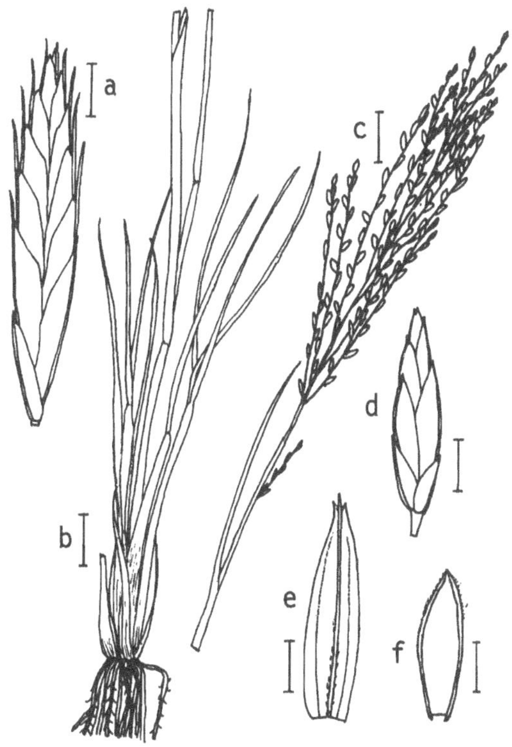

FIG. 297. *Diplachne fascicularis*: a, spikelet (1 mm). *D. fusca*: b, base of plant (2 cm); c, inflorescence (2 cm); d, spikelet (5 mm); e, lower lemma (1 mm); f, lower palea (1 mm).

Tufted annuals or stoloniferous perennials, very variable in size. Ligules absent or a fringe of hairs. Inflorescence of 1-sided, spike-like racemes along a central axis; rachis flattened or 3-angled, not ending in a bristle. Spikelets arranged in 2 or 4 rows, 2-flowered, terete, subsessile or in irregular clusters, disarticulating from the pedicel; lower floret sterile or male; upper floret bisexual. Glumes 2, unequal; lower glume less than half as long as the spikelet, 3- or 5-nerved, acute, cuspidate or short-awned, not clearly abaxial or adaxial; upper glume as long as the spikelet. Lemma of lower floret as long as the upper glume, flat or flattened on the back, cuspidate or awned (awn up to 5 cm long); lemma of upper floret rounded on the back with inrolled margins, not enclosing the apex of the palea. Caryopses fusiform, tightly enclosed by the hardened lemma and palea.

Annual or perennial. The caryopses of some are enclosed by an awned lemma, probably dispersed by adhesion to animals, others perhaps by water. Some are cultivated for grain or forage but the same or other species may be very serious weeds in rice fields and irrigated land. It is a taxonomically difficult genus, many species are variable and several species hybridize resulting in intermediate plants with many intergrading forms and local inbreeding races. Yabuno, T. Biosystematic studies of the genus *Echinochloa*. Japanese Journal of Botany 19(2): 277–323 (1966); Gould, F. W., Ali, M. A. and Fairbrothers, D. E. A revision of *Echinochloa* in the United States. American Midland Naturalist 87: 36–59 (1974).

Excluded: not found in wetlands: ***Echinochloa glabrescens*** Munro ex J. D. Hooker.

1. Ligule represented by a fringe of hairs, at least in the lower leaves; usually perennial
  2. Spikelets acute to cuspidate but never awned — ***E. pyramidalis***
  2. Spikelets with awns at least 2 mm long
    3. Spikelets (excluding the awn) up to 4 mm long — ***E. picta***
    3. Spikelets (excluding the awn) 5–7.5 mm long — ***E. stagnina***
1. Ligule absent; annual or rarely perennial
  4. Lemma of upper florets acuminate to awned; spikelets hispid; racemes untidily 2- to several-rowed, the longest 2–10 cm long
    5. Inflorescence axis usually drooping or at least curved, side branches soft, very densely covered in spikelets — ***E. crus-pavonis***
    5. Inflorescence axis erect or at most curved, side branches rough and hard, not densely covered in spikelets
      6. Lemma of upper floret 2–3 mm long; the longer racemes with secondary branchlets; spikelets very rarely more than 4 mm long — ***E. crus-galli***
      6. Lemma of upper floret 3.5–5 mm long; all racemes simple, unbranched; spikelets usually more than 4 mm long — ***E. oryzoides***
  4. Lemma of upper florets acute to cuspidate (the cusp not more than 1 mm long); spikelets hispid; racemes tidily 4- to several-rowed, the longest rarely exceeding 3 cm long
    7. Racemes not crowded, usually ± half their length apart and appressed to the axis or rarely spreading or subverticillate, ± in 4 rows; spikelets rarely more than 2.5 mm long — ***E. colona***
    7. Racemes crowded with at least some overlapping, not appressed to the axis, usually incurved at maturity, several-rowed; spikelets 3 mm or more long — ***E. frumentacea***

Echinochloa colona (Linnaeus) Link ['*colonum*' orth. mut., '*colonora*' orth. mut., *Panicum colonum* Linnaeus]: Fig. 298

Tufted. Culms erect or ascending, 10–100 cm long. Leaf sheaths glabrous, up to 6 cm long; ligules absent; blades (2–) 10–15 (–30) cm long, 2–8 mm wide, occasionally marked with purple bars. Inflorescence usually linear in outline, 1–15 cm long with

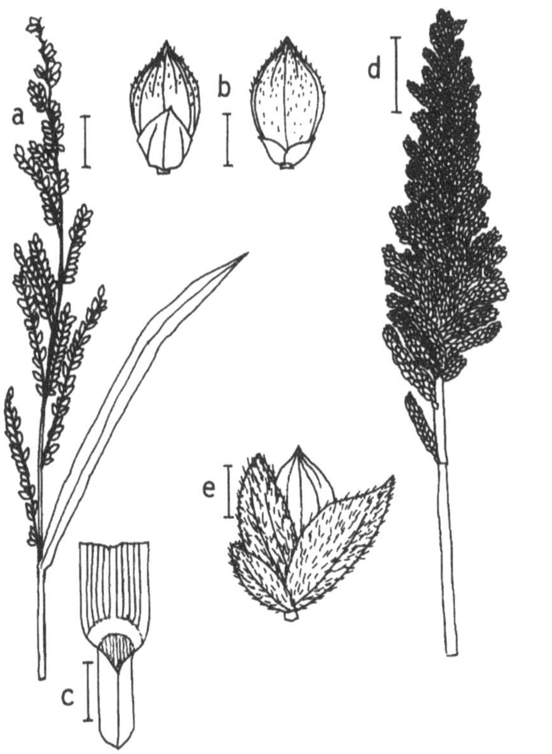

Fig. 298. *Echinochloa colona*: a, inflorescence (1 cm); b, spikelet viewed from the front and from the back (1 mm); c, ligule (5 mm). *E. frumentacea*: d, inflorescence (2 cm); e, spikelet viewed from the side (1 mm).

(3–) 8–10 racemes; racemes usually ± 4-rowed, 1–2 (–3) cm long, usually ± half their length apart and appressed to the axis or sometimes spreading and subverticillate. Spikelets ovoid-ellipsoid to subglobose, (1–) 2–2.5 (–3) mm long, pubescent. Lower glume ovate, 1–2 mm long, half as long as the spikelet, acuminate or shortly cuspidate but not awned, 5-nerved; upper glume ovate to boat-shaped, equalling the spikelet, acuminate, 7- to 11-nerved, hispid. Lemma of lower floret ovate, 2–4 mm long, acute or shortly cuspidate, the cusp not more than 1 mm long; lemma of upper floret 2–3 mm long. Anthers violet or purple.

Annual. Not confined to wetlands but often in wet and marshy places, in ditches, along canals, at the edges of ponds, sometimes a serious pest in rice fields. It grades into *E. frumentacea*. It is very good fodder both before and after flowering.

{⊙ Hel}. Tropics of Asia and Africa, introduced into America: Bur, Cey, Ind, Nep, Pak: <An, Ap, As, Dl, Gj, Go, Hp, Jk, Kl, Kt, Mh, Mp, Rj, Tn, Tp, Up, Wb. 0–2000 m.

Echinochloa crus-galli (Linnaeus) Palisot de Beauvois [?*E. hispidula* (Retzius) Nees ex Royle, *E. crus-galli* var. *breviseta* (Döll) Neilreich, *E. crus-galli* var. *longiseta* (Döll) Neilreich]: Fig. 299

Tufted. Culms erect or ascending, (15–) 25–100 (–150) cm long. Leaf sheaths glabrous or rarely with appressed hairs; ligules absent; blades 5–35 (–50) cm long, 4–20 mm wide. Inflorescence linear to ovate in outline, 6–20 (–25) cm long; racemes untidily 2- to several-rowed, the longest 2–10 cm long, usually with short secondary branchlets at the base; rachilla with some hairs as long as the spikelets. Spikelets ovoid-ellipsoid, (2.5–) 3–4 (–5) mm long (excluding cusp or awn), hispid, pale green or purple. Lower glume 1–3 mm long, one-third to two-fifths as long as the spikelet, 3- or 5-nerved; upper glume ovate-lanceolate, 3–5 mm long, as long as the spikelet, 3- or 5-nerved, with bristly hairs. Lemma of lower floret ovate-lanceolate, acuminate or with an awn up to 5 cm long; lemma of upper floret ovate, 2–4 mm long, including the short herbaceous tip, 2-keeled, 2-nerved. Anthers brown or yellow.

Annual perhaps sometimes perennating. Not confined to wetlands but often found in wet or marshy places. A common and troublesome weed in rice fields. It is good fodder before flowering; the grain is eaten in times of scarcity. It is autogamous or even cleistogamous and numerous segregates have been recognized.

{(⊙) Hel}. Warm and tropical, Asia, Africa and America: Bur, Cey, Ind, Pak: As, Dl, Go, Hp, Jk, Kl, Kt, Mg, Mh, Mp, Ng, Or, Pj, Rj, Tn, Up, Wb. 0–2000 m.

Echinochloa crus-pavonis (Humbolt, Bonpland et Kunth) Schultes: Fig. 299

Like *E. crus-galli* but inflorescence 10–30 (–40) cm long, soft, nodding, much more dense, with more numerous racemes and very crowded spikelets; lower racemes up to 14 cm long; rachilla without long hairs; spikelets mostly 3–4 mm long (excluding cusp or awn), usually awned, inconspicuously hairy; lemma of upper florets narrowly elliptical-ovate to elliptical, greyish.

Annual perhaps sometimes perennating. Usually found in water.

{(⊙) Hyp/Hel}. Tropical America and Africa, rare in Asia: Ind: An.

Echinochloa frumentacea (Roxburgh) Link [*E. colona* var. *frumentacea* (Roxburgh) C. E. C. Fischer, *E. crus-galli* var. *frumentacea* (Roxburgh) Wight, *E. utilis* Ohwi et Yabuno]: Fig. 298

Tufted. Culms erect, 30–150 cm long. Leaf sheaths glabrous and smooth, up to 12 cm long; ligules absent; blades linear-lanceolate, 5–30 (–60) cm long, 3–20 mm wide. Inflorescence lanceolate in outline, 6–20 cm long, with numerous racemes; racemes several-rowed with crowded spikelets, simple, unbranched, 1–3 cm

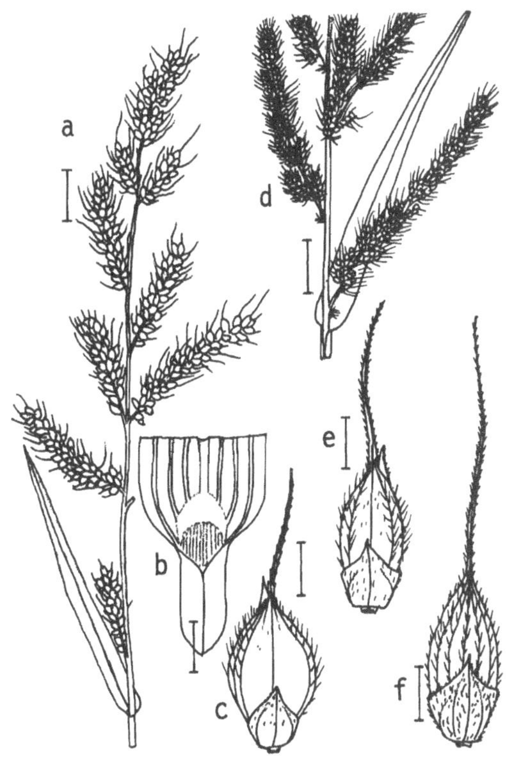

FIG. 299. *Echinochloa crus-galli*: a, inflorescence (1 cm); b, ligule (5 mm); c, spikelet (1 mm). *E. crus-pavonis*: d, lower part of inflorescence (1 cm); e, spikelet (1 mm). *E. oryzoides*: f, spikelet (1 mm).

long, closely spaced and sometimes overlapping, incurved at maturity; rachis and rachilla with long white hairs. Spikelets broadly ellipsoid to subglobose, 2.5–3.5 (–5) mm long, pubescent to hispid, pale green or purple, rather tardily deciduous, pale green or yellowish. Lower glume up to 1.5 mm long, 3-nerved; upper glume as long as the spikelet, with a short mucro at the tip, 3- or 7-nerved, with hairs on and between the nerves. Lemma of lower floret like upper glume, up to 2.5 mm long, acute, mucronulate, stiffly hispid; lemma of upper floret up to 2 mm long, with a short mucro. Anthers deep violet.

Annual. Not confined to wetlands but often found along rivers and round ponds as the water recedes. A common weed in rice fields. It is often cultivated as a dry season crop, particularly in river beds. It may be no more than a selected race of *E. crus-galli.* The cultivated race is sometimes known as *Echinochloa frumentacea* var. *utilis* (Ohwi et Yabuno) Tzvelev.

{⊙ Hel}. Tropical and warm Asia and Africa, naturalized in America, Ukraine and S Europe: Cey, Ind, Pak: Bh, Go, Jk, Kt, Mh, Mp, Sk, Tn.

Echinochloa oryzoides (Arduino) Fritsch [?*E. hispidula* (Retzius) Nees ex Royle, *E. phyllopogon* Stapf, *E. crus-galli* var. *oryzicola* (Vasinger-Alektorova) Ohwi, *E. macrocarpa* Vasinger-Alektorova]: Fig. 299

Tufted. Culms erect, 25–150 cm long. Leaf sheaths up to 20 cm long, smooth and glabrous; ligules absent; blades 7–35 cm long, 5–15 mm wide. Inflorescence linear to narrowly ovate in outline, 6–25 cm long; racemes irregularly 2-rowed, the longest 2–5 cm long, usually simple. Spikelets ovoid-ellipsoid, mostly (3.6–) 3.8–5 (–6.5) mm long, hispid. Lower glume 1–2 mm long, half to three-fifths as long as the spikelet, 3-nerved, with short mucro; upper glume ± as long as the spikelet, with awn up to 1.5 mm long. Lemma of lower floret up to 4 mm long, 5- or 7-nerved, acuminate or with an awn up to 2.5 cm long; lemma of upper floret 3.5–5 mm long, including the short herbaceous tip.

Annual. It tolerates water up to ± 25 cm deep. It is mostly found in rice fields but occasionally grows in inundated or marshy places. It is a common and troublesome weed in rice fields. It mimics rice and occasionally infection is so bad that cultivation of rice must be abandoned. In the New World it is exclusive to rice fields. It is perhaps best treated as a variety or ecotype of *E. crus-galli.*

{⊙ Hel}. Warm and tropical Asia, probably introduced in S Europe: Bhu, Ind, Nep, Pak: Kl, Sk, Tn.

Echinochloa picta (J. G. König) Michael: Fig. 300

Like *E. stagnina* but spikelets up to 4 mm long, with tubercule-base hairs; lower glume up to 1.5 mm long; upper glume up to 4 mm long; lemma of lower floret up to 3.5 mm long with an at least 2 mm long awn.

Perennial. In marshes or partly submerged, often mistaken for *E. stagnina* in India. It is a floating species with soft spongy culms and narrow spikelets tapering towards an awn at the tip.

{Hel}. SE Asia to New Guinea and Australia, also in E Africa: Ind: As, Tn (distribution very incomplete).

Echinochloa pyramidalis (Lamarck) Hitchcock et Chase [*E. senegalensis* Mez]: Fig. 300

Like *E. stagnina* but spikelets acute or shortly cuspidate but never awned, 3–4 mm long.

{Hel}. Tropical Africa and America, introduced in India: <Ind: <Kt. (distribution very incomplete).

Echinochloa stagnina (Retzius) Palisot de Beauvois: Fig. 300

Rhizomes stoloniferous, creeping, spongy when in water. Culms spongy below, prostrate or decumbent, rooting below, (20–) 40–150 cm long. Leaf sheaths

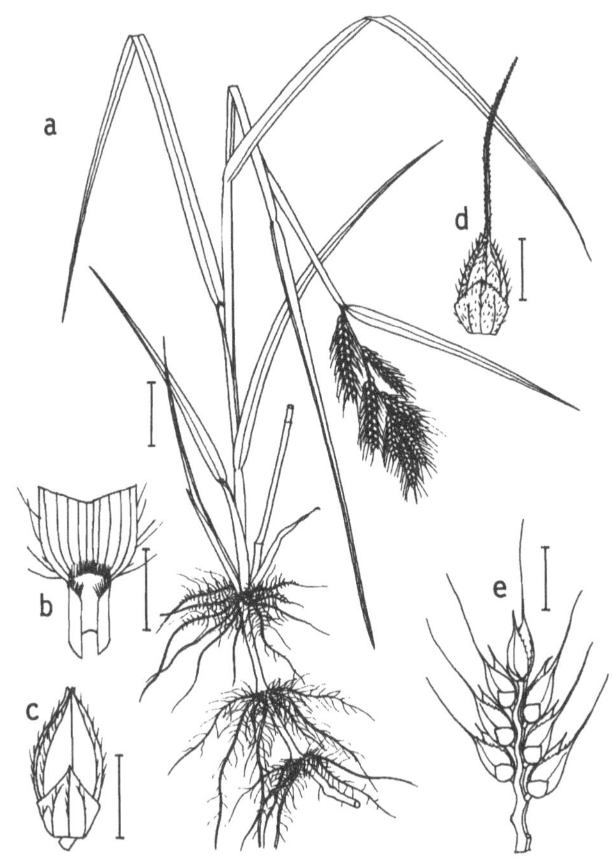

Fig. 300. *Echinochloa stagnina*: a, flowering plant (3 cm); e, rachilla with 7 spikelets (5 mm). *E. pyramidalis*: b, ligule (5 mm); c, spikelet (1 mm). *E. picta*: d, spikelet (2 mm).

glabrous; ligules a line of hairs; blades 8–50 cm long, 6–10 mm wide. Inflorescence narrowly lanceolate or rarely ovate in outline, 7–20 cm long, usually open with 5–15 racemes; racemes erect and usually spreading, 2–6.5 cm long. Spikelets ovoid to broadly ovoid or rounded, 3.5–6 mm long, pubescent to hispid. Lower glume ovate, 2–3 mm long, acuminate, 5-nerved; upper glume ovate-lanceolate, 5–6 mm long, acuminate, 7- or 9-nerved. Lemma of lower floret ovate-lanceolate, 4–6 mm long, 7-nerved, tapering into an awn; awn 5–10 mm long; lemma of upper floret 3–4 mm long. Anthers yellow.

Perennial or occasionally biennial or annual. Confined to wet places; found in marshes, along tanks, ditches and canals. It often forms floating mats of vegetation. It has rather distinct purple inflorescences. It is good fodder and the grains are eaten by humans in times of want. This species is often misapplied in Indian Floras.

{(⊙) Ple/Eph/Hel}. Tropical Asia: Bur, Ind, Pak: Ap, As, Kl, Kt, Mp, Rj, Tn, Wb. 0–1500 m.

Elytrophorus: 2 species: Old World tropics

Elytrophorus spicatus (Willdenow) A. Camus [*E. articulatus* auct. Ind., non Palisot de Beauvois]: Fig. 301

Culms branched from base, creeping below, 2–25 (–70) cm long. Leaf sheaths slightly keeled; ligules membranous, sometimes minutely hairy; leaf blades often overtopping the inflorescence. Inflorescence spike-like, 0.5–15 (–30) cm long, of dense, globose clusters of spikelets on a central axis, the lower clusters remote, the upper confluent and spike-like. Spikelets 4- to 6-flowered, 2–8 mm long (including awns), strongly flattened laterally; rachilla disarticulating between the florets. Glumes subequal, elliptical-lanceolate, the lower 1–2 mm long, cuspidate, the upper 2–3 mm long, cuspidate, 1-nerved. Lemma ovate-elliptical, 2–3 mm long, keeled, 3- or 5-nerved, midnerve prolonged into a 0.5–1 mm long awn. Paleas obovate, 1.5–2.5 mm long, 3- to 6-lobed at the apex, delicate. Caryopses small, ovate-elliptical to narrowly elliptical in outline.

Annual. Seasonally submerged, bottom-rooted or floating at edges of pools, ditches and lakes. Often gregarious, sometimes found in rice fields.

{⊙ Hel}. Old World tropics: Bur, Cey, Ind: Gj, Kl, Kt, Mh, Mp, Ng, Rj, Tn, Up, Wb.

Eragrostis: ± 350 species: warm and tropical regions

Annual or perennial. Inflorescence an open panicle. Spikelets bisexual, stalked, usually many-flowered (up to 45), laterally flattened; rachilla persistent. Glumes usually unequal (upper longer than lower), both longer than the lemma and paleas, usually 1-nerved. Lemmas glabrous, (1-) 3-nerved, awnless, falling together with the grain. Paleas membranous, hyaline, shorter than the lemmas, 2-keeled.

Excluded: many species grow in damp places but are not wetland plants: ***Eragrostis atrovirens*** (Desfontaines) Trinius ex Steudel; ***E. cilianensis*** (Allioni) Vignolo-Lutati [*Briza eragrostis* Linnaeus, *E. major* (Linnaeus) Host, *E. megastachya* (Koeler) Link]; ***E. gangetica*** (Roxburgh) Steudel [*E. stenophylla* Hochstetter]; ***E. japonica*** (Thunberg) Trinius [*E. diarrhena* (Schultes) Steudel]; ***E. pilosa*** (Linnaeus) Palisot de Beauvois [*Poa polymorpha* J. G. König ex Rottler]; ***E. riparia*** (Willdenow) Nees: ♦Ind; ***E. tenella*** (Linnaeus) Palisot de Beauvois; ***E. zeylanica*** Nees et Meyer.

Eragrostis unioloides (Retzius) Nees ex Steudel: Fig. 301

Culms tufted, 4.5–50 cm high, erect or creeping and rooting below. Ligule a narrow rim or thin membrane, ± 0.2 mm long; leaf blade (1–) 5–15 (–25) cm long, 1–8 mm wide, at the base rounded or shallowly cordate. Panicles up to 30 cm long, Spikelets ovate to ovate-oblong, 4–15 mm long, 1.5–4 mm wide, 6- to 40-flowered, usually tinged with purple. Lower glume oblong-lanceolate, 1–2 mm long, 1-nerved; upper

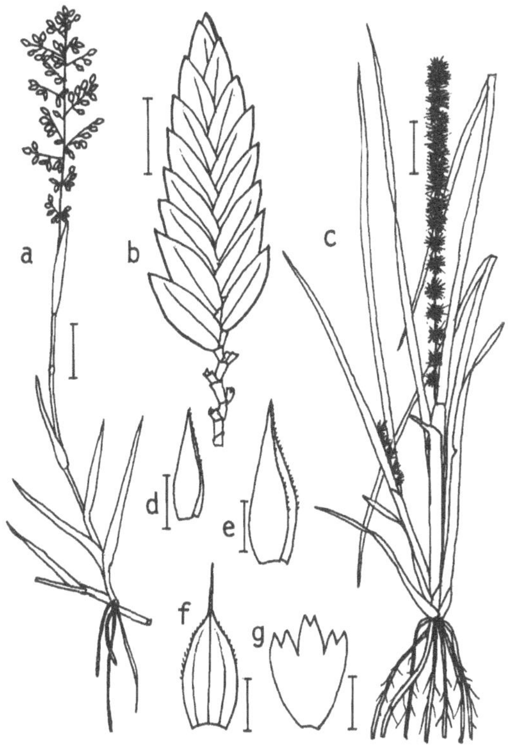

Fig. 301. *Eragrostis unioloides*: a, flowering culm (2 cm); b, spikelet (1 mm). *Elytrophorus spicatus*: c, flowering plant (2 cm); d, lower glume (1 mm); e, upper glume (1 mm); f, lemma (1 mm); g, palea (1 mm).

glume ovate-lanceolate, 1.5–2 mm long, 1-nerved. Lemma broadly ovate to ovate-lanceolate, 1–2.5 mm long, 3-nerved.

Annual or perennial. Very common, not confined to wetlands but often found in shallow water around pools, in and along ditches, in rice fields and in marshes. It is very variable in size. In water it sometimes develops runners or stolons.

{(⊙) Hel}. Almost cosmopolitan: Cey, Ind: An, Gj, Go, Hp, Kl, Kt, Mp, Rj, Tn, Tp, Up.

Eriochloa: ± 30 species

Eriochloa procera (Retzius) C. E. Hubbard [*E. polystachya* auct. Ind., non Humbolt, Bonpland et Knuth]: Fig. 302

Tufted. Culms sometimes shortly creeping at the base, erect, 30–80 (–150) cm long; nodes glabrous. Leaf sheaths slightly keeled; ligules membranous, fimbriate; blades linear-lanceolate or linear, 10–30 cm long, 2–8 mm wide, rounded at the base, acuminate. Inflorescence paniculate, 3–20 cm long, lax, purplish; racemes simple or branched, 2–10 cm long; rachis 3-sided, bearing paired spikelets. Spikelets elongate-ellipsoid, 3–4 mm long, with white silky hairs, green or purplish, supported on a ring-like swelling (lower glume and the top of the rachilla form a ring-like callus). Lower glume absent or reduced to a ring; upper glume ovate-lanceolate, 3–4 mm long, 1–1.5 mm wide, paper-like, 5-nerved, with white silky hairs. Lower floret sterile, without a palea. Upper floret bisexual. Lemma of lower floret resembling the upper glume, elliptical-lanceolate, 3–3.5 mm long, 1–1.5 mm wide, brittle, faintly 5-nerved, with white silky hairs; lemma of upper floret ovate-oblong, 2–2.5 mm long, ± 1 mm wide, mucronate, faintly 3- or 5-nerved. Paleas ovate-oblong, 2–2.5 mm long, ± 1 mm wide, brittle, 2-keeled, 2-nerved. Stamens 3, anthers ± 1 mm long, violet with yellow connectives.

Annual or perennial. A semi-aquatic grass found in wetlands, ditches, rice fields and along the banks of backwaters. Common and often abundant. Used as fodder. The ring-like growth below the spikelets is a good diagnostic character; it is sometimes called cup grass.

{(⊙) Hel}. Tropical Africa, S and SE Asia: Bur, Cey, Ind: Ap, Kl, Kt, Mh, Mp, Pj, Tn.

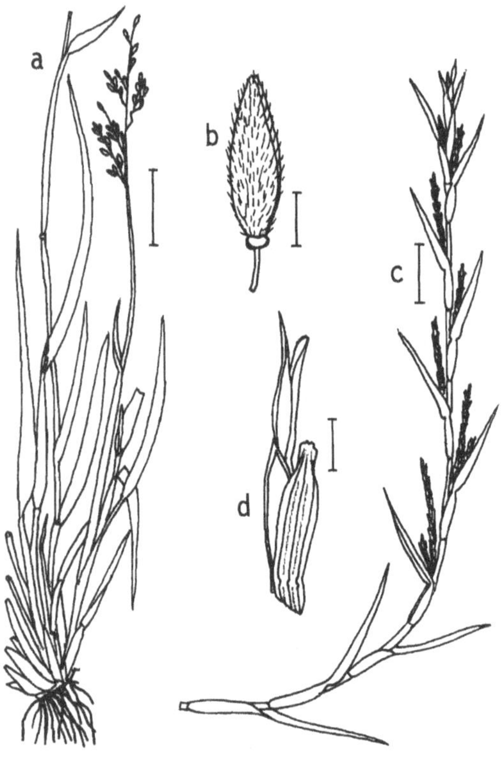

FIG. 302. *Eriochloa procera*: a, flowering plant (2 cm); b, spikelet (1 mm). *Hemarthria compressa*: c, flowering culm (3 cm); d, spikelet pair (1 mm).

Hemarthria: 12 species: Old World tropics and subtropics, perhaps also native in America

Excluded: found in damp or moist places but not wetland species: ***Hemarthria altissima*** (Poiret) Stapf et C. E. Hubbard, ***H. hamiltoniana*** Steudel; ***H. longifolia*** (J. D. Hooker) A. Camus; ***H. protensa*** Steudel.

Hemarthria compressa (Linnaeus fil.) R. Brown. [*Rottboellia glabra* Roxburgh, *H. coromandelina* Steudel, *H. laxa* Nees ex Steudel]: Fig. 302

Stolons creeping or floating, up to ± 8 m long. Culms floating, erect or ascending from a creeping base, 40–80 cm tall. Leaf sheaths hairy at base and throat; ligules very short, membranous, ciliate; leaf blades folded in bud, afterwards flat, 9–15 cm long, up to 8 mm wide, base rounded or subcordate, margins rough. Inflorescence of axillary, solitary or bunched, spike-like racemes; racemes 3–12 cm long, sometimes partly enclosed in leaf sheaths. Spikelets in opposite pairs; each pair with 1 pedicelled and 1 sessile spikelet (the pedicel joined to the internode of the rachis, thus each opposite pair is made up of a sessile spikelet and a pedicelled spikelet of the next lower node); rachis tough, disarticulating tardily. Spikelets 4–6 mm long, 2-flowered, awnless, flattened abaxially; the lower floret reduced to a sterile lemma; the upper floret bisexual. Lower glume lanceolate, ± 5 mm long leathery, usually gradually acuminate or obtuse, 5- or 7-nerved, indistinctly winged above; the upper glume ± 4 mm long, membranous except for a hardened tip. Sterile lemma ± 3 mm long, 2-nerved. Caryopses somewhat laterally flattened, oblong in outline, ± 1 mm long.

Perennial. Bottom-rooted or floating at edges of pools, lakes and rivers, also in damp places, ditches, banks and abandoned paddy fields. It can be a troublesome weed on irrigated land and in wet places.

{Hel/Eph/Ple}. Warm to tropical Asia, Australia and N Africa: Cey, Ind, Nep, Pak: Dl, Gj, Hp, Mp, Rj, Tn, Up, Wb.

Hemisorghum: 2 species: SE Asia

Excluded: ***Hemisorghum venustum*** (Thwaites) Clayton — not found in wetlands.

Hemisorghum mekongensis (A. Camus) C. E. Hubbard: Fig. 303

Culms very robust, tufted, tall. Ligules very short, truncate, densely short-ciliatee; leaf blades linear-lanceolate, flat. Inflorescence a large, loose, compound panicle, its primary branches subdivided, bearing long, loose racemes. Spikelets paired, 1 sessile the other pedicelled; sessile spikelets 2-flowered, the lower one sterile, the upper one bisexual, abaxially flattened, awnless, the callus obtuse with a ring of minute hairs. Glumes 2, unequal; lower glume thin, leathery, flat on the back, 8- to 11-nerved, 2-keeled for most of its length, the keels scabrid, the margins becoming sharply inflexed towards the base; upper glume rounded on the back, slightly keeled towards the apex. Lemma of lower floret slightly shorter than the glumes; lemma of upper floret nearly as long as the lower one, awnless or bidentate and awned. Pedicelled spikelets sterile, either well-developed with a 2-keeled lower glume or much reduced. Caryopsis oblong-obovoid, abaxially flattened, tipped with style base.

Perennial. Very robust, locally dominant on river banks.

{Hel}. Burma, Thailand and Laos: Bur.

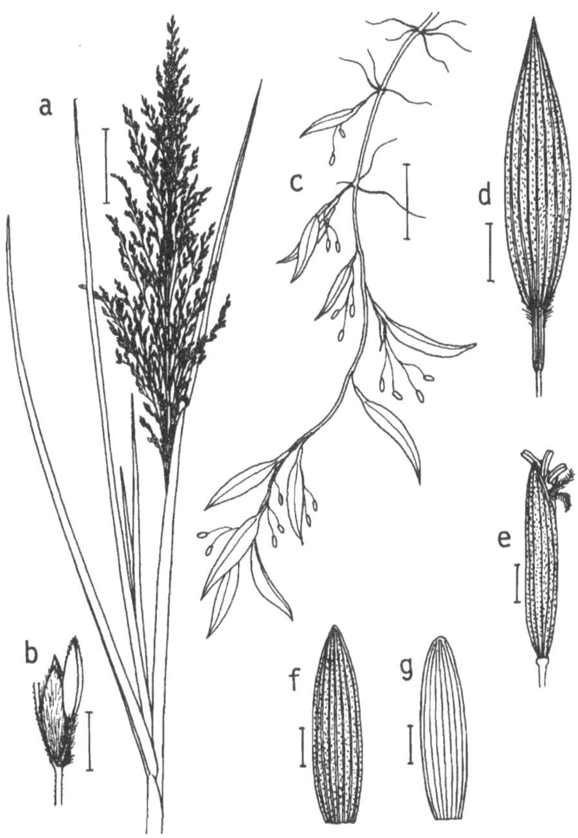

Fig. 303. *Hemisorghum mekongensis*: a, flowering culm (4 cm); b, two spikelets, one sessile and one stalked (2 mm). *Hubbardia heptoneuron*: c, hanging shoot (2 cm); d, leaf sheath and blade (5 mm); e, spikelet (0.5 mm); f, glume (0.5 mm); g, lemma (0.5 mm).

Hubbardia 1 genus

Hubbardia heptaneuron Bor: Fig. 303

Culms delicate, creeping on and hanging from rocks under waterfalls. Leaves arranged along the culms; leaf sheaths shorter than the internodes; ligules absent; leaf blades narrowly elliptical to oblong-elliptical, 1.5–4 cm long, 3–7 mm wide, very thin (only 2 layers of mesophyll cells), membranous, translucent, with slender microhairs on the adaxial surface. Inflorescence a scanty axillary panicle, terminating leafy branches. Spikelets all alike, ± 2.7 mm long, 2-flowered, without awns, terete or abaxially flattened, disarticulating above the glumes; rachilla without an extension. Florets similar but the lower empty, the other bisexual. Glumes persistent, membranous, 5- or 7-nerved, as long as the spikelet. Lemmas membranous, 7- or 9-nerved, rounded on the back; paleas absent. Caryopses spindle-shaped, 1.25–1.5 mm long, slightly flattened.

Probably annual. A remarkable *Elodea*-like plant found in water and on rocks in spray at the base of Jog (Gersoppa) Falls, Karnataka, probably extinct due to damming of the Sharavati River. Bor, N. L. A new genus of Indian grasses. Kew Bulletin 1950: 385–388 (1951)

{⊙? Vit/Hyp?}. Endemic to SW India: ♦Ind: Kt (Sharavati River, now probably extinct).

Hygroryza ['*Hygrorhyza*', orth. mut.]: 1 species

Hygroryza aristata (Retzius) Nees ex Wright et Arnott: Fig. 304

Culms creeping or trailing, floating, branching, ascending, rooting at internodes; nodes glabrous. Leaf sheaths open, strongly inflated, forming floats, partly emergent; ligules truncate, ± 0.8 mm long, membranous, unfringed; leaf blades emergent, ovate-lanceolate to elliptical, (1–) 2–5 (–8) cm long, (3–) 5–20 mm wide, glaucous beneath, somewhat cordate at base, without a prominent midrib. Inflorescence a pyramidal panicle,

FIG. 304. *Hygroriza aristata*: a, flowering plant (2 cm); b, spikelet (5 mm).

2–5 (–8) cm long. Spikelets solitary, 1-flowered, 6–8 mm long, slightly flattened laterally, long-pedicelled, disarticulating from the pedicel. Glumes and sterile lemmas absent. Lemma lanceolate, 6–8 mm long, papery, 5-nerved, narrowed into an awn; awn 0.5–1.7 cm long. Palea as long as the lemma, narrower, 3-nerved. Stamens 6. Caryopses narrowly oblong, ± 3.5 mm long.

Perennial or sometimes annual. Almost the habit of *Eichhornia crassipes* but with the floats developing from inflated leaf sheaths. A characteristic plant in mats of floating vegetation. It is common, often abundant and sometimes dominant. It is found in water reservoirs, pools, canals, backwaters and flooded rice fields. The caryopses are sharply awned, probably dispersed by animals. The grains are eaten by poor people; it makes good fodder for cattle.

{(⊙) Ple}. S and SE Asia: Ban, Cey, Ind, Pak: As, Bh, Kl, Mp, Rj, Up, Wb.

Hymenachne: 8 species: pantropics

Excluded: wetland species in the eastern Himalayas: ***Hymenachne assamica*** (J. D. Hooker) Hitchcock.

Hymenachne acutigluma (Steudel) Gilliland [*H. pseudo-interrupta* C. Mueller, *H. amplexicaulis* auct. Ind, non (Rudge) Nees, *Panicum myuros* auct. Ind., non Lamarck]: Fig. 305

Culms floating, creeping or ascending, often swollen and softly spongy (containing white, pithy aerenchyma), up to 1 m or more long, sparingly branched, with feather-like roots. Ligules membranous, 1–2.5 mm long; leaf blades linear to lanceolate, 7–35 cm long, 0.6–1.5 (–3.5) cm wide, rounded to amplexicaul at the base. Inflorescence a panicle, either spike-like and dense or long and narrow with more or less appressed spike-like branches towards the base, 10–40 cm long. Spikelets 2-flowered, 5–8 mm long, subterete; lower floret sterile, reduced to 1 lemma; upper floret bisexual. Glumes unequal, acuminate, the lower up to half as long as the spikelet, the upper as long as the spikelet. Lower (sterile) lemma as long as the spikelet, acute, acuminate or shortly awned; the lemma of upper floret membranous, acute, gripping the edges of the palea below but free towards the tip. Caryopses free, lemma and palea not hardened.

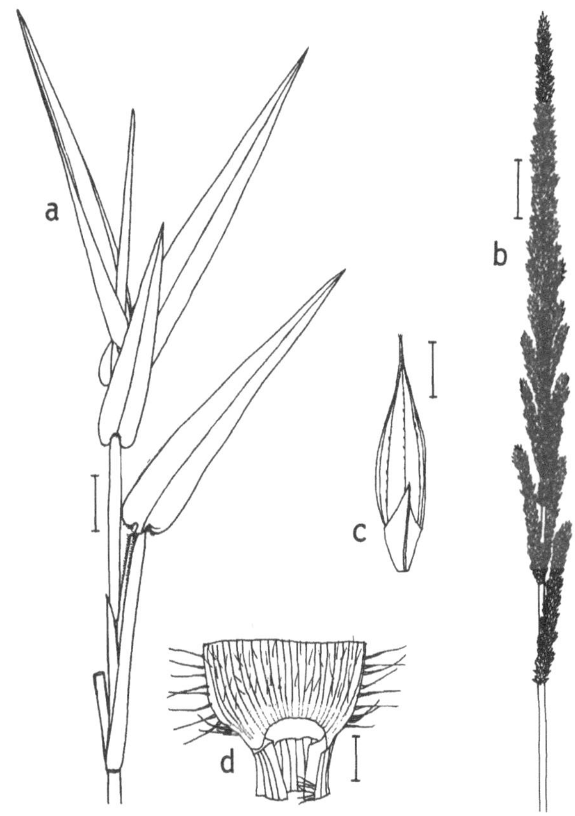

FIG. 305. *Hymenachne acutigluma*: a, sterile shoot (2 cm); b, inflorescence (5 mm); c, spikelet (1 mm); d, ligule (2 mm).

Perennial. Found in water in tanks, backwaters, swamps, irrigation ditches, streams and rice fields. Makes good fodder. It is superficially like *Sacciolepis interrupta* but can be distinguished by the presence of white pith in the floating culms (*Sacciolepis* has hollow culms); leaf blade bases rounded to amplexicaul; spikelets acuminate to shortly aristate.

{Hyp/Ple(Hel)}. S and SE Asia, Polynesia: Ban, Bur, Cey, Ind, Nep: As, Kl, Mp, Rj, Tn, Wb.

Isachne: ± 110 hardly separable species: tropical to warm regions, most in tropical Asia: 29 species in India most of which are local endemics in mountainous regions

Annual or perennial. Culms floating, ascending or erect, often rooting below. Leaves glabrous or with tubercule-based hairs; sheaths loose; blades mostly broad and flat, usually prominently nerved on the adaxial surface; ligule a fringe of hairs or absent. Inflorescence an open or somewhat contracted panicle or raceme. Spikelets 2-flowered, pedicelled, usually globose but sometimes ellipsoidal, gaping, awnless; both florets fertile, mostly bisexual or the lower sometimes male or rarely female. Glumes 2, alike or dissimilar, almost as long or longer than the spikelet. Lemmas alike or dissimilar. Paleas similar to the lemmas but usually flatter. Caryopses globose to fusiform, enclosed between lemma and palea.

There is much taxonomic confusion in *Isachne*. Some species such as *I. dispar*, *I. globosa*, *I. miliacea* and *I. pulchella* intergrade in parts of India and are sometimes difficult to distinguish, it is probably better to unite them under the name *I. globosa*. It is also difficult to find out which species grow in wetlands. Prakash, V. and Jain, S. K. Isachneae. Fascicles of Flora of India 14: 7–20 (1984).

Excluded: wetland species confined to the Himalayas: ***Isachne clarkei*** J. D. Hooker: ♦Ind: Mg, Ng, Sk; ***I. dispar*** Trinius: ♦Nep; ***I. himalaica*** J. D. Hooker.

Found in damp or moist places but not wetland species: ***I. angladei*** Fischer: ♦Ind: Tn; ***I. borii*** Hemadri: ♦Ind: Mh; ***I. deccanensis*** Bor: ♦Ind: Tn; ***I. henryi*** S. R. Srinivasan et P. V. Sreekumar: ♦Ind: Kl; ***I. kunthiana*** (Wight et Arnott ex Steudel) Miquel; ***I. multiflora*** (Thwaites) Ferguson: ♦Cey.

1. Lemmas in any single spikelet similar in outline and structure ***I. albens***

1. Lemmas in any single spikelet dissimilar in outline and structure

2. Glumes unequal, the lower shorter and narrower than the upper; spikelets (1.2–) 1.5–1.6 (–1.8) mm long; anthers of the lower floret 0.3–0.6 mm long ***I. pulchella***

2. Glumes subequal, the lower like the upper; spikelets 1.8–2.8 mm long; anthers of the lower floret 1–1.5 mm long

3. Leaf blades (1.5–) 3–12 cm long, mostly with cartilaginous and thickened margins; spikelets (1.8–) 2–2.8 mm long; nodes glabrous; ligule hairs (1.5–) 2–3.5 mm long ***I. globosa***

3. Leaf blades 1–2.5 (–3.5) cm long, mostly without cartilaginous and thickened margins; spikelets (1.2–) 1.5–2 (–2.8) mm long; nodes hairy or glabrous; ligule hairs 0.8–1.5 (–2) mm long ***I. miliacea***

Isachne albens Trinius: Fig. 306

Culms erect or ascending, (6–) 20–150 cm tall. Leaf sheaths glabrous; ligules hairs 1–2 (–3) mm long; leaf blades linear to broadly lanceolate or ovate, (1.5–) 3–12 cm long, 3–12 mm wide, glabrous; midnerve and 3–6 pairs of nerves very prominent on abaxial surface; margins cartilaginous. Inflorescence 7–40 cm long, spreading; branches secondarily divided, densely flowered, eglandular or with bands of glands. Spikelets

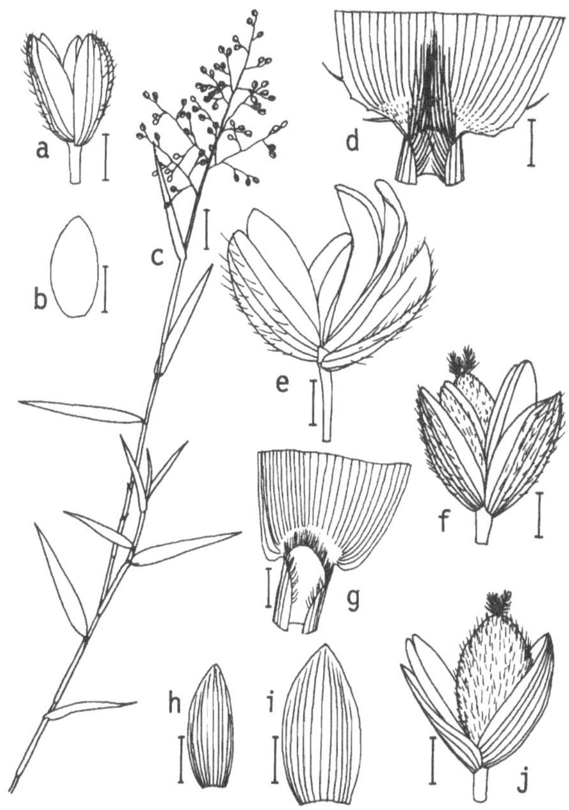

FIG. 306. *Isachne albescens*: a, spikelet (0.5 mm); b, lemma (0.5 mm). *I. globosa*: c, flowering shoot (1 cm); d, ligule (1 mm); e, spikelet (0.5 mm). *I. miliacea*: f, spikelet (0.5 cm); g, ligule (1 mm). *I. pulchella*: h, lower glume (0.5 mm); i, upper glume (0.5 mm); j, spikelet (0.5 mm).

globose or globose-obovoid, (0.8–) 1–2 (–3.2) mm long. Upper and lower florets almost alike in shape, size and texture, ellipsoid. Glumes subequal, 0.8–1.3 (–2) mm long. Lemmas broadly elliptical, concave, rather hard, with obscure nerves. Paleas like lemmas but flat with incurved margins, nerveless.

Perennial or perhaps sometimes annual. Not confined to wetlands but often found in wet places beside streams and in moist grassland. Readily eaten by cattle.

{(⊙) Hel}. SE Asia and Himalayas: Ban, Bhu, Bur, Ind, Nep: Ar, As, Hp, Kt, Mg, Mh, Mn, Ng, Sk, Up, Wb. 0–2000 m.

It is very variable and some variants from the western Ghats have been recognized at the rank of species.

Isachne bicolor V. N. Naik et B. W. Patunkar: in the original publication the description and illustration do not agree.

Like *I. albens* but leaf blades widest in the lower third; spikelets globose, (2–) 2.2–3.2 mm long, purple; panicle branches and pedicels with 1 or more yellowish glandular bands. ♦Ind; Mh (Aurangabad District).

Isachne elegans Dalzell ex J. D. Hooker.

Like *I. albens* but annual; culms delicate, ± 2.5 mm in diameter; spikelets (1.8–) 2–2.5 mm long; leaf blades linear (widest at the middle), not more than 15 cm long; lemmas pubescent and mostly purple on the upper half. ♦Ind: Kt? Mh.

Isachne meeboldii Fischer.

Like *I. albens* but culms robust, ± 5 mm in diameter; spikelets 2–2.5 mm long; leaf blades linear, (widest at the middle), 15–35 cm long, 2–10 mm wide; inflorescence branches and pedicels with reddish glandular bands. ♦ Ind: Kt, Mh?

Isachne veldkampii Bhatt et Nagendran.

Like *I. albens* but culms delicate, not more than 8.5 cm tall; leaf blades ovate (widest in the lower third), up to 1.5 cm long and 7 mm wide; inflorescence branches and pedicels with glandular bands; spikelets not more than 1.5 mm long. ♦Ind: Kt (South Kanara District).

Isachne globosa (Thunberg) O. Kuntze [*I. australis* R. Brown, *I. dispar* pro parte, auct. Ind., non Trinius, *Panicum violaceum* Klein ex Thiele, *P. lepidotum* Steudel, *P. nodibarbatum* Hochstetter ex Steudel]: Fig. 306

Culms tufted or spreading, often rooting below, (10–) 20–80 cm tall; nodes glabrous. Ligule hairs (1.5–) 2–3.5 mm long; leaf blades linear-lanceolate to ovate-lanceolate, (1.5–) 3–12 cm long, (3–) 6–15 mm wide, glabrous or hirsute, with cartilaginous margins. Inflorescence up to 18 cm long, compact or spreading; branches and pedicels with or without glandular bands. Spikelets globose or globose-obovoid, (1.8–) 2–2.5 (–3) mm long. Upper and lower florets dissimilar; the lower one longer and flatter. Glumes subequal, ± 2.5 mm long, 7- to 11-nerved. Lemma of the lower floret elliptical, ± 2.5 mm long, usually exceeding the lower glume, thin, paper-like, obscurely 5-nerved, glabrous. Lemma of the upper floret ovate-elliptical, 1.5–2 (–2.4) mm long, not exceeding the upper glume, concave, coriaceous, nerves obscure, glabrous or sparsely hairy. Palea of lower floret like lemma but narrower. Stamens 3; anthers 1–1.5 mm long. Lemma of upper floret and palea up to 2 mm long, glabrous, without nerves.

Annual or sometimes perennial. On mud or emergent in shallow water, often found in floating mats; common and occasionally very abundant. Sometimes becoming troublesome weeds in rice fields and irrigation ditches. Acceptable to grazing animals. Used as green manure. In many Indian Floras it is confused with *I. dispar*, a species endemic to Nepal.

{(⊙) Hel/Eph/(Hyp)}. S and SE Asia, extending to Japan and Australasia: Cey, Ind: Ap, As, Bh, Gj, Go, Kl, Kt, Mh, Mn, Mp, Or, Rj, Tn, Up, Wb. 0–2000 m.

Isachne miliacea Roth [*I. dispar* pro parte, auct. Ind., non Trinius, *I. geniculata* Griffith, *I. stigmatosa* Griffith]: Fig. 306

Culms erect or ascending, creeping and rooting below, 8–35 (–50) cm long; nodes hairy or glabrous. Ligule hairs 0.8–1.5 (–2) mm long; leaf blades lanceolate to ovate-lanceolate, 1–3 (–6) cm long, 2.5–6 (–10) mm wide, glabrous or sparsely hirsute, without cartilaginous margins. Inflorescence up to 6 (–10) cm long, spreading; branches and pedicels with or without glandular bands. Spikelets (1.2–) 1.5–2 (–2.8) mm long. Upper and lower florets dissimilar; the lower one longer and flatter. Glumes subequal 1.5–2 mm long, 7- to 11-nerved. Lemma of the lower floret elliptical, 1.8–2.5 mm long, slightly exceeding the lower glume, hard, glabrous. Lemma of the upper floret ovate-elliptical, 1.2–2 mm long, not exceeding the upper glume, concave, coriaceous.

Perennial or perhaps sometimes annual. Common and often abundant; found in wet places, flushes or hollows in grassland, sometimes turf-forming and sometimes forming floating mats. Considered to be good fodder. In many Indian Floras it is confused with *I. globosa* which is widespread and with *I. dispar*, a species endemic to Nepal. I suspect it is best united with *I. globosa*.

{(⊙) Hel/Eph/(Hyp)}. S and SE Asia extending to China: Bur, Cey, Ind: An, Ap, As, Go, Kl, Kt, Mg, Mh? Sk, Tn, Wb. 0–2000 m.

Isachne pulchella Roth ex Roemer et Schultes: Fig. 306

Like *I. globosa* and *I. miliacea* but spikelets (1.2–) 1.5–1.6 (–1.8) mm long; glumes distinctly unequal, the lower shorter and narrower than the upper; anthers of the lower floret 0.3–0.6 mm long.

{⊙ Hel?}. S and SE Asia: Bur, Ind: An, Ap, As, Mh, Mn, Mp, Ng, Or? Tn, Up, Wb. 0–± 1400 m.

Ischaemum: ± 70 species: warm and tropical regions, most in Asia

Culms creeping to erect, 10–350 cm tall, simple or branched. Ligule an unfringed membrane. Inflorescence terminal and axillary, of paired (rarely single or digitate) racemes, these generally 1-sided and interlocking back-to-back thus simulating a single spike, mostly exerted but sometimes enclosed in a leaf sheath. Spikelets 2-flowered, in pairs, 1 always pedicelled, the other usually sessile, disarticulating at the nodes; sessile or subsessile spikelet 5–6 (–8) mm long, abaxially flattened, awned or awnless. Lower floret male or sterile, the upper one bisexual. Pedicelled spikelet like sessile one but showing all stages in reduction, sometimes represented by no more than a small glume especially in the upper part of the raceme. Glumes 2, usually hard and rigid; lower glume awnless, often 2-winged near the margins; upper glume awned or awnless, 1-winged or keeled. Lemma of upper floret 2-lobed and awned from between the lobes; awn bent and twisted. Paleas as long as lemmas. Caryopses abaxially flattened.

Perennial with creeping rhizomes or sometimes annual. This genus is taxonomically rather confused with numerous local endemic species; few are found in wetlands, most grow in dry or, at most, damp places.

Excluded: the following may tolerate temporary flooding or waterlogging but are not wetland species: ***Ischaemum barbatum*** Retzius [*I. goebelii* Hackel]; ***I. bombaiense*** Bor: ♦Ind: Mh; ***I. commutatum*** Hackel; ***I. copeanum*** P. V. Sreekumar, V. J. Nair et N. C. Nair: ♦Ind: Kl; ***I. duthiei*** Stapf ex Bor: ♦Ind: Bh, Mp, Wb; ***I. flumineum*** Bor: ♦Ind: Kt Tn?; ***I. mangaluricum*** (Hackel) Stapf ex Fischer: ♦Cey, ♦Ind; ***I. nairii*** V. J. Nair et P. V. Sreekumar: ♦Ind: Kl; ***I. raizadae*** Billore et Hemadri: ♦Ind: Mh; ***I. santapauii*** Bor: ♦Ind: Mh; ***I. thomsonianum*** Stapf ex Fischer.

At anthesis the glumes are membranous with green nerves; in fruit they become hardened and the nerves are no longer visible. This key is based on fruiting plants.

1. Racemes 3–10, paniculate ***I. nilagiricum***
1. Racemes 2, rarely solitary or 3, digitate or nearly so
 2. Margins of lower glume of the sessile spikelet expanded below the middle
  3. Spikelets with projecting awns ***I. indicum***
  3. Spikelets without awns or awns rudimentary and hidden
   4. Racemes usually more than 4 cm long, borne well above the leaves; lower leaves narrowed and tapering towards the base; pedicels of the pedicelled spikelets ± 1 mm long ***I. jayachandranii***
   4. Racemes usually rarely more than 3.5 cm long, the lower part partly enclosed by sheathing leaves; lower leaves rounded or shallowly cordate at the base; pedicels of the pedicelled spikelets 3–5 mm long ***I. muticum***
 2. Margins of lower glume of the sessile spikelet narrowly and evenly inturned from base to apex
  5. Lower glumes of the sessile spikelets with white silky hairs, ± flat and almost smooth or with obscure nodules and furrows ***I. molle***
  5. Lower glumes of the sessile spikelets glabrous, curved, with distinct nodules along the lower margins
   6. Lower glumes of the sessile spikelets hardened below with sharply defined lateral furrows and ridges; pedicelled spikelets reduced, without flowers; lower leaf blades narrowed and tapered at the base ***I. rugosum***
   6. Lower glumes of the sessile spikelets leathery below with obscure lateral furrows and ridges; pedicelled spikelets well-developed and fertile; lower leaf blades rounded or shallowly cordate at the base
    7. Racemes 8–15 cm long; sessile spikelets usually without awns; lower glume of the pedicelled spikelet broadly winged on 1 margin ***I. travancorense***
    7. Racemes 5–8 cm long; sessile spikelets usually with awns; lower glume of the pedicelled spikelet narrowly winged on 1 margin ***I. vembanadense***

Ischaemum indicum (Houttuyn) Merrill [*I. indicum* var. *prorepens* Hackel, *I. indicum* var. *wallichii* (Hackel) Bor, *I. indicum* var. *longipilum* Hackel, *I. aristatum* auct. Ind., non Linnaeus, *I. ciliare* Retzius; *Andropogon malacophyllus* Hochstetter ex Steudel, *Spodiopogon scrobiculatus* Nees ex Steudel, *S. villosus* Nees]: Fig. 307

Culms tufted or creeping and rooting below, (10–) 20–80 (–100) cm tall; nodes glabrous or hairy. Leaf blades linear to lanceolate, (0.5–) 5–8 (–25) cm long, 2–8 (–15) mm wide, glabrous or hairy, at the base rounded or narrowed or toward the base of the culm usually developing a pseudopetiole, acute at tip. Racemes (1) 2 (3), (1–) 3–8 cm long. Sessile spikelets ovate-oblong, 3–6 mm long, distinctly awned; lower flower male; upper flower bisexual; lower glume oblong or ovate-oblong, 4–5.5 mm long, notched or acuminate at the tip, somewhat leathery below, paper-like above, faintly 5- to 9-nerved, the upper half broadly winged on both sides, the wings simple or ear-

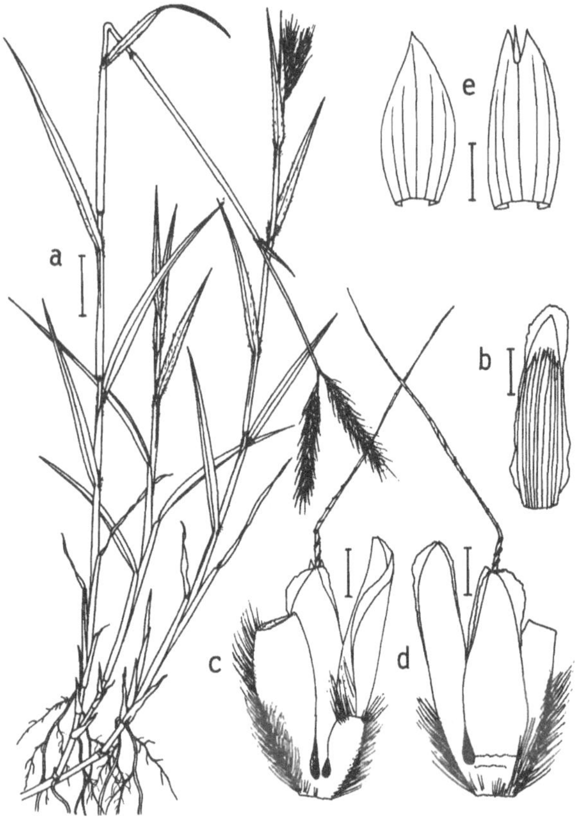

FIG. 307. *Ischaemum indicum*: a, flowering plant (2 cm); b, lower glume of sessile spikelet at anthesis (1 mm); c, spikelet viewed from the rachis side (1 mm); d, spikelet viewed from the outer side (1 mm). *I. jayachandranii*: e, lemmas, left without awn, right with rudimentary awn (2 mm).

like, margins ciliate below the wings; upper glume boat-shaped, 4–5 mm long, acute at tip or terminating in a 1–3 mm long awn, paper-like, 3- or 5-nerved, keeled on the back; lemma of the lower floret ovate-lanceolate, 3.5–4 mm long, paper-like, obscurely 3-nerved, glabrous; palea elliptical-lanceolate, 3–3.5 mm long; lemma of the upper flower 3.5–4 mm long, notched at tip, with awn, the awn 8–15 mm long. Pedicelled spikelet ovoid, 4–5 mm, awned; lower glume ovate-lanceolate, 3.5–4 mm long, winged on 1 margin, densely hairy on the upper half; upper glume and other floral parts like those in sessile spikelet.

Annual or perennial. Not confined to wetlands but often common in places subject to seasonal flooding, in rice fields and along ditches and canals. It is very variable in habit and hairiness, several intraspecific taxa have been described; some of which may be confined to wetlands.

{(⊙) Hel}. S and SE Asia, introduced into America: Ban, Bur, Cey, Ind: An, As, Gj, Go, Kl, Kt, Mh, Mp, Tp.

Ischaemum jayachandranii R. Ansari, V. S. Ramachandran et P. V. Sreekumar: Fig. 307

In habit like *I. indicum* but spikelets without awns or awns minute and concealed within the spikelet; lower glume of the sessile spikelets faintly 11- or 13-nerved, with narrowly inturned margins; pedicels of the pedicelled spikelets ± 1 mm long.

Perennial. Found in marshy places and rice fields.

{Hel}. Endemic to SE India: ♦Ind: Kl (Kannuur District).

Ischaemum molle J. D. Hooker: refer to Fig. 309

Like *I. rugosum* but usually more robust; the upper leaves pseudopetiolate, the base of the 'blade' part shallowly cordate; lower glume of the sessile spikelets densely covered with long white hairs, nodules and furrows obscure.

Perennial. In marshes often forming stands.

{Hel}. Endemic in India and Pakistan: ♦Ind, ♦Pak: Gj, Kl, Kt, Mh, Mp?

Ischaemum muticum Linnaeus [*I. repens* Roxburgh]: Fig. 308

In habit like *I. indicum* but racemes usually rarely more than 3.5 cm long, the lower part partly enclosed by the sheathing leaves; lower leaves rounded or shallowly cordate at the base; spikelets without awns; pedicels of the pedicelled spikelets 3–5 mm long.

Stoloniferous perennial. Not confined to wetlands but often found in standing water around pools and in seasonally flooded and marshy places, occasionally also growing floating. Collected for fodder in Sri Lanka. A smaller, prostrate race is found along the coast in Sri Lanka; it has been called *I. muticum* var. *srilankensis* Davidse.

{Hel(Hyp/Ple)}. Tropical to subtropical Old World and Pacific islands: Bur, Cey, Ind: Kl, Kt.

Ischaemum nilagiricum Hackel: Fig. 308

In habit like *I. indicum* but racemes paniculate, (3–) 5–10 (–16), each 2–10 cm long, clothed with silver hairs; leaf blades elliptical-lanceolate, 5–20 cm long, 5–15 mm wide; lower glume of the sessile spikelets not winged, densely hairy.

Perennial. Semi-aquatic, found in shallow water around lakes and ponds sometimes trailing over the surface, also found in rock crevices by running water.

{Hel/Hyp/(Ple)}. Endemic to S India: ♦Ind: Kl, Tn.

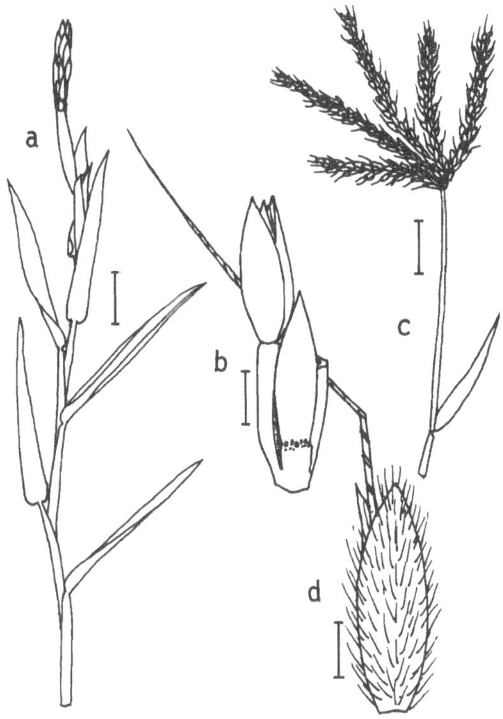

Fig. 308. *Ischaemum muticum*: a, flowering culm (1 cm); b, spikelet pair (5 mm). *I. nilagiricum*: c, flowering culm (2 cm); d, pedicelled spikelet (1 mm).

Ischaemum rugosum Salisbury: Fig. 309

Culms tufted or creeping and rooting below, 20–80 cm tall; nodes glabrous or hairy. Leaf blades linear to lanceolate, 4–15 cm long, 3–6 mm wide, at the base narrowed and tapering, acute at tip. Racemes (1) 2 (3), 3–8 cm long. Sessile spikelets ovate-oblong, 4–6 mm long, distinctly awned; lower flower male; upper flower bisexual; lower glume oblong-lanceolate, 4.5–5 mm long, becoming hard and shell-like with prominent side nodules connected by well-defined ridges and furrows, 13- or 15-nerved; upper glume boat-shaped, 4–6 mm long, somewhat leathery, 3- or 5-nerved, keeled on the back; lemma of the lower floret oblong-lanceolate, 4–5 mm long, paper-like, obscurely 3-nerved, glabrous; palea elliptical-lanceolate, 3–3.5 mm long; lemma of the upper flower 3–4 mm long, notched at tip, with awn, the awn 8–12 mm long. Pedicelled spikelet oblong, 4–5 mm, without an awn; pedicels ± 1 mm long, lower glume oblong, 4.5–5 mm long, hardened, broadly or narrowly winged on 1 margin, 11- to 15-nerved; upper glume and other floral parts like those in sessile spikelet.

Annual or perennial. It is not confined to wetlands but is common in wet places and found as a weed in rice fields, sometimes it grows in standing water. It is very variable in form. It is eaten by cattle when young, the grains are eaten in times of scarcity.

{(⊙) Hel(Hyp)}. S and SE Asia: Cey, Ind, Nep, Pak: An, Gj, Go, Hp, Kl, Kt, Mh, Mp, Pj, Rj, Tn, Tp, Up, Wb.

Ischaemum travancorense Stapf ex Fischer: Fig. 309

Like *I. rugosum* but racemes 8–15 cm long; leaf blades up to 30 cm long; sessile spikelets usually without awns; sessile spikelets usually with awns; lower glume of the sessile spikelet not or very narrowly winged on 1 margin at the tip, marginal nodules mostly joined by shallow and irregular transverse ridges; lower glume of the pedicelled spikelets with 1 broad and 1 narrow wing.

Perennial. Grows along streams, canals and backwaters, also in marshy places.

{Hel}. Endemic to S India: ♦Ind: Kl, Mh.

Ischaemum vembanadense R. B. Patil et R. D' Cruz: Fig. 309

Like *I. travancorense* but racemes 5–8 cm long; leaf blades up to 12 cm long; sessile spikelets usually with awns; lower glume of the pedicelled spikelet narrowly winged on 1 margin.

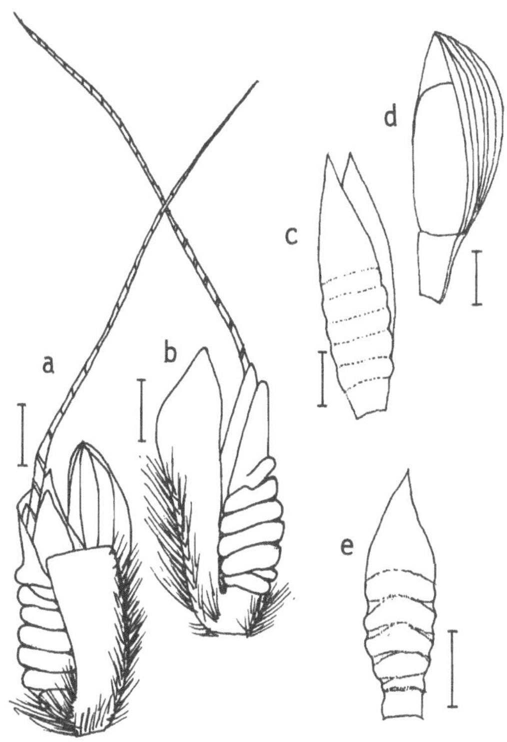

Fig. 309. *Ischaemum rugosum*: a, spikelet viewed from the rachis side (1 mm); b, spikelet viewed from the outer side (1 mm). *I. travancorense*: c, sessile spikelet (1 mm); d, pedicelled spikelet (1 mm). *I. vembanadense*: e, glume of sessile spikelet (2 mm).

Perennial. Grows along backwaters. Perhaps no more than an ecotype of *I. travancorense.*

{Hel}. Endemic to SW India: ♦Ind: Kl.

Leersia: 18 species: tropical and warm regions of the world

Excluded: ***Leersia hackelii*** Keng [*L. oryzoides* auct. Ind., non (Linnaeus) Swartz]; a wetland species from the NW Himalayas.

Leersia hexandra Swartz: Fig. 310

Stoloniferous. Culms ascending, 25–150 cm long or longer when floating in mats; nodes swollen, cuff-like, hairy. Ligules membranous, 4–9 mm long; leaf blades linear-lanceolate to lanceolate, 5–15 (–30) cm long, 2–6 (–12) mm wide. Inflorescence a lax panicle, 3–12 cm long, with flexuous branches. Spikelets 1-flowered, strongly flattened laterally, 3–4.5 mm long, yellowish-green, awnless, sessile or nearly so, pedicels not jointed, disarticulating below the glumes. Glumes reduced to an entire or 2-lobed and somewhat thickened rim. Lemmas all fertile, oblong, 3–4 mm long, leathery, 5-nerved. Paleas linear-oblong 2–3 mm long, 3-nerved, the edges clasping the inrolled margins of the lemma. Stamens 6 or sometimes less. Caryopses flat to somewhat laterally flattened.

FIG. 310. *Leersia hexandra*: a, flowering plant (3 cm); b, spikelet (2 mm); c, flower (2 mm).

Perennial. Common and widespread, found in or floating on water, in pools, streams and ditches, but also in marshes and rice fields. It is used as fodder and sometimes considered to be a weed in irrigation ditches and rice fields. The delimitation from *Oryza* is difficult, the most important character is the absence of sterile lemmas in *Leersia.*

{Hel/Ple/Eph}. Pantropics: Ban, Bur, Cey, Ind: As, Hp, Kl, Kt, Mn, Mp, Rj, Tn, Wb.

Leptochloa [excluding *Diplachne*]: 40 species: tropical to warm regions, most in New World

Excluded: sometimes found in damp or periodically flooded places but not wetland species: ***Leptochloa panicea*** (Retzius) Ohwi [*L. filiformis* auct. Ind., non (J. G. König) Roemer et Schultes; ***L. uniflora*** Hochstetter ex A. Richard.

Leptochloa chinensis (Linnaeus) Nees [*L. capillacea* Palisot de Beauvois]: Fig. 311

Culms usually erect, 10–80 (–150) cm tall or floating, smooth, hollow, leafy. Ligule a fringed membrane, 1.2–2 mm long; leaf blades linear 5–20 (–50) cm long, 3–7 (–10) mm wide, rounded at the base, acuminate at the tip. Inflorescence a panicle, 10–40 (–60) cm long, composed of numerous, slender, 1-sided, spike-like racemes (rachis visible over the whole length on one side). Spikelets, 3- to 9-flowered (mostly 5), oblong, 2–4 (–7) mm long, laterally flattened, purple, disarticulating above the glumes and between the florets. Florets laterally flattened. Glumes unequal to subequal, oblong-lanceolate to oblong, (the lower 0.6-2 cm long, the upper 1–2.5 mm long), papery, 1-nerved, with an acute or mucronate apex, usually shorter than the lowest lemma. Lemmas ovate-lanceolate, 1.5–2.7 mm long, 3-nerved, the nerves sometimes pubescent, keeled on the back. Paleas oblong, 1–1.5 mm long, 2-keeled, 2-nerved. Stamens 2 or 3. Caryopses ellipsoidal or obovoid, ± 1 mm long.

Annual or perhaps also perennial in permanent water. Not confined to wetlands but often found in shallow water around pools, in marshes and in rice fields. In aquatic habitats it is often luxuriant and makes excellent fodder.

{⊙ (⊙)? Hel/Hyp/Ple}. S and SE Asia: Ban, Bur, Cey, Ind, Nep?: Ap, Hp, Kl, Kt, Mp, Tn, Wb. 0–1000 m.

Limnopoa: 1 species

Limnopoa meeboldii (C. E. C. Fischer) C. E. Hubbard: Fig. 312

Culms delicate, trailing, submerged or partly floating, up to 150 cm or more long, usually leafless under water. Leaf sheaths 8–20 mm long, longer at the upper intern-

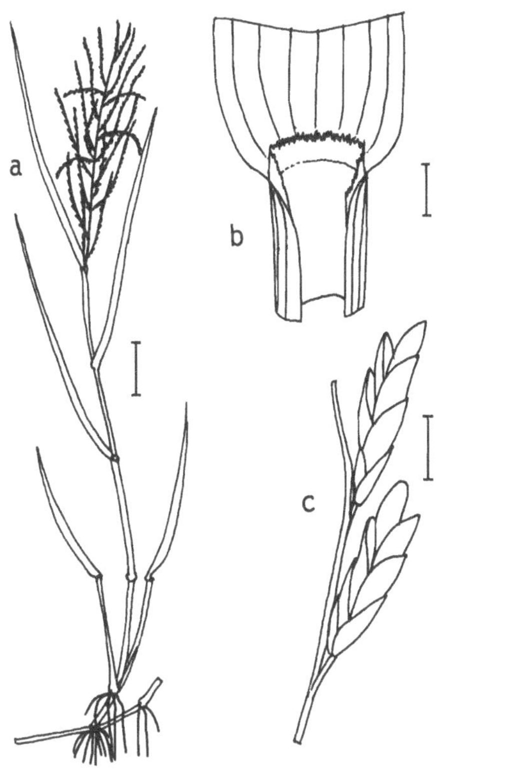

FIG. 311. *Leptochloa chinensis*: a, flowering plant (5 cm); b, ligule (0.5 mm); c, two spikelets (1 mm).

odes, the margins free; ligule a fringe of hairs, up to 0.3 mm long; leaf blades narrowly triangular, 4–18 mm long, up to 1.5 (–2.5) mm wide, reflexed on emergent parts of the culm. Inflorescence a solitary, unilateral, spike-like raceme, 1–3 cm long, emerging at maturity from the uppermost leaf sheath. Spikelets paired, alike but 1 sessile and the other pedicelled, 3–4.5 mm long, dorsoventrally somewhat flattened, awnless, 2-flowered; rachilla flattened and hollowed on one side, disarticulating above the glumes and between the florets. Florets 2, the lower male. Glumes subequal, 2.5–3.5 mm long, glabrous; lower glume lanceolate-oblong; the upper glume obovate to oblong. Lemma of lower floret as long as the spikelet, membranous, ovate-elliptical, glabrous, 5- to 7-nerved, palea almost as long as the lemma, 2-keeled; lemma of the upper floret densely and minutely pubescent, obscurely 5-nerved, elliptical, shorter than the lower one. Palea similar to the lemma. Caryopses ± 1.5–1.6 mm long, plano-convex, enclosed between the lemma and palea.

Annual or perhaps sometimes perennial. It grows bottom-rooted and floating in coastal lagoons that may become somewhat brackish after the monsoon. Sometimes locally dominant. The leaves die as they become submerged leaving long string-like stems trailing in the water. The caryopses are very hard and nut-like; they float and are probably dispersed by water but it could also be spread by aquatic animals.

{(⊙) Eph/Hyp}. Endemic to coastal SW India: ♦Ind: Kl, Kt. Sea level.

Oryza: 25 species: tropical to warm regions of the world

Annual or perennial with rhizomes or stolons. Culms emergent and erect, rarely more than 2 m long or floating and up to 6 m or more long. Leaf sheaths ± terete, with transverse veinlets; ligule membranous, well-developed; leaf blades linear to narrowly lanceolate. Inflorescence a panicle. Spikelets pedicellate, disarticulating above the glumes (cultivated races of *O. sativa* do not articulate), strongly flattened laterally, usually awned, 3-flowered; lower 2 florets reduced to sterile lemmas, the upper one bisexual. Glumes small and scale-like or reduced to a 2-lobed rim. Sterile lemmas 2, usually differing in shape, either scale-like or linear to ovate, mostly shorter than the fertile lemma; fertile lemma boat-shaped, keeled, prominently nerved, clasping the lateral nerves of the palea, apex usually prolonged into a long awn. Palea as long as but narrower than the fertile lemma, keeled, prominently nerved. Stamens 6. Caryopses enclosed within lemma and palea.

Humid forests and swamps. Juveniles of some are submerged, the adults emergent, floating or terrestrial. Some are autogamous. Caryopses dispersed by animals and perhaps also water. *Oryza sativa* (rice) is the most

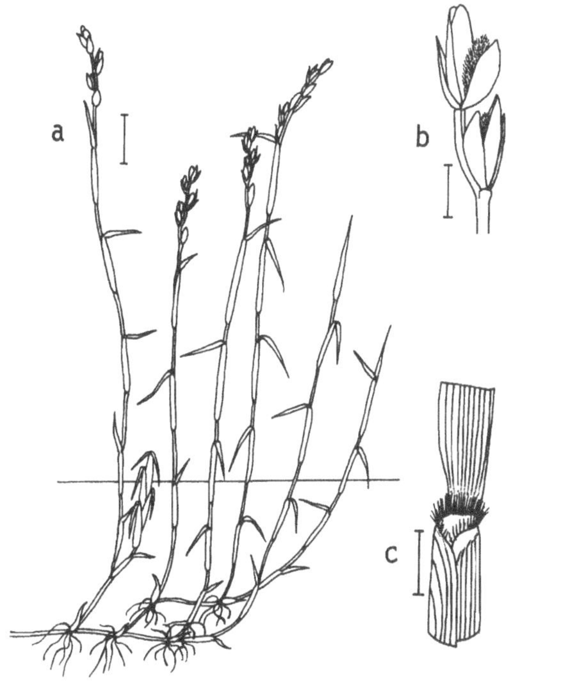

FIG. 312. *Limnopoa meeboldii*: a, flowering plant (3 cm); b, sessile and stalked spikelets (2 mm); c, ligule (1 mm).

important crop plant of the world; it has been in cultivation for at least 3000 years, there is evidence that the two major races from tropical India and cool-tolerant NE Asia, were domesticated independently. The domesticated *O. glaberrima* probably arose in Africa. Some wild species of rice are very troublesome weeds especially in rice fields. Duistermaat, H., A revision of *Oryza* in Malesia and Australasia. Blumea 32: 157–193 (1987); Oka, H.-I. Origin of cultivated rice. Developments in Crop Science. 14: 1–254 (1988).

Excluded: not found in wetlands: ***Oryza eichingeri*** A. Peter [*O. sativa* var. *collina* Trimen]; ***O. meyeriana*** (Zollinger et Moritzi) Baillon var. ***granulata*** (Nees et Arnott ex Watt) Duistermaat [*O. granulata* Nees et Arnott ex Watt, *O. filiformis* F. Hamilton, ?*O. jeyporensis* Govindaswamy et Krishnamurthy, *O. indandamanica* Ellis]; ***O. minuta*** J. S. Presl ex C. B. Presl [*O. officinalis* Wallich ex Watt, *O. malampuzhaensis* Krishnaswarmi et Chandrasekharan]; ***O. rhizomatis*** D. A. Vaughan: ♦Cey. Bor (1960) cites several introduced species which I have not been able to reconfirm.

1. Anthers (3.5–) 4–6.2 mm long, yellow or brown; spikelets deciduous when ripe, disarticulating above the glumes ***O. rufipogon***
1. Anthers 0.8–2 (–2.5) mm long, white or yellow; persisting when ripe ***O. sativa***

Oryza rufipogon Griffith [*O. fatua* J. G. König ex Trinius, *O. aquatica* Roschevicz]: Fig. 313

Tufted or stoloniferous. Culms decumbent and floating or ascending to erect, (30–) 70–90 (330) cm long, branching and rooting below. Leaf sheaths often inflated below, tight above; ligule triangular to narrowly triangular, 9–25 (–45) mm long; leaf blades linear, (10–) 27–60 cm long, 7–15 (–25) mm wide. Panicles loosely contracted, up to 30 cm long. Spikelets oblong to obovate-oblong, 6–11.5 mm long, deciduous, disarticulating above the glumes. Glumes ± 0.3 mm long. Sterile lemmas triangular to linear-lanceolate, 1.25–7.5 mm long. Fertile lemma obovate to lanceolate, 7–11 mm long; awn very variable, up to 110 mm long. Anthers (3.5–) 4–6.2 mm long, yellow or brown. Caryopses oblong, cylindrical, 5.2–6.7 mm long, reddish-brown.

Perennial or annual. In water emergent or sometimes forming floating mats or terrestrial in marshes or seasonally inundated places, also along water courses. Common in rice fields where it is a serious weed. Hybrids with cultivated rice are sometimes found. Some botanists recognize it as a variety of *O. sativa*. It is sometimes collected for fodder.

{(⊙) Hyp/Eph/Hel/(Ple)}. Warm to tropical regions of the world, often considered native to Asia: Ban, Bur, Cey, Ind, Nep: As, Kl, Kt, Mh, Mn, Mp, Rj, Up, Wb. It is much more widespread than my cited distribution — I have not been able to confirm more records.

Oryza sativa Linnaeus [*O. nivara* Sharma et Shastry]: Fig. 313

Tufted, without stolons. Culms erect to ascending, (30–) 50–150 (190) cm long, rooting below. Leaf sheaths often inflated below, tight above; ligule ovate-lanceolate to linear-lanceolate, (6–) 10–40 mm long; leaf blades linear, 20–60 cm long, 7–22 mm wide. Panicles loose, up to 30 cm long. Spikelets oblong to oblong-lanceolate, 5–12 mm long, persisting. Glumes ± 0.2 mm long. Sterile lemmas ovate-oblong to ovate-lanceolate, 1.5–4 (–10) mm long. Fertile lemma oblong to lanceolate, 6–11 mm long; awn very variable, up to 150 mm long. Anthers 0.8–2 (–2.5) mm long, white or yellow. Caryopses oblong, cylindrical, 5–8 mm long, whitish-yellow, brown to reddish-brown.

Annual or rarely perennial. Widely cultivated, occasionally escaping and found in a wide variety of wetland habitats. There are numerous races in cultiva-

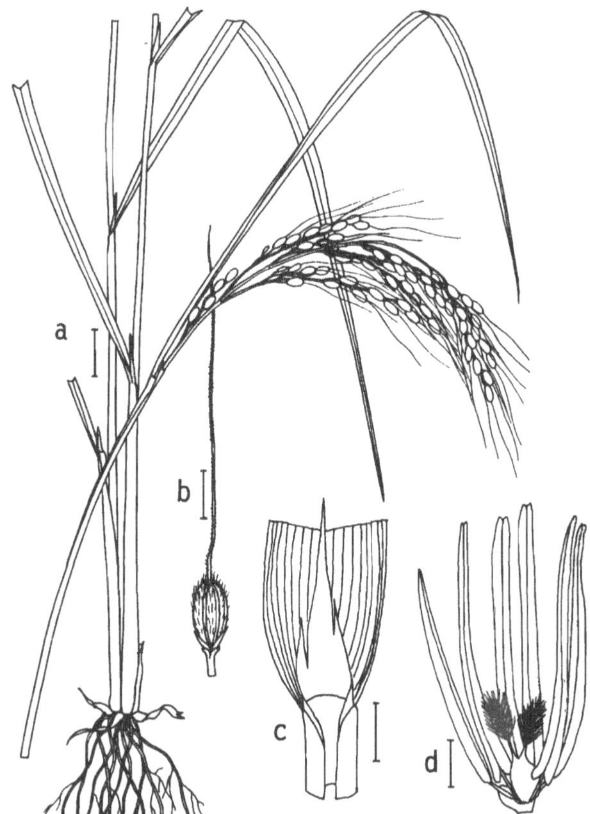

FIG. 313. *Oryza sativa*: a, flowering plant (2 cm); b, spikelet (4 mm). *O. rufipogon*: c, ligule (5 mm); d, flower (1 mm).

tion. Rice or paddy is thought to have originated in Asia independently in India and China from *O. rufipogon* and in Africa from *O. glaberrima* Steudel.

{⊙ (⊙) Hel/Eph}. Cultivated in warm to tropical regions of the world: probably throughout the subcontinent but rarely collected or recorded.

Panicum: ± 470 species: cosmopolitan except in cold regions

Tufted, stoloniferous or rhizomatous. Leaf blades cuneate or parallel, never amplexicaul at the base; ligules membranous or ciliate; leaf sheaths open, sometimes inflated. Inflorescence an open or closed panicle. All spikelets pedicelled, slightly flattened abaxially, disarticulating below the glumes, 2- or rarely 1-flowered, awnless; the lower floret usually reduced to a sterile lemma; the upper floret bisexual. Lower glume usually much shorter than the upper, sometimes reduced or absent; upper glume mostly equal to the sterile lemma and like it, the two usually completely enclosing the fertile floret. Fertile lemma hardened, margins overlapping the equally firm palea. Caryopses enclosed between the hardened lemma and palea, globose to ellipsoidal.

Annual or perennial. Terrestrial, floating or emergent, often covering relatively large areas: some are serious weeds in aquatic systems.

Excluded: found in moist or damp situations but not wetland species: ***Panicum auritum*** Presl [*Sacciolepis aurita* (Presl) A. Camus]; ***P. austroasiaticum*** Ohwi; ***P. bisulcum*** Thunberg; ***P. johnii*** S. M. Almeida: ♦Ind; ***P. trypheron*** Schultes [*P. neesianum* Wight et Arnott ex Steudel].

1. Tufted annual; lower glumes broadly ovate, apex acute; spikelets purple to red-purple ***P. psilopodium***
1. Stoloniferous perennial; lower glumes suborbicular, apex truncate; spikelets yellowish or green, at most tinged with purple
 2. Culms spongy at the base; stolons not irregularly thickened, rarely penetrating the substrate; lower glumes without nerves; usually developing grains ***P. paludosum***
 2. Culms not spongy; stolons irregularly thickened, deeply penetrating the substrate; lower glumes with 1 or 3 faint nerves; usually not developing grains ***P. repens***

Panicum paludosum Roxburgh [*P. natans* J. G. König ex Trinius, *P. proliferum* sensu J. D. Hooker non Lamarck]: Fig. 314

Like *P. repens* but aquatic or subaquatic with lower part of culms creeping, stolon-like, spongy and inflated, 30–100 cm long; leaf sheaths glabrous or pilose at the mouth only; leaf blades flat, bright green; spikelets narrowly lanceolate-ellipsoid, 3–4.5 mm long, appressed to the branches; lower glume without nerves; lemma of lower floret sterile; caryopses 1.5–2 mm long.

Perennial or annual. A gregarious and vigorous grass found growing in water. It roots in the substrate or forms floating mats on the surface. It is readily eaten by buffalo.

{(⊙) Eph/Hyp/Ple/(Hel)}. Tropical Asia and Australia: Ban, Bur? Cey, Ind, Nep? Pak: Go, Hp, Kl, Kt, Mp, Pj, Rj, Sk, Tn, Up, Wb.

Panicum psilopodium Trinius [*P. flexuosum* Retzius var. *glabrum* Retzius, *P. virgatum* Roxburgh ex Steudel]: Fig. 314

Tufted. Culms leafy, erect and rooting below, 15–40 cm tall. Ligules short, membranous, ± 0.5 mm high; leaf blades lanceolate to linear-lanceolate, flat or

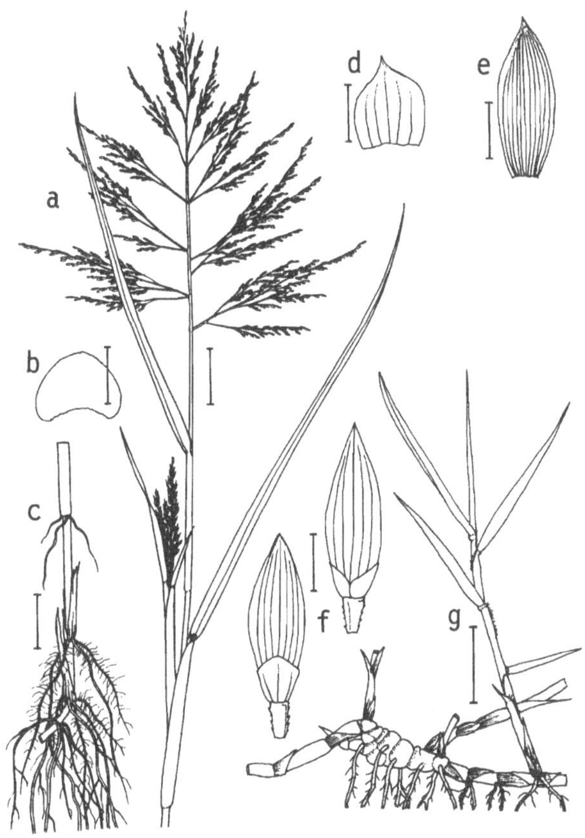

FIG. 314. *Panicum paludosum*: a, inflorescence (3 cm); b, lower glume (1 mm); c, culm base (3 cm). *P. psilopodium*: d, lower glume (0.8 mm); e, upper glume (1 mm). *P. repens*: f, spikelet viewed from opposite sides (1 mm); g, rhizome (2 cm).

folded, 4–15 cm long, 3–5 mm wide. Inflorescence open, 5–15 cm long, 3–10 cm wide, the branches spreading. Spikelets ovate to oblong, 2–3 mm long, usually purple or purple-red. Lower glume ovate to orbicular, ± 1 mm long, obtuse, with 5 faint nerves; upper glume as long as the spikelet, 11-nerved. Lemma of lower floret resembling the upper glume, its palea well-developed or absent; lemma of upper floret broadly ovate, ± 2 mm long, smooth and shining, hardened. Caryopses ovoid to globose, ± 1.5 mm long.

Annual or perennial. Not confined to wetlands but frequently found in marshes, along water courses and in seasonally inundated places.

{(⊙) Hel}. S and SE Asia: Bur, Cey, Ind, Pak:<Go, Kt, Mp, Pj, Rj, Tn.

Panicum repens Linnaeus: Fig. 314

Rhizomes deeply buried (up to 40 cm), white, here and there forming irregularly thickened parts from which new shoots arise. Culms leafy, erect and rooting below, 30–100 cm tall, with a large central cavity, the wall with many lacunae. Leaf sheaths hairy along the free margins; ligules short, membranous, 0.5 mm high, the upper margin white hairy; leaf blades linear, usually folded or inrolled, 4–30 cm long, 3–15 mm wide, blue-green (white waxy bloom), long white hairs behind the ligule otherwise almost glabrous, rounded at the base, acute at the tip, the midrib prominent and keeled. Inflorescence ovate to oblong in outline, 6–30 cm long, the branches ascending. Spikelets ovoid to oblong, 2.5–3.25 mm long, pale green to pale yellow, often with a purple tinge, glabrous, acute. Lower glume oblong-ovate, ±0.8 mm long, with depressed top, with 1 or 3 faint nerves; upper glume as long as the spikelet, 7- or 9-nerved. Lemma of lower floret resembling the upper glume, its palea well-developed or absent; lemma of upper floret smooth and shining, hardened, mucronatulate. Stamens and stigmas not exerted from the spikelet. Caryopses very rarely developing.

Perennial. It is not confined to wetlands but is often found in temporarily inundated places (it cannot withstand permanent flooding). It sometimes forms mats floating on the surface. Even in its native range it rarely develops caryopses. It is a serious weed in plantations and upland rice fields. Because of the deeply buried rhizomes it is difficult to eradicate. It is often called 'ginger grass' (the rhizomes can look a bit like ginger) or in the USA it is called 'torpedo grass' (the rhizomes spread under ground).

{Hel/(Eph/Hyp/Ple)}. Tropical and subtropical regions of both hemispheres, probably originating in Asia: Cey, Ind, <Pak: An, As, Bh, Kl, Kt, Mh, Mn, Mp, Rj, Tn, Wb.

Paspalidium: 40 species: tropics, most in Australia

Rhizomatous or with floating stolons. Ligule a fringe of short hairs; leaf blades linear, up to 30 cm long, up to 10 (–13) mm wide, flat or folded. Inflorescence up to 30 cm long, of short spike-like racemes, these overlapping or distant along a central axis and ending in a 1–2 mm long bristle, bearing the spikelets abaxially in 2 neat rows, each spikelet appressed to a shallow cavity formed by the keel of the rachis and by the disk- or cup-like pedicel. Spikelets subsessile, solitary, 2-flowered, up to 3 mm long, awnless, nearly flat abaxially, convex ventrally, disarticulating below the glumes. Lower floret male or sterile, the upper one bisexual. Lower glume abaxial, much shorter than the spikelet, truncate to obtuse; upper glume about as long as the spikelet, 3- to 7-nerved. Lemma of lower floret resembling the upper glume; lemma of upper floret embracing the palea. Caryopses tightly enclosed by hardened lemma and palea, plano-convex.

Few species grow in wetlands. The aquatic and wetland species are perennial; some are found as weeds in rice fields and irrigation channels.

1. Lowest raceme shorter than the internode; leaf blades abruptly obtuse or blunt and sometimes hooded at the tip; spikelets gibbous; inflorescence distinctly exserted over the subtending leaf sheath ***P. flavidum***

1. Lowest raceme as long or longer than the internode; leaf blades acute, acuminate or setaceous at the tip; spikelets not gibbous; inflorescence scarcely or very shortly exserted over the subtending leaf sheath

  2. Upper glume 7-nerved, almost as long as the spikelet; lemma of upper floret with a smooth, dull or shiny, grainy surface; lower floret with stamens ***P. geminatum***

  2. Upper glume 9- or 11-nerved, ± half as long as the spikelet; lemma of upper floret with a wrinkled surface; lower floret sterile ***P. punctatum***

Paspalidium flavidum (Retzius) A. Camus: Fig. 315

Tufted. Culms erect or ascending from a prostrate base, 10–80 (–120) cm long. Ligule hairs up to 1 mm long or sometimes absent; leaf blades 4–18 (–40) cm long, 2–7 (–7) mm wide, abruptly obtuse or blunt and sometimes hooded at the tip. Racemes (3–) 5–8 (–12), oblong, 1–2.5 (–3) cm long, distant by 2–4 times their own length; rachis narrowly winged, the wing 0.5–0.8 mm wide. Spikelets (4–) 15–25, swollen, globose, 2–3 mm long. Lower glume truncate or ± orbicular, one-third to half the length of the spikelet; upper glume 7-nerved, half to three-quarters as long as the spikelet. Lower floret with a palea, its lemma up to 2.2 mm long, as long as the spikelet, with nerves not raised; lemma of upper floret up to 3 mm long, granulose. Caryopses ovoid.

Perennial or sometimes annual. Roadside ditches and damp hollows, usually in rather shady places; it is not confined to wetlands and is also found in forest undergrowth. It is good fodder for horses and cattle; the grains are eaten by humans in times of want.

{(⨀) Eph/Ple/Hel}. Tropical Asia: Cey, Ind, Pak: Ap, Gj, Go, Hp, Kl, Kt, Mn, Mp, Rj, Tn, Up, Wb.

Paspalidium geminatum (Forsskål) Stapf [*P. fluitans* Retzius, *P. paspaloides* (Persoon) J. D. Hooker]: Fig. 315

Rhizomes creeping, spongy when in water and reaching 1 cm or more in diameter. Culms prostrate, rooting below, 10–50 (–150) cm long. Ligule hairs ± 1 mm long; leaf blades 10–20 (–30), 3–6 (–9) mm wide, acute to acuminate. Racemes 10–30, ovoid, 0.5–4 cm long, usually less than their own length apart; rachis winged, the wing 0.5–1 mm wide. Spikelets 10–30, ovoid, 1.6–2.6 (–3) mm long. Lower glume truncate, one-quarter to one-third the length of the spikelet; upper glume 7-nerved, two-thirds to four-fifths the length of the spikelet. Lower floret with a palea, its lemma as long as the spikelet, with nerves not raised; lemma of upper floret granulose. Caryopses ellipsoid.

Fig. 315. *Paspalidium flavidum*: a, spikelet viewed from opposite sides abaxial view above (1 mm); b, inflorescence (2 cm). *P. geminatum*: c, inflorescence (1 cm); d, spikelet viewed from opposite sides abaxial view above (1 mm); e, raceme viewed from above and below (2 mm). *P. punctatum*: f, inflorescence (1 cm); g, spikelet viewed from opposite sides abaxial view at left (1 mm).

Perennial. In permanently wet places such as marshes, borders of lakes and ponds and in seasonally inundated depressions.

{Eph/Ple/Hel}. Old World tropics: Cey, Ind, Pak: Go, Kt, Mp, Rj, Tn, Wb.

Paspalidium punctatum (N. L. Burman) A. Camus: Fig. 315

Rhizomes creeping, spongy when in water. Culms prostrate, rooting below, 30–100 (–200) cm long. Ligule hairs 1–2 mm long; leaf blades 10–25 cm or more long; 1.5–8 mm wide, acute to acuminate. Racemes 7–20, 1–6 cm long; rachis winged, the wing 0.5–1.5 mm wide. Spikelets 12–40, ovate-elliptical, 2–3.2 mm long. Lower glume truncate, one-quarter the length of the spikelet; upper glume 9- or 11-nerved, one-eighth to three-quarters the length of the spikelet. Lower floret sterile, without a palea, its lemma four-fifths to as long as the spikelet, often with nerves slightly raised; lemma of upper floret wrinkled.

Perennial. In marshes, along the edges of ponds, canals, streams and other wet places. A common weed in rice fields. It is excellent fodder.

{Eph/Ple/Hel}. E Africa and tropical Asia: Ban, Cey, Ind, Pak: Bh, Kl, Mp, Rj, Tn, Wb.

Paspalum [including *Reimaria*]: ± 330 species: tropical and warm regions

Annual or perennial. Leaf sheaths sometimes inflated when floating; ligules membranous or a fringe of hairs. Inflorescence of (1) 2 or many, digitate or scattered, 1-sided, spike-like racemes; rachis of the raceme flat, abaxially keeled, sometimes broadly winged, sometimes with a sterile tip or a terminal bristle 4–7 mm long. Spikelets alternate along the keel of the rachis, solitary or in unequally pedicelled pairs, disarticulating below the glume, awnless, abaxially flattened, plano-convex, orbicular to ovate. Lower glume absent or reduced to a minute scale; upper glume and lemma of lower floret as long as the spikelet. Upper (fertile) lemma leathery to hard, usually obtuse, obscurely 3- to 7-nerved, with inrolled margins embracing the palea, always awnless. Palea as long as lemma, 2-nerved, flat, membranous. Stamens 2 or 3. Caryopses enclosed by the hardened lemma, abaxially flattened, plano-convex.

Most species are not found in wetlands; a few grow emergent or seasonally submerged in fresh water, others in brackish water. Some are troublesome weeds in rice fields and other aquatic systems.

Excluded: found in moist or damp places but not wetland species: ***Paspalum dilatatum*** Poiret [*P. ovatum* Nees ex Trinius]; ***P. notatum*** Fluegge [*P. uruguayense* Arechavaleta]; ***P. plicatulum*** Michaux; ***P. urvillei*** Steudel.

1. Inflorescence of 3–15, erect or spreading, spike-like racemes; annual
 2. Leaf blades ovate-oblong to almost oblong-elliptical; spikelets 1–2 mm long ***P. canarae***
 2. Leaf blades linear to lanceolate; spikelets 2–2.5 mm or more long ***P. scrobiculatum***
1. Inflorescence of 2, spreading, spike-like racemes, rarely with a 3rd below; perennial or sometimes annual
 3. Spikelets with a fringe of long, fine, white hairs along the margins of the upper glume ***P. conjugatum***
 3. Spikelets without a fringe of long, fine, white hairs, glabrous or pubescent
  4. Glumes pubescent; (spikelets 2.5–3.5 mm long) ***P. distichum***
  4. Glumes glabrous
   5. Spikelets 2–2.5 mm long ***P. scrobiculatum***
   5. Spikelets 2.5–4.5 mm long ***P. vaginatum***

Paspalum canarae Veldkamp [*P. compactum* auct. Ind., non Roth, *P. canarae* var. *fimbriatum* (Bor) Veldkamp]: Fig. 316

Culms creeping and rooting below, prostrate, ascending or rarely erect, 10–40 cm long. Leaf sheaths densely or sparsely covered with tubercle-based hairs; ligules obscure; leaf blades ovate-oblong to oblong-elliptical, 1–5 cm long; 3–10 mm wide, glabrous or with tubercle-based hairs, the base slightly cordate or decurrent, the tip acute to acuminate. Racemes 3–16, usually spreading, 0.5–2 cm long. Spikelets broadly ovate to ovate-acute, 1–2 mm long. Upper glume obovate, 1–1.5 mm long, papery, 3- or 5-nerved, with or without tubercle-based hairs. Lemma of lower floret broadly ovate, 1–1.75 mm long, papery, 5-nerved; lemma of upper floret broadly ovate to orbicular, 1–1.25 mm long, hardened, faintly 3- or 5-nerved, with or without tubercle-based hairs. Anthers 0.3–0.5 mm long.

Annual or perhaps sometimes perennial. Grows in uplands. It is found floating on water, in marshy places or on wet sandy soil. Plants with tubercle-based hairs along the margins of the upper glume and lemma of lower floret have been recognized as var. *fimbriatum.*

{(⊙) Hel(Eph)}. Endemic to the western Ghats: ♦Ind: Mh, Kl, Kt, Mp? Tn.

Paspalum conjugatum Bergius: Fig. 317

Stolons up to 2 m long, flattened, with internodes up to 15 cm long. Culms erect, 20–50 (–100) cm long. Leaf sheaths flattened, glabrous except at the collar and the free margins which bear some hairs; ligules 0.4–1.5 mm long; leaf blades linear to lanceolate, (4–) 8–12 (–25) cm long; 3–15 mm wide, pubescent behind the ligule, glabrous or few tubercle-based hairs, the base rounded or narrowed, the tip acute to acuminate. Racemes 2, rarely with a 3rd below, widely spreading, each 4–16 cm long. Spikelets broadly elliptical to ovate, (1–) 1.4–1.8 (–2) mm long, conspicuously hairy, pale yellow-green to light brown, abruptly acute. Upper glume ovate to orbicular, as long as the spikelet, very thin, marginally 2-nerved, the midrib suppressed, with tubercle-based hairs. Lemma of lower floret ovate to orbicular, as long as the spikelet, papery, marginally 2-nerved, the midrib suppressed, glabrous; lemma of upper floret ovate to orbicular, 1–1.25 mm long, hardened, 5-keeled. Anthers 0.5–0.6 mm long.

Perennial. Common and widespread. It is not confined to wetlands but is frequently found on the banks of water courses and backwaters, in wet and seasonally flooded depressions, in marshes and in ditches, particularly those surrounding rice fields or draining plantations.

{Hel}. Pantropics but probably native to the New World: Bur, Cey, Ind: As, Kl, Kt, Tn.

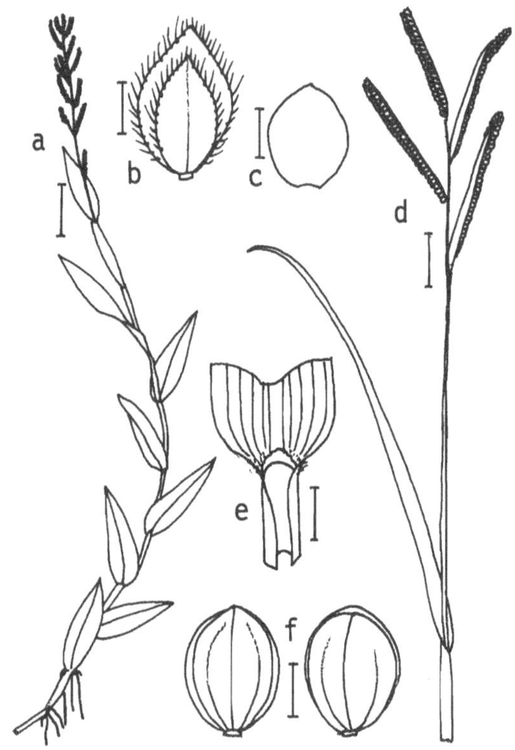

FIG. 316. *Paspalum canarae*: a, flowering culm, with hairs omitted (2 cm); b, spikelet showing upper glume and first lemma (0.5 mm); c, second lemma (0.5 mm). *P. scrobiculatum*: d, inflorescence (1 cm) e, ligula (5 mm); f, spikelet viewed from opposite sides abaxial view at left (1 mm).

Paspalum distichum Linnaeus [*P. paspalodes* (Michaux) Scribner, *'paspaloides'* orth. mut.]: Fig. 317

Stoloniferous or rhizomatous. Culms erect, 5–60 cm long. Leaf sheaths keeled towards the top, the collar glabrous; ligules 0.5–1.5 mm long; leaf blades linear to lanceolate, 3–12 cm long; 2–6 mm wide, usually glabrous or sometimes pubescent, the base slightly rounded and ciliate, the tip acute to acuminate. Racemes 2, rarely with a 3rd below, appressed or widely spreading and reflexed, each 1.5–8 cm long. Spikelets elliptical to ovate, 2.5–3.5 (–4) mm long, light green, acute. Upper glume broadly ovate to elliptical-lanceolate, ± as long as the lemma of lower floret, with short appressed hairs. Lemma of lower floret ovate-oblong to oblong-elliptical, papery; lemma of upper floret ovate-oblong to oblong-elliptical, 2–2.5 mm long, hardened, faintly 5- or 7-nerved. Anthers 1.3–1.7 mm long.

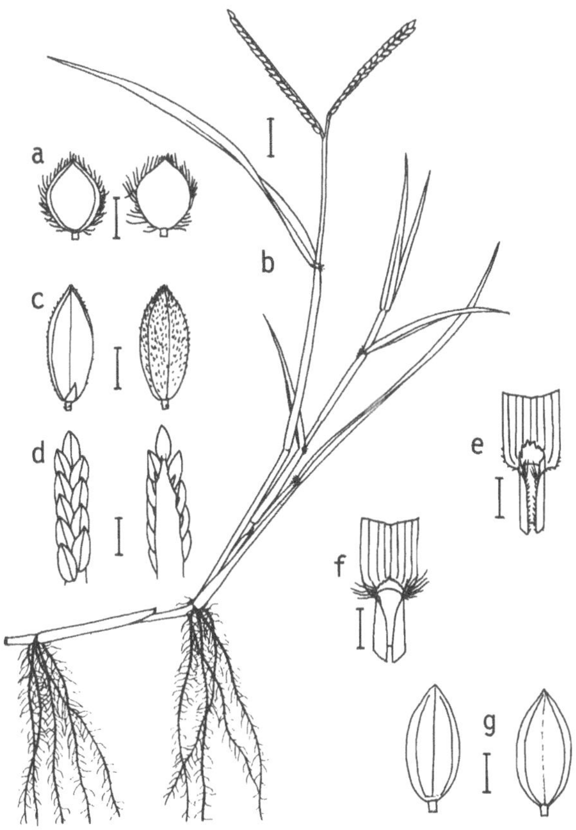

FIG. 317. *Paspalum conjugatum*: a, spikelet viewed from opposite sides abaxial view right side (1 mm). *P. distichum*: b, flowering culm (1 cm); c, spikelet viewed from opposite sides abaxial view right side (1 mm); d, raceme viewed from above and below (5 mm); e, ligule (5 mm). *P. vaginatum*: f, ligule (5 mm); g, spikelet viewed from opposite sides abaxial view at left (1 mm).

Perennial. Common and widespread, often forming large patches. It is not confined to wetlands but is usually in wet places, often found in or floating on water and in a wide variety of marshy habitats. It is reported to be a weed in rice fields. In the past it has often been confused with *P. vaginatum* a more tropical species mostly in coastal regions with narrowly ovate-elliptical spikelets, papery upper glumes and lemma of lower florets and glabrous upper glumes.

{Hel}. Tropics and subtropics of both hemispheres: Ban, Cey, Ind, Nep, Pak: An, Ap, Dl, Hp, Jk, Kl, Kt, Mh, Mp, Pj, Rj, Up, Wb.

Paspalum scrobiculatum Linnaeus [*P. cartilagineum* J. S. Presl et C. B. Presl, *P. orbiculare* G. Forster, *P. commersonii* Lamarck]: Fig. 316

Culms tufted, erect or decumbent at the base, rooting at the lower nodes, 10–120 cm long. Leaf sheaths slightly keeled towards the top, usually glabrous but sometimes pubescent; ligules 0.3–1.2 mm long; leaf blades linear to lanceolate, (2–) 5–30 (–40) cm long; 2–8 (–12) mm wide, usually glabrous or sometimes pubescent, the base rounded or shallowly cordate, the tip acute to acuminate. Racemes (1) 2–4 (5), ascending to widely spreading, each 1–8 (–10) cm long. Spikelets ovate to almost orbicular, (1.8–) 2–2.5 (–3) mm long, olive green to brown or glaucous green blotched with brown. Upper glume ovate to orbicular, ± as long as the lemma of lower floret, thin, 3- to 7-nerved. Lemma of lower floret ovate to orbiculate, papery; lemma of upper floret ovate, 2–2.5 mm long, hardened, faintly 3- to 7-nerved. Anthers 1–1.3 mm long.

Perennial or some races annual. Not confined to wetlands but often found in wet places such as in waterlogged soil along the margins of tanks and streams, in marshes, backwaters and rice fields. Some races are occasionally cultivated for grain.

{(⊙) Hel}. Old World tropics: Cey, Ind, Nep: &&An, Ap, As, Gj, Go, Hp, Kl, Kt, Mh, Mp, Rj, Tn, Up, Wb.

Phragmites: 2–4 species: cosmopolitan

Excluded: an aquatic and wetland reed-like grass found in the Himalayas but not extending into tropical India: ***Phragmites australis*** (Cavanilles) Trinius ex Steudel [*P. communis* Trinius].

Phragmites vallatoria (Linnaeus) Veldkamp [*P. karka* (Retzius) Steudel, *P. maxima* Blatter et McCann]: Fig. 318

Perennial reed with stoloniferous rhizomes. Culms 2–5 (–10) m high. Leaves cauline; ligule a very short membrane fringed with hairs; leaf blades 30–80 cm or more long, 1–4 cm wide, glabrous, the lower surface rough to the touch, the tips attenuate and stiff.

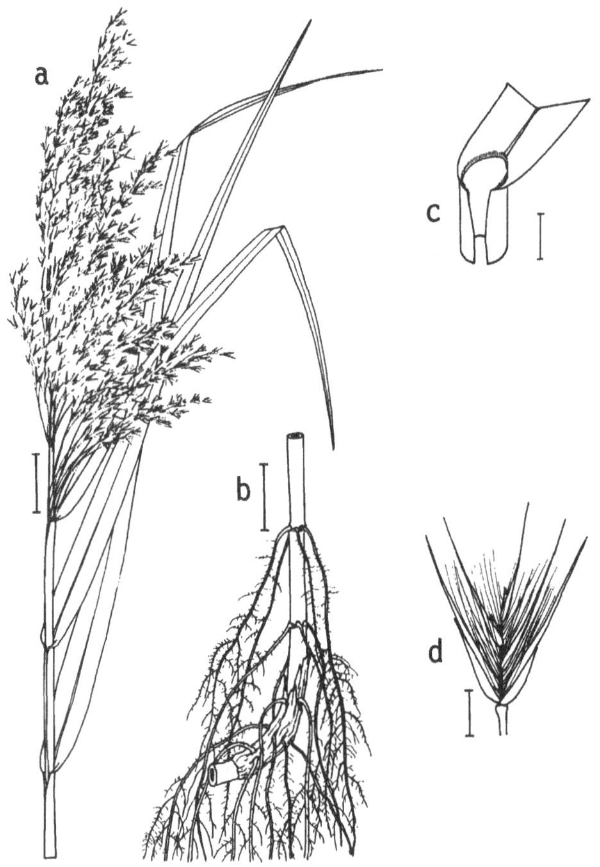

FIG. 318. *Phragmites vallatoria*: a, inflorescence (20 cm); b, base of culm (20 cm); c, ligule (1 cm); d, raceme (2 mm).

Inflorescence a feathery panicle up to 75 cm long. Spikelets solitary, 9–12 mm long, pedicelled, disarticulating above the glumes and between the florets, awned, subterete before anthesis becoming laterally flattened later, 3- to 7-flowered; lowest floret usually sterile or male, upper florets bisexual. Glumes 2, shorter than the lowest lemma, 3- or 5-nerved, with acute apex; lower glume just more than half as long as the upper; upper glume narrowly elliptical to very narrowly elliptical, 4–6 mm long. Lowermost lemmas narrowly elliptical, 7.5–12 mm long, awnless, 3-nerved; upper (fertile) lemmas becoming progressively shorter but awn tipped, 1- to 3-nerved, glabrous; rachilla with silky hairs 4–7 mm long. Stamens 2 or 3. Caryopses free, crowned by the style base.

Perennial reed. Often very abundant, found in swamps, in shallow water at the margins of lakes, ponds, in and along streams, ditches and irrigation canals. The caryopses are dispersed by wind. Often dominant over large areas. In stands it is a good soil binder and it is often planted. The young leaves can be used as fodder. The old culms are used for thatching, screens, baskets, paper and reeds for musical instruments. *Phragmites* stands are important refuges for wildlife but in irrigated land it is sometimes a major pest. Rodewald-Rudescu, L. Das Schilfrohr. Binnengewässer 27: 1–302 (1974).

It is rather similar to the temperate species *Phragmites australis* which differs in having leaf blades smooth on the lower surface, with filiform and flexuous tips; rhachilla hairs 6–10 mm long; upper glume 5–9 mm long.

{Hyp/(Eph)}. Tropical Africa, India, SE Asia and N Australia: Bur, Cey, Ind, Pak: An, Dl, Go, Hp, Hr, Jk, Kl, Kt, Mn, Mp, Pj, Rj, Tn, Up, Wb.

Pseudoraphis: 6 species: tropical and E Asia and Australia

Perennial. Ligules membranous, short or well-developed. Inflorescence of persistent racemes on a short common axis; the raceme axis slender, bearing 1–several distant spikelets, prolonged at the tip into bristle which extends beyond the uppermost spikelet. Spikelets acuminate to awned, terete or slightly flattened abaxially, 4–10 mm long, 2-flowered. Lower floret sterile or male, the upper floret bisexual or female. Lower glume very small, membranous, nerveless, with apex obtuse or appearing bitten off; upper glume as long as the spikelet, leathery, with an acuminate or awned apex, usually scabrid with short bristle-like hairs. Lemma of lower floret slightly shorter than the upper glume; lemma of upper floret much shorter than the lower one, thinly membranous, hyaline. Caryopses linear-oblong in outline, when ripe filling the spikelet.

In ditches and shallow pools, usually gregarious.

1. Lemma of the lower floret subacute; spikelets 2.5–4 mm long; inflorescence congested ***P. minuta***

1. Lemma of the lower floret acuminate-aristate; spikelets 4.5–10 mm long; inflorescence spreading

2. Spikelets 7.5–9 (–10) mm long; tip of the raceme axis extending 1–2.5 cm beyond the uppermost spikelet ***P. brunoniana***

2. Spikelets 4.5–6 mm long; tip of the raceme axis extending 0–1 cm beyond the uppermost spikelet ***P. spinescens***

Pseudoraphis brunoniana Griffith [*Chamaeraphis spinescens* Poiret var. *brunoniana* (Griffith) J. D. Hooker]: Fig. 319

Like *P. spinescens* but the tip of the raceme axis extending 1–2.5 cm beyond the uppermost spikelet; spikelets 7.5–9 (–10) mm long; lower glume oblong, 1 mm long; upper glume long-acuminate and aristate, 7.5–9 (–10) mm long.

Perennial and perhaps also annual. It is usually found floating on water in ditches, ponds and lakes but it may also grow as an emergent in shallow water.

{(⊙)? Eph/Hyp/Ple}. S and SE Asia: Ban, Bur, Cey? Ind: As, Mn, Mp, Or, Wb.

Pseudoraphis minuta (Mez) Pilger [*Chamaeraphis gracilis* Hackel]: Fig. 319

Like *P. spinescens* but inflorescences congested (dense panicles with small spikelets, reddish-green to deep purple); spikelets 2.5–4 mm long; lemma of the lower floret subacute, without an arista.

Perennial. Plants root at the edge of tanks and lakes and send out culms floating upon the water. It mostly forms large patches. It apparently tolerates brackish water.

{Eph/Hyp/Ple}. SE Asia and the northwestern part of the Indian subcontinent: Ban, Ind: As, Bh, Mg, Wb.

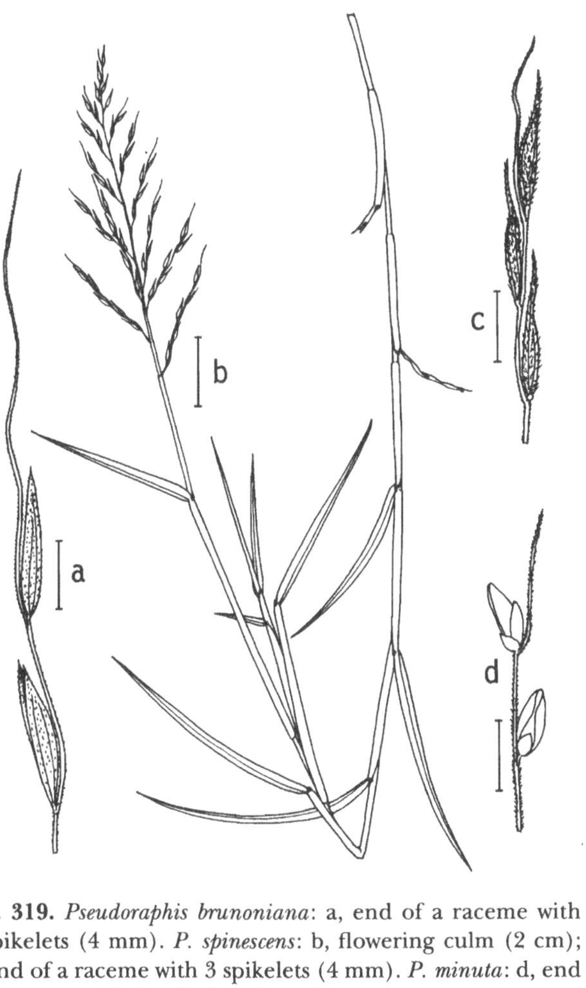

FIG. 319. *Pseudoraphis brunoniana*: a, end of a raceme with 2 spikelets (4 mm). *P. spinescens*: b, flowering culm (2 cm); c, end of a raceme with 3 spikelets (4 mm). *P. minuta*: d, end of a raceme with 2 spikelets (3 mm).

Pseudoraphis spinescens (R. Brown) Vickery [*P. aspera* auct. Ind., non (J. G. König) Pilger]: Fig. 319

Stoloniferous. Culms prostrate or floating, ascending to flower, rooting below, 15–30 (–60) cm tall. Leaf sheaths flattened laterally and keeled, tight but becoming loose with age; ligule membranous, ± 1 mm long, collar and orifice glabrous; leaf blades narrowly linear to lanceolate, (1–) 3–6 (–10) cm long, 2–5 mm wide, flat or folded, shortly hispid with tubercle-based hairs on the adaxial surface, rough to the touch. Inflorescences (3–) 6–15 cm long, up to 6 cm wide, with 9–12 lateral racemes, the bristle-like raceme prolongation extending 0–10 mm beyond the uppermost spikelet. Spikelets 4.5–6 mm long, lacking any subtending bristles of any sort. Lower glume broadly oblong, 0.5 mm long and wide, membranous; upper glume long-acuminate and aristate, 4.5–6 (–7) mm long, obscurely nerved. Lemma of lower floret aristate-acuminate, 4–6 mm long; lemma of upper floret 1.5–3 mm long, membranous, nerveless, glabrous, smooth, somewhat folded around its palea. Caryopsis obovate, ± 1 mm long.

Perennial or sometimes annual. An abundant and common aquatic grass found in flooded places after the monsoon rains. It roots on the banks of canals and ponds and spreads as a floating mat over the surface. It also grows in marshes and other wet places. It is gregarious and often locally dominant. It can be a troublesome weed in rice fields but is also good fodder. Is very variable; Bor (1960) recognizes 3 varieties.

{(⊙) Eph/Hel}. S and SE Asia and Australia: Bur, Cey, Ind: Go, Kl, Kt, Mh, Mp, Rj, Tn, Wb.

Saccharum [*Erianthus*]: ± 40 poorly defined species: tropics and warm regions

Whalen, M. D. Taxonomy of *Saccharum*. Baileya 23(3): 109–125 (1991)

Excluded: ***Saccharum ravennae*** (Linnaeus) Murray [*Erianthus elephantinus* J. D. Hooker]; a wetland species not extending south of the Himalayas.

Saccharum spontaneum Linnaeus: Fig. 320

Rhizomatous, forming clumps. Culms erect, 1–3 m tall, 5–15 mm in diameter, green or white, covered to the base with persistent leaf sheaths. Leaf sheaths with silky hairs (3–7 mm long) at the auricles and around the ligule; ligules membranous, ovate, 3–8 mm long; leaf blades linear-lanceolate, 30–150 (–200) cm long, 3–15 cm wide, glaucous, serrulate along the margins, with prominent midnerves. Inflorescence a large, feathery panicle, (15–) 40–70 cm long, bearing numerous spreading branches, all invested with silky hairs 2–8 mm long, these very prominent below the inflorescence. Spikelets linear-lanceolate, 3–5 mm long, with

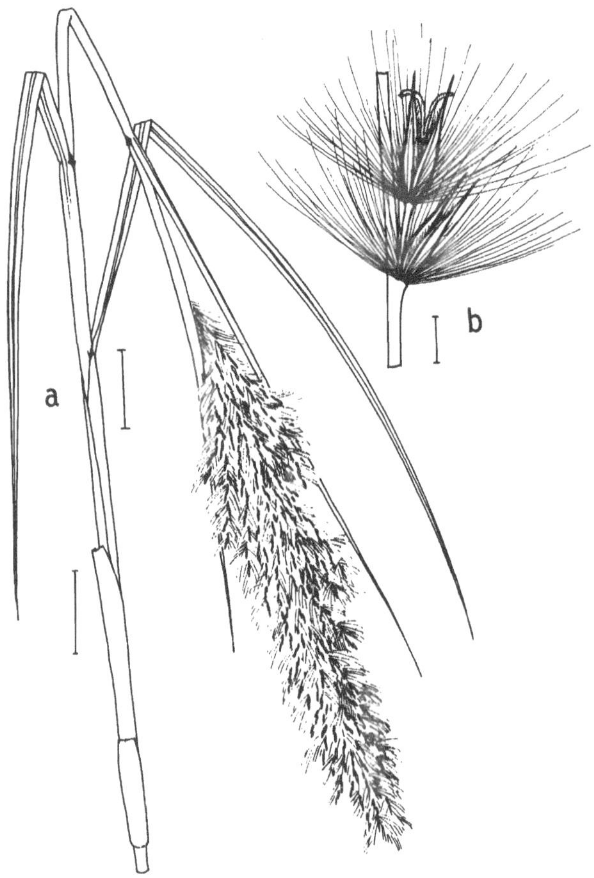

Fig. 320. *Saccharum spontaneum*: a, inflorescence (3 cm); b, spikelets pair (2 mm).

long silky hairs (8–14 mm long), in pairs, one sessile the other pedicellate; pedicelled spikelet disarticulating from the pedicel, the pedicel 0.5–3.2 mm long; the sessile spikelet disarticulating with the inferior internode and the pedicel of the pedicelled spikelet, with or without awns, 3.5–5.5 mm long, 2-flowered. Lower floret sterile, the upper one bisexual. Lower glumes ovate-lanceolate, 3–5 mm long, 2- or 3-nerved; upper glume 3–5, 1-nerved. Lemma of lower florets up to 4.5 mm long, nerveless or weakly 3-nerved, sometimes almost suppressed; lemma of upper floret strap-shaped, up to 4 mm long, very thin, sometimes almost suppressed. Caryopses oblong to globose.

Perennial. It is not confined to wetlands but is frequently found in seasonally inundated places, in swamps and along streams, rivers, backwaters and ditches. The caryopses are dispersed by wind. It usually forms deeply rooted clumps and is a useful soil binder but can become a serious weed in irrigation systems; it is used for decoration, thatching, making ropes and as fodder. In form it is highly variable. It does not produce sugar but is used in sugar cane breeding; hybrids between it and cultivated sugar cane (*Saccharum officinarum* Linnaeus) are sometimes found in the wild, which might, in part, explain the variability.

{Hel}. Warmer regions of the Old World, naturalized in the New World, it is often planted: Ban, Cey, Ind, Nep, Pak: An, Ap, As, Gj, <Go, Kl, Kt, Mh, Mn, Mp, Ng, Pj, Rj, Tn, Up, Wb. 0–1700 m.

Sacciolepis: 33 species: tropics and subtropics, most in Africa

Culms erect or when submerged or floating swollen and hollow (not spongy), rooting at internodes. Ligules membranous. Inflorescence a spike-like panicle. Spikelets pedicellate, 2-flowered, oblong, asymmetrical at base, awnless, readily disarticulating from the pedicel below the glumes. Lower floret male or sterile; upper floret bisexual. Glumes 2, unequal, prominently 5- to 11-nerved; lower glume one-quarter to three-quarters the length of the spikelet, the upper glume as long as the spikelet and inflated. Lemma of lower floret similar to the upper glume but almost straight on the back; lemma of upper floret shorter, papery, slightly hardened at maturity with opaque margins embracing the palea, closed at the apex. Caryopses enclosed between hardened lemma and palea but free within.

Annual or stoloniferous perennials. The disseminules are hardened and nut-like with a knob-like stipe, perhaps dispersed by animals or by water.

Excluded: not wetland species: ***Sacciolepis aurita*** (Presl) A. Camus [*Panicum auritum* Presl]; ***S. curvata*** (L.) Chase.

1. Spike-like panicle continuous, rarely more than 6 cm long; lower glume more than half as long as the spikelet; culms slender, never inflated ***S. indica***
1. Spike-like panicle interrupted below, mostly considerably longer than 6 cm; lower glume half or less as long as the spikelet; culms slender or inflated
  2. Spikelets (3–) 4–5 mm long; culms inflated, 1 cm or more thick ***S. interrupta***
  2. Spikelets 1–2 mm long; culms slender or if inflated then less than 1 cm thick ***S. myosuroides***

Sacciolepis indica (Linnaeus) A. Chase [*S. indica* var. *intermedia* S. M. Almeida]: Fig. 321

Tufted. Culms erect or spreading, slender, (5–) 20–70 (–100) cm tall, less than 1 cm thick. Ligules a ciliate membrane 0.5–1 mm long; leaf blades (0.5–) 2–26 cm long, 1–8 mm wide. Spike-like panicle cylindrical, 1–4 (–8) cm long, continuous. Spikelets 2.2–3.5 mm long, glabrous or hairy, green or with a purple tinge. Lower glume 0.5–2 mm long, more than half as long as the spikelet, 3- to 7-nerved; upper glume 1.5–3 mm long,

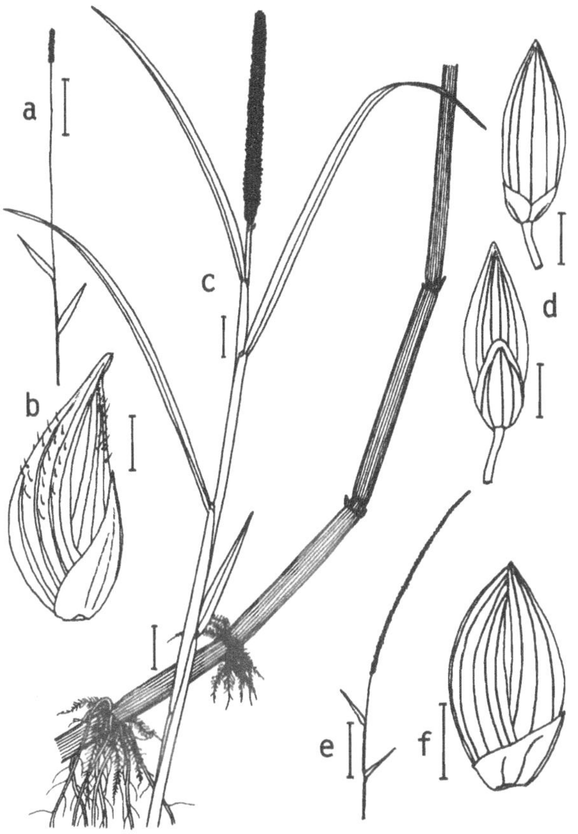

Fig. 321. *Sacciolepis indica*: a, flowering culm (10 cm); b, spikelet (0.5 mm). *S. interrupta*: c, young inflorescence before becoming interrupted, with inflated shoot in the background (2 cm); d, spikelet viewed from opposite sides (1 mm). *S. myosuroides*: e, inflorescence (10 cm); f, spikelet (1 mm).

5- to 11-nerved. Lemma of lower floret empty, 1–3 mm long, 5- to 11-nerved; lemma of upper floret 0.5–1.5, ± half as long as the spikelet, hard and shiny. Anthers purple or deep violet, 0.5–0.7 mm long. Caryopses ellipsoid, usually abaxially flattened, ± 0.8 mm long.

Annual. Not confined to wetlands but often found in marshes, seasonally flooded places, banks of rivers, often common in rice fields but not considered to be a serious pest.

{(⊙)? Hel}. India, SE Asia, Polynesia and Australia, naturalized in Africa and America: Cey, Ind, Nep: An, Gj, Kl, Kt, Mp, Mz, Tn, Up, Wb. 0–2100 m.

Sacciolepis interrupta (Willdenow) Stapf: Fig. 321

Rhizome and lower part of the culms creeping or floating, inflated but hollow, without spongy pith, usually more than 1 cm thick; culms 10–100 (–150) cm long, often purplish-red. Ligules truncate to ovate, 1.5–3 mm long; leaf blades linear to elliptical-lanceolate, 8–20 (–40) cm long, 3–8 (–20) mm wide, bluish-green. Spike-like panicle 15–30 (–40) cm long, interrupted below. Spikelets lanceolate-ellipsoid, (3–) 4–5 mm long, purplish or bluish. Lower glume broadly ovate, 1–2 mm long, less than half as long as the spikelet, 3- to 11-nerved; upper glume ovate-lanceolate, 3–5 mm long, 9- to 11-nerved. Lemma of lower floret empty, like the upper glume; lemma of upper floret ovate-oblong, 2–3 mm long, hardened, faintly 5-nerved. Anthers dark purple, 1–2 mm long. Caryopses obovate to elliptical, abaxially flattened, 2–2.5 mm long.

Perennial or perhaps sometimes also annual. Partly submerged or floating in stagnant water. Often dominant in recently flooded land or in lagoons with fluctuating water levels, sometimes found in rice fields and irrigation channels but not reported to be a serious weed. The inflated stems are eaten by buffaloes; emergent culms often gathered for fodder.

{(⊙)? Hyp/Ple/(Hel)}. Tropical Asia: Cey, Ind, Nep: Go, Kl, Kt, Mp, Tn, Wb.

Sacciolepis myosuroides (R. Brown) A. Camus: Fig. 321

Tufted. Culms erect or decumbent, 30–75 (–120) cm long, less than 1 cm thick. Ligules truncate, 0.5–1.5 mm long, ciliate; leaf blades linear to linear-lanceolate, 6–15 (–40) cm long, 2–4 (–6) mm wide. Spike-like panicle (2–) 5–15 (–35) cm long, 3–5 mm thick, densely flowered, narrow and whip-like, interrupted below. Spikelets ovoid-oblong to ellipsoid-subglobose, 1–2 mm long, dark-coloured. Lower glume ovate, ± half as long or shorter than the spikelet, 3- or 5-nerved; upper glume ovate to ovate-oblong, 1–2 mm long, 5- to 9-nerved. Lemma of lower floret empty, like the upper glume; lemma of upper floret elliptical, ± 1 mm long, hardened, faintly 5- or 7-nerved, shining. Anthers purple, ± 0.5 mm long. Caryopses ovoid, 1.5–2 mm long.

Annual. Not confined to wetlands but often found in seasonally inundated places, marshes, along water courses and in rice fields.

{⊙ Hyp/Hel}. Tropical Asia and Australia: Cey, Ind, Nep: Hp, Kt, Mp, Wb.

Sphaerocaryum [*Graya*]: 1 species

Sphaerocaryum malaccense (Trinius) Pilger [*S. elegans* (Steudel) Nees ex J. D. Hooker]: Fig. 322

Culms tufted or spreading, often rooting below, erect parts rarely more than 15 cm high. Leaf sheaths mostly 3–6 mm long, much shorter than the internodes; ligule a rim of hairs 1–1.5 (–2) mm long; leaf blades ovate-cordate, 1–2.5 (–5) cm long, (4–) 8–13 (20) mm wide, at base amplexicaul, at apex acute, glabrous or hairy. Inflorescence an open panicle, 2–5 (–10) cm long, 1–2.5 cm wide; the longer branches with 1–3 glandular bands; pedicels longer than the

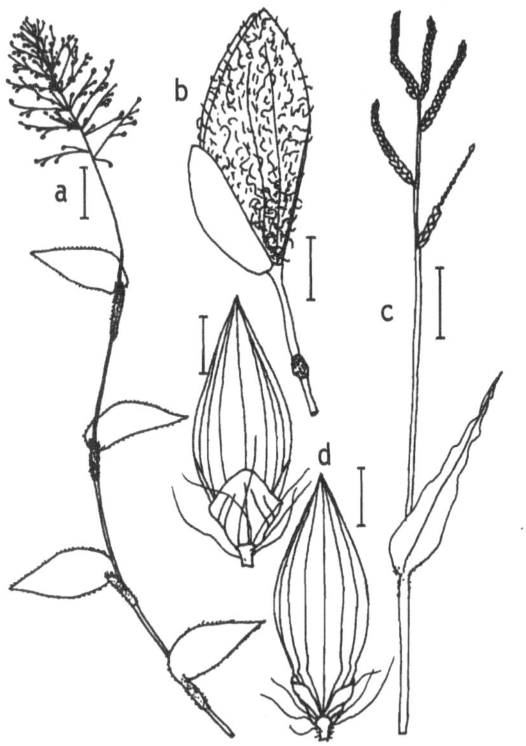

FIG. 322. *Sphaerocaryum malaccensis*: a, flowering culm (1 cm); b, spikelet (0.25 mm). *Urochloa panicoides*: c, inflorescence (2 cm); d, spikelet viewed from opposite sides (1 mm).

spikelets. Spikelets 1-flowered, globose, 0.7–0.9 (1.5) mm long. Glumes subequal, separately deciduous at an early stage, the floret disarticulating at maturity; lower glume thin, less than half as long as the spikelet; upper glume ± half as long as the spikelet. Lemma and palea alike, 0.7–0.9 mm long, membranous, 1-nerved, pubescent, obtuse. Caryopsis ± 0.6 mm long, with brownish longitudinal stripes.

Annual. On mud, in swamps or emergent in shallow water around pools and lakes and along streams. In the Flora of British India J. D. Hooker (1896) mistook this plant for *Isachne pulchella* with the result that it has probably been under-recorded in India.

{⊙ Hyp/Hel}. S and SE Asia: Ban, Bur, Cey, Ind: As, Mg, Mn. 0–1800 m.

Urochloa: 12 species: Old World Tropics, mainly in African savannas

Urochloa panicoides Palisot de Beauvois [*Panicum javanicum* auct. Ind., non Poiret, *P. urochloa* Desvaux, *U. pubescens* Kunth, *U. marathensis* Henrard, *U. marathensis* var. *velutia* Henrard]: Fig. 322

Culms (10–) 20–60 (–75) cm tall, ascending from a creeping base, rooting at lower nodes. Leaf sheaths hairy along the margins; ligule a row of hairs 1–2 mm long; leaf blades ovate-lanceolate to linear-lanceolate, (1–) 8–15 cm long, (3–) 8–20 mm wide, glabrous or somewhat hairy, rounded and cordate at the base. Inflorescence spreading, 8–10 cm long, of 2–7 (–12) spike-like racemes, each 1–6 cm long; rachis flattened, narrowly winged. Spikelets solitary or in groups of 2 or 3, borne in 2 rows, 3–5 mm long, glabrous or hairy. Lower glume turned away from the axis and facing the back of the lemma of upper floret, minute (0.3–1 mm long), 3- or 5-nerved; upper glume as long as the spikelet, 5- to 11-nerved. Lemma of lower floret ± like upper glume; lemma of upper floret slightly shorter than the spikelet, 2–4 mm long, with a short awn, hardened with transverse ridges. Palea hard like lemma of upper floret.

Annual. In rice fields, along ditches and canals, in marshy places and on wet sand along the coast. Not confined to wetlands and also found as a weed in cultivated fields. Generally considered to be excellent fodder. It is rather variable and several varieties have been described: see Basavaih T. C. S. Murthy. *Urochloa panicoides* P. Beauv. (Poaceae) in India. Journal of the Bombay Natural History Society 83: 687–689 (1986?).

{⊙ Hel}. Tropical Old World: Cey, Ind, Nep: Bh, Dl, Hp, Kl, Kt, Mh, Mn, Mp, Pj, Rj, Tn, Up. 0–1500 m.

Vetiveria: 10 species: Old World tropics

Vetiveria zizanioides (Linnaeus) Nash [*Andropogon squarrosus* auct Ind., non Linnaeus fil.]: Fig. 323

Rhizomatous but forming clumps. Culms up to 3 m tall. Leaf sheaths flattened, glabrous; ligule a ciliate membrane, the membrane 0.1–0.4 mm long, the cilia 0.2–1.4 mm long; leaf blades 25–90 (–130) cm long, 3–15 mm wide. Inflorescence a panicle, 15–40 cm long, of numerous whorls of simple or rarely branched racemes. Spikelets in pairs; sessile spikelet disarticulating with the pedicel of the pedicelled spikelet and sometimes the inferior internode, 3.6–5 (–7) mm long, sightly flattened laterally, 2-flowered; lower floret sterile; upper floret bisexual; lower glume convex, as long as the spikelet, keeled, with numerous upwardly directed spines, 7-nerved, with incurved (not inflexed) margins; upper glume like lower bur somewhat shorter, 3-nerved, with spines on midnerve only. Lemma of sterile floret as long as spikelet, thinly membranous, 2- or 3-nerved; lemma of upper floret, 2.8–3.9 mm long, shorter than the sterile one, 1- or 3-nerved, bearing a 0.4–1.8 mm long awn. Pedicelled spikelet 2.5–4.4 mm long, usually shorter than the sessile one, 2-flowered, the lower floret sterile , the upper one usually male; glumes thinner and not as spiny as the sessile spikelet; lemmas usually awnless. Caryopses oblong in outline, enclosed in spiny glumes.

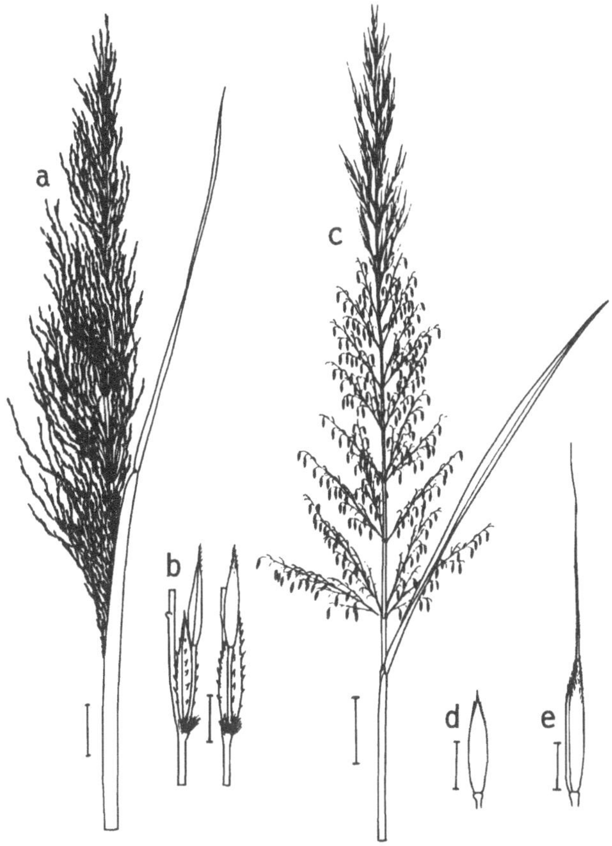

Fig. 323. *Vetiveria zizanioides*: a, inflorescence (3 cm); b, spikelet pair viewed from opposite sides (1 mm). *Zizania latifolia*; c, inflorescence (5 cm); d, male spikelet (4 mm); e, female spikelet (4 mm).

Perennial. Naturally growing in seasonally inundated places, around tanks and ponds and in ditches. Often cultivated as a source of vetiver oil, used in perfumes. The aromatic roots are woven into mats, rugs, fans and screens (khas-khas or tatties) which when wetted emit sweet fragrance. The caryopses are probably dispersed by animals.

{Hel}. Tropical Asia and Africa: Bur, Cey, Ind, Pak: Ap, Dl, <Go, Kl, Kt, <Mh, Mp, Rj, Tn, Up, Wb.

Vossia: 1 species

Vossia cuspidata (Roxburgh) Griffith: Fig. 324

Stolons submerged or floating. Culms up to 4.5 m long, spongy when in water. Ligules very short and densely hairy; leaf blades up to ± 90 cm long, up to ± 2.8 cm wide. Inflorescence terminal, of digitate (rarely single), 15–30 cm long, flattened, spike-like racemes; the internodes thickened towards the tip but hollowed at the tip. Sessile spikelets in cup-like cavities, appressed to rachis, disarticulating slowly together with the inferior internode of the rachis and the pedicel of the pedicellate spikelet, 2-flowered; lower floret male; upper floret bisexual. Pedicelled spikelets similar to sessile ones but usually slightly smaller and with both florets male. Lower glume ovate, leathery, narrowed into a flattened awn, the awn green with scabrous margins and up to 3 cm long and 2 mm wide at the base; upper glume ovate-oblong, 8 mm long, 5- to 9-nerved, mucronate. Lemmas membranous, of the lower floret 2-nerved, of the upper floret 3-nerved.

Floating or emergent perennial. Grows in swamps and tanks in up to 1 m deep water. The disseminules are awned and perhaps animal dispersed. It is readily eaten by buffaloes. It is highly gregarious and large and is often regarded as a pest in Africa where large populations are known to block waterways.

{Hyp/Ple}. Tropical Africa and India: Ban? Bur, Ind: As, Wb?

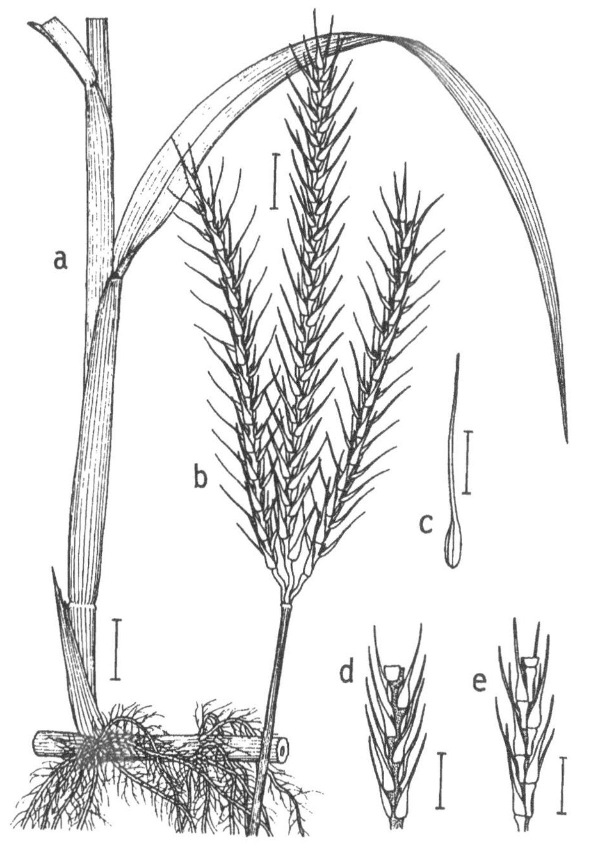

Fig. 324. *Vossia cuspidata*: a, part of a stolon with the base of a culm (2 cm); b, inflorescence (2 cm); c, spikelet (2 mm); d, adaxial side of the rachis (2 cm); e, abaxial side of the rachis (2 cm).

Zizania: 4 rather variable species: Asia and N America

Zizania latifolia Turczaninow [*Z. caduciflora* (Turczaninow) Handel-Mazzetti]: Fig. 323

Rhizomes up to 60 cm long, 1–2 cm thick (when infected with smut even thicker) or floating runners up to ± 3 m long or considerably more following flooding, hollow, with partitions at nodes. Culms usually ± erect, 2–3 m long, glabrous. Ligules membranous and prominent; leaf-blades linear-lanceolate, 1.5–2.5 cm wide. Inflorescence a panicle, up to ± 55 cm long; lower branches spreading or ascending, bearing pendulous male spikelets; upper branches ascending at anthesis becoming erect at maturity, bearing appressed female spikelets. Spikelets disarticulating from the pedicel, 1-flowered; glumes and sterile lemmas reduced to a small collar-like ridge at the tip of the pedicel. Male spikelet caducous, ± 1 cm long, not hardened, somewhat flattened laterally; lemma membranous, 5- nerved, acuminate to awn-tipped; palea about as long as the lemma; stamens 6. Female spikelet hardened, subterete but strongly angled, usually 1–1.8 cm long (3–4 cm or more long with awn); lemma 3-nerved, leathery, terminating in a ± 2 cm long awn; palea 2-nerved, closely clasped by the lemma. Caryopses cylindrical, ± 1.5 cm long.

Perennial. Partly emergent and partly floating, found in lakes and marshes, often dominant over large areas. The disseminules are awned and probably dispersed by animals, particularly water fowl. It makes excellent fodder. In China it is widely cultivated; the plants are infected with a smut (gausun) and the diseased rhizomes eaten, the leaves are used to make mats. The culms, grain and roots are widely used in Chinese medicine. There is some confusion concerning the correct name, I have adopted the widest used one.

{Hyp/Ple}. C and E Asia: Bur, Ind: Mn (Logtak Lake).

PODOSTEMACEAE [TRISTICHACEAE]: 47 genera, ± 268 species

Roots capillary, ribbon-shaped, thalloid or foliose, attached to but not penetrating hard objects in flowing water, usually photosynthetic. Shoots reduced and often only present when flowering or well-developed. Photosynthesis takes place under water, floral shoots develop as the water level drops. Leaves well-developed or reduced, extremely variable in size and shape. Anthesis takes place in air or flowers cleistogamous. Flowers solitary, completely enveloped in a membranous spathella or spathella absent. Perianth either present, 3-lobed, free or united at the base or reduced to linear or subulate, scale-like tepals; tepals 2 or 3, usually alternating with stamens. Stamens 1–3 rarely more, in 1 incomplete whorl or confined to one side of the flower, free or united, borne on a distinct stalk (andropodium); pollen in monads or diads. Ovary superior, 2- or 3-locular. Fruit a capsule, ribbed or smooth, with equal or unequal valves; stigmas 1–3, very variable in shape and size. Seeds axillary, (1) 2 to numerous.

Cosmopolitan in tropical and warm regions, extending into temperate NE America and E Asia. Most of the species are endemic to small geographical areas. The liverwort or alga-like plants live on rocks and sometimes on wood in rapids and waterfalls, usually in full sun. Different taxa are often found growing together. It is sometimes rather difficult to distinguish root from shoot. The term 'root' is used for the part adherent to the substrate, which is sometimes partly floating. 'Shoots' emerge from endogenous buds along the root. The pollination mechanism is unknown. They are essentially annual, dying as the water recedes but some may persist in permanent water. The floral development is very rapid; it may take only two or three days from anthesis to ripe fruit. The disseminules are small and mucilaginous seeds which stick to rocks, the dispersal mechanism is unknown. In India none seem to be of economic importance. Willis, J. C. Studies in the morphology and ecology of the Podostemaceae of Ceylon and India. Annals of the Royal Botanic Gardens Peradeniya 1: 181–250, 267–465 (1902); Cusset, G. and Cusset, C. Structures florales et végétatives des Tristichaceae. Bulletin du Muséum National d'Histoire Naturelle (Paris), 4e série, 10, section B, Adansonia, no. 2: 179–218 (1988); Cusset, C. Contribution à l'etude des Podostemaceae: 12. Les genres asiatiques. Bulletin du Muséum National d'Histoire Naturelle (Paris), 4e série, 14, section B, Adansonia, no. 1: 13–54 (1992). I would like to thank Ms Ursula M. Kradolfer and Dr R. Rutishauser for help with this family.

1. Perianth segments 3, ovate to oblong-obovate; young flowers enveloped by a few leaves but not enclosed in a membranous spathella; ovary 3-locular; stamens 3 or rarely 2, free

  2. Roots flattened, closely attached to substrate; flowering shoots simple, less than 2 cm long ***Dalzellia***

  2. Roots elongate, subterete, creeping and attached by disk-like holdfasts; flowering shoots branched, up to 60 cm long ***Indotristicha***

1. Perianth segments 2 or sometimes a third borne between the filaments, linear to subulate; young flowers totally enclosed in a membranous spathella; ovary 1- or 2-locular; stamens 1 or 2, when 2 then borne on an andropodium

3. Flowering shoots with leaves in 6 neat rows ***Willisia***
3. Flowering shoots with leaves in 2 rows or appearing irregular
4. Capsules indehiscent, containing 1 or 2 seeds ***Farmeria***
4. Capsules opening by valves, containing many more than 2 seeds
5. Capsule valves equal, either both persisting or both caducous
6. Roots foliose (lichen-like), closely attached to substrate; flowers subsessile, remaining within the leaves; spathella boat-shaped ***Hydrobryum***
6. Roots elongate (ribbon- or thread-like), with the upper part usually free; flowers pedicellate, held above the leaves ***Polypleurum***
5. Capsule valves unequal, the smaller one caducous
7. Capsules smooth or obscurely ribbed
8. Capsules pedicellate, held above the surrounding leaves; roots irregular in form, varying from star-shaped to ribbon-like ***Griffithella***
8. Capsules subsessile, remaining embedded in the surrounding leaves; roots ± regularly pinnately branched ***Hydrobryopsis***
7. Capsules with prominent, longitudinal ribs
9. Seeds (1) or 2–8 (–18) ***Farmeria***
9. Seeds more than 30
10. Spathellas funnel-shaped, opening irregularly at the tip; vegetative shoots without tufts of linear filaments (each leaf may terminate in a linear filament) ***Podostemum***
10. Spathellas boat-shaped, opening along the upper side by a longitudinal slit; vegetative shoots occasionally terminating in tufts of linear filaments ***Zeylanidium***

Dalzellia [*Lawia, Mnianthus, Terniola, Terniopsis, Tulasnea*]: 4 species: S and SE Asia

Dalzellia zeylanica (Gardner) Wight [*'ceylanica'* orth. mut., *Lawia zeylanica* (Gardner) Tulasne var. *gardneriana* Willis, var. *malabarica* Willis, var. *parkiniana* Willis]: Fig. 325

Roots frondose, flat, closely attached to rock, up to 35 cm in diameter, usually irregularly lobed, rather lichen-like, bright red or green. Shoots very short, simple. Leaves numerous, simple, occurring either on the upper surface and edges of the thalloid root or in closely packed rosettes on its older parts, those in the centre filiform to linear, those at the edges wider, becoming ovate or obovate; rosette leaves 6–10, 3–16 mm long. Flowers solitary or rarely in pairs, pedicillate, subtended by a small cup-like structure; spathella absent. Pedicel 2–3 mm long at anthesis becoming up to 10(–20) mm long in fruit. Perianth of 3 ± ovate lobes, 1.5–2.5 mm long, united below.

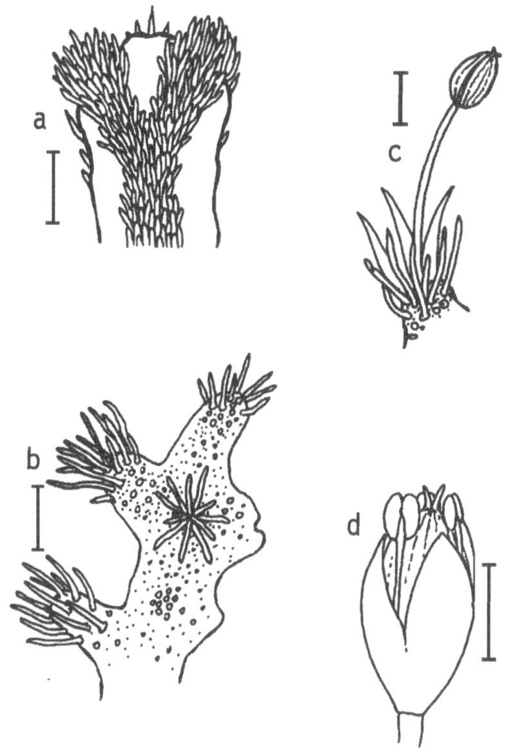

Fig. 325. *Dalzellia zeylanica*: a, young plant (5 mm); b, plant just before flowering (1 cm); c, young fruit (1 mm); d, flower (1 mm).

Stamens 3 or rarely 2, free; filaments 2–2.5 mm long; pollen in monads. Capsules ellipsoidal, 1–2.5 mm long, 9-ribbed, 3-locular, opening by 3 equal valves, remaining ± enclosed in the persistent perianth; stigmas 3, conical. Seeds ovoid, 220–270, ± 0.3 mm long, brown.

Annual perhaps perennial in deeper water. It is characteristically found on smooth rocks or wood in the slower flowing parts of lowland rivers. It is common along the Western Ghats. Flowering from September to February. Rather delicate plants with long leaves (up 16 mm long) and large capsules (2–2.5 mm long) have been recognized as *Dalzellia zeylanica* var. *konkanica* (Willis) Janardhanan [*Lawia zeylanica* (Gardner) Tulasne var. *konkanica* Willis]; it is doubtfully distinct.

{⊙/(⊙)? Hap}. Endemic to S India and Sri Lanka: ♦Cey, ♦Ind, Bh, Go, Kl, Kt, Mh, Or, Tn.

Farmeria [*Maferria*]: 2 species: Sri Lanka and S India

Roots ribbon- or thread-like, branched, forming entangled mats, closely attached to rocks. Vegetative shoots very short or absent. Leaves scattered in groups along the edge of the root, linear, up to 1 cm long, caducous. Floral shoots short, with usually 2–4 (–6) scale-like leaves, each leaf with linear and caducous tips. Flowers sessile; spathella boat-shaped, splitting at the tip and on the upper side. Tepals 2, linear, on each side of the stamen. Stamen 1; filament slightly flattened laterally, exceeding the ovary and stigmas. Capsules smooth or weakly (8)- 10- or 12-ribbed, 2-locular (locules very unequal, one aborting and the other 2- or 4- to 18-seeded), dehiscent or indehiscent; stigmas 2, linear.

1. Capsules shortly stalked, with 10 or 12 longitudinal ribs, 4- to 18-seeded, dehiscent *F. indica*
1. Capsules sessile, smooth, (1-) 2-seeded, indehiscent *F. metzgerioides*

Farmeria indica Willis: Fig. 326
Roots prostrate, thread-like and usually entangled, ± 1 mm in diameter, attached to the substratum by disk-like holdfasts. Leaves linear, sessile, 2–4.5 mm long, ± 0.5 mm wide, filiform tip caducous. Flowers sessile, hardly emerging from spathe. Tepals half as

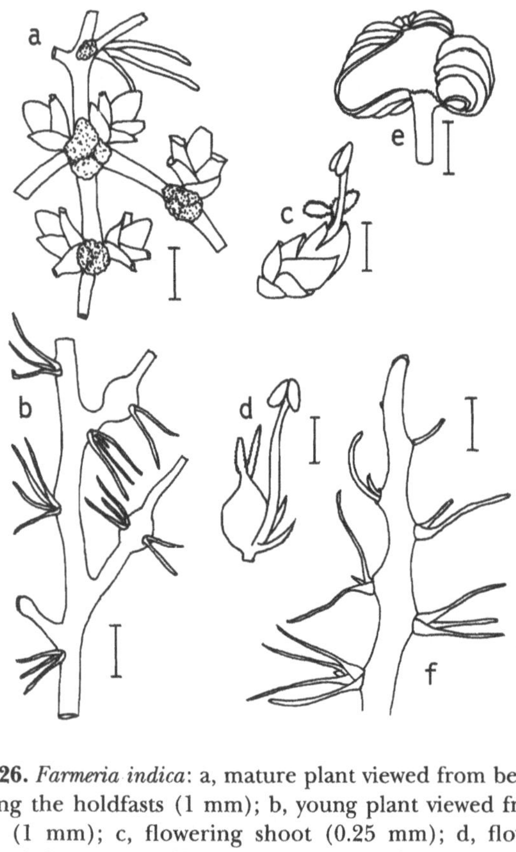

Fig. 326. *Farmeria indica*: a, mature plant viewed from below showing the holdfasts (1 mm); b, young plant viewed from above (1 mm); c, flowering shoot (0.25 mm); d, flower removed from spathella (0.25 mm); e, dehisced capsule (0.3 mm). *F. metzgerioides*: f, young plant (1 mm).

long as ovary. Stamen ± 1 mm long, exceeding ovary. Pedicel 0.5–1 mm long in fruit. Capsules obliquely ovoid, 0.5–1.2 mm long, ± 0.6 mm wide, (8-) 10- or 12-ribbed, dehiscent with 2 unequal lobes; stigmas 2 (3), subulate, unequal, minutely papillose. Seeds (4–) 8 (–18), ovoid, ± 0.5 mm long, ± 0.3 mm wide.

Ecology unknown — 'in a stream'. Arekal, G. D. and Nagendran, C. R. Additional notes on *Farmeria indica* Willis (Podostemaceae). Proceedings of the Indian Academy of Science, B, 80: 226–228 (1974). Cusset (1992) has transferred this species to a newly created monotypic genus *Maferria*; this decision does not seem to be justified.

{⊙ Hap}. Endemic to SW India: ♦Ind: Tn (Nellai Kattabomman District).

Farmeria metzgerioides (Trimen) Willis: Fig. 326
Like *F. indica*, but roots flattened, (1–) 2–4 mm wide, attached to the substratum by elongate rhizoids; tepals as long as or slightly shorter than the ovary; capsules smooth, sessile, indehiscent, ± 0.7 mm long, ± 0.4 mm wide; seeds (1) 2, often germinating *in situ*.

It grows on smooth rocks in rather shady places.

{⊙ Hap}. Endemic to S India and Sri Lanka: ♦Cey, ♦Ind: Kl.

Griffithella [*Hemidistichophyllum, Lawiella, Mniopsis* pro parte, excluded from *Cladopus*]: 1 species: S India
Griffithella hookeriana (Tulasne) Warming [*Griffithella hookeriana* var. *bombayensis* (Tulasne) Willis, *Cladopus hookerana* (Tulasne) C. Cusset [sic], in 1992 corrected to *'hookerianus'*]: Fig. 327
Roots star-shaped or ribbon-like, when ribbon-like then up 1 cm wide, red when young, usually closely attached to the rocks, but sometimes cup-like and then only attached by the central part. Floral shoots usually on the margins, with 2–6 distichous leaves. Leaves 3–4 mm long, at base ± hooded, tips deciduous. Flowers shortly pedicelled; spathella funnel-shaped, 2–3 mm long, toothed at apex and splitting irregularly at tip. Pedicel ± 3 mm at anthesis, 4 mm or more in fruit. Tepals 2, minute, one on each side of the andropodium. Stamens 2; andropodium 1.5–2 mm long, as long or longer than the ovary; filaments 0.5–0.7 mm long. Stigmas 2, linear, usually simple, sometimes ± toothed. Capsule smooth, 2.5 mm long, opening by 2, unequal valves, the larger valve persistent. Seeds many, ± 0.2 mm long.

Annual perhaps perennial. Found in deeper, swift flowing water in small rivers. Flowering October to February. *Griffithella hookeriana* var. *willisiana* (Warming) Willis [*Cladopus willisiana* (Warming) C. Cusset sic] has been described; it is like *G. hookeriana* but the roots are foliose; the spathella shorter and

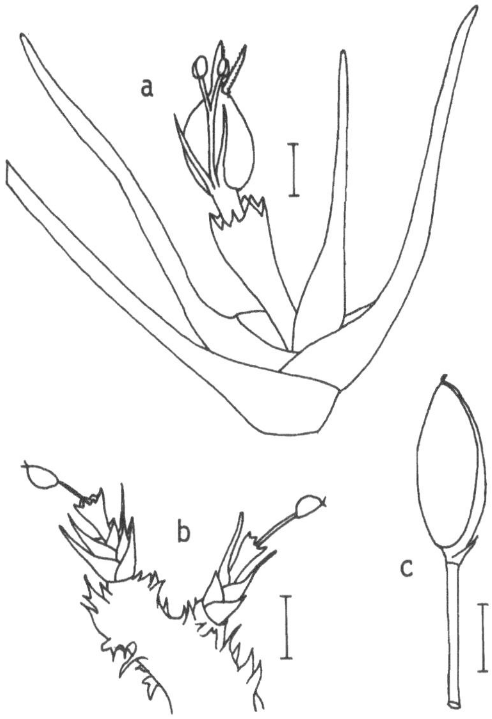

FIG. 327. *Griffithella hookeriana*: a, flowering shoot at anthesis (1 mm); b, plant with 2 young fruits (5 mm); c, dehisced capsule (1 mm).

broadly funnel-shaped, splitting into many teeth. This is probably no more than the environmentally induced cup-like form which appears when the plants grow crowded. Cusset, C. in Flore du Camboge, du Laos et du Viêt-Nam 14: 65–74 (1973) sunk the genus *Griffithella* into *Cladopus*; I do not see that this move was justified. Vidyashankari, B. and Mohan Ram, H. Y. *In vitro* germination and origin of thallus in *Griffithella hookeriana*. Aquatic Botany 28: 161–169 (1987).

{⊙/(⊙)? Hap}. Endemic to W Ghats of India: ♦Ind: Go, Kl, Kt, Mh.

Hydrobryopsis: 1 species: S India

Hydrobryopsis sessilis (Willis) Engler [*Hydrobryum sessile* Willis]: Fig. 328

Roots thalloid, irregular in outline, closely attached to rock, 3–10 mm wide, ± regularly pinnately branched, branches sometimes overlapping and giving the roots a foliose appearance. Floral shoots at edge of the root in branch 'axils', ± prostrate, each with 5–7 scale-like leaves. Leaves distichous, 1–3 mm long, the lower ones small, flat and apiculate at the apex, the upper ones larger and helmet-shaped. Flowers subsessile; spathella boat-shaped, remaining embedded in the surrounding leaves. Tepals 2, one on each side of the andropodium, linear, as long or longer than the ovary. Stamens 2; andropodium ± 1 mm long, as long or longer than the ovary; filaments up to 0.5 mm long; pollen in diads. Capsules globose, ± 1 mm in diameter, subsessile, remaining embedded within the surrounding leaves, smooth, unribbed or obscurely 8-ribbed, 2-locular, opening by 2 unequal valves (the larger one persistent); stigmas 2, linear. Seeds numerous, yellowish, ± 0.2 mm long.

Annual. Common in streams in the Western Ghats. Usually found on rounded boulders in very swift water. The thallus turns white when dry. This species may perhaps be better placed in the genus *Zeylanidium*.

{⊙ Hap}. Endemic to S India: ♦Ind: Ap, Go, Kl, Kt, Mh.

Hydrobryum [*Hydroanzia*]: ± 10 species: NE India, SE and E Asia

Hydrobryum griffithii (Wallich ex Griffith) Tulasne: Fig. 328

Roots hard, foliose, irregularly lobed, up to 30 cm in diameter. Vegetative shoots very short, simple or absent, the leaves linear, up to 12 mm long, scattered

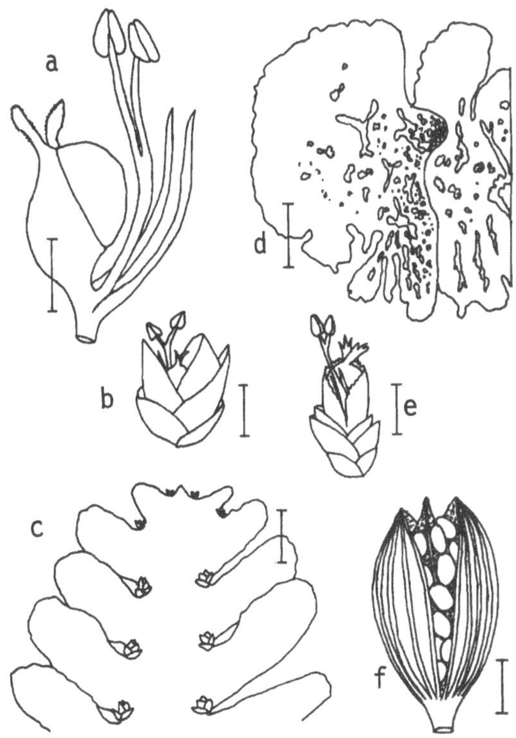

FIG. 328. *Hydrobryopsis sessilis*: a, flower removed from spathella (0.25 mm); b, flowering shoot (0.5 mm); c, plant viewed from above (1 cm). *Hydrobryum griffithii*: d, vegetative plant viewed from above (1 cm); e, flowering shoot (0.25 mm); f, capsule (0.5 mm).

in groups of 2–8 on the upper surface of the root. Floral shoots spread on the upper surface, simple and prostrate. Leaves of floral shoots scale-like, distichous, imbricate, enlarged at the base, tips caducous. Flowers subsessile, remaining within the leaves; spathella boat-shaped, splitting irregularly on the upper side and at the tip. Tepals 2 (rarely 1, 3 or 4), linear, 1–1.5 mm long. Stamens 2 (rarely 3), borne on an andropodium. Capsules shortly pedicelled, ellipsoidal, ± 2.2 mm long, ± 1 mm wide, (6-) 12- or 14-ribbed, 2-locular, opening by 2 equal valves; stigmas 2, linear to cuneate, entire, toothed or lobed. Seeds numerous, elliptical-patelliform, the surface granular.

Annual. On rocks, in rapid, but not deep water.

{⊙ Hap}. NE India, Indochina and N Vietnam: Bhu, Bur? Ind, Nep: As, Mg, Mn, Ng? Sk, Up, Wb.

Indotristicha: 2 species: S India

Roots long and slender, much branched, creeping. Vegetative shoots arising from the root, submerged, branched, bearing photosynthetic scales and filaments (ramuli). Fertile shoots densely clothed in ramuli and scales; scales and ramuli equal or unequal in length, usually overlapping, the uppermost ones united below and forming a cup-like structure at the base of the pedicel. Flowers terminal, solitary; spathella absent. Perianth segments 3, united below or free. Stamens 3, free; pollen panporate, in monads. Capsule 3-locular, 9-ribbed, opening by 3 equal or 2 unequal valves; stigmas 3, linear. Seeds numerous.

Cusset, C. and Cusset, G. Délimitations taxinomiques dans les Tristichaceae. Bulletin du Muséum National d'Histoire Naturelle (Paris), 4e série, 10, section B, Adansonia, no. 2: 149–177 (1988), Rutishauser, R. and Huber, K. A. The developmental morphology of *Indotristicha ramosissima* (Podostemaceae, Tristichoideae). Plant Systematics and Evolution 178: 195–223 (1991).

1. Ramuli feather-like; scale-like appendages scattered or spirally arranged, without midrib; perianth segments united in lower half ***I. ramosissima***
1. Ramuli not feather-like; scale-like appendages in 3 or 4 rows, each with midrib; perianth segments free ***I. tirunelveliana***

Indotristicha ramosissima (Wight) Van Royen [*Dalzellia ramosissima* Wight]: Fig. 329

Roots with disk-like and lobed holdfasts and adhesive rhizoids. Vegetative shoots arise from the holdfasts, up to 60 cm long, branched, bearing scales and ramuli; ramuli feather-like, up to 4 cm long, bearing many scales; the scales varying from leaf-like (triangular to lanceolate, 0.5–2.5 mm long) to filaments (0.5–1 mm long). Floral shoots abundant, each with 1–9 ramuli and scales, the uppermost 1–3 ramuli and 5–7 leaf-like scales form the cup-like structure at the base of the pedicel; pedicels 7–15 mm long in fruit. Perianth segments ovate to obovate, 2–3 mm long, membranous, reddish, united along the lower half, splitting later. Filaments 3–7 mm long. Capsule ellipsoidal, ± 2 mm long, opening by 2 or 3 slits. Seeds almost 100, bluntly ellipsoidal to oblong, ± 0.5 mm long.

Annual or sometimes perennial in deep permanent water. Usually found in the larger rivers, often very abundant. The seeds germinate during May and June.

{⊙/(⊙) Hap}. Endemic to W India: Cey? — record by Cusset and Cusset (1988) remains unconfirmed, ♦Ind: Kl, Kt, Tn. 100–1200 m.

Indotristicha tirunelveliana B. D. Sharma, S. Karthikeyan et B. V. Shetty: Fig. 329

Roots unknown. Shoots 5–25 cm long, densely covered with ramuli; vegetative ramuli 0.5–2 cm long; the scales leaf-like, overlapping, ovate to ovate-lanceo-

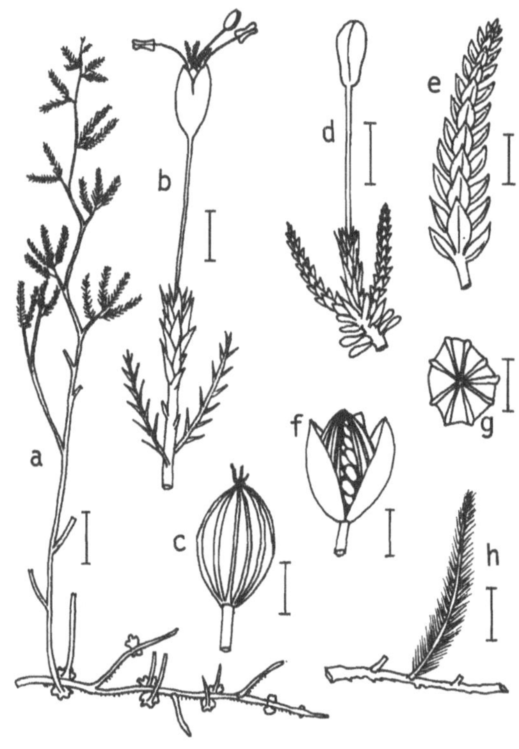

FIG. 329. *Indotristicha ramosissima*: a, vegetative shoot with several branches and ramuli removed (1 cm); b, fertile shoot (2 mm); c, capsule (1 mm). *I. tirunelveliana*: d, flowering shoot with ramuli (5 mm); e, ramulus with leaf-like scales (1 mm); f, capsule viewed from the side with perianth attached (1 mm); g, capsule viewed from above (1 mm); h, part of a vegetative shoot with ramuli (2 cm).

late, with a midrib, ± 1.5 mm long and 0.8 mm wide, arranged in 3 or 4 vertical rows. Floral shoots borne among the ramuli, clothed in leaf-like scales, the scales with elongate apices, the upper ones united at the base to form the cup-like structure. Pedicels ± 2 mm long at anthesis, ± 10 mm long in fruit. Perianth segments free, green, oblong-obovate, ± 2.5 mm long, ± 1.5 mm wide. Filaments ± 1 mm long. Capsule ellipsoidal, 2–3 mm long. Seeds, bluntly ellipsoidal, ± 0.5 mm long.

It is known only from the Thulukkamparai River near Tirukurungudi.

{⊙ Hap}. Endemic to S India: ♦Ind: Tn (Nellai Kattabomman District).

Podostemum [*'Podostemon'* orth. mut., *Zeylanidium* pro parte]: ± 10 species (generic limits somewhat uncertain): tropical to NE America and Asia

Roots elongate, branched, creeping. Shoots arising laterally from the root. Leaves in 2 rows, linear and entire (in Asian species only). Flowers borne in axils of leaves; spathella funnel-shaped, splitting irregularly from the top. Tepals 2 (3), one each side of the andropodium (and one on the back — abaxial side — of the andropodium or in the fork between the two filaments — in American species; abnormal flowers with proliferation of the tepals and stamens are not uncommon). Stamens 1 or 2, borne on an andropodium; pollen in diads. Capsules 8-ribbed, opening by 2 unequal valves, the larger valve persistent; stigmas 2, linear, equal or unequal.

Philbrick, C. T. Aspects of floral biology, breeding system and seed and seedling biology in *Podostemum ceratophyllum*. Systematic Botany 9(2): 166–174 (1984), Philbrick, C. T. and Bogle, A. L. A survey of floral variation in five populations of *Podostemum ceratophyllum*. Rhodora 90: 113–121 (1988). The Asian species are rather different from the New World species and the decision of Cusset (1992) to place the Indian species into the genus *Zeylanidium* is probably correct, at least, for *P. subulatus*.

1. Roots ribbon-like, not obscured by leaves when viewed from above; stamen 1; ovary with prominent almost wing-like ribs ***P. barberi***

1. Roots filiform, ± obscured by leaves when viewed from above; stamens 2; ovary ribbed but ribs rounded, not wing-like ***P. subulatus***

Podostemum barberi Willis [*Zeylanidium barberi* (Willis) C. Cusset]: Fig. 330

Roots ribbon-like, closely attached to rocks, ± 3 mm wide. Shoots 2–4 cm long. Leaves in 2 rows, linear, 5 cm long, not obscuring the root. Spathella elongate, splitting irregularly from the top. Tepals 2, minute. Stamen 1; filament as long or somewhat shorter than the ovary; pollen in diads. Capsule valve ribs usually prominent, almost winged.

{⊙/(⊙)? Hap}. Endemic to S India: ♦Ind: Kt, Kl, Mp, Tn.

Podostemum subulatus Gardner [*Podostemum subulatus* var. *mavaeliae* Willis, var. *sholaii* Willis, *Zeylanidium subulatus* (Gardner) C. Cusset]: Fig. 330

Like *P. barberi*, but roots filiform; leaves up to 30 cm long, very dense and obscuring the root; tepals 2, ± as long as the ovary; stamens 2, as long as the ovary; ribs of capsule valves rounded, not wing-like.

{⊙/(⊙)? Hap}. Endemic to Sri Lanka and S India: ♦Cey, ♦Ind: Kt, Mh, Tn.

Polypleurum [*Dicraea*]: 7 species: S Asia

Roots ribbon- or thread-like, closely attached to rock or the lower part closely attached to rock with the upper part free; free parts up to 50 cm long, rather tough. Specialized vegetative shoots absent. Floral shoots less than 0.5 mm long, developed at almost regular intervals along the margin of the root, bearing 2–8 or rarely more imbricate, distichous scale-like leaves. Flowers

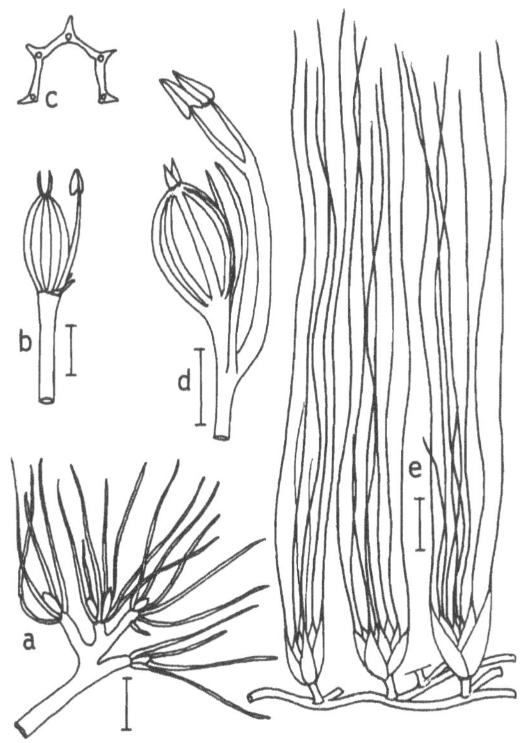

FIG. 330. *Podostemum barberi*: a, plant (1 cm); b, flower (1 mm); c, transverse section of the persistent capsule valve. *P. subulatus*: d, flower (1 mm); e, plant (1 cm).

pedicillate; spathella opening irregularly at the apex. Tepals 2, 1 each side of the andropodium. Stamens usually 2, borne on an andropodium. Capsules (6)- 8- or (10)-ribbed, opening by 2 equal and persistent valves; stigmas 2, ovate or linear.

Khosla, Charu and Mohan Ram, H. Y. Morphology of flower fruit and seed in *Polypleurum stylosum.* Aquatic Botany 46(3–4): 255–262 (1993); Sehgal, Anita, Mohan Ram, H. Y. and Bhatt, J. R. *In vitro* germination, morphogenesis and flowering of an aquatic angiosperm *Polypleurum stylosum* (Podostemaceae). Aquatic Botany 45(4): 269–283 (1993).

1. Root completely attached to substratum, foliose; shoots spread on upper surface of the root; upper pair of leaves of flowering shoot with 2, long, persistent apical filaments ***P. filifolium***

1. Roots attached to substrate at base only, the rest free; upper pair of leaves of each flowering shoot apiculate or with 1 caducous or persistent apical filament

- **2.** Free part of roots subterete, filiform (Sri Lanka) ***P. elongatum***
- **2.** Free part of roots distinctly flattened
  - **3.** Leaves of fertile shoots with persistent apical filaments ***P. munnarense***
  - **3.** Leaves of fertile shoots acuminate or tips caducous
    - **4.** Roots ± foliose, usually small, up to 10 mm wide, up to 10 cm long (NE India and Myanmar) ***P. wallichii***
    - **4.** Roots ribbon-like, much longer than wide (Peninsular India and Sri Lanka)
      - **5.** Free part of roots ± 6 mm wide, up to 30 cm long, not fleshy, sparingly branched above, often zigzagged towards the tip; floral shoots along the margins of the whole root; pedicel of fruit about 6 mm long ***P. dichotomum***
      - **5.** Free part of roots up to 10 mm wide, up to 50 cm long, fleshy, usually richly branched above, not zigzagged, floral shoots on the margins of the basal part of the root; pedicel of fruit 6–40 mm long ***P. stylosum***

Polypleurum dichotomum (Gardner) J. B. Hall: Fig. 331

Roots sparingly branched, ribbon-like, up to 6 mm wide, partly attached to rock and partly free; free parts 10–30 cm long. Floral shoots developed along the margin of the whole root at almost regular intervals, often on small projections making the root appear zigzagged, each bearing 2–5, imbricate, distichous scale-like leaves. Leaves with elongated tips, which sometimes persist or may fall off early. Pedicel 3–6 mm long in flower, elongating in fruit up to 6–8 mm; spathella opening irregularly at the apex. Stigmas 2, ovate to subulate.

Annual or perhaps sometimes perennial in deeper water. Found in streams and rivers.

{⊙/(⊙)? Hap}. Endemic to India: ♦Ind: Kl, Mh, Tn.

Polypleurum elongatum (Gardner) J. B. Hall: Fig. 331

Roots cylindrical, up to 3 mm in diameter, at base frequently branched and attached to rock by disk-like holdfasts, upper part free; free parts 50–60 cm long, sparingly branched; vegetative parts of the roots and the marginal parts of the fertile roots caducous. Floral shoots developed at almost regular intervals along the lower part of the free part of the root, each bearing (2–) 4 (5) imbricate, distichous scale-like leaves with caducous elongate tips. Pedicel 4–5 mm in flower, elongating in fruit up to 5–9 mm. Stigmas 2, subulate.

Annual or perhaps sometimes perennial in deeper water. On stones and rocks in rivers.

{⊙/(⊙)? Hap}. Endemic to Sri Lanka: ♦Cey.

Polypleurum filifolium (K. Ramamurthy et J. Joseph) Nagendran, Arekal et Subramanyam: Fig. 331

Like *P. dichotomum* but roots attached throughout their length without any free parts; shoots all on the upper surface of the root (never along the margins);

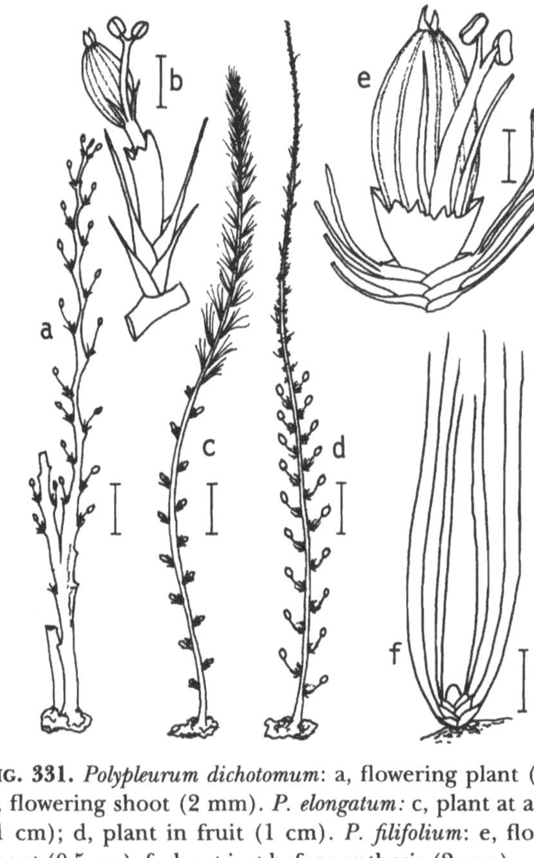

FIG. 331. *Polypleurum dichotomum*: a, flowering plant (1 cm); b, flowering shoot (2 mm). *P. elongatum:* c, plant at anthesis (1 cm); d, plant in fruit (1 cm). *P. filifolium*: e, flowering shoot (0.5 cm); f, shoot just before anthesis (2 mm).

leaf tips filiform, persistent, up to 20 mm long; the upper pair of leaves apparently each with two filiform appendages.

This species would, perhaps, be better placed in the genus *Farmeria*.

{⊙/(⊙)? Hap}. Endemic to S India: ♦Ind: Kl (Palakkad and Thrissur Districts). Gopal (1990) gives Or (Koraput), I have not been able to confirm it.

Polypleurum munnarense Nagendran et Arekal: Fig. 332

Like *P. dichotomum* but filiform leaf tips persisting during anthesis, up to 4 cm long; stigmas linear.

It is doubtful that this is a good species, future investigation will perhaps show that it belongs to *P. dichotomum*.

{⊙? Hap}. Endemic to India: ♦Ind: Kl (Munnar in Idukki district). 1820 m.

Polypleurum stylosum (Wight) J. B. Hall [*P. stylosum* var. *kumarensis* Nair, *Dicraea algaeformis* Warming, *D. stylosa* var. *bourdillonii* Willis, var. *fucoides* Willis, var. *kandrensis* Willis, var. *laciniata* Willis]: Fig. 332

Roots frequently branched, ribbon-like, up to 10 mm wide, fleshy, often crisped at the base, attached to rock by a stout foot and at other points by smaller disk-like holdfasts, partly free; free parts 5–50 cm long, irregularly branched but not zigzagged. Floral shoots developed at the base of the floating part, at almost regular intervals of 2–4 mm along the margin, each bearing (2–) 4–5 imbricate and distichous scale-like, helmet-shaped leaves. Flower buds usually as long as broad; pedicel 2–5 mm in flower, elongating in fruit up to 6–40 mm. Stamens usually 2, borne on an andropodium. Stigmas 2, linear.

Annual or perhaps perennial in deeper water. In rivers and streams. Sometimes in partial shade. The free roots are brown and leathery and closely resemble the brown alga *Fucus*.

{⊙/(⊙)? Hap}. Endemic to India and Sri Lanka: ♦Cey, ♦Ind: Go, Kl, Kt, Mh, Tn.

Polypleurum wallichii (R. Brown ex Griffith) Warming [*Dicraea minor* (Bentham et Hooker) Weddell, *D. pterophylla* Weddell, *Polypleurum agharkarii* (Nandi) Nagendran, Arekal et Subramanyan nom. illeg.]: Fig. 332

Roots ribbon-like, up to 10 mm wide, up to 10 cm long, branched and lobed, often frondose, with rather woody vascular tissue, at base attached to rock by a thick holdfast, at other points by disk-like or elongate holdfasts; vegetative parts caducous. Floral shoots develop at almost regular intervals of 2–3 mm along the upper edges of the margin, bearing (2–) 4 (–7) scale-like, helmet-shaped and keeled leaves. Flower buds usually as long as broad; pedicel 4–5 mm in

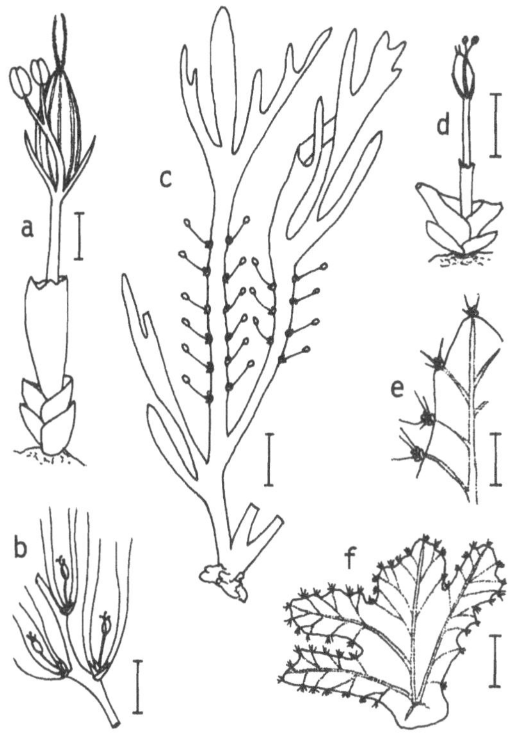

Fig. 332. *Polypleurum munnarense:* a, flowering shoot (5 mm). *P. stylosum*: b, flowering shoot (1 mm); c, plant in fruit (2 mm). *P. wallichii:* d, flowering shoot (2.5 mm); e, shoots just before anthesis (5 mm); f, root showing marginal shoots (1 cm).

flower, elongating to 8–25 mm in fruit; spathella opening irregularly at the apex. Capsule 8- or 10-ribbed. Stigmas 2, subulate.

Annual or perhaps perennial in deeper water. It is very close to *P. stylosum* – the best diagnostic character is the distribution.

{⊙/(⊙)? Hap}. Endemic to NE India and Myanmar: ♦Bur, ♦Ind: As, Mg.

Willisia: 1 species: SW India

Willisia selaginoides (Beddome) Warming ex Willis: Fig. 333

Roots foliose, closely attached to rock by an irregularly lobed, 2–3 cm in diameter holdfast. Vegetative shoots caducous as the water level falls, linear, simple or rarely branched, up to 50 cm long, almost leafless below, above with numerous, up to 8 cm long, filamentous leaves. Floral shoots densely crowded, simple, rigid, erect, 2–10 cm long, closely covered with leaves. Leaves of fertile shoots in 6 distinct rows, scale-like, triangular, rigid, 2 out of 6 rows terminating in a caducous, hair-like tip. Flowers sessile; spathella forked at the tip. Tepals 2. Stamens 2, borne on an andropodium.

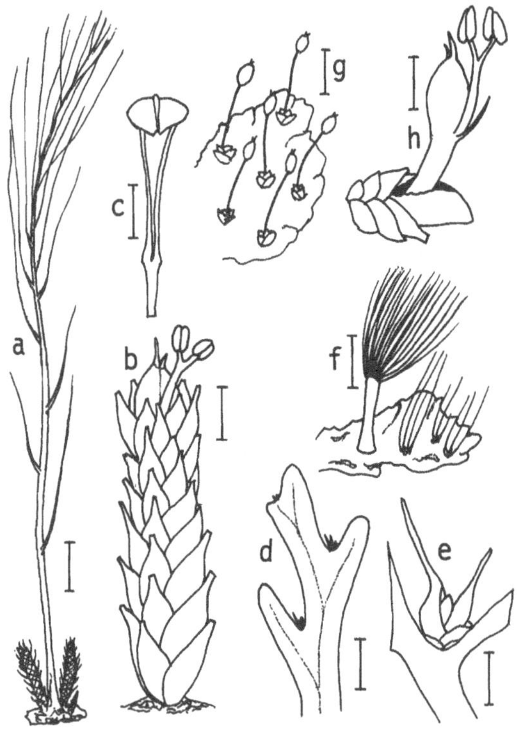

FIG. 333. *Willisia selaginoides*: a, plant before anthesis (2 cm); b, fertile shoot (1 mm); c, persistent capsule valve (1 mm). *Zeylanidium lichenoides*: d, plant before anthesis (5 mm); e, young fertile shoot (2 mm). *Z. olivaceum*: f, plant before anthesis (5 mm); g, plant in fruit (5 mm); h, flower at anthesis (2 mm).

Carpels smooth or the middle nerve of each valve ribbed, opening by 2 valves, the larger one persistent; stigmas 2, linear.

Annual or perhaps perennial in deeper water. On smooth rocks in torrents and waterfalls. Most accounts write that the leaves are in 4 rows; all the material we have seen has them in 6 rows.

{(⊙)? Hap}. Endemic to SW India: ♦Ind: Kl, Tn.

Zeylanidium [*Hydrobryum* pro parte, *Podostemum* pro parte]: 2 or 3 species: Sri Lanka, W, S and NE India and Myanmar

Roots thallose, ribbon-like or foliose, resembling lichens, up to 25 cm or more in diameter, closely attached to rock. Shoots short and simple, vegetative shoots terminating in tufts of linear filaments. Floral shoots rather prostrate, with distichous, bract-like leaves, each terminating in a caducous thread-like tip. Flowers pedicellate; spathella boat-shaped, opening at the upper side with a longitudinal slit. Tepals 2. Stamens 2 or rarely 1. Carpels 8-ribbed, opening by 2 unequal valves, the larger one persistent; stigmas 2, unequal, linear or slightly flattened.

As suggested by Cusset (1992) it may be better to place the Asian species of *Podostemum* in this genus.

1. Roots frondose, crustaceous, lichen-like, sometimes bearing stalked tufts of filamentous leaves; floral shoots scattered over the surface of the root ***Z. olivaceum***

1. Roots ribbon-like, pinnately branched, without stalked tufts of filamentous leaves; floral shoots arising along the margin of the root ***Z. lichenoides***

Zeylanidium lichenoides (S. Kurz) Engler [*Z. lichenoides* var. *bhorensis* (Willis) Santapau, *Z. lichenoides* var. *khandalense* (Willis) Santapau, Willis recognized an additional 10 varieties, *Podostemum acuminatum* Weddell, *P. microcarpus* Weddell, *P. kelensis* Willis]: Fig. 333

Roots ribbon-like (at least partly parallel-sided), closely attached to rock, at the periphery often pinnately branched. Vegetative shoots short, simple, with filiform leaves 3–6 mm long. Floral shoots ± prostrate with distichous, scale-like leaves, each terminating in a caducous thread-like tip. Flowers shortly pedicellate. Tepals 2, shorter than ovary. Stamens 2 or rarely 1, filaments as long or slightly longer than ovary. Stigmas 2, unequal, ovate to subulate, ± lobed or toothed.

Annual. It occurs on smooth rocks, in rapid currents. The vegetative form is very variable; Willis recognized 12 varieties. *Zeylanidium johnsonii* (Wight) Engler is like *Z. lichenoides* but it is reported to have stamen filaments (presumably including the andropodium) two and a half times as long as ovary.

{⊙ Hap}. Endemic to Sri Lanka, India and Myanmar: ♦Bur, ♦Cey, ♦Ind: Ap, As? Kl, Kt, Mg, Mh, Or, Tn. 1000–1300 m.

Zeylanidium olivaceum (Gardner) Engler [*Zeylanidium johnsonii* (Wight) Engler, *Hydrobryum olivaceum* var. *anamalaiensis* Willis, var. *grisum* Willis, var. *zeylanicum* Willis: Fig. 333

Roots frondose, closely attached to rock, ± circular outline, lobed, red when young. Shoots scattered on the surface, sometimes bearing stalked tufts of filamentous leaves, which are up to 10 cm long. Floral shoots rather prostrate with 4–8 distichous, boat-like leaves with caducous tips. Spathella ± papillose. Flowers pedicillate; pedicel 2–54 mm long. Tepals 2, one on each side of the andropodium. Stamens 2 or rarely 1, ± as long as ovary. Capsules 3–5 mm long, 8-ribbed, opening by 2 unequal valves, the larger one persistent; stigmas 2, unequal, subulate, ± lobed or toothed.

Annual. In rivers and small streams, usually in rapids with a strong current, mostly in shallow water. Towards the end of season the root decays and looses its shape. Willis (1902) describes *in situ* germination.

{⊙ Hap}. Endemic to Sri Lanka and India: ♦Cey, ♦Ind: Ap, Kl, Kt, Tn. 500–2000 m.

POLYGONACEAE: 51 genera

Leaves mostly alternate; stipules united into a membranous sheath (ochrea). Flowers bisexual or unisexual. Perianth 3- to 6-merous. Stamens 6–9. Ovary superior, 1-locular; ovule solitary, attached basally; styles 2–4. Fruit a nut.

Excluded: the following is recorded from wetland habitats in the Himalayas but does not extend further south in India: ***Rumex aquaticus*** Linnaeus.

Found in moist or damp habitats but not wetland species: ***R. dentatus*** Linnaeus; ***R. maritimus*** Linnaeus; ***R. nepalensis*** Sprengel.

Polygonum [including *Persicaria*]: 150 species: cosmopolitan

Ochreas inconspicuous and hyaline or conspicuous and membranous or paper-like, tubular or cleft, margins glabrous or variously ciliate or lacerate. Flowers aggregated in simple or branched spike-like racemes, in heads or in axillary clusters. Perianth segments (4) 5, subequal, free or united at the base, at least in part petaloid, not enlarging greatly in fruit. Stamens 4–8, alternating with glandular disk lobes. Ovary 3-angled or globose or obovoid; style arms 2 or 3, stigmas capitate; nut biconvex or 3-sided or rarely flattened and concave.

Annual or perennial. Found in a wide variety of habitats: submerged, floating, emergent, seasonally submerged or as helophytes. Pollination entomophilous, some species are distylic or functionally dioecious. The disseminules are nuts usually shed with the perianth attached; dispersed by animals or by water. There are good botanical grounds for placing nearly all the wetland species into the genus *Persicaria.* Nomenclatural stability is of utmost importance. I have, therefore, retained them in *Polygonum* but have included the full citation for the name in the genus *Persicaria* in the synonymy.

Excluded: the following are recorded from aquatic or wetland habitats in the Himalayas but do not extend further south in India: ***Polygonum amphibium*** Linnaeus [*Persicaria amphibia* (Linnaeus) S. F. Gray]; ***P. dolicopodum*** Ohki [*Persicaria dolicopoda* (Ohki) Sasaki]; ***P. perfoliatum*** Linnaeus [*Persicaria perfoliata* (Linnaeus) H. Gross]; ***P. posumbu*** F. Hamilton ex D. Don [*Persicaria posumbu* (F. Hamilton ex D. Don) H. Gross]; ***P. punctatum*** Elliot [*Persicaria punctata* (Elliot) Small; ***P. strigosum*** R. Brown [*Persicaria strigosa* (R. Brown) Nakai]; ***P. tenellum*** Blume [*Persicaria tenella* (Blume) H. Hara, *Persicaria kawagoneana* (Makino) Nakai].

Found in moist or damp habitats but not wetland species: ***P. limbatum*** Meisner [*Persicaria limbata* (Meisner) H. Hara]; ***P. minus*** Hudson [*Persicaria minor* (Hudson) Opiz]; ***P. mite*** Schrank [*Persicaria mitis* (Schrank) Dorstál]; ***P. orientale*** Linnaeus [*Persicaria orientalis* (Linnaeus) Assenov]. Indeterminate: ***P. praetermissum*** J. D. Hooker; ***P. strictum*** Allioni.

1. Leaf blades less than 4 mm wide, less than 1 cm long; flowers in axillary clusters; ochreas hyaline; ovary 3-angled ***P. plebeium***

1. Leaf blades more than 10 mm wide, more than 2 cm long; flowers in racemes or head-like clusters; ochreas membranous or paper-like; ovary globose or obovoid, not 3-angled

2. Flowers in remote, terminal and axillary head-like racemes; nodes with retrorse prickle-like hairs ***P. pedunculare***

2. Flowers in elongate, simple or panicled racemes; nodes glabrous or softly hairy

3. Ochrea not ciliate at the mouth (nuts flattened-biconvex; styles 2 or very rarely 3)

4. Perianth segments ± smooth, equal to or slightly longer than the fruit; bracts glabrous ***P. glabrum***

4. Perianth segments with raised anchor-shaped veins, clearly longer than the fruit; bracts hairy ***P. lapathifolium***

3. Ochrea ciliate at the mouth (nuts biconvex or 3-angled; styles 2 or 3)

5. Cilia at the top of the ochrea 5–20 mm long, nearly as long as the tube; perianth white (nuts 3-angled) ***P. barbatum***

5. Cilia at the top of the ochrea not more than 3 mm long, much shorter than the tube; perianth pink to red or very rarely white (nuts biconvex or 3-angled)

6. Racemes slender, lax, often hanging; bracts ± distant ***P. hydropiper***

6. Racemes dense and tight, usually erect; bracts overlapping ***P. pulchrum***

Polygonum barbatum Linnaeus sensu lato [*P. flaccidum* Roxburgh non Meisner, *P. rivulare* Roxburgh, *Persicaria barbata* (Linnaeus) H. Hara]: Fig. 334

Stems erect, sometimes creeping below, 40–80 (–150) cm tall. Ochreas membranous, tubular, truncate, 1.5–2.5 cm long, the numerous nerves extended into cilia, the cilia 0.5–2 cm long; leaf blades shortly petiolate (the petioles up to 1 cm long), lanceolate, (4–) 7–12 (–20) cm long, (0.5–) 1–3 cm wide, glabrous or hairy on the nerves and margins, lateral nerves 13–20 pairs, base cuneate to attenuate, tips attenuate, acute. Flowers in terminal and axillary, up to 7 cm long spike-like

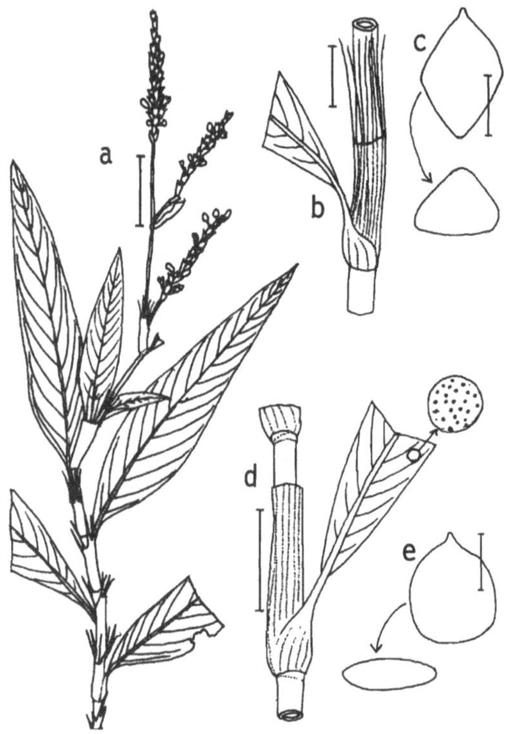

Fig. 334. *Polygonum barbatum*: a, flowering shoot (2 cm); b, ochrea (1 cm); c, nut with transverse section (inset) (1 mm). *P. glabrum*: d, ochrea (1.5 cm) with enlarged portion of the undersurface of the leaf blade (inset); e, nut with transverse section (inset) (1 mm).

racemes; bracts closely overlapping, oblong, 2.5–3 mm long, narrowed below, sparsely hairy; bracteoles up to 2 mm long. Perianth white, ± 2.5 mm in diameter; segments 5, subequal, ovate, ± 2 mm long, obtuse. Stamens 5–7; filaments subequal, ± 1.5 mm long; anthers ± 0.25 mm long, Styles (2) 3, ± 0.5 mm long. Nut rhomboid in outline, 3-sided or some on the same plant occasionally convex, 1.8–2 mm long, ± 1.5 mm wide.

Perennial but sometimes perhaps also annual. Gregarious and often forming distinct clumps. It is found by streams and rivers, in marshes or in shallow water in pools and ditches or in floating mats of vegetation. The roots are used in medicine as an astringent and cooling agent; the seeds are used in relieving colic. It is very variable; the following varieties are usually recognized in India but intermediate plants are relatively common.

Polygonum barbatum var. ***gracilis*** (Danser) Steward [*P. serrulatum* sensu J. D. Hooker, non Lagasca, *Persicaria barbata* subsp. *gracile* (Danser) H. Hara].

Like *P. barbatum* var. *barbatum* but bristles on the ochrea shorter; leaf blades narrower (7–12 mm broad), glabrous or scarcely setulose, the bases rounded; inflorescences narrower, 1–3 spike-like racemes borne at ends of branches.

Polygonum barbatum var. ***stagninum*** (F. Hamilton ex Meisner) Steward [*P. stagninum* F. Hamilton ex Meisner, *Persicaria barbata* var. *stagnina* (Meisner) Grierson]

Like *P. barbatum* var. *barbatum* but ochreas longer (± 2.5 cm long); leaf blades hispid, with greyish hairs.

{⊙? Hel/Hyp/Ple}. Tropical Old World and Australia. The following distribution refers to all three varieties: Ban, Bhu, Bur, Cey, Ind, Nep, Pak: <An, As, Bh, Hp, Hr, Jk, Kl, Kt, Mg, Mh, Mp, Or, Pj, Rj, Sk, Tn, Tp, Up, Wb.

Polygonum glabrum Willdenow, [*P. densiflorum* Meisner, *Persicaria glabra* (Willdenow) Gomes de la Maza, *Persicaria densiflora* (Meisner) Moldenke]. Fig. 335

Stems erect, shrubby, up to ± 2.5 m tall. Ochreas membranous, glabrous, tubular, truncate at the mouth, 2–4 cm long, the nerves not prolonged into cilia; leaf blades shortly petiolate (petioles up to 2 cm long); blades narrowly lanceolate, 10–25 cm long, 1.5–4 cm wide, gland-dotted, glabrous except for the midrib and some lateral nerves, lateral nerves 25–40 in

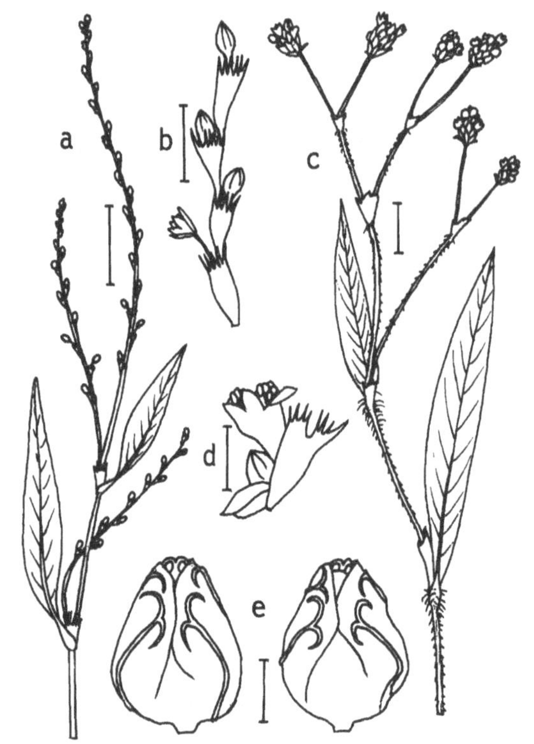

Fig. 335. *Polygonum hydropiper:* a, flowering shoot (2 cm); b, fragment of inflorescence (5 mm). *P. lapathifolium*: e, perianth enveloping the nut (1 mm). *P. pedunculare*: c, flowering shoot (2 cm); d, flower with bract (1 mm).

closely pinnate pairs, base cuneate to decurrent, tips gradually attenuate. Flowers in terminal and axillary, 7–10 cm long spike-like racemes, together forming a panicle; bracts oblong, ± 3 mm long, broadly triangular, margins scarious; bracteoles up to 2 mm long. Perianth rose or red, ± 3.5 mm in diameter; segments 5, subequal, oblong, ± 3 mm long, obtuse. Stamens 5; filaments ± 2 mm long; anthers ± 0.7 mm long. Styles 2, ± 0.8 mm long. Nut ± orbicular in outline, flattened, 2–3 mm long, dark brown and shiny.

Perennial or annual. Gregarious along water courses and tanks but is also found in marshes. It is often dominant along large rivers where it withstands considerable flooding and swift currents. Tender branches used as a vegetable in Rajasthan. It is also used for dispelling fever and colic.

{(⊙) Hel}. Old World tropics: Ban, Cey, Ind, Nep, Pak: Ap, As, Bh, Gj, Go, Kl, Kt, Mh, Mp, Or, Pj, Rj, Sk, Tn, Up, Wb. 0–2900 m.

Polygonum hydropiper Linnaeus [*P. hydropiper* subsp. *microcarpum* Danser, *Persicaria hydropiper* (Linnaeus) Spach]: Fig. 335

Stems erect, prostrate or creeping and rooting, 30–60 (–100) cm tall, glabrous. Ochreas membranous, glabrous or hairy, tubular, truncate, ± 2 cm long, nerves prolonged into cilia at the tips; leaf blades shortly petiolate (petioles up to ± 1 cm long), narrowly lanceolate, 3–6 (–15) cm long, 0.7–1.5 cm wide, glabrous, cuneate to decurrent, tips gradually attenuate. Flowers in terminal and axillary, lax spike-like racemes, 4–6 cm long, often hanging; bracts mostly distant and not overlapping. Perianth rose to white, with yellowish transparent glands; segments 4 (rarely 3 or 5), subequal, oblong, 3–4 mm long, obtuse. Stamens 6 (–8). Styles 2, ± 0.8 mm long. Nut biconvex or weakly 3-sided, 2.5–3.5 mm long, ± 2.8 mm wide.

Annual. On moist, rich soils which are regularly flooded, also found along the margins of ponds and in irrigation ditches. Frequently found as a weed in fields and along paths. It has a hot peppery taste; it is used as a fish poison and in medicine as a diuretic. It is very variable, several infraspecific classifications have been proposed. Tall, large-leaved plants with sharp pointed bristle-like hairs on the ochrea and undersurface of the leaf blades are often called *Polygonum hydropiper* var. *flaccidum* (Meisner) Steward [*P. pubescens* Blume: *P. flaccidum* sensu Meisner non Roxburgh, *P. donii* Wight]; these plants intergrade with *P. hydropiper*.

{⊙ Hel}. N Africa, temperate and tropical Eurasia, introduced in N America: Bhu, Ind: An, As, Kt, Mh, Mn, Mp, Or, Rj, Sk, Tn, Up, Wb. 0–2000 m.

Polygonum lapathifolium Linnaeus [*P. tomentosum* Schrank, *P. nodosum* Persoon, *Persicaria lapathifolia* (Linnaeus) S. F. Gray]: Fig. 335

Like *P. glabrum* but ochreas hairy; undersurface of leaf blades clothed in short white hairs; bracts hairy; perianth segments with raised anchor-shaped veins, pink to white or greenish, clearly longer than the fruit; nuts remaining enclosed in perianth.

Annual. It is not confined to wetlands but is sometimes found in water or in mats of floating vegetation. The aquatic plants usually have swollen internodes (up to 3 cm thick) which become purple. It is autogamous and has many local 'true breeding' races; the Indian ones tend to be hairy and are sometimes recognized as *P. lapathifolium* var. *lanatum* (Roxburgh) Steward. In India it seems to be often overlooked or incorrectly identified and is probably more widespread than some local Floras let us believe.

{⊙ Hel/Hyp/Ple}. Almost cosmopolitan, probably introduced in N America: Ban, Cey, Ind, Pak: Ap, As, Gj, Jk, Kt, Mg, Mh, Tn, Up.

Polygonum pedunculare Wallich ex Meisner [*P. pedunculare* var. *angustissimum* J. D. Hooker, *P. pedunculare* var. *robustum* J. D. Hooker, *Persicaria peduncularis* (Wallich ex Meisner) Nemoto]: Fig. 335

Stems erect, up to ± 50 cm tall, nodes with stout, retrorse, almost prickly hairs. Ochreas membranous, glabrous, tubular, obliquely cleft, up to 1.5 cm long, tips gradually acute; leaf blades not distinctly petiolate, linear-lanceolate, 5–9 cm long, 3–6 mm wide, glabrous except for the midrib, lateral nerves obscure, base attenuate, tips gradually acuminate. Flowers in remote, dense, terminal and axillary, up to 2 cm long, almost head-like racemes; peduncles glandular hairy; bracts ovate, 1–1.5 mm long, margins ciliate towards the tips; bracteoles up to 1.5 mm long. Perianth purplish, ± 3 mm in diameter; segments 5, subequal, oblong-concave, ± 2 mm long, obtuse. Stamens 5; filaments ±1.5 mm long; anthers ± 0.5 mm long, Styles 2, ± 0.5 mm long. Nut biconvex, almost circular, ± 2 mm in diameter.

Perennial. Prickly subshrubs not confined to wetlands but often found along rivers, in marshes and in seasonally inundated places. It is very variable and several varieties have been described. Should *Polygonum dichotomum* Blume [*Persicaria dichotoma* (Blume) Masamune] be shown to be conspecific then it is the earlier name.

{Hel}. Tropical Asia and Australia: Ban, Cey, Ind: As, Kt, Mg, Mp, Tn.

Polygonum plebeium R. Brown ['*plebejum*' orth. mut., *P. indicum* Heyne]: Fig. 336

Stems prostrate. Ochreas hyaline, obliquely tubular, ±3 mm long, the upper part lacerate; leaf blades sessile or nearly so, oblong to linear, 6–10 (–20) mm long, 1.5–3 mm wide, relatively thick, midnerve impressed above, prominent below, lateral nerves obscure, base attenuate to shortly decurrent, tips obtuse or acute. Flowers in axillary clusters; pedicel up to 2 mm long, jointed; bracts hyaline, 1.5 mm long; bracteoles up to 1 mm long. Perianth rose, ± 2 mm in diameter; segments 5, unequal; the outer 3 vertically folded and usually spotted; the inner 2 oblong, smooth, up to 2 mm long. Stamens 5 (6), filaments ± 0.8 mm long; anthers ± 0.2 mm long. Styles ± 0.3 mm long; stigmas 3, simple. Nut distinctly 3-sided, ± 2 mm long, ± 1 mm wide, with persistent style.

Annual or perhaps sometimes perennial, forming dense prostrate mats. Gregarious by rivers, canals and drying out pools; sometimes growing in rice fields but rarely considered to be troublesome. The seeds need to be immersed in water to leach out the germination suppressors; they often germinate under water. Used in medicine for pneumonia and bowel complaints. It is rather variable in habit and several varieties have been described.

{⊙ or (⊙)? Hel}. Old World tropics: Ban, Bhu, Bur, Ind, Nep, Pak: Ap, As, Bh, Dl, Gj, Go, Kl, Kt, Mh, Mp, Or, Pj, Rj, Sk, Tn, Up, Wb. 0–2500 m.

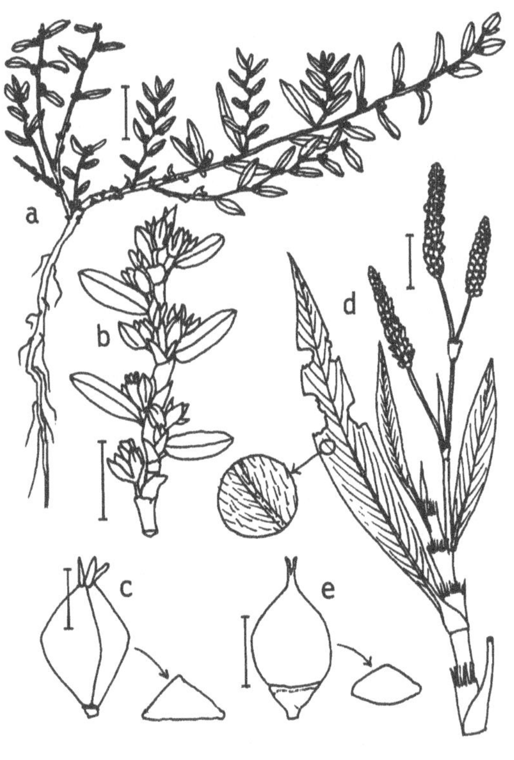

Fig. 336. *Polygonum plebeium:* a, flowering plant (1 cm); b, fragment of inflorescence (5 mm); c, nut with transverse section (inset) (1 mm). *P. pulchrum*: d, flowering shoot (2 cm); e, nut with transverse section (inset) (4 mm).

Polygonum pulchrum Blume [*P. tomentosum* Willdenow, *Persicaria pulchra* (Blume) Sasaki]: Fig. 336

Stems erect, shrubby, up to ± 1 m tall. Ochreas membranous, tubular, truncate, 2–3 cm long, softly ciliate; leaf blades petiolate (petiole up to 2 cm long); blades lanceolate to linear-obovate, 10–15 cm long, 2–4 cm wide, silky-hairy, lateral nerves 15–20 in closely pinnate pairs, base cuneate to decurrent, tips gradually acuminate. Flowers in terminal and axillary, up to ± 10 cm long spike-like racemes usually forming a panicle; bracts oblong, ± 4 mm long, ± 2 mm wide, sericeous, margins ciliate, up to 1.5 mm long; bracteoles up to 1 mm long. Perianth pinkish, ± 4 mm in diameter; segments 5, equal, ovate, ± 3.5 mm long, shortly united at the base, subacute. Stamens 5; filaments ± 2 mm long; anthers ± 0.5 mm long. Styles 2, ± 0.7 mm long. Nut biconvex, almost circular, ± 3 mm in diameter.

{Hel}. Tropical Old World: Ind: An, As, Bh, Kl, Kt, Mh, Or, Tn, Wb.

PONTEDERIACEAE: 9 genera

Stems rhizomatous, stoloniferous or erect, submerged, floating or creeping. Leaves sheathed and usually stipulate at the base, distichous or spirodistichous, linear or petiolate; blades linear to orbicular, cuneate to cordate at base. Flowers bisexual, often tristylic, zygomorphic to almost actinomorphic, solitary, paired or in spikes, umbel-like racemes or panicles, the inflorescence subtended by 2 spathes, one of which is sometimes leaf-like. Sepals and petals combined to a perianth. Perianth petaloid, tubular or lobed almost to the base, white, blue, yellow or blue with yellow flecks; lobes 6 or rarely 4 or 3. Stamens 6, 3 or 1, similar or dissimilar, inserted on the perianth. Ovary superior, 3-locular, occasionally only 1 locule fertile; style 1, elongate. Fruit a capsule or 1-seeded nut enclosed in a hardened perianth base (anthocarp). Seeds numerous or solitary.

Eckenwalder, J. E. and Barrett, S. C. H. Phylogenetic systematics of Pontederiaceae. Systematic Botany 11: 373–391 (1986); Rosatti, T. J. The genera of Pontederiaceae in the southeastern United States (literature survey). Journal of the Arnold Arboretum 68: 35–71 (1987).

The American species ***Heteranthera reniformis*** Ruiz et Pavon and ***H. rotundifolia*** (Kunth) Grisebach are spreading in rice fields in the Old World; neither has been reported in India. The genus *Heteranthera* is rather like *Monochoria* in having 6 almost equal perianth segments but it has 3 stamens, with 1 anther larger than the other 2.

1. Perianth lobed almost to the base, nearly actinomorphic; anthers unequal (1 large and blue, 5 small and yellow), basifixed ***Monochoria***

1. Perianth distinctly tubular at the base, distinctly zygomorphic; anthers subequal in length, dorsifixed

2. Mature plants floating with inflated petioles; fruit a 3-locular, many-seeded capsule; not enclosed by the base of a modified perianth tube ***Eichhornia***

2. Mature plants erect and bottom-rooted with narrow petioles; fruit a 1-seeded nut, enclosed by the hardened base of the perianth tube (anthocarp) ***Pontederia***

Eichhornia: 7 species: tropical America and Africa, through introductions now in most tropical and warm regions of the world

Eichhornia crassipes (Martius) Solms-Laubach: Fig. 337

Stems floating, stoloniferous or creeping. Leaves of juveniles submerged, cauline, linear; leaves of adults emergent, differentiated into petiole and blade; petiole up to ± 30 cm long, usually with a bulbous swelling below; blade broadly ovate to orbicular, 5–25 cm in diameter, at base subcordate, truncate or

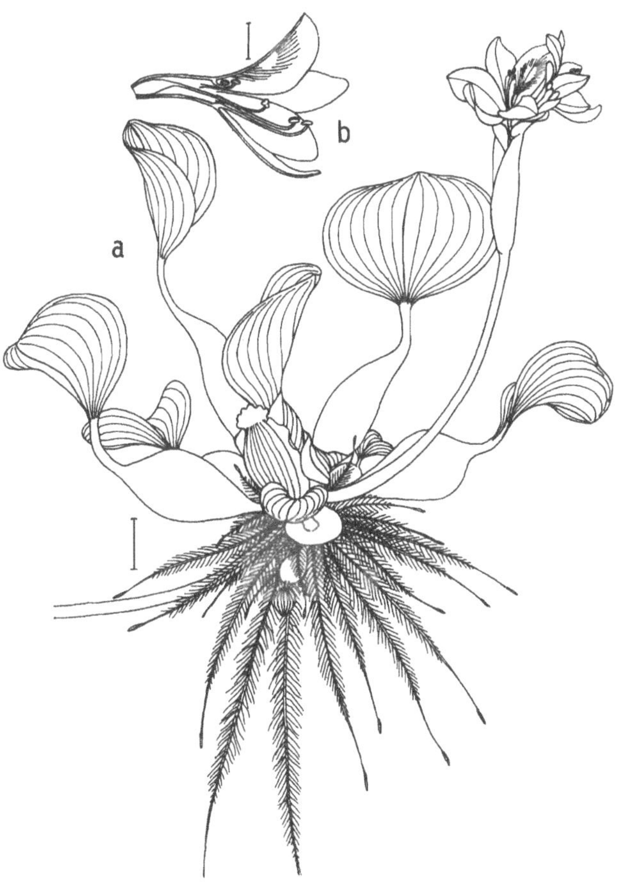

FIG. 337. *Eichhornia crassipes*: a, flowering plant (3 cm); b, longitudinal section of a mid-styled flower (1 cm).

rounded. Inflorescence a spike, up to 50 cm long; spathes dissimilar, the lower leaf-like, the upper scale-like. Perianth blue, tubular below, 6-lobed, the lobes unequal, the upper one larger, blue with a yellow fleck on the adaxial lobe, up to 4 cm long. Stamens 6, borne at 2 levels; filaments with glandular hairs; anthers subequal, dorsifixed. Fruit a 3-valved capsule. Seeds numerous, oblong, ± 1.5 mm long, with 8–10 longitudinal ribs.

Perennial or occasionally annual. The juveniles are submerged, the adults are free-floating. It is found in a large variety of aquatic habitats and is often dominant over large areas. Known as water hyacinth, it has become a very serious pest in many regions. It is utilized locally as fodder, manure, a source of methane and alcohol, for purifying water and also cultivated for decoration. It is effectively pollinated by insects. In India the long and middle style morphs are found. The seeds are usually dispersed in mud. Gopal, B. Water hyacinth. Aquatic Plant Studies 1: Elsevier, Amsterdam, i–xii, 1–471 (1987).

{(⊙) Ple/(Hyd)/(Ten)}. Introduced to the Indian subcontinent in about 1890 from tropical S America probably via Malaysia or Indonesia: <Ban, <Bhu 1992, <Bur ± 1904, <Cey 1904, <Ind, <Nep, <Pak: <Ap, <Ar, <As, <Bh, <Dl, <Gj, <Hp, <Hr, <Kl, <Kt 1928, Lk, <Mg, <Mh, <Mn, <Mp, <Mz, Ng, <Or, <Pj 1917, <Rj, <Sk, <Tn, <Tp, <Up, <Wb 1896.

Monochoria: 8 species: tropical to warm Africa, Asia and Australia, naturalized in Europe and the New World

Stems floating, creeping, corm-like or rhizomatous. Leaves radical, spirodistichous, submerged, floating or emergent, extremely variable as to size and shape, scale-like to linear or differentiated into petiole and blade; blade linear to ovate, sagittate or hastate; stipules sheathing and closed, usually with a ligule. Inflorescences paniculate or umbel-like, sometimes reduced to 2 flowers; spathes either similar and leaf-like or dissimilar, the lower leaf-like with a sheath enclosing the upper spathe; the upper sheathing and lacking a blade but sometimes hidden from view so the inflorescence appears to arise directly from the petiole of a leaf (the lower spathe). Perianth very shortly tubular, 6-lobed; lobes almost reaching the base, subequal, blue, pink or white, somewhat accrescent, persistent and enveloping the ripe fruit. Stamens 6, one larger than the others; filaments often with 1 or 2 obliquely erect teeth; anthers basifixed, unequal, 5 small and yellow, 1 large and blue. Fruit a 3-valved capsule, normally released as a unit and opening later; seeds numerous, with 8–14 longitudinal ribs or wings.

Annual or perennial. Usually found in shallow water but most are capable of accommodating to great changes in water level. The juveniles are submerged, the adults are floating, emergent or terrestrial. Pollinated by insects or autogamous, some with pre-anthesis cleistogamy. As the flowers open the blue anther takes a lateral position, the style lies in the other direction; within an inflorescence there are 'left' and 'right' handed flowers. The disseminules are capsules enclosed in the persistent perianth which opens later to release the seeds, dispersed in mud and by water. Leaves and inflorescences are eaten and also used locally in medicine. It is a minor weed in irrigation channels and rice fields. Cook, C. D. K. A revision of the genus *Monochoria.* in Tan Kit, Mill, R. R. and Elias, T. S. (eds.), Plant taxonomy, phytogeography and related subjects. The Davis and Hedge Festschrift. Edinburgh University Press 149–184 (1989).

1. Inflorescences elongate, overtopping the leaves; blades of the lower spathes elongate-sagittate or elongate-hastate **_M. elata_**

1. Inflorescences contracted or subumbellate, borne below the leaves; blades of the lower spathes linear to hastate

  2. Robust rhizomatous perennial; inflorescences usually with more than 20 flowers; blue anthers 6–10 mm long; styles more than 8 mm long **_M. hastata_**

  2. Annual or weakly rhizomatous perennial; inflorescences usually with less than 20 flowers; blue anthers (1.8–) 1.5–4 mm long; styles (2–) 4–5 mm long **_M. vaginalis_**

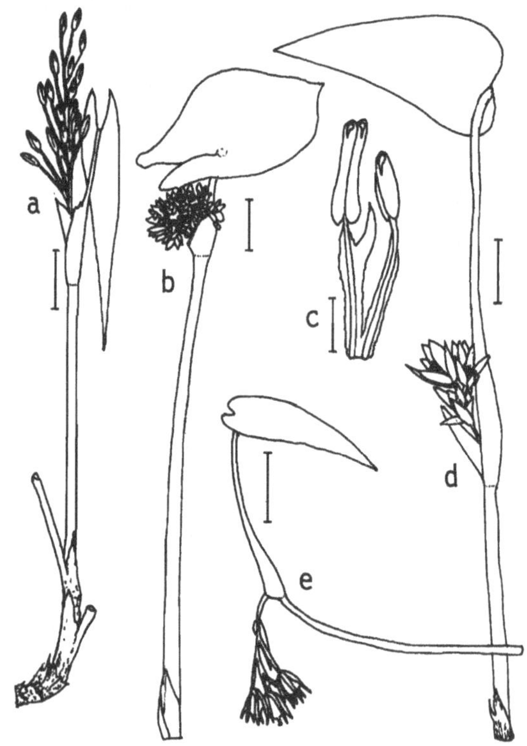

FIG. 338. *Monochoria elata*: a, flowering shoot (6 cm). *M. hastata*: b, flowering shoot (3 cm). *M. vaginalis*: c, one large and one small stamen (2 mm); d, flowering shoot (2 cm); e, fruiting shoot (2 cm).

Monochoria elata N. H. Ridley: Fig. 338

Blades of the lower spathes elongate-sagittate or elongate-hastate. Inflorescences elongate, erect, racemose, overtopping the leaves, with 20 or more flowers. Pedicels up to 3 cm or more long, simple. Blue anthers (5.3–) 5.5–6.5 mm long. Styles 5–6 mm long.

Large perennial with shoots up to 2 m or more tall. Ecology unknown but clearly a plant of wet places. Sometimes regarded as a variety of *M. hastata.*

{Hyp/Ten}. Tropical SE Asia. Bur.

Monochoria hastata (Linnaeus) Solms-Laubach: Fig. 338

Blades of the lower spathes widely sagittate or hastate. Inflorescences subumbellate, borne below the leaves, usually with more than 20 flowers. Pedicels mostly 1–2 cm long, the lower ones sometimes branched. Blue anthers 6–10 mm long. Styles more than 8 mm long. Seeds 0.8–1 mm long.

Perennial with a large and thick rhizome. Grows in clumps at the edges of pools, tanks and canals and in ditches. It attains a height of up to 2–3 m during the rains, adjusting its height with the rise in water level. During the winter the above-ground parts die down. The leaf stalks are used as a vegetable in Bengal; the rootstock is fed to livestock. In medicine it is used as a cooling agent.

{Hyp/Ten}. Tropical S and SE Asia: Ban, Bur, Cey, Ind, Nep: Ap, As, Bh, Kt? Mg? Mh, Mn, Mp, Mz? Ng? Or, Pj, Rj, Up, Wb.

Monochoria vaginalis (N. L. Burman) Kunth [*Pontederia plantaginea* Roxburgh]: Fig. 338

Blades of the lower spathes linear or lanceolate to ovate, the bases usually cuneate but sometimes truncate or even cordate. Inflorescences usually contracted-racemose but sometimes reduced to 1 or 2 flowers, strongly deflexed in fruit, borne below the leaves, with (1–) 2–7 (–20) flowers. Pedicels mostly simple, sometimes bearing 2 flowers, up to 3 cm long but usually very much less. Blue anthers (1.8–) 1.5–4 mm long. Styles (2–) 4–5 mm long. Seeds (0.5–) 0.7–0.8 (–1) mm long.

Annual or perhaps sometimes perennial. Often highly gregarious in inundated places, in swamps or at the edges of pools, ditches and canals. It is frequently found in rice fields but is rarely considered to be a serious pest. It is highly plastic and has stems from 3 cm to over 50 cm tall. The leaves and flowers are used as pot herbs and medicine; the roots are used for toothache.

{(⊙)? Hyp/Ten}. S, SE and E Asia, Australia and W Pacific, introduced in Hawaii, N America and Europe: Ban, Bhu, Bur, Cey, Ind, Nep, Pak: <An, Ap, As, Bh, Dl, Go, Hr, Jk, Kl, Kt, Mn, Mp, Or, Pj, Rj, Tn, Up, Wb. 0–1800 m.

Pontederia: 4 species: tropical to warm America, naturalized in the Old World

Pontederia cordata Linnaeus: Fig. 339

Stems stoloniferous or rhizomatous, floating, creeping or ascending, 30–80 (–100) cm tall. Leaves submerged, floating or emergent, linear or differentiated into petiole and blade; blade elliptical to ovate or sagittate with rounded lobes, up to 25 cm long, up to 6 cm wide (2–3 times as long as wide). Inflorescence a spike; spathes dissimilar, the lower leaf-like, the upper scale-like; spikes up to 15 cm long. Perianth hairy, ± 12 cm long, lower half tubular, 2-lipped above, 6-lobed, the 3 abaxial lobes partly united, the 2 adaxial ones subequal and free, blue to purple or whitish, often with a yellow fleck on the adaxial lobe. Stamens 6 at 2 levels; filaments with glandular hairs; anthers subequal, dorsifixed. Fruit a 1-seeded anthocarp (nut surrounded by a hardened perianth tube), with wing-like ridges.

Fig. 339. *Pontederia cordata*: a, sterile plant (1 cm); b, inflorescence (1 cm); c, fruit (2 mm); d, longitudinal section of a mid-styled flower (5 mm).

Perennial. Mostly found in shallow water, often gregarious, sometimes in brackish conditions. The juveniles are submerged; the adults are emergent. The flowers tristylic; pollinated by insects. The disseminules are winged nuts; dispersed by water and probably also by adhesion to animals. Cultivated for ornament. Lowden, R. M. A revision of the genus *Pontederia.* Rhodora 75: 426–487 (1973); Price, S. D. and Barrett, S. C. H. The function and adaptive significance of tristyly in *Pontederia cordata.* Biological Journal of the Linnean Society 21: 315–329 (1984).

{Ten/Hyp}. Atlantic N America to northern S America, naturalized in Europe, Australia and India: <Ind: <Mh 1986.

POTAMOGETONACEAE [excl. RUPPIACEAE]: 2 genera

Potamogeton: 80–90 species and ± 40 hybrids: cosmopolitan

Stems usually elongate, flexible, submerged or floating or stoloniferous and creeping, often developing specialized tuber-like buds (turions). Leaves distichous, alternate or occasionally some opposite on part of the shoot, stipulate, often heteromorphic; petioles present or absent; squamulae intravaginales present; stipules free or united with the leaf base, sometimes sheathing the stem or united at the margins to form a ligule; submerged leaves thin, translucent, capillary or linear or with broadened, sessile or petiolate blades; floating leaves often lacking, the blades leathery, opaque, narrowly to broadly elliptical-oblong. Flowers 4-merous, in terminal spikes. Perianth of 4, bract-like, clawed scales inserted opposite each stamen (the perianth is often considered to be an outgrowth of the stamen). Stamens 4, united with the perianth at the base; anthers sessile. Ovary superior, of 4, free or partly united carpels; style 1, short. Fruit a drupelet with fleshy exocarp and bony endocarp.

Perennial or sometimes annual. Found in a wide variety of aquatic habitats, totally submerged or with floating leaves, some temporarily emergent. Occasionally it grows in brackish water. Pollination by wind, epihydrophilous or some hypohydrophilous (submerged autogamy). The disseminules are drupelets with fleshy exocarp and woody endocarp, dispersed by waterfowl and also by water. Some species develop turions in conditions poor for growth. Most species are very important for wildlife both as food and as refuge. Some species are reported to be a nuisance in ditches and canals; some are cultivated for decoration. *Potamogeton* species are frequently wrongly named in some Floras; this account must be considered to be provisional as I have not seen sufficient herbarium material from India.

Excluded: the following aquatic species are found in the Himalayas but do not extend southwards into India: ***Potamogeton alpinus*** Balbis; ***P. berchtoldii*** Fieber; ***P. distinctus*** A. Bennett; ***P. filiformis*** Persoon; ***P. gramineus*** Linnaeus; ***P. lucens*** Linnaeus; ***P. natans*** Linnaeus; ***P. pusillus*** Linnaeus.

1. Leaves filiform; the leaf sheaths when young wrapped around the stem; peduncles long and flexible, usually floating; flowering spikes interrupted, with up to 5 pairs or whorls of flowers ***P. pectinatus***
1. Leaves flat; leaf sheaths absent or if present then not wrapped around the stem; peduncles rigid, usually held above the water surface; the flowering spikes continuous, densely covered in flowers
  2. Blades of submerged leaves clasping the stem (amplexicaul) at the base, sessile; floating leaves lacking
    3. Leaf margins clearly serrate (teeth visible with the naked eye) and wavy; leaves parallel-sided for most of their length; beak of the fruit ± 2 mm long; flowering spikes not more than 2 cm long ***P. crispus***
    3. Leaf margins entire or minutely serrate (teeth not visible with the naked eye) and flat; leaves lanceolate to ovate, not parallel-sided for most of their length; beak of the fruit not more than 1 mm long; flowering spikes usually more than 2 cm long ***P. perfoliatus***
  2. Blades of submerged leaves gradually narrowed at the base, never clasping the stem, petiolate or sessile; floating leaves present or absent
    4. Submerged leaves terminating in a sharp-pointed mucro or elongate appendage; floating leaves lacking ***P. malaianus***
    4. Submerged leaves rounded or acute at the apex, not terminating in a mucro or appendage; floating leaves usually present at time of flowering
      5. Blades of submerged leaves lanceolate (not parallel-sided); blades of floating leaves 5–10 (–20) cm long, with the petiole longer than the blade ***P. nodosus***
      5. Blades of submerged leaves linear (parallel-sided for most of their length); blades of floating leaves 1–4 cm long, with the petiole shorter than the blade ***P. octandrus***

Potamogeton crispus Linnaeus: Fig. 340

Stolons not more than 2 mm thick, much branched, reddish, lying flat just below the surface of the substrate, often developing turions (tubers) at the tips. Stems 10–70 (–200) cm long, somewhat flattened and angular, usually branched, reddish. Stipules very thin, deciduous, 4–15 mm long; submerged leaves translucent, linear with almost parallel sides (5–10 times as long as wide), (1–) 4–6 (–10) cm long, (3–) 5–10 (–15) mm wide, the margins serrate (teeth just visible to the naked eye) and usually wavy, rounded or sometimes acute at the tips, rounded at the base, 3- or 5-veined, midrib prominent; floating leaves absent. Peduncles rigid, erect, emergent, up to 10 cm long. Spikes with 2–10 flowers, dense, rarely more than 2 cm long. Drupelets globose, usually united below, 5–6 mm long including the ± 2 mm long curved beak.

Submerged perennial. It is found in still and flowing water and shows a preference for eutrophic conditions. Often gregarious, large populations may constitute a serious hazard. It is sometimes found in temporary pools where it survives the dry period as dormant tubers.

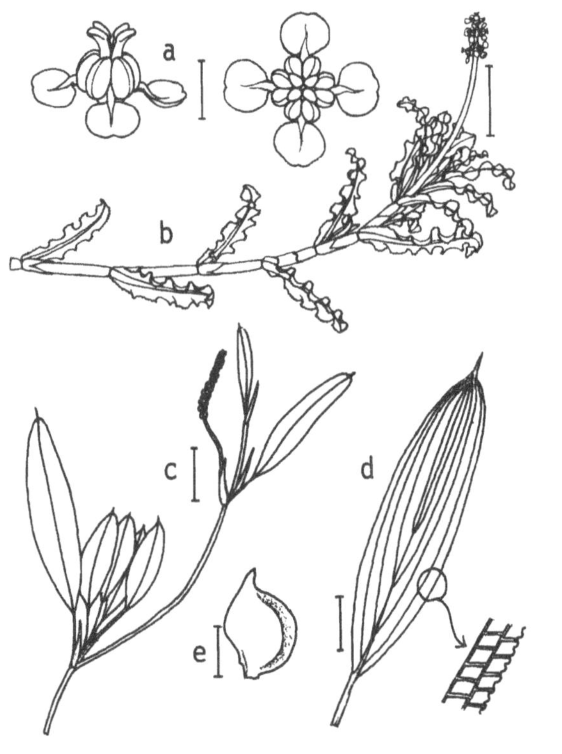

FIG. 340. *Potamogeton crispus*: a, flower viewed from the side and from above (1 mm); b, flowering shoot (2 cm). *P. malaianus*: c, flowering shoot (2 cm); d, leaf (1 cm) with enlargement of margin and cross nerves (inset); e, fruit (1.5 mm).

{Vit}. Temperate to tropical Europe, Africa, Asia, Australia and N America: Ban, Bhu, Ind, Nep, Pak: As, Bh, Dl, Go, Jk, Mh, Mn, Mp, Or, Rj, Tn, Up, Wb.

Potamogeton malaianus Miquel [*P. mucronatus* sensu Presl non Schrader ex Sonder]: Fig. 340

Stolons creeping, sparsely branched. Stems up to 2 m or more long, internodes 3-6 cm long, terete, sparsely branched. Submerged leaves present at time of flowering, distinctly petiolate; petioles 0.5–3 cm long; stipules very thin, amplexicaul, 2.5–3.5 cm long, usually persisting; leaf blades translucent, green or brownish, oblong-elliptical, (2–) 6–12 cm long, 1.5–2.5 cm wide, with 7–13 nerves, the innermost nerve pair forking from the midrib at about the middle, the margins minutely serrate (teeth not visible to the naked eye) and flat or minutely undulate, narrowed or cuneate at the base, at the tip terminating in a mucro or elongate appendage up to ± 1 cm long; floating leaves absent. Peduncles rigid, erect, emergent, up to 10 cm long. Spikes many-flowered, dense, 2–5 cm long, elongating in fruit. Drupelets almost globose, 3-keeled, the central keel larger, ± 3 mm long, the beak ± 1 mm long.

Perennial. In lakes and ponds in water 2 m or more deep.

{Vit}. Warm to tropical Asia, perhaps also Caribbean: Ban, Ind: Bh, Mp.

Potamogeton nodosus Poiret [*P. fluitans* Roth, *P. indicus* Roxburgh non Roth]: Fig. 341

Stolons rhizomatous, robust, creeping, yellowish-white. Stems up to 2 m long, terete, nodes often swollen, simple or branched. Stipules thin up to 6 cm long; submerged leaves usually present at time of flowering, translucent, lanceolate, 6–20 (–30) cm long, 2–3 cm wide (the lower ones usually smaller and wider), the margins entire and flat, acute at the tips, cuneate at the base and running into a distinct petiole, many-veined; floating leaves usually present, rather leathery, dark green to reddish-brown, ovate to elongate lanceolate, 5–10 (–20) cm long, 2–3.5 cm wide, the margins entire and flat, acute at the tips, rounded at the base and running into a distinct petiole 5–12 (–25) cm long, many-veined; some leaves intermediate between the submerged and floating ones usually develop. Peduncles rigid, erect, emergent, up to 12 cm long. Spikes many-flowered, dense, 3–4 (–5) cm long. Drupelets sharply keeled, 2–3.5 (–4) mm long; the beak less than 1 mm long.

Perennial. Usually found in flowing water and showing a preference for eutrophic conditions but also growing in lakes where it may build up large populations.

{Eph/(Vit)}. Warm and tropical regions of the world: Ban, Bur, Cey, Ind, Nep, Pak: An, Ap, Bh, Go, Hp, Kl, Kt, Mh, Mn, Mp, Or, Rj, Sk, Tn, Wb. ?–3000 m.

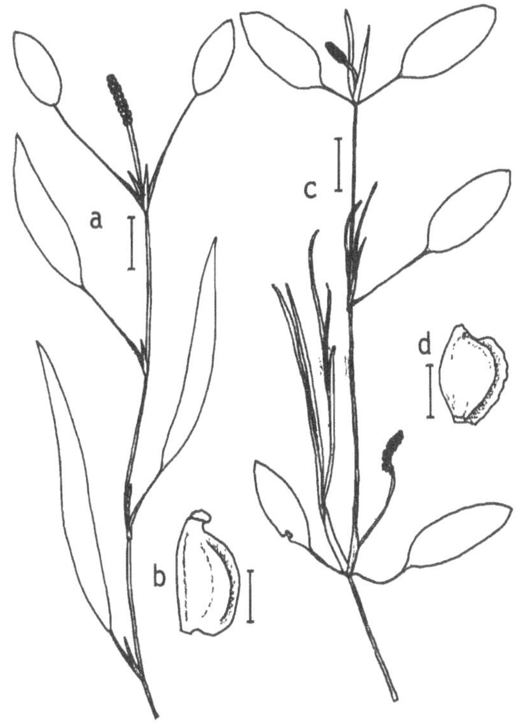

Fig. 341. *Potamogeton nodosus*: a, flowering shoot (2 mm); b, fruit (1.3 mm). *P. octandrus*: c, flowering shoot (1 cm); d, fruit (1 mm).

Potamogeton octandrus Poiret [*P. javanicus* Hasskarl]: Fig. 341

Like *P. nodosus* but generally smaller and more delicate; submerged and floating leaves clearly distinct from each other without any intermediate ones, both kinds sometimes arising side by side in the same cluster; submerged leaves linear, 3–6 cm long, 3–10 mm wide; floating leaf blades lanceolate, (1–) 1.5–3.5 (–4) cm long, 0.3–8 mm wide; the petiole shorter than the blade; peduncle 1–2 cm long; inflorescence 1–1.5 cm long; drupelets ± 2 mm long, 3-keeled, the central keel obscurely toothed.

Perennial. Submerged aquatic with floating leaves found in fresh and brackish water. It may be covered by the tide.

{Eph/(Vit)}. Tropical Africa, Asia and Australia: Ban, Ind, Nep, Pak?: Hp, Jk, Kl, Mg, Mp, Sk, Wb. 0–1000 m.

Potamogeton pectinatus Linnaeus: Fig. 342

Stolons not more than 1.5 mm thick, much branched, whitish, sometimes developing swollen turions. Stems (2–) 10–100 (–400) cm long, almost terete, ± 1 mm thick, usually repeatedly branched. Stipules arising from the leaf sheaths, transparent, up to 1 cm long; leaf sheaths wrapped around the stem, (–0.5) 2–5 (–6) long, 0.3–3 (–4) mm wide; submerged leaves filiform, (2–) 5–15 (–30) cm long, usually ± circular in transverse section but sometimes flattened, tapering to fine point at the tips, 3-veined but only the midrib prominent, thus appearing 1-veined; floating leaves absent. Peduncles flaccid, flexible, floating, up to 25 cm long. Spikes interrupted with up to 5 pairs or whorls of flowers. Drupelets asymmetrically ovoid, 2.5–4 (–5) mm long, 2–3 mm in diameter, the beak less than 1 mm long.

Perennial, persisting through unfavourable periods as turions or drupelets. It is found in still and flowing water in a very wide spectrum of aquatic habitats. It is mostly in nutrient-rich water. It is an important food plant for wildlife but is also considered to be an undesirable plant in small irrigation canals and ditches.

{Vit}. Almost cosmopolitan: Ban, Cey, Ind, Nep, Pak: Bh, Dl, Gj, Hp, Kt, Mh, Mp, Or, Rj, Tn.

Potamogeton perfoliatus Linnaeus: Fig. 342

Stolons rhizomatous, robust, creeping, sparsely branched. Stems up to ± 6 m long but usually less, terete, usually branched above. Stipules very thin,

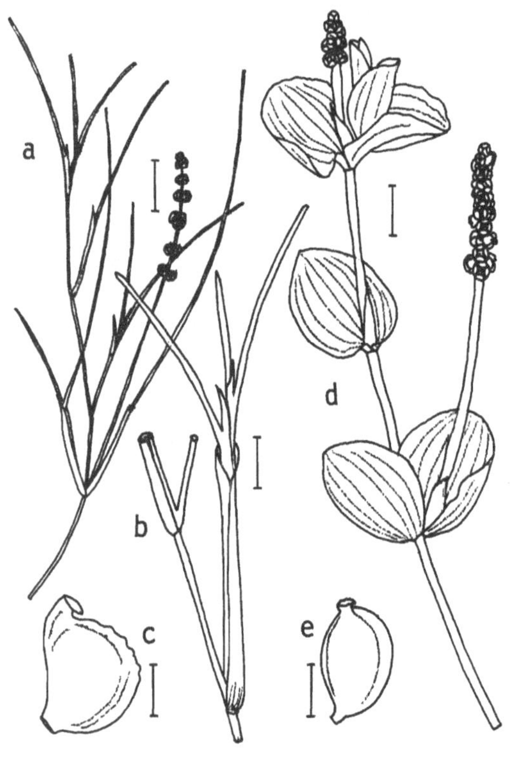

FIG. 342. *Potamogeton pectinatus*: a, flowering shoot (2 mm); b, detail of leaf sheaths (3 mm); c, fruit (1.7 mm). *P. perfoliatus*: d, flowering shoot (1 cm); e, fruit (1.7 mm).

membranous, deciduous, 1–2 cm long; submerged leaves present at time of flowering, sessile, translucent, pale green, elongate-ovate to lanceolate, 6–12 cm long, 3.5–6 cm wide, rounded to cordate at the base, the lobes clasping the stem (amplexicaul), 5–7 (–10) veins, midrib prominent, the margins minutely serrate (teeth not visible to the naked eye) and flat, the tips acute to rounded; floating leaves absent. Peduncles rigid, erect, emergent, 2–5 (–6) cm long. Spikes many-flowered, dense, up to 3 cm long, elongating in fruit. Drupelets obovoid, bluntly keeled, up to 3 mm long, the beak 1 mm long.

Perennial. Usually in 0.5–6 m deep, permanent, rather nutrient-rich, still or slowly flowing water. In lakes it often forms large stands in the deep sublittoral zone. An important plant for wildlife but sometimes regarded as a weed in irrigation canals.

{Vit}. Almost cosmopolitan: Ind, Pak: Hp, Kt, Mh, Mp, Rj, Tn, Up.

RANUNCULACEAE: 48 genera

Excluded: ***Caltha*** is found in the Himalayas but does not extend southwards to Peninsular India; the following species have been recorded: ***C. alba*** Jacquin ex Cambessèdes; ***C. govaniana*** Wallich ex Royle (perhaps no more than a synonym of *C. palustris*); ***C. palustris*** Linnaeus; ***C. palustris*** var. ***himalensis*** (D. Don) Mukerjee; ***C. scaposa*** J. D. Hooker et Thomson.

Ranunculus [*Batrachium*]: 400 species: cosmopolitan but most in temperate climates

Excluded: the following aquatic and wetland species are found in the Himalayas but do not extend south into India: ***Ranunculus emuricatus*** Majeed: ♦Ind; ***R. flavidus*** (Handel-Mazzetti) C. D. K. Cook; ***R. hyperboreus*** Rottbøll; ***R. lingua*** Linnaeus; ***R. natans*** C. A. A. Meyer; ***R. rionii*** Lagger [*Batrachium trichophyllum* subsp. *rionii* (Lagger) C. D. K. Cook]; ***R. sphaerospermus*** Boissier et Blanche [*Batrachium rhipiphyllum* subsp. *sphaerospermum* (Boissier et Blanche) C. D. K. Cook]; ***R. trichophyllus*** Chaix [*R. aquatilis* auct. Ind., non Linnaeus, *R. trichophyllus* subsp. *glaber* (Tamura) H. Hara, *Batrachium paucistamineum* (Tausch) F. Schultz, *B. drouetii* (F. Schultz) van den Bosch].

Ranunculus sceleratus Linnaeus: Fig. 343

Stems erect, (10–) 15–45 (–100) cm tall, furrowed, hollow, usually branched. Basal leaves in a rosette, floating or emergent; petioles 3–25 cm long, with wide sheaths; blades kidney-shaped in outline, simple, 1–6 cm long, 1–10 cm wide, 3-lobed, the lateral lobes often 2- or 3-lobed, at base cordate, margin toothed, the tips acute or obtuse; stem leaves alternate, at base of stem like rosette leaves, becoming petioleless with

3 lanceolate lobes above. Inflorescence cymose but flowers appearing solitary and opposed to leaf-like bracts. Flowers bisexual, radially symmetrical; pedicels elongating in fruit. Sepals 5, yellowish-green, ovate, (1–) 4–6 mm long, underside hairy, at anthesis reflexed, deciduous in fruit. Petals (morphologically nectar-leaves, often called honey-leaves) 5, broadly elliptical to narrowly ovate, 2–4 (–5) mm long, usually shorter than the sepals, pale yellow. Stamens numerous. Receptacle at anthesis 1–2 mm long, becoming 2.5–9 mm long in fruit. Carpels numerous, superior, free. Fruit a head of (40–) 70–100 (–300) 1-seeded nutlets. Nutlets oval, 0.8–1 mm in diameter, the sides smooth or laterally wrinkled, glabrous; style persistent and somewhat beak-like, 0.1–0.2 mm long.

Annual or sometimes biennial. In seasonally inundated places, marshes and along the edges of ponds, streams and rivers. It tolerates brackish water. Pollinated by insects or autogamous. The disseminules are nutlets, dispersed by water, in mud or perhaps also by animals. It contains protanemonin which may poison livestock.

{⊙ or biennial Ten/Eph}. Almost cosmopolitan, probably native in Eurasia: Ban, Ind, Nep, Pak: As, Dl, Jk, Mp, Pj, Rj, Up, Wb.

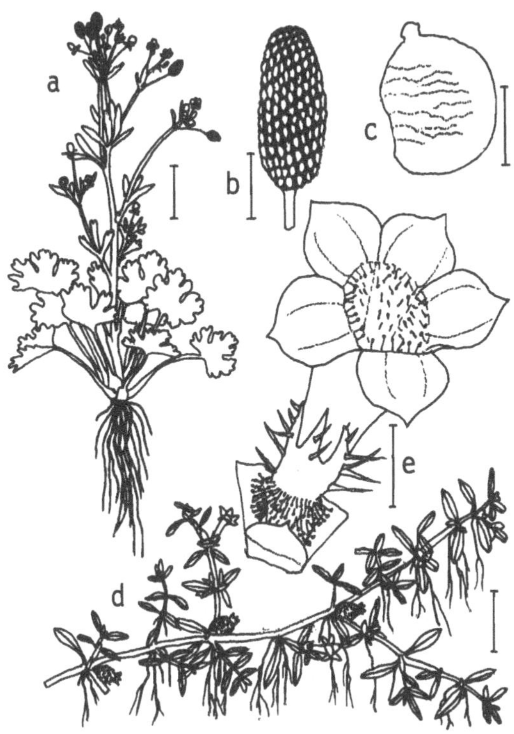

FIG. 343. *Ranunculus sceleratus*: a, flowering plant (2 cm); b, fruiting head (2 mm); c, nutlet (0.5 mm). *Dentella repens*: d, flowering shoot (1 cm); e, flower (2 mm).

RUBIACEAE: 637 genera

Trees, shrubs, lianas or herbs. Leaves opposite or whorled; stipules inter- or intrapetiolar; blades simple, usually entire. Inflorescences usually cymose, paniculate or in heads or flowers solitary. Flowers bisexual, radially symmetrical. Sepals 3- to 5-lobed. Petals united below, 3- to 5-lobed. Stamens as many as and alternating with the petal lobes. Ovary inferior, fused to and imbedded within the basal part of the sepals; carpels (1) 2 (–∞); ovules 1–∞. Fruits berries, drupes, dehiscent or indehiscent capsules or of 2 mericarps.

There is considerable generic confusion in this family. On a worldwide basis the genera *Hedyotis* and *Oldenlandia* run into each other. In India, however, they seem to be well separated. The wetland species of India, small herbs with thin-walled capsules, are placed in the genus *Oldenlandia.*

Excluded: wetland species not extending south of the Himalayas: ***Galium palustre*** Linnaeus.

The following grow in wet or damp places but do not really qualify as wetland plants; others are weedy annuals that may grow in places where water stood during the monsoon but has dried out (in the dry season they replace the wetland species in this kind of habitat): ***Kohautia aspera*** (Heyne ex Roth) Bremekamp [*Oldenlandia aspera* (Heyne ex Roth) A. P. de Candolle]; ***K. gracilis*** (Wallich) A. P. de Candolle [*Hedyotis gracilis* Wallich, *H. stricta* Wallich non Smith]; ***K. retrorsa*** Boise; ***Hedyotis hirsuta*** Linnaeus fil. [*H. japonica* Miquel, *H. lindleyana* J. D. Hooker ex Wight et Arnott, *H. stipulata* R. Brown ex J. D. Hooker]; ***H. pinifolia*** Wallich; ***Neanotis indica*** (A. P. de Candolle) W. H. Lewis [?*Hedyotis affinis* Wight et Arnott, *H. leschenaultiana* Wallich ex Wight et Arnott]; ***N. nummularia*** (Arnott) J. D. Hooker: ♦?Cey; ***N. oxyphylla*** (G. Don) W. H. Lewis: ♦?Ind; ***Spermacoce hispida*** Linnaeus [*Borreria articularis* (Linnaeus fil.) F. N. Williams]; ***S. pusilla*** (Wallich) A. P. de Candolle [*Bigelowia lasiocarpa* Wight et Arnott, *B. roxburghiana* Wight et Arnott; *B. kleinii* Wight et Arnott, *Borreria stricta* auct. Ind., non Linnaeus fil.].

1. Stipules entire; flowering nodes alternating with flowerless ones ***Dentella***
1. Stipules lobed or fimbriate; flowering nodes not alternating with flowerless ones, most nodes bearing flowers ***Oldenlandia***

Dentella ± 10 species: Indomalaysia and Australia

Dentella repens (Linnaeus) J. R. et J. G. A. Forster [*D. serpyllifolia* Wallich ex Craib, *Oldenlandia repens* Linnaeus]: Fig. 343

Stems creeping, rooting, much branched, cushion-forming. Leaves distichous in opposite pairs, shortly petiolate at base of shoot, sessile above; stipules interpetiolar, united below, transparent, entire, with few hairs; petiole 2–4 mm long; blades elliptical-oblong to oblong-obovate to rhomboid, 4–10 mm long, 2–5 mm wide, 1-nerved, base cuneate, apex acute to attenuate. Flowering nodes alternating with flowerless ones. Flowers solitary in the axils of leaf-like bracts or in branch axils, subsessile; pedicel 0.3–2 mm long. Sepal tube 5-ribbed, 2–3 mm long, glabrous or hairy; lobes 5, equal, valvate, narrowly triangular, ± 2 mm long, acute-acuminate. Petals funnel-shaped, 5–15 mm long, with club-shaped hairs inside; lobes 5, usually somewhat unequal, white with a purplish spot at the base, ± 4.5 mm long, membranous, subacute. Stamens 5, included within the petals, attached near to the base of the tube; filaments up to 0.5 mm long; anthers ± 1 mm long. Capsule 2-locular, indehiscent, subglobose, 3–4 mm in diameter, densely hairy with white somewhat swollen hairs or glabrous, crowned by the persistent sepals. Seeds ± 0.5 mm long, angular, dark brown, with a reticulate surface pattern.

Annual or perhaps sometimes perennial. Delicate herb forming dense mats. Creeping on drying out mud around pools and in rice fields, river banks and in marshy places. It flowers the whole year with a peak in winter. The disseminules are probably capsules that break up later. Plants with glabrous capsules are sometimes recognized as a distinct species called *Dentella serpyllifolia.*

{⊙ (⊙)? Hel}. S and SE Asia, Australia and Polynesia: Ban, Bur, Cey, Ind: <An, Ap, Bh, Gj, Go, Kl, Kt, Lk, Mh, Mp, Or, Rj, Tn, Tp, Up, Wb.

Oldenlandia [excluded from *Hedyotis*] ± 150 species (including *Hedyotis* 300–480 species): pantropics

Annual or perennial herbs. Leaves decussate; stipules develop between the leaf pairs, united below into a sheath, with triangular lobes above, margins usually fimbriate; leaf blades entire, sessile or shortly petiolate. Bracts leaf-like, scale-like or united below into a sheath. Flowers in cymes, pairs or solitary in axils. Sepals 4-lobed, usually persisting in fruit. Petals 4-lobed. Stamens 4, attached to the petal tube. Disk ring-like. Ovary globose or nearly so, 2-locular; style filiform; stigmas 2, linear. Fruit a capsule, thin-walled, opening above the sepal lobes by 2 valves. Seeds 2 to numerous.

Except for changing the generic name, I have accepted the treatment of Sivarajan, V. V. and Biju, S. D. Taxonomic and nomenclatural notes on the *Hedyotis corymbosa-diffusa* complex (Rubiaceae) in India. Taxon 39(4): 665-674 (1990) but am not convinced the treatment is correct; see also Halford, D. A. Review of the genus *Oldenlandia* L. and related genera in Australia. Austrobaileya 3(4): 683–722 (1992).

Excluded: species sometimes found moist or damp places but not wetland plants: ***Oldenlandia affinis*** (Roemer et Schultes) A. P. de Candolle [*Hedyotis affinis* Roemer et Schultes, *H. dichotoma* auct. Ind., non Heyne ex Roth]: ♦Ind; ***O. auriculata*** (Linnaeus) Schumacher [*Hedyotis auricularia* Linnaeus]; ***O. biflora*** Linnaeus [*O. paniculata* auct. Ind., non Linnaeus, *Oldenlandia biflora* sensu Roxburgh, non Linnaeus; *Hedyotis biflora* (Linnaeus) Lamarck, *H. racemosa* Lamarck]; ***O. erecta*** (K. S. Manilal et V. V. Sivarajan) ined. [*Hedyotis erecta* K. S. Manilal et V. V. Sivarajan, *H. corymbosa* var. *linearis* sensu Duatta, non *O. linearis* A. P. de Candolle]; ***O. herbacea*** (Linnaeus) Roxburgh [*O. heynei* G. Don, *Hedyotis herbacea* Linnaeus]; ***O. nitida*** (Wight et Arnott) Gamble [*Hedyotis nitida* Wight et Arnott, *H. glabella* R. Brown ex Beddome]: ♦Cey, ♦Ind; ***O. pumila*** (Linnaeus fil.) A. P. de Candolle [*O. crystallina* Roxburgh, *Hedyotis pumila* Linnaeus fil.]; ***O. trinervia*** Retzius [*Hedyotis trinervia* (Retzius) Roemer et Schultes]: ♦?Ind.

1. Stems 4-angled or 4-winged; petals with a ring of hairs at the throat; capsules spreading or erect ***O. corymbosa***

1. Stems terete or longitudinally ribbed but never 4-angled or 4-winged; petals glabrous, without a ring of hairs; at least some capsules nodding

2. Stems, peduncles and pedicels glabrous, without short bristle-like hairs; flowers not more than 2 in a cyme ***O. brachypoda***

2. Stems, peduncles and pedicels with short bristle-like hairs; at least some flowers in 3- or more-flowered cymes ***O. diffusa***

Oldenlandia brachypoda A. P. de Candolle [*O. diffusa* auct. Ind., pro parte, non Willdenow, *H. brachypoda* (A. P. de Candolle) V. V. Sivarajan et S. D. Biju]: Fig. 344

Stems floating, prostrate or erect, often rooting at nodes, branched, glabrous, terete, smooth or longitudinally ribbed but never 4-angled or 4-ribbed, 10–45 (–200) cm long. Leaf blades sessile, linear or linear-lanceolate, 1–3 (–5) cm long, 1–5 mm wide, base narrowed, margin usually inrolled, apex acute to apiculate. Flowers solitary, 2 at a node or in 2-flowered cymes; peduncles up to ± 5 mm long; pedicels 0–1 (–3) mm long at anthesis sometimes elongating later. Sepal tube 1–1.3 mm long, usually glabrous; lobes 4, equal, narrowly triangular, much shorter than the petal tube, often spreading in bud, 1–1.3 mm long in fruit, ciliate, acute or acuminate. Petal tube 1.25–2.5 mm long, glabrous; lobes 4, triangular, white or purplish, glabrous at the throat. Stamens 4, attached towards the top of the tube; anthers ±0.25 mm long. Capsule mostly nodding, globose to ovoid, somewhat compressed, 2–3 mm long, 2–3 mm wide, crown of capsule hardly raised, opening by slits which enlarge into a pore. Seeds numerous, irregularly angular, ± 0.3 mm long, brown, surface reticulate.

Annual. Weed in irrigated fields, temporarily inundated places and marshes. It sometimes grows in water 1 m or more deep. The submerged stems are usually leafless. It is used locally in medicine.

{(⊙)? Hel}. Warm and tropical Asia, naturalized in America: Ban, Bhu, Cey? Ind, Nep: An, Ap, Ar, As, Bh, Kl, Kt, Lk? Mg, Mn, Or, Sk, Tn, Tp, Up, Wb.

Oldenlandia corymbosa Linnaeus [*O. alsinifolia* G. Don, *O. capillaris* A. P. de Candolle, *Hedyotis corymbosa* (Linnaeus) Lamarck, *H. depressa* Roemer et Schultes, *H. intermedia* Wight et Arnott]: Fig. 344

Stems prostrate with ascending branches, 4-angled, 4-ribbed or 4-winged, usually branched from the base, (1.5–) 5–30 (–60) cm long. Leaf blades sessile, linear to narrowly elliptical, (0.6–) 1–3 (–5) cm long, (0.5–) 1.5–3 (–7) mm wide, glabrous, base narrowed, margin usually inrolled, apex acute to apiculate. Flowers homostylous, mostly in pedunculate, umbel-like, corymbs, of 3–8 flowers at a node, rarely in pedunculate pairs or solitary; peduncles (0–) 4–8 (–20) mm long; pedicels 1–7 (–20) mm long, very slender. Sepal tube subglobose, ± 1 mm in diameter; lobes 4 (5), equal, triangular, shorter than the petal tube, 0.5–2 mm long in fruit, green, margin bristly, acute or acuminate. Petal tube cylindrical, 0.5–1 (–3.5) mm long; lobes 4 (5), elliptical or ovate, 0.5–2 mm long, white, pale purple or pale pink, with a ring of rather long silky hairs at the throat. Stamens 4 (5), inserted at the base of the petal tube; filaments 0.2–0.5 mm long; anthers dark blue, ± 0.4 mm long, included in the petal tube. Style and stigma shorter than the sepals. Capsule erect, subglobose to obovoid, roundish in transverse section, somewhat flattened, 1.2–3 mm long, 1–3.5 mm wide, crown of capsule slightly raised, opening by slits which enlarge into a pore. Seeds numerous, obconic, laterally compressed, irregularly angular, ± 0.3 mm long, light brown, surface reticulate.

Annual. A common and widespread weedy plant, not confined to wetlands but frequently found in rice fields, flooded lake shores, pools on laterite and other seasonally inundated places, sometimes even in flowing water. It is very variable in form and some varieties have been described.

{⊙ Hel}. Tropical Africa, S and SE Asia, probably naturalized in tropical America and Australia: Ban? Cey, Ind, Nep, Pak: Ap, As, Bh, Go, Kl, Kt, Lk, Mh, Mp, Or, Pj, Rj, Tn, Tp, Up. 0–2400 m.

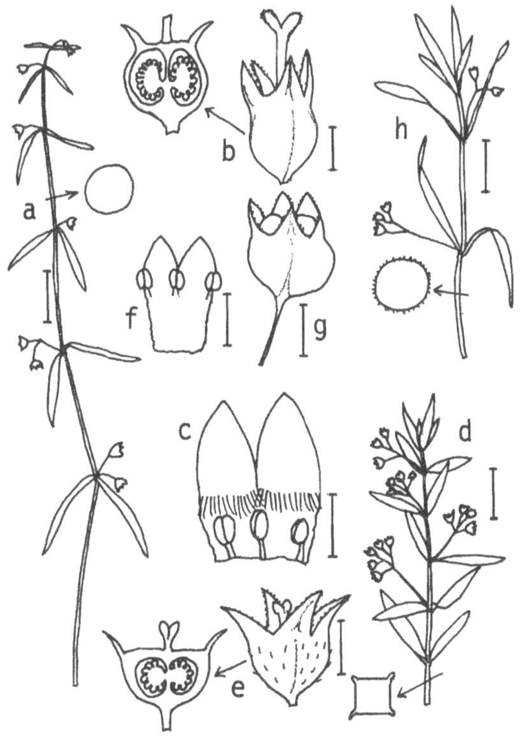

Fig. 344. *Oldenlandia brachypoda:* a, flowering shoot (1 cm) with transverse section of stem (inset); b, fruit with longitudinal section (inset) (1.5 mm). *O. corymbosa*: c, two petals viewed from inside (1 mm); d, flowering shoot (1 cm) with transverse section of stem (inset); e, fruit with longitudinal section (inset) (1.5 mm). *O. diffusa*: f, two petals viewed from inside (0.5 mm); g, fruit (1.5 mm); h, flowering shoot (1 cm) with transverse section of stem (inset).

Oldenlandia diffusa (Willdenow) Roxburgh [*O. pseudocorymbosa* (R. C. Bakhuizen van den Brink) C. R. Babu, *Hedyotis diffusa* Willdenow, *H. ramosissima* S. Kurz]: Fig. 344

Stems floating, prostrate or erect, often rooting at nodes, branched, with small bristle-like hairs, terete, smooth or longitudinally ribbed but never 4-angled or 4-ribbed, 3–10 (–45) cm long or in water 1 m or more long. Leaf blades sessile, linear-lanceolate, 1–3 (–5) cm long, 2–5 mm wide, base narrowed, margin usually inrolled, apex acute to apiculate. Flowers very rarely solitary, sometimes 2 at a node, usually in axillary 3- to 7-flowered cymes; peduncles up to ±1.5 cm long; pedicels 0.3–2 (–3) mm long at anthesis sometimes elongating later. Sepal tube ± 2 mm long, usually glabrous; lobes 4 (5), equal, narrowly triangular, shorter than the petal tube, 1–1.3 mm long in fruit, ciliate, acute or acuminate. Petal tube 0.9–1.3 mm long, glabrous; lobes 4 (5), triangular, white or purplish, glabrous at the throat. Stamens 4, attached towards the top of the tube; anthers ±0.25 mm long.

Capsule usually nodding, subglobose, somewhat flattened, 2–3 mm long, 2–3 mm wide, crown of capsule distinctly raised, opening by slits which enlarge into a pore. Seeds numerous, irregularly angular, ± 0.3 mm long, brown, surface reticulate.

Annual or perhaps also perennial in wet places. A weedy plant by no means confined to wetlands but common in rice fields and temporary pools.

{(⊙)? Hel}. Tropical and subtropical Asia: Ban, Bur, Cey, Ind: An, As, Go, Kl, Kt, Mg? Rj, Tn, Tp, Up, Wb. 0–1500 m.

RUPPIACEAE excluded from POTAMOGETONACEAE: 1 genus

Ruppia: 1–10 species: almost cosmopolitan

Ruppia maritima Linnaeus sensu lato [*R. rostellata* Koch, including *R. cirrhosa* (Petagna) Grande, *R. maritima* subsp. *spiralis* (Linnaeus ex Dumortier) Syme]: Fig. 345

Roots 1 or 2 at nodes. Stolons elongate often developing bulbil-like turions. Stems submerged, slender, very short or up to ± 2 m long. Leaves alternate or subopposite, loosely sheathed at the base but without a ligule, the blades capillary, 1-nerved, somewhat toothed at the tip. Inflorescence a short terminal spike at anthesis, appearing umbel-like in fruit. Flowers in pairs on slender peduncles which at first are short and surrounded by a spathe-like base of the subtending leaf; at anthesis the peduncles remain short or elongate; the elongated peduncles sometimes coil after pollination. Perianth vestigial or absent. Stamens 2. Fruits (2–) 4–12 (–26), at first sessile becoming stalked at maturity, nutlets or drupelets with spongy exocarp.

Perennial or annual. Totally submerged, mostly in brackish or alkaline water. In fresh water the plants are usually green but brown in brackish water. Pollinated at the water surface and perhaps also sometimes under water; the pollen grains are isobilateral and curved and often stick together in chains. The disseminules are nutlets or drupelets heavier than water, dispersed by waterfowl and also by water. It is an important food for wildfowl. It is taxonomically complex; the difficulties are well explained by Jacobs S. W. L. and Brock, M. A. A revision of the genus *Ruppia* in Australia. Aquatic Botany 14(4): 325–337 (1982). Recent work on DNA sequences by Les, D. H., Garvin, Denise, K. and Wimpee, C. F. in Molecular Phylogenetics and Evolution 2(4): 304–314 (1993) indicates that *Ruppia* is not very closely related to the Potamogetonaceae. I have seen no adequately collected material from India. I am therefore not sure which species are to be found; historically all was named *R. maritima*. My illustration shows a plant with long spirally wound fruit-bearing stalks which would usually be called *Ruppia cirrhosa.*

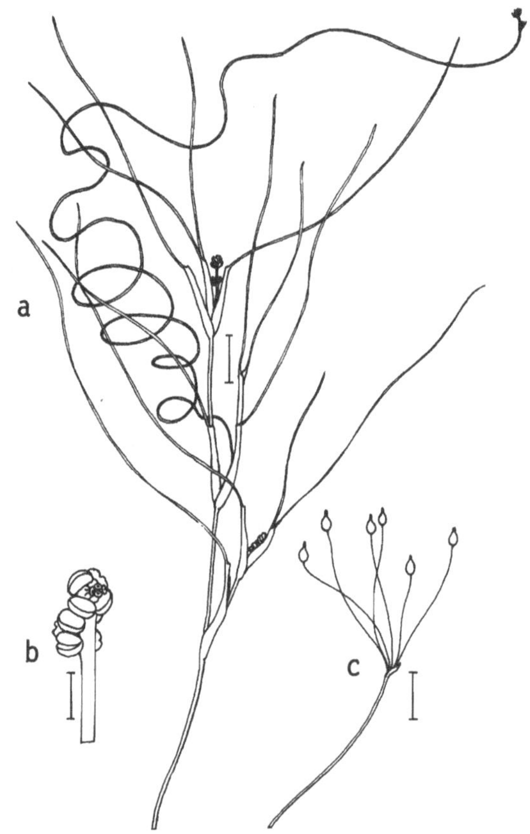

Fig. 345. *Ruppia cirrhosa*: a, flowering shoot (1 cm); b, inflorescence (2 mm); c, fruits (1 cm).

{(⊙) Vit}. Almost cosmopolitan: Ban, Cey, Ind: Ap, Gj, Kl, Mh, Or, Tn.

SCROPHULARIACEAE: 222 genera

Herbs. Leaves opposite or whorled. Flowers radially or bilaterally symmetrical, when bilateral then usually 2-lipped. Sepals tubular with 3–5 lobes. Stamens 4, 2 or rarely 5, staminodes sometimes present; anthers 2-locular, opening lengthwise or the locules joined at the tip and opening by 1 continuous slit. Ovary superior of 2 united carpels with a horizontal septum; ovules axile. Fruit a capsule. Seeds usually numerous with fleshy endosperm.

In this family it is difficult to draw the line between wetland plants and those which grow in moist or damp situations.

Excluded: aquatic or wetland species not extending south of the Himalayas in India: ***Gratiola griffithii*** J. D. Hooker; ***Limosella aquatica*** Linnaeus.

The following grow in wet or damp places but do not really qualify as wetland plants; others are weedy annuals that may grow in places where water stood during the monsoon but has dried out (in the dry season they replace the wetland species in this kind of habitat): ***Adenosma camphoratum*** (Vahl) J. D. Hooker: ♦Cey; ***A. indiana*** (Loureiro) Merrill; ***A. javanica*** (Blume) Koorders; ***A. subrepens*** (Thwaites) Bentham ex J. D. Hooker: ♦Cey, ♦Ind; ***Angelonia goyazensis*** Bentham [*A. salicariifolia* auct. Ind., non Humboldt et Bonpland]; ***Artanema longifolium*** (Linnaeus) Vatke; ***Legzpia polygonoides*** (Bentham) Yamazaki [*Torenia mucronulata* Bentham]; ***Craterostigma plantaginaeum*** Hochstetter; ***Mazus delavayi*** Bonati; ***M. pumilus*** (N. L. Burman) van Steenis; ***Mecardonia procumbens*** (Miller) Small [*Bacopa procumbens* (Miller) Greenman, *Lindernia dianthera* Swartz]; ***Scoparia dulcis*** Linnaeus [*Gratiola micrantha* Nuttall]; ***Sopubia delphinifolia*** (Linnaeus) G. Don; ***Stemodia viscosa*** Roxburgh; ***Sutera dissecta*** (Delpino) Walpers; ***Torenia bicolor*** Dalzell; ***T. cyanea*** Alston: ♦Cey; ***T. lindernioides*** Saldanha: ♦Ind: Kl.

Indeterminate: ***Bonnaya quinqueloba*** Blatter et Hallberg and ***Bonnayodes limnophiloides*** Blatter et Hallberg — their status and ecology are unknown; the original descriptions are inadequate. I have not been able to see any material of these two plants.

1. Petal tube very short, the lobes appearing free, flat and spreading; abaxial (lower) petal lobe smaller than the rest; stamens 2, spreading laterally ***Veronica***

1. Petal tube short or long, the lobes not appearing free; abaxial (lower) petal lobe or lobes not smaller than the rest; stamens 2, 4 or rarely 5, not spreading laterally

2. Abaxial filaments (of fertile or sterile stamens) appendaged (spur-like or elongate), swollen or knee-like at point of attachment to the petals ***Lindernia***

2. Abaxial filaments not appendaged at the point of attachment to the petals

3. Sepal lobes 3 or 4

4. Sepal lobes 4, subulate-lanceolate, much longer than the sepal tube; stigma subcapitate, ± straight; stamens 2; stems erect or floating ***Bythophyton***

4. Sepal lobes 3, broadly triangular, equal to or shorter than the sepal tube; stigma broadly spathulate, entire, curving over the stamens; stamens 2 or 4; minute; stems creeping ***Glossostigma***

3. Sepal lobes 5

5. Sepal and petal lobes ciliate (minute herbs) ***Microcarpaea***

5. Sepal and petal lobes without cilia

6. Sepal tube 5-angled; abaxial petal lip with 2, hairy patches at the base ***Mimulus***

6. Sepal tube not 5-angled, abaxial petal lip without 2, hairy patches

7. Sepals enlarged and globose in fruit; stamens 2 (small, creeping herbs) ***Peplidium***

7. Sepals not enlarged in fruit, not globose; stamens usually more than 2

8. Petal tube not strongly 2-lipped; petal lobes equal or subequal

9. Anther locules spurred; 1 loculus of each anther small and empty or aborted; roots orange (hemiparasitic herbs with entire leaves) ***Centranthera***

9. Anther locules not spurred; both locules of each anther fertile or both aborted; roots not orange (hemiparasitic herbs with entire leaves)

10. Stigmas simple; stamens each with a tuft of hairs at the base; petal tube long and slender, 5–10 times longer than the petal lobes (hemiparasitic annual herbs with leaves or their segments filiform) ***Rhamphicarpa***

10. Stamens without tufts of hairs at the base; petal tube not long and slender, rarely more than 5 times longer than the petal lobes ***Bacopa***

8. Petal tube 2-lipped; petal lobes unequal

11. Basal leaves in a rosette with stem leaves much reduced and scale-like ***Dopatrium***

11. Basal leaves not in a rosette

12. Flowers in terminal or axillary spikes or racemes ***Limnophila***

12. Flowers solitary or in pairs in leaf axils

13. Anther locules united, parallel ***Bacopa***

13. Anther locules free from each other, usually divergent ***Limnophila***

Bacopa [*Bramia, Herpestis, Mella, Monniera, Moniera* orth. mut. and perhaps *Sinobacopa*]: ± 70 species: warm regions of the world, most in America

Herbs sometimes aromatic when bruised. Stems creeping or erect. Leaves opposite or whorled; blades sessile. Flowers solitary or in pairs or clusters in the leaf axils; bracteoles if present just beneath the calyx tube. Sepal lobes 5, usually unequal. Petal lobes 5, subequal or 2-lipped, blue, purple or white. Stamens 2–5, all fertile; anther locules united and parallel. Stigmas 2, free or united below; capsules loculicidally or septicidally dehiscent.

Annual or perennial. Submerged, seasonally submerged or terrestrial. The disseminules are seeds, some dispersed in mud, some by water, others unknown. Several species are cultivated for decoration. ***Bacopa rotundifolia*** (Michaux) Wettstein, an American species, is spreading in Asia but not recorded from India; it differs from the Indian species in lacking bracteoles and having obovate to orbicular leaf blades.

1. Stems creeping with ascendent or erect branches, solid with narrow lacunae; pedicels 5–20 mm long; sepals glabrous **B. monnieri**
1. Stems erect, hollow, with a ring of large lacunae; pedicels 0–5 mm long; sepals hairy
  2. Flowers pedicellate, 1–3 in leaf axils **B. floribunda**
  2. Flowers sessile, solitary in leaf axils **B. hamiltoniana**

Bacopa floribunda (R. Brown) Wettstein: Fig. 346

Stems glabrous, 4-sided, erect, simple or branched from below, 10–30 (–65) cm tall, hollow with a ring of large lacunae. Leaves decussate, sessile or nearly so, linear-lanceolate, 1–3.5 (–7) cm long, 2–5 (–10) mm wide, margin slightly notched, apex acute. Flowers pedicellate, 1 or 2 (3) in leaf axils; bracteoles 2, filiform, 1.5–2 mm long; pedicels (0.5–) 1–3 (–5) mm long, hairy. Sepal lobes divided to the base, imbricate, membranous, unequal; the outer 2 larger, broadly oval, obtuse, ± 4 mm long increasing to ± 6 mm in fruit, 2–4 mm wide; the inner 2 and the median

FIG. 346. *Bacopa floribunda*: a, flowering shoot (2 cm); b, flower (2 mm). *B. hamiltoniana*: c, portion of a fruiting shoot (1 cm). *B. monnieri*: d, flowering shoot (1 cm); e, petal tube opened (2 mm).

3–4 mm long. Petals white usually with purple veins; petal tube ±1.5 mm long; petal lobes weakly 2-lipped; the adaxial lip ±2.5 mm long, ± 1.2 mm wide, glabrous; abaxial lip oblong to spathulate. Stamens 4, all fertile; filaments 2 ± 1 mm long, the other 2 ± 2.5 mm long; anther locules united and parallel, ± 1.5 mm long. Capsules ellipsoid-ovoid to oblong-globose, 3–5 mm long, ± 2.5 mm in diameter, 2- or 4-valved, septicidally or loculicidally and septicidally dehiscent, the placentas not separating and not winged. Seeds prismatic, 0.6–1 mm long, reticulate.

Annual. In wet places, in rice fields and often on wet sand around drying ponds. Locally frequent and often abundant.

{⊙ Hel}. Old World tropics: Cey (not found recently), Ind: <An, Go, Kl, Kt, Rj, Tn.

Bacopa hamiltoniana (Bentham) Wettstein: Fig. 346

Like *B. floribunda* but stems ± terete; flowers solitary and sessile in leaf axils; petals pink or bluish-pink, with purple veins; placentas winged and separating in fruit.

Annual. In wet places.

{⊙ Hel}. Old World tropics: Ban, Ind, Nep: As, Gj, Go, Kl, Kt, Mg, Mh, Mp, Pj, Rj, Wb.

Bacopa monnieri (Linnaeus) Pennell [*Bramia monniera* (Linnaeus) Wettstein, *Moniera cuneifolia* Michaux]: Fig. 346

Stems creeping with ascendent or erect branches, 5–50 cm long, indistinctly 4-sided, solid with narrow lacunae. Leaves decussate, sessile or nearly so, usually somewhat succulent, oblong-cuneate to obovate, 6–20 mm long, 3–10 mm wide, with sunken glandular dots, margin entire, apex obtuse. Flowers pedicellate, solitary in leaf axils; pedicels 5–15 (–20) cm long; bracteoles 2, linear, 2–3 (–8) mm long. Sepal lobes divided to the base, imbricate, somewhat succulent, glabrous, punctate, unequal; the outer 2 larger, ovate, 6–8 mm long, 3–4 mm wide; the inner 2 linear-lanceolate, 5–6 mm long, 0.7–1.5 mm wide; the median 1 oblong, ±5.5 mm long, ± 2 mm wide. Petals white to pink or pale violet, usually with green bands inside the throat, ± 8 mm in diameter; petal tube campanulate, 5–9 mm long; petal lobes subequal or weakly 2-lipped, 3–5 mm long. Stamens 4, all fertile; filaments one pair ± 1 mm long, the other pair ±2.5 mm long; anther locules united and parallel,± 1.5 mm long. Capsules oblong-globose, 3–5 mm long, ± 2.5 mm in diameter, 2- or 4-valved, septicidally or loculicidally and septicidally dehiscent. Seeds ovoid-oblong, 0.5–0.6 mm long, reticulate but also longitudinally striate.

Perennial or perhaps sometimes annual. Gregarious and often forming dense mats in marshy places, the banks of pools and along streams and ditches. It tolerates brackish conditions. Frequently found in rice fields but not considered harmful. It contains hersaponin which is used in medicine as a potent diuretic, cardiac tonic and tranquillizer. Plants boiled in water are used as purgative or skin lotion; also said to be a nerve tonic.

{(⊙)? Hel/(Ple)/(Hyd)}. Tropics and subtropics of the world: Cey, Ind, Nep, Pak: An, Ap, As, Bh, Dl, Gj, Go, Kl, Kt, Lk, Mh, Mp, Or, Pj, Rj, Tn, Wb.

Bythophyton: 1 species
Bythophyton indicum J. D. Hooker: Fig. 347

Stems submerged or emergent, flaccid, up to ± 8 cm long. Leaves opposite, subulate-lanceolate, entire. Flowers solitary, shortly pedicelled, in leaf axils, submerged flowers cleistogamous. Sepal lobes 4, united only at the base, equal, subulate-lanceolate, much longer than the sepal tube. Petal lobes 4, subequal, shorter than the tube. Stamens 2, the filaments very short, not appendaged. Style short, curved; stigma subcapitate. Capsules flattened laterally, broadly oblong, much shorter than the sepals.

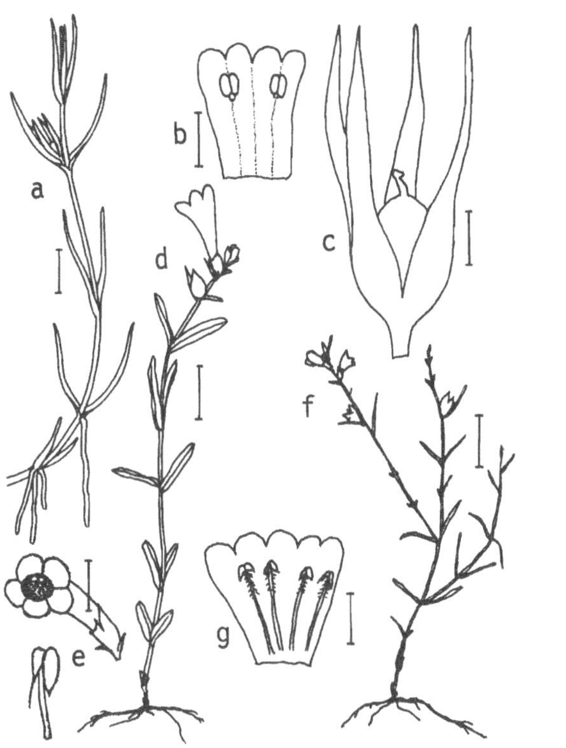

Fig. 347. *Bythophyton indicum*: a, flowering shoot (5 mm); b, petal tube opened (1 mm); c, immature capsule and sepals (1 mm). *Centranthera indica*: d, flowering plant with hairs omitted (1 cm); e, flower (1 cm) with stamen (inset). *C. tranquebarica*: f, flowering plant (2 cm); g, petal tube opened (2.5 mm).

Probably annual. A poorly known plant; the juveniles are submerged; the adults are submerged, partly emergent and perhaps also temporarily terrestrial. Perhaps pollinated by insects but most likely to be autogamous. The disseminules are seeds; the dispersal is unknown.

{⊙? Vit/Hyp/Ten?}. NE India and Philippines (Luzon): Ban? Ind: Mg (Khasi Hills).

Centranthera: 5 or 6 species

Annuals with orange roots. Stems prostrate or erect. Flowers solitary or in lax spikes or racemes; bracteoles 2, oblong. Sepals irregular, shortly 3- to 5-lobed. Petals pink to purple, the tube funnel-shaped and curved, the lobes ± equal or weakly 2-lipped, two adaxial (upper) lobes inside in bud. Stamens 4, included; anthers combined in pairs, the locules spurred or pointed at the base, one loculus of each anther small, empty or aborted. Stigma tongue-shaped. Capsules locicidicidally 2-valved.

Reported to be root parasites on grasses.

1. Blades of stem leaves oblong to ovate-oblong; petals more than 15 mm long, the lobes white or pink
C. indica

1. Blades of stem leaves linear to linear-filiform; petals less than 15 mm long, the lobes yellow
C. tranquebarica

Centranthera indica (Linnaeus) Gamble [*C. hispida* R. Brown, *C. nepalensis* D. Don, *C. procumbens* Bentham]: Fig. 347

Stems prostrate and branched from below or erect, up to 90 cm high, hispid. Leaf blades sessile, oblong to ovate-oblong, 0.9–2.8 cm long, ± 2 mm wide, hispid, at base truncate, at tip subacute. Bracts linear to linear-lanceolate, 4–7 mm long, 1.5–4 mm wide; pedicels 1–2 mm long; bracteoles oblong to ovate-oblong, 3–3.5 mm long, ± 1 mm wide. Sepals 5–19 mm long, the lobes lanceolate. Petals 18–31 mm long, the lobes white or pink. Capsules ovoid, 5–8 mm long, 4–5 mm wide, shorter than the sepals. Seeds oblong-cuneate, ± 1 mm long, striped.

Annual. Not confined to wetlands but often found in grass-dominated marshes and swamps.

{⊙ Hel/Ten?}S and SE Asia and Australia: Ban, Bur, Cey, Ind, Nep: Kl, Kt, Mh, Pj, Up.

Centranthera tranquebarica (Sprengel) Merrill [*C. humifusa* Wallich ex Bentham, *Torenia lepidota* Roth]: Fig. 347

Stems erect, 6–15 (–20) cm high, often branched from below, glabrous to sparsely hispid. Leaf blades sessile, at base of stem oblong to ovate-oblong, along the stem linear to linear-filiform, 5–25 (–30) mm long, 0.5–1.5 (–20) mm wide, glabrous to sparsely hispid, at base narrowed, at tip subobtuse. Bracts linear-lanceolate to ovate-lanceolate, 4–7 mm or more long; flowers subsessile; bracteoles 2, linear, 1–1.5 mm long. Sepals 3 (–5) mm long, inflated, the lobes subacute. Petals 10–11 mm long, the tube maroon, the lobes yellow. Capsules globose, ± 4 mm long and wide, slightly shorter than the sepals. Seeds oblong cuneate, 0.8–0.9 mm long, reticulate.

Annual. Usually found in rather wet places often near to the sea, in marshes, around ponds, along water courses and in rice fields.

{⊙ Hel/Ten?}. S and SE Asia: Ban, Cey, Ind: Bh, Gj, Go, Kl, Tn.

Dopatrium: 12 species: Old World tropics and subtropics, naturalized in N America

Stems erect, simple or sometimes branched below, usually somewhat succulent. Leaves opposite; the basal leaves often forming a loose rosette; upper leaves usually reduced and scale-like. Flowers solitary, in axils of bracts; submerged flowers cleistogamous. Sepal lobes 5, equal or subequal, longer than the tube. Petal tube 2-lipped; lobes 5, white, blue, purple or yellow. Stamens 4, the 2 adaxial ones fertile with loosely joined and parallel anther locules, the 2 abaxial ones staminodial. Stigma almost peltate with 2 flattened lobes. Capsules loculicidally dehiscent.

Annual. The juveniles are submerged; the adults are submerged, emergent or temporarily terrestrial. Pollinated by insects or autogamous (sometimes cleistogamous). The disseminules are seeds which are dispersed in mud, perhaps also by wind and water.

1. Capsules obovoid-oblong, apex acute; pedicel deflexed in fruit ***D. nudicaule***
1. Capsules globose, apex obtuse; pedicel erect or horizontal in fruit
   2. Lower flowers sessile or nearly so; pedicel erect in fruit; anthers unequal; petals pink to blue or purple ***D. junceum***
   2. Lower flowers pedicellate; pedicel horizontal in fruit; anthers equal; petals violet to deep blue ***D. lobelioides***

Dopatrium junceum (Roxburgh) F. Hamilton ex Bentham [*D. junceum* var. *multiloba* Blatter et Hallberg]: Fig. 348

Stems 10–25 (–40) cm tall. Blades of lower leaves oblong to lanceolate, 4–25 mm long, 1–7 mm wide, middle and upper leaves much smaller. Flowers in decussate pairs, the lower ones sessile or nearly so, the upper ones pedicellate; pedicel very slender, 1.5–8 (–12) mm long, erect in fruit. Sepal tubes campanulate, 1–1.5 mm long; lobes oval-lanceolate, 1–1.5 mm long, obtuse. Petals (of chasmogamous flowers) pink to blue

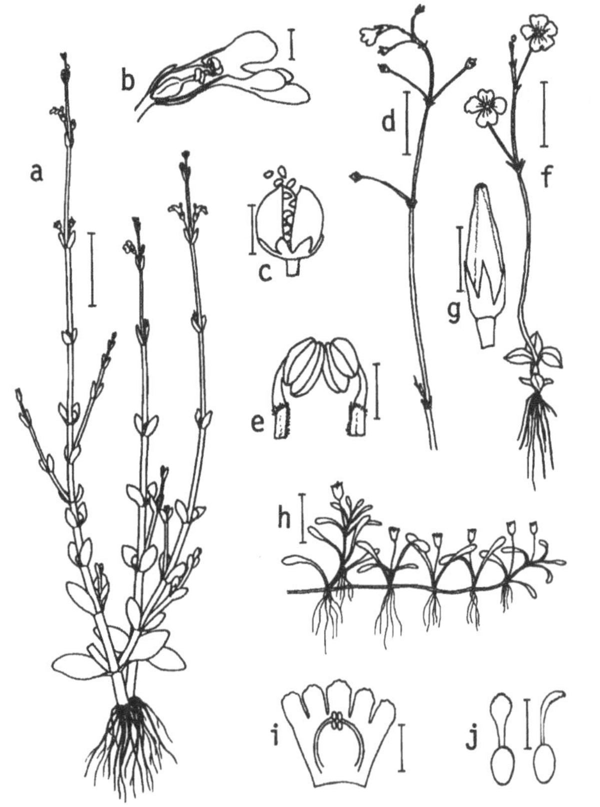

FIG. 348. *Dopatrium junceum*: a, flowering plant (2 cm); b, flower in longitudinal section (1 mm); c, mature capsule (2 mm). *D. lobelioides*: d, inflorescence (2 cm); e, stamens (1 mm). *D. nudicaule*: f, flowering plant (2 cm); g, sepals and capsule (2 mm). *Glossostigma diandrum*: h, flowering plant (1 cm); i, petal tube opened (1 mm); j, ovary with stigma viewed from two sides (1 mm).

or purple; petal tube 3.5–4 mm long; adaxial lip ± 4 mm long; abaxial lip ± 3 mm long, with a white blotch at the base of the middle lobe. Stamen filaments ± 0.5 mm long; anthers equal, ± 0.5 mm long, bearded. Capsules broadly ellipsoid to globose, 2–4 mm long and wide. Seeds cuneately ovoid, 0.5 mm long, longitudinally ribbed, transversely rugose between the ribs.

Annual. Germinating under water, flowering as the water recedes. It is often gregarious and locally abundant in post-monsoon pools, rice fields and perennially moist places. The lower flowers are usually cleistogamous.

{⊙ Ten}. Tropical Asia, Australia and Oceania, naturalized in N America: Ban, Cey, Ind, Nep: An, As, Gj, Go, Jk, Kl, Kt, Mp, Rj, Tn, Wb.

Dopatrium lobelioides (Retzius) Bentham: Fig. 348

Stems up to 80 cm tall, 4-winged, sometimes spongy and thickened below. Lower leaves crowded, oblong-lanceolate to oblong, 2.5–5.5 cm long, 1–1.7 cm wide, subacute at apex; stem leaves distant, 0.7–2 cm long, 0.5–0.9 cm wide, those subtending flowers scale-like. Flowers pedicellate, in decussate pairs; pedicel 7–17 mm long, horizontal in fruit. Sepal tube ± 1.5 mm long; lobes oblong, 1.5–2.5 mm long, obtuse. Petals violet to deep blue; petal tube 4–5 mm long; adaxial lip ± 10 mm long; abaxial lip ± 6 mm long. Stamen filaments ± 1 mm long; anthers equal, ± 0.7 mm long, bearded along the margin. Capsules globose, 1.5–4.5 mm in diameter. Seeds oblong-cuneiform, ± 0.6 mm long, longitudinally ribbed, transversely rugose between the ribs.

Annual. In shallow water in tanks and flooded rice fields. Sometimes gregarious. A common but rarely harmful weed in paddy fields.

{⊙ Ten}. Endemic to southern India and Sri Lanka: ♦Cey, ♦Ind: Kl, Tn.

Dopatrium nudicaule (Rottler) Bentham [*Gratiola cuneifolia* Roxburgh]: Fig. 348

Stems 9–12 (–25) cm tall. Lower leaves obovate-oblong, 2–7 mm long, 1–3.5 (–5) mm wide, apex obtuse. Flowers pedicellate, solitary or in decussate pairs; pedicels 7–15 mm long, elongating and deflexed in fruit; bracts linear-lanceolate, ± 1 mm long. Sepal tube ± 1 mm long; lobes up to 0.75 mm long, acute. Petals bluish; petal tube 3–4 mm long, dull blue; adaxial lip ± 3 mm long, 2-lobed, brilliant dark purple-blue; abaxial lip ± 3 mm long, pale bluish-purple, the middle lobe with a whitish blotch at the base. Stamen filaments ± 0.7 mm long; anthers unequal, ± 0.7 mm long, minutely bearded at the base. Capsules obovoid-oblong, 3–3.5 mm long, bluntly pointed at the apex. Seeds irregularly cuneiform, 0.3 mm long, rugulose.

Annual. Locally abundant, in post-monsoon pools on clayey soil and in rice fields.

{⊙ Ten}. S India, Sri Lanka and Philippines. Cey, Ind: Tn.

Glossostigma [excluded from *Microcarpaea*]: 5 species: Africa, India, Australia and New Zealand; 1 species in India

Glossostigma diandrum (Linnaeus) O. Kuntze [*Microcarpaea spathulata* J. D. Hooker]: Fig. 348

Stems capillary, creeping, rooting at most nodes, mat- or turf-forming. Leaves opposite or occasionally whorled, spathulate, sometimes somewhat succulent, entire, 0.5–1 cm long, up to 1.5 mm wide, nerves obscure, apex obtuse. Flowers minute, pedicellate, solitary, in leaf axils; submerged flowers cleistogamous; pedicel up to 1 cm long. Sepal tube campanulate, lobes 3, broadly triangular, ± 2 mm long, equal to or shorter than the sepal tube. Petal ± 1 mm long; lobes (4) 5, pinkish-white, adaxial lobes up to 0.4 mm long, abaxial lobes up to 0.7 mm long. Stamens 2; filaments filiform, 0.8 mm long; anther locules diverging at the base, confluent at the apex, 0.2 mm long. Style ± 1 mm long, elongate, flattened above into a broad spathulate stigma usually covering the stamens. Capsules globose; seeds ellipsoid.

Annual or when in permanent water perhaps perennial. It is seasonally submerged and flowers as the water recedes. It is gregarious and forms mats. Probably mostly autogamous and sometimes cleistogamous, but perhaps occasionally pollinated by insects. The disseminules are seeds, dispersed in mud. It is frequent in rice fields but not considered troublesome. It is cultivated in aquaria for decoration.

{(⊙) Ten/Hel}. Old World tropics: Ban? Ind: Ap, Bh, Dl, Kl, Kt, Mh, Mn, Mp, Rj, Tn, Up.

Limnophila [*Ambulia*]: 36 species: warm Old World, naturalized in N America

Frequently aromatic when bruised. Stems floating or emergent, erect, ascending or creeping. Leaves uniform or polymorphic: submerged leaves usually whorled, laciniate, pinnatifid to compound-pinnate with capillary lobes or decussate and entire; emergent leaves opposite or whorled, sessile or petiolate, entire to serrate, laciniate or pinnately divided. Flowers solitary in leaf axils or in loose or compact, terminal or axillary, spikes or racemes. Sepal lobes 5, subequal or the adaxial lobe enlarged. Petals tubular or funnel-shaped, blue, pink, yellow or white, the tube 2-lipped; the adaxial lip outside in bud, entire or 2-lobed; the abaxial lip 3-lobed and erect or spreading. Stamens (2) 4, all fertile, the adaxial pair shorter or sometimes absent; anther locules free, divergent. Styles capillary, deflexed at the tip; stigma 2-lipped. Capsules ellipsoidal to globose, septicidally 4-valved, the valves bifid. Seeds numerous.

Annual or perennial. Found in a variety of aquatic habitats, submerged, emergent or terrestrial. Often found in rice fields and irrigation ditches but rarely considered to be noxious. Pollinated by insects or autogamous and occasionally cleistogamous. The disseminules are seeds; dispersed in mud, by water and perhaps also by wind. Several species are cultivated for decoration; some are used locally in medicine and for flavouring. Philcox, D. A. A taxonomic revision of the genus *Limnophila*. Kew Bulletin 24: 101–170 (1970); Dutta Narendra Mohan. A revision of the genus *Limnophila* R. Br. of Eastern India. Bulletin of the Botanical Society Bengal 29(1): 1–7 (1975) both these works do not inspire confidence; Wannen, B. S. and Waterhouse, J. T. A taxonomic revision of the Australian species of *Limnophila*. Australian Journal of Botany 33: 367–380 (1985).

1. Leaves divided into linear segments present
2. Flowers distinctly pedicellate ***L. indica***
2. Flowers sessile or very nearly so
3. Upper aerial leaves and bracts finely divided or deeply lobed almost to the base ***L. sessiliflora***
3. Upper aerial leaves and bracts entire or shallowly toothed, never finely divided
4. Petals 10–14 mm long, violet (Bangladesh) ***L. cana***
4. Petals 5–6 mm long, white or pink
5. Inflorescence simple; flowers opposite and solitary in the axils of bracts; petals pale pink ***L. heterophylla***
5. Inflorescence much branched; flowers alternate in long, slender spikes; petals white or very pale pink ***L. polystachya***
1. Leaves all with flattened blades, none divided into linear segments
6. Mature calyx smooth or with not more than 5, prominent longitudinal veins
7. Upper leaves on flowering shoots divided to the base
8. Flowers distinctly pedicellate ***L. indica***
8. Flowers sessile or very nearly so ***L. sessiliflora***
7. Upper leaves on flowering shoots with flat, laminate blades not divided to the base
9. Leaf blades 1-nerved, pinnately nerved or pinnately lobed
10. Leaf blades sessile, linear-lanceolate to oval-oblong, 5–10 (–20) mm long, 2–4 (–7) mm wide; adaxial sepal lobe enlarged ***L. heterophylla***
10. Leaf blades petiolate (the petiole 5–30 mm long), ovate-lanceolate to oval-elliptical, 15–90 mm long, 7–50 mm wide; adaxial sepal lobe not enlarged ***L. rugosa***
9. Leaf blades with 3–7 parallel nerves
11. Flowers pedicellate ***L. indica***
11. Flowers sessile or nearly so
12. Bracteoles present ***L. connata***
12. Bracteoles absent
13. Inflorescence branched with terminal and axillary spikes; flowers alternate ***L. polystachya***
13. Inflorescence simple; flowers in decussate pairs, solitary in axils of leaf-like bracts
14. Petals 5–6 mm (–10) long, pink to pinkish-violet ***L. heterophylla***
14. Petals 12–13 long, blue to purple (Bangladesh) ***L. cana***
6. Mature calyx with more than 5, prominent longitudinal veins (veins sometimes obscured by indumentum)
15. Sepals 2.5–3 mm long in fruit; petals 2.5–4 (–5) mm long, glabrous inside (NW part of Indian subcontinent)
16. Flowers pedicillate and in axillary cymes ***L. helferi***
16. Flowers sessile or nearly so, solitary in the axils of leaf-like bracts
17. Stems glabrous; leaf blades linear-oblong to ovate, 3–14 mm long, 1–5 mm wide ***L. micrantha***
17. Stems with sessile or shortly stalked glands and eglandular hairs; leaf blades linear to linear-lanceolate, 3–12 mm long, 0.5–1.5 mm wide ***L. polyantha***
15. Sepals more than 3 mm long in fruit; petals more than–5 mm long, hairy inside
18. Petals 6–10 mm long
19. Stems rarely more than 5 cm long (Myanmar and NW India) ***L. pulcherrima***
19. Stems much more than 5 cm long
20. Bracteoles 3–5 mm long; stems, leaves and sepals densely covered in minute glands (Kerala) ***L. glandulifera***
20. Bracteoles 0.5–2 (–3.5) mm long; stems, leaves and sepals without or with few minute glands
21. Pedicels reflexed in fruit; stems glabrous or with short, appressed hairs ***L. erecta***
21. Pedicels erect in fruit; stems with spreading or upcurved hairs
22. Stems with moderately long, straight and spreading hairs particularly towards the apex; flowers usually solitary in axils of leaf-like bracts; pedicels (0.5) 3–5 (–7) mm long ***L. laxa***
22. Stems glabrous or with upcurved hairs; flowers usually in dense racemes; pedicels (0.5–) 1–2 (4) mm long ***L. repens***
18. Petals 10–18 mm long
23. Stems erect up to 1 m high; without simple hairs; bracteoles appressed ***L. aromatica***
23. Stems prostrate or decumbent below, erect parts rarely more than 20 cm high; with soft and flexuose simple hairs; bracteoles spreading ***L. chinensis***

Limnophila aromatica (Lamarck) Merrill [*L. gratissima* Blume, *L. punctata* Blume]: Fig. 349

Stems erect, simple or branched from the base, up to 1 m tall, glabrous to minutely glandular. Leaves decussate or in whorls of 3, simple; blades sessile, ovate-lanceolate to lanceolate-elliptical, 1–6 (–9) cm long, 0.3–2.2 cm wide, glabrous to densely and minutely glandular, punctate, pinnately nerved, base cuneate to subamplexicaul, margin rarely entire usually crenate-serrate or serrate, apex acute to occasionally obtuse. Flowers solitary and axillary or in many-flowered, terminal or axillary racemes, up to 15 cm long; pedicels slender, 0.3–2 (–3) cm long, remaining divergent-erect in fruit, glabrous to glandular; bracteoles 1.5–3 mm long, linear to linear-lanceolate, subciliate. Sepals 4–7 (–8) mm long, glabrous to

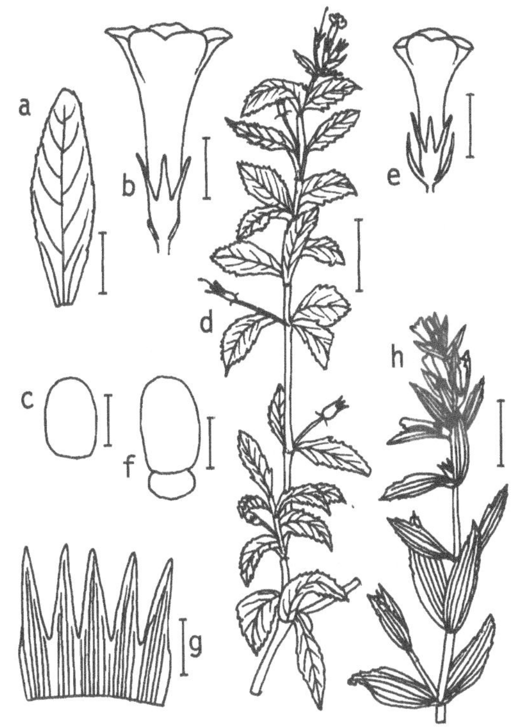

FIG. 349. *Limnophila aromatica*: a, leaf (1 cm); b, flower (5 mm); c, seed (0.2 mm). *L. chinensis*: d, flowering shoot (2 cm); e, flower (5 mm); f, seed with caruncle (0.2 mm); g, sepals opened with hairs omitted (5 mm). *L. connata*: h, flowering shoot with hairs omitted (2 cm).

minutely glandular, densely punctate, striate at maturity; sepal lobes subequal, 2.5–4.5 (–6.5) mm long, more than half as long as the tube, narrowly lanceolate-acuminate. Petals 10–18 mm long, pink to mauve or purple with a yellow throat, glabrous or finely and sparsely glandular outside, villous inside; tube 9-12 (–19) mm long; lobes oblong-orbicular. Capsules broadly ellipsoid, 4–6 mm long; style up to 6 mm long with 2 lateral processes at the apex below the stigma. Seeds cylindrical to oblong, 0.25–0.5 mm long, ± 0.2 mm in diameter, muricate.

Annual or perennial. In swamps, drains and moist habitats. Sometimes in flooded rice fields but not considered harmful. The whole plant smells of turpentine.

{(⊙) Hyp/Vit/Ten}. Tropical Asia and N Australia: Ban, Bur, Cey, Ind: As, Bh, Go, Kl, Kt, Mp, Tn, Wb.

Limnophila cana Griffith: refer to Fig. 349

Like *L. connata* but lacks glands and bracteoles and possesses pinnatisect to lacerate submerged leaves in whorls of up to 12.

Annual. 'Stagnant water'.

{⊙ Vit/Hyp/Ten?}. Endemic to Bangladesh, known from only 3 collections: ♦Ban.

Limnophila chinensis (Osbeck) Merrill [*Limnophila hirsuta* (Heyne ex Bentham) Bentham, *L. chinensis* var. *clarkei* (Haines) Murti et Panigrahi, *L. chinensis* var. *scaberrima* (S. Kurz) Thothathri]: Fig. 349

Stems simple or branched from the base, spongy, creeping and rooting below, 5–20 (–40) cm tall, the branchlets usually somewhat succulent and reddish, subglabrous to densely hairy, the hairs soft and flexuose. Leaves decussate or in whorls of 3–8, entire; blades sessile, ovate-lanceolate to elliptical-lanceolate, when submerged up to 7 cm long, up to 1 cm wide, when aerial (0.5–) 2–4 (–5) cm long, (0.2–) 0.7–1.5 cm wide, subglabrous to sparsely lanate, particularly on major nerves, obscurely punctate, pinnately nerved; margin serrulate to serrate; apex acute; base cuneate-decurrent to subamplexicaul. Flowers pedicellate, solitary and axillary or in many-flowered, terminal or axillary racemes; pedicels 2.5–15 mm long, remaining divergent-erect in fruit, hirsute; bracteoles 1.5–5 mm long. Sepals 4–7.5 mm long, hairy, striate at maturity; sepal lobes subequal, more than half as long as the tube, the adaxial one larger, (2.5–) 4–4.5 mm long, lanceolate, long acuminate. Petals (6–) 10–15 (–17) mm long, pink to blue or purple, occasionally white, finely pubescent outside, densely villous inside. Capsules ovoid to ellipsoid, 2.5–5 mm long, pale brown; style up to 5 mm long, with 2 lateral processes at the apex below the stigma. Seeds cylindrical to oblong, 0.5–0.7 mm long, with a caruncle.

Annual. Moist shady habitats, sometimes in water up to ±20 cm deep. Old plants usually have more hairs than young ones. The aerial leaves are very aromatic.

{⊙ Ten?/Hel}. Tropical Asia and Australia: Ban, Bur, Cey not found recently, Ind, Nep: An, As, Kl, Kt, Mg, Mn, Mp, Or, Sk, Tn, Wb.

Limnophila connata (F. Hamilton ex D. Don) Handel-Mazzetti [*L. hypericifolia* (Bentham) Bentham]: Fig. 349

Stems creeping and branching below, with erect flowering shoots, 20–50 cm tall, minutely glandular-puberulent. Leaves decussate; blades simple, sessile, oblong to ovate-oblong, (10–) 15–30 (–50) mm long, 5–14 (–22) mm wide, sparsely glandular-puberulent to almost glabrous, densely punctate, with 3–7 parallel nerves arising from the base, the base semi-amplexicaul, the apex acute or obtuse. Flowers sessile or nearly so, solitary, axillary, rarely in pedunculate spikes; bracteoles 3–5 mm long, linear, glandular-puberulent, apex obtuse. Sepals 5–6 (–7) mm long, glandular-puberulent, not striate at maturity; sepal lobes subequal, 3–3.5 (–5) mm long, lanceolate-acuminate. Petals 12–13 mm long, blue to purple, shortly and sparsely hairy outside, villous inside. Capsules 3–4 mm long, broadly ellipsoid, blackish-brown; style up to 4.5 mm long with 2 lateral processes at the apex below the stigma.

Annual or perhaps also perennial. Robust herb in swampy grassland. The leaf blades with 3–7 parallel nerves and clasping bases are distinctive for this species.

{(⊙)? Hel/Ten?}. Tropical and warm Asia: Bur, Ind, Nep: As, Kl, Kt, Mn, Tn. (up to 3000 m in Yunnan).

Limnophila erecta Bentham: Fig. 350

Like *L. repens* but stems with short, appressed hairs or sometime glabrous; pedicels reflexed in fruit, 2–4 mm long, with sparse short hairs; bracteoles linear, smaller (0.5–0.8 (–1.5) mm long).

{⊙ Hel/Ten?}. SE Asia: Bur, Ind?: Wb?

Limnophila glandulifera Philcox: Fig. 350

Like *L. aromatica* but plants covered in a dense covering of minute glands; pedicels shorter (1–6.5 mm long); bracteoles larger (3–5 mm long); petals smaller (8.5–9.5 mm long) and glabrous outside.

Annual. Known only from a single collection 'Travancore ... in locis humidis argillosis'.

{⊙ Hyp/Ten?/Hel?}. Endemic to S India: ♦Ind: Kl.

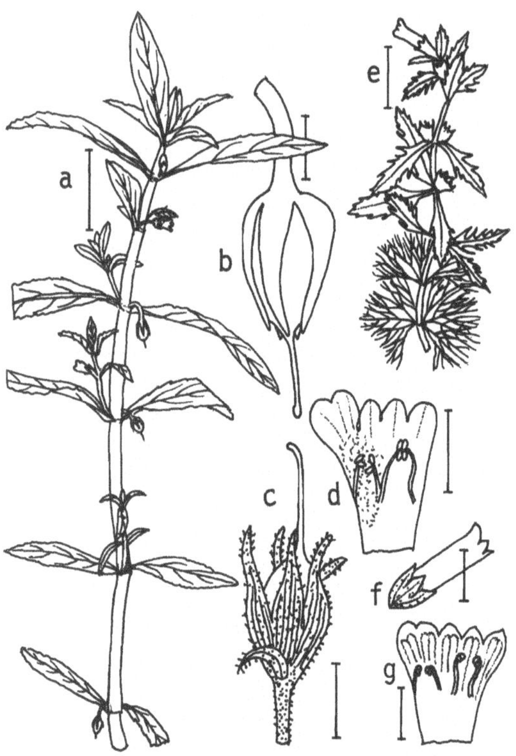

Fig. 350. *Limnophila erecta*: a, flowering shoot (2 cm); b, fruit (2 mm). *L. glandulifera*: c, fruit (2.5 mm); d, petals opened (5 mm). *L. heterophylla*: e, flowering shoot (1 cm); f, flower (4 mm); g, petals opened (4 mm).

Limnophila helferi J. D. Hooker: refer to Fig. 351

Like *L. micrantha* but flowers pedicellate and in axillary cymes (pedicels 0.5–1.5 (–2) mm long); leaves sometimes in whorls of 3–5; floral parts smaller (sepals (1.5–) 2–2.5 mm long, petals ± 2.5 mm long, bracteoles ± 1 mm long).

Annual. Known from a single gathering from Tenasserim, Myanmar.

{⊙ Ten?/Hel?}. Endemic to Myanmar: ♦Bur.

Limnophila heterophylla (Roxburgh) Bentham [*L. roxburghii* G. Don, *L. reflexa* Bentham]: Fig. 350

Submerged stems up to 80 cm long, spongy, glabrous; aerial stems erect, 5–10 cm high, subglabrous to shortly hirsute below, usually with sessile glands towards the apex. Submerged leaves in whorls of (3–) 8–12, compound-pinnate, 0.5–3 (–5.5) cm long; segments multifid, capillary; aerial leaves decussate or in whorls below, the blades simple, sessile, linear-lanceolate to oval-oblong, 5–10 (–20) mm long, 2–4 (–7) mm wide, glabrous, margin subcrenate to serrate, apex acute. Flowers solitary and axillary, sessile or nearly so, sometimes forming a lax terminal spike, cleistogamous flowers common; bracteoles absent. Sepal tube (2–) 3–4 mm long, without hairs but bearing sessile glands, not striate at maturity; sepal lobes subequal, 1.5–2 mm long, ± as long as the tube, with sessile glands, shortly acuminate. Petals 5–6 (–10) mm long, pale pink to pinkish-violet, glabrous outside. Capsules globular, 2.5–3 (–4) mm long; style up to 6 mm long, with 2 lateral processes at the apex below the stigma. Seeds cuneiform, ± 0.5 mm long, 4- or 5-angular, muricate.

Aquatic perennial or perhaps sometimes annual. It is usually found in fresh or brackish water in ponds and backwaters; sometimes also in rice fields. It is like *L. sessiliflora* but has the upper aerial leaves entire and flowers in a lax terminal spike. It is, perhaps, not specifically distinct.

{(⊙)? Vit/Hyp/Ten?}. Tropical Asia: Ban, Bur, Cey, Ind, Nep: As, Kl, Kt, Mn, Mp, Rj? Tn? Up, Wb.

Limnophila indica (Linnaeus) Druce [*L. aquatica* (Roxburgh) Alston, *L. gratioloides* auct. Ind., non R. Brown, *L. hyssopifolia* Roth, *L. racemosa* Bentham, *L. tillaeoides* J. D. Hooker]: Fig. 351

Submerged stems up to 1 m or more long, branching, glabrous or hairy; aerial stems erect or creeping below, simple or sparsely branched, up to 40 cm tall, with or without eglandular hairs, with sessile or stalked glands. Submerged leaves in whorls of (2–) 6–12, 0.7–3 (–6) cm long, compound-pinnate or pinnatisect, the lobes divided into capillary or linear and flattened segments; aerial leaves decussate and usually entire or whorled and variously dissected, 0.5–2 (–3 cm) long when dissected, up to ± 6.5 cm when entire; entire leaves with 1, 3 or rarely 5 nerves arising from the base, without hairs but often sessile glands, punctate above, margin finely spinulose-serrate. Flowers pedicellate, solitary, axillary, sometimes single, mostly in decussate pairs, all chasmogamous; bracts leaf-like (divided or entire) becoming reduced and scale-like towards the apex of old shoots; pedicels slender, (1–) 6–12 (–20) mm long, mostly exceeding the bracts, bearing sessile or stalked glands; bracteoles (0) 2, linear to linear-lanceolate, (1.5–) 3–4 mm long, simple or divided, glandular or not. Sepal tube 1.5–3 mm long, glabrous or hispid, bearing sessile glands, not striate at maturity, becoming subglobose in older flowers; sepal lobes subequal, linear-lanceolate, 2–3 mm long, acuminate, occasionally ciliate. Petals (6–) 8–15 mm long, entirely white, cream, pale yellow or blue to purple, with or without a yellow fleck at the top of the throat; petal tube usually greenish-white, glabrous or slightly pubescent outside, softly villous inside; petal lobes subequal, entire. Capsules subovoid, 3–3.5 (–5) mm long; style up to 4.5 mm long with 2 lateral processes

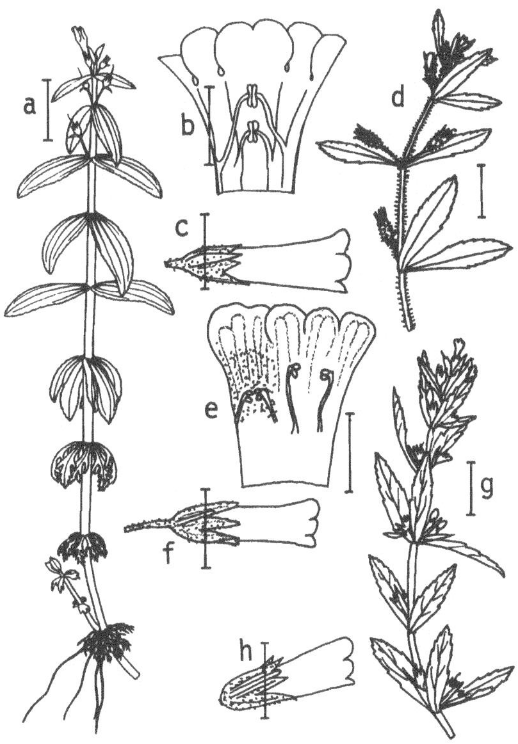

FIG. 351. *Limnophila indica*: a, flowering shoot (1 cm); b, petals opened (5 mm); c, flower (5 mm). *L. laxa:* d, flowering shoot (1 cm); e, petals opened (5 mm); f, flower (5 mm). *L. micrantha*: g, flowering shoot (1 cm); h, flower (5 mm).

± 0.2 mm wide at the apex below the stigma. Seeds cylindrical to oblong-cuneate, 0.5–0.7 mm long, ± 0.25 mm in diameter, 4- or 5-angled, muricate.

Annual or perennial. It is found in a wide variety of aquatic habitats. From the point of view of habit it is a very plastic species. In shallow water it has dissected leaves under water and normal ± entire ones above. In deep water the stems look like bottle brushes and very few laminate leaves develop (usually only those subtending flowers). On land, however, it develops normal laminate leaves and is branched forming dense mats. Flowers open by ± 10.00, are fully open by 11.30 and close ± 15.00. The submerged leaves close and wrap around the stem at night. It is used in medicine as an antiseptic and as a cure for dysentery.

It is highly variable with regard to leaf form and pubescence. Some of this variation is environmentally induced and some is genetically determined. Most authors recognize *L. aquatica* as a species distinct from *L. indica. Limnophila aquatica* is characterized by being annual, with hirsute submerged stems and entire aerial leaves in decussate pairs. In Sri Lanka and SW India it seems to be distinct but elsewhere there are so many intermediates that it is impossible to delimit them as two separate species. I feel it is wiser, pending experimental investigations, to combine them into one species. I rather suspect that *Bonnayodes limnophiloides* also belongs here.

Limnophila tillaeoides is known only from two gatherings from the banks of Kullung River, Assam; it is like *L. indica* but is smaller in all its parts (sepal tube 2.5–3.5 mm long, sepal lobes 1–1.5 mm long, petals 5–5.5 mm long, capsules 1.5–3.5 mm long, style 1.2–1.75 mm long) and is totally glabrous. It is probably no more than a depauperate state of *L. indica.*

Limnophila indica is known to hybridize with *L. sessiliflora.* This hybrid (*L.* × *ludoviciana* Thieret) is fertile; it is not recorded from India.

{(⊙) Vit/Hyp/(Ten)}. Tropical Africa, Asia and Australia: Ban, Bur, Cey, Ind, Nep? Pak: As, Go, Jk, Kl, Kt, Mh, Mn, Mp, Or, Rj, Sk, Tn, Up, Wb.

Limnophila laxa Bentham [*Stemodia diffusa* Bentham]: Fig. 351

Like *L. repens* but flowers mostly solitary in axils; pedicels longer ((0.5–) 3–5 (–7) mm long); main stems and branches usually hairy with moderately long, straight and spreading hairs, particularly towards the apex; sepals longer (4–8.5 mm long); petals puberulent outside, deep violet to white.

Annual. It is found at the edges of ponds, lakes and rivers. Sometimes it grows in brackish water.

{⊙ Hel/Ten?}. Sri Lanka and SE Asia from Myanmar to Sumatra: Bur, Cey unconfirmed.

Limnophila micrantha (Bentham) Bentham: Fig. 351

Stems suberect to creeping, profusely branched; branches, stout, erect, 3–15 (–20) cm high, glabrous. Leaves decussate or rarely in whorls of 3; blades simple, sessile, linear-oblong to ovate, 3–14 mm long, 1–5 mm wide, glabrous, densely punctate, pinnately nerved; margin entire to crenate-serrate. Flowers decussate, solitary in leaf axils or in heads of 2–4 flowers, sessile or nearly so; pedicels not exceeding 1 mm long; bracteoles 1.5–3 mm long, linear to filiform, glabrous. Sepals 2.5–3 mm long, glabrous to minutely pubescent at the apex, striate at maturity; sepal lobes subequal, 1.5–2 mm long, lanceolate to subulate, ciliate. Petals 2.5–4 mm long, white or mauve to purple-red, throat sparsely villous otherwise glabrous. Capsules 2–2.5 mm long, flattened-spherical to broadly ellipsoid, light purplish-brown; style up to 4.5 mm long with 2 lateral processes ± 0.2 mm wide at the apex below the stigma.

Annual. Open seasonally inundated places and rice fields.

{⊙ Hel/Ten?}. S and SE Asia: Ban, Bur, Ind: As? Mg? Mh? Wb.

Limnophila polystachya Bentham [*L. polystachyoides* Blatter]: Fig. 352

Submerged stems up to 50 cm long, sub-glabrous to densely covered with white hairs; aerial stems erect, up to 20 cm tall, glabrous or minutely hairy, particularly at the nodes. Submerged leaves in whorls of 3–12, much shorter than the aerial leaves, repeatedly divided into fine segments, the segments capillary or linear and flattened; aerial leaves decussate or in whorls of 3 or 4, the blades sessile, linear to elliptical-lanceolate, (some intermediate leaves pinnatifid below and becoming lacerate towards the apex), (6–) 10–40 mm long, 2–8 mm wide, glabrous, densely and minutely punctate, 3-nerved, the margin crenate to serrate, the base attenuate, the tip acute. Flowers alternate, sessile in long-peduncled, slender but compact spikes; upper bracts ± 3.5 mm long, oblong-lanceolate, crenate-serrate, minutely glandular. Sepals 2–3 mm long, bearing sessile glands, not striate at maturity; sepal lobes subequal, shortly ciliate. Petals ± 6.5 mm long. white to pinkish-white, villous within; adaxial lip almost entire. Capsules ± 2 mm long, globose, pale brown; style ± 3 mm long with 2 lateral processes at the apex below the stigma.

Annual. Forms extensive clumps in still water.

{⊙ Hyp/Vit/Hel?}. Endemic to S India and Sri Lanka: ♦Cey (record not confirmed), ♦Ind: Mh, Kl, Tn.

Limnophila polyantha S. Kurz ex J. D. Hooker: Fig. 352

Like *L. aromatica* but all green parts densely covered in sessile and short-stalked glands and eglandular hairs; leaf blades narrower (linear to linear-lanceolate, 3–12 mm long, 0.5–1.5 mm wide); floral parts smaller (sepals 2.5–3 mm long, petals 4–5 mm long, capsules ± 2 mm long).

Annual. Found emergent or floating on the surface of water.

{⊙ Hel?}. Endemic to NE India and Myanmar: ♦Bur, ♦Ind: Sk, Wb.

Limnophila pulcherrima (Griffith) J. D. Hooker [? *L. diffusa* sensu (Bentham) A. P. de Candolle, non (Roxburgh) G. Don, *L. pygmaea* J. D. Hooker, ? *L. cimicina* (Bentham ex Wallich) Narenda Mohan Dutta]: refer to Figs. 349 and 351

Like *L. chinensis* but green parts not bearing wool-like hairs; stems bearing minute glands; leaves sometimes shortly petiolate; leaf blades glabrous and smaller (3–12 (–17) mm long, 1–3 (–5) mm wide); flowers all solitary in axils and smaller (sepals 3–4 mm long, petals 6–8 mm long, capsules 2.5–3 mm long); petals glabrous outside.

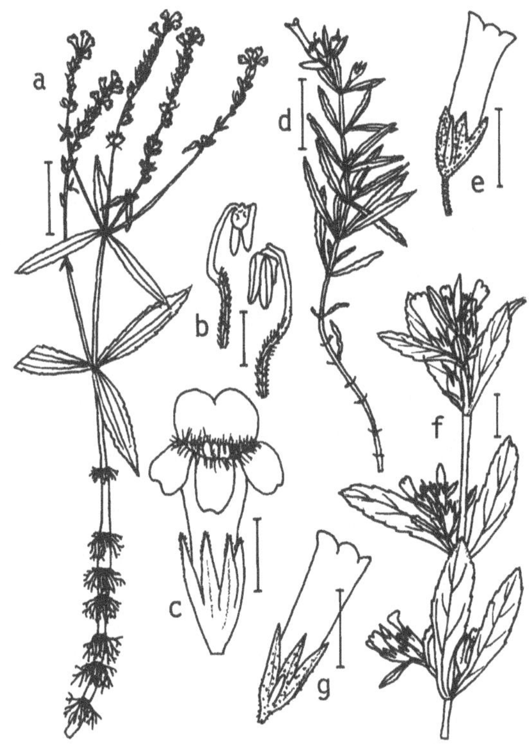

FIG. 352. *Limnophila polystachya*: a, flowering shoot (2 cm); b, the long stamen viewed from opposite sides (1 mm); c, flower (2 mm). *L. polyantha*: d, flowering shoot (2 cm); e, flower (5 mm). *L. repens*: f, flowering shoot (1 cm); g, flower (5 mm).

Annual. Ecology unknown but probably a plant from marshes. I am not sure that this is really a good species. Perhaps it should be combined with *L. laxa*.

{⊙ Hel?/Ten?}. Endemic to Myanmar, Bangladesh and West Bengal: ♦Ban, ♦Bur, ♦Ind: Wb.

Limnophila repens (Bentham) Bentham [*L. conferta* (Bentham) Bentham, *L. serrata* Thwaites]: Fig. 352

Stems decumbent, rooting at nodes or erect, simple or branched from or near the base; 5–40 cm high, glabrous or hispid on the upper part, the hairs mostly upcurved, not spreading. Leaves decussate; blades simple, sessile or sub-petiolate, oblong-lanceolate to oblong, (5–) 12–26 (–55) mm long, (2.5–) 4–8 (–15) mm wide, punctate, pinnately nerved, glabrous to hispid; the base cuneate to subamplexicaul; the margins crenate to serrate; the tip subacute. Flowers subsessile to shortly pedicellate, solitary in leaf axils or occasionally in short axillary spikes or racemes; pedicels (0.5–) 1–2 (–4) mm long, subglabrous to hispid, erect at maturity; bracteoles 1.5–2.5 (–3.5) mm long, linear-spathulate. Sepals 4–5 mm long; the tube ± 1.3 mm long, subglabrous to hirsute or hispid, striate at maturity; sepal lobes subequal, 2–4 mm long, narrowly lanceolate-acuminate, ciliate. Petals white, pink, blue or mauve (often variable within a population), glabrous to finely pubescent outside, villous inside; petal tube 6–8 mm long; abaxial petal lobe ± 4 mm long, sub-truncate. Capsules ovoid to ellipsoid, 2.5–3 (–4) mm long; style up to 4 mm long with 2 lateral processes at the apex below the stigma. Seeds cylindrical and bluntly angular, ± 0.3 mm long, ± 0.3 wide, brown and shining.

Annual. Widespread in marshy places and in rice fields particularly in low lying areas. It often develops quite large patches and may be locally dominant. The leaves smell and taste slightly like lemons; they are sometimes added to betel in paan.

{⊙ Hel/Ten?}. Tropical Asia: Ban, Bur, Cey, Ind, Nep: As, Kl, Kt, Mh? Or, Sk, Wb.

Limnophila rugosa (Heyne ex Roth) Merrill [*L. roxburghii* auct. Ind., non G. Don, ?*L. cimicina* (Bentham ex Wallich) Narenda Mohan Dutta, *Stemodia menthastrum* Bentham]: Fig. 353

Stems stout, simple or branched; 15–50 cm high, glabrous to densely hairy. Leaves decussate, petiolate; petioles 5–30 mm long, sparsely to densely hairy; leaf blades simple, ovate to oblong-ovate, 1.5–9 cm long, 7–50 mm wide, pinnately nerved, glabrous to scabrid adaxially, hirsute to scabrid on the major nerves of the abaxial surface, the base decurrent into the petiole, the margins crenate to crenate-dentate, the apex acute. Flowers sessile, solitary in leaf axils or in axillary, short or long pedunculate heads, (the flowers in heads or clusters subtended by a subspathulate to oblong, bract-like leaf, 11–15 mm long, 3–5 mm wide, 3 or 5-nerved from the base, glabrous or ciliate at the base); peduncles 2–35 mm long, glabrous to hirsute; bracteoles absent. Sepals 5–6 mm long, glandular, hirsute on the main nerves, not striate at maturity; sepal lobes unequal; adaxial lobe 8–11 long, ± 4 mm wide, broadly ovate-lanceolate, acute, ciliate; remaining sepal lobes 6–9 mm long, narrowly ovate-lanceolate, ciliate. Petals 13–16 mm long, pale blue to violet with yellow in the throat, pilose outside and inside; petal tube ± 7 mm long; abaxial petal lobe ± 6 mm. Capsules flattened-ovoid, 5–6.5 mm long; style up to 6 mm long, broadened above just below the stigma. Seeds cylindrical to oblong, ± 0.5 mm long, ± 0.25 mm in diameter.

Annual. Marshy places, streams and river banks. It seems to prefer shady habitats. In permanently wet habitats it flowers throughout the year.

{⊙ Hel/Ten?}. Tropical Asia: Ban, Bur, Cey, Ind, Nep: Ap, As, Bh, Kl, Mg? Mn, Mp, Or, Sk, Tp, Up, Wb.

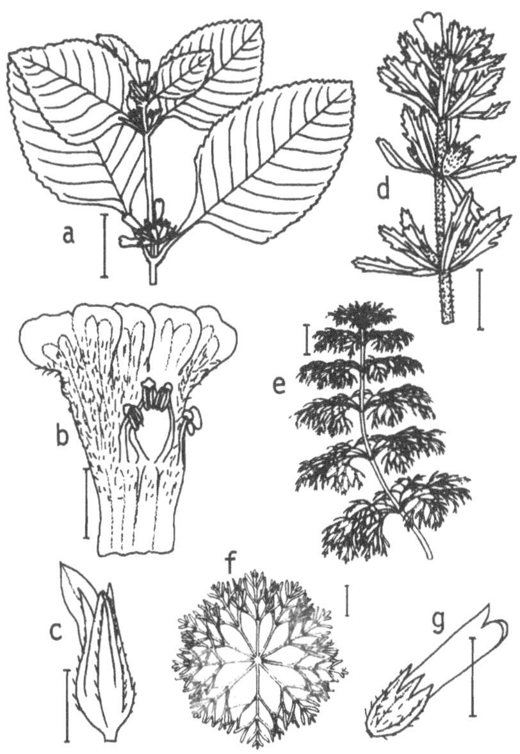

FIG. 353. *Limnophila rugosa*: a, tip of a flowering shoot (2 cm); b, petals opened (4 mm); c, sepals in fruit (4 mm). *L. sessiliflora*: d, tip of a flowering shoot (1 cm); e, submerged shoot (2 mm); f, whorl of submerged leaves (2 mm); g, flower (5 mm).

Limnophila sessiliflora (Vahl) Blume: Fig. 353

Submerged stems 10–60 cm long, glabrous to subglabrous; aerial stems erect or creeping, simple or branched, 5–20 cm high, subglabrous to hairy. Submerged leaves in whorls of 5–12, (5–) 10–20 (–30) mm long, glabrous, compound-pinnatisect; lobes divided into capillary or linear and flattened segments; aerial leaves in whorls of 3–8, usually dissected or rarely entire, 4–12 (–20) mm long, glabrous, when laminate densely punctate, 1- or 3-nerved. Flowers solitary in the axils of leaves, sessile or very nearly so, frequently cleistogamous; bracteoles usually absent or if present then not more than 1.5 mm long and entire. Sepals 3–4 at anthesis becoming 5–7 mm long in fruit, densely to sparsely hirsute, not striate at maturity; sepal lobes subequal, lanceolate, 2–3.5 (–4) mm long, acuminate. Petals (5–) 8–10.5 mm long, blue to violet or purple, glabrous outside. Capsules flattened-globose, 3–4 (–5.5) mm long; style ± 3 mm long with 2 lateral processes at the apex below the stigma. Seeds cylindrical to ovate-oblong, ± 0.5 mm long, ± 0.2 mm in diameter, 5-angular, muricate.

Annual or perennial. Found in permanent and temporary water. It is very like *L. heterophylla* but differs in being usually glabrous and lacks finely divided aerial leaves and bracts. These characters are readily modified by environmental factors and perhaps *L. heterophylla* and *L. sessiliflora* should be united. Further investigations are needed. *Limnophila sessiliflora* is also like *L. indica* but the flowers are usually sessile (if pedicellate then the pedicels not exceeding 1.5 mm long); submerged stems glabrous; emergent stems and sepals with patent white hairs; bracteoles absent or minute; cleistogamous flowers common. It is known to hybridize with *L. indica.* This hybrid (*L.* × *ludoviciana* Thieret) is fertile; it is not recorded from India.

{(⊙) Vit/Hyp/Ten?}. Tropical and E Asia: Ban, Bur, Cey, Ind, Nep: As, Mg, Mp, Sk, Wb.

Lindernia [*Bonnaya, Ilysanthes, Vandellia*]: ± 80 species: warm regions of the world

Stems erect ascending or creeping, usually quadrangular in transverse section. Leaves opposite, simple; sessile or shortly petiolate, entire, crenate or toothed. Flowers solitary in bract axils or in terminal or axillary racemes or umbel-like clusters; bracts leaf-like or scale-like. Sepal lobes 5, shallow or almost reaching the base of the tube; sepal tube 5-nerved, each nerve with an obscure or distinct rib or wing. Petals tubular, 2-lipped; adaxial lip 3-lobed, spreading, longer than the abaxial lip. Stamens either 4 and all with fertile anthers or with 2 fertile and 2 sterile; filaments of the adaxial pair frequently bear spur-like or elongate appendages at or near the base; anther locules divergent, free or joined. Style capillary, erect; stigma 2-lipped. Capsules globose to narrowly cylindrical, with septicidal dehiscence, usually leaving the entire septum as a median plate.

Annuals or short-lived perennials. Most species grow in wet places. They may be seasonally submerged or partly emergent but none are strictly aquatic. Pollination is by insects or some are autogamous or even cleistogamous. The disseminules are seeds; dispersed in mud or otherwise. Several have been reported as weeds in aquatic systems but they are rarely considered to be troublesome. Mukherjee, S. K. Revision of the Indo-Burmese species of *Lindernia* Allioni. Journal of the Indian Botanical Society 24: 127–134 (1945); Philcox, D. A. Revision of the Malesian species of *Lindernia.* Kew Bulletin 22: 1–72 (1968); Yamazaki, T. Revision of the Indo-Chinese species of *Lindernia* All. Journal of the Faculty of Science, University of Tokyo, Section III, Botany 13(1): 1–64 (1981); Sivarajan, V. V. and Mathew, P. The genus *Lindernia* All. (Scrophulariaceae) in India. Journal of the Bombay Natural History Society 80: 131–140 (1983). For this account I have mostly relied on the work of Takasi Yamazaki.

1. Sepals united to above the middle
- **2.** Capsules linear-cylindrical, ± twice as long as the sepals
  - **3.** Leaf blades elliptical-ovate to oblong, the teeth along the margins terminating in spines ***L. ciliata***
  - **3.** Leaf blades ovate-orbicular to almost orbicular, the teeth along the margins not terminating in spines ***L. nummulariifolia***
- **2.** Capsules ellipsoid to oblong-ovoid, not or slightly longer than the sepals
  - **4.** Petals 7–11 mm long; all flowers solitary and borne in the axils of leaf-like bracts ***L. crustacea***
  - **4.** Petals 4–5 mm long; the uppermost flowers forming umbel-like clusters and borne in the axils of scale-like bracts ***L. molluginoides***

1. Sepals free almost to the base or if united then to below the middle
- **5.** Stamens 4, all bearing well-developed anthers
  - **6.** Capsules much longer than the sepals ***L. anagallis***
  - **6.** Capsules equalling or shorter than the sepals
    - **7.** At least the uppermost flowers in terminal racemes with each flower subtended by a scale-like bract
      - **8.** Petals 9–12 mm long; stems robust, erect; leaf blades 3–13 cm long, 2–5 cm wide ***L. latifolia***
      - **8.** Petals less than 6 mm long; stems slender, usually creeping below; leaf blades less than 3.5 cm long, 0.3–2 cm wide

9. Bracts gradually decreasing in size towards the top of the inflorescence; capsules ellipsoid, 3–4 mm long, 1.5–2 mm wide ***L. elata***
9. Bracts suddenly decreasing in size towards the top of the inflorescence; capsules globose, 2–2.5 mm long and wide ***L. viscosa***
7. All flowers solitary in the axils of leaf-like bracts
10. Leaf blades palmately 3- or 5-nerved (appearing parallel-nerved from the base to the tip of the blade)
11. Capsules cylindrical, 9–13 mm long, more than twice as long as the sepals ***L. micrantha***
11. Capsules ellipsoid, 3–4 mm long, ± as long or slightly longer than the sepals ***L. procumbens***
10. Leaf blades pinnately nerved or 1-nerved
12. Capsules obovoid to globose, 2–4 mm long and wide; adaxial petal lip brown ***L. pusilla***
12. Capsules ellipsoid, 4–5 mm long, 25–3 mm wide; adaxial petal lip white to pale violet or purple
13. Anthers of abaxial stamens much smaller than the adaxial ones; petioles not more than 0.5 cm long; adaxial anthers with tails ***L. hookeri***
13. Anthers of abaxial stamens ± the same size as the adaxial ones; petioles 1.5–4 cm long; adaxial anthers without tails ***L. mollis***
5. Stamens 2 or 4, when 4 then 2 staminodial (not bearing fertile anthers)
14. Leaf blades palmately 3- or 5-nerved (appearing parallel-nerved from the base to the tip of the blade)
15. Capsules ellipsoidal, 5–7 mm long, ± twice as long as the sepals
16. Leaf blades linear-lanceolate to oblong, 5–30 cm long, 1–6 mm wide ***L. hyssopioides***
16. Leaf blades elliptical to ovate, up to 10 mm long, up to 6 mm wide ***L. manilaliana***
15. Capsules ± globose, 2.5–4 mm long, ± equal to or shorter than the sepals
17. Stems creeping or decumbent; all flowers subtended by leaf-like bracts; adaxial petal lip with entire margins (South India and Sri Lanka) ***L. rotundifolia***
17. Stems erect; upper flowers subtended by scale-like bracts; adaxial petal lip with saw-like margins (Sri Lanka) ***L. srilankana***
14. Leaf blades pinnately nerved or 1-nerved
18. Most flowers alternate, leaf opposed (1 flower only at each node)
19. Capsules oblong-globose ***L. parviflora***
19. Capsules ellipsoid to cylindrical (at least twice as long as the sepals)
20. Bracts ± as long as the pedicels (pedicels up to 7 mm long); sepals up to 6 mm long; petals white; staminodes absent (SW India) ***L. estaminodiosa***
20. Bracts much shorter than the pedicels; sepals in fruit 4 mm or less long; petals pink blue or purple; staminodes present (SW India)
21. Capsules ovoid-ellipsoid, ± twice as long as the sepals; pedicels (0.5–) 2–5 cm long; sepals 3–4 mm long ***L. hyssopoides***
21. Capsules cylindrical, 3 to 4 times as long as the sepals; pedicels 0.2–1 (–1.7) cm long; sepals 1.5–2 (–4) mm long ***L. tenuifolia***
18. Most flowers in opposite pairs
22. Capsules shorter than sepals, ellipsoid to almost globose, not more than 5 mm long
23. Stems and leaves hairy, creeping below, with more than 3 pairs of leaves (N India) ***L. hookeri***
23. Stems and leaves glabrous or nearly so, erect with rarely more than 3 pairs of leaves (S India) ***L. minima***
22. Capsules much longer than the sepals, ellipsoid to almost globose, not more than 5 mm long
24. Leaf blade margins bluntly and shallowly toothed (teeth ± 3 mm apart) or almost entire (W India) ***L. estaminodiosa***
24. Leaf blade margins saw-like (with sharp teeth, less than 3 mm apart)
25. Leaf blades linear to oblong, the margins never saw-like ***L. oppositifolia***
25. Leaf blades broadly ovate to elliptical, the margins saw-like
26. Leaves petiolate (petioles 4–20 mm long) ***L. ruelloides***
26. Leaf blades sessile or very nearly so
27. Teeth along the leaf blade margin not terminating in spines; staminodes ± 3 mm long ***L. antipoda***
27. Each tooth on the leaf blade margin terminating in a spine; staminodes ± 1.5 mm long ***L. ciliata***

Lindernia anagallis (N. L. Burman) Pennell [*Vandellia cordifolia* (Colsmann) G. Don, *V. angustifolia* auct. Ind., non Bentham, *V. pedunculata* Bentham]: Fig. 354

Stems glabrous, spreading, creeping and rooting at nodes, (3–) 20–60 cm long. Leaves glabrous, sessile or shortly petiolate; petiole up to ± 5 mm long; blades ovate to elliptical-deltate, 0.5–2.5 cm long, 0.3–1.8 cm wide, pinnately nerved, base subcordate to truncate, margins appressed crenate to subentire, tips subacute. Flowers solitary in the axils of leaf-like bracts; pedicels 0.6–5 cm long, stiff, glabrous. Sepals deeply 5-lobed almost to the base, 4–6 mm long; lobes narrowly lanceolate, acuminate, glabrous. Petals 10–16 mm long, ± 5 mm in diameter, white to pink or purple, with or without white flashes on the lip; petal tube ± 6 mm long; adaxial lip ± 3.5 mm long, entire; abaxial lip

± 3 mm long. Stamens 4, all with fertile anthers; adaxial pair with ± 1.8 mm long filaments, the anthers ± 1 mm long and spurred at both ends; abaxial pair with 3.5 mm long filaments, each with a rounded, spur-like, appendage, 0.4–1 mm long, arising near the base, the anthers each with a basal spur as long as the anther itself. Capsules cylindrical, acuminate, 0.8–1.2 cm long, 1.5–2 mm in diameter, more than twice the length of the sepals, arching upwards when ripe. Seeds shortly cylindrical, 0.3–0.5 mm long, ± 0.15 mm in diameter, longitudinally grooved, minutely pitted.

Annual. Gregarious and locally abundant in marshy places, banks of ponds, riversides and wet rice fields. Used as a remedy for gonorrhoea

{⊙ Hel}. Tropical Asia and Australia: Cey, Ind, Nep: As, Bh, Gj, Go, Kl, Kt, Mh, Mn, Mp, Or, Rj, Sk, Tn, Tp, Up, Wb. 0–1400 m.

Lindernia antipoda (Linnaeus) Alston [*Ilysanthes verbenaefolia* (Colsmann) Kerr, *Bonnaya veronicifolia* (Retzius) Sprengel, *B. grandiflora* (Retzius) Sprengel]: Fig. 354

Stems glabrous, prostrate and rooting below, spreading above, 5–20 (–40) cm long. Leaves sessile, oblanceolate to obovate-oblong, 0.6–4.5 cm long, 0.2–1.4 cm wide, pinnately nerved, base narrowed, margins shallowly saw-like, tips acute. Flowers solitary in the axils of leaf-like bracts or in lax racemes, borne singly or in pairs; bracts oblong-lanceolate to linear, 3–7 mm long, margins saw-like; pedicel stout, 2–6 mm long at anthesis, becoming 4–17 mm long in fruit, glabrous. Sepals deeply 5-lobed almost to the base, 3–5 mm long at anthesis, 5–6 mm long in fruit; lobes linear-lanceolate, minutely scabrid and minutely toothed. Petals 8–10 mm long, ± 6 mm in diameter, white or blue to purple, usually yellow at the throat; petal tube ± 5 mm long; adaxial lip 3–4 mm long, entire; abaxial lip 5–6 mm long, with 2 yellow tongues below. Stamens 4, 2 with fertile anthers; filaments ± 2 mm long, anthers ± 1 mm long; staminodes 2, ± 3 mm long, yellow, hooked at the tips, glabrescent, the hooks convergent. Capsules cylindrical, apiculate, 10–16 mm long, 1–1.5 mm in diameter, 2 to 3 times as long as the sepals, spreading horizontally or slightly deflexed when ripe. Seeds shortly cylindrical, 0.3–0.5 mm long, ± 0.2 mm in diameter, striate, minutely pitted.

Annual. Along the banks of rivers, tanks and ponds, on banks, in rice fields, in short grass and in exposed humus-filled rocky hollows. It is rather variable in form, size and flower colour.

{⊙ Hel}. Temperate and tropical Asia, Australia and Polynesia: Ban? Bur, Cey, Ind, Nep: An, As, Bh, Kl, Kt, Mn, Mp, Or, Rj, Tn, Tp, Up, Wb. 0–2300 m.

FIG. 354. *Lindernia anagallis*: a, flowering shoot (2 cm); b, petal tube opened (2 mm); c, seed (0.15 mm). *L. antipoda*: d, flowering shoot (2 cm); e, petal tube opened (2.5 mm). *L. ciliata*: f, flowering shoot (2 cm); g, petal tube opened (1.7 mm).

Lindernia ciliata (Colsmann) Pennell [*Ilysanthes serrata* (Roxburgh) Urban, *Bonnaya brachiata* Link et Otto, *B. bracteoides* Blatter et Hallberg]: Fig. 354

Stems glabrous or sparsely hairy at the nodes, laxly branched, erect or ascendent, 5–20 cm tall. Leaves glabrous, sessile, elliptical-obovate to oblong, 0.6–3.5 cm long, 0.2–1.2 cm wide, pinnately nerved, base narrowed and semi-amplexicaul, margins saw-like, the teeth very sharp and each terminating in a spine, the tips terminating in spines. Flowers in up to 10 cm long, terminal, laxly-flowered racemes, borne in opposite pairs; bracts reduced, scale-like, linear, 2–4 mm long; pedicels 1–5 mm long at anthesis, 3–15 mm long in fruit. Sepals glabrous, when young tubular almost to the tip, usually becoming deeply 5-lobed almost to the base later, 3–5 mm long; lobes lanceolate to subulate, equal or 1 smaller. Petals 6–8 mm long, up to 1 cm in diameter, petal tube ± 3.5 mm long, whitish to pink or blue below, blue or purple in the throat; adaxial lip 2–6 mm long, oblong; abaxial lip 3–4 mm long. Stamens 4, 2 with fertile anthers; filaments ± 1.5 mm long, anthers unequal, ± 1 mm long; staminodes ± 1.5 mm long, the tips divergent. Capsules cylindrical, acuminate, 7–17 mm long, 2–3 times as long as the sepals, arching upwards when ripe. Seeds ellipsoid with truncate ends, ± 0.3 mm long, ± 0.15 mm in diameter, pitted.

Annual. Often gregarious, found álong the banks of post-monsoon streams, in rice fields, moist ground, damp rocks, lateritic slopes and in wet grassland.

{⊙ Hel}. Tropical Asia and Australia: Bur, Cey, Ind, Nep: Ap, As, Bh, Dl, Gj, Go, Kl, Kt, Mh, Mn, Mp, Or, Pj, Rj, Sk, Tn, Up, Wb. 0–1500 m.

Lindernia crustacea (Linnaeus) F. von Mueller [*L. crustacea* var. *clarkei* Giri et Majumdar, *Vandellia racemosa* Bonati, *V. gracilis* Bonati]: Fig. 355

Stems 4-angled, glabrous or sparsely pubescent on the angles, laxly branched, prostrate or erect, with ascending branches, occasionally rooting below, 2–20 cm long. Leaves petiolate becoming sessile towards the tips; petiole up to 1 cm long, glabrous or pubescent; blades ovate to almost rhomboid or suborbicular, 0.5–1.7 cm long, ± 0.4–1.5 cm wide, pinnately nerved, glabrous to sparsely hispid on the margins, base truncate to cordate, margins coarsely toothed with rounded or pointed teeth and ciliate, tips subacute. Flowers solitary in the axils of leaf-like bracts, borne in opposite pairs; bracts becoming smaller above; pedicel 0.4–2.5 cm long, sometimes forming lax terminal racemes. Sepals tubular, united from above the middle, often red, 3–4 mm long at anthesis becoming 4–6 mm long in fruit; tube 5-ribbed, ridged or slightly winged, sparsely hispid on the main nerves; lobes short, acute. Petals 7–11 mm long, ± 4 mm in diameter, white or pink to blue or purple or white with blue flecks on the margins; petal tube 4–7 mm long, sometimes pale yellow at the base; adaxial lip 2–3 mm long, 2-lobed; abaxial lip 2–4 mm long. Stamens 4, all with fertile anthers; adaxial pair with ± 1.5 mm long filaments, the anthers ± 0.7 mm long; abaxial pair with ± 3.5 mm long filaments, each with a subulate, purplish-blue blunt appendage 0.4–0.5 (–0.9) mm long, near the base, the anthers like the abaxial ones. Capsules oblong-ovoid, 3–5 mm long, 2–3 mm in diameter, not or slightly exceeding the sepals, patent when ripe. Seeds oval, 0.3–0.5 mm long, ± 0.25 mm in diameter, pitted.

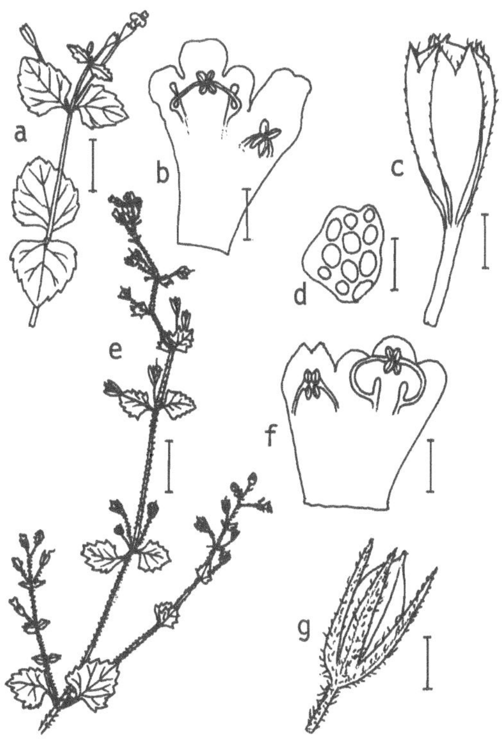

Fig. 355. *Lindernia crustacea*: a, flowering shoot (1 cm); b, petal tube opened (1.6 mm); c, sepals with capsule (1.6 mm); d, seed (0.25 mm). *L. elata*: e, flowering shoot (1 cm); f, petal tube opened (1 mm); g, sepals with capsule (1.6 mm).

Annual. Low lying pastures, river banks, rice fields after the harvest and drying out tanks. Sometimes it grows near the sea shore. Mostly common and gregarious, the small leaves and large flowers make it a elegant plant. Used in medicine for biliousness, dysentery, in poultices for boils and against ringworm.

{⊙ Hel}. Warm and tropical Old World, naturalized in the New World: Cey, Ind, Nep: An, Ap, As, Dl, Gj, Go, Kl, Kt, Mh, Mn, Mp, Mz, Or, Rj, Tn, Tp, Up, Wb. 0–1800 m.

Lindernia elata (Bentham) Wettstein [*Vandellia urticifolia* Hance]: Fig. 355

Stems laxly branched, creeping and rooting at nodes or suberect, 10–50 cm long, sparsely to densely hispid with spreading long white hairs intermixed with glandular hairs. Leaves sessile or shortly petiolate; petioles 2–13 mm long, with long white hairs; blades ovate to ovate-deltate, 4–20 mm long, 3–15 mm wide, pinnately nerved, with long white hairs intermixed with glandular hairs on both surfaces, the base truncate, margins sharp toothed, the tip acute. Flowers either solitary or in clusters of 4–6 flowers at nodes; pedicels slender, 2–10 mm long, with long white hairs. Sepals 3–5 mm long, cleft almost to the base, with spreading white hairs; lobes linear-lanceolate, acuminate. Petals 4–5 mm long, white to bluish, fragrant; adaxial (upper) lip 2-lobed; abaxial (lower) lip 3-lobed, longer than the adaxial lip, the central lobe larger, 3–4 mm long. Stamens 4, all with fertile anthers; abaxial (lower) pair each with a rounded, knee-like appendage ± 0.4 mm long. Capsules ellipsoid, 3–4 mm long,

1.5–2 mm wide, shorter than the sepals, spreading or deflexed when ripe. Seeds shortly cylindrical, ± 0.35 mm long, ± 0.25 mm wide, minutely pitted.

Annual. Mostly in dry sandy fields and meadows but it also grows in rice fields and marshes.

{⊙ Hel}. Tropical Asia: Bur, Ind: An.

Lindernia estaminodiosa (Blatter et Hallberg) Mukherjee: refer to Fig. 362

Like *L. tenuifolia* but larger, leaf blades oblong to oblanceolate, 15–30 mm long, the margins with shallowly saw-like teeth; flowers in lax terminal and lateral racemes; bracts ± as long as the pedicels; pedicels up to 7 mm long; sepals up to 6 mm long in fruit; petals white; staminodes absent; flowers in terminal racemes.

Annual. Wet lowlands and backwaters. The description and orthography of the name in Sivarajan and Mathew (1983) do not agree with the original description of Blatter and Hallberg — I have seen no herbarium material and have followed the original description.

{⊙ Hel}. Endemic to SW India: ♦Ind: Kl, Kt, Mh.

Lindernia hookeri (C. B. Clarke) Wettstein [*L. hookeri* var. *kumaunensis* Pennell, *Vandellia stemonoides* Haines]: Fig. 356

Like *L. anagallis* but capsules shorter than sepals, ellipsoid, ± 5 mm long; pedicels not exceeding 15 mm long; the abaxial stamen pair sterile, bearing a ± 0.3 mm long spur near the middle and rudimentary anthers.

Annual. In moist and wet places, also in running water on hill slopes.

{⊙ Hel}. Himalayas and Indo-China (India, Nepal, Sikkim), India: Bur, Ind, Nep: Ap? Ar, As, Bh, Mg, Mp, Or, Up.

Lindernia hyssopoides (Linnaeus) Haines ['*hyssopioides*' orth. mut.]: Fig. 356

Stems glabrous, simple or diffusely branched, erect or ascending, rather brittle, 5–30 cm or more tall. Leaf blades sessile, linear-lanceolate to oblong, 5–30 mm long, 1–6 mm wide, glabrous faintly 3-nerved from the base or 1-nerved, base narrowed, margins entire or minutely dentate, tips acute. Flowers solitary in the axils of leaf-like bracts, borne singly or very rarely in pairs; bracts linear, reduced and scale-like above; pedicel (0.5–) 2–5 cm long. Sepals 3–4 mm long, deeply lobed almost to base; lobes linear-lanceolate. Petals (7–) 10–15 mm long (mostly more than 10 mm long), ± 8 mm in diameter pale to deep blue, to purple at the throat, with whitish lobes; petal tube 6–9 mm long; adaxial lip ± 4 mm long, slightly notched, with dark blue lines; abaxial lip ± 5 mm long, obtuse, with dark blue blotches at the bases of the lobes. Stamens 4, the adaxial 2 with fertile anthers; filaments 2–2.5 mm long, anthers ± 0.7 mm long; staminodes blunt, ± 0.7 mm long, with a linear and glandular hairy basal appendage. Capsules ovoid-ellipsoid, 5–7 mm long, 2–3 mm in diameter, almost 2 times as long as the sepals, deflexed when ripe. Seeds oblong, ± 0.4 mm long, ± 0.15 mm in diameter, longitudinally ridged, reticulate.

FIG. 356. *Lindernia hookeri*: a, flowering shoot (1 cm); b, petal tube opened (2 mm); c, sepals with capsule (2.5 mm). *L. hyssopoides*: d, flowering shoot (2 cm); e, petal tube opened (4 mm); f, sepals with capsule (4 mm).

Annual. Common and abundant, often forming dense mats in perennially moist places and harvested rice fields. A very variable species which sometimes intergrades with *L. parviflora* and *L. manilaliana*.

{⊙ Hel}. SE Asia: Bur? Cey, Ind: As, Bh, Kl, Kt, Mg, Mn, Mp, Or, Rj, Tn.

Lindernia latifolia (Blume) Koorders [*Vandellia foliosa* Bonati]: Fig. 357

Stems erect, 20–50 cm high; branches spreading. Leaves petiolate becoming almost sessile above; petioles up to 2.5 cm long; blades elliptical, ovate to obovate, (2.5–) 7.5–15 cm long, (1–) 2.5–6 cm wide, pinnately nerved, hairy on both surfaces, particularly along the nerves on the abaxial surface, base cuneate, margins saw-like, tips obtuse. Flowers in terminal or axillary racemes; racemes with up to ± 50 flowers;

pedicels slender, 10–25 mm long, with glandular hairs; bracts subulate, 1–1.5 mm long. Sepals 6–8 mm long at anthesis becoming 10–13 mm long in fruit, cleft almost to the base, with spreading hairs and glands; lobes linear-lanceolate, acute, ciliate. Petals 8.5–12 mm long, white to pink with or without a yellow throat; adaxial lip broadly ovate, 2-lobed; abaxial lip longer than the adaxial lip. Stamens 4, all with fertile anthers; adaxial pair with ± 1 mm long filaments; abaxial pair each with a rounded, knee-like appendage ± 0.7 mm wide, the filaments ± 3 mm long. Capsules ellipsoid to globose, 4–6 mm long, shorter than sepals, remaining erect when ripe. Seeds ovoid to spherical, ± 0.5 mm in diameter, pitted.

Annual. A plant found on moist ground, rarely growing in wetlands.

{⊙ Hel}. Tropical SE Asia. Bur.

Lindernia manilaliana V. V. Sivarajan: Fig. 357

Like *L. parviflora* but stems rather succulent and stout; leaf blades always sessile, ovate to elliptical; pedicels not more than 1 cm long, remaining erect in fruit; bracts all leaf-like; sepal lobes oblong-obtuse; petals 6–7 mm long; staminodes rather smaller and simple.

Annual. In swampy lowlands. It grades into *L. parviflora* which, in turn, grades into *L. hyssopiodes.* I suspect it is a late developing rather xeromorphic terrestrial ecad of *L. parviflora.*

{⊙ Hel}. Endemic to S India: ♦Ind: Kl (Calicut). August–December.

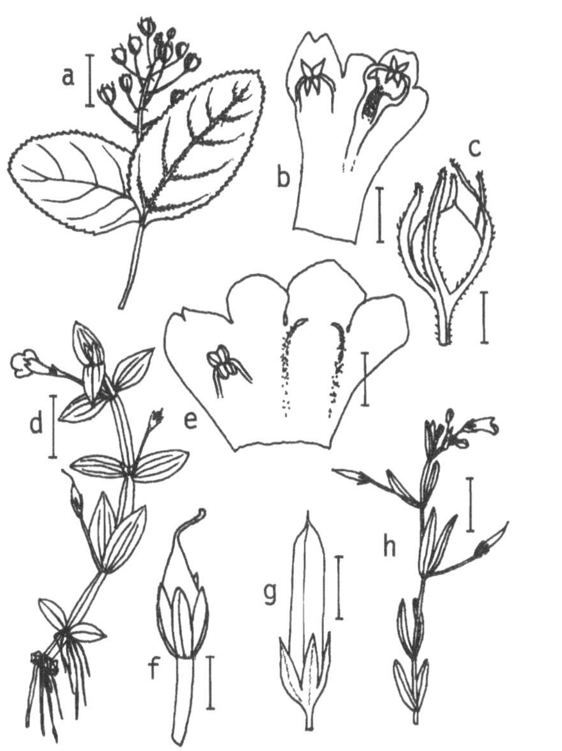

Fig. 357. *Lindernia latifolia*: a, tip of a flowering shoot (6 mm); b, petal tube opened (1.5 mm); c, sepals with capsule (2.5 mm). *L. manilaliana*: d, flowering shoot (1 cm); e, petal tube opened (2.5 mm); f, sepals with capsule (4 mm). *L. micrantha*: g, sepals with capsule (3 mm); h, flowering shoot (1 cm).

Lindernia micrantha D. Don [*Vandellia angustifolia* Bentham]: Fig. 357

Stems glabrous, laxly branched, erect or ascendent, usually not rooting at nodes, 7–25 cm tall. Leaves glabrous, sessile, linear to linear-lanceolate, 1–2 (–4) cm long, 0.3–6 mm wide, pinnately nerved, base attenuate, margins entire or sparsely and obscurely crenate, tips obtuse. Flowers solitary in the axils of leaf-like bracts; pedicels 0.5–2 (–3) cm long, glabrous. Sepals deeply 5-lobed almost to the base, 2–4.5 mm long; lobes linear-lanceolate, acuminate, glabrous. Petals 8–9 mm long, pale purple; adaxial lip ± 2 mm long, emarginate to shallowly 2-lobed; abaxial lip ± 2 mm long. Stamens 4, all with fertile anthers; anthers ± 1 mm long and spurred at both ends; abaxial filaments each with a rounded appendage, 0.8–1 mm long arising near the base, the anthers each with a basal spur as long as the anther itself. Capsules cylindrical, acuminate, 0.9–1.3 cm long, 1–1.5 mm in diameter, more than twice the length of the sepals, arching upwards or becoming deflexed when ripe. Seeds shortly cylindrical, ± 0.3 mm long, ± 0.15 mm in diameter, minutely pitted.

Annual. Usually found in wet rice fields. It often grows with *L. anagallis* which it closely resembles in floral structure. It is almost certainly under-recorded in India.

{⊙ Hel}. Tropical and E Asia: Cey, Ind, Nep: Kl, Rj.

Lindernia minima (Bentham) Mukherjee: Fig. 358

Like *L. rotundifolia* but stems erect, simple below, 2.5–7.5 cm long; leaves in 2–3 pairs; leaf blades 2.5–6.5 mm long, the tip obtuse, the margin entire; pedicels 6.4–17.5 mm long; petals 6.4–8.3 mm long; capsules subglobose, ± 2.5 mm in diameter, ± equalling sepals.

{⊙ Hel}. Endemic to S India: ♦Ind: Tn (Palayankottai).

Lindernia mollis (Bentham) Wettstein [*L. nutans* (Yamazaki) Philcox, *Vandellia montana* (Blume) Koorders]: Fig. 358

Stems creeping and rooting at nodes, 20–50 cm long, sparsely to densely covered with spreading long white hairs. Leaves subsessile or shortly petiolate; peti-

Fig. 358. *Lindernia minima*: a, flowering plant (1 cm); b, petal tube opened (1.4 mm). *L. mollis*: c, flowering shoot (1.3 cm); d, petal tube opened (2.5 mm); e, sepals with capsule (2.5 mm). *L. molluginoides*: f, flowering shoot (1.3 cm); g, petal tube opened (1 mm) with one enlarged adaxial stamen (inset); h, sepals enclosing the capsule (1 mm).

oles 0.5–13 mm long, with long white hairs; blades ovate to ovate-lanceolate, 1.5–4 cm long, 0.8–3 cm wide, pinnately nerved, with long white hairs both surfaces, the base narrowed to truncate, margins with rounded or pointed teeth, the tip subacute to obtuse. Flowers either in subumbels of 4–6 flowers, borne at the ends of sympodially branched stems or without peduncles and 1- or 2-flowered; pedicels slender, 1–2 cm long, with long white hairs; bracts linear-lanceolate to linear, 1–3 mm long. Sepals 4–7 mm long, cleft almost to the base, with spreading white hairs; lobes linear-lanceolate, acute, ciliate, the abaxial one usually shorter than the others. Petals 8–11 mm long, white to pale violet, fragrant; adaxial (upper) lip 2-lobed; abaxial (lower) lip 3-lobed, longer than the adaxial lip, the central lobe larger 3–4 mm long. Stamens 4, all with fertile anthers; abaxial (lower) pair each with a rounded, knee-like appendage ± 0.4 mm wide at the base. Capsules ellipsoidal, 4–5 mm long, 2.5–3 mm wide in diameter, shorter than the sepals, spreading or deflexed when ripe. Seeds shortly cylindrical, ± 0.3 mm long, ± 0.2 mm wide, pitted.

Annual. Marshy land and seasonally inundated places on clayey soil. Glabrescent plants are sometimes called *Lindernia nutans.*

{⊙ Hel}. S and SE Asia: Ban, Bur, Ind, Pak: As, Mg, Ng, Sk, Wb.

Lindernia molluginoides (Bentham) Wettstein: Fig. 358

Like *L. crustacea* but flowers 2–10 in umbel-like clusters or rarely solitary-axillary, smaller (sepals 3–4 mm long, petals 4–5 mm long); leaf blades lanceolate to oblong-ovate; capsules ellipsoid, ± 0.25 mm long, ± 0.2 mm in diameter, shorter than the sepals, deflexed when ripe; filament appendages clavate.

Annual. Sandy ground on river beds or in open wet meadows.

{⊙ Hel}. India, Myanmar and Thailand: Bur, Ind: Kt. At low altitudes.

Lindernia nummulariifolia (D. Don) Wettstein ['*nummularifolia*' orth. mut., *Vandellia sessiliflora* Bentham]: Fig. 359

Stems 4-angled, with short hairs on the angles, simple or laxly branched, erect, 4–17 cm tall, occasionally decumbent and rooting at lower nodes. Leaves subsessile to petiolate; petioles up to 3 mm long; blades suborbicular to ovate-orbicular, 1–2.5 cm in diameter, pinnately nerved but looking palmately 3- or 5-nerved as the lateral nerves arise near the base, glabrous adaxially, sparsely hirsute on the nerves of the abaxial surface, the base truncate, margins with sharp saw-like teeth and ciliolate, the tip obtuse. Flowers solitary in the axils of leaf-like bracts or in lax racemes; bracts scale-like. 0.7–1 mm long; pedicels 0.3–1.5 cm long, glabrous below sparsely hispid above. Sepals campanulate, ± 1.5 mm long at anthesis, becoming 2–3 mm long in fruit; lobes subequal, about as long as the tube, sparsely pubescent on the nerves and margins. Petals 7–10 mm long, white and pink or violet to purple with a yellow fleck at the throat; adaxial lip ± 3 mm long, pale violet-brown, emarginate; abaxial lip ± 4 mm long, middle lobe with a pale yellow blotch at the base. Stamens 4, all with fertile anthers; abaxial filaments each with a rounded appendage, ± 0.15 mm thick, arising near the base. Capsules linear-cylindrical, acuminate, 5–11 mm long, ± 2 mm in diameter, the wall almost transparent, 1.5- to 2-times as long as the sepals, arching upwards when ripe. Seeds ellipsoidal to oblong, 0.4–0.5 mm long, ± 0.3 mm in diameter, obscurely longitudinally ribbed, minutely pitted.

Annual or perhaps sometimes perennial. Sandy and open ground in evergreen forests, along mountain rides but usually in wet places, particularly in rice fields. The epithet '*sessiliflora*' refers to cleistogamous

FIG. 359. *Lindernia nummulariifolia*: a, flowering shoot (1 cm); b, petal tube opened (1.6 mm); c, sepals with capsule (2.5 mm). *L. oppositifolia*: d, flowering shoot (2 cm); e, sepals (2 mm); f, petal tube opened (2 mm).

plants which have very short pedicels; they are particularly common during the monsoon.

{(⊙) Hel}. African mountains and Asia from Kashmir to W and C China: Bur, Cey, Ind, Nep: As, Bh, Hp, Jk, Kt, Mn, Mp, Ng, Sk, Tn, Up, Wb. 0–2900 m.

Lindernia oppositifolia (Retzius) Mukherjee [*Gratiola oppositifolia* Retzius, non G. Don, *Bonnaya minima* G. Don]: Fig. 359

Stems erect, up to 30 cm tall. Leaves sessile or nearly so, linear-lanceolate to oblong-lanceolate, 2–3.5 cm long, 0.4–0.7 cm wide, pinnately nerved, base narrowed, margins saw-like, tips acute. Flowers axillary or leaf-opposed, solitary or in racemes, borne singly or in opposite pairs; bracts leaf-like below becoming linear and reduced above; pedicel up to ± 1 cm long. Sepals 3–6 mm long; lobes lanceolate, acute. Petals 2.5–3 mm long, ± 2.5 mm in diameter; pinkish below, with blue to whitish lobes, with or without a yellow fleck at the throat; petal tube ± 3 mm long; adaxial lip ± 2.5 mm long, entire, obtuse; abaxial lip ± 3 mm long. Stamens 2; filaments ± 1.5 mm long, anthers unequal, ± 0.5 mm long; staminodes linear, ± 0.6 mm long, acute. Capsules lanceolate, ± 0.7 cm long, much exceeding the sepals, deflexed.

Annual. Wet lowlands, scrub jungle and pasture land. Often found in rice fields after the harvest. It is often gregarious and is easily seen as it has very lush green leaves. It is used locally in medicine.

{⊙ Hel}. S and SW Asia: Ind, Nep: Bh, Gj, Go, Kl, Kt, Or, Tn, Up, Wb. 0–1400 m.

Lindernia parviflora (Roxburgh) Haines: Fig. 360

Stems erect or ascending, simple or branched from below, slender, 7–15 (–30) cm tall, glabrous. Leaves sessile; blades linear-lanceolate to oblong-lanceolate; the lower ones 3–10 (–15) mm long, 1.5–3 mm wide, 3- or sometimes 5-nerved from the base, base rounded, margins entire, tips acute; the upper ones smaller and narrower, often 1-nerved, often minutely 1- to 2-dentate. Flowers axillary, mostly borne singly and alternating or sometimes in opposite pairs; bracts leaf-like below becoming smaller above and sometimes scale-like; pedicels slender, 3–10 (–15) mm long, often reflexed in fruit. Sepals 1.5–2 (–3) mm long, deeply

FIG. 360. *Lindernia parvifolia*: a, flowering shoot (1 cm); b, sepals with capsule (2 mm); c, petal tube opened (2 mm); d, seed (0.2 mm). *L. procumbens*: e, flowering shoot (1 cm); f, petal tube opened (1.6 mm); g, sepals with capsule (1.6 mm); h, seed (0.13 mm).

lobed; lobes linear-lanceolate, glabrous, acute. Petals 4–6 mm long, ± 2.5 mm in diameter, white or pinkish-white, with 2 yellow dots at the throat; adaxial lip ovate, notched; abaxial lip with 3 oblong lobes. Stamens 4, the adaxial pair with fertile anthers; filaments ± 1 mm long, anthers ± 0.4 mm long; staminodes clavate, ± 0.5 mm long. Capsules oblong-globose, slightly oblique at the tips, 3–4.5 mm long, 1.5–2.5 mm in diameter, 1.5–2 times as long as the persistent sepals. Seeds narrowly ellipsoidal, ± 0.2 mm long, ± 0.12 mm in diameter, longitudinally ridged, reticulate.

Annual. Moisture loving, in humid sandy places and rice fields. Gregarious and locally abundant in perennially moist areas and along river banks. The flowers opening after 10 a.m. A very variable species that sometimes intergrades with *L. hyssopoides.*

{⊙ Hel}. Old World tropics: Cey, Ind, Nep: Ap, As, Bh, Dl, Gj, Kl, Kt, Mg, Mh, Mn, Rj, Tn, Up, Wb. 0–1800 m.

Lindernia procumbens (Krocker) Borbás [*L. pyxidaria* Linnaeus, *Gratiola integrifolia* Roxburgh, *Vandellia erecta* Bentham]: Fig. 360

Stems erect or ascending, 5–20 (–30) cm tall, glabrous. Leaves sessile; blades elliptical to ovate-elliptical, glabrous, 0.7–2.5 cm long, 0.3–1 cm wide, 3- or sometimes 5-nerved from the base, base rounded, margins entire, tips obtuse. Flowers solitary; pedicel 0.5–2 cm long. Sepals lobed almost to the base, 2.5–3 mm long; lobes linear-lanceolate, acute. Petals 4–5 mm long, pale pink; adaxial lip ovate, notched; abaxial lip with 3 rounded lobes. Stamens 4, all with fertile anthers; adaxial filaments each with a clavate to linear spur 0.4–1 mm long arising near the base, anthers notched at the ends. Capsules ellipsoid, 3–4 mm long, 2.5–3 mm in diameter, subequalling or slightly longer than the persistent sepals. Seeds oblong, 3–4 mm long, ± 1.5 mm in diameter, longitudinally ridged, reticulate.

Annual. Rice fields, river beds and other moist and muddy habitats.

{⊙ Hel}. Temperate to tropical Eurasia: Ind, Nep: Bh, Jk, Mp, Or, Pj, Rj, Sk, Up, Wb. 0–2400 m.

Lindernia pusilla (Willdenow) Boldingh [*L. caespitosa* (Blume) Panigrahi, *L. laxa* (Bentham) Mukherjee, *Vandellia scabra* Bentham, *Torenia hirta* Chamisso et Schlechtendal]: Fig. 360

Stems hispid with spreading long hairs, prostrate, ascending or erect, 5–15 (–25) cm tall; branches widely spreading from the root stock. Leaves sessile or shortly petiolate; petiole up to 3 mm long; blades ovate to suborbicular, 5–12 mm long, 4–14 mm wide, pinnately nerved, pubescent above, hirsute beneath, base truncate, margins obscurely toothed, tips acute. Flowers solitary or in clusters of up to 4 in the axils of leaf-like bracts, borne singly or in opposite pairs; pedicels 3–15 (–30) mm long, hispid. Sepals 2–3 mm long, cleft almost to the base, glandular hairy; lobes lanceolate, acute, ciliate. Petals 5–8.5 mm long, ± 7 mm in diameter, violet, yellow and/or white at the throat or white, usually with a brown adaxial lip; petal tube 4–5 mm long; adaxial lip erect, 2.5–3 mm long, entire or 2-lobed; abaxial lip longer than the adaxial lip, ± 3 mm long. Stamens 4, all with fertile anthers; adaxial pair with ± 1.5 mm long filaments, the anthers ± 0.7 mm long, with broad connectives; abaxial pair with 3 mm long filaments, each with a rounded, knee-like appendage ± 0.2 mm wide, the anthers like the abaxial ones. Capsules obovoid or globose, 2–4 mm in diameter, the septum winged, shorter or equalling the sepals, spreading or deflexed when ripe. Seeds shortly cylindrical, 0.3–0.5 mm long, ± 0.2 mm wide, truncate at both ends, minutely pitted.

Annual. Borders of ponds or tanks, in marshes, in wet grassy places and in rice fields. Often locally very abundant.

{⊙ Hel/Ten?}. Old World tropics: Cey, Ind, Nep: As, Bh, Go, Kl, Kt, Mp, Ng, Or, Sk, Tn, Tp, Wb.

Lindernia rotundifolia (Linnaeus) Alston: Fig. 361

Stems 4-angled, with short hairs on the angles, decumbent and rooting at lower nodes, with branches ascending, up to 20 cm tall. Leaves sessile or subsessile; blades ovate to orbicular, 5–13 mm long, 4–12 mm wide, 4- or 5-nerved from the base, base truncate, glabrous, margins minutely toothed towards the tips, tips obtuse. Flowers solitary in the axils of leaf-like bracts; pedicels 0.5–1.5 cm long, glabrous. Sepals divided to the base; lobes ± equal, lanceolate, 2.5–3 mm long; ± 1 mm wide, minutely glandular hispid. Petals (9–) 11–14 mm long, bluish-white; tube white at base; adaxial lip 2-lobed, usually with a yellow patch inside; abaxial lip 3-lobed, white with purplish-blue blotches at the bases of the lobes. Stamens 4, with 2 bearing fertile anthers; filaments 1.5 mm long; staminodes linear to club-shaped, curved outwards at the tips, dark blue with age. Capsules ovoid-globose, 2.5–3.5 mm long, equal to or slightly shorter than the sepals. Seeds oblong, ± 0.5 mm long, with transverse bars.

Annual or perhaps sometimes perennial. In rice fields, particularly after harvesting, along ditches and the edges of ponds and tanks.

{(⊙) Hel}. Mauritius, Madagascar, W and S India and Sri Lanka: Cey, Ind: Kl, Kt, Tn.

Lindernia ruelloides (Colsmann) Pennell ['*ruellioides*' orth. mut., *Bonnaya reptans* (Roxburgh) Sprengel]: Fig. 361

Stems sparsely pubescent, prostrate, branching and rooting below. Leaves petiolate; petioles 4–20 mm

Fig. 361. *Lindernia rotundifolia*: a, flowering shoot (1 cm); b, petal tube opened (2 mm); c, sepals with capsule (1 mm). *L. ruellioides*: d, petal tube opened (2.5 mm); e, flowering shoot (1 cm); f, sepals with capsule (3.3 mm).

long; blades elliptical to orbicular-elliptical; 1–5 cm long, 0.5–2.5 cm wide, pinnately-nerved, sparsely pubescent, base narrowed, margins closely dentate with the teeth inwardly curved and acute at the apices, tips acute. Flowers axillary, in a lax raceme, mostly borne in opposite pairs; bracts linear, 4–6 mm long; pedicels 3–5 mm long at anthesis, 5–20 mm long in fruit, glabrous. Sepals deeply 5-lobed almost to the base, 7–9 mm long; lobes linear-lanceolate, acuminate, glabrous to ciliate towards the tip. Petals 12–15 mm long, 9 mm in diameter, pinkish to lilac, purple or violet; adaxial lip entire; abaxial lip 3-lobed. Stamens 4, 2 with fertile anthers; filaments ± 2 mm long, anthers ± 0.7 mm long, connectives broad; staminodes 2, 2–3 mm long, hooked, pubescent; hooks divergent. Capsules cylindrical, acute, 10–20 mm long, 1.2–2 mm in diameter, 2–2.5 times as long as the sepals, arching upwards when ripe. Seeds ellipsoid, ± 0.35 mm long, ± 0.25 mm in diameter, minutely pitted.

Annual. Rice fields, riversides, stream sides and swampy grasslands.

{⊙ Hel}. Tropical and subtropical Asia: Bur, Ind, Nep: As, Kl, Mg, Ng, Sk, Tn.

Lindernia srilankana Cramer et Philcox: refer to Fig. 360

Like *L. pusilla* but leaves 5-nerved from the base; stamens 2 with fertile anthers and 2 staminodial; the staminodes horn-like, bending outwards.

Annual. In open grassy localities. Flowering in February.

{⊙ Hel?}. Endemic to Sri Lanka: ♦Cey.

Lindernia tenuifolia (Colsmann) Alston [*L. tenuifolia* var. *pygmaea* Sivarajan et Philip Mathew]: Fig. 362

Stems glabrous, simple or laxly branched, erect or ascendent, 5–20 (–40) cm tall. Leaves sessile or nearly so, glabrous, linear to linear-lanceolate or almost filiform, usually somewhat succulent, 0.5–3 (–5) cm long, 0.5–5 (–9) mm wide, 1-nerved (lateral nerves obscure), base decurrent, margins entire to sparingly crenate, tips obtuse or subacute. Flowers solitary, usually leaf-opposed and subtended by a small bract; bracts scale-like, linear, 0.5–2.5 mm long; pedicels stout, 2–10 (–17) mm long. Sepals deeply 5-lobed almost to the base, 1.5–2 (–4) mm long; lobes

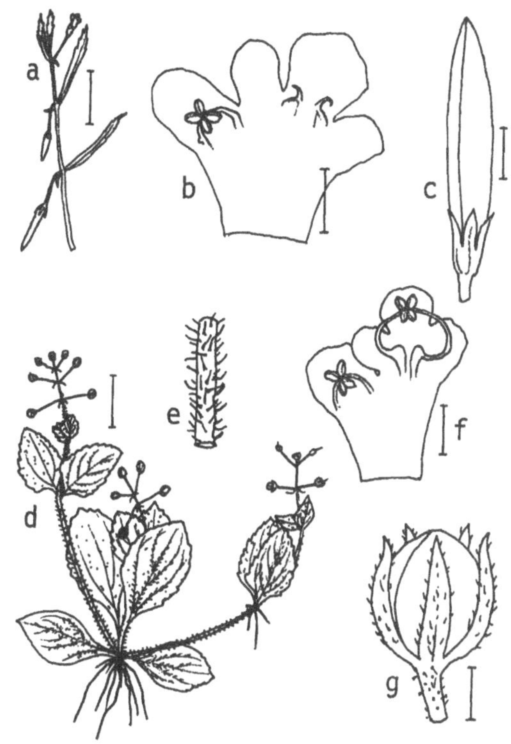

Fig. 362. *Lindernia tenuifolia*: a, tip of a flowering shoot (1 cm); b, petal tube opened (1 mm); c, sepals with capsule (1.6 mm). *L. viscosa*: d, flowering plant (2 cm); e, enlargement of the stem; f, petal tube opened (1.6 mm); g, sepals with capsule (1.6 mm).

lanceolate, obtuse. Petals (2.5–) 3–4 mm long, 2–5 mm in diameter, pinkish-white to blue or purple. Stamens 4, the adaxial 2 with fertile anthers; adaxial filaments ± 0.7 mm long, anthers unequal, ± 0.5 mm long; staminodes 2, filiform, 0.3–0.5 (–1.5) mm long. Capsules cylindrical, acute, 4–8 (–12) mm long, 0.7–1 mm in diameter, 3–4 times as long as the sepals, deflexed when ripe. Seeds ellipsoid 0.5–0.7 mm long, minutely pitted.

Annual. A slender herb, often gregarious in marshy ground, along backwaters and in rice fields after the harvest. Sometimes it is found close to the sea shore. On dry sand or in gravelly soil it is sometimes minute with stems ± 2 cm tall, with minute flowers and almost filiform leaves ± 3 mm long; these plants have been named var. *pygmaea* Sivarajan et Philip Mathew.

{⊙ Hel}. S and SE Asia: Bur, Cey, Ind: Go, Kl, Tn.

Lindernia viscosa (Hornemann) Boldingh [*Vandellia hirsuta* F. Hamilton ex Bentham, *V. multiflora* (Roxburgh) G. Don]: Fig. 362

Stems sparsely to densely hispid with spreading long hairs, laxly branched, erect or ascending, 5–15 cm tall; branches widely spreading from the root stock. Leaves sessile or petiolate; petioles 3–8 mm long; blades ovate to suborbicular, 1.5–3.5 cm long, 1–2 cm wide, pinnately nerved, hirsute on both surfaces, base truncate, margins undulate or crenate-dentate, tips acute. Flowers in lax racemes borne at the ends of branches; pedicels 4–8 mm long, glandular pubescent; bracts lanceolate to subulate, 1–2 mm long. Sepals 2–3 mm long, cleft almost to the base, with spreading hairs and glandular; lobes linear-lanceolate, acute, ciliate. Petals ± 5 mm long, ± 7 mm in diameter, white to pink with or without a yellow throat; adaxial lip broadly ovate, 2-lobed; abaxial lip longer than the adaxial lip, ± 3 mm long. Stamens 4, all with fertile anthers; adaxial pair with ± 1.5 mm long filaments, the anthers ± 0.7 mm long, with broad connectives; abaxial pair each with a rounded, knee-like appendage ± 0.15 mm wide, the anthers like the abaxial ones. Capsules globose, 2–2.5 mm in diameter, shorter or equalling the sepals, spreading or deflexed when ripe. Seeds shortly cylindrical, ± 0.3 mm long, ± 0.2 mm wide, minutely pitted.

Annual. Waysides, edges of forests, swamps and moist shady places. It is rarely found in wetlands. Glabrescent plants are sometimes recognized as a separate species — *Lindernia multiflora* (Roxburgh) Mukherjee.

{⊙ Hel?}. Tropical Asia: Ban, Bur, Cey not found recently, Ind, Nep: An, Bh, Go, Kl, Mg, Mp, Pj, Rj, Sk, Up, Wb.

Microcarpaea [excluded from *Peplidium*]: 1 species

Microcarpaea minima (J. G. König ex Retzius) Merrill [*M. muscosa* R. Brown, *Ammannia dentelloides* S. Kurz]: Fig. 363

Stems 1–15 cm long, capillary, creeping, rooting at the base ascending at the apex. Leaves decussate, subsessile; blades simple, linear to oblong-linear, 2–5.5 mm long, 0.5–1 (–2) mm wide, glabrous, 1-nerved, base decurrent, margin entire, tip obtuse. Flowers minute, sessile or nearly so, solitary in leaf axils; submerged flowers cleistogamous; bracteoles absent. Sepals 1.5–2.5 (–3) mm long; lobes 5, slightly shorter than the tube, recurved, margins ciliate, acute. Petals 2–3 mm long pink or light purple; tube ± 1–1.5 mm long; lobes 5, ciliate at the apex, the adaxial lobe longest (± 1 mm long) and exceeding the sepals. Stamens 2, adaxial; filaments filiform, ± 1 mm long; anthers exerted, appearing 1-locular at maturity, 0.2 mm long. Style linear, ± 1 mm long; stigma capitate. Capsules ellipsoidal to globose, 1.2–2 mm long, loculicidally dehiscent. Seeds ellipsoid, ± 0.4 mm long.

Annual. Small creeping plants forming mats of vegetation in seasonally inundated areas and along marshy

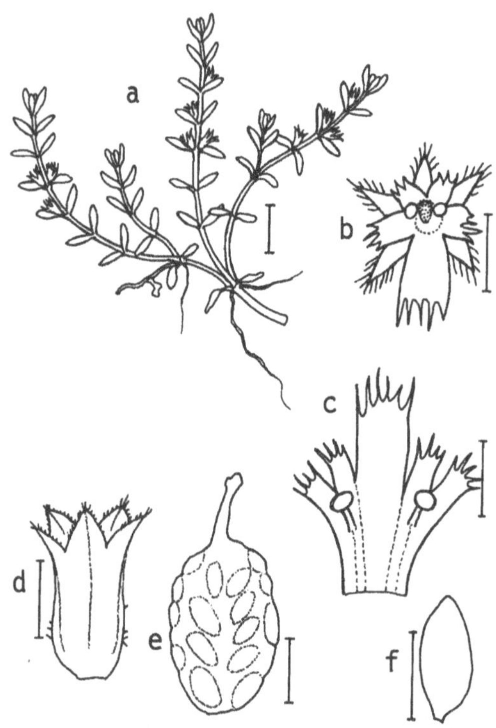

FIG. 363. *Microcarpaea minima*: a, flowering plant (3 mm); b, flower viewed from the front (1 mm); c, petal tube opened (1 mm); d, sepals (1 mm); e, capsule (0.5 mm); f, seed (0.25 mm).

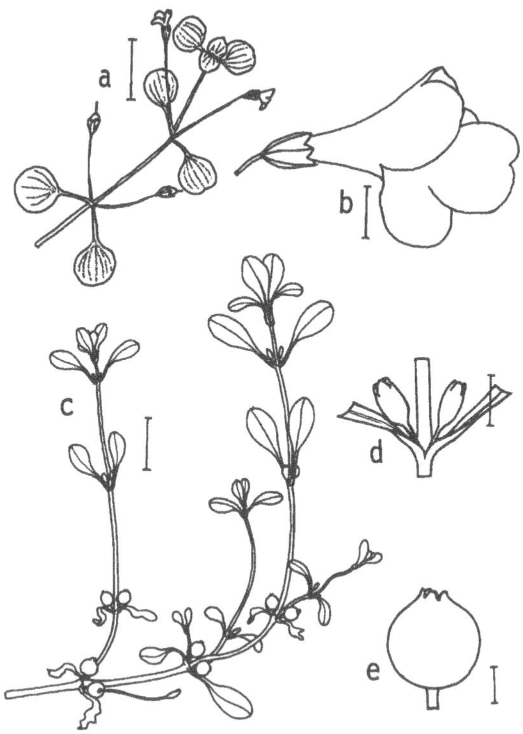

FIG. 364. *Mimulus orbicularis*: a, tip of a flowering shoot (2 cm); b, flower (2 mm). *Peplidium maritimum*: c, flowering plant (1 cm); d, two cleistogamous flowers at anthesis (2 mm); e, sepals enclosing capsule (1 mm).

banks of streams and ponds. Autogamous or perhaps sometimes entomophilous. The disseminules are seeds; their dispersal is unknown, probably carried in mud. Frequently found in rice fields but not considered to be troublesome.

{⊙ Ten/Hel}. Tropical Asia and Australia: Ban, Bur, Cey, Ind, Nep: Kl, Kt, Mp, Rj, Tn.

Mimulus: 150 species: cosmopolitan, mainly New World

Excluded: wetland plant confined to the Himalayas: ***Mimulus nepalensis*** Bentham.

Found in wet or damp places but not in wetlands: ***Mimulus strictus*** Bentham [*M. gracilis* auct. Ind., non R. Brown, *M. pusillus* Bentham].

Mimulus orbicularis Wallich ex Bentham: Fig. 364

Stems creeping and branched, up to 50 (–100) cm long, rooting at nodes. Leaves opposite; petioles 0.2–3 cm long; leaf blades suborbicular, 0.9–3 cm long, up to ± 3 cm wide, attenuate at the base, entire, rounded or emarginate at the apex. Flowers solitary or paired in leaf axils; pedicels 1.5–4 cm long, as long or longer than the leaves. Sepals bell-shaped, 3.5–4.5 mm long, 5-angled with 5 longitudinal ribs, 5-toothed, the teeth deltate. Petals 12–15 mm long, tubular, 2-lipped, slightly constricted towards the throat, bluish-white to purple with a yellow spot; adaxial lip exterior in bud, 2-lobed; abaxial lip shorter, 3-lobed, with middle lip 2-lobed. Stamens 4; filaments fused to the lower part of the petal tube, 6–7 mm long; anther locules joined at the apex, more or less spreading at the base. Style ± 4.5 mm long; stigma 2-lobed, with sensitive plate-like lobes 1–3 mm in diameter. Capsules ellipsoid, 5–8 mm long, 3–4 mm wide, somewhat angular, with loculicidal dehiscence and a persistent style. Seeds narrowly cylindrical, ± 0.6 mm long, ± 0.2 mm wide

Perennial perhaps sometimes annual. Often found floating on water or rooted in shallow water in lakes, ditches and fallow rice fields. It can withstand brackish conditions and it often grows in estuaries and lagoons by the coast. Banerjee, L. K. and Maity, G. G. Observations on the distribution and effect of salinity on a little known plant, *Mimulus orbicularis* Wall. ex Benth. (Scrophulariaceae). Bulletin of the Botanical Survey of India 32(1–4): 135–140 (dated 1990, published 1992).

{(⊙)? Hyp/Eph/Ple}. Tropical SE Asia: Ban, Bur, Ind: Or, Wb. May–December.

Peplidium [excluding *Microcarpaea*]: 2 species: tropical and subtropical Old World

Peplidium maritimum (Linnaeus fil.) Ascherson [*P. humifusum* Delile, *Microcarpaea cochlearifolia* J. G. König ex J. E. Smith]: Fig. 364

Stems much branched, creeping, rooting at most nodes. Leaves decussate, subsessile; blades orbicular to obovate or spathulate, somewhat succulent, 5–10 mm long, 4–6 mm wide, glabrous, pinnately nerved, base decurrent, margin entire, apex obtuse. Flowers minute, usually solitary in leaf axils but sometimes 2 or 3, sessile or shortly pedicellate; submerged flowers cleistogamous; pedicels up to 1 mm long. Sepals tubular, 1.7–2 mm long, 5-ribbed, bluntly 5-toothed, enlarged in fruit. Petals ± 2 mm in diameter, cream or pinkish-white; tube ± 2 mm long; lobes 5, subequal, ± 1 mm long, undulate, obtuse. Stamens 2, adaxial; filaments flattened, curved and narrowed above, facing each other, ± 0.8 mm long; anthers 1-locular, reniform, 0.2 mm long. Style linear, ± 0.5 mm long; stigma 2-lobed. Capsules globose, 2.5–3.5 mm in diameter, dehiscing irregularly or regularly by 2 or 4 valves. Seeds oblong to cuneate, ± 0.5 mm long, longitudinally striate.

Annual. A small, mat-forming plant growing in seasonally inundated places, flowering as the water recedes. It tolerates saline water and is sometimes found in salt marshes along the coast. Autogamous or perhaps sometimes entomophilous. The disseminules are small seeds; the dispersal is unknown but probably carried in mud.

{⊙ Ten/Hel}. Egypt, tropical Asia and Australia: Cey, Ind: Rj, Tn.

Ramphicarpa: 6 species: warm to tropical Old World

Hansen, O. J. The genus *Ramphicarpa* Benth. emend. Engl. A taxonomic revision. Botanisk Tidsskrift 70 (2–3): 103–125 (1975).

Ramphicarpa longiflora Wight ex Bentham: Fig. 365

Stems slender, erect, 2–30 cm high. Leaves cauline, opposite, rarely simple, usually pinnately divided into filiform segments, (0.5–) 1–4 (–7) cm long. Flowers solitary in leaf axils, relatively large; pedicels 0–5 (–7) mm long; bracteoles 0.5–5 mm long. Sepals bell-shaped; tube 0.5–2 mm long; lobes 5, narrowly triangular, 4–8 (–10) mm long. Petals ± equally 5-lobed, white, purple or yellow; tube straight, 1.5–3 (–3.5) cm long and slender; lobes suborbicular to spathulate, 5–10 mm long and broad. Stamens 4, all alike, with a tuft of hairs at the base of each filament; filaments 0.8–1.5 mm long; anthers 2–3 mm long. Capsules globose to ovoid, slightly flattened, 4–10 mm long, 3–6 mm broad, obliquely beaked, dehiscence mostly along the adaxial suture. Seeds numerous, cylindrical, ±1 mm long, reticulate.

Fig. 365. *Ramphicarpa longiflora*: a, flowering plant (5 cm); b, stamen (1 mm); c, sepals and capsule (5 mm); d, flower (1 cm).

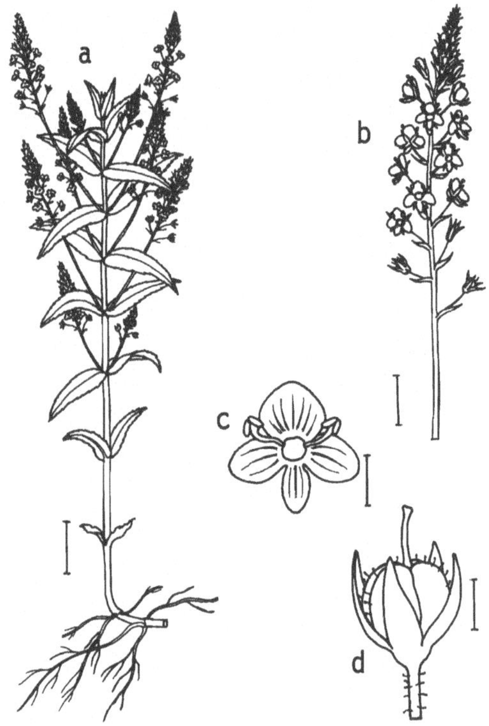

Fig. 366. *Veronica anagallis-aquatica*: a, flowering plant (3 cm); b, inflorescence (1 cm); c, flower viewed from the front (2 mm); d, sepals and capsule (2 mm).

Annual. It is probably hemiparasitic and is often found growing in marshes and places which are seasonally inundated. It is particulary common in rice fields where it might be a parasite of rice. Pollination is by insects. The disseminules are seeds; their dispersal is unknown.

{⊙ Hel}. Endemic to W India: ♦Ind: Go, Kl, Kt, Mp?

Veronica: 250 species: almost cosmopolitan

Excluded: wetland species found in the Himalayas but not extending southwards in India: ***Veronica beccabunga*** Linnaeus; ***V. oxycarpa*** Boissier; ***V. undulata*** Wallich [*V. anagallis* auct Ind., non Linnaeus].

Veronica anagallis-aquatica Linnaeus: Fig. 366

Stems creeping, ascending or erect (3–) 30–60 (–200) cm high, 3–17 mm thick, weakly 4-angled, pith green. Leaves opposite; blades lanceolate to oblong, (2–) 5–12 (–15) cm long, (0.8–) 1.5–4 (–5.5) cm wide, base cordate or cuneate, margins slightly to distinctly toothed. Flowers in opposite, axillary racemes, with 10–60 flowers pedicels 4–7 mm long, usually exceeding the bracts. Sepals 4, practically free. Petals very shortly tubular, 4-lobed, (5–) 6 (–8) mm in diameter; lobes appearing free, blue, violet, pinkish or white, flat and spreading, the adaxial lobe the largest, the abaxial lobe the smallest. Stamens 2; filaments diverging laterally, bearing the anthers beyond the petals. Style elongate, 1.5–2 (–3) mm long; stigma capitate. Capsules ovate, flattened laterally, 2.5–3.5 mm long, 2–3 mm wide, loculicidally dehiscent at first later also septicidally dehiscent. Seeds oval, flattened, 0.6–0.7 mm long.

Annual or short-lived perennial (usually dying after flowering). Found in a variety of aquatic habitats, seasonally submerged, floating or emergent. Pollinated by insects. The disseminules are flattened seeds; dispersed by adhesion to animals or by water. Used locally in medicine, the leaves are used as a vegetable and it is cultivated for decoration. I have seen no material from lowland India but it is frequently recorded.

{(⊙) Hel/Ple/Vit)}. Almost cosmopolitan: Bhu, Ind, Nep: Dl, Jk, Mp, Pj, Rj, Up, Wb. ?–4700 m.

SPHENOCLEACEAE [sometimes united with CAMPANULACEAE]: 1 genus

Sphenoclea: 2 species: 1 in W Africa the other pantropical but perhaps naturalized in America and SE Asia

Sphenoclea zeylanica Gaertner: Fig. 367

Stems erect or decumbent, usually swollen and pithy at the base, up to ± 150 cm high. Leaves alternate; stipules absent; petioles up to 1.3 cm long; blades lanceolate, up to 10 cm long, up to 3 cm broad, glabrous, entire, tips ± acute. Inflorescence a dense terminal spike, up to 7 cm long, ±1 cm in diameter. Flowers bisexual, radially symmetrical. Sepals united to the ovary, with 5 ovate-rounded, imbricate lobes. Petals bell-shaped, perigynous; lobes 5, as long as the sepals, greenish-white. Stamens 5, adnate to the petals, alternating with the petal lobes; filaments very short. Ovary semi-inferior, 2-locular; style 1, short; stigma very shortly forked. Fruit a capsule, subglobose, apically flattened, opening along a horizontal line; the valve coming off like a lid; seeds minute, numerous, attached to a large and spongy axile placenta.

Annual. Seasonally submerged, emergent or temporarily terrestrial, often found in swampy areas, along the banks of water courses and in rice fields. It can tolerate brackish water. It is autogamous but perhaps also sometimes pollinated by insects. The disseminules are small seeds, dispersed in mud and perhaps also by other means. It is often gregarious becoming a troublesome weed in rice fields but it is also cultivated and eaten as a vegetable. Monod, T. A propos du *Sphenoclea zeylanica*. Adansonia sér. 2, 20(2): 147–164 (1980).

{⊙ Hel/Hyp}. Almost pantropical: Ban, Cey, Ind, Nep, Pak: An, As, Dl, Go, Kl, Mn, Pj, Rj, Up, Wb.

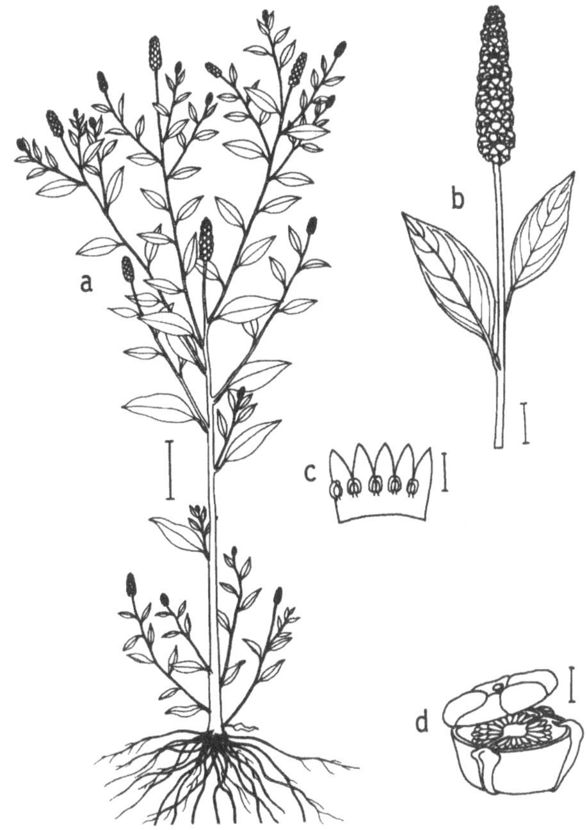

FIG. 367. *Sphenoclea zeylanica*: a, flowering plant (6 cm); b, inflorescence (1 cm); c, petal tube opened (2 mm); d, capsule with detached lid (1 mm).

STERCULIACEAE excluded see p. 12.

TILIACEAE excluded see p. 12.

TRAPACEAE: 1 genus

Trapa: 1 polymorphic or up to ± 20 species: warm to tropical Old World, naturalized in N America

This genus is very variable and without a satisfactory taxonomic treatment. In India it is most variable in the Himalayas where the following occur: ***Trapa natans*** var. ***natans*** [*T. quadrispinosa* Roxburgh] and ***T. maximowiczii*** Korschinsky [*T. incisa* Siebold et Zuccarini] (a much smaller plant with smaller incise-dentate leaves and much smaller fruits with 4 long and thin spines). Kadono, Y. A preliminary study on the variation of *Trapa* in Japan. Acta Phytotaxonomica et Geobotanica (Kyoto) 38: 199–210 (1987).

Trapa natans Linnaeus var. ***bispinosa*** (Roxburgh) Makino [*T. bispinosa* Roxburgh]: Fig. 368

Stems elongate, mostly simple, flexible, submerged, bottom-rooted. Leaves opposite below, alternate above,

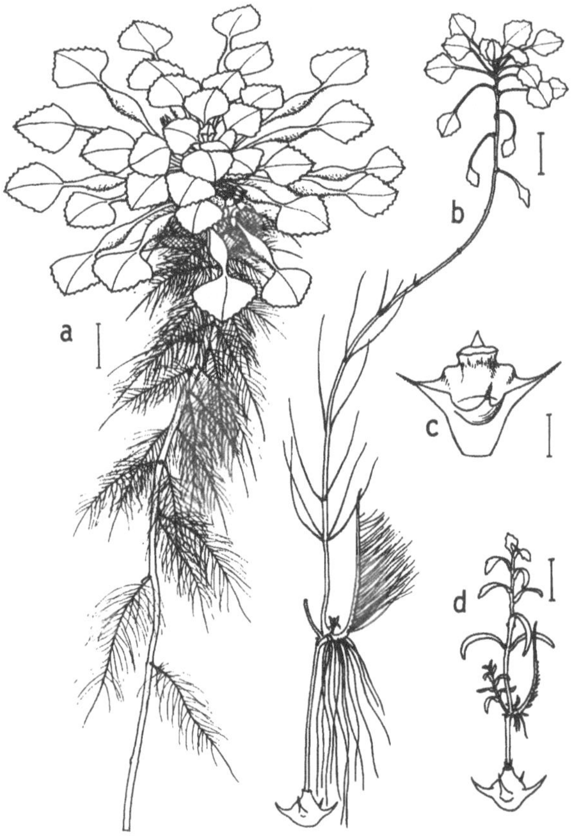

Fig. 368. *Trapa natans*: a, floating rosette showing pinnately branched roots (2 cm); b, young plant with linear submerged leaves (2 cm); c, nut (1 cm); d, seedling (2 cm).

dimorphic; submerged leaves sessile, linear, entire, caducous (after falling off green, pinnately branched, adventitious roots develop from the leaf scars, these roots are often mistaken for leaves in the literature); floating leaves in terminal rosettes, the blades rhombic with toothed margins and broadly cuneate to truncate bases, the petiole often with an ellipsoidal and spongy swelling. Flowers bisexual, radially symmetrical, solitary, shortly pedicelled, often cleistogamous. Sepals 4, triangular, united to the ovary, persistent, developing into 2, 3 or 4 hard spines or horns in fruit. Petals 4, white or lilac, caducous. Stamens 4. Ovary half-inferior, 2-locular; ovules pendent, 1 in each locule, but only 1 develops. Fruit a large, woody or bony, very polymorphic, variously sculptured, spinose nut, mostly with 2 large and 2 small (often rudimentary) spines; the spines with or without barbed tips.

Annual or perennial. Found in still or slowly flowing water, the seedlings are submerged, the adult leaves float in rosettes. Pollination autogamous or sometimes entomophilous. The disseminules are large spiny nuts, dispersed by animals and also by water. It is frequently cultivated in Asia (singhara in India) and elsewhere where it is a staple food. The fruits are eaten raw or cooked; in China they are also used for making flour. There are many cultivars; some of them are without spines. The cultivated races are annual or cultivated as such. Large populations may be very troublesome to navigation, fisheries and recreation.

{(⊙) Eph}. Old World, it is widely cultivated and it is difficult to say today where it is native, probably mostly introduced in tropical India: Ban, Bur, Cey, Ind, Pak: As, Bh, Dl, <Go, Hr, <Kl, <Kt, Mh, Mp, <?Rj, Up, Wb.

TYPHACEAE [excluding SPARGANIACEAE]: 1 genus

Typha: ± 8 species: cosmopolitan

Stems erect and corm-like, connected by stolons or rhizomes. Leaves distichous, sheathing below, not differentiated into petiole and blade, scale-like on underground parts otherwise linear; linear leaves mostly radical and equitant, emergent or rarely submerged. Flowers arranged in superposed, cylindrical, unisexual spikes, the female below the males. Perianth segments scale- or hair-like, simple or branched, in female flowers modified to numerous hairs. Male flowers ephemeral, tightly packed and hardly definable as individual flowers; filaments usually united below; pollen single or in tetrads. Female flowers with elongate and accrescent gynophores, each with 1 or more whorls of perianth hairs; the gynoecia of 2 different kinds: one with ovules the other without; ovulate gynoecia with solitary, uniovulate carpels, with simple styles that elongate and persist in fruit; non-ovulate gynoecia (pistillodia) with terminal, club-shaped carpels tipped with mucros; terminal pistillodia sometimes lack perianth hairs. Fruit stalked, opening by a single valve, a 1-seeded capsule (or follicle) or indehiscent. Seeds solitary, cylindrical, the chalazal end sharply pointed.

Perennial. Found in a wide variety of aquatic habitats, the juveniles are submerged, the adults emergent or sometimes terrestrial. It is usually gregarious and often dominant over large areas. Pollinated by wind. The disseminules are 1-seeded capsules or follicles, dispersed by wind, releasing the seeds in water where they are further transported by water or animals (mostly fish). The rhizomes and pollen are gathered by humans and eaten locally. The leaves are used for weaving mats and roofs. Large stands make a valuable refuge for wildlife. In irrigation systems it is often a serious pest and can even block navigable water. The literature is enormous, the following reference is a short summary, Smith S. G. *Typha*: its taxonomy and the ecological significance of hybrids. Archiv für

Hydrobiologie: Beiheft (Ergebn. Limnol.) 27: 129–138 (1987).

Excluded: the following are temperate species found in the W Himalayas but not extending further south in India: ***Typha latifolia*** Linnaeus; ***T. laxmannii*** Lepechin; ***T. minima*** Funck ex Hoppe.

1. Leaves in transverse section lens-shaped, lacking a keel, with 1 central row of gas chambers; female heads in fruit light cinnamon brown; hairs in male heads forked; pollen grains single ***T. domingensis***
1. Leaves in transverse section 3-angled, distinctly keeled below, with more than 1 row of gas chambers; female heads in fruit dark reddish-brown; hairs in male heads simple; pollen in tetrads ***T. elephantina***

Typha domingensis Persoon [*T. angustifolia* auct. Ind., non Linnaeus, *T. angustata* Chaubert et Bory]: Fig. 369

Rhizomes robust up to 1.5 cm in diameter, superficial (not deeply rooted), up to 50 cm or more between aerial shoots, internodes 2.5–6.6 cm long, the cortex with spongy aerenchyma and some, irregularly arranged gas chambers. Scale leaves nearly surrounding the nodes. Photosynthetic leaves (0.3–) 1–3 (–4) m long, (3–) 5–10 (–16) mm wide, abaxially convex, adaxially flat or weakly convex, with a single and central row of gas chambers (usually 10–12 large and 2–4 small); sheaths usually decurrent or uppermost cauline leaves with shoulders or small auricles. Inflorescence stalks 1.0–2.5 m tall, longer than or slightly shorter than the subtending leaves. Female inflorescences in fruit (5–) 7–20 (–35) cm long, 1.4–2.5 cm in diameter and light cinnamon brown. Gap between male and female inflorescences (0–) 1–3 (–9) cm long. Male inflorescences up to ± 35 cm long, about as long as the female inflorescence, interrupted by 4–9 caducous spathes. Male flowers: hairs simple or branched (forked, anchor- or antler-like), at tip lanceolate or irregularly laciniate; stamens in groups of (1–) 2 (–6); filaments usually simply forked; anthers after anthesis (1–) 1.4–1.8 (–2.2) mm long; connective appendages 0.3–0.8 mm long; pollen single, 20–31 μm in diameter. Female flowers with bracts; bracts exceeding the perianth hairs, at base linear, at tip lanceolate to ovate, light brown, the apex distinctly acuminate or mucronate, 4–8 cells across, the acumen or mucro usually white (contrasting with the rest); perianth hairs 20–41, arising from about 3, loose whorls occupying the basal ± 0.6 mm of the gynophore, linear or enlarged at tip, white except for isolated brown cells near the apex; stigmas linear, filamentous, ± 0.5–1.0 mm long, partly borne above the perianth hairs, at anthesis green, later becoming dark brown but in fruit becoming light cinnamon brown. Pistillodia club-shaped, mucronate at the apex, at maturity as long as the perianth hairs, visible from the surface. Ramuli in fruit with a 0.7–1.0 mm high basal part; terminal part 2–2.4 mm long, mostly bearing 1 or 2 (3) pistillodia and a terminal tuft of (2) 3 (–5) bracts. Fruits 1–1.5 mm long. Seeds (0.6–) 0.8–1.1 (–1.2) mm long, 0.21–0.29 mm wide.

Perennial. Often dominant in fresh and brackish marshes, backwaters, lagoons, pools and along water courses. It is less common in the humid tropical parts of India. It is widely used for weaving, particularly in mat making. The rhizome is astringent and diuretic. The immature inflorescence is used as food.

{Hyp}. Warm and tropical regions of the world: Ban, Bur, Ind, Pak: Dl, Hp, Jk, Kl, Kt, Mh, Mn, Mp, Rj, Tn, Up, Wb.

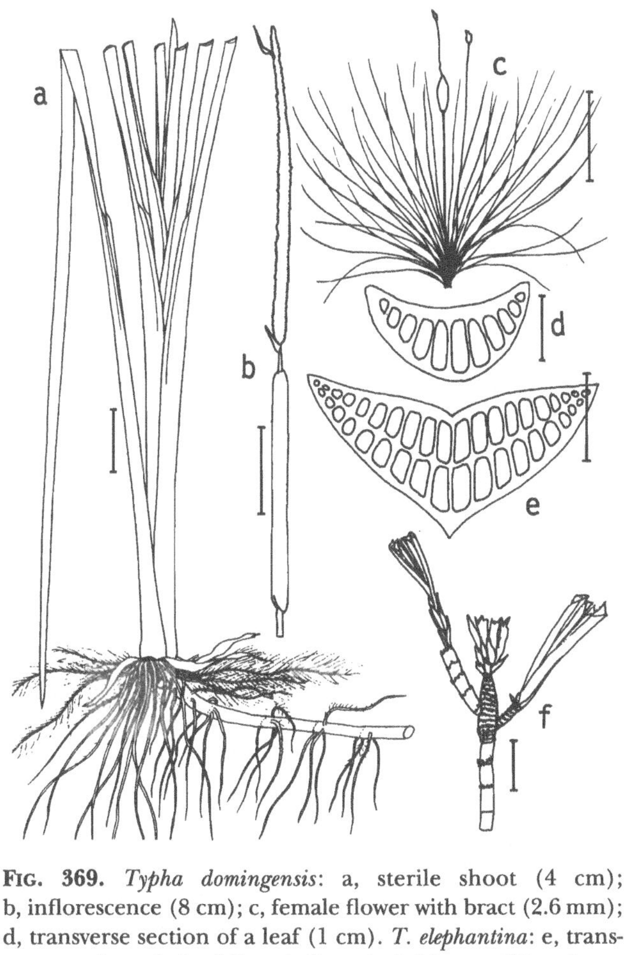

Fig. 369. *Typha domingensis*: a, sterile shoot (4 cm); b, inflorescence (8 cm); c, female flower with bract (2.6 mm); d, transverse section of a leaf (1 cm). *T. elephantina*: e, transverse section of a leaf (1 cm); f, vertical rhizome (15 cm).

Typha elephantina Roxburgh [*T. latifolia* auct. Ind., non Linnaeus]: Fig. 369

Rhizomes very robust, up to 5 cm in diameter, with internodes up to 8 cm long, the cortex with many gas chambers arranged regularly in several rows, usually dimorphic with part horizontal and part vertical; horizontal rhizomes usually deep, 0.8–2 m below the soil surface; vertical rhizomes terminate in aerial shoots crowned with persistent leaf bases. Scale leaves appressed to but not surrounding and enclosing the internodes. Photosynthetic leaves robust and stiff with rather sharp margins, up to 4 m or more long, (14–) 17–22 (–40) mm wide, with a prominent adaxial ridge from base to apex, in transverse section distinctly 3-angled and containing (1) 2 (3) rows of gas chambers; sheaths decurrent. Inflorescence stalks up to 3 m or more tall, subtended by numerous leaves which about equal the flowering stems in length. Female inflorescence in fruit 10–25 cm long, dark reddish-brown. Gap between male and female inflorescences 2–10 cm or more long. Male flowers: hairs simple, with spathulate or club-shaped tips; stamens mostly single; filaments mostly simple, up to 2 mm long; anthers after anthesis ± 2.5–5 mm long; pollen in tetrads, individual grains (16–) 20–26 μm in diameter. Female flowers subtended by simple bracts with dark brown to almost black, linear to slightly club-shaped tips; perianth hairs 15–21 (–28), slightly swollen at the tip, arising from the base of the gynophore; stigmas linear-lanceolate to spathulate, up to 1 mm long, distinctly exceeding the perianth hairs. Pistillodia narrowly club-shaped with mucronate tips, overtopped by perianth hairs. Ramuli in fruit with 0.3–0.5 mm tall basal part; terminal part 0.3–1.5 mm or perhaps more long, terminating in a naked pistillodium or a tuft of hairs or bracts. Seeds 1.1–1.3 mm long, ± 0.2–0.25 mm in diameter.

Found in a wide variety of wetland habitats, including alkaline and brackish conditions. The rhizomes readily spread from wet places into dryer ones; it can thus invade fields used for cultivating wheat, maize or soya. It is difficult to eradicate. It is used in mat and paper making.

{Hyp/(Hel)}. Warm to tropical Asia and perhaps also Africa: Ban, Ind, Nep, Pak: As? Rj, Wb. This species is under recorded.

UMBELLIFERAE see APIACEAE

VERBENACEAE ± 91 genera

Trees, shrubs, lianas or herbs. Leaves usually opposite, without stipules. Flowers usually bisexual. Sepals tubular below with (4) 5 (–8) lobes above. Petals tubular below, ± irregular to 2-lipped. Stamens 2 or 4; filaments inserted on the petal tube. Ovary of (1) 2 carpels. Fruit a head of 1-seeded mericarps, a drupe or a 2- or 4-valved capsule. Seeds with a straight, oily embryo.

Phyla [*Lippia*]: ± 11 species

Phyla nodiflora (Linnaeus) E. Greene [*Lippia nodiflora* (Linnaeus) Michaux]: Fig. 370

Stems creeping and rooting at nodes, (5–) 30–95 cm long, pubescent with centrally attached and appressed white hairs; flowering shoots sometimes ascending, very rarely more than 10 cm tall. Leaves in opposite pairs, sessile or tapered with short subpetiolate bases, oblanceolate, spathulate or obovate, (0.7–) 1–5 (–7) cm long, (2.5–) 3–7 (–10) mm wide, apically rounded or obtuse, rarely acute, marginally toothed from the middle or just above; teeth sharp, (3–) 5 (–7) on each side. Inflorescence a very dense, many-flowered, cylindrical spike with flowers borne in the axils of tightly overlapping bracts; bracts cuneate-obovate, 2.5–3 mm long, with ciliate margins, somewhat apiculate. Sepal tube somewhat flattened, with 2 narrow pubescent keels each running into an erect triangular to subulate lobe; lobes 1.5–2 mm long, split to more than half their length, extending beyond the fruit. Petals white to pale rose-purple, tubular, 2-lipped, 4- or 5-lobed. Stamens 4, in 2 pairs; anthers not exerted beyond the petals. Fruit ellipsoid-globose, separating into 2 mericarps, each truncate and 1–2 mm long.

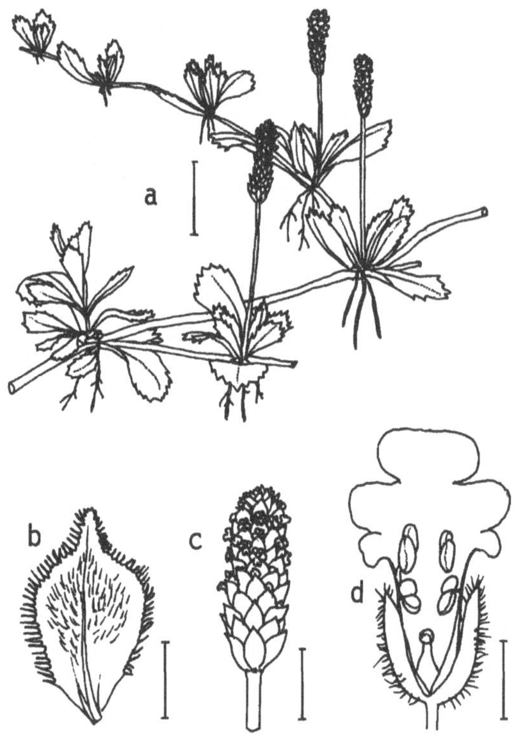

Fig. 370. *Phyla nodiflora*: a, flowering plant (2 cm); b, bract (1 mm); c, inflorescence (1 cm); c, petal tube opened from adaxial side (1 mm).

Perennial or annual. Found in regions regularly flooded, flowering as the water recedes, particularly common on the banks of irrigation ditches and ponds but also found in moist grassland and regularly watered lawns. It forms a dense turf and is sometimes planted as ground cover. Used locally in medicine. It is very variable in form and some varieties have been described. A second, probably not wetland species, *Phyla canescens* (Kunth) E. Greene, has been reported from S India; it is like *P. nodiflora* but it is covered with soft greyish hairs; leaves entire or bluntly toothed; mature spikes ovoid-globose; bracts elliptical-ovate; sepal lobes less than half as long as the tube; petals lilac or pink.

{(⊙)? Hel}. Cosmopolitan in tropical to warm regions: Ind, Nep, Pak: <An, Ap, As, Bh, Dl, Gj, Go, Kt, Lk, Mh, Mp, Pj, Rj, Tn, Tp, Up, Wb.

VAHLIACEAE see SAXIFRAGACEAE excluded see p. 11.

XYRIDACEAE: 5 genera

Xyris: 250 species: almost cosmopolitan

Stems reduced, corm-like. Leaves distichous, mostly radical, simple, linear or capillary, sheathing at the base. Inflorescence a simple, terminal, bracteate head, borne on an elongate culm. Bracts overlapping, persistent. Flowers bisexual, tightly packed between bracts. Sepals 3, 2 lateral, keeled, persistent, the median large, membranous and hooded, covering the petals in bud. Petals 3, almost free, yellow or rarely white, withering before noon. Stamens 3, inserted on the petals. Staminodes 3, filiform, forked, each arm brush-like, with numerous hairs. Pollen grains smooth. Ovary superior, 1-locular, with 3 parietal placentas; style without appendages; stigmas 3, linear. Fruit a capsule. Seeds numerous, ellipsoid-fusiform, with dark-coloured longitudinal ribs.

Perennial or annual. Most species are found in marshy or wet areas and may grow seasonally submerged. Pollinated by insects, the petals usually wither by midday. The disseminules are seeds; dispersal unknown. Some are used locally in medicine and some are cultivated for decoration; the culms are used to make figures for Hindu temples. Hansen, B. Xyridaceae in Flora of Thailand 5(1): 130–140 (1987).

1. Bracts in the middle of the inflorescence with distinctly notched tips, margin otherwise entire (Myanmar) ***X. lobbii***

1. Bracts in the middle of the inflorescence with entire tips, although sometimes irregularly torn, if notched then the margins ciliate

2. Bracts in the middle of the inflorescence with ciliate margins at least towards their tips; culms flattened above with 2 strong ridges; leaf margins rough to touch ***X. complanata***

2. Bracts in the middle of the inflorescence without cilia on the margins; culms terete or if flattened then without 2 strong ridges; leaf margins smooth or rough to touch

3. Leaf blades with short transverse ribs connecting the longitudinal ones; the margins smooth; culms terete (circular or nearly so in transverse section), with 6–15, low, longitudinal ridges ***X. indica***

3. Leaf blades without transverse ribs; the margins smooth or rough; culms terete or somewhat flattened toward the tip, smooth or with 1–7, longitudinal ridges

4. Bracts in the middle of the inflorescence without a greenish-brown or grey field below the tip, at most a longitudinal keel on the upper third of the abaxial surface ***X. capensis***

4. Bracts in the middle of the inflorescence with a greenish-brown or grey field below the tip on the abaxial surface

5. Greenish-brown or grey field below the tip on the abaxial surface oblong-triangular, 3–10 times longer than wide, not protruding upwards; leaf sheath ciliate below (NE India and Mynamar) ***X. wallichii***

5. Greenish-brown or grey field below the tip on the abaxial surface triangular, at most 2.5 times longer than wide, protruding or not protruding upwards; leaf sheath not ciliate

6. Leaves more than 3.5 mm wide, when dry smooth; greenish-brown or grey field below the tip on the abaxial surface not protruding upwards; plants robust (Myanmar) ***X. intersita***

6. Leaves rarely more than 3 mm wide, when dry rough due to protruding cell walls; greenish-brown or grey field below the tip on the abaxial surface ridge-like and protruding upwards towards the tip of the bract; plants slender (widespread) ***X. paucifolia***

Xyris capensis Thunberg [*X. capensis* Thunberg var. *schoenoides* (Martius) Nilsson, *X. schoenoides* Martius, *X. nilagarensis* Steudel]: Fig. 371

Leaves with and without ligules; blades flat, (5–) 10–25 (–35) cm long, (1.8–) 2–3 (–4) mm wide, the margins slightly thickened. Culms terete, (13–) 25–60 (–70) cm long, 1–2 mm wide, with 1 or 2 ridges. Inflorescence globose, 6–9 mm long, 5–11 mm wide. Bracts in the middle of the inflorescence blackish-brown to light brown, smooth except for a longitudinal keel on the upper third of the abaxial surface, margins entire. Lateral sepals brownish with an entire and smooth crest.

Perennial. In open wet places in hills.

{Hel}. S and SE Asia, S Africa and S America: Ind, Nep: As, Mg, Tn. 900–2700 m.

Xyris complanata R. Brown [*X. anceps* auct. Ind., non Lamarck]: Fig. 371

Leaves with ligules; sheaths dark brown to black; blades flat, (6–) 25–60 cm long, (0.8) 1–4 mm wide, surface smooth, the margins thickened and rough (with rows of backwardly pointing cells). Culms flattened in transverse section or subterete, occasionally 4-angled, (10–) 70–80 (–90) cm long, (0.7–) 1.5–2 (–2.8) mm wide, with 1–5 longitudinal ridges. Inflorescence ellipsoid to ovoid or cylindrical, 9–31 (–39) mm long, (2.5–) 6–8 mm wide. Bracts in the middle of the inflorescence greenish-brown to light brown, darkening with age, margins ciliate towards the tip and entire below, numerous, slightly notched at the tip; greenish-brown or grey field on the abaxial surface below the tip elliptical-rhomboid, 2.5–3 mm long, 1.5–2 wide, with some scattered light spots. Lateral sepals brownish with a ciliate crest. Capsules 2.4–4.5 mm long. Seeds ovoid, reddish-brown, 0.35–0.4 mm long, with ± 14 dark brown ridges.

Perennial. In marshes, moist depressions, drainage ditches, along streams and other wet places. The petals may persist until well into the afternoon in this species.

{Hel}. S and SE Asia and Australia: Bur, Cey, Ind: Kl. 0–1600 m.

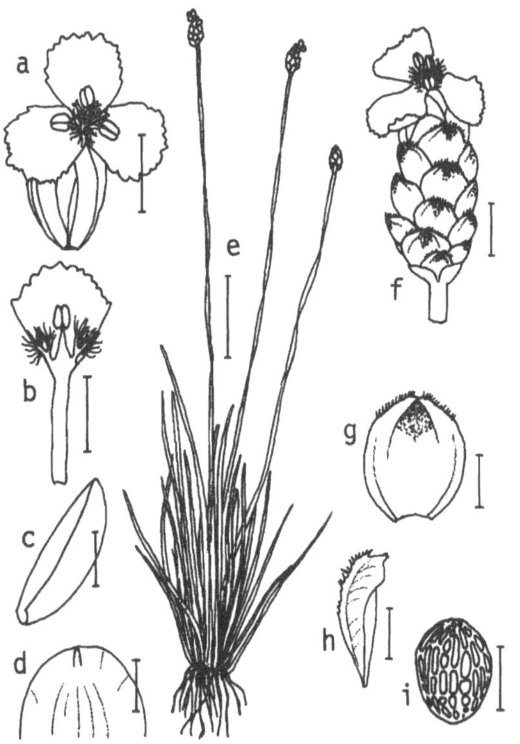

Fig. 371. *Xyris capensis*: a, flower (4 mm); b, a petal with 1 stamen and 2 staminodes (4 mm); c, lateral sepal (1.5 mm); d, upper part of bract (1.5 mm). *X. complanata*: e, flowering plant (8 cm); f, spike with 1 flower (4 mm); g, bract (3 mm); h, lateral sepal (4 mm); i, seed (0.8 mm).

Xyris indica Linnaeus [*X. robusta* Martius, ?*X. coronata* Haines]: Fig. 372

Leaves without ligules; sheaths reddish-green to reddish-brown; blades unifacial, flat, (10–) 30–50 (–70) cm long, (1.5–) 3–6 (–13) mm wide, the margins not thickened, smooth or sparsely papillate, with conspicuous, short transverse ribs connecting the longitudinal nerves. Culms terete, (5–) 30–70 (–110) cm long, (1.2–) 2–3 (–4.5) mm wide, with 6–15 longitudinal ridges. Inflorescence globose to ovoid or cylindrical, (4) 10–15 (35) mm long, (3–) 8–14 mm wide. Bracts in the middle of the inflorescence dark or light brown, margins entire, the tip entire or slightly notched; greenish-brown or grey field on the abaxial surface below the tip triangular-rhombic, 1–2 mm long, 0.8–2 mm wide, not protruding upwards. Lateral sepals light brown to almost hyaline, with a coarsely saw-like crest. Capsules 2.9–4.2 mm long. Seeds ovoid, 0.5–0.6 mm long, yellow, with 14–18 longitudinal ridges.

Perennial and perhaps sometimes annual. In wet sand or muddy soil in swampy places, sometimes in shallow water in streams, pools, ditches and canals, also in rice fields. It is used as cure for ringworm, itch and leprosy.

{(⊙) Hel/Hyp}. S and SE Asia and Australia: Ban, Bur, Cey, Ind: As, Go, Kl, Kt, Mg, Mh, Mn, Mp, Sk, Wb.

Xyris intersita Malme: Fig. 372

Leaves without ligules; blades flat, (7) 10–25 (–38) cm long, (3.5–) 4–6 (–7.5) mm wide, the margins widened but not thickened, smooth. Culms terete, (6–) 10–35 (–48) cm long, (0.7–) 1.2–1.7 (–2) mm wide, without longitudinal ridges. Inflorescence ovoid to globose, (6–) 7–1.1 (–14) mm long, (0.5–) 7–10 (–11) mm wide. Bracts in the middle of the inflorescence light brown, with some scattered light spots scattered below the grey-brown field, margins entire or minutely toothed towards the tip, sometimes torn at the tip; grey-brown field on the abaxial surface below the tip triangular, 1–2.5 mm long, 1–1.5 mm wide. Lateral sepals light brown to almost hyaline, with a thin entire crest.

Perennial. On moist sandy soil.

{Hel}. Indo-China, Thailand and Myanmar: Bur.

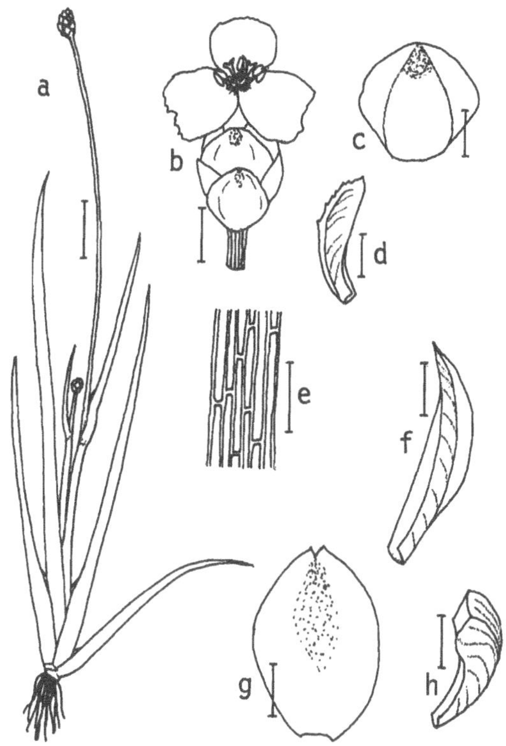

Fig. 372. *Xyris indica*: a, flowering plant (4 cm); b, spike with 1 flower (4 mm); c, bract (2 mm); d, lateral sepal (2 mm); e, portion of leaf surface (4 mm). *X. intersita*; f, lateral sepal (3 mm). *X. lobbii*: g, bract (1.5 mm); h, lateral sepal (1.5 mm).

Xyris lobbii Rendle: Fig. 372

Leaves without ligules; blades flat, 20–40 cm long, 2–5 mm wide, the margins not thickened. Culms terete, (35–) 40–60 (–65) cm long, 1.5–2.5 mm wide, without ridges. Inflorescence subglobose, 8–15 mm long, 6–1.2 mm wide. Bracts in the middle of the inflorescence light brown, with many light spots scattered below the grey-brown field, margins smooth, with a deep notch at the tip; grey-brown field on the abaxial surface below the tip triangular, ± 2 mm long, ± 1.2 wide. Lateral sepals brown, firm, widening upwards, with an entire or minutely toothed crest.

Perennial. Wet places in forests.

{Hel}. Vietnam, Thailand and Myanmar: Bur. 1000–2150 m.

Xyris pauciflora Willdenow: Fig. 373

Leaves with or without ligules; sheaths red to reddish-green; blades subterete, (2.5–) 5–20 (–33) cm long, 0.5–2 (–3.7) mm wide, surface rough (papillate-scabrid) the margins sometimes thickened, rough (epidermis tuberculate). Culms terete or nearly so, (3–) 10–30 (–44) cm long, 0.3–1.2 (–1.8) mm wide, rarely with longitudinal ridges, mostly unridged. Inflorescence ellipsoid to ± hemispherical, 3.5–10 (–14) mm long, 2–7 (–12) mm wide. Bracts in the middle of the inflorescence light or dark brown to blackish, with some light spots scattered below the grey-brown field, margins entire, tip entire; grey-brown field on the abaxial surface below the tip triangular, 0.7–2.2 mm long, 0.4–1.5 mm wide, sometimes tuberculate in the upper half, usually protruding towards the tip of the bract. Lateral sepals light brown to almost hyaline, with a coarsely saw-like to entire crest. Capsules 2.5–3 mm long. Seeds ellipsoid, 0.3–0.4 mm long, yellowish to reddish-brown, with 12–14 longitudinal ridges.

Annual or perhaps sometimes perennial. In open wet places, swamps, the margins of pools and rice fields. Flowering during and after the monsoon.

{(⊙)? Hel}. S and SE Asia and Australia: Ban, Bur, Cey, Ind, Nep: Ap, Go, Kl, Kt, Tn, Wb. 0–1500 m.

Xyris wallichii Kunth: Fig. 373

Leaves without ligules; sheaths with ciliate margins; blades flat, 5–30 cm long, 1.5–4.6 mm wide, the margins not thickened, smooth. Culms terete or

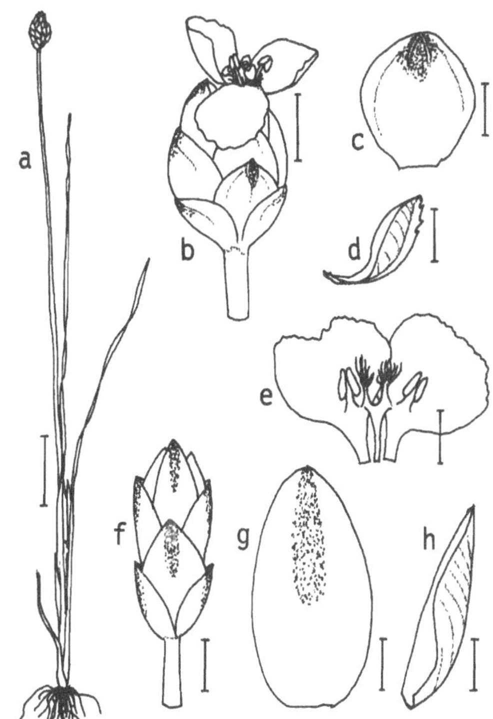

Fig. 373. *Xyris pauciflora*: a, flowering plant (2 cm); b, spike with 1 flower (3 mm); c, bract (1.5 mm); d, lateral sepal (1.5 mm); e, two petals with stamens and staminodes (2 mm). *X. wallichii*: f, spike (5 mm); g, bract (1.5 mm); h, lateral sepal (1.5 mm).

slightly compressed above, 5–40 cm long, 0.3–1.4 mm wide, with or without longitudinal ridges, the ridges 1 or 2 and shallow. Inflorescence ellipsoid, 4–11 mm long, 2–6 mm wide. Bracts in the middle of the inflorescence light brown to brown, margins entire, tip entire; grey-brown field on the abaxial surface below the tip triangular, 1.7–3.2 mm long, 0.3–1.2 mm wide, 3–10 times longer than wide, with a narrowly triangular hyaline area on each side widening but not protruding upwards. Lateral sepals light brown (never hyaline), with an entire crest.

Perennial. In wet places, on dripping rocks.

{Hel}. NE India, Myanmar, Vietnam and the Malay Peninsula: Ban? Bur, Ind: Mg. 1200–1400 m.

ZANNICHELLIACEAE: 4 genera, 1 in India

Zannichellia: ± 6 or 1 polymorphic species

Zannichellia palustris Linnaeus: Fig. 374

Roots not branched. Stems erect or creeping, much branched, sometimes stoloniferous, with erect, filamentous, leafy stems. Leaves mostly opposite or the distal ones clustered in pseudowhorls and enveloped by a membranous prophyll; squamulae intravaginales filiform, usually 2 at the base of each leaf; basal leaf sheaths (or stipules) either absent or free from the leaf and ligule-like; leaf blades linear, flat, 1-nerved, up to 5 cm long, usually less than 1 mm wide. Flowers unisexual, solitary or clustered, in leaf axils. Male flowers naked; stamens 1 (or 2 united according to interpretation); connective small, prolonged into a blunt knob; anther with (2) 4–8 locules; pollen grains globose. Female flowers borne in cup-shaped sheaths (spathe or perianth depending on the morphological interpretation); carpels (1–) 4–5 (–9); style elongate, persisting in fruit; stigma oblique-peltate, asymmetrically funnel-shaped. Fruit a pithy drupe, obliquely oblong, flattened, 3–6 mm long, with irregular peg-like teeth and/or warty ridges on one or both sides.

Annual or perennial. Totally submerged, in still or slowly flowing, fresh or brackish water. Gregarious and often locally dominant. It is pollinated under water

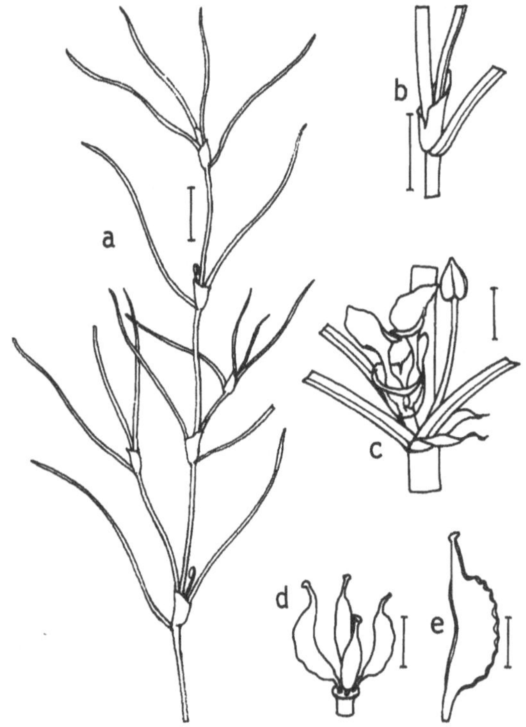

Fig. 374. *Zannichellia palustris*: a, flowering shoot (1 cm); b, node with sheathing prophyll or stipule (2 mm); c, inflorescence with 1 male and 1 female flower (1 mm); d, head of fruits (4 mm); e, ripe fruit (2 mm).

(hypohydrophilous) Guo, You-Hao, Sperry, Rebecca, Cook, C. D. K. and Cox, P. A. The pollination ecology of *Zannichellia palustris* L. (Zannichelliaceae). Aquatic Botany, 38: 341–356 (1990). The disseminules are pithy drupes with spiny endocarps, dispersed by water and probably also by animals or in mud. In India the fruits are often stalked and could be assigned to *Z. palustris* var. *pedicillata* Wahlenberg et Rosén. It is useful as food and as a refuge for wildlife.

{(⊙) Vit}. Almost cosmopolitan: Ban, Ind, Nep, Pak: Bh, Dl, Hp, Mh, Mp, Rj, Up.

Index to plant names

Abildgaardia 96, 132
densa 96
eragrostis 132
ovata 145
triflora 149
ACANTHACEAE 30
Acanthospermum hispidum 66
Acanthus ilicifolius 30
Achyranthes 41
aquatica 43
prostrata 42
Acmella 73
calva 73
ACORACEAE 50
Acorus 50
calamus 51
gramineus 50
Acroceras 283
munroanum 283
zizanioides 283
Acrostichum 27
aureum 27
Actinoscirpus 174
grossus 177
Adenosma 30
camphoratum 339
indiana 339
javanica 339
subrepens 339
thymus 31
uliginosa 32
Adenostemma angustifolium 66
Adhatoda 36
Aeschynomene 206
americana 207
aspera 207
indica 208
paludosa 211
agarbotti 124
Agasicles hydrophila 42
Ageratum conyzoides 66
Agrostis semiverticillata 281
stolonifera 281
verticillata 281
AIZOACEAE 11
Ajuga bracteosa 224
macrosperma 224
Aldrovanda 182
vesiculosa 182
Alisma 37
glandulosum 37
gramineum 37
lanceolata 37
orientale 37
plantago-aquatica 37
ALISMATACEAE 37
Allmania longipedunculata 41
nodiflora 41
Alternanthera 41
ficoidea 41
nodiflora 41
paronychioides 41
philoxeroides 42
pungens 41
sessilis 42
tenella 41
triandra 42
Alysicarpus hamosus 206
vaginalis 206
AMARANTHACEAE 41
AMARYLLIDACEAE 43
Amaryllis vivipara 43
Ambulia 343
Ameletia 252
polystachya 257
uliginosa 257
Amischophacelus 87
Ammannia 248
auriculata 249
baccifera 249
baccifera subsp. aegyptiaca 249
cordata 252
dentelloides 360
desertora 250
multiflora 250
nagpurensis 250
nana 257
octandra 250
repens 257
salicifolia 249, 251
senegalensis 249, 251
verticillata 251
Anabaena azollae 22
Anagallis pumila 11
Andropogon malacophyllus 299
squarrosus 314
Aneilema 89
hamiltonianum 89
hispida 88
lineolatum 89
nanum 91
nudiflorum var. terminale 90
paniculatum 90
radicans 90
sinicum 89
Angelonia goyazensis 339
salicariifolia 339
ANGIOSPERMAE 30
Aniseia 92
martinicensis 92
APIACEAE 44
Aponogeton 46
appendiculatus 47
bruggenii 47
crispus 48
echinatum 48
indicus 48
jacobsenii 48
lakhonensis 48
microphyllum 49
monostachyon 48
natans 48
rigidifolius 49
satarensis 49
undulatus 49
APONOGETONACEAE 46
ARACEAE 50
ARECACEAE 12
Artanema longifolium 339
Arthraxon breviaristatus 281
castratus 281
jubatus 281
lanceolatus 281
nudus 281
quartinianus 281
Arum colocasia 51
Arundo 284
donax 284
ASCLEPIADACEAE 66
Ascolepis tenuior 160
Asteracantha 30
longifolia 35
ASTERACEAE 66
Asystasia coromandelina 30
gangetica 30
Azolla 22
imbricata 22
pinnata 22
pinnata subsp. asiatica 22
pinnata var. imbricata 22
AZOLLACEAE 22

Bacopa 339
floribunda 340
hamiltoniana 340
monnieri 340
procumbens 339
rotundifolia 339
Baeothryon 182
BALSAMINACEAE 74
Barclaya 273
longifolia 273
BARCLAYACEAE 273
Barringtonia acutangula 12
BARRINGTONIACEAE 12
Batrachium 334
drouetii 334
paucistamineum 334
rhipiphyllum subsp. sphaerospermum 334
trichophyllum subsp. rionii 334
Baumea 160
crassa 160

Bergia 184
aestivosa 184
ammannioides 184
ammannioides var. trimera 186
capensis 185
odorata 184
polyantha 184
suffruticosa 184
trimera 186
verticillata 185
Berula erecta 44
Bidens bipinnata 66
biternata 66
cernua 67
foliosa 66
pilosa 66
radiata 66
tripartita 67
Bigelowia kleinii 335
lasiocarpa 335
roxburghiana 335
Bischofia javanica 12
BISCHOFIACEAE 12
Bistella 11
Blainvillea 73
acmella 66
rhomboidea 66
Blastus cochinchensis 12
Blumea bifoliata 66
mollis 66
oligocephala 66
Blyxa 216
aubertii var. aubertii 216
aubertii var. echinosperma 217
japonica var. japonica 217
octandra 218
quadricostata 218
Bobartia indica 104
Boerhavia erecta 11
Bolboschoenus 96
maritimus 96
Bonnaya 350
brachiata 352
bracteoides 352
grandiflora 352
minima 357
quinqueloba 339
reptans 358
veronicifolia 352
Bonnayodes limnophiloides 339, 347
BORAGINACEAE 75
Borreria articularis 335
stricta 335
Bothriochloa pertusa 281
Bothriospermum tenellum 75
Brachiaria 284
decumbens 285
deflexa 285
distachya 285
eruciformis 285
munae 285
mutica 285
nilgirica 285
paspaloides 285
ramosa 286
remota 285
reptans 286
seminuda 285
setigera 285
villosa 285
Bramia 339
monniera 340
Brasenia 80
schreberi 80
BRASSICACEAE 78
Briza eragrostis 293
Bulbostylis 96
capillaris 87
capillaris var. trifida 96
densa 96
Burmannia 79
azurea 79
caelestis 79
coelestis 79
distachya 79
disticha 79
pusilla 80
sumatrana 79
triflora 79
BURMANNIACEAE 79
BUTOMACEAE 11
Butomopsis 247
lanceolata 247
latifolia 247
Butomus umbellatus 11
Bythophyton 341
indicum 341

Cabomba 81
aquatica 81
australis 81
caroliniana var. caroliniana 81
CABOMBACEAE 80
Caesulia 67
axillaris 67
Caladium nymphaeaefolia 51
Calamus erectus 12
tenuis 12
Caldesia 37
grandis 37
oligococca 37
parnassifolia 38
parnassiifolia var. major 38
reniformis 38
CALLITRICHACEAE 81
Callitriche 81
stagnalis 81
wightiana 81
Caltha 334
alba 334
govaniana 334
palustris 334
palustris var. himalensis 334
scaposa 334
CAMPANULACEAE 82
Canna 83
edulis 83
× generalis 83
indica 83
× orchidioides 83
speciosa 83
CANNACEAE 83
Canscora decurrens 211
decussata 211
diffusa 211
CAPPARIDACEAE 11
Cardamine 78
Cardanthera 30
avana 31
balsamica 31
griffithii 31
pinnatifida 33
thwaitesii 31
triflora 32
uliginosa 32
verticillata 31
Carex 97
heterostachya 97
lenticularis 98
maculata 98
ovata 145
phacota 98
CARYOPHYLLACEAE 11
Cassia alata 12
tora 206
Catabrosa aquatica 281
Celosia argentea 41
Centella 44
asiatica 44
Centipeda 68
minima 68
orbicularis 68
Centranthera 341
hispida 341
humifusa 341
indica 341
nepalensis 341
procumbens 341
tranquebarica 341
Centrostachys 43
aquatica 43
Cephalanthus occidentalis 12
CERATOPHYLLACEAE 84
Ceratophyllum 84
demersum 84
muricatum subsp. muricatum 85
submersum 84
submersum subsp. muricatum 85
Ceratopteris 27
cornuta 28
pteridoides 28
thalictroides 29
Chamaeraphis gracilis 311
spinescens var. brunoniana 310
Chamissoa albida 41
aspera 41
dichotoma 41
CHENOPODIACEAE 11
Chenopodium ambrosioides 11
Chinese water-chestnut 126, 128
Chrysopogon aciculatus 281
Cicuta virosa 44
Cipura paludosa 11
Cladium glomeratum 160
jamaicense subsp. chinense 94
mariscus subsp. jamaicensis 94
riparium 160
Cladopus 319
hookerana 318
hookerianus 318
willisiana 318
Cleome chelidonii 11
Clerodendron indicum 12
Clinogyne 12
Cochlearia cochlearioides 78
flava 78

Coelachne 286
ghatica 287
minuta 287
perpusilla 287
perpusilla var. nilagirica 287
pulchella 287
simpliciuscula 287
Coix 287
aquatica 288
gigantea 288
lachryma-jobi 288
lingulata 288
puellarum 288
Coldenia 75
procumbens 75
Colocasia 51
antiquorum 51
esculenta 51
COMBRETACEAE 12
Commelina 85
aquatica 86
albescens 85
attenuata 85
benghalensis 86
clavata 85
diffusa 86
erecta 85
hasskarlii 86
kurzii 85
longicaulis 86
longifolia 85
nudiflora 86
obliqua 85
paludosa 85
salicifolia 85
subulata 87
undulata 85
COMMELINACEAE 85
COMPOSITAE 66
CONVOLVULACEAE 91
Convolvulus reniformis 93
uniflorus 92
Corchorus aestuans 12
capsularis 12
CORDIACEAE 75
Coricarpha aphylla 157
Cotula anthemoides 66
bicolor 66
conica 73
coronopifolia 66
hemisphaerica 66
Courtoisia 162
cyperoides 164
Courtoisinia 162
Crassocephalum crepidioides 66
Crassula schimperi var. schimperi 11
CRASSULACEAE 11
Craterostigma plantaginaeum 339
Crinum 43
asiaticum 43
defixum 43
ensifolium 43
viviparum 43
CRUCIFERAE 78
Crypsis schoenoides 281
Cryptocoryne 52
alba 53
albida 53
axelrodii 58
balansae 56
beckettii 54
bogneri 54
burmensis 56
ciliata 54
cognata 55
cognatoides 57
consobrina 55
costata 53
crispatula var. flaccidifolia 56
cruddasiana 56
dalzellii 57
gomezii 61
huegelii 57
korthausae 53
legroi 58
lucens 57
lutea 58
nevillii 56
parva 56
petchii 54
retrospiralis 57
retrospiralis var. albida 53
retrospiralis var. crispatula 56
roxburghiana 57
roxburghii 57
sinensis 56
spiralis 57
spiralis var. cognatoides 58
thwaitesii 58
tonkinensis 56
tortuosa 57
undulata 58
uniloculare 57
walkeri 58
wendtii 59
wightii 55
× willisii 57
Cyanotis 87
axillaris 87
cristata 87
cucullata 88
fasciculata 87
Cyathocline 68
flava 68
jacquemontii 68
lutea 68
lyrata 68
purpurea 68
Cymodocea rotundata 11
serrulata 11
CYMODOCEACEAE 11
Cynodon arcuatus 281
dactylon 281
intermedius 281
CYPERACEAE 94
Cyperus 99
albidus 171
alopecuroides 102
alternifolius subsp. flabelliformis 113
altus 111
alulatus 103
amabilis 104
angulatus 173
angustifolius 108
arenarius 104
aristatus 166
articulatus 105
atkinsonii 99
baccha 171
bifax 105
brachiatus 106
brevifolius 156
bromoides 173
bulbosus 99
capillaris 168
carnosus 121
castaneus 105
cephalotes 106
circinatus 99
clarkei 162
compactus 163
compressus 106
compressus subsp. micranthus 106
compressus var. capillaceus 106
conglomeratus 99
conglomeratus var. arenarius 104
corymbosus 107
corymbosus var. longispiculatus 107
corymbosus var. scariosus 124
cruentus 171
cuspidatus 108
cyperinus 162
cyperoides 164
dehiscens 118
denudatus 113
diaphanus 168
difformis 109
diffusus 99
diffusus subsp. macrostachyus 99
digitatus 109
dilutus 163
disruptus 105
distans 110
distans var. major 117
distans var. pseudonutans 110
donianus 119
dubius 164
elatus 110
eleusinoides 117
eragrostis 171
esculentus 111
exaltatus 111
exaltatus var. serpens 111
flabelliformis 113
flavidus 125, 168
fuscus 99
glaber 99
globosus 168
glomeratus 99
haenkianus 163
halpan 112
haspan subsp. haspan 112
haspan subsp. juncoides 113
hookeri 109
hookerianus 106
hyalinus 155
imbricatus 113
involucratus 113
iria 114
iria var. rectangularis 103
juncoides 113
kyllingaeoides 164
kyllingia 155
laevigatus 114
latespicatus 168
latespicatus var. gracilescens 168
leucocephalus 99, 122

Cyperus (*cont.*):
locuples 99
longus 115
macrostachyos 169
maderaspatanus 165
malabaricus 99
malaccensis 115
marginellus 119
meeboldii 116
melanospermus 156
membranaceus 170
meyenii 106
meyenii subsp. micranthus var. capillaceus 106
michelianus subsp. pygmaeus 122
mitis 124
monti 99
mysurensis 117
nigroviridis 99
nilagiricus 168
niveus 99
nutans var. eleusinoides 117
nutans var. nutans 117
obliquus 119
odoratus 170
ohwii 117
pachyrrhizus 99
pallidus 120
pangorei 118
paniculatus 119, 170
papyrus 118
pectiniformis 106
pennatus 162
pentabracteatus 99
pilosus 119
platyphyllus 120
platystylis 120
polyanthelus 99
polystachyos 170
procerus 121
pseudobromoides 168
pseudokyllingioides 164
pubisquama 99
pulchellus 122
pulcherrimus 99
pulvinatus 170
pumilus 155, 170
punctatus 170
puncticulatus 171
pygmaeus 122
racemosus 110
radians 99
radiatus 113
ramosii 99
retzii 105
rotundus 123
rotundus subsp. retzii 105
roxburghii 120
rubicundus 99
sanguinolentus 171
scariosus 124
serotinus 99
setaceus 130
sphacelatus 99
spinulosus 163
squarrosus 166
steadii 162
stenostachys var. indica 124
stoloniferus 99
strictus 168
subcapitatus 124
substramineus 172
tegetiformis 107
tegetum 118
teneriffae 99
tenuiculmis 124
tenuispica 125
triceps 156
tuberosus 105, 123
uncinatus 108
unioloides 173
venustus 111
zollingeri 99, 124
Cyrtobagous singularis 30

Dalzellia 317
ceylanica 317
ramosissima 320
zeylanica 317
zeylanica var. konkanica 317
Damsonium flava 248
Dentella 335
repens 335
serpyllifolia 335, 336
Desmodium ferrugineum 206
heterophyllum 206
laxiflorum 206
rufescens 206
triflorum 206
Dichaespermum 89
Dichanthium annulatum 281
Dichoespermum 89
juncoides 90
repens 89
Dichostylis 99
Dichrocephala integifolia 66
minutifolia 68
schmidtii 68
Dicraea 321
algaeformis 323
minor 323
pterophylla 323
stylosa var. bourdillonii 323
stylosa var. fucoides 323
stylosa var. kandrensis 323
stylosa var. laciniata 323
Digera alternifolia 41
arvensis 41
muricata 41
Digitaria ciliaris 281
marginata 281
setigera 281
stricta 281
Dillenia indica 12
DILLENIACEAE 12
Diplachne 289
fascicularis 289
fusca 289
Diplacrum 125
caricinum 125
tridentatum 125
zeylanicum 125
Donax arundastrum 12
Dopatrium 342
junceum 342
junceum var. multiloba 342
lobelioides 342
nudicaule 343
Drosera 183
burmannii 183
indica 183
peltata 183
peltata var. lunata 183
DROSERACEAE 182
Dyschoriste 30
erecta 31
Dysophylla 224
benthamiana 225
crassicaulis 225
gracilis 224
kachinensis 224
stellata var. gracilis 224
tetraphylla 224
tomentosa 224
verticillata 225

Ecbolium 36
Ecbolium procumbens 36
Echinochloa 289
colona 290
colona var. frumentacea 291
colonora 290
colonum 290
crus-galli 291
crus-galli var. breviseta 291
crus-galli var. frumentacea 291
crus-galli var. longiseta 291
crus-galli var. oryzicola 292
crus-pavonis 291
frumentacea 291
frumentacea var. utilis 292
glabrescens 290
hispidula 291, 292
macrocarpa 292
oryzoides 292
phyllopogon 292
picta 292
pyramidalis 292
senegalensis 292
stagnina 292
utilis 291
Eclipta 69
alba 69
latifolia 66
prostrata 69
Egeria densa 215
EHRETIACEAE 75
Eichhornia 329
crassipes 329
ELATINACEAE: 184
Elatine 186
ambigua 186
americana 186
gracilis 186
triandra 186
Eleocharis 123, 126, 182
acicularis 126
actangula 127
acutangula 127
andamanensis 128
atropurpurea 127
capitata 129
chaetaria 130
confervoides 182
congesta 127

Eleocharis (*cont.*):
congesta var. japonica 128
dulcis 128
equisetina 128
fistulosa 127
geniculata 129
japonica 127
lankana 129
mitracarpa 126
obscura 126
ochrostachys 129
palustris 126, 128
parishii 126
pauciflora 126
pellucida 127
plantaginea 128
quinqueflora 126
retroflexa subsp. chaetaria 130
sphacelata 128
spiralis 130
swamyii 131
tetraquetra 131
thomsonii 126
tuberculosa 126
tuberosa 128
uniglumis 126
variegata var. laxiflora 129
Eleogiton 131
curvulus 131
fluitans 131
Elytrophorus 293
articulatus 293
spicatus 293
Enhalus 215
acoroides 11
Enhydra 69
Enicostema axillare 211
littorale 211
Enydra 69
fluctuans 69
Epaltes 70
divaricata 70
pygmea 70
Epilobium 276
EQUISETACEAE 23
Equisetum 23
debile 23
diffusum 23
fluviatile 23
limosum 23
palustre 23
ramosissimum 23
ramosissimum subsp. debile 23
ramosissimum subsp. ramosissimum 23
Eragrostis 293
atrovirens 293
cilianensis 293
diarrhena 293
gangetica 293
japonica 293
major 293
megastachya 293
pilosa 293
riparia 293
stenophylla 293
tenella 293
unioloides 293
zeylanica 293
Erianthus 311
elephantinus 311
Erigeron molle 66
ERIOCAULACEAE 187
Eriocaulon 187
achiton 187
alpestre 187
argenteum 199
atratum 187
atratum var. major 187
barba-caprae 187
barbeyanum 190
bhutanicum 187
bifistulosum 201
bombayanum 194
bracteosum 188
breviscapon 190
breviscapum 190
brownianum 187, 188
brownianum var. latifolium 188
capillus-naiadis 201
catopsioides 187
caulescens 187
ceylanicum 187
ceylanicum var. subcaulescens 187
cherrapunjianum 187
christopheri 187
cinereum 191
colletii 187
collinum 187
collinum var. nanum 187
conicum 188
cristatum 187
cristatum var. mackii 187
cuspidatum 191
cuspidatum var. bracteatum 195
dalzellii 192
dalzellii var. glabratum 192
dianae 196
dianae var. conicum 188
dianae var. longibracteatum 196
dianae var. parviflorum 198
dianae var. richardianum 199
dianae var. triloboides 198
duthiei 193
echinulatum 193
edwardii 187
edwardii var. clarkei 187
elenorae 188
ensiforme 188
eurypeplon 194
fergusonii 194
fluitans 201
fluviatile 194
fysonii 195
gamblei 187
geoffreyi 188
glaucum 188
gracile 187
gracile var. kurzii 187
gregatum 187
hamiltonianum 187
hamiltonianum var. minimum 188
hamiltonianum var. minor 197
hanningtonii 196
helferi 195
heterolepis 196
horsley-kondae 188
horsley-kondae var. megalocephalum 188
horsley-kundae 188
humile 188
indicum 194
infirmum 187
infirmum var. kurzii 187
intermedium 201
kasaragodense 196
katmanduense 187
keralense 194
lanceolatum 196
lanceolatum var. pilosum 196
leucomelas 188
limosum 201
longicuspe 187
longicuspe var. polycephalum 187
longicuspis 187
longifolium 204
luzulaefolium 187
luzulifolium 187
margaretae 188
mariae 187
martianum 197
meeboldii 188
melaleucum 188
melanocephalum 201
minimum 197
minutum 188
miserum 187
miserum var. mitophyllum 187
mitophyllum 187
myriophyllum 201
mysorensis 199
nairii 188
neesianum 187
nepalense 187
nilagirense 188
nilagirense var. parvifolium 188
odoratum 187
oliveri 187
oryzetorum 197, 198
palghatense 188
panagudianum 188
parviflorum 198
pectinatum 188
philippo-coburgi 187
polycephalum 187
psammophilum 199
pseudoquinquangulare 187
pumilio 187
pygmaeum 205
quinquangulare 199
quinquangulare forma viviparum 199, 201
quinquangulare var. martianum 197
quinquangulare var. walkeri 204
rajendrababui 188
ramnadense 188
redactum 191
reductum 191
rhodae 188
richardianum 199
ritchieanum 188
rivulare 192
robusto-brownianum 199
robustum 188
robustum var. caulescens 187
roseum 199
rouxianum 188
santapaui 200
schweinfurthii 201

Eriocaulon (*cont.*):
sedgwickii 201
setaceum 201
sexangulare 201
sexangulare var. longifolium 204
sharmae 188
sieboldianum 191
sieboldtianum 191
sivarajanii 202
sollyanum 203
staintonii 187
stellulatum 202
subcaulescens 187
subglaucum 187
talbotii 188
thomasii 194
thwaitesii 187
transvaalicum 196
trilobum 203
trimeni 187
trisectoides 187
truncatum 204
truncatum var. disepalum 204
truncatum var. malaccense 204
tuberiferum 188
vanheurckii 194
vanheurckii forma minimum 194
vasudevanii 188
viride 187
walkeri 204
wallichianum 201
wightianum 195, 204
wightianum forma viviparum 195
wightianum var. helferi 195
willdenovianum 204
willdenovianum var. fergusonii 194
wrightianum 199
xeranthemum 205
zeylanicum 187
Eriochloa 294
polystachya 294
procera 294
Eriophorum 94
Ethulia divaricata 70
Euphorbia prostrata 11
thymifolia 11
EUPHORBIACEAE 11, 12
Euryale 274
ferox 274
Eusteralis 224
Exacum hyssopifolium 211
pedunculatum 211
pumilum 212
sessile 212
tetragonum 212

FABACEAE 12, 206
Farmeria 317
indica 318
metzgerioides 318
Fimbristylis 132
acuminata 135
aestivalis 135
aestivalis subsp. major 135
aestivalis var. esquarrosa 133
aestivalis var. macrostachya 135
aestivalis var. squarrosa 148
aestivalis var. trichopoda 135
affinis 149
aggregata 132
alboviridis 136
amplocarpa 132
annua forma tomentosa 149
annua var. diphylla 141
aphylla 136
argentea 137
arnottii 149
asperrima 132
assamica 150
autumnalis 138
autumnalis var. complanata 138
autumnalis var. microcarya 144
bisumbellata 137
bisumbellata var. hirtistyla 133
carpopoda 132
cinnamometorum 138
circumciliata 132
comata 148
complanata 138
complanata var. kraussiana 139
complanata var. microcarya 144
compressa 145
connectens 139
consanguinea 139
contorta 132
crystallina 140
cylindrocarpa 149
cymosa 140
cyperoides var. cinnamometorum 138
dauciformis 140
densa 96
dichotoma 137, 141
dichotoma subsp. podocarpa 149
dichotoma var. nilgaririca 133
dichotoma var. villosa 133
diphylla 141
diphylla var. pluristriata 149
dipsacea 141
disticha 142
dura 132
eligulata 132
eragrostis 132, 142
falcata 132
ferruginea 142
ferruginea subsp. sieberiana 148
fimbristyloides 132
fusca subsp. fulvescens 132
fusca subsp. fusca 132
glauca 141
globulosa 150
glomerata 140
griffithii 142
haspaniformis 146
hirsutifolia 143
insignis 132
intonsa 132
junciformis 132
littoralis 143
longistigmata 132
merguensis 132
microcarya 144, 151
miliacea 143, 144
monospicula 132
monostachyos 145
monticola 132
multinervia 132
nayaranii 132
nigro-brunnea 132
nutans 144
ovata 145
pallescens 137
paludosa 146
pentagona 154
pentaptera 147
podocarpa 149
polytrichoides 145
propinqua 133
pubisquama 145
pustulosa 132
quinquangularis 144
quinquangularis var. crassa 136
rectifolia 132
rigidiuscula 133
rugosa 133
salbundia subsp. pentaptera 147
salbundia subsp. salbundia 146
scaberrima 147
scabrisquama 133
schoenoides 147
semi-disticha 133
sieberiana 148
sincta 141
spathacea 140
squarrosa 148
squarrosa var. esquarrosa 133
stolonifera 133
strigosa 133
swamyii 133
tenuicula 149
tetragona 149
thomsonii 133
tomentosa 149
tortifolia 133
triflora 149
tristachya 133
uliginosa 133
umbellaris 150
unispicularis 133
utilis 150
velata 133
wightiana 140
woodrowii 151
zeylanica 151
Fissendocarpa 276
Floscopa 88
hamiltonii 88
scandens 88
Fuirena 151
capitata 152
ciliaris 153
glomerata 153
pubescens 153
rottboellii 153
rubiginosa 160
swamyii 154
trilobites 154
tuwensis 154
umbellata 154
uncinata 152
wallichiana 155

Galium palustre 335
exaristata 281
Garnotia tenella 281
gausun 316

Geissaspis 208
cristata 208
tenella 209
tenella var. malabarica 209
Geniosporum parviflorum 224
GENTIANACEAE 211
ginger grass 306
Glaux maritima 11
Glinus lotoides 11
oppositifolius 11
Glossostigma 343
diandrum 343
Glyceria plicata 281
tonglensis 281
Gnaphalium affine 66
hypoleucum 66
indicum 66
luteo-album 66
polycaulon 66
pulvinatum 66
Gomphrena 41
celosioides 41
GRAMINEAE 280
Grangea 70
latifolia 66
maderaspatana 70
Gratiola cuneifolia 343
griffithii 338
integrifolia 358
micrantha 339
oppositifolia 357
Graya 313
Griffithella 318
hookeriana 318
hookeriana subsp. bombayensis 318
hookeriana var. willisiana 318

Halodule uninervis 11
Halophila 215
beccarii 11
decipiens 11
ovalis 11
ovata 11
stipulacea 11
HALORAGACEAE 212
Haloragis isomera 212
philipinensis 212
Hanguana 214
malayana 214
HANGUANACEAE 214
Hapalosa loeflingiae 11
Hedyotis 335
affinis 335, 336
auricularia 336
biflora 336
brachypoda 336
corymbosa 337
corymbosa var. linearis 336
depressa 337
dichotoma 336
diffusa 337
erecta 336
glabella 336
gracilis 335
herbacea 336
hirsuta 335
intermedia 337
japonica 335
leschenaultiana 335
lindleyana 335
nitida 336
pinifolia 335
pumila 336
racemosa 336
ramosissima 337
stipulata 335
stricta 335
trinervia 336
Hedysarum neli-tali 208
Heliotropium 76
coromandelianum 77
curassavicum 76
indicum 76
keralense 77
malabaricum 77
ovalifolium 77
supinum 77
Hemarthria 294
altissima 294
compressa 294
coromandelina 294
hamiltoniana 294
laxa 294
longifolia 294
protensa 294
Hemicarpha 158
isolepis 158
Hemidelphis 30
polysperma 33
polysperma var. joshianus 33
Hemidistichophyllum 318
Hemigraphis dura 30
hirta 30
Hemisorghum 295
mekongensis 295
venustum 295
Herpestis 339
Heteranthera reniformis 328
rotundifolia 328
HIPPURIDACEAE 11
Hippuris vulgaris 11
Hoppea 212
dichotoma 212
fastigiata 212
Hormonia retusa 12
riparia 12
Hoshiarpuria minutiflora 259
Houttuynia cordata 11
Hubbardia 295
heptaneuron 295
Hybanthus suffruticosa 12
HYDATELLACEAE 215
Hydrilla 218
verticillata 218
Hydroanzia 319
Hydrobryopsis 319
sessilis 319
Hydrobryum 319, 324
griffithii 319
olivaceum var. anamalaiensis 324
olivaceum var. grisum 324
olivaceum var. zeylanicum 324
sessile 319
Hydrocera 75
triflora 75
Hydrocharis dubia 215
morsus-ranae 215
HYDROCHARITACEAE 11, 215
Hydrocotyle 45
nitidula 45
perexqua 45
puncticulata 45
ranunculoides var. incisa 45
rotundifolia 45
sibthorpioides 45
tenella 45
zollingeri 45
Hydrolea 221
Hydrolea spinosa 222
Hydrolea zeylanica 222
HYDROPHYLLACEAE 221
Hydrostemma 273
Hygrophila 30
angustifolia 34
auriculata 35
balsamica 31
biplicata 31
difformis 32
erecta 31
griffithii 31
heinei 32
heloides 32
incana 31
obovata 34
phlomoides 31
pinnatifida 33
polysperma 33
pusilla 31
quadrivalvis 34
ringens 34
salicifolia 34
schulli 35
serphyllum 35
serpylla 35
serpylla var. hookeriana 36
spinosa 35
stocksii 35
thwaitesii 31
Hygrorhyza 295
Hygroryza 295
aristata 295
Hymenachne 296
acutigluma 296
amplexicaulis 296
assamica 296
pseudointerrupta 296
Hypaelyptum argenteum 158
HYPERICACEAE 11
Hypericum laxum 11
Hypolytrum latifolium 94
nemorum 94
wrightianum 94

Ilysanthes 350
serrata 352
verbenaefolia 352
Impatiens 74
angustifolia 75
natans 75
Imperata cylindrica 281
Indigofera cordifolia 206
Indocourtoisia 162
cyperoides 162
Indotristicha 320
ramosissima 320

Indotristicha (*cont.*):
tirunelveliana 320
Ipomoea 92
aquatica 92
asarifolia 92
beladamboe 92
carnea subsp. fistulosa 93
crassicaulis 93
fistulosa 93
repens 92
reptans 92
rugosa 92
IRIDACEAE 11
Isachne 297
albens 297
angladei 297
australis 298
bicolor 298
borii 297
clarkei 297
deccanensis 297
dispar 297, 298
elegans 298
geniculata 298
globosa 298
henryi 297
himalaica 297
kunthiana 297
meeboldii 298
miliacea 298
multiflora 297
pulchella 298
stigmatosa 298
veldkampii 298
Ischaemum 299
aristatum 299
barbatum 299
bombaiense 299
ciliare 299
commutatum 299
copeanum 299
duthiei 299
flumineum 299
goebelii 299
indicum 299
indicum var. longipilum 299
indicum var. prorepens 299
indicum var. wallichii 299
jayachandranii 300
mangaluricum 299
molle 300
muticum 300
muticum var. srilankensis 300
nairii 299
nilagiricum 300
raizadae 299
repens 300
rugosum 301
santapauii 299
thomsonianum 299
travancorense 301
vembanadense 301
Iseilema anthephoroides 281
laxum 281
prostratum 281
ISOETACEAE 23
Isoetes 23
bilaspurensis 24
brachyglossa 23
coromandelina 23
dixitei 24
indica 23, 25
mahadevensis 24
mirzapurensis 25
panchananii 24
panchganiensis 24
panchganiensis subsp. kemangundiensis 24
pantii 24
rajasthanensis 24
reticulata 24
sahyadriensis 24
sahyadrii 24
sampathkumaranii 25
tuberculata 25
unilocularis 25
Isolepis 131, 158, 174
bellula 158
corymbosus 176
kernii 158
oryzetorum 178
prolongata 175
roylei 179
senegalensis 179
squarrosa 160
trifida 96
uninodis 178

Java fern 27
JUNCACEAE 222
JUNCAGINACEAE 11
Juncella 215
Juncellus 99
alopecuroides 102
laevigatus 114
pygmaeus 122
serotinus 99
Juncus 222
bufonius 223
maritimus 223
prismatocarpus 223
Jussiaea 276
fissendocarpa 277
repens 276
speciosa 278
suffruticosa 277
Justicia 36
betonica 36
glabra 36
peploides 36
quinqueangularis 36
quinqueangularis var. peploides 36
simplex 36

Kedarnatha sanctuarii 44
Kohautia aspera 335
gracilis 335
retrorsa 335
Kyllinga 155
aromatica 155
bifolia 155
brevifolia 156
brevifolia var. stellulata 155
bulbosa 156
cyperina 162
cyperoides 164
hyalina 155
melanosperma 156
melanosperma subsp. bifolia 156
metzii 155
monocephala 155
nemoralis 155
odorata subsp. cylindrica 155
panicea 165
polyphylla 155
squamulata 155
sumatrensis 167
tenuifolia 156
triceps 156
umbellata 167

LABIATAE 223
Lagarosiphon 219
alternifolia 219
Lagenandra 59
blassii 62
bogneri 60
dewitii 60
erosa 61
gomezii 61
insignis 63
jacobsenii 61
koenigii 62
lancifolia 62
meeboldii 62
nairii 63
ovata 63
praetermissa 63
schultzei 63
thwaitesii 63
toxicaria 63, 64
undulata 64
LAMIACEAE 223
Lasia 64
aculeata 64
heterophylla 64
spinosa 64
Laurembergia coccinea 212
Laurentia longiflora 82
Lawia 317
zeylanica var. gardneriana 317
zeylanica var. konkanica 317
zeylanica var. malabarica 317
zeylanica var. parkiniana 317
Lawiella 318
LECYTHIDACEAE 12
Leersia 302
hackelii 302
hexandra 302
oryzoides 302
LEGUMINOSAE 12, 206
Legzpia polygonoides 339
Lemna 226
aequinoctialis 227
blatteri 227
cruciata 228
eleonorae 227
gibba 227
minima 227
minor 228
minor auct. Ind. 227
orbiculata 229
paucicostata 227
perpusilla 227

Lemna (*cont.*):
tenera 228
trisulca 228
turionifera 228
LEMNACEAE 226
LENTIBULARIACEAE 231
Lepironia 157
articulata 157
compressa 157
conifera 157
mucronata 157
zeylanica 161
Leptochloa 302
capillacea 302
chinensis 302
filiformis 302
malabarica 289
panicea 302
uniflora 302
LILIACEAE 43
Limnanthemum biflorum 264
kleinianum 265
moonii 266
nymphoides 263
wrightianum 265
Limnocharis 248
flava 248
LIMNOCHARITACEAE 247
Limnochloa 126
media 127
Limnophila 343
aquatica 346
aromatica 344
cana 345
chinensis 345
chinensis var. clarkei 345
chinensis var. scaberrima 345
cimicina 348, 349
conferta 349
connata 345
diffusa 348
erecta 346
glandulifera 346
gratioloides 346
gratissima 344
helferi 346
heterophylla 346
hirsuta 344
hypericifolia 345
hyssopifolia 346
indica 346
laxa 347
× ludoviciana 347, 350
micrantha 348
polyantha 348
polystachya 348
polystachyoides 348
pulcherrima 348
punctata 344
pygmaea 348
racemosa 346
reflexa 346
repens 349
roxburghii 346, 349
rugosa 349
serrata 349
sessiliflora 350
tillaeoides 346
Limnophyton 38
obtusifolium 38
Limnopoa 302
meeboldii 302
Limosella aquatica 338
Lindernia 350
anagallis 351
antipoda 352
caespitosa 358
ciliata 352
crustacea 353
crustacea var. clarkei 353
dianthera 339
elata 353
estaminodiosa 354
hookeri 354
hookeri var. kumaunensis 354
hyssopioides 354
hyssopoides 354
latifolia 354
laxa 358
manilaliana 355
micrantha 355
minima 355
mollis 355
molluginoides 356
multiflora 360
nummularifolia 356
nummulariifolia 356
nutans 355
oppositifolia 357
parviflora 357
procumbens 358
pusilla 358
pyxidaria 358
rotundifolia 358
ruellioides 358
ruelloides 358
srilankana 359
tenuifolia 359
tenuifolia var. pygmaea 359
viscosa 360
Lipocarpha 158
chinensis 158
hemisphaerica 158
kernii 158
pygmaea 158
raynalena 159
raynaliana 159
reddyi 159
sphacelata 159
squarrosa 160
triceps 159
Lippia 366
nodiflora 366
Lobelia 82
affinis 83
alsinoides 82
dichotoma 82
dopatrioides 82
griffithii 82
heyneana 83
microcarpa 82
stipularis 82
subincisa 83
succulenta 83
terminalis 82
trialata 83
trigona 82
zeylanica 83
LOBELIACEAE 82
LOGANIACEAE 11
Lophotocarpus 38
guayanensis 39
lotos 272
lotus 272
Ludwigia 276
adscendens 276
hyssopifolia 277
linifolia 277
natans 276
octovalvis 277
octovalvis subsp. sessiliflora 277
parviflora 278
perennis 278
peruviana 278
prostrata 279
repens 276
Lycopus europaeus 223
exaltatus 223
Lysimachia candida 11
lobeliodes 11
obovata 11
LYTHRACEAE 248
salicaria 248
tribracteum 248

Machaerina 160
crassa 160
rubiginosa 160
rubiginosa subsp. crassa 161
Maferria 317
MAGNOLIOPSIDA 30
Manisuris myuros 281
Mapania 161
andamanica 161
cuspidata 161
cuspidata var. angustifolia 161
cuspidata var. petiolata 161
enodis 161
humilus 161
immersa 161
palustris 161
silhetensis 161
zeylanica 161
MARANTACEAE 12
Mariscopsis suaveolens 155
Mariscus 162
albescens 162
blatteri 164, 165
bulbosus 162
clarkei 162
compactus 163
cyperinus subsp. cyperinus 162
cyperinus subsp. laxtus 162
cyperoides 164
dregeanus 164
dubius 164
javanicus 162
konkanensis 162
maderaspatanus 165
microcephalus 163
paniceus 165
paniceus var. roxburgianus 165
pedunculatus 162
pictus 162
sieberianus 167
sieberianus var. subcompositus 162

Mariscus (*cont.*):
squarrosus 166
sumatrensis 167
tenuifolius 162
Marsilea 25
aegyptiaca 25
ballardii 25
brachycarpa 26
brachypus 26
condensata 25
coromandeliana 26
coromandelica 26
coromandelina 26
gracilenta 27
maheshwari 27
minuta 26
poonensis 26
quadrifolia 25
rajasthanensis 25
MARSILEACEAE 25
Mazus delavayi 339
pumilus 339
Mecardonia procumbens 339
Melastoma malabarthricum 12
MELASTOMACEAE 12
Mella 339
Melochia cochorifolia 12
concatenata 12
Mentha aquatica 223
pulegium 223
stellata 224
verticillata 225
MENYANTHACEAE 263
Menyanthes trifoliata 263
Merremia 93
emarginata 93
gangetica 93
Mesona palustris 224
Meyera orientalis 67
Microcarpaea 360
cochlearifolia 361
minima 360
muscosa 360
spathulata 343
Microsorium 27
Microsorum 27
pteropus 27
pteropus var. minor 27
Mimosa adenanthera 210
natans 209
Mimulus 361
gracilis 361
nepalensis 361
orbicularis 361
pusillus 361
strictus 361
Mitrasacme indica 11
alsinoides 11
Mnianthus 317
Mniopsis 318
MOLLUGINACEAE 11
Mollugo disticha 11
hirta 11
pentaphylla 11
Moniera 339
cuneifolia 340
Monniera 339
Monochoria 329
elata 330
hastata 330
vaginalis 330
morlo 305
Murdannia 89
blumei 89
crocea subsp. ochracea 89
elata 89
esculenta 89
keisak 89
loriformis 90
malabarica 90
nudiflora 90
ochracea 89
pauciflora 90
semiteres 90
simplex 89
spirata 91
vaginata 91
wightii 90
Myriophyllum 213
aquaticum 213
brasiliense 213
indicum 213
intermedium 213, 214
oliganthum 214
proserpinacioides 213
spathulatum 214
spicatum 213
tetrandrum 214
tuberculatum 214
verticillatum 213
Myriostachya wightiana 281
MYRTACEAE 12

NAJADACEAE 267
Najas 267
falciculata 269
foveolata 267
foveolata var. minor 269
graminea 268
graminea var. angustifolia 270
graminea var. minor 270
grossareolata 269
heteromorpha 269
indica 269
kingii 270
kurziana 270
lacerata 269
malesiana 270
marina 270
marina subsp. armata 271
marina subsp. intermedia 271
marina subsp. marina var. kashmirensis 271
minor 271
minor var. spinosa 269
oguraensis 271
tenuis 271
Nasturtium 78
microphyllum 78
officinale 78
Neanotis indica 335
nummularia 335
oxyphylla 335
Nechamandra 219
alternifolia 219
Nelumbium speciosum 272
Nelumbo 272
lutea 272
nucifera 272
NELUMBONACEAE 272
Nepeta calminthoides 224
hindostana 224
ruderalis 224
Neptunia 209
acinaciformis 209
oleracea 209
plena 210
prostrata 209
triquetra 209
Nesaea 252
brevipes 252
lanceolata 252
prostrata 252
Neyraudia arundinacea 281
Nicoteba 36
Nomaphila 30
parishii 31
pinnatifida 33
stricta 31
Nothosaerva brachiata 41
Nuphar 273
NYCTAGINACEAE 11
Nymphaea 274
alba 274
caerulea 275
candida 274
× daubeniana 275
lotus 275
micrantha 275
nouchali 274, 275
odorata 274
pubescens 275
rubra 275
stellata 274
tetragona 274
tuberosa 274
NYMPHAEACEAE 273
Nymphoides 263
aurantiaca 264
cristata 264
hydrophylla 264
indica 265
krishnakesara 265
macrosperma 265
parvifolia 266
peltata 263
sivarajanii 267

Oenanthe 45
bengalensis 45
hookeri 46
javanica 46
linearis 46
rivularis 46
schlechteri 46
stolonifera 46
thomsonii 46
Oldenlandia 336
affinis 336
alsinifolia 337
aspera 335
auriculata 336
biflora 336
brachypoda 336
capillaris 337

Oldenlandia (*cont.*):
corymbosa 337
crystallina 336
diffusa 336, 337
erecta 336
herbacea 336
heynei 336
linearis 336
nitida 336
paniculata 336
pentandra 11
pseudocorymbosa 337
pumila 336
repens 335
trinervia 336
Oligmeris linifolia 11
ONAGRACEAE 276
ORCHIDACEAE 279
Oryza 303
aquatica 304
eichingeri 304
fatua 304
filiformis 304
glaberrima 304, 305
granulata 304
indandamanica 304
jeyporensis 304
malampuzhaensis 304
meyeriana var. granulata 304
minuta 304
nivara 304
officinalis 304
rhizomatis 304
rufipogon 304, 305
sativa 304
sativa var. collina 304
triticoides 281
Osbeckia rostrata 12
Ottelia 220
alismoides 220
cordata 220
Oxystelma 66
esculentum 66
secamone 66

paddy 305
padupai korai 111
PALMAE 12
PANDANACEAE 279
Pandanophyllum 161
immersum 161
Pandanus 279
fasicularis 280
odoratissimus 280
tectorius 280
pangorai 108, 118
Panicum 289, 305
adscendens 281
amphibium 285
auritum 305, 312
austroasiaticum 305
bisulcum 305
colonum 290
flexuosum var. glabrum 305
isachne 285
javanicum 314
johnii 305
latifolium 283
lepidotum 298
myuros 296
natans 305
neesianum 305
nodibarbatum 298
paludosum 305
procumbens 286
proliferum 305
prostratum 286
psilopodium 305
purpurescens 285
repens 306
setigerum 285
supervacuum 286
trypheron 305
umbrosum 286
urochloa 314
violaceum 298
virgatum 305
Papyrus latifolia 120
paragrass 286
PARKERIACEAE 27
Parnassia laxmannii 11
nubicola 11
palustris 11
PARNASSIACEAE 11
Paspalidium 306
flavidum 306
fluitans 307
geminatum 307
paspaloides 307
punctatum 307
Paspalum 307
canarae 308
canarae var. fimbriatum 308
cartilagineum 309
commersonii 309
compactum 308
conjugatum 308
dilatatum 308
distichum 309
notatum 308
orbiculare 309
ovatum 308
paspalodes 309
paspaloides 309
plicatulum 308
scrobiculatum 309
uruguayense 308
urvillei 308
vaginatum 309
Pennisetum hordeoides 281
purpureum 281
Pentapetes phoenicea 12
Peplidium 361
humifusum 361
maritimum 361
Peristylus lawii 279
Persicaria 325
amphibia 325
barbata 325
barbata subsp. gracile 326
barbata var. stagnina 326
densiflora 326
dichotoma 327
dolicopoda 325
glabra 326
hydropiper 327
kawagoneana 325
lapathifolia 327
limbata 325
minor 325
mitis 325
orientalis 325
peduncularis 327
perfoliata 325
posumbu 325
pulchra 328
punctata 325
strigosa 325
tenella 325
Phalaris aquatica 281
arundinacea 281
nodosa 281
tuberosa 281
Pharnaceum depressum 11
PHILYDRACEAE 280
Philydrum 280
laniginosum 280
Phragmites 309
australis 309
communis 309
karka 309
maxima 309
vallatoria 309
Phrynuim 12
Phyla 366
canescens 367
nodiflora 366
Physichilus 30
Physichilus serpyllum 35
Pistia 65
stratiotes 65
Plaesiantha 30
Plaesianthera 30
thwaitesii 31
Pleiogyne cardiosperma 66
Poa palustris 281
polymorpha 293
POACEAE 280
podma 272
PODOSTEMACEAE 316
Podostemon 321
Podostemum 321, 324
acuminatum 324
barberi 321
kelensis 324
microcarpus 324
subulatus 321
subulatus var. mavaeliae 321
subulatus var. sholaii 321
Pogostemon 224
auricularius 224
cruciatus 224
deccanensis 224
erectum 224
gracilis 224
griffithii 225
helferi 224
linearis 224
paludosus 224
peguanus 224
pumilus 225
quadrifolius 224
salicifolius 224
stellatus 225
stocksii 224
tomentosa 224

Pogostemon (*cont.*)
wightii 224
Polycarpon prostratum 11
tetraphyllum 11
Polygala elongata 11
POLYGALACEAE 11
POLYGONACEAE 325
Polygonum 325
amphibium 325
barbatum 325
barbatum var. gracilis 326
barbatum var. stagninum 326
densiflorum 326
dichotomum 327
dolicopodum 325
donii 327
flaccidum 325, 327
glabrum 326
hydropiper 327
hydropiper subsp. microcarpum 327
hydropiper var. flaccidum 327
indicum 327
lapathifolium 327
lapathifolium var. lanatum 327
limbatum 325
minus 325
mite 325
nodosum 327
orientale 325
pedunculare 327
pedunculare var. angustissimum 327
pedunculare var. robustum 327
perfoliatum 325
plebeium 327
plebejum 327
posumbu 325
praetermissum 325
pubescens 327
pulchrum 328
punctatum 325
rivulare 325
serrulatum 326
stagninum 326
strictum 325
strigosum 325
tenellum 325
tomentosum 327, 328
Polypleurum 321
agharkarii 323
dichotomum 322
elongatum 322
filifolium 322
munnarense 323
stylosum 323
stylosum var. kumarensis 323
wallichii 323
POLYPODIACEAE 27
Polypodium tridactylon 27
Polypogon littoralis var. mutica 281
monspeliensis 281
viridis 281
Pontederia 331
cordata 331
plantaginea 330
PONTEDERIACEAE 328
Porteresia coarctata 281
Potamogeton 331
alpinus 332
berchtoldii 332
crispus 332
distinctus 332
filiformis 332
fluitans 333
gramineus 332
indicus 333
javanicus 333
lucens 332
malaianus 333
mucronatus 330
natans 332
nodosus 333
octandrus 333
pectinatus 333
perfoliatus 334
pusillus 332
POTAMOGETONACEAE 331
Potentilla supina 11
Pratia 82
PRIMULACEAE 11
Pseudognaphalium luteo-album 66
Pseudoraphis 310
aspera 311
brunoniana 310
minuta 311
spinescens 311
Psilotrichum ferrugineum 41
PTERIDACEAE 27
PTERIDOPHYTINA 22
Pulicaria foliosa 66
Pycreus 167
albomarginatus 169
atroglumosa 168
bolei 168
capillaris 168
curvibracreatus 168
decumbens 168
diaphanus 168
diaphanus var. laxus 168
dwarkensis 168
flavescens 168
flavidus 168
globosus 168
govindarajalui 168
hyalinus 155
lancelotii 122, 168
latovaginatus 168
lurida 168
macrostachyos 169
malabaricus 168
nitens 170
odoratus 170
plumbeonuceus 168
plurinodosus 168
polystachus 170
polystachyos 170
polystachyos var. circinatus 170
polystachyos var. laxiflorus 170
pseudobromoides 168
pumilus 170
pumilus subsp. membranaceus 171
pumilus var. patens 171
puncticulatus 171
rubiglumosus 168
sanguinolentus 171
stramineus 172
stricticulmis 168
sulcinux 168
unioloides 173

Queenslandiella 155
hyalina 155

Ramphicarpa 362
longiflora 362
RANUNCULACEAE 334
Ranunculus 334
aquatilis 334
emuricatus 334
flavidus 334
hyperboreus 334
lingua 334
natans 334
rionii 334
sceleratus 334
sphaerospermus 334
trichophyllus 334
trichophyllus subsp. glaber 334
Reimaria 307
RESEDACEAE 11
Rhabdia 78
fluviatilis 78
lycioides 78
Rhaphidospora 36
abyssinica 36
Rhynchospora 173
aurea 173
ceylonica 173
chinensis 173
corymbosa 173
glauca var. condensata 174
laxa 174
rubra 173
rugosa subsp. brownii 174
ruppioides 182
triflora 173
wallichiana 173
wightiana 173
rice 303, 305
Rikliella 158
kernii 158
squarrosa 160
Rorippa 78
benghalensis 78
dubia 78
indica 78
madagascariensis 78
micrantha 78
montanum 78
nasturtium-aquaticum 78
palustris 78
pseudoislandica 78
ROSACEAE 11
Rostellaria 36
peploides 36
Rostellularia 36
procumbens var. simplex 36
quinqueangularis var. peploides 36
Rotala 252
alata 261
andamanensis 257
cookii 254
cordata 255

Rotala (*cont.*):
densiflora 255
diversifolia 255
fimbriata 255
floribunda 255
fysonii 256
hexandra 256
illecebroides 257
indica 257
leptopetala 260
macrandra 258
malabarica 258
malampuzhensis 258
mexicana 259
occultiflora 259
pentandra 260
pygmaea 259
repens 256
ritchiei 259
rosea 260
rotundifolia 260
rubra 261
serpyllifolia 261
simpliciuscula 261
subrotunda 262
tenuis 261
vasudevanii 254
verticillaris 262
wallichii 262
Rottboellia exaltata 281
glabra 294
Rotula 78
aquatica 78
RUBIACEAE 12, 335
Ruellia 30
erecta 31, 34
ringens 34
Rumex aquaticus 325
dentatus 325
maritimus 325
nepalensis 325
Rungia repens 30
Ruppia 338
cirrhosa 338
maritima 338
maritima subsp. spiralis 338
rostellata 338
RUPPIACEAE 338

Saccharum 311
officinarum 312
ravennae 311
spontaneum 311
Sacciolepis 312
aurita 305, 312
curvata 312
indica 312
indica var. intermedia 312
interrupta 313
myosuroides 313
Sagittaria 38
chinensis 40
edulis 40
greggii 39
guayanensis subsp. lappula 39
guyanensis 39
hastata 40
latifolia 39
longiloba 39
montevidensis subsp. montevidensis 39
pygmaea 39
sagittifolia 39
sagittifolia subsp. leucopetala 40
sinensis 40
tengtsungensis 39
triflora 40
SALICACEAE 12
tetrasperma 12
Salomonia ciliata 11
Salvia plebeia 224
Salvinia 29
cucullata 29
molesta 30
natans 29
SALVINIACEAE 29
Samolus valerandi 11
Sarcostemma esculentum 66
SAURURACEAE 11
Saururus chinensis 11
SAXIFRAGACEAE 11
Schoenoplectus 174
articulatus 175
corymbosus 176
fuscorubens 174
grossus 177
juncoides 177
lacustris var. tabernaemontani 174
lateriflorus 178
litoralis 180
maritimus 96
mucronatus 178
roylei 179
senegalensis 179
subulatus 180
supinus 175
supinus subsp. lateriflorus 178
tabernaemontani 174
triqueter 174
wallichii 181
Schoenus 182
cyperoides 149
tuberosus 123
Schumanianthus dichotoma 12
Scirpidiella 131
Scirpus 126, 131, 132, 174, 181, 182
affinis 96
asiaticus 94
bispicatus 147
brachyceras 176
brevifolius 141
chinensis 94
compactus 96
confervoides 182
coronarius 99
corymbosus 176
cyperoides 167
erectus 177
grossus 177
inclinatus 176
incurvatus 175
jacobii 179
javanus 178
junciformis 177
juncoides 177
lacustris subsp. glauca 174
lateralis 178
lateriflorus 178
laxiflorus 129
litoralis 180
littoralis 180
macrostachys 96
maritimus 96
maximus 177
mucronatus 178
muticus 178
occultus 122
plantaginoides 128
quinquefarius 179
subarticulatus 175
submersus 182
subulatus 180
supinus 175, 178
ternatanus 94
thermalis 180
triangulatus 178
tuberosus 96
tumidus 128
wallichii 181
wichurai 94
Scleria 181
biflora 181
capitata 181
caricina 125
fenestrata 181
flaccida 181
hebecarpa 181
lateriflora 181
levis 181
neesii 181
oryzoides 181
parvula 181
pergracilis 181
poaeformis 181
rugosa 181
steudeliana 181
stricta 181
sumatrensis 181
swamyii 181
tessellata 181
uliginosa 181
Scoparia dulcis 339
SCROPHULARIACEAE 338
Scutellaria galericulata 224
Serpicula brevipes 212
hirsuta 212
indica 212
zeylanica 212
Sesbania 210
aculeata 210
aculeata var. paludosa 211
aegyptiaca 210
bispinosa 210
cannabina 210
concolor 210
javanica 211
procumbens 211
roxburghii 211
sericea 211
sesban 210
uliginosa 211
Setaria intermedia 281
pallide-fusca 281
paniculifera 281

Setaria intermedia (*cont.*)
pumila 281
tomentosa 281
singhara 364
Sinobacopa 339
Sium laciniatum 46
latijugum 44
Smithia aspera 208
bigemina 206
conferta 206
sensitiva 206
setulosa 206
Sonchus brachyotus 66
Sopubia delphinifolia 339
Sorghum halepense 281
Sorostachys 99
kyllingioides 122
SPARGANIACEAE 11
Sparganium emersum 11
erectum subsp. stoloniferum 11
Spermacoce hispida 335
pusilla 335
SPERMATOPHYTINA 30
Sphaeranthus 71
africanus 71
amaranthoides 72
hirtus 72
indicus 72
microcephalus 71
peguensis 73
senegalensis 73
zeylandicus 72
Sphaerocaryum 313
elegans 313
malaccense 313
Sphenoclea 363
zeylanica 363
SPHENOCLEACEAE 363
SPIGELIACEAE 11
Spilanthes 73
acmella 66
acmella var. calva 73
acmella var. paniculata 74
calva 73
ciliata 74
iabadicensis 74
oleracea 73
paniculata 74
pseudo-acmella 73
radicans 73
uliginosa 74
Spirodela 229
intermedia 229
maxima 229
oligorhiza 230
polyrhiza 229
polyrrhiza 229
punctata 230
Spodiopogon cotulifer 281
scrobiculatus 299
villosus 299
Sporobolus diander 281
tremulus 281
virginicus 281
Stemodia diffusa 347
menthastrum 349
viscosa 339
STERCULIACEAE 12
Steris aquatica 222
Struchium sparganophorum 66
sugar cane 312
Sutera dissecta 339
Swertia corymbosa 212
minor 212
Synedrella nodiflora 67
Synnema 30
triflora 32
Syringodium isoetifolium 11
Syzygium cumini 12

TAMARICACEAE 12
Tamarix dioica 12
ericoides 12
troupii 12
Tanacetum viscosum 68
taro 51
Telanthera 41
Tenagocharis 247
latifolia 247
Terminalia arjuna 12
Terniola 317
Terniopsis 317
Teucrium scordium 224
Thalassia 215
hemprichtii 11
Thyrocephalon nemoralis 155
Thysanolaena maxima 281
TILIACEAE 12
Tillaea pentandra 11
Tonningia 87
Torenia bicolor 339
cyanea 339
hirta 358
lepidota 341
lindernioides 339
mucronulata 339
torpedo grass 306
Tradescantia paniculata 88
Trapa 363
bispinosa 626
incisa 363
maximowiczii 363
natans var. bispinosa 363
natans var. natans 363
quadrispinosa 363
TRAPACEAE 363
Trichelostylis 132
chaetorrhiza 132
latifolia 132
torta 132
Triglochin maritima 11
palustre 11
Trigonella occulta 206
TRISTICHACEAE 316
Trithuria 215
konkanensis 215
Tulasnea 317
Tungia laevigata 158
Typha 364
angustata 365
angustifolia 365
domingensis 365
elephantina 365
latifolia 365
laxmannii 365
minima 365
TYPHACEAE 364
UMBELLIFERAE 44
Urochloa 314
marathensis 314
marathensis var. velutia 314
panicoides 314
pubescens 314
Utricularia 231
acuta 239
affinis 246
alata 236
albiflora 237
albocaerulea 233
antirrhinoides 236
arcuata 234
arenaria 234
aurea 235
australis 236
bifida 236
bifida var. wallichiana 243
biflora 239
brachiata 232
brachypoda 246
caerulea 237
caerulea var. graminifolia 239
caerulea var. smithiana 244
capillaris 237
cecilii 238
christopheri 232
coerula 237
conferta 239
confervifolia 235
corynephora 232
decipiens 246
elachista 246
equiseticaulis 239
exoleta 239
fasciculata 235
filicaulis 237
flexuosa 235
forestii 232
foveolata 238
furcellata 232
gibba 239
glochidiata 245
graminifolia 239
hirta 240
humilis 236
inflexa 240
involens 232
khasiana 239
kumaonensis 232
lazulina 240
liliacina 246
macrocarpa 244
macrolepis 243
minor 232
minutissima 241
moliniformis 241
multicaulis 232
nayarii 246
nepalensis 232
nivea 237
ogmosperma 234
orbiculata 245
parviflora 239
paucifolia 237
pedicellata 239
polygaloides 242
praeterita 242

Utricularia (*cont.*):
pubescens 242
punctata 232
purpurescens 239
pusilla 245
racemosa 237
ramosa 236, 237
recta 232
reticulata 242
rogersiana 232
rosea 237
roseopurpurea 237
roxburghii 239
salwinensis 232
sampathii 237
scandens 243
smithiana 244
squamosa 247
stellaris 244
striatula 245
stricticaulis 242
suberecta 239
subulata 246
tayloriana 240
uligenoides 239
uliginoides 239
uliginosa 246
volubilis 243
vulgaris 236
wallichiana 236, 243
wallichii 243
wightiana 247

Vahlia 11
dichotoma 11
digyna 11
oldenlandioides 11
viscosa 11
VAHLIACEAE 11
Vallisneria 221
denseserrulata 221
minor 221
natans 221
spiralis 221
spiraloides 221
Vandellia 350
angustifolia 351, 355
cordifolia 351
erecta 358
foliosa 354
gracilis 353
hirsuta 360
montana 355
multiflora 360
pedunculata 351
racemosa 353
scabra 358
sessiliflora 356
stemonoides 354
urticifolia 353
VERBENACEAE 12, 366
Verbesina acmella 66
alba 69
Veronica 362
anagallis 362
anagallis-aquatica 362
beccabunga 362
oxycarpa 362
undulata 362
Vetiveria 314
zizanioides 314
Victoria 273
VIOLACEAE 12
Vitex negundo 12
Vogtia malloi 42
Vossia 315
cuspidata 315

Wahlenbergia erecta 82
gracilis 82
marginata 82
water hyacinth 329
watercress 78
Websteria 182
confervoides 182
Wedelia calendulacea 67
chinensis 67
Wiesneria 40
triandra 40
Willisia 323
selaginoides 323
Wolffia 230
angusta 230
arrhiza 230
delilii var. schleideni 231
globosa 231
microscopica 231
neglecta 231
Wolfiella 231
hyalina 231

Xanthium indicum 67
strumarium 67
XYRIDACEAE 367
Xyris 367
anceps 368
capensis 367
capensis var. schoenoides 367
complanata 368
coronata 368
indica 368
intersita 368
lobbii 369
nilagarensis 367
pauciflora 369
robusta 368
schoenoides 367
wallichii 369

Zannichellia 370
palustris 370
palustris var. pedicillata 370
ZANNICHELLIACEAE 370
Zeuxine 279
strateumatica 279
sulcata 279
Zeylanidium 321, 324
barberi 321
johnsonii 324
lichenoides 324
lichenoides var. bhorensis 324
lichenoides var. khandalense 324
olivaceum 324
subulatus 321
Zizania 316
caduciflora 316
latifolia 316
Zygomenes 87

The manufacturer's authorised representative in the EU for product safety is Oxford University Press España S.A. of El Parque Empresarial San Fernando de Henares, Avenida de Castilla, 2 - 28830 Madrid (www.oup.es/en or product.safety@oup.com). OUP España S.A. also acts as importer into Spain of products made by the manufacturer.
Printed and bound by CPI Group (UK) Ltd, Croydon, CR0 4YY
14/09/2026
02216298-0002